Game Theory for
Wireless Communications
and Networking

Game Theory for Wireless Communications and Networking

Edited by
Yan Zhang and Mohsen Guizani

CRC Press
Taylor & Francis Group
Boca Raton London New York

CRC Press is an imprint of the
Taylor & Francis Group, an **informa** business

AN AUERBACH BOOK

CRC Press
Taylor & Francis Group
6000 Broken Sound Parkway NW, Suite 300
Boca Raton, FL 33487-2742

© 2011 by Taylor and Francis Group, LLC
CRC Press is an imprint of Taylor & Francis Group, an Informa business

No claim to original U.S. Government works

Printed in the United States of America on acid-free paper
10 9 8 7 6 5 4 3 2 1

International Standard Book Number: 978-1-4398-0889-4 (Hardback)

Library of Congress Cataloging-in-Publication Data

Game theory for wireless communications and networking / editors, Yan Zhang and Mohsen Guizani.
     p. cm. -- (Wireless networks and mobile communications)
  "A CRC title."
  Includes bibliographical references and index.
  ISBN 978-1-4398-0889-4 (alk. paper)
  1. Radio resource management (Wireless communications) 2. Game theory. I. Zhang, Yan, 1977-
II. Guizani, Mohsen. III. Title. IV. Series.

TK5103.4873.G36 2011
621.38201'5193--dc22
                                                            2011000132

Visit the Taylor & Francis Web site at
http://www.taylorandfrancis.com

and the CRC Press Web site at
http://www.crcpress.com

Contents

Preface

It is envisioned that game theory will be a fundamental technique in the field of wireless communications and networks. Its application also prompts interdiscipline between economics and wireless engineering. Game theory was originally invented to explain complicated economic behavior. With its effectiveness in studying complex dynamics among players, game theory has been widely applied in politics, philosophy, military, sociology, telecommunications, and logistics. In these fields, game theory has been employed to achieve socially optimal equilibrium.

A wide spectrum of issues can be efficiently addressed in recent efforts in applying game theory in wireless communications and networking. The applications are mainly motivated by the inherent characteristics of wireless systems. First, the wireless systems become more complicated with respect to network size, protocol heterogeneity, multimedia service variations, dynamic interactions, and unknown attacks. These challenges have made it difficult to model, analyze, predict, and optimize wireless network performance. In such contexts, game theory is able to significantly help better understand the complex interactions, and design more efficient, scalable, and robust protocols. Second, there are very limited resources in wireless communications, e.g., bandwidth, power, and capacity. With the increasing number of wireless access terminals, resource scarcity and hence competition among mobile users becomes more severe. This competition for limited resources closely matches with game theory rationale. Third, from the point of view of the whole wireless industry, more and more wireless technologies are emerging to offer high-speed, high-capacity, and high-coverage services. This demands intelligent spectrum management and marketing, which may well be solved within the game theoretical framework. Game theory can be employed in both infrastructure-based wireless networks and multi-hop networks, e.g., wireless ad hoc networks, mesh networks, and sensor networks. Depending on the interested environments, the objectives may include power consumption reduction, system capacity improvement, packet loss decrease, and network resilience enhancement. The game theoretic model is able to address efficiently resource allocation, congestion control, attack, routing, energy management, packet forwarding, and medium access control (MAC).

This book systematically introduces and explains the application of game theory in wireless communications and networking. It provides a comprehensive technical guide covering introductory concepts, fundamental techniques, recent advances, and open issues in related topics. It also contains illustrations and provides complete cross-referencing.

This book is organized into four parts:

- Part I: Fundamentals
- Part II: Power Control Games
- Part III: Economic Approaches
- Part IV: Resource Management

Part I introduces the fundamental issues and solutions in applying different games in different wireless domains, including wireless sensor networks, vehicular networks, and OFDM-based wireless systems. Part II introduces issues and solutions in power control games. Power control has been a long-term issue in using game theory. Part III explores applications of different economic approaches, including bargaining and auction-based approaches. Part IV offers the game theoretic approach to address radio resource management issues.

This book has the following salient features:

- Presents game theory basics, models, and applications in wireless domain
- Serves as a comprehensive reference on using game theory in wireless communications and networking
- Covers a broad range of topics and future development directions
- Introduces architectures, protocols, security, and applications
- Assists professionals, engineers, students, and researchers

The book can serve as an essential reference for students, educators, research strategists, scientists, researchers, and engineers in the field of wireless communications and networking. In particular, it will be of interest to students, researchers, developers, and consultants in optimizing future generation wireless systems and networks. The content provided in this book should help readers understand the necessary background, concepts, and principles in using game theory approaches in wireless systems.

We would like to acknowledge the effort and time invested by all the contributors for their excellent work. They have been extremely professional and cooperative. Special thanks go to Richard O'Hanley, Stephanie Morkert, and Joette Lynch of Taylor & Francis Group for their support, patience, and professionalism from the beginning to the final stages. We are very grateful to Suganthi Thirunavukarasu for her assistance during the typesetting period. Last but not least, a special thank you to our families and friends for their constant encouragement, patience, and understanding throughout this project.

For MATLAB® and Simulink® product information, please contact

The MathWorks, Inc.
3 Apple Hill Drive
Natick, MA, 01760-2098 USA
Tel: 508-647-7000
Fax: 508-647-7001
E-mail: info@mathworks.com
Web: www.mathworks.com

Editors

Yan Zhang is currently heading the Wireless Networks research group at Simula Research Laboratory, Norway. He is also an adjunct associate professor in the Department of Informatics, the University of Oslo, Norway. He received his PhD in the School of Electrical and Electronics Engineering, Nanyang Technological University, Singapore. Since August 2006, he has been with Simula Research Laboratory, Norway (http://www.simula.no/).

Dr. Zhang is an associate editor and serves on the editorial board of the *International Journal of Communication Systems* (*IJCS*—Wiley), *Wireless Communications and Mobile Computing* (*WCMC*—Wiley), and *Security and Communication Networks* (Wiley). He is currently serving as the book series editor for the book series on Wireless Networks and Mobile Communications (Auerbach Publications, CRC Press, Taylor & Francis Group). He also serves as a guest coeditor for the Wiley *WCMC* special issue for best papers from the conference *IWCMC 2009*; ACM/Springer *Multimedia Systems Journal* special issue on Wireless Multimedia Transmission Technology and Application; Springer *Journal of Wireless Personal Communications* special issue on Cognitive Radio Networks and Communications; Interscience *International Journal of Autonomous and Adaptive Communications Systems* (*IJAACS*) special issue on Ubiquitous/Pervasive Services and Applications; *EURASIP Journal on Wireless Communications and Networking* (*JWCN*) special issue on Broadband Wireless Access; *IEEE Intelligent Systems* special issue on Context-Aware Middleware and Intelligent Agents for Smart Environments; and Wiley *Security and Communication Networks* special issue on Secure Multimedia Communication. He is also a guest coeditor for Elsevier's *Computer Communications* special issue on Adaptive Multicarrier Communications and Networks; Interscience *IJAACS* special issue on Cognitive Radio Systems; *The Journal of Universal Computer Science* (*JUCS*) special issue on Multimedia Security in Communication; Springer *Journal of Cluster Computing* special issue on Algorithm and Distributed Computing in Wireless Sensor Networks; EURASIP *Journal on Wireless Communications and Networking* (*JWCN*) special issue on OFDMA Architectures, Protocols, and Applications; and Springer *Journal of Wireless Personal Communications* special issue on Security and Multimodality in Pervasive Environments.

Zhang is serving as coeditor for the following books: *Resource, Mobility and Security Management in Wireless Networks and Mobile Communications*; *Wireless Mesh Networking: Architectures, Protocols and Standards*; *Millimeter-Wave Technology in Wireless PAN, LAN and MAN*; *Distributed Antenna Systems: Open Architecture for Future Wireless Communications*; *Security in Wireless Mesh Networks*; *Mobile WiMAX: Toward Broadband Wireless Metropolitan Area Networks*; *Wireless Quality-of-Service: Techniques, Standards and Applications*; *Broadband Mobile Multimedia: Techniques and Applications*; *Internet of Things: From RFID to the Next-Generation Pervasive Networked Systems*; *Unlicensed Mobile Access Technology: Protocols, Architectures, Security, Standards and Applications*;

Cooperative Wireless Communications; *WiMAX Network Planning and Optimization*; *RFID Security: Techniques, Protocols and System-on-Chip Design*; *Autonomic Computing and Networking*; *Security in RFID and Sensor Networks*; *Handbook of Research on Wireless Security*; *Handbook of Research on Secure Multimedia Distribution*; *RFID and Sensor Networks*; *Cognitive Radio Networks*; *Wireless Technologies for Intelligent Transportation Systems*; *Vehicular Networks: Techniques, Standards and Applications*; *Orthogonal Frequency Division Multiple Access (OFDMA)*; *Game Theory for Wireless Communications and Networking*; and *Delay Tolerant Networks: Protocols and Applications*.

He also serves as organizing committee chairs and on technical program committees for many international conferences. He received the Best Paper Award at IEEE AINA 2007. His research interests include resources, mobility, spectrum, data, energy, and security management in wireless networks and mobile computing. He is a senior member of IEEE and IEEE ComSoc.

Mohsen Guizani is currently a professor and the associate dean of academic affairs at Kuwait University. He was the chair of the Computer Science Department at Western Michigan University from 2002 to 2006 and chair of the Computer Science Department at the University of West Florida from 1999 to 2002. He also served in academic positions at the University of Missouri-Kansas City, University of Colorado-Denver, and Syracuse University. He received his BS (with distinction) and MS in electrical engineering, and his MS and PhD in computer engineering in 1984, 1986, 1987, and 1990, respectively, from Syracuse University, Syracuse, New York.

Dr. Guizani's research interests include computer networks, wireless communications and mobile computing, and optical networking. He currently serves on the editorial boards of six technical journals and is the founder and editor-in-chief (EIC) of *Wireless Communications and Mobile Computing* journal published by John Wiley (http://www.interscience.wiley.com/jpages/1530-8669/) and the *Journal of Computer Systems, Networks and Communications* (http://www.hindawi.com/journals/) published by Hindawi. He is also the founder and the steering committee chair of the *Annual International Conference of Wireless Communications and Mobile Computing (IWCMC)*. He is the author of 7 books and more than 270 publications in refereed journals and conferences and has guest edited a number of special issues in IEEE journals and magazines. He has also served as member, chair, and general chair of a number of conferences.

Dr. Guizani served as the chair of IEEE ComSoc WTC and chair of TAOS ComSoc Technical Committees. He was an IEEE Computer Society Distinguished Lecturer from 2003 to 2005. He is also an IEEE fellow and a senior member of ACM.

Contributors

Virgilio A.F. Almeida
Universidade Federal de Minas
Gerais, Brazil

Josephina Antoniou
Department of Computer Science
University of Cyprus
Nicosia, Cyprus

Elena Veronica Belmega
Laboratoire des Signaux et Systèmes
Supélec
Gif-sur-Yvette, France

Mehdi Bennis
Centre for Wireless Communications
University of Oulu
Oulu, Finland

Bir Bhanu
Center for Research in Intelligent Systems
Bourns College of Engineering
University of California at Riverside
Riverside, California

Hanna Bogucka
Department of Wireless Communications
Poznan University of Technology
Poznan, Poland

Dimitris E. Charilas
School of Electrical and Computer Engineering
National Technical University of Athens
Athens, Greece

Bin Chen
Institute for Infocomm Research
Singapore, Singapore

Philip Constantinou
School of Electrical and Computer Engineering
National Technical University of Athens
Athens, Greece

P.G. Cottis
Department of Electrical and Computer
  Engineering
National Technical University of Athens
Athens, Greece

Gabriele D'Angelo
Department of Computer Science
University of Bologna
Bologna, Italy

Luiz A. DaSilva
Centre for Telecommunications Value-Chain
  Research
Trinity College Dublin
Dublin, Ireland

and

Department of Electrical and Computer
  Engineering
Virginia Tech
Blacksburg, Virginia

Mérouane Debbah
Supélec
Gif-sur-Yvette, France

Manos Dramitinos
Stormrider Informatics Ltd.
Heraklion, Greece

and

Department of Informatics
Athens University of Economics and Business
Athens, Greece

Cesar Fernandes
Universidade Federal de Minas
Gerais, Brazil

Stefano Ferretti
Department of Computer Science
University of Bologna
Bologna, Italy

Jie Gao
Department of Electrical and Computer
  Engineering
University of Alberta
Edmonton, Alberta, Canada

Savo Glisic
Centre for Wireless Communications
University of Oulu
Oulu, Finland

Zhu Han
Department of Electrical and Computer
  Engineering
University of Houston
Houston, Texas

Mahbub Hassan
School of Computer Science and Engineering
University of New South Wales
Kensington, New South Wales, Australia

Gaoning He
Supélec
Gif-sur-Yvette, France

Anh Tuan Hoang
Institute for Infocomm Research
Singapore, Singapore

Ekram Hossain
Department of Electrical and Computer
  Engineering
University of Manitoba
Winnipeg, Manitoba, Canada

Jia Hu
Department of Computer Science
Liverpool Hope University
Liverpool, United Kingdom

Jane Wei Huang
Electrical Computer Engineering Department
University of British Columbia
Vancouver, British Columbia, Canada

Weijia Jia
Department of Computer Science
City University of Hong Kong
Kowloon, Hong Kong

Hai Jiang
Department of Electrical and Computer
  Engineering
University of Alberta
Edmonton, Alberta, Canada

Timotheos Kastrinogiannis
School of Electrical and Computer Engineering
National Technical University of Athens
Athens, Greece

Manzoor Ahmed Khan
Distributed Artificial Intelligence Laboratory
Technische Universität Berlin
Berlin, Germany

Zaheer Khan
Centre for Wireless Communications
University of Oulu
Oulu, Finland

Zhen Kong
Department of Electrical and Computer
  Engineering
Wayne State University
Detroit, Michigan

Vikram Krishnamurthy
Electrical Computer Engineering Department
University of British Columbia
Vancouver, British Columbia, Canada

Yu-Kwong Kwok
Department of Electrical and Electronic
   Engineering
The University of Hong Kong
Pokfulam, Hong Kong

Samson Lasaulce
Laboratoire des Signaux et Systèmes
Supélec
Gif-sur-Yvette, France

Isabelle Guérin Lassous
Centre National de la Recherche Scientifique
Institut National de Recherche en
   Informatique et en Automatique
L'Université Claude Bernard Lyon 1
Lyon, France

Yiming Li
Center for Research in Intelligent Systems
Bourns College of Engineering
University of California at Riverside
Riverside, California

Ying-Chang Liang
Institute for Infocomm Research
Singapore, Singapore

Antonio A.F. Loureiro
Universidade Federal de Minas
Gerais, Brazil

Geyong Min
Department of Computing
School of Computing, Informatics and Media
University of Bradford
Bradford, United Kingdom

Raquel A.F. Mini
Pontificia Universidade Católica de Minas
Gerais, Brazil

Dusit Niyato
School of Computer Engineering
Nanyang Technological University
Singapore, Singapore

Athanasios D. Panagopoulos
School of Electrical and Computer Engineering
National Technical University of Athens
Athens, Greece

Vicky Papadopoulou
Department of Computer Science
European University Cyprus
Nicosia, Cyprus

Symeon Papavassiliou
School of Electrical and Computer Engineering
National Technical University of Athens
Athens, Greece

Andreas Pitsillides
Department of Computer Science
University of Cyprus
Nicosia, Cyprus

Rajatha Raghavendra
Department of Electrical and Computer
   Engineering
University of Houston
Houston, Texas

Dwayne Rosenburgh
U.S. Department of Defense
University of Maryland Institute for Advanced
   Computer Studies
College Park, Maryland

Giovanni Rossi
Department of Computer Science
University of Bologna
Bologna, Italy

Fikret Sivrikaya
Distributed Artificial Intelligence Laboratory
Technische Universität Berlin
Berlin, Germany

Hamidou Tembine
Computer Science Laboratory
University of Avignon
Avignon, France

Eirini-Eleni Tsiropoulou
School of Electrical and Computer Engineering
National Technical University of Athens
Athens, Greece

Rémi Vannier
Centre National de la Recherche Scientifique
Institut National de Recherche en
 Informatique et en Automatique
L'Université Claude Bernard Lyon 1
Lyon, France

Athanasios V. Vasilakos
Department of Electrical and Computer
 Engineering
National Technical University of Athens
Athens, Greece

Stavroula G. Vassaki
School of Electrical and Computer Engineering
National Technical University of Athens
Athens, Greece

Vasos Vassiliou
Department of Computer Science
University of Cyprus
Nicosia, Cyprus

Pedro O.S. Vaz de Melo
Universidade Federal de Minas
Gerais, Brazil

Sergiy A. Vorobyov
Department of Electrical and Computer
 Engineering
University of Alberta
Edmonton, Alberta, Canada

Artemis C. Voulkidis
Department of Electrical and Computer
 Engineering
National Technical University of Athens
Athens, Greece

Mike E. Woodward
Department of Computing
School of Computing, Informatics and Media
University of Bradford
Bradford, United Kingdom

FUNDAMENTALS

Chapter 1

Game Theory in Multiuser Wireless Communications

Jie Gao, Sergiy A. Vorobyov, and Hai Jiang

Contents

Wireless communications are always resource limited. In multiuser wireless systems, all users compete for resources and can interfere with each other. The conflicting objectives of users make it highly unlikely for any user to gain more profit without harming other users. The possibility of exploiting interactions among wireless users was not considered in traditional information-theoretic studies of the multiuser systems. Therefore, the results obtained from the information-theoretic perspectives might lead to unstable or even infeasible solutions for multiuser systems when the selfish nature of the users is taken into account. Indeed, it is reasonable to assume that all users compete for the maximum achievable benefit at all time. Then, the competition among the users should be analyzed and regulated to improve the overall performance of the whole multiuser system.

For a multiuser system, the resource-sharing problem can be investigated from a game-theoretic perspective. Without coordination among users, the existence of stable outcomes, corresponding to the so-called *Nash Equilibria* (NE), can be analyzed. On the other hand, if there is a voluntary cooperation among users, extra benefits for all users can be gained and optimally distributed among users. In both the noncooperative and cooperative cases, the efficiency of resource utilization can be boosted and the system stability can be guaranteed.

Due to the aforementioned advantages, there is an increasing amount of research efforts studying multiuser wireless systems from the game-theoretic perspective [1–6]. In the literature, game theory has been exploited in various systems such as code division multiple access (CDMA) [7,8], orthogonal frequency-division multiplexing (OFDM) systems [9,10], ad hoc [11,12], and cognitive radio networks [13–15]. The purpose of this chapter is to discuss applications of game theory for designing multiuser strategies in wireless communications. A brief introduction to the basics of game theory will be given first. Power allocation, beamforming, and precoding problems will then be discussed. These problems will be formulated as scalar-, vector-, and matrix-valued games, respectively, and analyzed from a game-theoretic perspective. Both the noncooperative and cooperative cases will be considered. Performance metrics such as efficiency, fairness, uniqueness of the solution, and complexity will be compared and discussed. One conclusion of the studies will be the demonstration of the advantages associated to the cooperative strategies over the noncooperative ones. The cooperative strategies, however, require coordination among users and, thus, can lead to an increase of system overhead.

1.1 A Brief Survey: Multiuser Games in Wireless Communications

Mathematically, an M-player game can be modeled as

$$\Gamma = \left\{ \Omega, \{s_i | i \in \Omega\}, \{u_i | i \in \Omega\} \right\} \tag{1.1}$$

where

$\Omega = \{1, 2, \ldots, M\}$ is the set of all players

s_i is the strategy of player i

u_i is the utility (payoff) for player i as a function of $\{s_1, s_2, \ldots, s_M\}$

Depending on whether players collaborate or not, a game can be cooperative or noncooperative. In the following sections, some basic concepts of noncooperative and cooperative games will be reviewed, and examples based on simplified wireless systems will be given.

1.1.1 Noncooperative Games and Nash Equilibria

In a noncooperative game, there is no collaboration among users and the existence of an *equilibrium* is the main concern. An equilibrium $\{s_i | i = 1, \ldots, M\}$ is the strategy set composed of such strategies from which none of the players wants to deviate [16]. The most popular example of an equilibrium is the *Nash equilibrium* (NE), which can be mathematically expressed as

$$u_i \left(s_i^{\mathrm{NE}}, s_{\tilde{i}}^{\mathrm{NE}} \right) \geq u_i \left(s_i', s_{\tilde{i}}^{\mathrm{NE}} \right) \quad \forall s_i' \tag{1.2}$$

where

s_i^{NE} is the strategy of player i in the NE

$s_{\tilde{i}}^{\mathrm{NE}}$ are the corresponding strategies of all players but player i in the NE

s_i' stands for any possible strategy but s_i^{NE} for player i

An NE is a stable combination of all users' strategies such that no player can increase its utility by deviating from his current strategy given that other players do not deviate as well.

Generally, there are two typical issues with NE solutions. First, more than one NE may exist for a game, which renders difficulty in predicting the final outcome of the game. Second, an NE can lead and even usually leads to an inefficient outcome for all the players. In order to show it, the following two-user two-channel communication system is explored as an example.*

Example 1.1

Assume that there are two users and two communication channels, and each user has to make a decision about which channel to transmit on. First consider a simplified case: each user has a fixed power budget and is only allowed to transmit on one channel. Let c_1 denote the strategy that a user transmits on channel 1 and let c_2 denote the strategy that a user transmits on channel 2, respectively. Thus, the players are the two users and the strategy space is $\{c_1, c_2\}$. The information rates of the players are chosen to be their utilities. The combination of users' strategies and corresponding utilities are organized in Table 1.1.

In the table, the utilities of the two users are ordered as (u_1, u_2) with u_i representing the utility of user i, and $\hat{r}_{ij}$ and r_{ij} represent the information rate user i can obtain on channel j with and without interference from the other user, respectively (obviously $\hat{r}_{ij} < r_{ij}$). Although it is a simple two-user game, the uniqueness of NE[†] depends on the channel conditions (including desired channels and interference channels) and the noise power on each channel for each user. Let us assume that the noise power is identical on each channel for each user and study the impact of channel conditions on the game. We stress on the three following possible cases:

---

* The discussion of NE in this chapter is limited to pure strategy NE only.

† The abbreviation NE is used for both Nash equilibrium and Nash equilibria.

Table 1.1 Simplified Two-User Two-Channel Game

		User 1	
		c_1	c_2
User 2	c_1	$(\hat{r}_{11}, \hat{r}_{21})$	(r_{12}, r_{21})
	c_2	(r_{11}, r_{22})	$(\hat{r}_{12}, \hat{r}_{22})$

- *Case 1—Symmetric channels.* If the desired channels are identical for each user ($r_{i1} = r_{i2}, \forall i \in \{1, 2\}$), then the game has two NEs, which are $\{s_1 = c_1, s_2 = c_2\}$ and $\{s_1 = c_2, s_2 = c_1\}$. Note that in this case, the interference channels do not have to be identical.
- *Case 2—Asymmetric channels and high interference.* If channel 1 is better for both users than channel 2 ($r_{i1} > r_{i2}, \forall i \in \{1, 2\}$), and the interference channels are strong for user 1 but weak for user 2 such that $\hat{r}_{11} < r_{12}, \hat{r}_{21} > r_{22}$, then the game has a unique NE, which is $\{s_1 = c_2, s_2 = c_1\}$.
- *Case 3—Asymmetric channels and low interference.* If channel 1 is better for both users than channel 2 ($r_{i1} > r_{i2}, \forall i \in \{1, 2\}$), and the interference channels are weak such that $\hat{r}_{11} > r_{12}, \hat{r}_{21} > r_{22}$, then the game has a unique NE, which is $\{s_1 = c_1, s_2 = c_1\}$.

There are some other cases to discuss, which are left to the readers.

There is a special case of NE, which can be shown in the generalized two-user two-channel game. Assume that the number of channels that each user can transmit on is not limited. However, the power budget on each channel is fixed for each user (power is nontransferable between users or channels). Then the game can be described as in Table 1.2, where strategy 0 corresponds to allocating no power on any channel and strategy $c_1 + c_2$ corresponds to allocating power on both channels.

One feature of this game is that there is a unique NE regardless of the channel conditions. Indeed, given any choice that one player makes, the other player will choose the strategy of using all the channels, that is, it has a fixed best strategy. Such an NE is recognized as a dominant strategy equilibrium and can be formally described as

$$u_i \left(s_i^{\mathrm{D}}, s_i \right) > u_i \left(s_i', s_i \right) \quad \forall s_i \quad \forall s_i' \neq s_i^{\mathrm{D}} \tag{1.3}$$

Table 1.2 Generalized Two-User Two-Channel Game

		User 1			
		0	c_1	c_2	$c_1 + c_2$
User 2	0	$(0, 0)$	$(r_{11}, 0)$	$(r_{12}, 0)$	$(r_{11} + r_{12}, 0)$
	c_1	$(0, r_{21})$	$(\hat{r}_{11}, \hat{r}_{21})$	(r_{12}, r_{21})	$(\hat{r}_{11} + r_{12}, \hat{r}_{21})$
	c_2	$(0, r_{22})$	(r_{11}, r_{22})	$(\hat{r}_{12}, \hat{r}_{22})$	$(r_{11} + \hat{r}_{12}, \hat{r}_{22})$
	$c_1 + c_2$	$(0, r_{21} + r_{22})$	$(\hat{r}_{11}, \hat{r}_{21} + r_{22})$	$(\hat{r}_{12}, r_{21} + \hat{r}_{22})$	$(\hat{r}_{11} + \hat{r}_{12}, \hat{r}_{21} + \hat{r}_{22})$

A dominant strategy equilibrium, if exists, is the unique NE in the game. However, the utilities resulted for the players may not be "dominant." Consider the same example as in Table 1.2. If the interference between users are very strong on both channels such that ($\hat{r}_{11} + \hat{r}_{12} < r_{11}$ and $\hat{r}_{21} + \hat{r}_{22} < r_{22}$), the strategy set $\{s_1 = c_1 + c_2, s_2 = c_1 + c_2\}$ will be inferior to the strategy set $\{s_1 = c_1, s_2 = c_2\}$.

The inefficiency of the NE is due to the fact that there is no cooperation among the players. The lack of cooperation usually leads to inefficient resource allocation in multiuser systems. We will next introduce the so-called Nash bargaining (NB) games in multiuser systems in order to improve the system performance.

1.1.2 Cooperative Games: Nash Bargaining and Other Bargaining Solutions

To achieve better payoffs, users may resort to cooperation. By sharing some information, players can determine whether there are potentially extra utilities for everyone if they cooperate. If there are such extra utilities, players may bargain with each other to decide how to share them. Otherwise, they come back to the noncooperative state.

For cooperative games, the Nash axiomatic bargaining theory states that, in a convex utility space, there is a unique point that satisfies four specific axioms and maximizes the Nash function defined as

$$F = \prod_{i \in \Omega} \left(u_i - u_i^{NC} \right) \tag{1.4}$$

where
u_i^{NC} is the utility that user i obtains in the noncooperative case
the point $\left(u_1^{NC}, \ldots, u_M^{NC} \right)$ is known as the disagreement point with M being the number of players

Readers are referred to [17] for details of the four axioms.

Nash axiomatic bargaining focuses on describing the properties of the final solution of a cooperative game. However, the manner of cooperation based on which the users cooperate to reach the solution is not specified. Thus, the NB solution in a specific game depends on the manner of cooperation. For example, wireless users may perform time division multiplexing (TDM) or frequency division multiplexing (FDM) in a cooperative game.

Example 1.2

Assume that there is one communication channel and two users, and denote the information rate of user i (user i's utility) that can be achieved by using the channel exclusively as R_i. Now let the users cooperate by using the channel alternatively in time. User 1 uses a fraction α of the time and obtains information rate αR_1 as its utility, while user 2 uses a fraction $1 - \alpha$ of the time and obtains information rate $(1 - \alpha)R_2$ as its utility. There are three following possible cases.

- *Case 1.* If the disagreement point is $(0, 0)$, then the maximization of the Nash function (1.4) satisfies the so-called proportional fairness principle [18].
- *Case 2.* If the disagreement point is (R_1', R_2') and there exists $0 \leq \alpha \leq 1$ such that $\alpha R_1 > R_1'$ and $(1 - \alpha)R_2 > R_2'$, then the game has an NB solution.

Table 1.3 The "Fairness"of NB Solution

(R_1, R_2)	Disagreement Point	α in NB Solution (%)
(10,10)	(3,3)	50
(10,8)	(3,2)	52.5
(10,5)	(3,1)	55

- *Case 3.* If the disagreement point is (R_1', R_2') and there does not exist $0 \leq \alpha \leq 1$ such that $\alpha R_1 > R_1'$ and $(1-\alpha)R_2 > R_2'$, then the game does not have an NB solution if the TDM manner of cooperation is assumed.

Case 1 shows that the NB is related to a certain fairness principle, which justifies the necessity of cooperation among users. Case 2 states that all users should be able to improve their utilities to achieve a cooperative solution. Case 3 illustrates that the existence of the NB solution is not guaranteed in a cooperative game under a specific manner of cooperation.

Note that the proportional fairness of the NB solution does not mean that it must be always fair. In Table 1.3, the parameters in Example 1.2 are given different values and it is shown that the NB solution sometimes favors the user with better utility in the disagreement point.

One main limitation of the NB approach is that it requires convex utility spaces. In multiuser wireless systems, information rates are usually chosen as users' utilities. However, the interference among the users always renders the utility space (rate region) non-convex. The most popular approach to transfer the non-convex utility space to a convex one in multiuser wireless communication systems is to use orthogonal signaling such as time division multiple access (TDMA), frequency division multiple access (FDMA), or both. The efficiency of such methods will be discussed later.

There are several other important results on cooperative games, such as Kalai–Smorodinsky and Egalitarian solutions, which also deal only with convex games [19]. In the literature, there are limited research efforts that extend the Nash axiomatic bargaining, Kalai–Smorodinsky, and Egalitarian solutions to certain non-convex games [20,21].

1.1.3 Multiuser Systems: Generalized Signal Model and Game Model

In this section, a generalized model for many games played in multiuser systems is given. It will be specified and investigated in details in subsequent sections. Our focuses are the resource-sharing games, which are well studied in the literature in wireless communications. The following assumptions are adopted:

1. Players are the wireless users in a communication system. Codebook at each transmitter is assumed to be Gaussian codebook, and the achievable information rate between each transceiver pair is selected as the user's utility. Players' strategies depend on specific setups of the problem.
2. Users have to interact with each other due to the existence of interference among them, which makes game theory applicable. Interference perceived by a user at its receiver side is treated as noise. No interference canceling decoding is adopted for neither noncooperative games nor cooperative games.

3. The transmission channels of each user are known at both receiver and transmitter sides, that is, receivers are able to transmit the channel information back to their transmitters without errors.

Generalized Signal Model

Assume that there are M wireless users in a communication system. We have the following model:

$$\mathcal{Y}_i = \mathcal{H}_{ii}\mathcal{Q}_i\mathcal{S}_i + \sum_{j\neq i}\mathcal{H}_{ji}\mathcal{Q}_j\mathcal{S}_j + \mathcal{N}_i \quad \forall i \in \Omega = \{1, 2, \ldots, M\} \tag{1.5}$$

where

$\mathcal{Y}_i$ is the received symbol (or symbol vector) for user i

$\mathcal{S}_i$ is the information symbol (or symbol vector) to be transmitted for user i

$\mathcal{Q}_i$ can be the power allocation parameter/beamforming vector/precoding matrix of user i depending on the setup of the problem

$\mathcal{H}_{ji}$ is the channel between transmitter of user j and receiver of user i

$\mathcal{N}_i$ is the additive white Gaussian noise

It is further assumed that $E\left\{\mathcal{S}_i\mathcal{S}_i^H\right\} = \mathbf{I}$ (information symbols are uncorrelated and have unit-energy) and $E\left\{\mathcal{N}_i\mathcal{N}_i^H\right\} = \sigma_i^2\mathbf{I}$ (noise is white with variance σ_i^2), where $\mathbf{I}$ and $(\cdot)^H$ stand for the identity matrix and the Hermitian transpose, respectively. The dimensions of $\mathcal{Y}_i, \mathcal{S}_i, \mathcal{Q}_i, \mathcal{H}_{ji}$, and $\mathcal{N}_i$ depend on the specific problem setup, such as the channel fading conditions or number of antennas at the transceiver pairs. These parameters will be specified later in each example considered.

The corresponding game model for the generalized signal model is as follows:

Generalized Game Model

$$\Gamma = \left\{\Omega = \{1, 2, \ldots, M\}, \{\mathcal{Q}_i | i \in \Omega\}, \{R_i | i \in \Omega\}\right\} \tag{1.6}$$

where R_i is the achievable information rate for user i under the strategy set $\{\mathcal{Q}_i | i \in \Omega\}$.

The generalized signal and game models provide a unified structure for many games in wireless systems. In subsequent sections, the generalized models (1.5) and (1.6) will be specified to investigate the power allocation, beamforming, precoding games, etc. We start from the most basic games—power allocation games.

1.2 Power Allocation Games: Competition versus Cooperation

Power allocation games are the most fundamental and well-studied games in wireless communications. Games involving more complex signal-processing techniques can be transferred to power allocation games, as we will show later. In this section, several power allocation games are studied for both flat fading and frequency selective fading channels. Both the noncooperative and cooperative cases are covered.

1.2.1 Power Allocation on Flat Fading Channels

Assume that there are M wireless users sharing a channel with bandwidth W, which is flat fading for each user. The channel gain from transmitter j to receiver i is h_{ji}, and the power budget for user i is p_i. Then the generalized signal model (1.5) in this case can be specified as follows:

Signal Model for the M-User Flat Fading Channel

$$y_i = h_{ii}\sqrt{p_i}s_i + \sum_{j \neq i} h_{ji}\sqrt{p_j}s_j + n_i \quad \forall i \in \Omega \tag{1.7}$$

Note that the users' strategies are not shown in the signal model. The strategy is characterized by the bandwidth w_i occupied by each user. The corresponding game model is as follows.

Game Model for the M-User Flat Fading Channel

$$\Gamma = \left\{\Omega = \{1, 2, \ldots, M\}, \{w_i | 0 < w_i \leq W\}, \{R_i\}\right\} \tag{1.8}$$

Unlike the discrete games in the brief survey, this game is a continuous game. First, consider the game in the noncooperative case. Similar to the discussion in the preceding section, the result of the noncooperative game depends on the channel parameters. Following cases summarize some typical results. The detailed proof is omitted and the readers are referred to [22–24].

- *Case 1*. If the interference channels h_{ji} are weak, there exists a unique NE in which user i's strategy is $w_i = W, \forall i \in \Omega$. The utility for user i in this case is

$$R_i = \frac{W}{2} \log_2 \left(1 + \frac{|h_{ii}|^2 p_i}{W\sigma^2 + \sum_{j \neq i} |h_{ji}|^2 p_j}\right) \tag{1.9}$$

- *Case 2*. When the interference channels become stronger, there may exist more than one NE. As the interference keeps increasing, the competitive solution converges to FDM.

It is straightforward to see that the NE in Case 1 is actually a dominant strategy equilibrium. The intuition behind these two cases is that it is more beneficial for the players to use a larger bandwidth when the interference from other users is weak. On the other hand, avoiding the interference leads to a better payoff when the potential interference is very high.

Next, we consider the cooperative case. A simple manner for the users to cooperate is FDM [24]. The assumption of FDM adds additional constraints to the utility set of the game. Practically, the users should use nonoverlapping frequency bands and $\sum_i w_i \leq 1$. Equivalently, a user's strategy can be defined as the portion of the whole bandwidth that the user obtains. If user i obtains a fraction α_i of W, then $R_i(\alpha_i) = (\alpha_i W/2) \log_2(1 + |h_{ii}|^2 p_i / \alpha_i W\sigma^2)$ in the cooperative game case.

In this specific game, the Nash function is

$$F(\boldsymbol{\alpha}) = \prod_i \left(R_i(\alpha_i) - R_i^{\text{NC}}\right) \tag{1.10}$$

where $\boldsymbol{\alpha} = [\alpha_1, \ldots, \alpha_M]$ and R_i^{NC} is decided by the disagreement point. The NB solution can be derived by solving the following optimization problem:

$$\max_{\boldsymbol{\alpha}} \quad \prod_i \left(R_i(\alpha_i) - R_i^{\text{NC}}\right)$$

$$\text{s.t.} \quad 0 < \alpha_i < 1 \; \forall i \tag{1.11}$$

$$\sum_i \alpha_i = 1 \; \forall i$$

Table 1.4 Users' Utilities (R_1, R_2) in Two-User Flat Fading Channel Games

	$h = 0.1$	$h = 0.3675$	$h = 1$
$w_1 = w_2 = 1$	(2.0715, 2.0715)	(1.3394, 1.3394)	(0.4826, 0.4826)
$w_1 = w_2 = 0.5$		(1.3394, 1.3394)	

Note that the equality in the second constraint follows from the fact that the objective $F(\alpha)$ can be further increased if $\sum_i \alpha_i < 1$.

The NB solution exists in this game if and only if there exists $\alpha > 0$ such that $\sum_i \alpha_i \le 1$ and $R_i(\alpha_i) > R_i^{NC}$, where $>$ denotes "larger than" in element-wise comparison. In other words, the NB solution exists if and only if all users are able to improve their utilities using FDM.

It may appear contradictory that the previously discussed noncooperative NE solution may converge to FDM, which is assumed in the cooperative NB game. However, unlike the cooperative case, the users are driven to FDM in their pursuits of individual benefits when the interference is high in the noncooperative game. Moreover, generally the partitions of the whole bandwidth are different for the FDM-based NB and the NE in FDM.

The efficiency of the NB solution depends on the manner of cooperation assumed. For example, FDM is highly inefficient in a low-interference multiuser system.

Example 1.3

A comparison between the strategy in which both users choose to use the whole bandwidth and interfere with each other and the strategy in which the users share the bandwidth using FDM is listed in Table 1.4. Particularly, assuming a symmetric system in which $M = 2$, $W = 1$, $p_1 = p_2 = 10$, $\sigma^2 = 1$, $|h_{11}| = |h_{22}| = 1$, $|h_{12}| = |h_{21}| = h$, the users' utilities under different strategy set $\{w_i\}$ are shown in Table 1.4.

In the table, the noncooperative strategy $w_1 = w_2 = 1$ is the dominant strategy equilibrium when $h = 0.1$. The FDM-based NB solution does not exist when the interference channels are very weak (e.g., when (2.0715, 2.0715) is the noncooperative NE solution for the case $h = 0.1$). The FDM-based cooperative solution becomes comparatively more and more efficient when the interference channel becomes stronger. For the case when the interference between the users is strong enough, for example, $h = 1$, the NB solution of the FDM cooperative game exists under strategies $w_1 = w_2 = 0.5$.

The fundamental reason behind the fact that FDM (and some other manners of cooperation) can be inefficient is that it reduces the utility space of the game while making it convex and, thus, limits the solution to belong to only a subset of the original utility space. We will introduce a different manner of cooperation, which enlarges the utility space and produces a "convex hull" in the subsequent sections.

1.2.2 Power Allocation on Frequency Selective Fading Channels

On frequency selective fading channels or equivalently inter-symbol interference (ISI) channels, multiple users have to spread their power over frequency bins of the wideband channel. The optimal power allocation scheme on a frequency selective fading channel for a single user can be derived using the well-known water-filling algorithm. However, the competition for resource arises when there are more than one user in the system.

Assume that there are M users, and the frequency selective fading channel can be decoupled into N frequency bins, each of which is flat fading for all users. The signal model (1.5) in this case can be specified as follows:

Signal Model for the M-User Frequency Selective Fading Channel

$$\mathbf{y}_i = \mathbf{H}_{ii}\sqrt{\mathbf{p}_i} \odot \mathbf{s}_i + \sum_{j \neq i} \mathbf{H}_{ji}\sqrt{\mathbf{p}_j} \odot \mathbf{s}_j + \mathbf{n}_i, \quad \forall i \in \Omega \qquad (1.12)$$

where

$\mathbf{y}_i = [y_i(1), \dots, y_i(N)]$ is the received symbol vector for user i on the frequency bins 1 to N

$\mathbf{s}_i = [s_i(1), \dots, s_i(N)]$ is the information symbol vector transmitted by user i

$\mathbf{p}_i = [p_{i1}, \dots, p_{iN}]$ is the power allocation vector of user i on the N frequency bins

$\mathbf{H}_{ji}$ is the $N \times N$ diagonal channel matrix with its kth diagonal element $h_{ji}(k)$ denoting the sampled channel gain of the kth frequency bin between transmitter of user j and receiver of user i

$\mathbf{n}_i$ is the additive white Gaussian noise for user i

$\odot$ denotes the Hadamard product

Note that the assumptions made on $\mathbf{s}_i$ and $\mathbf{n}_i$ in the generalized signal model in Section 1.1.3 are inherited here.

Power Allocation Game under Spectral Mask Constraints

On frequency selective fading channels, the so-called spectral mask (also known as power spectral density [PSD] mask) constraints are typically adopted. These constraints can be written as

$$\mathbf{p}_i \preceq \mathbf{p}_i^{\max} \quad \forall i \in \Omega \qquad (1.13)$$

where

$\mathbf{p}_i^{\max} = \left[p_{i1}^{\max}, \dots, p_{iN}^{\max} \right]$ is the spectral mask for user i

$\preceq$ denotes "less than or equal to" in element-wise comparison

In the noncooperative case, the game model corresponding to the signal model under the spectral mask constraints is as follows:

Game Model (Noncooperative, Spectral Mask) for the M-User Frequency Selective Fading Channel

$$\Gamma = \left\{ \Omega = \{1, 2, \dots, M\}, \left\{ \mathbf{p}_i \middle| 0 \preceq \mathbf{p}_i \preceq \mathbf{p}_i^{\max} \right\}, \{R_i\} \right\} \qquad (1.14)$$

where $R_i = \sum_k \log_2(1 + (|h_{ii}(k)|^2 p_{ik})/(\sigma_i^2(k) + \sum_{j \neq i} |h_{ji}(k)|^2 p_{jk}))$ and $\sigma_i(k)^2$ is the noise power for user i on frequency bin k.

It is straightforward to see that this noncooperative game has a dominant strategy equilibrium, which is $\mathbf{p}_i = \mathbf{p}_i^{\max}$. The proof is based on the fact that if one user does not use maximum power on all the frequency bins, it can always improve its utility by changing to the strategy of using maximum power on all the frequency bins. The details of the proof are left to the readers.

In the cooperative case, the users can cooperate by adopting the joint TDM/FDM scheme [25]. In the joint TDM/FDM scheme

1. All frequency bins are shared among the users in time domain, that is, each frequency bin can be used by different users over different time intervals (TDM part).
2. All frequency bins are partitioned and allocated to the users, while only one user is allowed on any frequency bin at any given time (FDM part).

The joint TDM/FDM scheme generates a convex utility space for all users to perform NB. Since the users' utilities depend only on the lengths of time durations which the users have obtained on the frequency bins, the users' strategies can be defined in terms of their TDM/FDM coefficients as

$$\boldsymbol{\alpha}_i = [\alpha_{i1}, \dots, \alpha_{iN}] \quad \forall i \in \Omega \tag{1.15}$$

where α_{ik} is the proportion of time that is allocated to user i on frequency bin k.

The cooperative game model can be, then, given as follows:

Game Model (Cooperative, Spectral Mask) for the M-User Frequency Selective Fading Channel

$$\Gamma = \left\{ \Omega = \{1, 2, \dots, M\}, \left\{ \boldsymbol{\alpha}_i \middle| \boldsymbol{\alpha}_i \succ 0, \sum_i \boldsymbol{\alpha}_i \preceq 1 \right\}, \{R_i(\boldsymbol{\alpha}_i)\} \right\} \tag{1.16}$$

where $R_i(\boldsymbol{\alpha}_i) = \sum_k \alpha_{ik} \log_2(1 + |h_{ii}(k)|^2 p_{ik}^{\max}/\sigma_i^2(k))$.

The corresponding NB solution can be obtained by solving the following convex optimization problem:

$$\begin{aligned} \max_{\{\boldsymbol{\alpha}_i\}} \quad & \sum_i \log\left(R_i(\boldsymbol{\alpha}_i) - R_i^{\text{NC}}\right) \\ \text{s.t.} \quad & \boldsymbol{\alpha}_i \succ 0 \quad \forall i \\ & \sum_i \boldsymbol{\alpha}_i = 1 \quad \forall i \\ & R_{i(\boldsymbol{\alpha}_i)} > R_i^{\text{NC}} \quad \forall i \end{aligned} \tag{1.17}$$

The objective function in (1.17) is the logarithm of the Nash function. The first two constraints are the requirements of the joint TDM/FDM, while the third constraint states that the NB solution exists if and only if all users can benefit from the joint TDM/FDM–based cooperation.

Power Allocation Game under Total Power Constraints

The games under total power constraints are different and, in fact, more complex for both noncooperative and cooperative cases. It is because the available power for all users is bounded by the total power constraints in this case.

The noncooperative game model under total power constraints for the same signal model (1.12) can be expressed as follows:

Game Model (Noncooperative, Total Power) for the M-User Frequency Selective Fading Channel

$$\Gamma = \left\{ \Omega = \{1, 2, \ldots, M\}, \left\{ \mathbf{p}_i \,\middle|\, \sum_k p_{ik} \leq P_i^{\max} \right\}, \{R_i\} \right\} \tag{1.18}$$

where
   $P_i^{\max}$ is the maximum power that user i can use in total
   R_i is the same as in (1.14)

Fortunately, this game is a convex game, and, therefore, it has at least one NE [26]. Thus, the existence of an NE is guaranteed. Practically, iterative water-filling (including sequential water-filling and parallel water-filling) algorithms work well for finding NE of such games. However, there is no guarantee that the water-filling algorithms can converge. There are examples in which both sequential water-filling and parallel water-filling may diverge while searching for an NE of a simple game with a unique NE [25].

The cooperative case is even more complex. Assume again that the users cooperate according to the joint TDM/FDM scheme. Then the users' strategies are composed of both the TDM/FDM coefficients and the power allocation coefficients on the frequency bins due to the total power constraints. The cooperative game model under total power constraints in this case is as follows:

Game Model (Cooperative, Total Power) for the M-User Frequency Selective Fading Channel

$$\Gamma = \left\{ \Omega = \{1, 2, \ldots, M\}, \left\{ \mathbf{p}_i, \boldsymbol{\alpha}_i \,\middle|\, \sum_k \alpha_{ik} p_{ik} \leq P_i^{\max} \right\}, \{R_i\} \right\} \tag{1.19}$$

where
   $\boldsymbol{\alpha}_i$ has the same meaning as in the cooperative game under spectral mask constraints
   $R_i(\boldsymbol{\alpha}_i) = \sum_k \alpha_{ik} \log_2(1 + |h_{ii}(k)|^2 p_{ik}/\sigma_i^2(k))$

The challenge in this problem lies in the fact that the maximization of the Nash function is a non-convex optimization problem. A water-filling-based algorithm can still be used to search for the NB solution in the simplified two-user version of this model [27]. The algorithm bargains in many different convex subspaces of the original utility space and obtains one NB solution in each subspace. Then the largest of the NB solutions is selected as the final NB solution of the game. However, the complexity of such an algorithm is high even for the two-user case and the algorithm cannot be extended to the M-user ($M > 2$) games.

Similar to the FDM-based cooperative game on the flat fading channel, the joint TDM/FDM–based cooperation can be inefficient when the interference among users is weak.

There are some other models for cooperative power allocation games. For example, in [28], a cooperative game similar to the game on the frequency selective fading channel with total power constraints is studied. In [28], neither TDM nor FDM is assumed. Instead, the users maximize the Nash function iteratively given the strategies of other users. The algorithm is proved to converge, however, not necessarily to the global optimal point.

Power allocation games are basic, since many complex games can be transformed to equivalent power allocation games. The power allocation games introduced in this section pave the way to the higher-level games in the following two sections. A power allocation cooperative game under both spectral mask and total power constraints is considered in [29].

1.3 Beamforming Games on MISO Channels

The games in the preceding section are all games on single-input single-output (SISO) channels. Note that games involving multiple antennas have also been studied in the literature. The focus in this section is to investigate the games played on multi-input single-output (MISO) channels where the users' strategies are defined as their beamforming vectors.

1.3.1 Beamforming on the MISO Channel: Noncooperative Games

Assume that there are M wireless users in a system, all communication channels are flat fading, and each user has N antennas at the transmitter side and a single antenna at the receiver side. All users interfere with each other if they communicate simultaneously, and this setup constitutes an MISO interference channel. The signal transmission model in this case is as follows [30].

Signal Model for the Beamforming on M-User MISO Channel

$$y_i = \mathbf{h}_{ii}^T \mathbf{w}_i s_i + \sum_{j \neq i} \mathbf{h}_{ji}^T \mathbf{w}_j s_j + n_i \quad \forall i \in \Omega \tag{1.20}$$

where
  y_i is the received symbol for user i
  s_i is the information symbol transmitted by user i
  $\mathbf{w}_i$ is the beamforming vector of user i
  $\mathbf{h}_{ji} = [h_{ji}(1), \ldots, h_{ji}(N)]$ is the $N \times 1$ channel vector between transmitter of user j and receiver
    of user i
  n_i is the additive white Gaussian noise for user i
  $(\cdot)^T$ stands for transpose

The assumptions made on $\mathbf{s}_i$ and n_i in the generalized signal model in Section 1.1.3 are also inherited here.

First, consider a two-user beamforming game and assume that the number of transmit antennas is only two (this is a practically important case). Moreover, let all users be subject to the total power constraint $\|\mathbf{w}_i\|^2 \leq P, \forall i \in \Omega$. The game model in this case can be written as follows:

Game Model (Noncooperative) for the Beamforming on Two-User MISO Channel

$$\Gamma = \left\{ \Omega = \{1, 2\}, \{\mathbf{w}_i | \ \|\mathbf{w}_i\|^2 \leq P\}, \{R_i\} \right\} \tag{1.21}$$

where $R_i = \log_2(1 + |\mathbf{w}_i^T \mathbf{h}_{ii}|^2 / (|\mathbf{w}_j^T \mathbf{h}_{ji}|^2 + \sigma^2))$, $\{j\} = \Omega \setminus \{i\}$.

It can be shown that this game has a unique NE (which is actually a dominant strategy equilibrium), and user i's strategy in the NE is $\mathbf{w}_i^{\text{NE}} = \mathbf{h}_{ii}^* / \|\mathbf{h}_{ii}\|$ where $(\cdot)^*$ denotes complex conjugate [30]. The proof here is similar to that in the power allocation game on the frequency selective fading channel with total power constraints in the previous section.

The same result can be derived in much more general cases, for example, in the cases when

1. The number of users is arbitrary.
2. The number of antennas at each transmitter is arbitrary and can be different for different users.
3. The power constraints are different for different users.

Note that NE may be a poor solution for all users, especially when the interference among users is strong. Thus, cooperation is a better choice for both users if the communication system can afford the related overhead.

1.3.2 Cooperative Beamforming: Time Sharing, Convex Hull, and Nash Bargaining

Recall that the NB solutions are only defined for games with convex utility spaces. In the cooperative power allocation games, FDM or joint TDM/FDM schemes are used to generate convex subspaces of the original utility spaces in order to perform NB. The inefficiency is noticed when the interference among users is weak, or equivalently, when the convex sub-spaces are quite small compared to the original utility spaces. Unlike the FDM or join TDM/FDM, here a new manner of cooperation is introduced in [30], which enlarges the original utility space by generating a *convex hull* of the original space to perform bargaining.

Denote the original utility space (rate region) of the two-user game with constraints $\|\mathbf{w}_i\|^2 \leq P$, $\forall i \in \Omega$ as $\mathcal{R} = \{R_1, R_2\}$. The convex hull of the game can be obtained by performing time sharing

$$\tilde{\mathcal{R}} = \left\{ \beta R_1^1 + (1 - \beta)R_1^2, \beta R_2^1 + (1 - \beta)R_2^2 \right\} \tag{1.22}$$

where
(R_1^1, R_2^1) and (R_1^2, R_2^2) are two points in $\mathcal{R}$
β is the time-sharing coefficient

In the first β portion of time, both users choose their strategies $\mathbf{w}_i^1$, $\forall i \in \Omega$ and get payoff βR_i^1, $\forall i \in \Omega$. Then in the remaining $1 - \beta$ portion of time, the users choose strategies $\mathbf{w}_i^2$, $\forall i \in \Omega$ and get payoff $(1 - \beta)R_i^2$, $\forall i \in \Omega$. Thus, the overall payoff for user i is $\tilde{R}_i = \beta R_i^1 + (1 - \beta)R_i^2$, $\forall i \in \Omega$.

It can be proved that $\tilde{\mathcal{R}}$ is a convex set, which includes $\mathcal{R}$ as its subset. Therefore, the bargaining can be performed in a larger space. However, there is a price paid for generating the convex hull using time sharing. Indeed, the users have to determine their strategies twice along with the time-sharing coefficient β. Thus, the number of variables of the NB problem to be solved is increased.

The model for the two-user cooperative time-sharing game is as follows.

Game Model (Cooperative, Time Sharing) for the Beamforming on Two-User MISO Channel

$$\Gamma = \left\{ \Omega = \{1, 2\}, \left\{ \mathbf{w}_i^l; \beta \left| \|\mathbf{w}_i^l\|^2 \leq P, l = 1, 2; 0 \leq \beta \leq 1 \right\}, \{\tilde{R}_i\} \right\} \tag{1.23}$$

Then, the NB solution can be obtained from solving the following optimization problem:

$$
\max_{\{\mathbf{w}_i^l\},\beta} \quad \sum_i \log\left(\tilde{R}_i - R_i^{\mathrm{NC}}\right)
$$

$$
\text{s.t.} \quad \left\|\mathbf{w}_i^l\right\|^2 \le P \quad \forall i \quad \forall l \tag{1.24}
$$

$$
0 \le \beta \le 1
$$

$$
\tilde{R}_i > R_i^{\mathrm{NC}} \quad \forall i
$$

Cooperative solutions on MISO channels have also been studied using other bargaining theories. For example, Nokleby and Swindlehurst [31] investigate the cooperative bargaining game on the MISO channel and derive the Kalai–Smorodinsky solution.

1.3.3 Pareto-Optimality: Competition and Cooperation

One property of the NB solution is its Pareto-optimality. However, it is not the only Pareto-optimal point in the utility space. In fact, all points in the Pareto boundary of the corresponding utility space are Pareto-optimal. For multiuser games in wireless systems, the points on the Pareto boundary generally represent efficient allocations of communication resources. For the above studied games on the two-user MISO channels, an interesting result is that any Pareto-optimal point can be realized through certain balance of users' competition and cooperation [32,33]. To explain the proposition, we start by considering the NE and the zero-forcing (ZF) beamforming strategies.

Recall that the NE for the two-user noncooperative beamforming game is $\mathbf{w}_i^{\mathrm{NE}} = \mathbf{h}_{ii}^*/\|\mathbf{h}_{ii}\|$ [30]. The NE strategy can be viewed as completely competitive and selfish because the user that uses this strategy aims at maximizing its own payoff only. On the other hand, a user is considered as altruistic if it adopts the strategy that generates no interference to other users. Such a strategy is known as the ZF strategy. Note that the ZF strategy for the two-user beamforming game can be expressed as [32]

$$
\mathbf{w}_i^{\mathrm{ZF}} = \frac{\mathbf{\Pi}_{\mathbf{h}_{ij}^*}^{\perp}\mathbf{h}_{ii}^*}{\left\|\mathbf{\Pi}_{\mathbf{h}_{ij}^*}^{\perp}\mathbf{h}_{ii}^*\right\|} \quad \forall i \in \Omega, \{j\} = \Omega\backslash\{i\} \tag{1.25}
$$

where $\mathbf{\Pi}_{\mathbf{A}}^{\perp} = \mathbf{I} - \mathbf{A}(\mathbf{A}^H\mathbf{A})^{-1}\mathbf{A}^H$ is the orthogonal projection onto the orthogonal complement of the column space of $\mathbf{A}$.

Indeed, if both users choose the ZF strategies, they obtain nonoptimal payoffs for themselves yet avoid interfering with each other. Thus, the ZF strategies can be considered as completely cooperative and altruistic strategies. Note that in general, the NE and ZF strategies do not lead to the Pareto-optimal solutions. However, it is proved that any point on the Pareto-optimal boundary can be achieved using a certain combination of the NE and the ZF. This combination can be formally expressed as [32]

$$
\mathbf{w}_i(\lambda_i) = \frac{\lambda_i\mathbf{w}_i^{\mathrm{NE}} + (1 - \lambda_i)\mathbf{w}_i^{\mathrm{ZF}}}{\left\|\lambda_i\mathbf{w}_i^{\mathrm{NE}} + (1 - \lambda_i)\mathbf{w}_i^{\mathrm{ZF}}\right\|} \quad \forall i \in \Omega \tag{1.26}
$$

where λ_i is a parameter satisfying $0 \le \lambda_i \le 1$.

The detailed proof is omitted here. However, intuitively, one can understand it in the following way. To achieve global efficiency, each user has to combine its own utilities and all other users' utilities in its objective to be maximized. Considering NB as an example, the NB solution is Pareto-optimal in a convex utility space because every user maximizes the Nash function, which combines all players' utilities. In fact, if each user maximizes the weighted Nash function defined as

$$F = \prod_i \left(R_i - R_i^{\mathrm{NC}} \right)^{\gamma_i} \qquad (1.27)$$

where $\gamma_i > 0, \forall i$, then any point on the Pareto-optimal boundary can be achieved using the weighted NB by varying γ_i's. Using the ZF strategy, a user actually aims at maximizing the achievable utility of other users, while using the NE strategy, a user aims at maximizing its own utility only. Thus, the combination of the ZF and the NE strategies covers the utilities of all players in the game and can lead to Pareto-optimal solutions.

1.4 Matrix Games: Precoding Games and Others

In the previous sections, we have introduced power allocation games on SISO channels and beamforming games on MISO channels. Next, we proceed to the games on multiple-input multiple-output (MIMO) channels. It will be shown that such games can be transformed into power allocation games under the assumption of orthogonal signaling and, thus, can be solved using the aforementioned methods.

1.4.1 Precoding Games on the MIMO Channels

Consider an M-user communication system based on OFDM. The signal transmission model can be written as follows [34,35].

Signal Model for the Precoding on M-User MIMO Channel

$$\mathbf{y}_i = \mathbf{H}_{ii}\mathbf{F}_i\mathbf{s}_i + \sum_{j \neq i} \mathbf{H}_{ji}\mathbf{F}_j\mathbf{s}_j + \mathbf{n}_i \quad \forall i \in \Omega \qquad (1.28)$$

where
  $\mathbf{y}_i$ is the received symbol block for user i
  $\mathbf{s}_i$ is the information symbol block transmitted by user i
  $\mathbf{F}_i$ is the precoding matrix of user i
  $\mathbf{H}_{ji}$ is the $N \times N$ channel matrix between transmitter of user j and receiver of user i
  $\mathbf{n}_i$ is the additive white Gaussian noise for user i

Note that the assumptions made on $\mathbf{s}_i$ and $\mathbf{n}_i$ in the generalized signal model are inherited here.

In an OFDM system, cyclic prefixes (CP) are used to cancel the ISI and decouple the wideband frequency selective fading channel into a number of flat fading frequency bins. The channel matrix $\mathbf{H}_{ji}$ can be diagonalized as $\mathbf{H}_{ji} = \mathbf{W}\mathbf{D}_{ji}\mathbf{W}^H$ due to the CP insertion, where $\mathbf{W}$ is the $N \times N$ discrete Fourier transform (DFT) matrix and $\mathbf{D}_{ji}$ is the diagonal sampled channel matrix between transmitter of user j and receiver of user i.

The transmitters in the system are subject to both total power and spectral mask constraints, that is,

1. *Total power constraints:* $Tr\left(\mathbf{F}_i\mathbf{F}_i^H\right) \leq NP_i, \forall i \in \Omega$, where $Tr(\cdot)$ stands for the trace operator and P_i is the maximum available power for the transmission of a signal symbol.

2. *Spectral mask constraints:* $E\{|[\mathbf{W}^H\mathbf{F}_i\mathbf{s}_i]_k|^2\} = [\mathbf{W}^H\mathbf{F}_i\mathbf{F}_i^H\mathbf{W}]_{kk} \leq p_i^{max}(k), \forall i \in \Omega, \forall k \in \{1, \ldots, N\}$, where $E\{\cdot\}$ stands for the expectation and $p_i^{max}(k)$ is the maximum power that user i can allocate on frequency bin k.

Let $\mathfrak{F}_i$ be the strategy space of user i subject to the above constraints, then the noncooperative game model of this signal model can be written as follows:

Game Model (Noncooperative) for the Precoding on M-User MIMO Channel

$$\Gamma = \left\{\Omega = \{1, 2, \ldots, M\}, \{\mathbf{F}_i|\mathbf{F}_i \in \mathfrak{F}_i\}, \{R_i\}\right\} \qquad (1.29)$$

where $R_i = (1/N)\log_2\left(\left|\mathbf{I} + \mathbf{F}_i^H\mathbf{H}_{ii}^H\mathbf{R}_i^1\mathbf{H}_{ii}\mathbf{F}_i\right|\right)$ is the information rate of user i with $\mathbf{R}_i = \sigma^2\mathbf{I} + \sum_{j\neq i}\mathbf{H}_{ji}\mathbf{F}_j\mathbf{F}_j^H\mathbf{H}_{ji}^H$ representing the noise plus interference for user i.

The above game is a matrix game. However, it is proposed that this game can be transformed into a power allocation game due to the CP insertion. Practically, it is proved that the NE of the precoding game can be achieved using the following precoding strategies [34]:

$$\mathbf{F}_i = \mathbf{W}\sqrt{diag(\mathbf{p}_i)} \quad \forall i \in \Omega \qquad (1.30)$$

where $diag(\mathbf{a})$ is a square diagonal matrix with its kth diagonal elements as $[\mathbf{a}]_k$ and $\mathbf{p}_i$ can be found by solving the following power allocation game.

Game Model (Noncooperative) for the Precoding on M-User MIMO Channel, Equivalent Power Allocation Game

$$\Gamma = \left\{\Omega = \{1, 2, \ldots, M\}, \left\{\mathbf{p}_i\middle|\mathbf{p}_i \preceq \mathbf{p}_i^{max}, \sum_k \mathbf{p}_i(k) \leq NP_i\right\}, \{R_i\}\right\} \qquad (1.31)$$

where $\mathbf{p}_i^{max} = [p_i^{max}(1), \ldots, p_i^{max}(N)]$ is the spectral mask vector. The expressions for R_i's can also be simplified. However, we skip it here for the sake of brevity.

The equivalent game (1.31) is similar to the noncooperative power allocation game on the frequency selective fading channel (1.18). One difference is that this game considers both spectral mask and total power constraints. However, the water-filling-based algorithms are still applicable [35].

In what follows, for the cooperative case of the precoding game, only spectral mask constraints are considered for simplicity because the total power constraints will render the problem non-convex [36,37]. Adopting again the joint TDM/FDM scheme as the manner of cooperation among users, it can be observed that the diagonal structure as in (1.30) also applies to the cooperative case, and thus, the cooperative precoding game can be transformed to a cooperative power allocation game, which can be written as follows:

Game Model (Cooperative) for the Precoding on M-User MIMO Channel, Equivalent Power Allocation Game

$$\Gamma = \left\{\Omega = \{1, 2, \ldots, M\}, \left\{\mathbf{p}_i, \alpha_i\middle|\mathbf{p}_i \preceq \mathbf{p}_i^{max}, 0 \leq \alpha_i \leq 1, \sum_i \alpha_i \leq 1\right\}, \{R_i\}\right\} \qquad (1.32)$$

where $\boldsymbol{\alpha}_i = [\alpha_i(1), \ldots, \alpha_i(N)]$ is the TDM/FDM coefficient vector of user i as defined in (1.15).

The above game is similar to the cooperative power allocation game on the frequency selective fading channel (1.16). The difference is that this game is not a two-player game and the algorithm used for solving the two-player game cannot be extended to the multiuser case. This equivalent cooperative game can be solved in a distributed manner using a dual decomposition–based algorithm [37]. A non-convex game with both spectral mask and total power constraints is considered in [29].

1.4.2 Other Matrix Games

Besides the precoding games, some other matrix games have also been studied [38,39]. Specifically, games can be played on MIMO channels with the strategies defined as the users' signal covariance matrices subject to total power constraints [38]. For the noncooperative case, the MIMO water-filling algorithm based on singular value decomposition of the channel matrices and signal covariance matrices is adopted to obtain the NE of the game [38]. It is noticed that the algorithm, which works well practically, is not guaranteed to converge. Note here the similarity with the noncooperative power allocation game on the frequency selective fading channel under total power constraints. For the cooperative case, the gradient projection method is proposed for finding the NB solution of the game. However, the convergence can be only guaranteed to a local optimal point.

A game can also be played to allocate communication resources among the secondary users in an MIMO cognitive radio system [39]. The constraints can be set on each user's total power and the maximum interference that it can generate. It is observed that this matrix game can be transformed to a power allocation game while the interference constraints can be transformed to the rotations of the channel matrices. Moreover, the NE solution for the covariance matrices adopts the diagonal structure similar to (1.30). Again the MIMO water-filling-based algorithm can be used to derive the NE with the original channel matrices substituted by the rotated channel matrices.

1.5 Further Discussions

1.5.1 Efficiency and Fairness

The efficiency of the NE and NB solutions has been mentioned and compared many times in preceding sections. The examples lead to the conclusion that the FDM or joint TDM/FDM–based NB solutions are efficient when the potential interference among users is strong and vice versa. We have two supplementary remarks here.

First, the above conclusion is based on the assumption that the noise power is fixed for all users. Actually the efficiency of the FDM and joint TDM/FDM-based NB solutions depends on the interference to noise ratio (INR). If the interference power dominates the noise power, the FDM and joint TDM/FDM–based cooperations are efficient. Otherwise, they are inefficient because the FDM and joint TDM/FDM schemes cannot decrease the noise PSD.

Secondly, the efficiency of the FDM or joint TDM/FDM–based NB solutions is not equivalent to the efficiency of the NB solution. Indeed, the NB solution depends on the manner of cooperation assumed. For the same multiuser wireless system, different manners of cooperation can generate different NB solutions. For example, time sharing (recall the cooperative beamforming game on the MISO channel) can achieve better efficiency than joint TDM/FDM at the price of increasing the number of parameters in the corresponding optimization problem.

As for the fairness, it has been mentioned that the NB solution is related to the so-called proportional fairness. However, our previous example also shows that the NB solution favors the user with better utility in the disagreement point. On the other hand, NE are considered as completely selfish strategies. However, NE are not necessarily unfair. The fairness of NE, if any, depends on the system setup (wireless users' desired channels, interference channels, and the noise power). For example, in a symmetric system, an NE will be completely fair for all users. However, unlike the NB solution, NE may not be unique, and the discussion on fairness would be obviously meaningless in this case.

1.5.2 Uniqueness and Complexity

It is well known that in a convex utility space only one point maximizes the Nash function, while there may be multiple NE regardless of the convexity of the utility space. However, it is more accurate to conclude that the NB solution is unique if it exists. In fact, the NB solution may not exist even in a convex utility space. Again, the existence of the NB solution depends on the specific manner of cooperation and the problem setup. If all users are able to benefit from cooperation, there is a unique NB solution in the corresponding cooperative game. A detailed study can be found in [40,41]. The condition for the uniqueness of NE is much more complex, and we refer the readers to the references of noncooperative games in preceding sections.

Therefore, one difficulty in finding NE is related to the fact that it may not be unique, which adds to the complexity of the game. In many cases, the algorithms designed for finding NE are not guaranteed to converge. However, for NB, determining the manner of cooperation, which renders the corresponding utility space convex and as large as possible contributes to the main complexity.

1.5.3 Implementation: Centralized versus Distributed Structure

An NE solution has an advantage over the NB solution in developing distributed implementations. It is because in an NE each player only considers its own utility, which usually requires local information only. However, to find the NB solution, all users need to cooperate with each other and information exchange is inevitable. For example, in the joint TDM/FDM–based cooperation scheme, the TDM/FDM coefficients of all users need to be collected and broadcasted to the users. Thus, a completely distributed algorithm is not possible for most cooperative bargaining games. However, a distributed structure with a coordinator is still possible for cooperative games. Two examples are developed in [37,42]. In these examples, the original bargaining problems are decoupled into two-level problems. Then in the corresponding two-level structure, the higher level problem is solved by the coordinator using the information collected from all users, while the lower level problems are solved in parallel by individual users using local information and the broadcasted information from the coordinator. Note that the coordinator may be selected from one of the users or performed by the users in a round-robin manner.

1.6 Open Issues

1.6.1 On the User Number

Currently two-user games have been extensively studied in the literature. Generalizations to the multiuser cases, however, are still open problems. Algorithms for two-user games are not always applicable for multiuser cases. For example, the algorithm used in [24] in order to find the NB solution cannot be extended for solving the problem with the number of players $M > 2$. However, this problem exists only in cooperative games. For most noncooperative games, the water-filling-based solutions are applicable to both two-user and multiuser games.

1.6.2 On the Constraints

This applies mostly to the cooperative games due to the requirement of convex utility spaces for the NB. As mentioned before, maximization of the Nash function can be a non-convex optimization problem under certain constraints. One example is the cooperative power allocation game with total power constraints. Moreover, many games in the literature deal with other power-related constraints. Games that incorporate multiple constraints related to different practical requirements are still open for research.

1.6.3 On the Strategies

Most of the works on applications of game theory in wireless communications consider pure strategies only. Mixed strategies are seldom studied in multiuser wireless systems. Thus, the investigation of noncooperative and cooperative games based on mixed strategies remains an open problem. Moreover, players of the games introduced in this chapter have their strategies defined on a single target, for example, power. In a multiuser system, however, power allocation is usually not the only issue that needs to be solved to optimize the performance of the system. Admission control, scheduling, and others may also need to be taken into account. Thus, the users may have joint strategies defined on power, scheduling, and/or other parameters. In many applications, such games are still yet to be investigated.

1.7 Conclusion

This chapter reviews the applications of game theory for the multiuser wireless systems. Different games on different channels are considered, for both noncooperative and cooperative cases. The power allocation games are reviewed in details. These power allocation games construct the basis for higher-level beamforming and precoding games. The focuses of the discussion are on the comparison between noncooperative and cooperative games, as well as on games under different constraints. A number of similarities and differences between different games are emphasized and different performance metrics for cooperative and noncooperative games are analyzed. Throughout this chapter, we show that game theory is a powerful tool for solving many different problems in the multiuser wireless communications.

References

1. A. B. MacKenzie and S. B. Wicker. Game theory in communications: Motivation, explanation, and application to power control. In *Proceedings of IEEE Global Communications Conference*, pp. 821–826, San Antonio, TX, November 2001.
2. F. Meshkati, H. V. Poor, and S. C. Schwartz. Energy-efficient resource allocation in wireless networks. *IEEE Signal Processing Magazine*, 24(3):58–68, May 2007.

3. N. Dusit and H. Ekram. Radio resource management games in wireless networks: An approach to bandwidth allocation and admission control for polling service in IEEE 802.16. *IEEE Wireless Communications*, 14(1):27–35, February 2007.

4. A. Agustin, O. Muhoz, and J. Vidal. A game theoretic approach for cooperative MIMO schemes with cellular reuse of the relay slot. In *Proceedings of IEEE International Conference on Acoustics, Speech, and Signal Processing*, pp. 581–584, Montreal, Canada, May 2004.

5. J. O. Neel, J. H. Reed, and R. P. Gilles. Convergence of cognitive radio networks. In *Proceedings of IEEE Wireless Communications and Networking Conference*, pp. 2250–2255, Atlanta, GA, March 2004.

6. J. Virapanicharoen and W. Benjapolakul. Fair-efficient guard bandwidth coefficients selection in call admission control for mobile multimedia communications using game theoretic framework. In *Proceedings of IEEE International Conference on Communications*, pp. 80–84, Paris, France, June 2004.

7. S. Gunturi and F. Paganini. Game theoretic approach to power control in cellular CDMA. In *Proceedings of IEEE Vehicular Technology Conference*, pp. 2362–2366, Orlando, FL, October 2003.

8. S. M. Betz and H. V. Poor. Energy efficient communications in CDMA networks: A game theoretic analysis considering operating costs. *IEEE Transactions on Signal Processing*, 56(10):5181–5190, October 2008.

9. Z. Han, Z. Ji, and K. J. R. Liu. Power minimization for multi-cell OFDM networks using distributed non-cooperative game approach. In *Proceedings of IEEE Global Communications Conference*, pp. 3742–3747, Dallas, TX, November 2004.

10. H. Bogucka. Game theoretic model for the OFDM water-filling algorithm with imperfect channel state information. In *Proceedings of IEEE International Conference on Communications*, pp. 3814–3818, Beijing, China, May 2008.

11. Y. Wei and K. J. R. Liu. Game theoretic analysis of cooperation stimulation and security in autonomous mobile ad hoc networks. *IEEE Transactions on Mobile Computing*, 6(5):507–521, 2007.

12. O. Ileri, M. Siun-Chuon, and N. B. Mandayam. Pricing for enabling forwarding in self-configuring ad hoc networks. *IEEE Journal of Selected Areas in Communications*, 23(1):151–162, 2005.

13. M. Maskery, V. Krishnamurthy, and Q. Zhao. Decentralized dynamic spectrum access for cognitive radios: Cooperative design of a non-cooperative game. *IEEE Transactions on Communications*, 57(2):459–469, February 2009.

14. D. Niyato and E. Hossain. Competitive spectrum sharing in cognitive radio networks: A dynamic game approach. *IEEE Transactions on Wirless Communications*, 7(7):2651–2660, July 2008.

15. A. Attar, M. R. Nakhai, and A. H. Aghvami. Cognitive radio game for secondary spectrum access problem. *IEEE Transactions on Wirless Communications*, 8(4):2121–2131, April 2009.

16. D. Fudenberg and J. Tirole. *Game Theory*. MIT Press, Cambridge, MA, 1991.

17. J. Nash. Two-person cooperative games. *Econometrica*, 21(1):128–140, January 1953.

18. Z. Han, Z. J. Ji, and K. J. R. Liu. Fair multiuser channel allocation for OFDMA networks using Nash bargaining solutions and coalitions. *IEEE Transactions on Communications*, 53(8):1366–1376, August 2005.

19. W. Thomson. Cooperative models of bargaining. In *Handbook of Game Theory with Economic Applications*, R. J. Aumann and S. Hart, (Eds.), vol. 2, pp. 1237–1284, Elsevier Science B.V., New York, 1994.

20. M. J. Herrero. The Nash program: Non-convex bargaining problems. *Journal of Economic Theory*, 49:266–277, 1989.

21. T. C. A. Anant and B. Mukherji. Bargaining without convexity: Generalizing the Kalai-Smorodinsky solution. *Economics Letters*, 33:115–119, 1990.

22. Z. Q. Luo and J. S. Pang. Analysis of iterative waterfilling algorithm for multiuser power control in digital subscriber lines. *EURASIP Journal on Applied Signal Processing*, 2006 (24012), 2006. Available at: http://www.hindawi.com/journals/asp/2006/024012.abs.html

23. A. Leshem and E. Zehavi. Bargaining over the interference channel. In *Proceedings of IEEE International Symposium on Information Theory*, pp. 2225–2229, Seattle, WA, July 2006.

24. A. Leshem and E. Zehavi. Cooperative game theory and the gaussian interference channel. *IEEE Journal of Selected Areas in Communications*, 26(7):1078–1088, September 2008.

25. A. Leshem and E. Zehavi. Game theory and the frequency selective interference channel. *IEEE Signal Processing Magazine*, 26(5):28–40, September 2009.

26. T. Basar and G. Olsder. *Dynamic Non-Cooperative Game Theory*. pp. 447–551, Academic Press, London, 1982.

27. E. Zehavi and A. Leshem. Bargaining over the interference channel with total power constraints. In *Proceedings of International Conference on Game Theory for Networks*, pp. 447–451, Istanbul, Turkey, May 2009.

28. J. E. Juan, L. A. DaSilva, Z. Han and A. B. MacKenzie. Cooperative game theory for distributed spectrum sharing. In *Proceedings of IEEE International Conference on Communications*, pp. 5282–5287, Glasgow, Scotland, June 2007.

29. J. Gao, S. A. Vorobyov, and H. Jiang. Cooperative resource allocation games under spectral mask and total power constraints. *IEEE Transactions on Signal Processing*, 58(8):4379–4395, August 2010.

30. E. G. Larsson and E. A. Jorswieck. Competition versus cooperation on the MISO interference channel. *IEEE Journal of Selected Areas in Communications*, 26(7):1059–1069, September 2008.

31. M. Nokleby and A. L. Swindlehurst. Bargaining and the MISO interference channel. *EURASIP Journal on Advances in Signal Processing*, 2009 (368547), 2009. Available at: http://www.hindawi.com/journals/asp/2009/368547.html

32. E. A. Jorswieck and E. G. Larsson. The MISO interference channel from a game theoretic perspective: A combination of selfishness and altruism achieves pareto optimality. In *Proceedings of IEEE International Conference on Acoustics, Speech, and Signal Processing*, pp. 5364–5367, Las Vegas, NV, March 2008.

33. E. A. Jorswieck, E. G. Larsson, and D. Danev. Complete characterization of the Pareto boundary for the MISO interference channel. *IEEE Transactions on Signal Processing*, 56(10):5292–5296, October 2008.

34. G. Scutari, D. P. Palomar, and S. Barbarossa. Optimal linear precoding strategies for wideband non-cooperative systems based on game theory—part I: Nash equilibria. *IEEE Transactions on Signal Processing*, 56(3):1230–1249, March 2008.

35. G. Scutari, D. P. Palomar, and S. Barbarossa. Optimal linear precoding strategies for wideband non-cooperative systems based on game theory—Part II: Algorithms. *IEEE Transactions on Signal Processing*, 56(3):1250–1267, March 2008.

36. J. Gao, S. A. Vorobyov, and H. Jiang. Game theoretic solutions for precoding strategies over the interference channel. In *Proceedings of IEEE Global Communications Conference*, New Orleans, LA, November 2008.

37. J. Gao, S. A. Vorobyov, and H. Jiang. Game theory for precoding in a multi-user system: Bargaining for overall benefits. In *Proceedings of IEEE International Conference on Acoustics, Speech, and Signal Processing*, pp. 2361–2364, Taipei, Taiwan, April 2009.

38. M. Nokleby, A. L. Swindlehurst, Y. Rong, and Y. Hua. Cooperative power scheduling for wireless MIMO networks. In *Proceedings of IEEE Global Communications Conference*, pp. 2982–2986, Washington, DC, November 2007.

39. G. Scutari, D. P. Palomar, and S. Barbarossa. Cognitive MIMO radio: A competitive optimality design based on subspace projections. *IEEE Signal Processing Magazine*, 25(6):46–59, November 2008.

40. Z. Chen, S. A. Vorobyov, C.-X. Wang, and J. Thompson. Nash bargaining over MIMO interference systems. In *Proceedings of IEEE International Conference on Communications*, pp. 1–5, Dresden, Germany, June 2009.

41. Z. Chen, S. A. Vorobyov, C.-X. Wang, and J. Thompson. Pareto region characterization for rate control in multi-user systems and Nash bargaining. *IEEE Transactions on Automatic Control*. Submitted. (Arxiv: 1006.1380).
42. H. Yaiche, R. R. Mazumdar, and C. Rosenberg. A game theoretic framework for bandwidth allocation and pricing in broadband networks. *IEEE/ACM Transactions on Networking*, 8(5):667–678, October 2000.

Chapter 2

Decision Theory with Its Applications in Wireless Communication

Dwayne Rosenburgh

Contents

The purpose of this chapter is to provide an introduction to decision theory before the topic of game theory is discussed. Since game theory is essentially multiagent decision theory, it is probably beneficial to first devote some time to the study of classical decision theory. Relevant decision theory concepts will be introduced, and, where appropriate, analogies will be made to wireless communications.

In general, decision theory is concerned with understanding, describing, and quantifying decision processes and methods, and their outcomes. In other words, decision theory provides a rational framework for choosing between alternative courses. In this chapter, we will treat decision activities as goal-oriented behavior in the presence of alternatives. Almost all human activity involves making a decision. Now, we require our machines (e.g., cognitive radios) to make decisions. It is reasonable to expect that a human decision and a machine decision may be similar.

As the chapter proceeds, we will draw on economic theory to establish the bases and conditions for making decisions. We will examine the concepts of a best decision, a correct decision, and an optimal decision and then see how those concepts can be applied to physical layer or network layer preferences in wireless communications.

2.1 Introduction

Before proceeding into game theory, this chapter will present an overview of decision theory and provide insight into how it may be applied to wireless communications. The relationship between decision theory and game theory becomes obvious when one considers that game theory is essentially multiagent decision theory. It is important to remember that, while decision theory is focused on a single-agent decision maker, the decision maker does not have to be a singular (or individual) entity. In other words, several entities that have the same goal (or goals) can be considered a single-agent decision maker. Therefore, multiagent decision theory (i.e., game theory) comes into reality when there are several, and usually competing, goals.

Decision theory is a body of knowledge and analytic techniques that is focused on quantifying the process of making choices between alternatives. Decision theory is rooted in statistical theory and is fundamental to economics. Originally, decision theory dealt primarily with human decision making. Now that humans are creating more sophisticated and complicated machines that are capable of selecting alternatives, it is reasonable to examine the processes of human and machine decision making in a uniformed manner. One may argue that there is, or should be, no difference between human and machine decision making. After all, humans create the algorithms for machine-based decision making. In addition, the result or product of a machine's decision process will eventually have some type of impact on a human. In this chapter and throughout this book, the machines that we are most interested in are wireless communication and networking devices.

One thing that makes decisions and decision theory interesting is that the decision maker usually must act with an incomplete knowledge of the consequences that will result from the action. Confronting uncertainty is not easy or comfortable. In order to deal with those issues on a rational basis, the structure on which decisions are made must include uncertainty. Instead of examining decisions from a purely scientific viewpoint, we will take the engineering viewpoint. Therefore, our concern and focus will be on actually making decisions, that is, making a choice between alternatives.

Decision theory can be normative or descriptive. Normative decision theory focuses on how decisions *should be* made if the desired outcome is to maximize an expected utility (EU). Descriptive decision theory refers to theories about how decisions are actually made. In this year of 2010, we are still at a relatively early stage of applying decision making (or even game theory) to wireless communications. Therefore, the majority of the research and literature on decision theory and wireless communications is normative.

Before proceeding, let us examine what is meant by "utility." Utility, or utility function, comes from the utility theory that was formulated by von Neumann and Morgenstern [1]. Those researchers developed preference relationships that assumed, amongst other things,

1. That given two alternatives, a decision maker either prefers one to the other or is indifferent between them.

2. There are well-defined chance events, having probabilities associated with them, which can be manipulated according to the rules of probability.

We will return to the discussion of utility later, after we establish what is meant by a preference relationship. For now, we can summarize an EU as being the value that a decision maker places on her decision.

As we proceed, our setting will be the preferences of the users of wireless communication devices. The preferences of the users (or their devices) may include things like battery life, transmitter power, frequency use, data routes, signal-to-interference-and-noise ratio (SINR), quality of service (QoS), etc.

The main components of decision theory are probability and utility. If we are to use a decision-theoretic approach in wireless communications, we must somehow determine the probabilities and utilities [2].

2.2 Preference Relationships

Earlier, we mentioned a preference relationship. The simplest preference relationship to consider is a binary one. If we let **C** be a nonempty set of alternatives, which contains at a minimum x and y, and we let $\succeq$ be a weak (and binary) relation on **C**, then $x \succeq y$ denotes that x is preferred to y or x is indifferent to y. (An equivalent statement is that y is not preferred to x.) Informally, the weak relation says that one alternative is at least as good as the other alternative.

The weak preference relationship $\succeq$ is said to be complete if, for all x and y, either $x \succeq y$ OR $y \succeq x$. Note that "OR" is Boolean, that is, "OR" does not preclude both statements from being true. A relationship is defined as transitive if $x \succeq y$ and $y \succeq z$ infer that $x \succeq z$.

Axiom 2.1   A binary relation is a weak relation if it is complete and transitive.

We will let the above axiom stand by itself for now. Later, we will examine some potential problems with that statement.

The strong relation $x \succ y$ denotes that x is *always* preferred over y (or x is better than y). When alternatives are always equal to each other, that is, there is no preference, the notation $x \equiv y$ is used.

Now, we will explore the connection between preferences and wireless communications. At a high layer, or application layer, a user may prefer a high QoS to a low QoS if she is streaming video. If the user is browsing the Internet, then low latency is probably preferred over high latency. However, if the user is sending or receiving a small email, then she may be indifferent to the latency. Those preferences will be, or should be, translated into low layer, or physical layer, preferences. Therefore, lower layer algorithms must also determine preference relationships for the SINR, the bit rate, the bit error rate (BER), etc. In advanced communications, such as cognitive radio, the lower layer alternatives will include additional features, such as, frequency, type of modulation, etc.

Let us now return to Axiom 2.1. The first problem with that axiom—completeness—can be easily corrected. The property of completeness requires that there exists a preference between any two elements in set of outcomes, even if they are unrelated. For example, completeness requires a user to express a preference between, say, high QoS or transmitter frequency, if they comprise a two-element set. Clearly, expressing such a preference is almost as meaningless as selecting between listening to jazz music and eating a banana. Therefore, we need to define completeness as *connectedness*, thereby requiring completeness to be defined for a relation and its domain. More formally, a weak relation is complete if and only if, for any elements x and y of its domain, either $x \succeq y$ OR $y \succeq x$. Note that we now require the elements to be connected to the domain. That means that we no longer express a preference between unrelated choices.

The second problem with Axiom 2.1 is transitivity. Strict transitivity requires that we make fine distinctions. For example, if we prefer a higher data throughput over a lower throughput, then transitivity requires that we prefer 10.00012–10.00011 Mbps, although such a distinction is probably senseless. One obvious solution is to establish a meaningful threshold, beyond which the alternatives are considered equal. In that case, one would express no preference amongst the choices. Since indifference is transitive, Axiom 2.1 is not violated, and we do not have to be concerned with expressing a preference amongst almost infinitesimal choices.

In decision making, preference relations are used to find the best alternative. The best decision (or alternative) is the one that is better than all other available alternatives, in light of the information that is available. This definition recognizes that often the information that is available may be incomplete, and thus the best alternative, relative to one information set, may not be the best alternative in view of a more complete (or different) set of information. However, since we can only be concerned with the information that is available, the guiding rule should be that if there is a uniquely best (i.e., best of the best) alternative, then select it. More generally, an alternative is amongst the best if and only if it is at least as good as all other alternatives. If all of the alternatives are equally good, then select one of them. If the alternatives are truly equally best alternatives, then the human designer decides the overall "character" of the system. If predictability and repeatability are important, then the selected alternative is hard coded. An example of hard coding is always to select the alternative that appears first. However, by allowing a random selection when alternatives are equal, it may be possible to arrive at nonintuitive solutions. That is especially true in complex situations when one alternative is the input to a separate algorithm that will produce another alternative (and so on).

2.3 Utility and Classification

Numerical values, or utilities, are frequently used in decision making and decision theory. The basic decision rule is simple and obvious—select the alternative with the highest utility. In decision theory, the goal is to develop a numerical representation of a (weak) preference. This brings us to another axiom:

Axiom 2.2   A binary relation $\succeq$ on the finite set $\mathbf{C}$ is a preference relation if and only if there exists a utility function $\boldsymbol{u}$ that represents $\succeq$.

Axiom 2.2 can be proved by restricting the numerical values to real numbers. Since the real numbers order, we can write $x > y$. Therefore, the set $\mathbf{C}$ is complete. Similarly, a set of real numbers that is complete is also transitive. Therefore, $\succeq$ is complete and transitive, making it a preference relation. If $\boldsymbol{u}$ is a function that represents the number of elements in the set, and if $x \succeq y$ implies $\boldsymbol{u}(x) \geq \boldsymbol{u}(y)$, then $x \succeq y$ if and only if $\boldsymbol{u}(x) \geq \boldsymbol{u}(y)$. Therefore, every preference relationship on a finite set $\mathbf{C}$ can be represented with a utility function $\boldsymbol{u}$. Utility functions are not unique. Therefore, if more than one alternative have equally high utilities, then select one of them (it does not matter which).

Since utilities are nonunique and subjective, obtaining them is usually problematic. Consider, for example, intersubjective utility: one user's utility of, say, 97 may not be equal to another user's utility of 97. In decision theory, these issues can be made less significant by assuming that there is just one user or decision maker. The same single-user approach can be used, sparingly, in game theory. In game theory, amalgamating various utilities is often a problem. Therefore, it would probably not be of much practical use to replace multiple utilities with a single utility, consistently.

In a wireless scenario, we can consider the issue of network usage and data throughput. The decision maker must decide what the relative utilities of blocking or dropping some users should be when attempting to achieve a certain throughput for another user. There are many pieces of information that would go into determining the utility. For example, what other frequency resources are available; what are the adaptive capabilities of the users; etc.? If users cannot be moved to different frequencies, or modulation types, etc., then other issues such as the loss of profit from dropping a user will become important. As one can imagine, the set of inputs to a utility function can become large.

Typically, decisions are classified based on what is known (or believed) about the results. It is customary to classify that knowledge into three categories, which range from complete knowledge to ignorance. The categories are

- Certainty
- Risk
- Uncertainty

Decision making under certainty assumes that the decision maker has complete knowledge available; therefore, the outcome is deterministic. The decision maker is considered a perfect predictor because it is assumed that there is only outcome for each alternative.

Decision making under certainty is usually appropriate in relatively simple situations, or when there are a small number of outcomes. One simple scenario is in the cognitive radio situation where primary and secondary users share a channel. Here we assume that the simple goal is for multiple users to share the channel with minimum amount (ideally zero) interference between the users. If the radios are truly cognitive, then we assume that there is some cooperation between them. Therefore, at least one of the radios has full a priori knowledge of the channel (often called the cognitive channel). The radio that has full knowledge and uses that knowledge in its decision-making algorithm meets the definition for decision making under certainty.

Decision making under risk (i.e., known probabilities) is also referred to as EU. In the classical literature of decision theory, EU hypothesis asserts that the utility of a lottery is equal to the EU of its component prizes [3]. Therefore, EU could be referred to as "probability-weighted utility theory." In EU theory, each alternative is assigned a weight, where the weights are probabilities of the alternative being the "winning" alternative. The basic expected utility (EU) function is represented by Equation 2.1:

$$EU = \sum_{i=1}^{N} w_i a_i \tag{2.1}$$

where the weights, w_i, applied to each alternative, a_i, represent importance, or preference. Therefore, a *utility function* is a numerical representation of an entity's preferences when there is the possibility of a set of specific outcomes, each outcome occurring with a known probability.

In the published literature, a linear-logarithmic utility function is often used in order to avoid the problem of too often finding the optimums at the extreme edges. Equation 2.2 (the Cobb–Douglas utility function) adds a degree of convexity to the optimization curve. In economics, q_i is the quantity consumed; for our purposes, it is the objective. The weights w_i in Equation 2.1 are replaced by the adjustment coefficient β_i

$$\ln(EU) = \sum_{i=1}^{N} \beta_i \ln q_i \qquad (2.2)$$

When multiple outcomes are possible, and each of the outcomes may be acceptable in different environments or scenarios, decision making under risk may be most appropriate. For example, consider a wireless network with multiple users. The users may have different requirements for QoS (in this example, QoS is defined in terms of latency) and data throughput. Let us assume that there are 10 users on the network—9 users require a medium QoS, 1 user requires a high QoS, half of the users require a high data throughput, and half of the users are indifferent to throughput. In that scenario, establishing a utility function for the users and the network becomes a stochastic problem that matches well with decision making under risk.

Decision making under uncertainty considers situations in which several outcomes are possible for each course of action. The difference between uncertainty and risk is that under uncertainty the decision maker does not know, or cannot estimate, the probability of occurrence of the possible outcomes. Decision making under uncertainty is more difficult to do because of insufficient information. Consider the case when a cognitive radio is first trying to determine whether a frequency is available for use. Initially, there may be no way of determining if any energy that is present at the frequency of interest is noise or data. Similarly, if the energy is data, it is initially difficult to determine if the transmitter is a primary or secondary user. Consequently, the problem needs to be approached as decision making under uncertainty.

A different approach to decision making under uncertainty is to attempt to reduce ignorance to risk. This is done by the use of the *principle of insufficient reason*, which was formulated by Jacques Bernoulli (1654–1705). This principle states that if there is no reason to expect that one event is more likely to occur than another is, then the events should be assigned equal probabilities. The principle is intended for use in situations where there is an exhaustive list of alternatives, all of which are mutually exclusive. This approach may be more reasonable to the readers who prefer Bayesian decision theories.

Bayesian decision making is always decision making under certainty or risk, never under uncertainty. The distinction between risk and uncertainty is not meaningful in Bayesian decision making. Bayesian decision making requires the outcome with the highest EU to be selected.

2.4 Behavior

Now, we will examine decision making and behavior, in a wireless sense. In much of game theory, which is a central theme of this book, decisions and behavior are intertwined. In [4], Devroye et al.* used Figure 2.1 to illustrate competitive, cognitive, and cooperative behaviors in wireless environments:

(a) *Competitive behavior*: The two transmitters transmit independent messages. There is no cooperation in sending the messages, and thus the two users compete for the channel. An example of this behavior is carrier sense multiple access (CSMA) schemes commonly used in wired and wireless networks.

---

* Permission to use Figure 2.1 and related description is granted under the Creative Commons Attribution License.

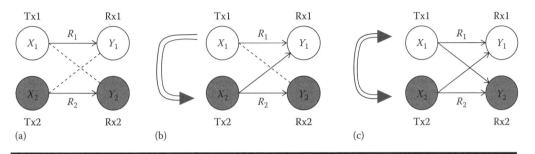

Figure 2.1 **(a) Competitive behavior: the interference channel. The transmitters may not cooperate. (b) Cognitive behavior: the cognitive channel. Asymmetric transmitter cooperation is possible. (c) Cooperative behavior: the two Tx antenna broadcast channel. The transmitters, but not the receivers, may cooperate fully and symmetrically. In these figures, solid lines indicate desired signal paths, while dashed lines indicate undesired (or interfering) signal paths.)**

(b) *Cognitive behavior*: Asymmetric cooperation is possible between the transmitters. This asymmetric cooperation is a result of Tx2 knowing Tx1 message, but not vice versa, and is indicated by the one-way double arrow between Tx1 and Tx2. We idealize the concept of message knowledge: whenever the cognitive node Tx2 is able to hear and decode the message of the primary node Tx1, we assume it has full a priori (i.e., noncausal) knowledge. We use the term cognitive behavior, or cognition, to emphasize the need for Tx2 to be a smart device capable of altering its transmission strategy according to the message of the primary user.

(c) *Cooperative behavior*: The two transmitters know each other's messages (two-way double arrows) and can thus cooperate fully and symmetrically in their transmission. The channel pictured in Figure 2.1c may be thought of as a 2×2 multiple-input–multiple-output (MIMO) channel.

The work of Devroye et al. is important because they showed that through cooperation, the data rates of the primary and secondary users might be significantly improved. Those results should lead to better decision making in cognitive systems.

When there is a single user who is making a decision between two alternatives, behavior is not usually an issue. For example, suppose that User A wants to transmit a message to User B; there are two available frequencies from which to select. All that is required is for User A to select from the frequencies from which User B has provided information. There is little, if any, concern for behavior in this situation.

Now consider the situation where there are, say, three users who desire to communicate, and the users have different priorities (e.g., message length, power usage, etc.). There are two reasonable options to consider: (1) one user becomes the master (or dictator) and makes the decisions and (2) the users attempt to reach mutually agreed-upon decisions (i.e., group decisions). The easiest choice is number 1; that is the approach used in the *Bluetooth*® wireless protocol.* In his seminal work in economic theory, Arrow [5] showed the difficulty in implementing number 2.

---

* The *Bluetooth* word mark is owned by the Bluetooth SIG.

Following is a summary of Arrow's research, which led to his famous *impossibility theory*. In establishing his theorem, Arrow proposed five conditions:

Condition 2.1   There are at least two individuals, and at least three alternatives.

Condition 2.2   (Positive association of individual and group preferences) If alternative $x \succeq y$ for a given set of individual preferences, then that preference is maintained when the set is modified in x's favor.

Condition 2.3   (Independence of irrelevant alternatives) If the group's preference is $x \succeq y$, and some of the individual preferences between alternatives *other than* x and y are modified, but the individual preferences between x and y remain unchanged, then the original group preference $x \succeq y$ is maintained.

Condition 2.4   (Citizen's sovereignty) For each pair of alternatives x and y, there is some set of individual orderings for the group, such that $x \succeq y$.

Condition 2.5   (Non-dictatorship) There is no individual with the property that whenever he or she expresses the preference $x \succeq y$, the group does likewise, regardless of the preferences of other individuals.

One can argue that the above axioms are reasonable and not too restrictive. Condition 2.2 states that in a situation where the group prefers x over y, if some of the members maintains or increase their enthusiasm for x via some other alternative(s) and none of the members lessen their enthusiasm for x via any other alternative, then the group maintains its preference for x over y.

Condition 2.3 says that a group's preference of x over y should depend only on how individuals rank x and y and not on how individuals rank some irrelevant alternative (say, "alternative z") relative to x and y. Satisfying Condition 2.3 requires that the group selects x as long as the majority of the members prefer x to all other alternatives.

Condition 2.4 states that the collective opinions of the group have some relevance to the group decision. Thus, for any pair of alternatives x and y, there is some set of individual preferences, which could lead to a group preference of x over y. If Condition 2.4 does not hold true, then the group does not have the power to establish preferences between all of the alternatives.

Condition 2.5 states that there is no dictator in the group, that is, there is no individual whose preference between x and y will dominate the group decision even if all other members of the group have the opposite preference structure.

Arrow's impossibility theorem states that Conditions 2.1 through 2.5 are not logically consistent. That is, there is not a group decision process, or method, that possesses the properties demanded by all of these conditions. In other words, if a function satisfies Conditions 2.1 through 2.4, then Condition 2.5 cannot hold.

Arrow provided a basis for an evaluation of choice and decision making. In his research, Arrow established several apparently reasonable conditions for the method of group decision making (or social choice); these require that a decision be responsive to the desires of individuals and that, among any given set of possible alternatives, a decision depends only on the preferences of the individuals among the members of that set. He then showed that it is not possible to meet the complete set of conditions. We will not duplicate the proof here, but the reader is referred to [5] for a complete proof.

Other researchers, such as Farris and Sage [6] did additional work on Arrow's theory to address some of the logical inconsistencies. In general, researchers have found that by relaxing some of Arrow's conditions, the inconsistencies may be resolved. As we have already established, group decision making is a foundation of game theory. Therefore, in some sense, the remainder of this book deals with how to ensure that when one is applying game theory to wireless scenarios, one does not reach a logically inconsistent state.

2.5 Decision Efficacy

In most situations, after a decision is made, it is probably valuable to determine the effectiveness or value of that decision. Let us establish a few definitions that should be helpful when discussing decisions. We use the term *best* in a restricted way. The best decision (or alternative) is the one that is supported by the preponderance of the information that is available to the decision maker. This definition recognizes that often the information that is available may be incomplete, and thus the best alternative, relative to the information available at the time, may not be the best alternative in view of a more complete set of information.

A *correct* decision may, or may not, be supported by the preponderance of available information. However, it is a decision that is determined by a knowledgeable and authoritative entity (other than the decision maker) to be the most accurate, desirable, valid, or significant decision.

We use the term *optimal* to refer to Pareto optimality (named after Vilfredo Pareto [1848–1923]). Pareto optimality is meaningful in multiagent or group settings; it is a topic discussed in game theory. For now, we can generally state that a Pareto optimal condition occurs when an attempt to improve the results for one entity will mean that at least one other entity experiences detriment.

The preceding three definitions mean that in decision theory, we can strive for a best decision OR a correct decision. In game theory, we can strive for a best decision, OR a correct decision, OR an optimal decision.

We have seen that quality or efficacy of a decision is relative and may not be static under all circumstances. In addition, as we shall see in the next few paragraphs, there are probably no decisions that are so undesirable that they are always avoided. Likewise, there are probably no decisions that are so attractive that they are always selected.

If we consider that outcomes are usually expressed as a finite set, and there is typically a numeric value associated with the outcomes, then it is reasonable to say that decisions are normally represented by an ordered set. Therefore, the axioms of continuity should be applicable. In particular, the Archimedean axiom is applicable; this brings us to our third axiom:

Axiom 2.3   For all x, y, $z \in \mathbf{C}$ such that $x \succ y \succ z$, there exist a, $b \in (0, 1)$ such that $ax + (1 - a)z \succ y \succ bx + (1 - b)z$.

Axiom 2.3 (the Archimedean axiom) can be interpreted to mean that there are no extremely bad or extremely good outcomes. That means that regardless of how bad z is, if we strongly prefer x to y, then we will accept a gamble that assigns a large probability to x and a small probability to z over one that give us y with certainty. Similarly, regardless of how good x is, if we strongly prefer *y* to *z*, then we will accept *y* with certainty over a gamble that assigns a small probability to *x* and a large probability to *z*.

You may ask yourself, "Do people really behave in accordance with the Axiom 2.3?" The answer is "yes." For example, consider the use of mobile phones while driving an automobile. People who use the phone while driving are expressing a strong preference for communicating now and hoping that they do not cause an accident while talking on the phone (outcome *x*), over the option for communicating later and not causing an accident while talking on the phone (outcome *y*). It is probably fair to say that the people who are driving while using a phone are aware of outcome *z*—talking on the phone while driving could cause an accident that could lead to someone's death. Regardless of how bad outcome *z* is, many people accept the gamble and risks associated with alternatives *x* and *z*, over the certainty and safety of alternative *y*. It is worthwhile to mention that it is rare to have extreme alternatives or outcomes present. Aside from the example provided above, one is not likely to encounter many decisions in wireless communications where an alternative is death.

2.6 Open Issues

In doing a search of the literature on decision theory and wireless communications, one will find that there is not a great deal of scholarly writings. Additionally, one is likely to find that most of what has been written was done relatively recently (roughly, since the year 2000). This is not too surprising since we are just at the point where it is possible to have a (portable) wireless device that can efficiently implement sophisticated decision algorithms. Examples of relevant and recent published literature include [2,4,7].

If we remember that decision theory is most appropriate when there is a single decision maker (otherwise, it is a game-theoretic situation), then probably, the best application for decision theory and wireless communications is cognitive radio. It is likely that most cognitive radio-based decisions will be for a single user. If there are multiple users, then we have seen that cooperation rather than competition should produce better outcomes.

What may be the biggest open issue (i.e., concern) is that more research is needed on applying decision theory to, and developing decision theory for, wireless communications. Some additional research will occur naturally, because of the increasing interest in cognitive radios. However, more proactive research in decision theory and wireless communications may hasten development and progress in cognitive radio.

2.7 Conclusion

In this chapter, we have explored some of the basics of decision theory. We have established that a decision maker analyzes possible outcomes in two dimensions—utility theory and probability of occurrence. A decision maker cannot guarantee that her outcome will be best in all environments; but she attempts to make the best decision based on preferences and available information. We have seen how decision theory can be applied to situations that may be encountered in wireless communications. Now that we know what it means for an entity to have a utility function, and we can differentiate between certainty, risk, and uncertainty, and we can understand the difference between a best decision, correct decision, and optimal decision, we can proceed into game theory.

References

1. J. von Neumann and O. Morgenstern, *Theory of Games and Economic Behavior*, 2nd edition, Princeton, NJ: Princeton University Press, 1947.
2. Z. Haas, J. Halpern, L. Li, and S. Wicker, A decision-theoretic approach to resource allocation in wireless multimedia networks, in *Proceedings of the 4th International Workshop on Discrete Algorithms and Methods for Mobile Computing and Communications*, Boston, MA, pp. 86–95, June 2000.
3. R. Luce and H. Raiffa, *Games and Decisions—Introduction and Critical Survey*, New York: Dover Publications, 1989.
4. N. Devroye, M. Vu, and V. Tarokh, Achievable rates and scaling laws for cognitive radio channels, *EURASIP Journal on Wireless Communications and Networking*, 2008, 13–24, New York: Hindawi Publishing, 2008.
5. K. Arrow, *Social Choice and Individual Values*, 2nd edition, New Haven, CT: Yale University Press, 1963.
6. D. Farris and A. Sage, Introduction and survey of group decision making with applications to worth assessment, *IEEE Transactions on Systems, Man and Cybernetics*, SMC-5(3), 346–358, 1975.
7. T. Rondeau and C. Bostian, *Artificial Intelligence for Wireless Communications*, Norwood, MA: Artech House, 2007.

Chapter 3

Game Theory in Wireless Sensor Networks

Pedro O.S. Vaz de Melo, Cesar Fernandes, Raquel A.F. Mini,
Antonio A.F. Loureiro, and Virgilio A.F. Almeida

Contents

3.1 Introduction

A wireless sensor network (WSN) consists of spatially distributed autonomous devices that cooperatively monitor physical or environmental conditions, such as temperature, sound, luminosity, vibration, pressure, motion, and pollutants. The potential for observation and control of the real world allows WSNs to present themselves as a solution to a variety of monitoring and control applications, such as environment monitoring, biotechnology, industrial monitoring and control, public safety, transportation, and medical control. A WSN comprises compact and autonomous devices called sensor nodes, composed of processor, memory, battery, sensor devices, and transceiver. A sensor network is the result of the fast convergence of three key technologies: microelectronics, wireless communication, and MEMS (micro electro mechanical systems). Sensor nodes are energy-constrained devices, and, thus, algorithms and protocols designed for WSNs should consider the energy consumption in their conception.

In this chapter, we describe how different interactions in WSNs can be modeled as a game [26]. Game theory is based on models that express the interaction among players, in this case nodes, by modeling them as elements of a social network in such a way that they act to maximize their own utility. This allows the analysis of existing algorithms and protocols for WSNs, as well as the design of equilibrium-inducing mechanisms that provide incentives for individual nodes to behave in socially constructive ways. Game theory defines mechanisms for players to choose the best available action, but at the same time it provides a scenario where other players' utilities can be maximized.

This chapter aims to present some of the problems of WSNs addressed with the help of game-theoretic models. Some of the issues studied that aim to achieve an energy-efficient solution are scheduling sensor nodes to switch between energy-conserving modes of operation, efficient routing algorithms, clustering and density control, and information fusion. Another important problem is security in WSNs. Sensor nodes tend to be deployed in different geographical environments and, thus, are more prone to failure from hostile environmental conditions besides being vulnerable to network attacks. Furthermore, very interesting questions arise when we combine these two issues that have the potential to lead to novel results. For instance, how do we route/forward packets efficiently in terms of energy in a WSN when there are misbehaving nodes?

Emerging research in game theory applied to WSNs shows much promise to help understand the complex interactions between sensor nodes in this highly dynamic and distributed environment.

3.2 Taxonomy

In this section, we show the possibilities of applying game theory to WSNs. Differently from general ad hoc networks, a WSN is controlled by a single authority, which is responsible for programming all its network devices to respond indiscriminately to its commands, without any rational decision making. Thus, in a single WSN, the unique rational agent that can make strategic decisions is the authority that governs the network. This implies that game theory, at first, can only be naturally applied when there is a conflict involving this authority.

A possible conflicting scenario involving the authority that governs the WSN is when one or more WSNs are deployed near each other and they can interact among themselves. In this case, WSNs may cooperate to improve their functionalities and extend their lifetime by trading routing favors [5], or increase the information quality by using a common data aggregation technique. The problem with this cooperation is that the owner of the sensor networks is only concerned with its network operability, and will only cooperate if this brings clear benefits to the network.

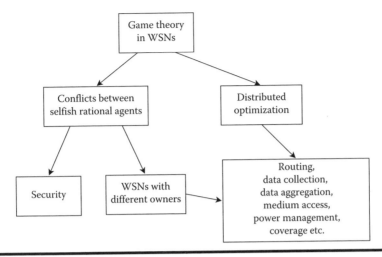

Figure 3.1   Taxonomy of possible applications of game theory to WSNs.

The owner must ensure that its network capability will be higher in a cooperative scenario than in a non-cooperative one.

Besides the application to scenarios that present conflicts among rational agents, game theory may be also applied to distributed decision making in WSNs. Since the number of sensor nodes in WSNs can reach a large amount such as hundreds of thousands, its decision making should be local, without any assistance of a central authority. Thus, game theory may be used in the design of utility functions that aim to optimize a global network metric based on local decisions of the sensor nodes. Figure 3.1 depicts possible applications of game theory to WSNs, which will are detailed in this chapter.

3.3 Conflict among Selfish Rational Agents

3.3.1 Cooperation among Different Networks

In the future, we can expect to have WSNs deployed in different places, such as forests, volcanos, seas, cities, and deserts. Furthermore, we can expect to have distinct WSNs owned by different authorities working at the same place, supporting a wide variety of applications. In a rain forest, for example, we could have a WSN, owned by the government, to detect a fire, and another WSN, owned by a private company, to detect the presence of specific species.

When two WSNs, installed at the same place, share their sensor nodes in the execution of one or more activities in a profitable way, both networks may improve their capabilities and perform their activities in a more efficient way. Despite being obvious and simple, this idea brings several implications that hinder cooperation among the networks. Whereas a WSN has a rational and selfish character, it will cooperate with another WSN only if this association provides services that justify the cooperation. Thus, to model this scenario, game theory is used to assess the conflict among the authorities responsible for managing the networks, which are modeled as the players. Each authority wishes to maximize both the lifetime and the quality of service of its network and will only cooperate with another network if this cooperation brings benefits. Felegyhazi et al. [9] consider that different WSNs may cooperate forwarding packets toward the sink node to save their energies. They investigate whether cooperation can exist based solely on the self-interest of the authorities that control the networks, without any cooperation enforcement mechanism. In this way, a sensor node that receives a packet from another network may drop it for no particular reason.

In the game-theoretic model, described in the following, the time is divided into slots and the sensor nodes send their data packets, once per time slot, to their sink nodes. In this model, the authorities that control the networks are the players of the game and the strategy of an authority i at time slot t is to define a move $m_i(t)$, characterized by the combination of two actions: (i) let their nodes forward other network packets and/or (ii) ask the other network to forward its packets. For each time slot t, it is also defined as a variable $\xi(t) \in \{true, false\}$ that tells if the data collection at that time slot was satisfactory. This is measured by comparing the number of collected data received $\rho_i(t)$ at time slot t with the number of collected data requested SR_i, defined by the application. If $\rho_i(t) \geq SR_i$, then $\xi(t) = true$, otherwise, $\xi(t) = false$. If $\xi(t) = true$, player i receives a gain $g_i(t) = G_i$, otherwise it receives $g_i(t) = 0$. Thus, the payoff $\pi_i(t)$ of player i at time slot t is $\pi_i(t) = g_i(t) - c_i(t)$, where $c_i(t)$ is the total transmission and reception cost of all sensors that belong to authority i for all packets, considering that $G_i \gg c_i(t)$ for every time slot t. This assumption is commonly used when applying game theory to WSNs, since it is rational to think that it is worth sending packets toward the sink when these packets are relevant to the application.

Thus, at every time slot, each player updates its strategy, by altering its move $m_i(t + 1)$ in function of $\xi(t)$. To do this, every sensor node receives at each time slot t a bit informing the result of $\xi(t - 1)$. The utility U_i of each player i is the cumulative payoff for each time slot t until T (time slot in which the network controlled by authority i becomes inactive). The utility function of player i is then defined as $U_i = \sum_{t=0}^{T} \pi_i(t)$ and the objective is to maximize U_i, that is, report data collections, successfully as many times as possible, while minimizing the energy consumption.

Felegyhazi et al. [9] show that the networks converge basically into two states of equilibria: noncooperative—no node provides and asks for services to nodes from another network and cooperative—all nodes provide and ask for services to nodes from another network. Moreover, it was found that when the environment is very dense, the reception cost dominates the transmission cost, making the algorithm to converge to the noncooperative equilibrium. It was also shown that when the environment is hostile, that is, the value of path loss exponent is high, there are strong incentives for cooperation.

In contrast to a practical approach, we can understand the theoretical issues of providing strategies that allow cooperation among different networks. Wu and Shu [27] propose an *InterSensorNet* scheme, which is a federation of multiple sensor networks that is built depending on whether a node is willing to cooperate with nodes in a foreign network. In this case, each player of the game is a sensor node that has a type t_i responsible for defining its marginal cost c_i, associated with a given activity. Player i's strategy is to reveal a cost a_i to perform its strategy, not necessarily equal to c_i. Given the declared costs of every sensor, a player can decide whether it will contract its services or not. Every player that provides a service receives a payment in the form of credits according to a_i.

Given this formulation, Wu and Shu [27] propose a mechanism for data collection considering the cooperation among different WSNs. In this problem, the cooperation should be a choice whenever a sensor node cannot be reached by its local network and nodes from a foreign network can offer assistance in its routing. Moreover, cooperation is beneficial when nodes from a foreign network can assist routing traffic via a less expensive path. Thus, in this mechanism designed for data collection, if an agent reveals its real cost for routing a packet, its benefits will be maximized. In Figure 3.2a, we can observe a sensor map in which two WSNs are deployed. Each node has a different identification, and, in particular, the sink node receives the number 0. The number on each edge represents the physical distance between a pair of nodes. Network E_1 comprises nodes

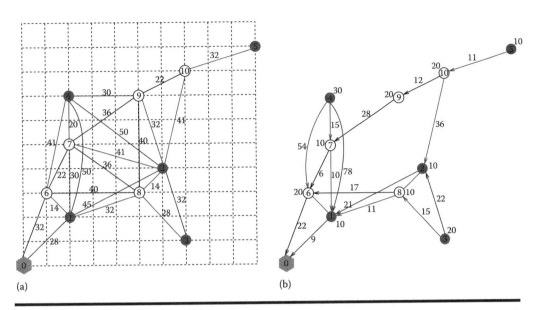

Figure 3.2 *InterSensorNet* for data collection. (a) A sensor network map. (b) A sensor network graph. (From Wu, M.-Y. and Shu, W., InterSensorNet: Strategic routin and aggregation. In: *Proceedings of the IEEE Global Telecommunications Conference (GLOBECOM'05)*, St. Louis, MO, 5 pp., November–December 2005. With permission.)

$\{n_1, n_2, n_3, n_4, n_5\}$, whereas the network E_2 comprises nodes $\{n_6, n_7, n_8, n_9, n_{10}\}$. The network graph is shown in Figure 3.2b, where the reception costs are marked next to the nodes and the transmission costs are marked on the edges. Only edges that are used in communication are shown. We observe that several sensor nodes from a network can use sensor nodes from the other network to reduce its communication costs. For instance, node $n_7 \in E_2$, without cooperation, routes its packets through $n_6 \in E_2$ with a path cost of 48 and, with cooperation, it may route its packets using $n_1 \in E_1$ and reduce the path cost to 29.

Crosby and Pissinou [8] propose another approach to study the problem of cooperation among different WSNs using evolutionary game theory. The problem is modeled as a noncooperative iterated N-player game $g = (P, S, U)$, where P denotes the set of players, S is the set of strategies, and U is the set of utility functions that define the payoff. The players are, again, the authorities that control the networks, and their strategies are to define if all their sensor nodes will cooperate or not with sensor nodes of a foreign network. The time is divided into slots and, once per time slot, the sensors of each WSN send packets to be forwarded by a node from a foreign network, which follows its own strategy. This game was assumed to be infinite, since there is a very small probability that the game ends in any time slot, that is, the players are unaware of the ending of the game.

The modeling of this game is identical to the Iterated Prisoner Dilemma game, which has a payoff matrix shown in Figure 3.3. In this model, the networks have a disincentive in the amount β to not expend energy and an incentive in the amount γ for cooperation, following the inequality $\gamma > \beta > 0$. While the so-called "reward" $R = \gamma - \beta$ represents the payoff for mutual cooperation, the payoff for defection on both sides is the "punishment" $P = 0$. When a network unilaterally defects, it gets the "temptation" payoff $T = \gamma$, while the cooperating network gets the "sucker's payoff" $S = -\beta$. Given the inequality $T > R > P > S$ and the assumption $2R > T + S$, mutual cooperation returns the highest collective payoff in repeated encounters.

$$N_q$$

		Cooperate	Not cooperate
N_p	Cooperate	P, P	S, T
	Not cooperate	T, S	R, R

Figure 3.3 Payoff matrix of the "Prisoner's Dilemma" game.

The cooperation is not evolutionary stable when the networks are playing the iterated N-player prisoner's dilemma [8]. Moreover, in the case of stationary classes, there is some possibility for cooperation to emerge without any incentive. For this case, Crosby and Pissinou [8] present a localized distributed and scalable algorithm, called Patient Grim Strategy, that tells a sensor node to cooperate and continue cooperating until the other player defects n times ($n > 0$) and then the node defects forever. This protocol is proved to be a Nash equilibrium of the modeled problem.

Miller et al. [20] consider the possibility of different WSNs to exchange different favors such as routing, sensing, processing, and data storage. They consider in the problem formulation two sponsoring organizations $i \in \{A, B\}$, which are the same authorities that control the networks, each one with K sensors $s_{i1}, \ldots, s_{iK}$. Each sensor has a probability λ of providing a favor at time slot t, with the payoff of sensor node s_{ik} at time t being given by $u_{ik}(t) = \alpha R_t - \beta P_t - \gamma C_t$, where α, β, and γ are positive parameters, R_t is the number of favors received by s_{ik}, P_t is the number of favors provided by s_{ik}, and C_t is the number of transmissions made by s_{ik}. Given this formulation, the utility of a network i in the entire game is

$$U_i = \sum_{t=1}^{T} \sum_{k=1}^{K} u_{ik}(t) + \tau_i,$$

where τ_i is the monetary transfer received by the authority i after the end of period T. The monetary transfers should be bounded above, that is, $\tau < \infty$, and zero sum, that is, the transfers τ_A received by authority A should be equal to the transfers τ_B given by authority B. Moreover, given the assumption that $\gamma / \lambda < \alpha - \beta$, the networks can gain by cooperating. Based on these constraints, Miller et al. [20] show that a stable and beneficial solution to both systems is only feasible if the owners of the networks sign a financial contract before the network deployment.

It is important to emphasize that the problem of cooperation among different WSNs involves several parameters that can significantly influence the establishment of cooperation. Vaz de Melo et al. [25] list and explain these parameters and, through simulation, discuss the benefits of cooperation. That work shows that the network configuration impacts the benefits that cooperation can bring, and different parameters, such as the path loss exponent, the network density, the data collection rate, and the routing algorithm, must be carefully considered when this problem is addressed.

3.3.2 Security

WSNs have severe constraints regarding energy consumption and processing capacity. Due to these characteristics, most of the well-known techniques to ensure security cannot be directly applied to WSNs. Therefore, new strategies need to be devised to achieve good security levels while satisfying these constraints [23].

3.3.2.1 Message Forwarding

An important problem during the message forwarding is to determine which nodes to trust in this process. Agah et al. [1] propose a game-theoretic technique to deal with this problem in WSNs. The approach takes into account reputation, cooperation, quality of security, and distance. The distance is taken into account because the nodes spend more energy sending long-range signals than sending short-range ones. Hence, choosing a closer hop saves energy.

In this game, the nodes are the players and a utility function is defined for each pair of nodes, which considers the reputation, cooperation, and distance between them. The Nash equilibrium is presented, and a pair cooperates with another one if the reputation of the other node is enough, there is a good history of joint operations, and they are close enough. Based on this cooperation, clusters of cooperative nodes emerge.

Agah et al. [1] developed a simulator to evaluate the proposed solution. In the simulation, nodes can move and, thus, the distance between them can change and its utility as well. The proposed scheme is compared with the strategy that only considers proximity. Over time, the average value of the payoff functions and the number of clusters drop. The number of clusters is considerably lower than with the distance-based algorithm. Also, the number of messages passing per node, on average, is lower in the proposed method if compared with the distance-based one.

The proposal presented in Agah et al. [1] helps to decide whether to trust or not another node. That decision is done locally within each node. However, sometimes, it is crucial to flag the untrustworthy nodes and, ultimately, exclude them from the network. Thus, the network would achieve a higher quality.

3.3.2.2 Message Dropping Attacks

Agah et al. [2] address the detection of passive Denial of Service (DoS) at the routing layer. DoS occurs when, to save energy, malicious nodes refuse to deliver packets from other nodes. This behavior degrades the quality of the results achieved within the network. The problem is modeled as a game, where the players are the intrusion detection systems (IDSs), which monitor the nodes. The game is repeated several times. During each round, the IDS plays against each node. Based on the node behavior during the last round, the IDS decides whether it is malicious or not, based on a utility function devised to model the payoffs and depicted in Figure 3.4.

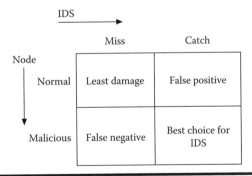

Figure 3.4 Payoff matrix. (From Agah, A. et al., Prevention of DoS attack in sensor networks using repeated game theory. In: *Proceedings of the International Conference on Wireless Networks (ICWN'06)***, Las Vegas, NV, pp. 29–36, September 2006. With permission.)**

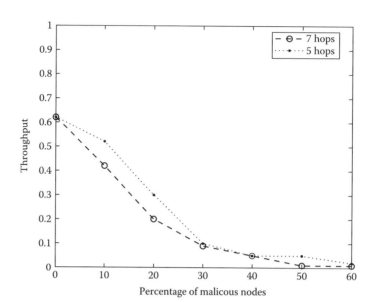

Figure 3.5 Impact of the fraction of malicious nodes. (From Agah, A. et al., Prevention of DoS attack in sensor networks using repeated game theory. In: *Proceedings of the International Conference on Wireless Networks (ICWN'06)***, Las Vegas, NV, pp. 29–36, September 2006. With permission.)**

Nodes flagged as malicious will behave maliciously for the rest of the game, regardless if the classification was correct. Also, catching a bad node has a high payoff. On the other hand, each node has to decide to play cooperatively or not. To promote cooperative actions, a scheme of reputation was created. The impact in the reputation is considered, along with the energy consumption, to decide whether to forward a packet.

Agah et al. [2] propose a technique to define a route to send a packet. This route has the highest payoff among all possible paths. The payoffs are calculated as the difference between the reputations of each node in the path and the power consumption of these nodes. Since malicious nodes have low reputation, usually, the chosen routes exclude them. The protocol is simulated to evaluate the resilience in the increase of malicious nodes and the accuracy of detections of the IDS. The increase in the number of malicious nodes has a high impact on the system, which is depicted in Figure 3.5. With 10% of malicious nodes, the throughput dropped to 52%, and with 20% of bad nodes, the throughput dropped to 35%. Along the time, the algorithm improves its performance by isolating the malicious nodes, preventing them from participating in transmissions, as depicted in Figure 3.6.

Another security breach occurs when malicious nodes refuse to forward broadcasts, which is called denial of message. This happens when broadcasted messages, which may be control messages, are not forwarded. McCune et al. [19] deal with this issue. They analyze the problem using a game-theoretic approach and suggest countermeasures.

The most common countermeasure to an attack is for every node to reply to a broadcast with an ACK (acknowledgment message). However, this overloads the network. McCune et al. [19] propose a protocol called Secure Implicit Sampling, which reduces the amount of ACKs sent. This protocol chooses a subset of nodes that are required to send an ACK, which is unknown to the attacker. Also, since the model takes into account the possibility of network failures, ACKs may be lost despite the action of malicious nodes. The system is trained to define the threshold to set apart expected failures from attacks.

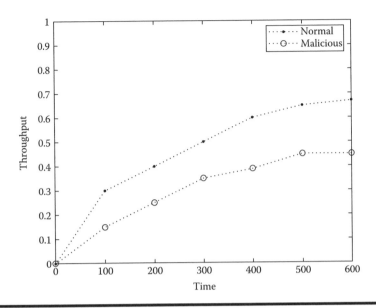

Figure 3.6 **Behavior over time. (From Agah, A. et al., Prevention of DoS attack in sensor networks using repeated game theory. In:** *Proceedings of the International Conference on Wireless Networks (ICWN'06)*, **Las Vegas, NV, pp. 29–36, September 2006. With permission.)**

The attacker is modeled to maximize its payoff over time. The increase on the number of compromised nodes also increases the probability of detection. Therefore, the utility function of the attacker takes into account this probability and models a trade-off between the detection probability and nodes prevented from receiving the broadcast. Also, theoretical forecasts about the accuracy of the model are given. A study is conducted to assess how well the proposed model predicts the simulation results. The forecasts are shown to be quite close to the simulated results.

McCune et al. [19] also present the optimum parameters for their strategy on the evaluated topology as depicted in Figure 3.7. The x-axis shows the fraction of the ACKs and the y-axis shows the payoff of the attacker. In a scenario with no natural packet loss, increasing the number of ACKs also decreases the attacker's payoff. However, surprisingly indeed, when there are losses due to natural causes, increasing the number of ACKs is only helpful till a certain extent. After this threshold point, increasing the number of ACKs decreases the quality of the system, increasing the gain of the attacker.

3.3.2.3 Generic Intrusion Detection

There are several other possible attacks to sensor networks besides the problem of dropping malicious messages. To address the intrusion detection issue without focusing on a specific attack, some studies proposed generic frameworks, as described in the following.

Agah et al. [3] divide the nodes inside clusters with the objective to choose a cluster to protect at each period of time. Due to limitations, the IDS is modeled as capable of protecting only one node at a time. To meet this requirement, each cluster has a cluster head, which is the node to be protected if the cluster is chosen to be kept safe.

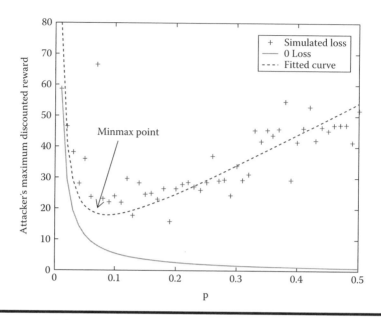

Figure 3.7   Impact of the fraction of ACKs. (From McCune, J.M. et al., Detection of denial-of-message attacks on sensor network broadcasts. In: *Proceedings of the IEEE Symposium on Security and Privacy (S & P'05)*, pp. 64–78, Oakland, CA, May 2005. With permission.)

Three techniques are presented and evaluated to decide the cluster to be protected. The first technique is based on game theory. The modeled game is a two-player noncooperative, where the players are the IDS and the attacker. The network is divided into clusters, each one containing a node that would be protected. Therefore, the attacker has three options: attack a cluster k, attack a different cluster, or not attack at all. The IDS has two options: defend cluster k, or defend another cluster. That work assumes that the choice of the node to be protected is perfect; hence, whenever an attack occurs to cluster k, the IDS is capable of detecting it by monitoring the chosen node from this cluster. The Nash equilibrium is given, and the IDS needs to protect the cluster k, while the attacker would always attack cluster k. The cluster k to be defended is chosen to be the cluster with the highest utility of all available clusters. The second strategy is based on a Markov decision process. It is designed to, through a learning process, assess the cluster that will be attacked, and, therefore, needs to be protected. The third strategy protects the cluster with the highest traffic.

Those three schemes are evaluated through simulation. The strategy of protecting only the most valuable cluster achieved the highest overall quality among all three strategies.

Alpcan and Basar [4] define several virtual nodes that correspond to software-monitoring agents. They are responsible for monitoring the system and signaling whenever they detect a suspicious behavior. These nodes report to the IDS, which decides if an attack is occurring. Therefore, differently from [3], where the IDS actively protects a node, in the proposal of Alpcan and Basar [4], there are several monitoring nodes that report a suspicious behavior to the IDS. Although that work does not address this problem in the context of WSNs, the proposed solution can be directly applied to a WSN.

The game is modeled involving the IDS and the attackers. The attacker can choose the type of attack, and the IDS can choose, based on the information received, the response to the signals of the monitoring nodes. The sensor network is modeled as a third player, which, according to a specific attack, has a probability of correctly alerting the IDS. Alpcan and Basar [4] conclude that using a pure strategy is not adequate for either of the players. Therefore, each player needs to randomize its action. Also, they present the Nash equilibrium for the intrusion detection problem.

3.3.2.4 Jamming Attack

A jamming attack occurs when malicious nodes prevent other nodes from communicating, which can be done by interfering on the transmission frequency or by corrupting packets. Also, jamming can occur naturally, since packets may be lost or corrupted without the action of an attacker. Li et al. [17] address the issue of jamming attacks in WSNs, in particular, attacks based on packet collision. The attacker is modeled as having a probability of attacking, and the attack consists of sending a packet with the intention of causing a collision. The success of the attack depends on the probability of attacking and the probability of a packet being sent during the attack. Also, the attacker faces the trade-off: attacking more frequently increases the short-term payoff and the probability of being caught, and attacking less frequently increases the payoff on the long run. The defense mechanism has a fixed number of nodes, which are trained to assess the probability of random and natural collisions. Therefore, they signal an attack whenever the perceived rate of collisions is expected.

The first evaluated scenario is the jamming with constant power and one monitor node. Initially, both sides have full knowledge about the probability of an access (probability of a transmission by a legitimate node) and the probability of jamming. Therefore, this is a zero-sum game where the attacker tries to maximize the time before being caught while damaging the network by choosing its probability of jamming, and the network tries to minimize this time, by choosing its access probability. Li et al. [17] conclude that the attacker needs to choose the minimum probability of jamming that does enough damage to the network to stay undetected. The "amount" of damage corresponds to positive values for the utility function of the attacker that takes into account the damage and the consumed energy. For the network, it is shown that there is an optimum value of access probability. However, the solution needs to be the closest to the optimum that satisfies the energy constraints. Still in this scenario, both sides are analyzed considering no knowledge about the probability of the adversary. Thus, the game is analyzed as a minimax, where each party defines the best possible response considering that the other party has chosen the probability that achieves its highest payoff.

The second scenario considers a constant jamming power and several monitors. In this case, the attacker needs to remain undetected by all monitors, while damaging the network. The presented solution indicates that the attacker needs to choose the lower probability of attacking while still doing some damage.

The last scenario is a controllable jamming power and several monitors. In this case, the attacker can choose how far its jamming signal can reach. Therefore, it needs to choose the correct power that maximizes its payoff. That work presents an algorithm for the attacker that calculates the payoffs for both the shortest and longest jamming. After this evaluation, it decides whether increasing the power also increases the payoff.

3.4 Distributed Optimization

In a WSN, the cost of data communication is a major constraint [24] and, ideally, a transmission must occur whenever its utility compensates its cost. An interesting challenge in WSNs is to develop solutions able to identify sensor activities essential for a given scenario. However, because of the distributed nature and large scale of WSNs, the main challenge for the sensor nodes is to perform a local decision that does not depend on a central entity. Decentralized and/or distributed control systems can be viewed in mathematical terms as "games of identical interests" in which the player status is attributed to system components with decision-making or control authority [7].

Game theory proposes interesting concepts that can aid in the design of distributed solutions for WSNs that have two important properties. The first one is the "sensor-centric" paradigm [14], that is, to maximize the network utilization and information viability in a WSN; sensors should cooperate to achieve network wide objectives while maximizing their individual lifetime, since the longer a node survives, the better it is for the network. The second property is the fact of treating sensor nodes as "selfish." Using appropriate utility functions, it is possible to design distributed algorithms for optimizing the network performance as a "whole" [21].

3.4.1 Data Collection

Byers and Nasser [6] propose a model in which the sensor nodes adapt themselves locally to improve the user's utility. The utility is measured by a real value, which is mapped by a monotone utility function $U : S \rightarrow [0 : 1]$ that, for a network graph $G = (V, E)$, defines the real value from the sensing subset $S \subset V$, that is, the set of all nodes in the graph that are sensing. The strategy of a sensor node is to define locally which type of activity it will perform not only to increase the function U but also to save its energy. This strategy change is called *node specialization* and it defines four types of operation modes for a sensor node: idle, routing, sensing, and routing/sensing. There are costs c_s, c_t, c_r, c_a associated with the four operations a node may perform: sensing, transmitting and receiving a fixed size message, and aggregating sensory data, respectively.

In that work, Byers and Nasser consider two scenarios: the all-or-nothing case, where the output of the network is either totally useful or useless, and a more forgiving scenario, where the utility of the output varies inelastically. The utility function for both scenarios is illustrated in Figure 3.8. In the first scenario, the utility function is on the left side, and a useful data fusion is only possible if, and only if, the number of nodes participating in the sensing operation is at least as large as the threshold set by the function. This scenario characterizes applications in which the user needs, for instance, the complete information of the environment where the WSN is deployed. If the user can not have the complete information, all the information is useless. In the

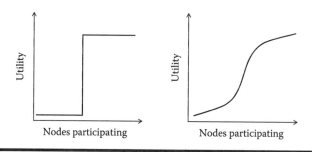

Figure 3.8   Utility functions described in. (From Byers, J. and Nasser, G. Utility-based decision-making in wireless sensor network. In: *Proceedings of the First Annual Workshop on Mobile Ad Hoc Networking and Computing (MobiHoc'00)***, pp. 143–144, Boston, MA, 2000. With permission.)**

second scenario, the utility function is illustrated on the right side of Figure 3.8, where there is some freedom in tuning the number of participating nodes in the sensing activity. Notice that the utility function may be modeled using other types of curves, since it will depend on the WSN application.

Byers and Nasser [6] also propose an objective function that reflects a natural goal for the WSN, which is the maximization of the total aggregated network utility along the time. The proposed objective function takes into account the costs incurred by each possible activity of the sensor node that was previously modeled and has two constraints: first, a node can not consume more energy than its capacity, and, second, the data collected by all nodes participating in the sensing subset S at time t have to be routed to the sink node. This is only possible through a combination of a careful power management solution combined with distributed coordination techniques among the nodes by choosing their roles accordingly along the time.

3.4.2 Data Communication

Utility functions are also employed to design routing solutions in WSNs, in particular for the construction of a data-collecting tree. Each sensor node must decide locally which node will be its ancestor and the nodes that will be its descendants in the tree. Different criteria can be used such as reliability, energy, and delay.

Singh et al. [21] apply techniques from mechanism design and game theory to facilitate the design of decentralized algorithms for building a data-collecting tree that aims to optimize data gathering in sensor networks. They propose the design of suitable local utility functions such that each sensor node optimizes simultaneously its own local utility function and the desired global objective. They consider the problem of constructing a load-balanced spanning tree rooted at the sink node for continuous data-gathering applications.

Basically, the algorithm initiates with a flooding at the sink node. Each sensor marks its level, which is the distance in hops to the sink node. After this, a load-balanced tree is built choosing an ancestor node of level i by its descendants of level $i + 1$. In this problem, Singh et al. [21] consider an iterative game, that is, an iteration is defined as a round in which parents in the collecting tree announce their bandwidth guarantees and the children decide the parent they want to attach based on the maximum bandwidth guarantee they will receive. The bandwidth offered by the ancestors varies as the descendants are being chosen. A descendant node may disconnect itself from an ancestor node in case it considers that the association with another ancestor is more beneficial to it. The process finishes when no descendant has the intention to connect to another ancestor node.

Besides distributing the data collection fairly among the sensor nodes, it is also crucial to the WSN operation that the routing protocol employs a solution that tries to distribute evenly the energy consumption among the nodes. When a group of sensor nodes run out of energy, the network may have a partition, which reduces its lifetime and compromises its functioning. Thus, it is important to have both a load- and energy-balanced data-collecting tree. For the construction of such a tree, the utility function considers the residual energy e_i of the sensor node and of the sensor nodes at level i in the subtree rooted at the sink node. Thus, a sensor node tends to route its data to the sink over the shortest path using $1/e_i$ as the edge length metric, resulting in the desired energy-balanced data aggregation tree.

Zeng et al. [29] also propose a game-theoretic energy balance routing algorithm to even the energy consumption among the sensor nodes. The players are the sensor nodes and their strategies are to relay or not a packet toward the sink node. Every relay course is considered as a game and

		Other nodes	
		Relay	Silent
Node i	Relay	$a, *$	b, c
	Silent	$c, *$	d, d

Figure 3.9 Payoff matrix of the routing game. (From Zeng, J. et al., Game theoretic distributed energy control in sensor networks. In: *Proceedings of the Seventh IEEE International Conference on Computer and Information Technology (CIT'07)*, pp. 1015–1019, Fukushima, Japan, October 2007. With permission.)

the payoff matrix of the nodes is illustrated in Figure 3.9. In this game, a node is playing against its neighbors, in a way that if it relays the message and another neighbor also relays it, the nodes that relayed the message receive a benefit of a, proportional to the number of nodes that relayed the message, and the other ones that did not relay the packet receive a benefit of $c = 0$. If the node relays the message and the other nodes do not, that is, they keep silent, the relay node receives a benefit of $b > a$ and the other nodes receive a benefit of c. If all nodes do not relay the message, they receive a benefit of $d < 0$.

To support the decision making of the sensor nodes, Zeng et al. [29] define probabilities of a node relaying a message and keeping silent. Also, they define a probability that all nodes keep silent at a given time. These probabilities are defined as a function of the residual energy of the nodes, which is done to fairly distribute the energy consumption in relaying the messages among the sensor nodes. Moreover, in simulation results, it was shown that the proposed algorithm has a desirable convergence and good performance.

The above solutions consider that the energy management is the only and primary objective of the routing algorithm. On the other hand, we should consider that WSNs can be deployed in hazardous and hostile environments in which they can fail to operate or be destroyed. To operate efficiently, such networks should consider other QoS metrics rather than long-term survivability, which depends on energy economy. Those networks should consider that the information they are routing is somewhat essential and/or time-critical and the sensor nodes should route them via the most reliable paths available.

Kannan et al. [15] consider this scenario, which adds more constraints to sensor nodes that should make their decisions without the assistance of a central authority. They consider that the probability of a sensor node s_i to fail is p_i, given the inhospitality of the environment where the network is deployed. Thus, in this scenario, the sensor nodes should route over the most reliable paths while minimizing their own energy consumption rather than some aggregate energy criterion. Kannan et al. claim that this model of reliable energy-constrained routing has the advantage of allowing a node to choose its cheapest next hop in the path using only its local information, saving its energy, and prolonging its lifetime.

In this game-theoretic formulation, the players of the routing game are the sensor nodes $S = \{s_1, \ldots, s_n\}$ and each sensor s_i has a piece of information with value $v_i \in \mathbb{R}^+$ that represents an abstract quantification of the value of the event sensed at node s_i. The value 0 is assigned to those nodes whose sensed information does not satisfy the specified attributes of the query. The data are routed to the sink node s_q through an optimal chosen set $S' \subseteq S$ of intermediate nodes. The formation of links between node s_i and s_j is built rationally by each node, taking into account the reliability of the path from s_j and the energy cost $c_{i,j}$ of link i, j. Thus, the strategy of each node s_i is to choose

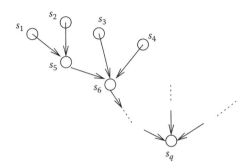

Figure 3.10 Abstract view of a vehicular area network. (From Kannan, R. et al., Sensor-centric quality of routing in sensor networks. In: *Proceedings of the 22nd Annual Joint Conference of the IEEE Computer and Communications Societies (INFOCOM'03),* **volume 1, pp. 692–701, San Francisco, CA, March–April 2003. With permission.)**

locally the node to propagate its message. The payoff $\Pi_i(l)$ of sensor node s_i is a function of the strategy profile l, that is, the routing tree T formed in the network, and it is given by

$$\Pi_i(l) = \begin{cases} V_i R_i - c_{i,j}, & \text{if } s_i \in T, \\ 0, & \text{otherwise,} \end{cases}$$

where

R_i denotes the path reliability from s_i onward to s_q

V_i denotes the expected value of the data at node s_i given the set of its parents F_i, where

$V_i = v_i + \sum_{j \in F_i} p_j V_j$

Thus, s_i gets a piece of information from its parents only if it survives with the given probabilities. In Figure 3.10, the payoff of node s_5 is $\Pi_5 = R_5(v_5 + p_1 v_1 + p_2 v_2) - c_{5,6}$.

Kannan et al. [15] also show that the Nash Equilibrium for this game is the tree generated when all nodes are playing their best response to the other nodes' strategies, that is, every node is receiving its optimal payoff. However, finding the strategy that leads to the optimal payoff requires each node to determine the optimal tree formed by each of its possible successors on receiving its data, which is a NP-Hard problem. Thus, Kannan et al. propose a distributed algorithm, so the nodes compromise to maximize their least possible payoff rather than selecting a neighbor to maximize their individual payoff. This solution was compared with other centralized solutions and, given the proposed scenario, it found reasonable good paths. Other results for this problem are described in [12–14].

Game theory can also be used to define the cluster heads of hierarchical routing algorithms. Since the number of sensor nodes in WSNs can extend to large values such as hundreds of thousands, clustering is a grouping technique that poses an efficient method to reduce the energy consumption. In that technique, a group of sensor nodes are grouped and elect one of them as their cluster head that is responsible for gathering the data from all sensor nodes in its group and forwarding it to the sink node. One of the main challenges is to efficiently distribute the energy consumption among the nodes in a group, since the energy expenditure at the cluster head is significantly higher. Thus, game theory is used to solve the conflict of selecting the cluster head, since no node, if selfish, would naturally declare itself as a cluster head.

		s_2	
		D	ND
s_1 D		$v-c, v-c$	$v-c, v$
ND		$v, v-c$	$0, 0$

Figure 3.11   Payoff matrix of the two-player clustering game.

Koltsidas and Pavlidou [16] study the problem of choosing a single cluster head among the population of N nodes using a game-theoretic approach. A clustering game $CG = \langle I, S, U \rangle$ is defined, where I is the set of players, $S = \{S_i\}$ is the set of the available strategies, and $U = \{U_i\}$ is the set of utility functions for each node. The players are the sensor nodes and they have two possible pure strategies $\{D, ND\}$: to declare (D) itself as cluster head or not (ND).

The payoff of the players is again modeled considering the abstract value v of the data that should be transmitted to the sink and the cost involved in this transmission. If a player declares itself as a cluster head, its payoff is $c - v$, where v is the abstract value for successfully delivering the data with value v, and c is the cost of becoming a cluster head, which involves all the energy consumption associated with this role, such as the reception of all data collected by the sensor nodes of its group and the transmission of this data to the sink. If a player does not declare itself as a cluster head and no other node becomes a cluster head either, its payoff will be 0, as the player will be unable to send its data toward the sink. On the other hand, if at least one node from its group declares itself as a cluster head, then its payoff will be solely v, which is the highest possible for this game.

For the two-player clustering game, the payoff matrix is illustrated in Figure 3.11. In this two-player symmetric game, the symmetrical strategies (D, D) and (ND, ND) are not Nash equilibrium strategies, since, in the first one, a player can improve its payoff by changing its strategy to ND and, in the second one, a player does it by changing its strategy to D. Thus, the Nash equilibrium strategies of this game are all asymmetrical, that is, the (D, ND) and (ND, D) strategies.

In the N player clustering game, the results are basically the same, where no symmetric pure strategy leads to a Nash equilibrium. The only Nash equilibrium is the asymmetrical strategy profile, which all players play ND and only one plays D. Because there is no symmetrical equilibrium in this game, the payoff of the nodes is unbalanced, which causes an unfair energy consumption among the nodes and, consequently, an early degradation of the network functionality. In order to solve this problem, Koltsidas and Pavlidou [16] allow the players to play mixed strategies, that is, players choose their strategies randomly following a probability distribution. In this game, the probability

$$p = 1 - \left(\frac{c}{v}\right)^{(1/N-1)}$$

leads to a mixed strategy Nash equilibrium.

From the above-described games and results, Koltsidas and Pavlidou [16] develop a more realistic energy consumption model and, then, propose and evaluate an algorithm to select the cluster head of a group of sensor nodes. They compare their algorithm with the LEACH protocol [10]. The proposed method provides a higher network lifetime for most of the evaluated cases.

3.4.3 Cross-Layer Optimization

While general-purpose networks are designed to support a wide variety of applications, a WSN is a special case of an ad hoc network designed for a specific application with specific constraints. Given these characteristics, it is possible to design a WSN that does not follow a traditional network architecture such as the OSI model. In this case, we can conceive protocols that rely on interactions between different layers, a method called cross-layer design. Proposals in this group aimed to achieve performance improvements and optimize network functionalities, and should be applied carefully. A cross-layer design is particularly important for any network using wireless technologies, since the state of the physical medium can significantly change over time [18].

In a WSN, the application layer is responsible for obtaining information from the underlying physical phenomenon as accurately as possible while the other layers operate under network resource limitations. Yuan and Yu [28] formulate a network optimization problem, in which the objective is to the minimize the overall distortion at the application layer considering the resource constraints of a WSN. They formulate a source coding game at the application layer and a power control game at the physical layer, both implemented in a distributed fashion.

In the power control game, the players are the communication links formed by the sensor nodes and their strategies are to increase or decrease their power transmission. The main problem of the power control game is the transmission interference among nearby sensors. In order to intelligently avoid interference, Yuan and Yu introduce a tax mechanism such that the higher the power a link uses, the more interference it will produce to others and more tax it has to pay. Each link player maximizes a payoff function and Yuan and Yu prove that at a certain defined condition, the game is ensured to converge to a unique and stable Nash equilibrium.

In the application layer, the source coding game characterizes the interaction among sensor rates and estimation distortion. The underlying physical phenomenon is denoted as θ, which is a vector of M-independent random variables. Each sensor node i deployed in the field makes a local observation θ_j of θ, being corrupted by an independent observation noise n_i. The observation channel is characterized by a matrix H. At each sensor i, the noisy observation y_i, that is a result of n_i over θ_j, is quantized into a codeword u_i. The quantized information from all sensors is transmitted through the network to the sink node (Yuan and Yu [28] refer to the sink node as the remote central office—CEO) with source rates $(s_1, \ldots, s_N)$. At the sink node, the decoder first jointly decodes the codewords u and then estimates the source $\hat{\theta}$. Thus, the performance criterion is to minimize the mean squared error

$$D(\hat{\theta}, \theta) = \left\| \hat{\theta} - \theta \right\|^2$$

considering that smaller distortions came from higher source rates, which imply higher energy consumptions. In order to control this, Yuan and Yu introduce a payoff function with a price mechanism such that it is easy for nodes to make an appropriate trade-off between energy consumption and distortion in a distributed manner, reaching for a defined condition, a unique and stable Nash equilibrium. This process is depicted in Figure 3.12.

For both games, that is, power control and source coding, Yuan and Yu [28] propose distributed algorithms. Then, they present a distributed primal-dual algorithm that iteratively executes these algorithms and updates shadow prices. This cross-layer design, incorporated with game-theoretic concepts, allows the overall network optimization problem to be solved approximately in a distributed manner, which is a fundamental issue for WSNs.

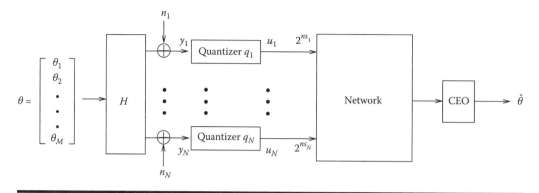

Figure 3.12 Distributed source coding. (From Yuan, J. and Yu, W., Distributed cross-layer optimization of wireless sensor networks: A game theoretic approach. In: *Proceedings of the IEEE Global Telecommunications Conference (GLOBECOM'06)*, **San Francisco, CA, 5 pp., November–December 2006. With permission.)**

3.4.4 Power Management

A good strategy to save energy in a WSN is to turn off parts of a sensor node that are not needed at a given moment for a period of time. For instance, if there will be no communication nor sensed data, a sensor node can turn off its radio or its sensing device, respectively. Hill et al. [11] state that a WSN should embrace the philosophy of getting the work done as quick as possible and going to sleep, to save the network energy and extend its lifetime. The problem is the trade-off that exists between the energy saving and the network operability. An efficient power management design for a WSN should save the energy resources of the sensor nodes without compromising its network operability.

In this direction, Niyato et al. [22] propose a game-theoretic approach to optimal energy management in a WSN. Basically, while the radio is turned on, a sensor node can relay its data more efficiently but the energy consumption is higher. On the other hand, if a sensor node spends too much time sleeping, it will save its energy at the cost of possibly missing important data messages. Niyato et al. define three states of node operation: active, listen, and sleep. In the active mode, a sensor node can receive and transmit packets. In the listen mode, the transmitter circuitry is turned off, making the node not able to transmit packets. In the sleep mode, both the receiver and transmitter are turned off.

Niyato et al. [22] propose a bargaining game to obtain an equilibrium strategy for the energy-saving mechanism at a tagged sensor node and the other nodes that are relaying packets to this tagged node, in a way that all nodes be satisfied with the perceived QoS performance. There are two players in this game: the first one is the entity that governs the packet arrival at the queue of the sensor node, and the second one is the sensor node itself. The packet arrival at a sensor node comes from other sensor nodes that are relaying their packets, and from the sensor device within the tagged node. The strategy for the first player is to select the wake up probability $P_{active,\,sleep}$ when the sensor node is in sleep mode. The strategy for the second player is to select the wake up probability $P_{active,\,listen}$ when the sensor node is in listen mode. The time a sensor node stays in sleep mode defines a packet blocking probability P_{block} that an incoming packet is blocked. The time a sensor node stays in listen mode defines a packet dropping probability P_{drop} that a received/generated packet is dropped due to the lack of buffer space, making the energy used in receiving/generating this packet to be wasted. These probabilities define the payoff of both players, in a way that the payoff of the first player is $U_1 = 1 - P_{block}$ and the payoff of the second player is $U_2 = 1 - P_{drop}$.

In this two-person bargaining game, the players try to come to an agreement on trading/sharing a limited amount of resources and cooperating to gain a higher benefit than that they could have obtained by playing the game without cooperation. While the first player wants the second one to be active and to receive the incoming packets, the second player wants to save its energy and be in a sleep state. On the other hand, the second player wants the first one to transmit its packets that are in the buffer, but the first player wants to transmit only when the second player is in the listen mode. Then, both players change their strategies in the direction of making an agreement. If an agreement between both players cannot be reached, the utility they receive is 0. Based on the Nash equilibrium, the optimal values of the $P_{active,\,sleep}$ and $P_{active,\,listen}$ probabilities can be obtained so that the throughput of a sensor node is maximized while the power supply constraint is met.

Another concern in an efficient WSN power management is whether a sensor node should keep its sensing devices turned on or off. Considering that each sensor has a sensing coverage, if a large region is being covered by several sensor nodes, probably a subset of them should turn off its sensing devices to save its energy. In this direction, Campos-Nañes et al. [7] propose a game-theoretic distributed scheme to efficiently manage the sensing coverage in a WSN. They model a game in which the players are the n sensor nodes and the strategy a_i of player i ($i = 1, \ldots, n$) is to define whether its sensor device is turned on ($a_i = 1$) or off ($a_i = 0$). They also define a coverage function based on the network operational configuration $a = (a_1, \ldots, a_n)$:

$$K(a) = \sum_{i=1}^{n} \mathbf{1}_i(a),$$

where

$$\mathbf{1}_i(a) = \begin{cases} 1, & \text{if sensor } i \text{ is covered under action profile } a, \\ 0, & \text{otherwise,} \end{cases}$$

and a sensor is considered covered if all points within its sensing radius are covered by the sensor itself or by other active sensors. To quantify the trade-off between energy usage and sensing coverage of the network, Campos-Nañes et al. [7] modeled a payoff function that is common to all sensor nodes:

$$U(a) = K(a) - c \sum_{i=1}^{n} a_i,$$

where c is the cost of maintaining a sensor device turned on. This payoff function measures how effective the sensing coverage of the network is and an optimal sensor coverage configuration is given by the solution of

$$\max_{a \in A} \{U(a)\},$$

where $A = \{0, 1\}^n$.

The solution of this problem can be solved in a centralized and off-line mode for reasonable sized instances of n, but since scalability is a major concern in WSNs, Campos-Nañes et al. propose a distributed online algorithm that finds suboptimal solutions and converges to a Nash equilibrium state. They also show good experimental performance results on a MicaZ testbed and on large-scale topologies.

3.5 Open Issues

Game theory is used to model and solve problems related to conflicts of interests among selfish rational agents. In a WSN, the existence of natural selfish rational agents is more restricted than in traditional ad hoc networks, since in a WSN an authority controls all the nodes in the network, while in traditional ad hoc networks each node is responsible for just itself. Thus, game theory naturally emerges in a WSN when there is a conflict of interests between two or more authorities that control a WSN. As discussed in Section 3.3, this problem is studied in the literature but is not solved yet. We believe that proper solutions for this problem are well-defined protocols that allow every sensor node to communicate with each other when it is desired. Based on this communication, sensor nodes should have means to provide and ask for resources that nodes of a foreign network do not have or have plenty. These protocols should be distributed and robust about cheating, guaranteeing that no malicious sensor node will make use of this cooperation to its benefit only.

In this direction, game theory can be naturally applied to model security games, where a malicious external agent tries to jeopardize the network operability or even tries to access confidential information. Because of the severe energy and processing constraints of sensors, most of the well-known techniques to ensure security cannot be applied to a WSN. Therefore, new strategies need to be devised to achieve good security levels while satisfying these constraints. A possible future direction is to address the problem of security modeling and comparing different detection techniques for the same security threat. This would allow the choice of the most reliable and robust strategy given the power consumption limitations. Moreover, the design and analysis of punishment strategies to apply, after an attacker has been caught, can result in a valuable contribution.

Finally, game theory can be applied to the design of distributed solutions for WSNs. Since the number of sensor nodes in a WSN can be very large, its decision making should be local, without any assistance of a central authority. Thus, game theory may be used in the design of utility functions that aim to optimize a global network metric from local decisions of the sensor nodes. As discussed in Section 3.4, there are several solutions that use this concept. Every part of the network operability is subject to a conflict between the quality of the network operation and the energy consumption to realize it. In the MAC layer, for instance, the more the node hears the channel to avoid collisions and to reduce the latency, the more energy is consumed and the shorter is its lifetime. The design of intelligent utility functions that balance the use of network resources and the quality of its operability is still an open challenge in the WSN literature.

References

1. A. Agah, S. K. Das, and K. Basu. A game theory based approach for security in wireless sensor networks. In *Proceedings of the 23rd IEEE International Conference on Performance, Computing, and Communications (IPCCC'04)*, pp. 259–263, Phoenix, AZ, April 2004.
2. A. Agah, M. Asadi, and S. K. Das. Prevention of DoS attack in sensor networks using repeated game theory. In *Proceedings of the International Conference on Wireless Networks (ICWN'06)*, pp. 29–36, Las Vegas, NV, September 2006.
3. A. Agah, S. K. Das, K. Basu, and M. Asadi. Intrusion detection in sensor networks: A non-cooperative game approach. In *Proceedings of the Third IEEE International Symposium on Network Computing and Applications (NCA'04)*, pp. 343–346, Boston, MA, August–September 2004.
4. T. Alpcan and T. Basar. A game theoretic analysis of intrusion detection in access control systems. In *Proceedings of the 43rd IEEE Conference on Decision and Control (CDC'04)*, volume 2, pp. 1568–1573, Atlantis, December 2004.

5. C. Busch, R. Kannan, and A. V. Vasilakos. Quality of routing congestion games in wireless sensor networks. In *Proceedings of the fourth Annual International Conference on Wireless Internet (WICON'08)*, Maui, HI, November 2008. 6 pp.

6. J. Byers and G. Nasser. Utility-based decision-making in wireless sensor networks. In *Proceedings of the First Annual Workshop on Mobile Ad Hoc Networking & Computing (MobiHoc'00)*, Boston, MA, pp. 143–144, 2000.

7. E. Campos-Nañez, A. Garcia, and C. Li. A game-theoretic approach to efficient power management in sensor networks. *Operations Research*, 56(3):552–561, May 2008.

8. G. V. Crosby and N. Pissinou. Evolution of cooperation in multi-class wireless sensor networks. In *Proceedings of the 32nd IEEE Conference on Local Computer Networks (LCN'07)*, Dublin, Ireland, pp. 489–495, October 2007.

9. M. Felegyhazi, L. Buttyan, and J. P. Hubaux. Cooperative packet forwarding in multi-domain sensor networks. In *Proceedings of the Workshop on Sensor Networks and Systems for Pervasive Computing (PerSeNS'05)*, Kauai Island, HI, pp. 345–349, March 2005.

10. W. R. Heinzelman, A. Chandrakasan, and H. Balakrishnan. Energy-efficient communication protocol for wireless microsensor networks. In *Proceedings of the 33rd Hawaii International Conference on System Sciences (HICSS'00)*, Maui, HI, volume 2, January 2000. 10 pp.

11. J. Hill, R. Szewczyk, A. Woo, S. Hollar, D. Culler, and K. Pister. System architecture directions for networked sensors. In *Proceedings of the Ninth International Conference on Architectural Support for Programming Languages and Operating Systems (ASPLOS-IX)*, Cambridge, MA, pp. 93–104, November 2000.

12. R. Kannan and S. S. Iyengar. Game-theoretic models for reliable path-length and energy-constrained routing with data aggregation in wireless sensor networks. *IEEE Journal on Selected Areas in Communications*, 22(6):1141–1150, August 2004.

13. R. Kannan, L. Ray, R. Kalidindi, and S. S. Iyengar. Max-min length-energy-constrained routing in wireless sensor networks. In *Proceedings of the First European Workshop on Wireless Sensor Networks (EWSN'04)*, pp. 234–249. Volume 2920 of *Lecture Notes in Computer Science*, Berlin, Germany, January 2004.

14. R. Kannan, S. Sarangi, and S. S. Iyengar. Sensor-centric energy-constrained reliable query routing for wireless sensor networks. *Journal of Parallel and Distributed Computing*, 64(7):839–852, July 2004.

15. R. Kannan, S. Sarangi, S. S. Iyengar, and L. Ray. Sensor-centric quality of routing in sensor networks. In *Proceedings of the 22nd Annual Joint Conference of the IEEE Computer and Communications Societies (INFOCOM'03)*, San Francisco, CA, volume 1, pp. 692–701, March–April 2003.

16. G. Koltsidas and F.-N. Pavlidou. Towards a game theoretic formulation of clustering routing in wireless sensor networks. In *Proceedings of the Second International Workshop on Game Theory in Communication Networks (GameComm'08)*, Athens, Greece, October 2008. 9 pp.

17. M. Li, I. Koutsopoulos, and R. Poovendran. Optimal jamming attacks and network defense policies in wireless sensor networks. In *Proceedings of the 26th IEEE International Conference on Computer Communications (INFOCOM'07)*, pp. 1307–1315, Anchorage, AK, May 2007.

18. X. Lin, N. B. Shroff, and R. Srikant. A tutorial on cross-layer optimization in wireless networks. *IEEE Journal on Selected Areas in Communications*, 24(8):1452–1463, August 2006.

19. J. M. McCune, E. Shi, A. Perrig, and M. K. Reiter. Detection of denial-of-message attacks on sensor network broadcasts. In *Proceedings of the IEEE Symposium on Security and Privacy (S&P'05)*, Oakland, CA, pp. 64–78, May 2005.

20. D. A. Miller, S. Tilak, and T. Fountain. "Token" equilibria in sensor networks with multiple sponsors. In *Proceedings of the Workshop on Stochasticity in Distributed Systems (StoDiS'05)*, San Jose, CA, December 2005.

21. M. Singh, N. Sadagopan, and B. Krishnamachari. Decentralized utility based sensor network design. *Mobile Networks and Applications*, 11(3):341–350, June 2006.

22. D. Niyato, E. Hossain, M. M. Rashid, and V. K. Bhargava. Wireless sensor networks with energy harvesting technologies: A game-theoretic approach to optimal energy management. *IEEE Wireless Communications*, 14(4):90–96, August 2007.

23. A. Perrig, J. Stankovic, and D. Wagner. Security in wireless sensor networks. *Communications of the ACM*, 47(6):53–57, June 2004.

24. G. J. Pottie and W. J. Kaiser. Embedding the internet wireless integrated network sensors. *Communications of the ACM*, 43(5):51–58, May 2000.

25. P. O. S. Vaz de Melo, F. D. da Cunha, J. M. Almeida, A. A. F. Loureiro, and R. A. F. Mini. The problem of cooperation among different wireless sensor networks. In *Proceedings of the 11th International Symposium on Modeling, Analysis and Simulation of Wireless and Mobile Systems (MSWiM'08)*, Vancouver, Canada, pp. 86–91, October 2008.

26. A. C. Voulkidis, A. V. Vasilakos, and P. G. Cottis. Evolutionary game theory and wireless networks. In Y. Zhang and M. Guizani (Eds.), *Game Theory for Wireless Communications and Networking*, CRC Press, 2011.

27. M.-Y. Wu and W. Shu. InterSensorNet: Strategic routing and aggregation. In *Proceedings of the IEEE Global Telecommunications Conference (GLOBECOM'05)*, St. Louis, MO, November–December 2005. 5 pp.

28. J. Yuan and W. Yu. Distributed cross-layer optimization of wireless sensor networks: A game theoretic approach. In *Proceedings of the IEEE Global Telecommunications Conference (GLOBECOM'06)*, San Francisco, CA, November–December 2006. 5 pp.

29. J. Zeng, C. Mu, and M. Jiang. Game theoretic distributed energy control in sensor networks. In *Proceedings of the Seventh IEEE International Conference on Computer and Information Technology (CIT'07)*, pp. 1015–1019, Fukushima, Japan, October 2007.

Chapter 4

Game-Theoretic Models for Vehicular Networks

Dusit Niyato, Ekram Hossain, and Mahbub Hassan

Contents

4.1 Introduction

An intelligent transportation system (ITS) integrates information, computer, and telecommunication technology to enhance the safety, efficiency, and pleasantness of road transportation. ITS applications include both safety-related (e.g., emergency message communication) and non-safety-related (e.g., road traffic condition, pollution monitoring, remote vehicle diagnostics, and infotainment) applications. These applications can be supported through vehicle-to-infrastructure (V2I) and/or vehicle-to-vehicle (V2V) communications based on different wireless technologies. In a V2I communication scenario, the onboard unit (OBU) in a vehicle communicates with the infrastructure (i.e., roadside base station or RBS). In a V2V communication scenario, OBUs in the vehicles communicate with each other directly. Also, V2I and V2V communications can be integrated to improve the efficiency and flexibility of data transmission in a vehicular networking environment. Different wireless technologies such as the 3G cellular wireless, mobile broadband wireless access (MBWA), wireless local area network (WLAN), and dedicated short-range communication (DSRC) technologies can be used in a vehicular network. Wireless access protocols for these different technologies, however, need to be optimized for a vehicular network considering its unique characteristics (e.g., high mobility of vehicle) and specific quality of service (QoS) requirements of ITS applications.

For wireless access by OBUs, many conflicting situations arise in a vehicular network. For example, several OBUs may competitively access the radio channel to connect to an RBS. The OBUs are generally rational to maximize their own benefits. However, the benefit of one OBU depends not only on its own action (i.e., strategy) but also on the actions of other OBUs. Game theory is a set of mathematical tools used to analyze the conflicting situations involving multiple agents (i.e., players). The players are rational (or of self-interest) to strategically maximize their own benefits (i.e., payoffs). The players decide to perform actions according to their received payoffs. The payoff of one player is a function of its own action and the actions of other players. To obtain the optimal strategy of each OBU in a vehicular wireless access environment, noncooperative game theory can therefore be applied. Besides, there are situations where the vehicles may cooperate with each other for information sharing through V2V communications. Cooperative game models can be used to model such V2V communication scenarios and obtain the optimal wireless access methods for the OBUs. In this chapter, we present several game-theoretic models that can be used as a basis to design optimal wireless access methods for V2I and/or V2V communications in a vehicular networking environment. For each game model, the motivation, system model, formulation, and some selected numerical examples are presented. At the beginning, we provide an overview of the different ITS applications that can be supported by vehicular networks and the basics of V2I and V2V communications. A brief introduction to the different game models considered here followed by the different conflicting situations, which are modeled by these games, is then presented.

The rest of this chapter is organized as follows: Section 4.2 presents an overview of the different ITS applications and V2I/V2V communications scenarios in vehicular networks. Section 4.3 introduces several game models and the different conflicting situations that can be modeled by using these games. A noncooperative game model is presented in Section 4.4 for truthful dissemination of road traffic information by OBUs in a V2V communications scenario. For V2I communications, a game model is presented in Section 4.5 for bandwidth auction among OBUs at an RBS. A stochastic game model for competitive wireless access for streaming data through V2I communications is presented in Section 4.6. A game model is presented in Section 4.7 for transmission rate control of traffic sources to the mobile routers (i.e., vehicles) for delay-tolerant vehicular telematic applications using V2I communications. Again, for a V2V scenario, a bargaining game model is discussed in Section 4.8 for peer-to-peer (P2P)-based data transfer. For applications involving both V2V and V2I communications, a hierarchical game model is presented in Section 4.9 where the vehicles form clusters and the limited radio bandwidth is shared among the vehicles in a heterogeneous wireless access environment. Extensions of these game models are discussed in Section 4.10. Finally, Section 4.11 concludes the chapter.

4.2 ITS Applications and Vehicular Networks

4.2.1 ITS Applications

Vehicular networks support data transfer among moving vehicles and fixed infrastructure for various ITS applications as follows:

- *Public safety applications:* These ITS applications aim to increase the safety in transportation systems, for example, to reduce the number of vehicle collisions and accidents. The collision warning system [1–4] can avoid vehicle crashes. For example, a vehicle detecting obstacle on the road can broadcast this information to other vehicles so that drivers can react properly and timely.
- *Traffic management applications:* These applications aim to improve the traffic flow on the roads, which can reduce the travel time, congestion, transportation cost, and accident. A vehicle or an RBS can monitor the local traffic conditions. This information is then passed to other vehicles so that the optimal route can be chosen to minimize the travel time [5,6]. Also, traffic lights can be scheduled according to the traffic load condition to minimize congestion.
- *Driver support applications:* These applications aim to provide useful information including traffic, road, and weather condition to the drivers [7]. Drivers can retrieve these information from the RBSs. The road conditions (e.g., water on the road, repairing bridge, or bumps) can

also be proactively reported to the drivers in advance. This type of application is referred to as vehicular telematics applications [8,9]. Also, they include applications to collect highway tolls and parking payments automatically. In other applications, the repair and maintenance data can be collected and remotely reported to the vehicle service centers.

■ *Infotainment applications:* These applications aim to enhance the pleasantness of the road journey for passengers. Voice and instant messaging can be communicated among moving vehicles. Web access, video, and multimedia streaming can be provided through the RBSs [10,11].

4.2.2 Communication Scenarios in Vehicular Networks

Different types of communication scenarios, namely, V2I, V2V, and hybrid communications, can be identified in vehicular networks to support different types of ITS applications (Figure 4.1). In V2I communications, data are transferred between vehicles and fixed infrastructure (e.g., RBS or gateway), which is connected to the public network (e.g., Internet). Alternatively, the vehicles can communicate with each other directly. This is referred to as V2V communications. V2V communications can involve either single-hop or multihop transmissions. In single-top communication, the OBU in a vehicle transmits data directly to the other vehicles. On the other hand, in multihop communication, data from the source vehicle can be relayed by other vehicles to the destination. In a hybrid scenario, V2V and V2I communications are integrated (Figure 4.2). In one such scenario, data from a vehicle are transmitted to an RBS. Then, the RBS relays these data to the destination vehicle [12]. This is referred to as infrastructure-based relay approach. In another scenario, data from the source vehicle are relayed through multiple vehicles to the infrastructure (i.e., RBS) [13]. This is referred to as vehicle-to-vehicle-to-infrastructure (V2V2I) [14] or vehicle-based relay approach.

Two major communication scenarios in vehicular networks, that is, V2I and V2V scenarios, are discussed next.

4.2.3 Vehicle-to-Infrastructure Communications

In V2I communications, vehicles transmit and/or receive data from the infrastructure (i.e., RBS, access point, or gateway). For traffic management and driver support applications, a vehicle can monitor the road and traffic condition, and then report the information to the ITS servers residing in the external network connected with the RBS. Also, a vehicle can download the road and traffic

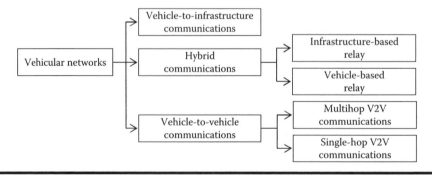

Figure 4.1   Communication scenarios in vehicular networks.

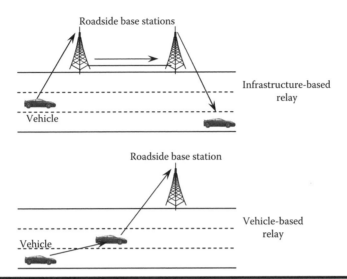

Figure 4.2 Infrastructure-based and vehicle-based relay communications in a vehicular network.

information from the ITS servers via roadside access point. V2I communication can also be used for infotainment applications. Passengers can access Internet and download video and multimedia streaming data during the journey.

V2I communications can be based on the traditional cellular wireless technology [15] and/or mobile broadband wireless access (MBWA) such as IEEE 802.16e technology [16]. For short-range V2I communications (e.g., from a vehicle to an RSB installed in traffic light post), IEEE 802.11 WLAN technology can also be used [17]. A feasibility study and comparison between using IEEE 802.16 and IEEE 802.11 technologies for V2I communications were presented in Chou et al. [18]. It was observed that although IEEE 802.16 can provide larger coverage for wireless access, the delay could be higher due to the complicated network structure and access protocol.

However, since the IEEE 802.11 standard was not designed for high mobility wireless access environment, it lacks of the proper mobility management functionalities. Therefore, DSRC standard/IEEE 802.11p has been introduced specifically for vehicular networks, which has similar physical and medium access control (MAC) layer specifications as the IEEE 802.11 [19]. An experimental study on using IEEE 802.11p for V2I communications was presented in Xiang et al. [20]. DSRC can be used in many ITS applications such as for transferring real-time safety-critical data [21], exchanging logistic information [22], and performing traffic control [23,24]. In Jonsson and Bohm [25], a collision-free MAC mechanism was introduced as an enhancement to the IEEE 802.11p standard for safety-critical and real-time V2I communications. To ensure that the delay-sensitive data are transmitted before deadline, a prioritization mechanism based on the vehicle's position and road traffic intensity was presented. With this protocol, the QoS requirements for the application can be met, while at the same time, the network throughput can be improved. In Jhang and Liao [26], the IEEE 802.11 DCF (distributed coordination function) protocol was enhanced for V2I communications to support collaborative and opportunistic data forwarding between vehicles. A new transmission technique based on cooperative diversity was applied in Ilhan et al. [27] for V2I communications to gain higher transmission rate.

Wireless access by the vehicles in a V2I communication scenario depends on the bandwidth demand of the vehicles, the availability of the bandwidth at the corresponding RBSs, locations of the vehicles, and vehicle mobility. The limited available bandwidth of the V2I link to an RBS needs to be shared among multiple vehicles. The bandwidth demand of a vehicle can be determined from the application's QoS requirement and the mobility of the vehicle. For example, for infotainment type of applications, a vehicle can download data in advance (i.e., pre-fetch) when wireless connectivity is available to RBSs to avoid service outage due to buffer underrun effect while the vehicle is on the road. The bandwidth sharing becomes a challenging issue when each vehicle has self-interest to maximize its own benefit. In a competitive wireless access environment, each of the vehicles will compete to obtain as much bandwidth as possible to meet its communication requirement in a short period of time (e.g., due to fast mobility). Game theory can be applied to analyze competitions among vehicles.

4.2.4 Vehicle-to-Vehicle Communications

In V2V communication scenarios, vehicles transmit data to each other directly without the use of any infrastructure. Due to direct transmission, the communication delay is much smaller than that for V2I communications. Therefore, V2V communication is suitable for real-time ITS applications such as collision warning and avoidance applications. For traffic management and driver support applications, road traffic information can be forwarded to the far-away vehicles to avoid congestion and to let the drivers be prepared for the road and weather conditions. For infotainment applications, voice and video conferences among users in moving vehicles can be supported using V2V communications [28]. Also, P2P applications such as file exchange/sharing can be supported in vehicular environment through V2V communications [29].

V2V communications form the basis of a vehicular ad hoc network (VANET), which is a special class of mobile ad hoc network (MANET) [30]. The mobility of vehicular nodes in a VANET is based on vehicle movements [31] rather than the random way point mobility model as adopted for many MANETs [32]. For V2V communications in a VANET, the specific requirements of ITS applications have to be taken into account.

DSRC can be used for V2V communications. In Kukshya and Krishnan [33] an experimental study was conducted to measure the performance of this protocol. In a multihop V2V communication scenario, routing is important for relaying data over multiple vehicles from source to destination. In Chen et al. [34] a routing protocol was proposed, which takes the driving information of the vehicles into account. For data forwarding, the routing metric combines hop count with vehicle speed, which results in smaller delay. In Kwon et al. [35] an on-demand unicast routing protocol was proposed. This protocol is composed of route search, establishment, and maintenance mechanisms optimized to achieve the best performance in terms of reachability while minimizing the signaling overhead.

Note that a comprehensive survey on inter-vehicle communication protocols and related ITS applications specifically based on V2V communications can be found in Willke et al. [36].

4.3 Game Theory for Designing Vehicular Networking Protocols

4.3.1 Game Models

The following game models can be used to model and analyze different conflicting situations in a vehicular network.

- *Noncooperative game:* In a noncooperative game [37], the players make decisions and perform actions independently to maximize their own payoffs. In general, a noncooperative game model is described by a set of players, a set of actions for each player, and the corresponding payoff, which is a function of actions of all players. A noncooperative game can be played by the players with either complete or incomplete information. In a complete information game, each player knows the available actions and preferences (i.e., *types*) of all other players. On the other hand, in an incomplete information game, each player knows only the probability distribution of the *type* of other players [37]. A player can decide and perform its action deterministically or randomly. The former is referred to as a pure strategy in which one specific action is chosen. The latter is referred to as a mixed strategy in which the probability distribution of all available actions is determined. For a noncooperative game, Nash equilibrium is one of the most widely used solutions. This Nash equilibrium ensures that none of players will change its action to achieve a better payoff given that all other players stick to using their Nash equilibrium actions. In short, at the Nash equilibrium, none of the players has any motivation to deviate.
- *Stochastic game:* Similar to a noncooperative game, in a stochastic game [38], players are noncooperative and they select actions such that their payoffs are maximized. However, in a stochastic game, the players are characterized by their states (e.g., payoff of a player is a function of its state and action). The state transitions are random and may possess the Markov property. Therefore, each player optimizes its policy (i.e., mapping of local state to the action) given the policies of other players. The solution of a stochastic game is generally referred to as constrained Nash equilibrium.
- *Evolutionary game:* An evolutionary game models the decision-making process of a population (i.e., group) of players [39]. Unlike a noncooperative game, in an evolutionary game, these players possess the property of bounded rationality in which the rationality of a player is limited by the available information. Therefore, a player may not be able to make decision to maximize its payoff directly. Alternatively, a player will evolve over time by gradually adjusting its action so that the payoff is maximized. The typical solution of an evolutionary game is the evolutionary equilibrium. At this equilibrium, the population stops evolving.
- *Cooperative game:* Players in a game can cooperate to achieve better payoffs compared to that with noncooperation. If the players are cooperative, a cooperative game model can be used to obtain an efficient and fair solution for all players. The most common type of cooperative game is the bargaining game [40] in which the players can negotiate on the solution. The most popular solution concept for a bargaining game is the Nash bargaining solution (NBS) which provides both fairness and efficiency (due to Pareto optimality). Two other solution concepts are Kalai-Smorodinsky solution (KSS) and Egalitarian solution (ES) [41].

4.3.2 Conflicts in Vehicular Networks

In a vehicular network, different entities (e.g., vehicles and roadside base stations) have different objectives that could be conflicting with each other. A few examples of these conflicting situations are provided in the following.

- *Broadcasting true road traffic information:* Road traffic information can be exchanged among vehicles so that they can choose their optimal routes to the destination. However, a selfish vehicle can minimize its travel time by propagating false traffic information. If other vehicles believe this false information, they will deviate from using the same route as that of the selfish vehicle. A conflicting situation arises here between the selfish and the non-selfish vehicles. A selfish vehicle has to decide whether it should broadcast true traffic information or not, and a regular vehicle should decide whether it should trust the received traffic information or not. A noncooperative game can be formulated to obtain the Nash equilibrium strategies for the vehicles [42].

■ *Bandwidth allocation among vehicles by the RBS:* RBSs are typically deployed at selected locations (e.g., bus stops or traffic lights) to provide wireless access to the passengers in the vehicles (e.g., buses). When a vehicle moves into the coverage area of an RBS, the vehicles/users download and cache data. When the vehicle moves out of the coverage area of the RBS, the cached data is used by the users, for example, to playout streaming data. When there are multiple vehicles connected to an RBS, the OBUs in these vehicles compete for bandwidth to download data from the RBS. An auction mechanism can be used for bandwidth allocation by the RBS among vehicles. A competitive bidding strategy has to be determined for each vehicle (i.e., OBU) so that its utility is maximized. This situation can be analyzed by using a noncooperative game model for which the Nash equilibrium can be obtained as the solution of the bidding prices of the OBUs [43].

■ *Location-aware wireless access for data streaming:* Vehicular users with streaming applications (e.g., infotainment applications) download and cache data from RBSs deployed at the different locations. The network service provider (NSP) can adjust the price of wireless connectivity for vehicles based on the total demand at the different locations of the RBSs. Therefore, each vehicle has to determine the optimal location-aware wireless access policy such that the application QoS requirement (i.e., buffer underrun probability for a streaming application) is met while the cost of wireless connectivity is minimized. With the buffer state and location of each vehicle, a stochastic game model can be formulated to obtain the constrained Nash equilibrium for data downloading policy [44].

■ *Data transfer from sources (e.g., telematic sensors) to mobile routers:* For delay-tolerant vehicular telematic applications, vehicles can be used as mobile routers to transfer data from the telematic sensors (i.e., data sources) to the destination (i.e., data sink) connected with the RBSs. However, due to the limited buffer space, a mobile router selectively receives data from different sources. Therefore, the sources have to compete with each other to upload data to the mobile routers. In this case, the sources can optimize their transmission rates such that their utility (which is a function of the end-to-end throughput and the cost of wireless connectivity) is maximized. A noncooperative game can be formulated to obtain the Nash equilibrium of transmission rates of data sources [45].

■ *P2P data transfer in vehicular networks:* Peer-to-peer (P2P) file sharing protocols can be used in vehicular networks to transfer large amount of data among vehicles through V2V communications. In such a case, each vehicle has an objective to maximize the amount of data received from other vehicles within a limited connection time. To reach a fair and efficient solution of data exchange among moving vehicles, a bargaining game can be formulated [29].

■ *V2I and V2V communications in cluster-based heterogeneous vehicular networks:* In a heterogeneous vehicular network, vehicular nodes (i.e., OBUs) can use different wireless technologies to communicate with the RBSs and other nodes. A vehicular node has to determine whether it should become a cluster head with a direct connection to RBS or become a cluster member and let the cluster head relay the data to the RBS or another vehicle (in the same cluster or in a different cluster). If the vehicular node chooses to become a cluster head, it has to determine the competitive price to be charged to its cluster members. On the other hand, if the vehicular node chooses to become a cluster member, it has to select the cluster head to relay its data to the RBS or to another cluster. For cluster heads, a noncooperative game can be formulated and solved to obtain the Nash equilibrium price to be charged to the cluster members. For the cluster members, an evolutionary game can be formulated to obtain the equilibrium solution on cluster head selection [46].

In the following, the details of the game models will be presented, which have been developed to analyze the aforementioned conflicting situations.

4.4 Game-Theoretic Modeling of Selfish Behavior in Vehicular Networks

Vehicles can exchange road traffic information among each other to help the drivers to identify road congestion [47], obtain optimal driving route [48], and improve traffic safety [49]. This is known as cooperative driving. To support cooperative driving, the OBU in a vehicle collects and propagates collected local road traffic information (e.g., location and speed of vehicle) to other vehicles using V2V communications based on WiFi radio. However, since the vehicles want to reach the destination as fast as possible, the OBUs can be programmed to broadcast false information about road traffic (i.e., cheating). In [42], this selfish behavior of vehicles in broadcasting road traffic information was studied. A noncooperative game model was formulated and the Nash equilibrium of the game was obtained.

4.4.1 Selfish Behavior of an OBU to Maximize Its Utility

The objective of a selfish OBU is to maximize its own utility, which is defined as the average duration of journey on the road. Since this objective can be achieved by removing congestion on the road, the OBU can broadcast false information in the network. With falsified information about high congestion condition in a certain route, the other vehicles will refrain from taking the same route.

While a regular OBU (i.e., naive OBU) will collect road traffic information (e.g., from local measurement or from other OBUs) and send it to other OBUs, the selfish OBU will perform differently. First, a selfish OBU calculates the shortest path from the origin to the destination. Then, for any road that is not in the route of the selfish OBU, the true information will be broadcast. However, for all roads that are in the route of the selfish OBU and have not been traveled by this selfish vehicle yet, a high congestion condition will be broadcast.

Simulation results show that a selfish vehicle can achieve a smaller delay compared to that of a regular OBU using the same or a different route (Figure 4.3). Note that a selfish vehicle gains benefit from reporting false information only after the second round of journey. Since in the first round, the regular OBUs have not received the false information yet, they do not deviate from the old route as the selfish vehicle wants.

A noncooperative game can be formulated to obtain the equilibrium strategies for selfish vehicles and regular vehicles. In this game, the *players* are selfish OBUs and regular OBUs. A selfish OBU broadcasts the duration it takes to travel on a certain road. Since this duration information could be false, the selfish OBU randomly chooses the value between the thresholds T_{slf} and T_{max} from an inverse geometric distribution, where T_{max} is the maximum duration. This threshold can be chosen to be $T_{avg} \leq T_{slf} \leq T_{max}$, where T_{avg} is the true average duration. The *payoff* of the selfish

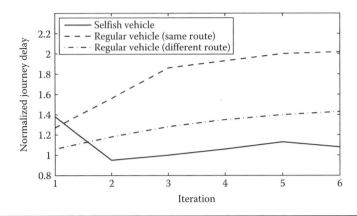

Figure 4.3 Normalized average delay for different vehicles.

OBU is the time duration of journey from origin to destination. Since the regular OBU knows that there could be false information in the network, it will ignore the received information if the broadcast information on duration of travel on a road is higher than the threshold T_{ign} for $T_{avg} \leq T_{ign} \leq T_{max}$. The *strategies* of the selfish and regular OBUs are the values of thresholds T_{slf} and T_{ign}, respectively.

If the duration information broadcast by the selfish OBU is higher than or equal to threshold T_{ign} of regular OBU, it will not gain any benefit, and its payoff is zero. In contrast, if it is lower than the threshold T_{ign}, the selfish OBU will gain a positive benefit. For a regular OBU, if the false duration broadcast by the selfish OBU is accepted, the payoff of a regular OBU will be negative. Otherwise, the regular OBU will gain positive benefit. Therefore, the payoffs of selfish and regular OBUs (i.e., U_{slf} and U_{reg}, respectively) can be defined as follows:

$$U_{slf}(T_{slf}, T_{ign}) = \begin{cases} 0, & T_{slf} \geq T_{ign} \\ T_{slf} - T_{avg} + 1, & T_{slf} < T_{ign} \end{cases} \tag{4.1}$$

$$U_{reg}(T_{ign}, T_{slf}) = \begin{cases} T_{ign} - T_{avg}, & T_{slf} \geq T_{ign} \\ T_{slf} - T_{ign}, & T_{slf} < T_{ign}. \end{cases} \tag{4.2}$$

The Nash equilibrium for selfish and regular OBUs is defined, respectively, by T_{slf}^* and T_{ign}^*, where $U_{slf}(T_{slf}^*, T_{ign}^*) \geq U_{slf}(T_{slf}, T_{ign}^*)$ and $U_{reg}(T_{ign}^*, T_{slf}^*) \geq U_{reg}(T_{ign}, T_{slf}^*)$. This Nash equilibrium is found to be $(T_{slf}^*, T_{ign}^*) = (T_{avg}, T_{avg})$ for which none of the selfish and regular OBUs will change its strategy to improve its payoff. This Nash equilibrium can be proved to be unique. However, the Nash equilibrium solution is inefficient since the selfish OBU cannot gain any benefit (i.e., payoff is zero as $T_{slf}^* = T_{ign}^* = T_{avg}$). Also, since the regular OBU will ignore all the information with the duration larger than or equal to the average duration, it is unable to detect the false information in the network.

In summary, for exchanging road traffic information, a vehicular network is vulnerable to the misbehavior of a selfish vehicle. The selfish behavior of vehicles can degrade the performance of the network. Consequently, game-theoretic incentive mechanisms can be designed to prevent such misbehaviors.

4.5 Bandwidth Auction Mechanism for Wireless Access for V2I Communications

For infotainment applications, OBUs can pre-fetch data from RBSs [50,51] which are deployed at the selected locations (e.g., bus stops in Figure 4.4). Due to the sporadic wireless connectivity along the route, the OBUs have to ensure that sufficient amount of data is cached, which can be used when wireless connectivity is not available (i.e., when the vehicle is on the road). When there are multiple vehicles at the same location, an auction mechanism can be used for bandwidth sharing among vehicles [43]. In Akkarajitsakul and Hossain [43], a game-theoretic bandwidth auction mechanism was developed to solve the conflict among vehicles to share the wireless bandwidth to connect to the RBS. In this model, the auctioneer and the bidders are the RBS and the vehicles, respectively. A vehicle can adjust its bidding strategy such that its utility is maximized.

4.5.1 Model for Bandwidth Auction

Let us consider one RBS that is connected with the Internet (Figure 4.4). When a vehicle moves into the coverage area of the RBS, the OBU can connect to the RBS and download data from the Internet. When the vehicle moves out of the coverage area of the RBS, the OBU uses the cached data in the buffer. The total available bandwidth to connect to RBS is denoted by B, which is shared among N vehicles. Vehicle i submits the bid defined as $d_i = (q_i, p_i)$ to the RBS, where q_i is the total required amount of bandwidth, and p_i is price per unit of downloaded data [52]. Given the bids from all vehicles, the amount of allocated bandwidth to vehicle i is obtained from

$$b_i = \min\left(q_i, \frac{p_i}{\sum_{j=1}^{N} p_j} B\right). \tag{4.3}$$

4.5.2 Game Formulation of Bidding Strategy

The bandwidth allocated to one vehicle will depend not only on its own bid but also on the bids from other vehicles. Therefore, each vehicle will optimize its bid such that its utility is maximized. Given this conflicting situation, a noncooperative game for sharing the bandwidth at the RBS can be formulated as follows: The *players* of this game are the vehicles connecting to the RBS. The *strategy* is the bidding price, that is, p_i. The *payoff* is the utility, which is defined as follows:

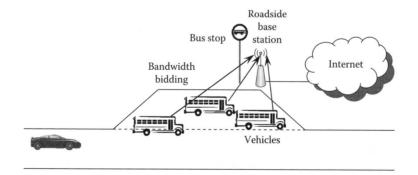

Figure 4.4 System model of bandwidth auction for bandwidth allocation among vehicles.

$$\mathscr{U}_i(p_i, \mathbf{p}_{-i}) = \mathscr{V}_i(b_i(\mathbf{b})) - \mathscr{C}_i(b_i(\mathbf{b}), p_i) \tag{4.4}$$

where

$\mathbf{p}_{-i}$ is the vector denoting the bidding prices of all vehicles except vehicle i

$\mathbf{b}$ is the vector of bids from all vehicles

$\mathscr{V}_i(b_i(\mathbf{b}))$ is the valuation function

$\mathscr{C}_i(b_i(\mathbf{b}), p_i)$ is the cost function defined as $\mathscr{C}_i(b_i(\mathbf{b}), p_i) = b_i p_i$

The valuation function is defined as follows:

$$\mathscr{V}_i(b_i) = t_{on,i}\alpha \log(1 + \gamma b_i) + \mathscr{S}(t_{out,i}) \tag{4.5}$$

where

$t_{on,i}$ is the time interval during which the vehicle is connected to the RBS

α and γ are the constants of the logarithmic utility function of allocated bandwidth b_i

$\mathscr{S}(t_{out,i})$ is the user's satisfaction as a function of the time interval t_{out} during which there is no cached data for the user (i.e., service is interrupted)

In other words, t_{out} is the time duration during which there is not enough cached data in the buffer. This satisfaction is defined as follows:

$$\mathscr{S}(t_{out,i}) = 1 - \frac{1}{1 + \exp(-\mu(t_{out,i} - \beta))} \tag{4.6}$$

where μ and β are constants. The best response of vehicle i can be defined as follows:

$$\mathscr{B}_i(\mathbf{p}_{-i}) = \arg\max_{p_i} \mathscr{U}_i(p_i, \mathbf{p}_{-i}). \tag{4.7}$$

Then, the Nash equilibrium is given by

$$p_i^* = \mathscr{B}_i\left(\mathbf{p}_{-i}^*\right) \tag{4.8}$$

where $\mathbf{p}_{-i}^*$ is the vector of Nash equilibrium of bidding prices of all vehicles except vehicle i.

For two vehicles, Figure 4.5 shows the utility (i.e., payoff) of vehicle 1 given the varying bidding prices. When the bidding price of this vehicle increases, the payoff first increases, since the RBS allocates more bandwidth to this user. However, at a certain point, the utility decreases since the cost becomes higher than the benefit. Therefore, an optimal bidding price can be obtained such that the maximum utility is achieved. This is referred to as the best response of the vehicle. Also, when the other vehicle changes its bidding price (e.g., from $p_2 = 1$ to $p_2 = 2$), this best response changes accordingly.

Figure 4.6 shows the best response of two vehicles under different bidding prices. When one vehicle increases its bidding price, the best response for the other vehicle will also be to increase the bidding price. The Nash equilibrium is located at the point where the best responses intersect.

In summary, an auction mechanism can be applied to determine the allocated bandwidth among multiple vehicles according to their demands. Since the vehicles are rational to maximize their own payoffs, the Nash equilibrium of the bidding strategy (i.e., bidding price) can be obtained from a noncooperative game formulation.

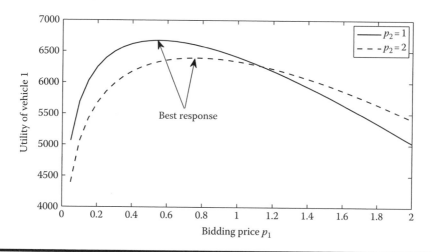

Figure 4.5 Utility of a vehicle under different bidding prices.

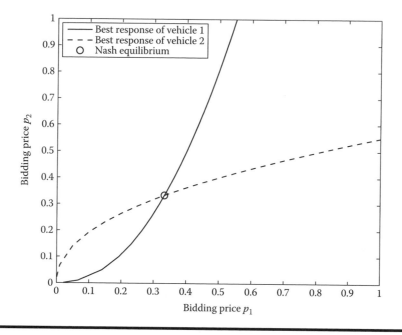

Figure 4.6 Best responses of two vehicles under different bidding prices.

4.6 Stochastic Game Model of Location-Aware Competitive Wireless Access for Data Streaming over V2I Communications

Data streaming is one of the major ITS applications, which relies on V2I communication to download data from Internet servers to the users in moving vehicles. In Niyato et al. [44] the problem of competitive wireless access for data streaming over V2I communications was modeled considering multiple RBSs, locations of those RBSs, mobility of the vehicles, QoS requirements for

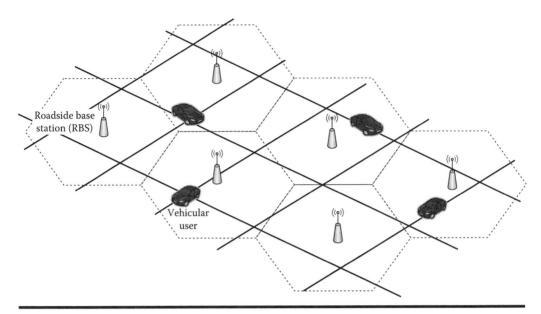

Figure 4.7   Example of service area of competitive wireless access for data streaming over V2I communications.

the streaming application (e.g., buffer underrun probability), and the cost of wireless connectivity. This buffer underrun probability denotes the probability that the data in the streaming buffer in the OBU of a vehicle are less than the demand from streaming application.

A streaming application scenario similar to that in Mancuso and Bianchi [53] is considered for a particular service area (Figure 4.7). This service area is composed of a set of locations denoted by $\mathbb{L}$ where wireless access is available through RBS $l \in \mathbb{L}$. All RBSs are assumed to belong to the same network service provider (NSP). The cost of wireless access through an RBS in a particular location depends on the total demand from all vehicles served by that RBS in that location. There are N vehicles in this service area. For vehicle $i \in \{1, \ldots, N\}$, the mobility is modeled by a transition matrix $\mathbf{M}_i$. The element $M_i(l, l')$ of this matrix $\mathbf{M}_i$ represents the probability that the vehicle changes its location from l to l'.

The vehicle has a finite proxy buffer of length X packets to store the downloaded streaming data. The packet retrieval process from the proxy buffer is modeled by a batch Markovain process. The transition probability matrix for the packet departure process from the proxy buffer is given by $\mathbf{D}_i^{(d)}$ for $d \in \{0, 1, \ldots, D\}$ departing packets, where D is the maximum batch size. In this case, if the number of packets in the proxy buffer is less than the demand from the streaming application, the playout of streaming data will be interrupted. The corresponding performance measure is the buffer underrun probability, which has to be maintained below a threshold $E_{\max}$. The OBU in vehicle i obtains wireless access to the RBS at location l using bandwidth u_i. A connection fee or price $\mathscr{P}(\bar{u})$ is charged to the vehicle per unit of bandwidth, which is a function of total bandwidth demand from all vehicles, where $\bar{u} = \lim_{t \to \infty} \sup (1/t) \sum_{t'=1}^{t} \sum_{i=1}^{N} u_i(t')$, and $u_i(t')$ is the bandwidth used by vehicle i at time t'.

4.6.1 Stochastic Game Formulation

Given the states of the vehicles (i.e., location, buffer size, and packet departure rate), a stochastic game model can be formulated [54]. The *players* of this game are vehicles. The *state space* of player

i is $\mathbb{S}_i$, which is defined as follows:

$$\mathbb{S}_i = \{(\mathcal{L}, \mathcal{X}, \mathcal{D}); \mathcal{L} \in \mathbb{L}, \mathcal{X} \in \{0, 1, \ldots, X\}, \mathcal{D} \in \{1, \ldots, H\}\} \tag{4.9}$$

where
- $\mathcal{L}$ is the location of the vehicle
- $\mathcal{X}$ is the number of packets in the buffer
- X is the maximum buffer size
- $\mathcal{D}$ is the phase of packet departure (i.e., due to playout of streaming data)
- H is the maximum number of phases

Let $\mathbb{S} = \prod_{n=1}^{N} \mathbb{S}_i$ denote the global state space (i.e., state space of all players), where $\prod$ is the Cartesian product. The *strategy* of each player is the action to request u_i units of bandwidth from RBS at each location. The action space of vehicle i is defined as $\mathbb{U}_i(s_i) = \{0, 1, \ldots, U\}$, where $s_i \in \mathbb{S}_i$. An action $u_i \in \mathbb{U}_i(s_i)$ corresponds to the amount of bandwidth to be used by vehicle i. The state transition probability matrix of player i is denoted by $\mathbf{P}_i(u_i)$, which depends on action u_i. The *payoff* of a player is the long-term average cost of wireless access. The constraint for a player (i.e., a vehicle) is to maintain the buffer underrun probability below or equal to threshold $E_{\max}$.

The strategy of a vehicle can be expressed as the wireless access policy. The stationary policy π_i of vehicle i defines the probability of performing an action in a given state. Let $\nu(\mathbb{U}_i(s_i))$ denote the probability distribution for a discrete set of actions $\mathbb{U}_i(s_i)$ given state $s_i \in \mathbb{S}_i$. The stationary policy for state s_i is defined as $\pi_i(\cdot|s_i) \in \nu(\mathbb{U}_i(s_i))$. We can also define $\boldsymbol{\pi}_i$ as a set of stationary policies for vehicle i, that is, $\pi_i \in \boldsymbol{\pi}_i$. In short, vehicle i will choose action u_i with probability $\pi_i(u_i|s_i)$ if the state of the vehicle is s_i. Then $\boldsymbol{\pi} = \prod_{i=1}^{N} \boldsymbol{\pi}_i$ denotes a set of stationary multi-policies, that is, the policies of all vehicles.

The policy of a vehicle is to be optimized such that the cost of wireless access is minimized and the QoS constraint is met. For this, the long-term average cost $\mathscr{J}_{C,i}(\pi)$ and long-term QoS performance measure (i.e., buffer underrun probability) $\mathscr{J}_{E,i}(\pi)$ for vehicle i are defined as follows:

$$\mathscr{J}_{C,i}(\pi) = \lim_{t \to \infty} \sup \frac{1}{t} \sum_{t'=1}^{t} E_\pi \left(\mathscr{C}_i(\mathcal{S}_{t'}, \mathcal{U}_{t'}, \pi) \right)$$

$$\mathscr{J}_{E,i}(\pi) = \lim_{t \to \infty} \sup \frac{1}{t} \sum_{t'=1}^{t} E_\pi \left(\mathscr{E}_i(\mathcal{S}_{t'}, \mathcal{U}_{t'}, \pi) \right). \tag{4.10}$$

$\mathcal{S}_{t'} \in \mathbb{S}$ is the global state, and $\mathcal{U}_{t'} \in \mathbb{U}$ is the actions of all vehicles at time t', where $\mathbb{U} = \prod_{i=1}^{N} \mathbb{U}_i$. $E_\pi(\cdot)$ in (4.10) denotes an expectation over the stationary multi-policy $\pi \in \boldsymbol{\pi}$. These long-term cost and buffer underrun probability measures are defined as functions of the stationary multi-policy π. $\mathscr{C}_i(s_i, u_i, \pi)$ and $\mathscr{E}_i(s_i, u_i, \pi)$ for $s_i \in \mathbb{S}_i$ and $u_i \in \mathbb{U}_i$ are the immediate cost and immediate buffer underrun probability functions, respectively, which are functions of local state s_i. The immediate cost function is defined as follows:

$$\mathscr{C}_i(s_i, u_i, \pi) = \omega \mathscr{E}_i(s_i, u_i, \pi) + \mathscr{P}(\bar{u})u_i. \tag{4.11}$$

A constrained Markov decision process (CMDP) can be formulated for each vehicle as follows:

$$\text{Minimize:} \quad \mathscr{J}_{C,i}(\pi) \tag{4.12}$$

$$\text{Subject to:} \quad \mathscr{J}_{E,i}(\pi) \leq E_{\max} \tag{4.13}$$

where

multi-policy π can be defined as $\pi = (\pi_{-i}|\pi_i)$

π_{-i} is the multi-policy of all vehicular users except vehicular user i

In this optimization problem, each vehicle can optimize its own policy π_i. The policy of one vehicle will affect the cost of other vehicles.

The constrained Nash equilibrium is considered to be the solution of this game. To obtain this constrained Nash equilibrium, first the feasibility condition for a multi-policy is defined. The multi-policy π is feasible if π satisfies $\mathscr{J}_{E,i}(\pi) \leq E_{\max}$. The multi-policy π is feasible if π is feasible for all $i = \{1, \ldots, N\}$. Then multi-policy π^* is the constrained Nash equilibrium if for each vehicle $i = 1, \ldots, N$ and for any $\tilde{\pi}_i$, the condition

$$\mathscr{J}_{C,i}(\pi^*) \leq \mathscr{J}_{C,i}(\tilde{\pi}) \tag{4.14}$$

is satisfied for feasible multi-policy $\tilde{\pi} = (\pi_{-i}|\tilde{\pi}_i)$.

The optimization problem defined in (4.12) and (4.13) can be solved to obtain the *best response policy* π_n^* of vehicle i given the multi-policy π_{-i} of other vehicles. This best response policy can be obtained by formulating and solving an equivalent linear programming (LP) problem. Let $\phi(s_i, u_i)$ denote the stationary probability that the vehicle takes action u_i when the local state is s_i. The LP problem corresponding to the optimization formulation in (4.12) and (4.13) can be expressed as follows:

$$\text{Minimize} \sum_{(s_i, u_i) \in \mathbb{K}_i} \mathscr{C}_i(s_i, u_i, \pi)\phi(s_i, u_i) \tag{4.15}$$

$$\text{Subject to} \sum_{(s_i, u_i) \in \mathbb{K}_i} \mathscr{E}(s_i, u_i)\phi(s_i, u_i) \leq E_{\max} \tag{4.16}$$

$$\sum_{u_i \in \mathbb{U}_i} \phi\left(s_i', u_i\right) = \sum_{(s_i, u_i) \in \mathbb{K}_i} P\left(s_i'|s_i, u_i\right)\phi(s_i, u_i) \tag{4.17}$$

$$\sum_{(s_i, u_i) \in \mathbb{K}_i} \phi(s_i, u_i) = 1, \quad \phi(s_i, u_i) \geq 0 \tag{4.18}$$

for $s_i' \in \mathbb{S}_i$, where $P\left(s_i'|s_i, u_i\right)$ is the probability that the state changes from s_i to s_i' when action u_i is taken. This probability is the element of matrix $\mathbf{P}_i(u_i)$. $\mathbb{K}_i = \{(s_i, u_i); s_i \in \mathbb{S}_i, u_i \in \mathbb{U}_i(s_i)\}$ is the local set of state-action pairs for vehicle i. The objective and the constraint defined in (4.15) and (4.16) correspond to those in (4.12) and (4.13), respectively.

Let $\phi^*(s_i, u_i)$ denote the optimal solution of the LP problem defined in (4.15) through (4.18). The best response policy π_i^* is a randomized policy, which can be uniquely mapped from the optimal solution of the LP problem as follows:

$$\pi_i^\star(u_i|s_i) = \frac{\phi^*(s_i, u_i)}{\sum_{u_i' \in \mathbb{U}_i} \phi^*\left(s_i, u_i'\right)} \tag{4.19}$$

for $\sum_{u_i' \in \mathbb{U}_i} \phi^*\left(s_i, u_i'\right) > 0$. Otherwise, the action $u_i = 0$ is chosen. The optimal solution $\phi^*(s_i, u_i)$ can be obtained by using a standard method for solving LP.

4.6.2 Constrained Nash Equilibrium

To obtain the solution (i.e., constrained Nash equilibrium), the stationary probabilities for the different states are required. The stationary probability for the vehicle to be in state s_i is denoted by $q_i^\pi(s_i)$ for $s_i \in \mathbb{S}_i$. This probability can be obtained by solving the following set of equations:

$$\left(\vec{\mathbf{q}}_i^\pi\right)^T \mathbf{P}_i^\pi(\cdot) = \left(\vec{\mathbf{q}}_i^\pi\right)^T, \quad \left(\vec{\mathbf{q}}_i^\pi\right)^T \vec{\mathbf{1}} = 1 \tag{4.20}$$

where

$$\vec{\mathbf{q}}_i^\pi = \left[\cdots \; q_i^\pi(s_i) \; \cdots \right]^T$$

$\vec{\mathbf{1}}$ is a vector of ones

$\mathbf{P}_i^\pi(\cdot)$ is the transition probability matrix for vehicle i when the multi-policy π is applied

The following iterative algorithm can be used to obtain the constrained Nash equilibrium.

1: Initialize multi-policy π
2: **repeat**
3:   **for** $i = 1, \ldots, N$ **do**
4:     $J_{C,i} = \mathscr{J}_{C,i}(\cdot, \pi)$ {Compute the stationary cost}
5:     Given multi-policy π, the stationary probability vector $\vec{\mathbf{q}}_i^\pi$ is obtained from (4.20)
6:     Obtain the best response policy $\pi_i^\star$ by solving the LP problem defined in (4.15)–(4.18) given multi-policy π and stationary probability $\vec{\mathbf{q}}_i^\pi$
7:     Update the multi-policy $\pi = \left(\ldots, \pi_i^\star, \ldots\right)$
8:   **end for**
9: **until** $\max \left|J_{C,i} - \mathscr{J}_{C,i}(\cdot, \pi)\right| < \epsilon$.

ϵ is the threshold used in the termination criterion of the algorithm (e.g., $\epsilon = 10^{-6}$). The constrained Nash equilibrium is obtained as $\pi^* = \left(\pi_1^\star, \ldots, \pi_i^\star, \ldots, \pi_N^\star\right)$.

4.6.3 Numerical Examples

Figure 4.8 shows the convergence of the iterative algorithm to obtain the constrained Nash equilibrium policy for two vehicles. In particular, the amount of bandwidth allocated to two vehicles

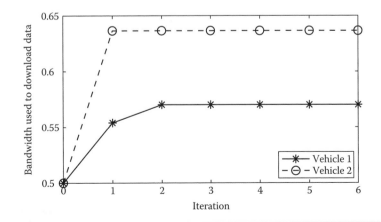

Figure 4.8 Convergence of the optimal wireless access policy adopted by a vehicle.

in each iteration is shown. Clearly, from an initial policy, the algorithm converges rapidly to the equilibrium (e.g., in less than five iterations).

Figure 4.9a and b show the amount of bandwidth and buffer underrun probability under different number of vehicles competing in a service area. When the number of vehicles increases, a larger bandwidth demand results in a higher price per unit of bandwidth. In this case, the cost of wireless access for a vehicle can be minimized by reducing the amount of allocated bandwidth (Figure 4.9a), which results in a higher buffer underrun probability. Nonetheless, due to the QoS requirement considered in the stochastic game formulation, the buffer underrun probability is bounded at $E_{max} = 0.05$ even with increasing number of vehicles (Figure 4.9b). Similarly, the amount of allocated bandwidth reaches a constant value, which is sufficient to maintain the buffer underrun probability at the threshold E_{max}.

In summary, if the price of wireless access through RBSs is adjusted dynamically by the NSP, the vehicles have to optimize their wireless access policy given its buffer state and location. This situation can be modeled as a stochastic game in which the states of the vehicles are independent of each other. Given the QoS requirement (i.e., maximum buffer underrun probability), the constrained Nash equilibrium can be obtained as the solution of this stochastic game.

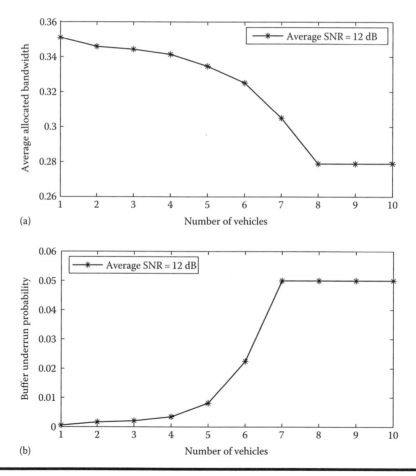

Figure 4.9 **(a) Amount of bandwidth and (b) buffer underrun probability under different number of vehicles.**

4.7 Rate Control in Vehicular Delay-Tolerant Networks

In a delay (or disruption)-tolerant vehicular network (VDTN), the telematic sensors (or data sources) rely on the mobility of the vehicular nodes (i.e., mobile routers) to carry the data and forward it to the sink [45] (Figure 4.10). Each traffic source and sink is connected to an RBS. When the mobile router (i.e., vehicle) moves into the transmission range of the RBS to which the traffic source is connected to, the traffic source transmits the data packet to the mobile router. The data are stored in the buffer of the mobile router when it travels. Once the vehicle moves into the transmission range of the RBS the sink is connected to, the mobile router transmits data in its buffer to this RBS. However, since the resources in a mobile router are limited (i.e., buffer size is finite), which are shared among multiple traffic sources, the performance (i.e., end-to-end throughput) of one traffic source depends not only on its own transmission rate but also on that of each of the other sources. For example, if one source transmits at a high data rate, the buffer of the mobile router will be full, and, hence, when it moves to other sources, data from the other sources cannot be stored and consequently their performances will degrade. A noncooperative game can be developed to obtain the equilibrium transmission rates of the traffic sources to the mobile router.

4.7.1 System Model of a VDTN

A set of locations, denoted by $\mathbb{L}$, is considered. At each location, there is a stationary node, which acts as either a traffic source or a sink (Figure 4.10). The data packets at traffic source i need to be delivered to sink i'. There is no direct connection between any traffic source and data sink. Therefore, the packets from traffic sources are delivered to the sink with the help of mobile routers. A mobile router may visit different locations in $\mathbb{L}$ randomly at a random speed. When the mobile router visits the location of traffic source i, this traffic source transmits the data packet (destined to sink i') to the mobile router. However, the mobile router can decide to accept or reject the incoming packet from the traffic source. This decision of the mobile router depends on the current buffer status and the importance of the packet. Once the mobile router travels and visits the corresponding

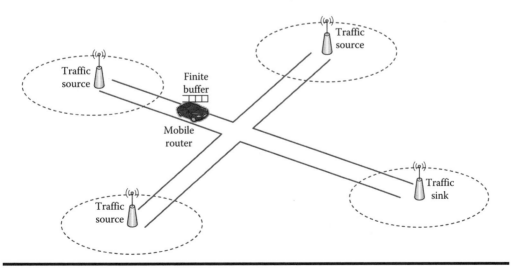

Figure 4.10 Example of a vehicular delay-tolerant network with three traffic sources and one sink.

sink i', the packet is transmitted by the mobile router to the sink. There is no strict delay constraint for packet delivery from a traffic source to the sink in this network.

Let the transmission rate of source i to sink i' be denoted by λ_i. The end-to-end throughput of source i received from the mobile router is denoted by $\tau_i(\lambda)$, where λ is a vector of transmission rates of all sources. The end-to-end throughput of each traffic source is a nondecreasing function of the transmission rate of the corresponding source.

4.7.2 Noncooperative Game Formulation for Rate Control of Traffic Sources in a VDTN

A noncooperative game model for rate control of the traffic sources in a VDTN can be formulated as follows: The *players* of this game are the traffic sources. The *strategy* of each traffic source is the transmission rate denoted as λ_i for source i. The *payoff* is the utility defined as follows:

$$\mathcal{U}_i(\lambda_i, \lambda_{-i}) = w_i \tau_i(\lambda) - c_i \lambda_i \tag{4.21}$$

where

w_i and c_i are the weight of the throughput and the cost of transmission rate, respectively

λ_{-i} is a vector of the transmission rates of all sources except source i

The Nash equilibrium of the noncooperative game is a set of strategies λ_i^* with the property that no traffic source can increase its payoff by choosing a different transmission rate, given other sources' transmission rates λ_{-i}^*. That is,

$$\mathcal{U}_i\left(\lambda_i^*, \lambda_{-i}^*\right) \geq \mathcal{U}_i\left(\lambda_i, \lambda_{-i}^*\right) \quad \forall i. \tag{4.22}$$

The Nash equilibrium can be obtained from the best response of each traffic source. This best response is defined as an optimal set of strategies of a particular source given the strategies of other sources. The best response of traffic source i, as defined in the following, can be obtained numerically.

$$\lambda_i^* = \mathcal{B}_i(\lambda_{-i}) = \arg\max_{\lambda_i} \mathcal{U}_i(\lambda_i, \lambda_{-i}). \tag{4.23}$$

The Nash equilibrium is considered to be the solution of the following optimization formulation, which minimizes the difference between the strategy of traffic source i and its best response:

$$\text{Minimize} \sum_{i=1}^{N} \left| \lambda_i - \mathcal{B}_i\left(\lambda_{-i}\right) \right|. \tag{4.24}$$

The Nash equilibrium is located at the point where the objective function is zero.

In a centralized decision-making scenario, the traffic sources can be cooperative to maximize the total utility, which is defined as follows:

$$\mathcal{T}(\lambda) = \sum_{i=1}^{N} \mathcal{U}_i(\lambda_i, \lambda_{-i}) \tag{4.25}$$

An optimization problem can be formulated to obtain the optimal strategy as follows:

$$\lambda^* = \arg \max_{\lambda} \mathcal{T}(\lambda). \tag{4.26}$$

Again, the optimal strategy can be obtained by using numerical method.

4.7.3 Numerical Examples

The payoffs (i.e., utility) of two traffic sources, which compete to transmit their packets to the sink through the mobile router, are shown in Figure 4.11. As the packet transmission rate increases, the utility first increases. However, at a high transmission rate, the buffer at the mobile router becomes congested. Therefore, the packet blocking probability increases, which results in a lower utility. The packet transmission rate of each source that yields the highest utility is defined as the best response given the packet transmission rates of other sources.

The best responses of sources 1 and 2 are shown in Figure 4.12. As one source increases its packet transmission rate, the other source has to decrease its transmission rate in order to achieve the highest utility. Since the buffer of the mobile router becomes congested when one source increases its packet transmission rate, the other source has to reduce its packet transmission rate to avoid high cost due to packet blocking. Therefore, the best response of each source decreases as the packet transmission rate of other source increases. For noncooperative sources, the point at which the best responses of all sources intersect is the Nash equilibrium.

Figure 4.13 shows the total utility of two traffic sources. As expected, as both sources increase their packet transmission rates, the total utility first increases and then decreases at a certain point due to the congestion at the mobile router. The point that yields the maximum total utility is the optimal strategy of both sources. This optimal strategy (which is obtained by global optimization) is different from the Nash equilibrium. This result indicates that the Nash equilibrium of the packet transmission rates cannot maximize the total utility.

In summary, in a VDTN, there is no end-to-end path from traffic sources to the sinks, and the data are transferred by the mobile routers. Since the buffer size at a mobile router is limited, the traffic sources have to optimize the data transmission rates to the mobile router. A noncooperative game can be formulated to obtain the Nash equilibrium of the transmission rates.

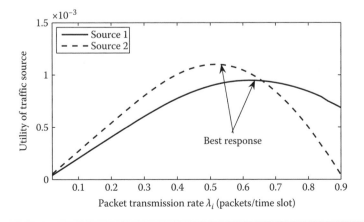

Figure 4.11   Utility of the traffic sources under different packet transmission rates.

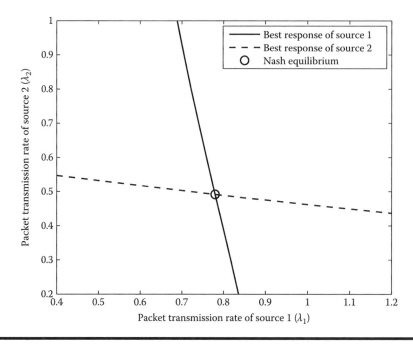

Figure 4.12 Best responses and Nash equilibrium transmission rates of traffic source 1 and source 2.

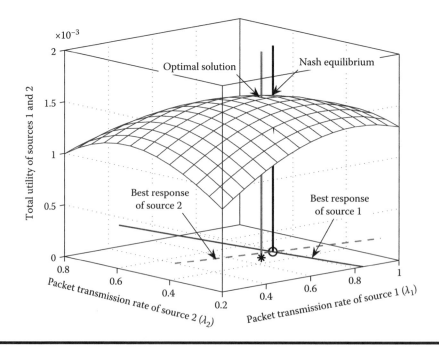

Figure 4.13 Total utility of traffic source 1 and source 2, and locations of Nash equilibrium strategy and optimal strategy.

4.8 Bargaining of P2P-Based Data Transfer in Vehicular Networks

P2P is a distributed network architecture in which resources (e.g., data) can be shared among multiple peers efficiently. In P2P file sharing, a file is fragmented into multiple chunks. These chunks are distributed and stored on a number of nodes in the network. When a user requests for the file, instead of downloading from a single server, the user can download the data chunks from multiple sources. Therefore, P2P file sharing can improve not only the speed but also the reliability (due to the redundant copies of the file). The concept of P2P file sharing (e.g., BitTorrent) has been applied to bulk data transfer in vehicular networks [55,56].

4.8.1 System Model of P2P-Based Data Transfer in Vehicular Networks

In Shrestha et al. [29] a P2P-based data transfer model over vehicular networks was considered. An example scenario is shown in Figure 4.14, where RBS 1 wants to transfer data to the vehicles in lane B, and RBS 2 wants to transfer data to the vehicles in lane A. However, since the file size is large and the duration during which the vehicle will be in the transmission range of an RBS is small, RBS 1 and RBS 2 cannot transmit the entire file at once to the vehicles in lanes B and A, respectively. Therefore, the RBSs transmit the data chunks to the vehicles in the opposite lanes. In this case, RBS 1 and RBS 2 first transmit the data chunks to the vehicles in lanes A and B, respectively. Then, on the road, these vehicles in lanes A and B exchange the data when they pass each other. In this way, a large file can be distributed to multiple vehicles efficiently.

In Shrestha et al. [29] wireless access in vehicular environment (WAVE) technology (i.e., IEEE 802.11p) was considered for data exchange among passing vehicles. The wireless propagation was modeled by a two-ray ground reflection model. Adaptive modulation was used for data transmission among vehicles. The file to be shared is divided into L chunks with the same size, and a vehicle needs to receive all chunks. The weight of chunk k is denoted by $w(k)$. The number of chunks c that can be transmitted within t_0 is defined as $c \leq Bt_0/M$, where B is the transmission rate.

The utility of vehicle i is defined as the sum of weights for the set $\mathcal{I}_i$ of chunks that vehicle i currently has, that is,

$$U_i = \sum_{k \in \mathcal{I}_i} w_i(k). \tag{4.27}$$

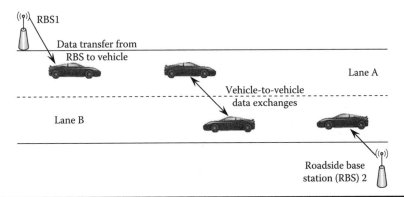

Figure 4.14 System model of P2P-based data transfer in a vehicular network.

This utility function corresponds to the user's satisfaction gained from an application-specific data chunk. For vehicle i and vehicle j, if each has some chunks that the other does not have, they will exchange these data chunks. For two vehicles, the bargaining problem can be stated as follows:

$$\text{Maximize} \quad \mathscr{F}(U_i, U_j) \tag{4.28}$$

$$\text{Subject to} \quad \sum_{k \notin \mathcal{I}_i, k \in \mathcal{I}_j} 1 + \sum_{l \notin \mathcal{I}_j, l \in \mathcal{I}_i} 1 \leq n_{i,j} \tag{4.29}$$

where

$n_{i,j}$ is the maximum number of data chunks that can be exchanged within the time period t_0
This time period t_0 is the duration when two vehicles are in the transmission range of each other
$\mathscr{F}(\cdot, \cdot)$ is a function that represents the social welfare

4.8.2 Bargaining Game among Vehicles

In [29], three fairness criteria for bargaining among vehicles were considered. First, the NBS [41] for a two-player game was considered. The definition of NBS is given in the following.

Definition 4.1 Nash Bargaining Solution   Define $\mathcal{U}$ as the feasible region, $\mathbf{U}$ as the utility vector after users' bargaining, and $\mathbf{U}^0$ as the utility vector before the negotiation (i.e., disagreement point). $\phi(\mathcal{U}, \mathbf{U}^0)$ is the NBS that maximizes the product of utility from both players as follows:

$$\phi(\mathcal{U}, \mathbf{U}^0) = \arg \max_{\mathbf{U} \geq \mathbf{U}^0, \mathbf{U} \in \mathcal{U}} \prod_{i=1}^{2} (U_i - U_i^0). \tag{4.30}$$

Under six general conditions shown in [41], the NBS is a unique solution.

Then, the Kalai-Smorodinsky solution (KSS) and the Egalitarian solution (ES) are considered. These solutions require having the restricted monotonicity property defined as follows:

Definition 4.2 Restricted Monotonicity   If $\mathcal{V} \subset \mathcal{U}$ and $H(\mathcal{U}, \mathbf{U}^0) = H(\mathcal{V}, \mathbf{U}^0)$ then $\phi(\mathcal{U}, \mathbf{U}^0) \geq \phi(\mathcal{V}, \mathbf{U}^0)$, where $H(\mathcal{U}, \mathbf{U}^0)$, called the *utopia point*, is defined as

$$H(\mathcal{U}, \mathbf{U}^0) = \left[\max_{\mathbf{U} > \mathbf{U}^0} U_1(\mathbf{U}) \quad \max_{\mathbf{U} > \mathbf{U}^0} U_2(\mathbf{U}) \right] \tag{4.31}$$

Definition 4.3 Kalai-Smorodinsky Solution   Let Λ be a set of points on the line containing $\mathbf{U}^0$ and $H(\mathcal{U}, \mathbf{U}^0)$. $\phi(\mathcal{U}, \mathbf{U}^0)$ is the KSS, which can be expressed as follows:

$$\phi(\mathcal{U}, \mathbf{U}^0) = \max \left\{ \mathbf{U} > \mathbf{U}^0 \,\middle|\, \frac{1}{\theta_1}(U_1 - U_1^0) = \frac{1}{\theta_2}(U_2 - U_2^0) \right\} \tag{4.32}$$

where $\theta_i = H_i(\mathcal{U}, \mathbf{U}^0) - U_i^0$. The solution is in Λ.

Definition 4.4 Egalitarian Solution $\phi(\mathcal{U}, \mathbf{U}^0)$ is the ES, which can be expressed as follows:

$$\phi(\mathcal{U}, \mathbf{U}^0) = \max\left\{\mathbf{U} > \mathbf{U}^0 \,\middle|\, U_1 - U_1^0 = U_2 - U_2^0\right\}. \tag{4.33}$$

From (4.32), the KSS assigns as the bargaining solution the point in the boundary of a feasible set that intersects the line connecting the disagreement point and the utopia point. From (4.33), the ES assigns as the bargaining solution the point in the feasible set where all players achieve maximal equal increase in utility relative to the disagreement point.

4.8.3 Data Exchange Algorithm

To exchange data among vehicles, the following algorithm is executed.

1: **repeat**
2:    *Neighbor Discovery*: Investigate the neighbor in the transmission range with the best channel and with the packets which are most beneficial.
3:    *Negotiation*: Vehicles exchange information of available data packets and their weights.

     ■ The expected number of transmitted packets $c_{i,j}$ between vehicle i and vehicle j is computed for a certain transmission duration t_0.
     ■ Assume that vehicle i initiates the negotiation by sending a message containing information about its available packets to vehicle j.
     ■ After receiving this information, vehicle j checks whether it has data packets of interest in vehicle i or not.
     ■ Vehicle j replies with a message containing information about the needed packets from vehicle i and their weights.
     ■ Information about the data packets available at vehicle j is piggybacked with this message and sent back to vehicle i.

4:    *Bargaining*: The solution of the bargaining game is obtained.
5:    *Data Transmission*: Exchange packets to the other vehicle.
6:    *Adaptation*: Monitor the channel and adjust modulation and coding rate accordingly.
7: **until** Both vehicles have similar packets or the channel becomes bad.

In Step 4, the bargaining solution is obtained from the following algorithm.

1: Determine weights of available packets k from vehicles $i \in \{1, 2\}$ (i.e., $w_i(k) \in \mathcal{I}_i$), transmission rate between vehicles i and j (i.e., $c_{i,j}/t_0$) where $i \neq j$.
2: Sort packets according to their weights, i.e., $w_i(1) > \cdots > w_i(k) > \cdots > w_i(\langle \mathcal{I}_i \rangle)$, where $\langle \mathcal{I}_i \rangle$ gives the number of elements in set $\mathcal{I}_i$.
3: Define the set of number of transmitted packets by vehicles i and j as $\{(c_i, c_j) : c_i = \{0, \dots, c_{i,j}\}, c_j = c_{ij} - c_i\}$. $U_i(n)$ can be obtained based on (4.27), i.e., $U_i(n) = \sum_{k=1}^{n} w_i(k)$.
4: **if** Nash solution **then**
5:    Obtain solution in terms of $\left(c_i^*, c_j^*\right) = \arg\max_{(c_i, c_j)} \left(U_i(c_i) - U_i^0\right) \times \left(U_j(c_j) - U_j^0\right)$.
6: **else if** Kalai-Smorodinsky solution **then**
7:    Define normalized utility $\hat{U}_i(c_i) = \frac{1}{\theta_i}\left(U_i(c_i) - U_i^0\right)$, where $\theta_i = \max_{c_i\{0,\dots,c_{i,j}\}} U_i(c_i) - U_i^0$.
8:    $\left(c_i^*, c_j^*\right) = \arg\min_{(c_i, c_j)} \left|\hat{U}_i(c_i) - \hat{U}_j(c_j)\right|$.

9:   **else if** Egalitarian solution **then**

10:      The solution is obtained from $\left(c_i^*, c_j^*\right) = \arg\min_{(c_i,c_j)} \left| \left(U_i(c_i) - U_i^0\right) - \left(U_j(c_j) - U_j^0\right)\right|$.

11:   **end if**

12:   $\phi\left(\mathcal{U}, \mathbf{U}^0\right) = \left(U_i\left(c_i^*\right), U_j\left(c_j^*\right)\right)$

13:   The number of packets to be transmitted by vehicles i and j is given by $\left(c_i^*, c_j^*\right)$, respectively.

4.8.4 Numerical Examples

Figure 4.15 shows the transmission rate (i.e., packets/second) between two vehicles using the IEEE 802.11p radio. As two vehicles approach each other (Figure 4.14), the transmission rate becomes higher due to the shorter distance and hence closer transmission range. Note that the flat line on the top of the curve occurs when the highest transmission rate of IEEE 802.11p is used (i.e., transmission mode 11 with 64QAM and coding rate 7/8 is used). We observe that the vehicles with slower speed have longer duration for data transmission.

The different solutions of the bargaining game (i.e., Nash, KSS, and ES) are shown in Figure 4.16. Here, U_1 and U_2 denote the utilities of vehicles in the different lanes. The three solutions along with the Pareto optimal solutions under different transmission rates between two vehicles are shown. The Nash solution is located where $\max(U_1 \times U_2)/U_1$ intersects the Pareto-optimal rates (which is concave). Here, Nash solution, ES, and KSS are located at the different points on the Pareto-optimal utility line.

The rate of vehicles entering the highway is varied, and the utility of the vehicles under different bargaining solutions is shown in Figure 4.17. As expected, when the traffic intensity increases, the utility of the vehicles increases, since there is higher probability that most of data chunks will be exchanged. Different bargaining solutions yield different average utility.

In summary, with P2P-based data transfer in a vehicular network, moving vehicles exchange and collect data chunks from each other. With the different weights of the data chunks and limited connectivity period between vehicles, the exchange has to be performed in a fair and efficient fashion. A bargaining game model can be used to model this situation and various solution concepts (i.e., Nash, KSS, and ES) can be applied.

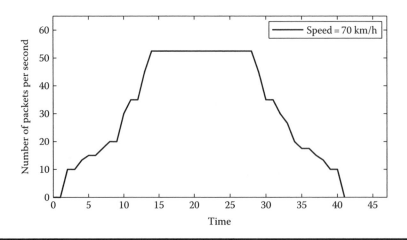

Figure 4.15   Transmission rate between two vehicles.

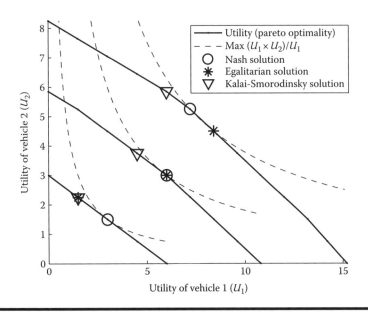

Figure 4.16 Utility, Pareto optimality, and bargaining solutions.

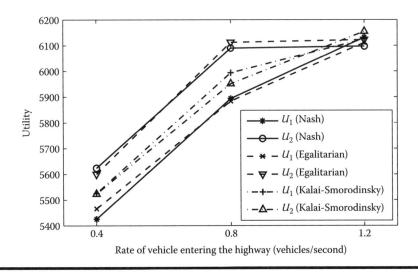

Figure 4.17 Average utility of vehicles under different bargaining solutions.

4.9 V2I and V2V Communications in Cluster-Based Heterogeneous Vehicular Networks

Heterogeneity will be one of the most important features in the next generation mobile communication networks. In such a network, multiple wireless access technologies will be integrated to provide seamless and high-speed wireless connectivity to the mobile users. A heterogeneous wireless access system based on the integration of IEEE 802.11-based WLAN and IEEE 802.16- or WiMAX-based MBWA can be used in a clustered vehicular network to support V2I and V2V communications

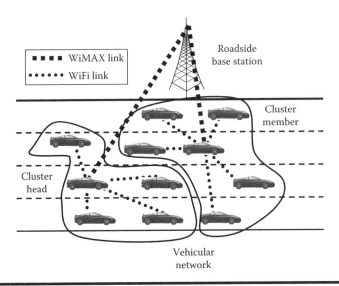

Figure 4.18 Heterogeneous vehicular network with integrated WiFi–WiMAX access.

efficiently [42]. With the clustering structure, IEEE 802.11-based WLAN is used for communications among vehicles (i.e., cluster head and cluster member) while IEEE 802.16-based MBWA is used for communications between vehicle (i.e., cluster head) and the RBS (Figure 4.18). In this section, a game model for such heterogeneous vehicular network proposed in [46] is discussed.

A clustered vehicular network is considered with rational vehicular nodes that have self-interest to maximize their own utilities [42]. A vehicular node has to decide whether to become a cluster head or a cluster member. A cluster head connects to the RBS using WiMAX radio interface. It receives data from cluster members using the WiFi radio interface. A cluster head charges a price to a cluster member to relay its data to the RBS. If a vehicular node decides to become a cluster member, it has to select the cluster head to transmit data to. If a vehicular node decides to become a cluster head, this node has to choose a price to be charged to the cluster member. To obtain these decisions, a hierarchical game model can be formulated by incorporating both networking and economic (i.e., pricing) aspects.

4.9.1 System Model of a Clustered Heterogeneous Vehicular Network

Consider a cluster-based transmission strategy in a vehicular network with N vehicular nodes moving in the same direction (Figure 4.18). Each of these N nodes is equipped with a dual-mode WiFi/WiMAX transceiver. All vehicular nodes need to communicate through the RBSs. Vehicular nodes can form clusters consisting of cluster heads and cluster members. Let n_g denote the number of cluster heads (i.e., number of clusters). The number of cluster members associated with cluster head i is denoted by $n_{c,i}$. The bandwidth on a WiMAX link is B_b bps. This link is shared among $1 + n_{c,i}$ nodes (a cluster head and its members). Each of the vehicular nodes is allocated with a logical WiMAX link with bandwidth $b_i = B_b/(1 + n_{c,i})$ bps. Let B_c denote the aggregated bandwidth on all WiFi links associated with a cluster head. Therefore, the end-to-end bandwidth is $b_i^{(ee)} = \min(B_b/(1 + n_{c,i}), B_c/n_{c,i})$ bps. The RBS charges P_b monetary units (MUs) to cluster

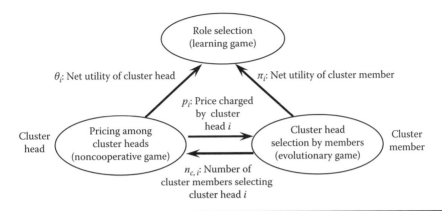

Figure 4.19 A hierarchical game model for vehicular nodes in a heterogeneous vehicular network.

head for WiMAX link with bandwidth B_b bps. Cluster head i offers relay service to the associated cluster members and charges each of its members price $p_i < P_b$ MUs.

Each vehicular node needs to make a two-level decision (Figure 4.19). In the first level, a vehicular node decides whether to be a cluster member or a cluster head. In the second level, a cluster member determines its cluster head, and the cluster head determines its competitive price. Every vehicular node is interested in maximizing its own net utility, which is defined as follows:

$$\mathcal{N}(b, p, r) = \mathcal{U}(b) - p + r \tag{4.34}$$

where

$\mathcal{U}(b)$ is the utility for bandwidth b

p is cost

r is revenue (for cluster head)

The utility is given by the following concave logarithmic function:

$$\mathcal{U}(b) = u_1 \log(1 + u_2 b) \tag{4.35}$$

where u_1 and u_2 are the parameters of the function [57]. This utility function is used to compute the benefit (i.e., net utility) of a vehicular node to become a cluster head or a cluster member.

4.9.2 Decision Making by a Vehicular Node

Given the hierarchical game model shown in Figure 4.19, in the first level, each vehicular node applies a *learning game* to determine its role (as a cluster member or as a cluster head). In the second level, a cluster head applies a *noncooperative game* to obtain the competitive price, while a cluster member uses an *evolutionary game* for cluster head selection. These game models can be analyzed by using backward induction.

4.9.2.1 Cluster Member Decision—Select a Cluster Head

For a cluster member, to select a cluster head, an evolutionary game [58] is applied. Here, each cluster member observes the prices broadcast by the cluster heads periodically. Each cluster member computes the expected net utility and selects the cluster head that yields the highest net utility. The net utility depends on both price and bandwidth in which this end-to-end bandwidth depends on the decision of other cluster members selecting the same cluster head. Therefore, all cluster members iteratively choose a cluster head. After reaching the equilibrium, the net utility will remain unchanged over the rest of the adaptation interval.

An evolutionary game for cluster head selection is formulated as follows: The *players* are the cluster members. The *population* is the group of all cluster members. The *strategy* of a cluster member is the selection of a cluster head. The *payoff* of a cluster member is the net utility. Each cluster member decides to join one of n_g clusters [59] which maximizes its net utility. In this evolutionary game, let x_i denote the proportion of cluster members selecting a cluster head i, where $\sum_{i=1}^{n_g} x_i = 1$. The evolutionary equilibrium is defined as a point where no strategy can lead to a change in the proportion of cluster members x_i. This evolutionary equilibrium satisfies the following condition:

$$\dot{x}_i = x_i \left(\pi_i - \overline{\pi} \right) = 0 \tag{4.36}$$

where

$\pi_i = \mathcal{N}(b_i, p_i, 0)$ denotes the payoff of each cluster member selecting cluster head i
$\overline{\pi}$ denotes the average payoff of n_c cluster members

Since the proportion x_i ceases to change at the evolutionary equilibrium, the number of cluster members associated with cluster head i, $n_{c,i} = x_i(N - n_g)$, ceases to change. At this point, the net utility of each cluster member becomes constant.

4.9.2.2 Cluster Head Decision—Set Competitive Price

A rational cluster head aims at setting price to attract cluster members and maximize its net utility. While a high price repels cluster members, leading to low revenue, a low price attracts cluster members, which may cause congestion. A noncooperative game is formulated to obtain this competitive price. This game can be described as follows: The *players* of this game are cluster heads. The *strategy* of a player is the price. The *payoff* of a cluster head is the net utility defined as

$$\theta_i = \mathcal{N}(b_i, P_b, n_{c,i} p_i) \tag{4.37}$$

where $n_{c,i}$ is the total number of cluster members selecting cluster head i. The Nash equilibrium is considered to be the solution of this noncooperative game at which none of the cluster heads would unilaterally change its strategy to improve its payoff θ_i.

4.9.2.3 Role Selection for Vehicular Nodes

Then, the decision of a vehicular node to choose its role (i.e., as cluster member or as cluster head) is analyzed. Since becoming a cluster member or a cluster head will yield different payoffs (π_i and θ_i, respectively), a vehicular node chooses its role so that its net utility is maximized. The role selection takes the following factors into account. A cluster head pays expensive WiMAX service fee; however, it achieves high data rate and earns revenue from traffic relaying for other nodes. A cluster member pays a cheap traffic relaying fee, but it cannot earn revenue from reselling the service. If there are few cluster heads in the network, each cluster member will obtain only small fraction of end-to-end bandwidth. In this case, the cluster member will switch its role to become a cluster head and earn revenue by relaying traffic. Since the net utility of each node is a private information, vehicular nodes choose their role by learning. The algorithm for role selection is as follows:

1: A vehicular node randomly chooses to become a cluster head or a cluster member
2: **loop**
3:     A vehicular node observes its net utility (i.e., θ_i or π_i if the node decides to be cluster head or cluster member, respectively).
4:     **if** node is a cluster head **then**
5:         Update $\tilde{\theta}_i = (1 - \alpha_i)\tilde{\theta}_i + \alpha_i\theta_i$
6:         **if** $\tilde{\theta}_i < \tilde{\pi}_i$ **then**
7:             Cluster head switches back to become a cluster member. {Becoming a cluster member yields a higher net utility}
8:         **else**
9:             Cluster head randomly becomes a cluster member with small rate ρ (e.g., $\rho = 10^{-3}$) {Learns by trying}
10:        **end if**
11:    **else**
12:        Update $\tilde{\pi}_i = (1 - \alpha_i)\tilde{\pi}_i + \alpha_i\pi_i$
13:        **if** $\tilde{\pi}_i < \tilde{\theta}_i$ **then**
14:            Cluster member switches to cluster head {Becoming a cluster head yields a higher net utility}
15:        **else**
16:            Cluster member randomly becomes a cluster head with small rate ρ. {Learns by trying}
17:        **end if**
18:    **end if**
19: **end loop**

Note that $0 < \alpha_i < 1$ is the learning rate of node i.

This role selection algorithm can be modeled as a learning game as follows: The *players* are the vehicular nodes. The *strategy* of a player is to become either a cluster head or a cluster member. The *payoff* is the net utility of a node. The solution of this learning game is an equilibrium where none of the vehicular nodes has any motivation (i.e., no improvement in its payoff) to switch its role. This equilibrium can be obtained analytically by formulating a finite discrete-state and continuous-time Markov chain.

4.9.3 Numerical Examples

Figure 4.20 shows the number of cluster members selecting different cluster heads at the evolutionary equilibrium. With $n_g = 3$ and $n_c = 30$, the prices of cluster heads 2 and 3 are $p_2 = 1$ and $p_3 = 1.5$, respectively, while that of cluster head 1 is varied. When price p_1 of cluster head 1 increases, the number of cluster members selecting this cluster head 1 decreases. At the same time, the number of cluster members selecting cluster heads 2 and 3 increases.

Figure 4.21 shows the net utility of two cluster heads under different prices (i.e., strategies). Clearly, there is an optimal price to maximize the net utility of each cluster head. This price is referred to as the best response. The Nash equilibrium prices of all cluster heads are located where all best responses intersect. At this Nash equilibrium, none of the cluster heads will change its price to improve its net utility.

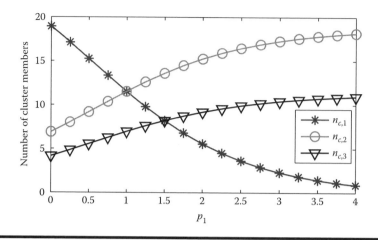

Figure 4.20 Number of cluster members selecting three cluster heads under different price.

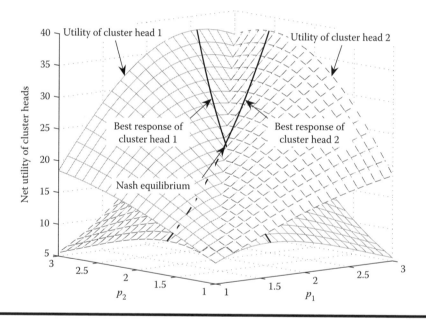

Figure 4.21 Net utility of cluster heads under varied price p_1.

Figure 4.22 shows the net utility of cluster head and cluster member given different number of available cluster heads in the vehicular network. As the number of cluster heads increases, the net utility of a cluster head decreases, since the number of cluster members as well as revenue per cluster head decreases. Conversely, the net utility of a cluster member increases as the number of cluster heads increases—this is due to the larger share of end-to-end bandwidth and lower relaying fee charged by the cluster heads (due to higher degree of competition). It is observed that there is a point where the net utilities of cluster head and cluster member are identical. This is the equilibrium of the learning game since none of the vehicular nodes can improve the net utility by changing the role. This equilibrium can be reached by the role selection algorithm.

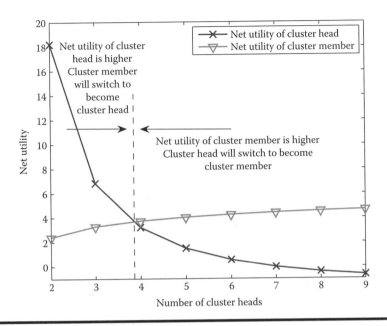

Figure 4.22 Net utility of cluster head and cluster member.

In summary, with heterogeneous wireless access in a vehicular network, different wireless technologies (e.g., IEEE 802.11 WiFi and IEEE 802.16 WiMAX) can be integrated to improve the efficiency of data transfer through V2I and V2V communications. Also, the vehicles can form clusters for better network scalability. For such a clustered vehicular network, a distributed decision-making framework would be required. A hierarchical game model can be formulated to realize this distributed decision framework to obtain decisions on role selection between cluster head and cluster member, the competitive price of a cluster head, and cluster head selection by a cluster member.

4.10 Extensions of the Game Models and Related Research Issues

The different game models described earlier in the context of different V2I and V2V communication scenarios can be extended considering several practical network design issues as described in the following.

- *Decision making of vehicular nodes under incomplete information as Bayesian games:* In a vehicular network, complete information to select a strategy or action may not be available to a vehicular node. For example, the bandwidth demand of multiple vehicles sharing the same RBS is a private information. In a P2P scenario, the weight of the data packets to be exchanged among vehicular nodes may not be known in advance. Therefore, to achieve the optimal strategy in such an environment, game models based on incomplete information (e.g., Bayesian noncooperative game) need to be formulated and solved (e.g., to obtain Bayesian Nash equilibrium).
- *Decision making of vehicular nodes in a dynamic environment and repeated games:* It may be common that the vehicles competing or cooperating for wireless access in a vehicular

network will interact repeatedly. For example, vehicles can bid for the bandwidth from different RBSs at the different points in time. In this case, the dynamics of decision making has to be modeled through dynamic and repeated games with an objective to achieve the optimal long-term decision (e.g., subgame perfect equilibrium).

- *Hierarchical decision making of vehicular nodes and leader–follower games:* In a vehicular network, there could be an entity (i.e., leader) that makes a decision and executes the corresponding action before others (i.e., followers). For example, the transport service provider (e.g., bus operator) reserves the bandwidth from network service provider (NSP). Then, the vehicles bid for the available bandwidth from the transport service provider. In this case, the leader will make decision to maximize its payoff given that the followers make their best decisions. A hierarchical game model such as a Stackelberg game can be used to obtain the solution (e.g., Stackelberg equilibrium).

- *Coalition among vehicular users and coalitional games:* To efficiently utilize the radio resources in a vehicular network, coalitions can be formed among vehicles. For example, given different mobility patterns, the vehicles can share the wireless access at the different RBSs efficiently by forming coalitions such that the cost for wireless access is minimized while the QoS requirements are satisfied. A coalition game model [60] can be applied whose solution is given by the *core*. This core ensures that none of the vehicles will deviate from the current coalition to improve its payoff.

4.11 Conclusion

In this chapter, applications of several game theory models to solve decision-making problems in vehicular networks have been discussed. First, an overview of the different communication scenarios in a vehicular network and several ITS applications, which can be supported through vehicular networks, have been presented. Examples of several different conflicting situations that arise for wireless access in a vehicular networking environment have been discussed, and the game theory tools to model these conflicts have been introduced. The details of these game models from the perspectives of the different communication scenarios have been presented along with representative numerical illustrations. To this end, several extensions of these game models have been outlined.

Acknowledgment

This work was done in the Centre for Multimedia and Network Technology (CeMNet) at the School of Computer Engineering, Nanyang Technological University, Singapore. It was supported in part by the AUTO21 NCE research grant for the project F303-FVT, in part by ITAP AUTO21, and in part by a Visiting Fellowship offered to E. Hossain by the School of Computer Science and Engineering, The University of New South Wales (UNSW).

References

1. T. ElBatt, S. K. Goel, G. Holland, H. Krishnan, and J. Parikh, Cooperative collision warning using dedicated short range wireless communications, in *Proceedings of International Workshop on Vehicular Ad Hoc Networks*, Los Angeles, CA, pp. 1–9, 2006.

2. X. Yang, J. Liu, F. Zhao, and N. H. Vaidya, A vehicle-to-vehicle communication protocol for cooperative collision warning, in *Proceedings of International Conference on Mobile and Ubiquitous Systems (MOBIQUITOUS)*, Boston, MA, pp. 114–123, August 2004.

3. J. Yin, T. ElBatt, G. Yeung, B. Ryu, S. Habermas, H. Krishnan, and T. Talty, Performance evaluation of safety applications over DSRC vehicular ad hoc networks, in *Proceedings of International Workshop on Vehicular Ad Hoc Networks*, Philadelphia, PA, pp. 1–9, 2004.

4. S. U. Rahman and U. Hengartner, Secure crash reporting in vehicular Ad hoc networks, in *Proceedings of International Conference on Security and Privacy in Communications Networks and the Workshops (SecureComm)*, Nice, France, pp. 443–452, September 2007.

5. T. Nadeem, S. Dashtinezhad, C. Liao, and L. Iftode, TrafficView: Traffic data dissemination using car-to-car communication, *ACM SIGMOBILE Mobile Computing and Communications Review*, 8(3), 6–19, July 2004.

6. L. Wischhof, A. Ebner, and H. Rohling, Information dissemination in self-organizing intervehicle networks, *IEEE Transactions on Intelligent Transportation Systems*, 6(1), 90–101, March 2005.

7. H.-Y. Kung and H.-H. Ku, A real-time mobile multimedia communication system for the prevention and alert of debris-flow disaster, in *Proceedings of IEEE Vehicular Technology Conference (VTC)-Fall*, Orlando, FL, vol. 5, pp. 3400–3404, October 2003.

8. J. Lee, Design of a network coverage analyzer for roadside-to-vehicle telematics networks, in *Proceedings of ACIS International Conference on Software Engineering, Artificial Intelligence, Networking, and Parallel/Distributed Computing (SNPD)*, Phuket, Thailand, pp. 201–205, August 2008.

9. M. Garanayak and A. Tripathy, A design framework for analysis of connectivity issues for vehicular adhoc network in a geographical information system, in *Proceedings of IEEE International Advance Computing Conference (IACC)*, Patiala, India, pp. 1031–1036, March 2009.

10. R. Baroody, A. Rashid, N. Al-Holou, and S. Hariri, Next generation vehicle network (NGVN): Internet access utilizing dynamic discovery protocols, in *Proceedings of IEEE/ACS International Conference on Pervasive Services (ICPS)*, Beirut, Lebanon, pp. 81–88, July 2004.

11. F. Xie, K. A. Hua, W. Wang, and Y. H. Ho, Performance study of live video streaming over highway vehicular ad hoc networks, in *Proceedings of Vehicular Technology Conference (VTC)-Fall*, Dallas, TX, pp. 2121–2125, September–October 2005.

12. S. Y. Wang, The potential of using inter-vehicle communication to extend the coverage area of roadside wireless access points on highways, in *Proceedings of IEEE International Conference on Communications (ICC)*, Glasgow, Scotland, U.K., pp. 6123–6128, June 2007.

13. K. Yang, S. Ou, H.-H. Chen, and J. He, A multihop peer-communication protocol with fairness guarantee for IEEE 802.16-based vehicular networks, *IEEE Transactions on Vehicular Technology*, 56(6), 3358–3370, November 2007.

14. J. Miller, Vehicle-to-vehicle-to-infrastructure (V2V2I) intelligent transportation system architecture, in *Proceedings of IEEE Intelligent Vehicles Symposium*, Eindhoven, the Netherlands, pp. 715–720, June 2008.

15. J. Santa, A. Moragon, and A. F. Gomez-Skarmeta, Experimental evaluation of a novel vehicular communication paradigm based on cellular networks, in *Proceedings of IEEE Intelligent Vehicles Symposium*, Eindhoven, the Netherlands, pp. 198–203, June 2008.

16. M. Aguado, J. Matias, E. Jacob, and M. Berbineau, The WiMAX ASN Network in the V2I Scenario, *IEEE Vehicular Technology Conference (VTC)-Fall*, Calgary, Canada, pp. 1–5, September 2008.

17. J. Ott and D. Kutscher, Drive-thru Internet: IEEE 802.11b for automobile users, in *Proceedings of Joint Conference of the IEEE Computer and Communications Societies (INFOCOM)*, Hong Kong, vol. 1, March 2004.

18. C.-M. Chou, C.-Y. Li, W.-M. Chien, and K.-C. Lan, A feasibility study on vehicle-to-infrastructure Communication: WiFi vs. WiMAX, in *Proceedings of International Conference on Mobile Data Management: Systems, Services and Middleware (MDM)*, Taipei, Taiwan, pp. 397–398, May 2009.

19. J. Zhu and S. Roy, MAC for dedicated short range communications in intelligent transport system, *IEEE Communications Magazine*, 41(12), 60–67, December 2003.

20. W. Xiang, Y. Huang, and S. Majhi, The design of a wireless access for vehicular environment (WAVE) prototype for intelligent transportation system (ITS) and vehicular infrastructure integration (VII), in *Proceedings of Vehicular Technology Conference (VTC)-Fall*, Calgary, Canada, pp. 1–2, September 2008.

21. A. Bohm and M. Jonsson, Supporting real-time data traffic in safety-critical vehicle-to-infrastructure communication, in *Proceedings of IEEE Conference on Local Computer Networks (LCN)*, Montreal, Canada, pp. 614–621, October 2008.

22. A. E. Coronado, C. S. Lalwani, E. S. Coronado, and S. Cherkaoui, Wireless vehicular networks to support road haulage and port operations in a multimodal logistics environment, in *Proceedings of IEEE/INFORMS International Conference on Service Operations, Logistics and Informatics (SOLI)*, Chicago, IL, pp. 62–67, July 2009.

23. C.-M. Huang, C.-C. Yang, C.-Y. Tseng, and C.-H. Chou, A centralized traffic control mechanism for evacuation of emergency vehicles using the DSRC protocol, in *Proceedings of International Symposium on Wireless Pervasive Computing (ISWPC)*, Melbourne, Australia, pp. 1–5, February 2009.

24. R. M. Daoud, M. A. El-Dakroury, H. M. Elsayed, H. H. Amer, M. El-Soudani, and Y. Sallez, Wireless vehicle communication for traffic control in urban areas, in *Proceedings of IEEE Conference on Industrial Electronics (IECON)*, Paris, France, pp. 748–753, November 2006.

25. M. Jonsson and A. Bohm, Position-based data traffic prioritization in safety-critical, real-time vehicle-to-infrastructure communication, in *Proceedings of IEEE Vehicular Networking and Applications Workshop (VehiMobil 2009)* in conjunction with IEEE ICC'09, Dresden, Germany, pp. 1–6, June 14, 2009.

26. M.-F. Jhang and W. Liao, On cooperative and opportunistic channel access for vehicle to roadside (V2R) communications, in *Proceedings of IEEE Global Telecommunications Conference (GLOBECOM)*, New Orleans, LA, pp. 1–5, November–December 2008.

27. H. Ilhan, M. Uysal, and I. Altunbas, Cooperative diversity for intervehicular communication: Performance analysis and optimization, *IEEE Transactions on Vehicular Technology*, 58(7), 3301–3310, September 2009.

28. M. Guo, M. H. Ammar, and E. W. Zegura, V3: A vehicle-to-vehicle live video streaming architecture, in *Proceedings of IEEE International Conference on Pervasive Computing and Communications (PerCom)*, Kauai, HI, pp. 171–180, March 2005.

29. B. Shrestha, D. Niyato, Z. Han, and E. Hossain, Wireless access in vehicular environments using BitTorrent and bargaining, in *Proceedings of IEEE Global Telecommunications Conference (GLOBECOM)*, New Orleans, LA, November–December 2008.

30. I. Chlamtac, M. Conti, and J. J.-N. Liu, Mobile ad hoc networking: Imperatives and challenges, *Ad Hoc Networks*, 1(1), 13–64, July 2003.

31. H. Conceicao, M. Ferreira, and J. Barros, A cautionary view of mobility and connectivity modeling in vehicular ad-hoc networks, in *Proceedings of IEEE Vehicular Technology Conference (VTC)-Spring*, Barcelona, Spain, pp. 1–5, April 2009.

32. T. Camp, J. Boleng, and V. Davies, A survey of mobility models for ad hoc network research, *Wireless Communications and Mobile Computing*, 2(5), 483–502, September 2002.

33. V. Kukshya and H. Krishnan, Experimental measurements and modeling for vehicle-to-vehicle dedicated short range communication (DSRC) wireless channels, in *Proceedings of IEEE Vehicular Technology Conference (VTC)-Fall*, Montreal, Canada, pp. 1–5, September 2006.

34. Y. Chen, Z. Xiang, W. Jian, and W. Jiang, An improved AOMDV routing protocol for V2V communication, in *Proceedings of IEEE Intelligent Vehicles Symposium*, Xian, China, pp. 1115–1120, June 2009.

35. T. J. Kwon, W. Chen, R. Onishi, and T. Hikita, Unicast routing among local peer group (LPG)-based VANETs, in *Proceedings of IEEE GLOBECOM Workshops*, New Orleans, LA, pp. 1–5, November–December 2008.

36. T. L. Willke, P. Tientrakool, and N. F. Maxemchuk, A survey of inter-vehicle communication protocols and their applications, *IEEE Communications Surveys and Tutorials*, 11(2), 3–20, Second Quarter 2009.

37. M. J. Osborne, *An Introduction to Game Theory*, Oxford University Press, Oxford, U.K., 2003.

38. J. Filar and K. Vrieze, *Competitive Markov Decision Processes*, Springer, New York, November 1996.

39. J. Hofbauer and K. Sigmund, *Evolutionary Games and Population Dynamics*, Cambridge University Press, Cambridge, U.K., June 1998.

40. A. Muthoo, *Bargaining Theory with Applications*, Cambridge University Press, Cambridge, U.K., September 1999.

41. D. Fudenberg and J. Tirole, *Game Theory*, MIT Press, Cambridge, MA, 1991.

42. L. Raz, K. Sarit, and S. Yuval, On the benefits of cheating by self-interested agents in vehicular networks, in *Proceedings of 6th International Joint Conference on Autonomous Agents and Multi Agent Systems (AAMAS'07)*, Honolulu, HI, pp. 1–8, 2007.

43. K. Akkarajitsakul and E. Hossain, An auction mechanism for channel access in vehicle-to-roadside communications, in *Proceedings of International Workshop on Multiple Access Communications (MACOM)*, Dresden, Germany, June 2009.

44. D. Niyato, E. Hossain, and P. Wang, Competitive wireless access for data streaming over vehicle-to-roadside communications, in *Proceedings of IEEE Global Communications Conference (GLOBECOM)*, Honolulu, HI, November–December 2009.

45. D. Niyato and P. Wang, Optimization of the mobile router and traffic sources in vehicular delay tolerant network, *IEEE Transactions on Vehicular Technology*, 58(9), 5095–5104, November 2009.

46. D. Niyato, E. Hossain, and T. Issariyakul, An adaptive WiFi/WiMAX networking platform for vehicle-roadside communications, Technical Report, Nanyang Technological University (NTU), Singapore, 2009.

47. A. Bejan and R. Lawrence, Peer-to-peer cooperative driving, in *Proceedings of International Symposium on Computer and Information Sciences (ISCIS)*, Orlando, FL, pp. 259–264, October 2002.

48. Y. Shavitt and A. Shay, Optimal routing in gossip networks, *IEEE Transactions on Vehicular Technology*, 54(4), 1473–1487, July 2005.

49. I. Chisalita and N. Shahmehri, A novel architecture for supporting vehicular communication, in *Proceedings of IEEE Vehicular Technology Conference (VTC)-Fall*, Vancouver, BC, Canada, vol. 2, pp. 1002–1006, 2002.

50. Y.-B. Lin, W.-R. Lai, and J.-J. Chen, Effects of cache mechanism on wireless data access, *IEEE Transactions on Wireless Communications*, 2(6), 1247–1258, November 2003.

51. K.-C. Lan, C.-M. Huang, and C.-Z. Tsai, On the locality of vehicle movement for vehicle-infrastructure communication, in *Proceedings of International Conference on ITS Telecommunications (ITST)*, Hilton Phuket, Thailand, pp. 116–120, October 2008.

52. M. Blomgren and J. Hultell, Demand-responsive pricing in open wireless access markets, in *Proceedings of IEEE 65th Vehicular Technology Conference (VTC)-Spring*, Dublin, Iveland, pp. 2990–2995, April 2007.

53. V. Mancuso and G. Bianchi, Streaming for vehicular users via elastic proxy buffer management, *IEEE Communications Magazine*, 42(11), pp. 144–152, November 2004.

54. E. Altman, K. Avrachenkov, N. Bonneau, M. Debbah, R. El-Azouzi, and D. S. Menasche, Constrained cost-coupled stochastic games with independent state processes, *Operations Research Letters*, 36(2), pp. 160–164, March 2008.

55. L. Chisalita and N. Shahmehri, A peer-to-peer approach to vehicular communication for the support of traffic safety applications, in *Proceedings of IEEE International Conference on Intelligent Transportation Systems*, Singapore, pp. 336–341, 2002.

56. V. Prinz, R. Eigner, and W. Woerndl, Cars communicating over publish/subscribe in a peer-to-peer vehicular network, in *Proceedings of International Conference on Wireless Communications and Mobile Computing (IWCMC)*, Leipzig, Germany, pp. 431–436, July 2009.

57. Z. Jiang, Y. Ge, and Y. Li, Max-utility wireless resource management for best-effort traffic, *IEEE Transactions on Wireless Communications*, 4(1), 100–111, January 2005.

58. T. L. Vincent and J. S. Brown, *Evolutionary Game Theory, Natural Selection, and Darwinian Dynamics*, Cambridge University Press, Cambridge, U.K., July 2005.

59. M. Khabazian and M. Ali, A performance modeling of connectivity in vehicular ad hoc networks, *IEEE Transactions on Vehicular Technology*, 57(4), pp. 2440–2450, July 2008.

60. D. Ray, *A Game-Theoretic Perspective on Coalition Formation*, Oxford University Press, Oxford, U.K., January 2008.

Chapter 5

EGT in Wireless Communications and Networking

Artemis C. Voulkidis, Athanasios V. Vasilakos, and P.G. Cottis

Contents

As wireless communication networks are getting more and more popular, new network architectures arise in an attempt to satisfy the increasing demand for various new services. However, installation, surveillance, and maintenance of these networks are growing harder. Different network characteristics, swarming wireless node implementations, as well as rapid changes in service demand and network conditions constitute the timely human intervention difficult and error prone. This chapter discusses how evolutionary game theory (EGT) may deal with these problems, transforming static wireless networks into adaptive, autonomic wireless networks. A short tutorial examining how EGT can be applied in network optimization is presented, followed by a number of relevant application examples found in the literature. This chapter identifies some open research issues concerning the application of EGT in real wireless networks.

5.1 Introduction

Multimodal, ubiquitous computing should be present anywhere, anytime, and anyway providing high quality services to anyone in need. Handheld communication devices such as personal digital assistants (PDAs) or cell phones with video and even GPS capabilities have become quite popular. The need to provide differentiated services with QoS assurance has led network designers to introduce new types of wireless networks based on the general ad hoc paradigm such as wireless sensor networks (WSNs), delay tolerant networks (DTNs), mobile ad hoc networks (MANETs), wireless multimedia networks (WMNs), heterogeneous Networks, etc. Each of these network architectures responds to different service requirements and faces different limitations and restrictions.

Despite the specific properties of such networks, they all exhibit many common characteristics due to their operation in an ad hoc manner. Routing, medium access control (MAC), connection admission control (CAC), and congestion control are some of the procedures that must be accomplished. Usually, the modules constituting these next-generation networks (NGNs) are mobile, subjected to low computational and power limitations. It is argued that the achievable QoS in a communication network depends on the predictability of the underlying network [1]. Wireless networks exhibit rapid variations due to mobility, network congestion, and varying propagation conditions. This makes traditional network management methods inadequate to provide QoS services. Taking into consideration the emergence of NGNs, it is evident that including human intervention is not efficient for network design, monitoring, and maintenance. On the contrary, networks should be autonomic, continuously evolving, and adaptive to the variations of the network environment. Moreover, the protocols developed should be *scalable* to operate in extreme cases of node population or network traffic, *self-healing*, that is, to recover from network breakdowns, *self-optimizing* to choose the best operating mode under any possible situation, and *self-stabilizing* to ensure that the network will always operate in stable state [2].

EGT was originally invented to introduce classical game theory (GT) into biology and has recently been used in network design and economics [3]. Its basic difference from classical GT is that it does not rely on players rationality; instead, it considers populations of decision-making entities, the players, that interact at random during the game. The popularity of each decision strategy among the players may change over time in response to the decisions taken by all the players of the population. Each member of this population evolves along with the rest of the population, each time trying to take the best decision, in an attempt to maximize its personal benefit when interacting with the other players [4]. Another critical point differentiating classical GT and EGT is EGT makes use of procedures appearing in biological systems such as *reproduction*, *selection, mutation,* and *death.* As it will be shown later, the use of EGT may render a wireless communication network to a fully autonomic network.

In the following sections, a formal introduction to EGT is given, followed by a tutorial on the generic application of replicator dynamics to NGNs. Next, representative papers are discussed dealing with the application of EGT in wireless networks. The chapter concludes marking out some open research issues in the field of evolutionary protocols for smart, autonomic wireless networks.

5.2 Evolutionary Game Theory and Wireless Networks

In this section, an introduction to the basic principles of EGT combined with a mini tutorial of how they can be related to the basic functions of a wireless network is provided. It is also described how evolutionary logic can be infused into network design. It can be used as a starter to a large variety of applications that need to incorporate replicator dynamics.

5.2.1 Evolutionary Game Theory Basics

As aforementioned, EGT makes use of procedures appearing in biological systems such as *reproduction, selection, mutation,* and *death.* The use of these procedures renders EGT very different from, e.g., repeated games. The evolutionary approach usually considers a set of players that interact within a game and then die, giving birth to a new generation player that fully inherits its ancestor's knowledge. The new player strategy is evaluated according to the one of its ancestors and its current environmental surroundings. Also, through mutation, a slightly different strategy may be adopted, possibly leading to better payoffs. Next, the player competes with the other players in the evolutionary game, while the selection process ensures that "the fittest will survive": strategies with high payoffs will survive as more players will tend to choose them, while weak strategies will eventually disappear.

Formally, consider an infinite population of individuals that react to changes of their environment using a finite set of n pure strategies $S = \{s_1, s_2, \ldots, s_n\}$. A *population profile* $\mathbf{x} = \{x_1, x_2, \ldots, x_n\}$ is a vector that denotes the popularity of each strategy $s_i \in S$ among the nodes. That is, x_i is the probability that a strategy s_i is played by the individuals. As to non-pure strategies, the population profile can also be interpreted as the proportion of pure strategies included in the mixed strategy followed by a random node; then, x is defined as the set of mixed strategies.

5.2.1.1 Replicator Dynamics

Consider an individual in a population with profile x. Its expected payoff when choosing to play strategy s_i is given by $f(s_i, x)$. In a two-player game, if an individual chooses strategy s_i and its opponent responds with choosing strategy s_j, the payoff of player 1 can be determined as $f(s_i, s_j)$. The expected payoff of strategy s_i can be evaluated as $f_i = \sum_{j=1}^{n} x_j \cdot f(s_i, s_j)$, whereas the average payoff is $f_x = \sum_{i=1}^{n} x_i \cdot f_i$.

The replicator dynamics is a differential equation that describes the dynamics of an evolutionary game. According to this differential equation, the rate of growth of a specific strategy is proportional to the difference between the expected payoff of the strategy and the overall average payoff of the population, that is

$$\dot{x} = x_i \cdot (f_i - f_x) \tag{5.1}$$

(5.1) describes mathematically the physical process of selection without mutation [4]. So, if a strategy yields a much better payoff than the average, the number of individuals that tend to use it increases. On the contrary, a strategy yielding lower payoffs than average is preferred less and eventually disappears. In games with two available strategies, the replicator dynamics implies that a strategy s_1 with payoff higher than the average will always increase in popularity, ultimately extincting the opponent strategy s_2. It is then said that strategy s_1 is the *dominant* strategy whereas s_2 is the *dominated* one.

5.2.1.2 Evolutionary Stable Strategy

Imagine that a small group of mutants $\epsilon \in (0, 1)$ with a profile $x' \neq x$ invades the previous population. The new profile of the newly formed population is $x_{final} = \epsilon \cdot x' + (1 - \epsilon) \cdot x$. Hence, the average payoff of nonmutants will be $f_x^{non\text{-}mutant} = f(x, x_{final}) = \sum_{j=1}^{n} x_j \cdot f(j, x_{final})$ and the average payoff of any mutant would be $f_{x_{final}}^{mutant} = f(x', x_{final}) = \sum_{j=1}^{n} x'_j \cdot f(j, x_{final})$.

A strategy x is called *evolutionary stable strategy* (*ESS*) if for any $x' \neq x$, $\epsilon_{mut} \in (0, 1)$ exists such that for all $\epsilon \in (0, \epsilon_{mut})$, the following equation holds:

$$f_{x_{final}}^{non\text{-}mutant} > f_{x_{final}}^{mutant} \tag{5.2}$$

(5.2) can be rewritten as

$$\sum_{j=1}^{n} \left(x_j - x'_j \right) \cdot f(j, x_{final}) > 0 \tag{5.3}$$

More formally, for a pairwise contest, if σ^* is an ESS, then $\forall \sigma \neq \sigma^*$ is either of the following:

- $f(\sigma^*, \sigma^*) > f(\sigma, \sigma^*)$
- $f(\sigma^*, \sigma^*) = f(\sigma, \sigma^*)$ and $f(\sigma^*, \sigma) = f(\sigma, \sigma)$

Conversely, if either (1) or (2) holds for any $\sigma \neq \sigma^*$, then σ^* is an ESS. The proof of this can be found in [3]. Note that if (5.2) and (5.3) are not strict, σ^* is called *neutral stable strategy*. Considering the above and the definition of a Nash equilibrium (NE), one could claim that if a strategy is a strict NE, it is also an ESS while if a strategy is an ESS, it is also an NE.

Concluding, it is claimed that when an ESS is reached, the population proportions distributed over the strategies do not change and the population is immune from being invaded by other groups with different population profiles. More information on the various stable strategies in EGT can be found in [3–5].

5.2.2 Application on Wireless Networks

After having introduced some of the basic principles of EGT, it is interesting to examine how these ideas may be applied in the design of wireless ad hoc networks. Consider a network consisting of a sufficiently large number of smart nodes. By the term "smart," basic intelligence concerning the evaluation of the other nodes' behavior is meant. As node transactions are usually twofold, a two-player game is considered. For simplicity, a simple two-strategy game is outlined, so that nodes

**Table 5.1 Payoff
Table for the Two
Players**

	C	D
C	(c, c)	(b, e)
D	(e, b)	(d, d)

should consider; cooperating (C) or defecting (D). The game to be analyzed is symmetric, as it is usually the case in wireless communication networks.

After having defined the network characteristic to optimize, the various payoffs should be set. Consider a node that needs the help of a random opponent node that lies within its neighborhood. Also suppose that when a node decides to cooperate (C) with a cooperative opponent, both have a payoff c. When both nodes decide to deny the cooperation request, their common payoff is d. When only one of the two cooperates, it has a benefit of b while the other one that does not cooperate has a payoff of e. The payoffs of this game are summarized in Table 5.1.

Considering a population profile $x = (x_1, x_2)$, the average payoff for a cooperative node is $f_c = x_1 \cdot c + x_2 \cdot b$ whereas a denying one would expect a payoff $f_{nc} = x_1 \cdot e + x_2 \cdot d$. The average payoff of the population is $f_x = x_1 \cdot f_c + x_2 \cdot f_{nc}$. According to the replicator dynamics, the regeneration rate of cooperators and defectors is $\dot{x}_c = x_1 \cdot (f_c - f_x)$ and $\dot{x}_{nc} = x_2 \cdot (f_{nc} - f_x)$, respectively. Therefore, if $f_c > f_{nc}$ cooperators will be favored, as more and more nodes will decide to cooperate in order to maximize their payoff.

An other important issue is that according to [3], all *symmetric, two action, pairwise* games have an ESS. Of course, not all evolutionary games are symmetric and two action. Another important note is that gains c, d, b, and e are not necessarily stationary. They could as well be and usually are functions of some other network or node metrics, such as the current network traffic, the node resilient power levels, the node neighboring count, etc.

After proving the existence of an ESS, the next step is to find the ESS. This can be done by appropriately solving or estimating the solution of the replicator dynamics defined by (5.1).

5.3 Relative Existing Bibliography

EGT may be successfully applied to model a variety of network problems. Also, due to its universal nature, it is suitable for almost any kind of wireless networks including WSNs, DTNs, WMNs, etc. Next, a quick preview of the existing literature concerning the applications of EGT to these kinds of networks will be given, classified according to the network type and problem addressed.

5.3.1 Wireless Sensor Networks

WSNs consist of extremely large numbers of nodes with low computational power and very limited energy resources. So in many cases, the minimization of the number of transmissions is studied since it is the main source of energy consumption [6]. Aggregation, packet forwarding, congestion control, and call admission control are some of the techniques used to eliminate unnecessary radio transmissions.

A special category of WSNs is considered in [7], where the authors study the evolution of cooperation among nodes in underwater sensor networks governed by different authorities that

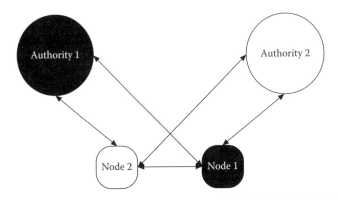

Figure 5.1   Simple scenario featuring two nodes answering different authorities.

act as multiple sinks. In these networks, the design aims at maximizing the network life; sensor nodes are uniformly scattered into the ocean and have no means of energy replenishment, battery replacement being almost infeasible. Moreover, in underwater environments, energy consumption due to radio activity is even more intense [8]. The approach followed to resolve this issue is based on node cooperation on packet forwarding.

Consider the scenario depicted in Figure 5.1, where the two colored nodes are governed by the authorities of the same color. In the majority of cases, the energy required to send a message from the source (node) directly to the sink (authority) and vice versa is greater than the total energy required by an equivalent hop-by-hop transmission. Hence, if the two nodes cooperate, the total energy spent will be significantly less. One or more forwarding nodes, however, may decide to deny cooperation and act as defectors (see Section 5.2.2). Having this in mind, the authors propose an evolutionary game among the nodes, considering the possible behaviors of the nodes in terms of direct and hop-by-hop transmission as strategies. The payoffs of the various strategies are evaluated in terms of energy consumption, giving no further incentives for cooperation to the nodes; the smallest energy consumption corresponds to the highest payoff.

After a series of simulations under various mean cooperation levels, it was shown that spontaneous cooperation may emerge, without any extra incentive mechanism to motivate cooperation. Furthermore, it is argued that in nonreactive strategies, equilibrium leads to full cooperation or full defection, the former requiring high QoS conditions. Reactive strategies, though, seem to outperform nonreactive ones. As to the equilibria achieved, *weak ESS* is achieved with the probability of successful invasion lying below 20% in almost every case. The authors conclude that EGT can assist the coexistence of different authorities to increase network lifetime.

Cooperation in WSNs under different authorities is also studied in [9]. A multi-class sensor network is considered, where each class is governed by a different authority. In contrast with [7], the authors consider the case of a wireless network where an incentive mechanism promotes cooperation among nodes for packet forwarding by enforcing node cooperation through punishment. An interesting aspect of this approach is that cooperation among nodes of the same class is taken for granted, so the games played concern neighboring nodes of different classes. Originally, the evolutionary game considers the two pure strategies, so that sensor nodes can either cooperate or defect. Besides the two pure strategies, the authors propose a hybrid "patient grim strategy" (PGS), where a node functions as a cooperator and continues to always cooperate with the opponent node, until the latter defects more than a threshold of times. In this case, the former node becomes a defector and never cooperates with this node again. Under certain restrictions, it is also proved that the proposed solution leads to an NE.

In the simulations, three scenarios are examined. In the first scenario, a sensor network with mobile classes was simulated. In this case, defectors seem to behave more efficiently as they dominate over cooperators. In the second scenario, a network characterized by spatially dispersed stationary classes was considered, where the equilibrium achieved consisted of both cooperators and defectors, with an estimated population profile being $x = (0.75, 0.25)$. In these cases, the proposed algorithm (PGS) was not tested. The third scenario tested was identical to the second one except for adopting PGS. A series of simulations demonstrate that the hybrid strategy performs better than the pure ones, leading to an equilibrium where the nodes adopting the PGS get a significantly better average payoff.

EGT has been used to eliminate interference and increase data rate in impulse radio ultra wideband (IR-UWB) sensor networks [10]. IR-UWB technology allows multiple, simultaneous transmissions suffering at the same time from excessive, impulsive, non-Gaussian interference [11]. Although many approaches—even game-theoretic ones—exist to address the data rate maximization problem, they usually require knowledge of the network condition as well as a substantial amount of computing power, which is normally not available [12].

EGT is presented as a means to create a robust, but light in computer power requirements, protocol. Taking into consideration that nodes can adjust the average pulse repetition frequency (RPF) per link—controlling thus the interference and the expected BER of each link—the authors form a game where the payoff of each strategy is calculated as a function of the link quality. Particularly, each node calculates the BER of its opponent and adaptively sets its RPF, reacting, thus, to the current network changes.

A series of simulations under various network configurations demonstrate that the proposed protocol can substantially improve the performance of low mobility nodes under normal data flow. However, when the node environment is unstable (due to increased node mobility or bursty traffic), the nodes fail to properly respond to the network changes and do not converge to an ESS.

5.3.2 Delay Tolerant Networks

DTNs are mobile ad hoc networks characterized by very frequent disconnections and transmission disruptions [13,14]. Due to these adverse inherent characteristics, transmission links between nodes have a rather opportunistic nature, and end-to-end node connectivity is rarely achieved.

Data aggregation techniques have been successfully used in other network types (mainly in WSNs) to facilitate the interconnection of dispersed or clustered nodes. In [15], an EGT approach to the realization of self-organized data aggregation schemes in DTN is presented. Based on the work presented in [16,17], Kabir et al. propose a dynamic, autonomic, decentralized scheme aiming at load balancing in opportunistic networks. The evolutionary game assumed is based on spontaneous rather than incentive-based cooperation.

Consider the topology depicted in Figure 5.2, where the various nodes are structured in clusters. Moving nodes, called *ferries*, travel across the network searching for nodes having packets to forward. Note that a node is visited by a ferry if it has an adequate number of aggregated messages to deliver. The nodes aggregating data packets within a cluster are called *aggregators*. On visit, the ferry provides the aggregator with a certain amount of energy (larger than the amount spent for data gathering and processing). Thus, a network node has two possible strategies to follow: become an aggregator, or just stay a sender, feeding aggregators with packets to forward. Energy is definitely the trade-off: an aggregator has to be more active in order to gather and

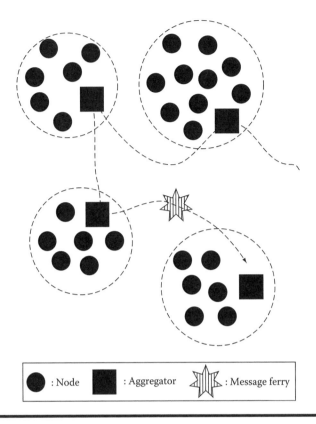

Figure 5.2 Message ferry visits the aggregators of each cluster to deliver the collected packet bundles to the sink.

process data packets, facing the possibility of not gathering enough to be visited by a ferry. On the other hand, being a sender means less aggressive energy consumption and no energy replenishment either.

Borrowing notions from evolutionary graph theory [18,19], the authors form a distributed protocol based on a game, where the strategy payoffs are determined in terms of energy consumption and recharge. The numerical and analytical results presented show that both aggregating and sending strategies may coexist in a stable way. Moreover, their proportion can be properly tuned by adjusting a set of variables such as the energy delivered by the ferries to the aggregators. Moreover, the protocol exhibits a satisfactory behavior even under heavy load conditions, balancing successfully the offered load using exclusively local information.

5.3.3 *Peer-to-Peer Networks*

The term peer to peer (P2P) refers to self-organized distributed networks, mostly for file-sharing purpose. Usually they are not wireless, but their distributed, self-organized nature resembles many wireless network architectures. In P2P networks, nodes (users) usually behave selfishly, trying only to maximize their own benefit. This selfish behavior usually leads to the *free riding problem* where most system users do not contribute anything to the rest of the users and the system operates due to a small number of cooperative, data-contributing nodes.

In [20], the authors present an EGT approach toward the enforcement of incentive-based cooperation in P2P systems. The proposed protocol assumes only fundamental node rationality rendering it lightweight with regard to both computer and network resources. Following the classical approach where each node has the option to decide either to cooperate (download and contribute at the same time) or to defect (download but do not share any file), a two-player, two-strategy game is formed. The payoffs used are measured in terms of data download or upload units, along with certain additional, virtual parameters that model the cost of each data packet transmitted. Interestingly, the authors also consider a migration mechanism, where nodes from other networks— sharing different strategy population profiles—invade the network under consideration, in an attempt to assure network stability in terms of data contribution. The stability of the suggested scheme is proved with the help of differential analysis. The respective simulations carried out also showed that the evolutionary approach adopted leads to an equilibrium where users tend to share their data packets with the rest of the nodes, even in the case of node migration, that was earlier mentioned.

P2P networks are also the target networks of the work filed in [21]. The authors consider a normal file-sharing network, in which nodes perform file and chunk caching, in an attempt to increase data availability and decrease latency across the network. Taking into consideration though that chunk caching requires both computational power and storage capacity, a node may decide not to cooperate and thus deny to perform the caching. This problem is also addressed with the help of EGT.

In detail, the authors describe a distributed algorithm to enable node cooperation, assuming that when a node chooses to cache, data packets receive a benefit b as a result of the increase in chunk availability within the network. Note that b actually refers to the popularity of the file or chunk to be cached. On the other hand, two kinds of costs for caching are investigated: computational power and storage capacity. The game is formed using these variables as payoffs for the two available strategies of the nodes. After solving the respective replicator dynamics equation, the authors prove that in a stable network state, the ratio of cooperators increases with the increase of the file demand b. In addition to that, when the demand for a file is relatively low, then the shared chunks of it tend to disappear from the network. Using a similar approach, agent-based dynamics are studied to examine the scalability of the proposed architecture. Through both mathematical analysis and simulation results, the effect of the presented scheme is shown to be beneficial in terms of file availability and scalability.

Sasabe et al. also consider P2P file-sharing systems in [22]. Extending their work presented in [21], the authors argue that EGT is beneficial not only in ensuring the availability of less popular file chunks but also in enhancing the overall network performance. A simple generic evolutionary model is initially proposed to approach the behavior of the nodes participating in the P2P network, very similar to the one proposed in [21]. Next, through a theoretic analysis based on agent-based dynamics, the authors review the relationship binding file availability, which is directly analogous to global node cooperation level, with processing load and storage capacity. Simulation results show that the proposed model performs well and scales under almost any network structure, even if nodes behave selfishly.

An incentive-based mechanism toward cooperation enforcement among users of P2P networks is the case of [23]. Arguing that the majority of P2P network users actually free ride, the authors present a model that could increase user cooperation and consequently file availability. They consider three major user behaviors, always defect (ALLD), always cooperate (ALLC) and reciprocate (R), and consider that reciprocation costs are not negligible. Initially, they show that, with small reciprocation costs and no mutation probability, ALLD is the only Nash equilibrium. However, when mutation probability is existent, the behavior population is proved to oscillate among the tree available strategies.

5.3.4 Wireless Networks in General

The literature relevant to EGT applications in generic wireless network architectures is presented in this section.

In [24], the authors adopt a game-theoretic approach in noncooperative wireless networks to study the evolution of congestion control protocols in a cooperation-hostile environment. Using the well-known Hawk and Dove game, they formulate a framework aiming at modeling and possibly adjusting network behavior and achieving node behavior convergence, in the presence of congestion control algorithms. The authors argue that network stability depends on time delay and also that there is always a trade-off in wireless networks between network stability and convergence speed, greater time delay resulting in slower convergence as stale information does not properly reflect the actual network state. Also, there is a trade-off between the higher gains in congestion control schemes and the respective stability.

In contrast with the rest of the studies already presented, a time delay version of replicator dynamics is considered. *Delayed evolutionary game dynamics* have also been studied in [25,26]. According to this realistic approach, the payoff acquired at some time will affect the rate of strategy population growth τ time units later. In a real wireless network, this is usually a realistic scenario compared to the approaches that do not consider this time delay [25], as information transmission and processing sometimes take substantial amount of time. A crucial characteristic of this study is that players have the option to choose among N available strategies.

The payoffs of the opponents in the games studied are evaluated with regard to available and reserved bandwidth, link transmission rate, and available throughput. A linear gain parameter that controls the congestion control protocol gain is also introduced. After proper mathematical analysis, the authors argue that this variable in combination with the mean time delay considered greatly affects system's stability and convergence speed: if the product of the time delay and this parameter surpasses a threshold, the network state cannot converge to an equilibrium. This conclusion is critical when designing a new congestion control protocol for wireless networks and may act as a guideline for successful development of wireless network types.

Another application of EGT in wireless networks is presented in [27,28], where high-frequency networks operating under IEEE 802.16 Wireless MAN specifications are investigated. Since these networks operate in the frequency range 10–66 GHz, they are very sensitive to rain [29]. The authors propose an adaptive, cross layer approach for routing under heavy rain attenuation. Anyway, this approach can easily be extended to any wireless network suffering from any kind of data flow distortion, that is, excessive traffic, link breakage, etc. The adaptability and stability of the proposed protocol are studied based on EGT principles.

The proposed protocol is based on *agent* evolution where traffic is determined by a large number of agents, each responsible for a infinitesimally small amount of traffic. Their aim is to selfishly maximize their payoff, namely, increasing the probability to successfully carry and deliver data packets, without considering the consequences of their choices to the rest of the nodes and the overall host network. The agent strategies are continuously revised giving rise to an evolutionary game where packet error rate (PER) is the payoff variable; high PER means low payoff and vice versa. In a multipath environment as the one studied in [27], where the available agent strategies are actually the set of available routes, the link PER determines the link reliability and, indirectly, the possible payoff of selecting this strategy.

EGT agents have the option to revise their preferred routing strategy, each time choosing the route with the lower PER between two randomly sampled paths. The mathematical analysis using linear programming and Lyapunov's first method has proven that system stability and convergence are guaranteed. Extensive simulations verified the mathematical results showing that the proposed protocol is stable under almost any climatic conditions. In addition to the basic protocol just described, an improved algorithm for faster system convergence was also given, further enhancing the total performance. According to the improved algorithm, the probability of sampling a specific link increases when route PER is lower, eventually leading to faster convergence toward the equilibrium.

Adaptive, reliable routing is also the subject of [30]. The case of *stale* information routing is considered, where agents reroute traffic based on passed information. As usual, the agents try to maximize their benefit from each transaction. The benefit of each agent, though, is considered as a function of the latency that data routing may suffer, the higher the latency the lower the benefit. It is argued that a game where nodes share timely information and every agent chooses to route data from slower to faster routing paths has a well-known equilibrium called "Wardrop equilibrium" [5]. However, when nodes deliver delayed information, this equilibrium cannot be achieved. Employing EGT, the authors in [30] attempt to reduce the unwanted effects of stale information routing. According to their approach, every agent samples a random route and chooses its favorable strategy from a set of disjoint routes. Note that the sampling performed is not uniform, varying according to the Poisson distribution. The game includes payoffs based on latencies and on link data rate. It also includes the probabilities of sampling and strategy migration, which are analytically evaluated. Through both mathematical analysis (using Lyapunov's second method and standard convex programming) and simulation results, it is shown that EGT can successfully lead the population in a steady state, which is a typical Wardrop equilibrium. However, it is shown that under certain conditions, and when nodes always choose the best possible strategy (best response), a significant oscillation around this equilibrium may arise, since it is possible that multiple nodes choose the same path, leading to node and link overloading, congestion, increased packet loss, etc. This problem can be dealt with employing *smooth adaptation policies*, where each node chooses to play the best response with a certain probability. Thus, oscillation effects are avoided.

An interesting point in the proposed protocol is the existence of a *bulletin board*, similar to the one described in [31]. This bulletin board functions as a server collecting information about link latencies, offering this information to nodes and, thus, agents. It is also shown that the protocol performance is dependent on how frequently this information is updated.

An application of EGT for resource allocation in service provisioning in cloud environments is presented in [32]. The authors argue that traditional system-centric resource management protocols are inadequate in the case of a cloud market. Instead, market-oriented approaches seem to be more appropriate for these systems, being distributed and scalable. According to their approach, the assumed market consists of selfish service sellers and buyers that interact within the context of a simple trading protocol: sellers announce their prices first, and then buyers determine the set of resources (services) to buy. It is argued that the market equilibrium only depends on sellers' pricing strategies, as buyers' optimal actions can be directly interpreted into seller strategies. Using a simple four-step algorithm based on sequential Monte Carlo methods, the authors form an evolutionary game among seller pricing strategies. The simulations performed showed that the proposed approach leads to ESS and that social welfare increases over time. Moreover, the scale of resource allocation problem does not seem to affect the convergence speed of the evolutionary dynamics.

In [33], the authors present an evolutionary adaptation framework for network applications, where agents are used to monitor and properly adjust applications' behavior to the current network conditions. Borrowing notions from the human immune system, the authors consider network surroundings as antigens and agents' behavior as antibodies. According to the proposed scheme called iNet-EGT, application behavior selection is based on EGT. Specifically, after agent initialization, two randomly selected behaviors (agents) repeatedly play games in the population. The winning behavior replicates itself, while the losing one vanishes. Then, eventually, one behavior will dominate over the others to define the application behavior. Note that a mutating mechanism is also considered, ensuring that the applications will have the chance to reconsider their behavior and timely adjust to the network environment changes. The stability of the proposed protocol is mathematically proven using classic theorems of EGT. Simulation results also show that in iNet-EGT, applications quickly adjust their behavior according to the network changes, achieving higher throughput and lower response times than when not using an EGT-enabled scheme. An extension of this framework is presented in [34] where iNet-EGT/C (iNet-EGT, coalitional edition) is presented. Borrowing notions from cooperative game theory, where agents behave as parts of larger groups (coalitions), C. Lee et al. propose a scheme where interactions occur among agent coalitions. Compared to iNet-EGT, iNet-EGT/C exhibits faster agent behavior convergence in the case of quick network changes, simultaneously preserving the core advantages of the former.

Wang et al. [35] study the effects of the application of EGT in cognitive networks [36]. The authors consider a cognitive network where multiple secondary users (players) occupy different sub-bands of one primary user that can be used when the primary user is inactive. According to their approach, when a secondary user senses the presence of the primary one, she announces the sensing results to the other secondary users through a narrow-band channel. However, sensing tasks are accompanied by both energy consumption and lower throughput. Thus, secondary users may consider not to cooperate in spectrum sensing; they might simply overhear the sensing outcome of other users and only use their resources for their own data transmissions. If no user senses the primary user, then all secondary users get a very low payoff. The payoff functions of the proposed evolutionary game are expressed in terms of throughput utilization and false alarm probability. Through mathematical analysis and simulation results, the authors argue that their evolutionary scheme can help secondary users to approach their optimal strategy only knowing their own payoff history and that, as far as network resource utilization is concerned, a global optimum is achieved.

In [37], an evolutionary framework for TCP throughput adaptation in WiMax networks is presented. In WiMax, the TCP protocol is mainly conditioned by wireless channel impairments. Adaptive coding and modulation (ACM) and forward error correction (FEC) techniques are widely used to cope with these impairments. However, it has been shown [38] that under different rain attenuation conditions, different ACM and FEC schemes present throughput optimality. The authors in [37] consider a scheme where TCP connections adaptively choose ACM strategies in order to maximize their throughput. Following the notions of agent-based dynamics, TCP traffic is seen as a total of data packets–agents that can continuously update their ACM strategies by sampling all the alternative ACM schemes, taking information from the physical layer. All utility functions and payoffs are expressed in terms of PER and throughput and the stability of the proposed scheme is mathematically proven based on evolutionary dynamics and Lyapunov's first method while the convergence analysis shows that the scheme quickly converges to the optimal ACM strategy under any channel state. Moreover, it is shown that convergence speed may be increased if more information from the physical layer is available. Furthermore, the authors consider an improved sampling algorithm to accelerate convergence. This algorithm takes advantage of Markovian modeling of rain attenuation in wireless channels operating at frequencies above 10 GHz quickly eliminating the obviously nonoptimal ACM schemes. Convergence speed is then shown to be significantly improved.

The case of heterogeneous 4G networks is considered in [39]. The authors combine notions from EGT with the basics of reinforcement learning, forming a robust framework aiming at efficient user service in both WLAN and LTE networks. They consider a realistic, dynamic network environment where the cost and delay for service or operator switching are nontrivial. According to the proposed scheme, only numerical values about payoffs achieved by previous strategic decisions are known by both the users and the operators. Moreover, an evolutionary reinforcement learning algorithm for strategy evaluation and optimal behavior determination is suggested, which renders the latter dynamic and distributed. The stability and convergence of the scheme are proven with the help of evolutionary dynamics.

5.4 Open Issues

EGT has been successfully applied to model the behavior of NGN nodes in a variety of network challenges such as routing, resource allocation, etc. However, the existing approaches only consider one level of evolutionary interaction among the various network entities. In real life, evolution takes place in more than one levels. Human body is a great example of a multilevel approach as human body evolves both separately (organs) and as a whole not only because of the interaction with other human beings but also because of the environmental changes. Moreover, human behavior changes in accordance with the evolution of the body. Thus, more than one levels of evolution exist in human nature. This multilevel approach is highly expected to further improve the efficiency of evolutionary protocols.

Based on a theory initially developed to model the evolution of species through time, EGT can effectively model a wide range of adaptive networking challenges. Through the processes of reproduction, selection, mutation, migration, and death, EGT can be successfully applied to model node behavior in dynamic and rapidly changing networking environments. Networks offering cloud services can greatly benefit from EGT in both pricing and resource allocation. Due to the opportunistic nature of service demand in these networks, static node behavior may lead to resource misuse and ultimately profit decrease. EGT can capture and handle service demand wobbles and accordingly optimize network parameters and node behavior. Networks modeled as immune systems as well as gene regulatory networks can also use EGT to further extend and improve their biology analogies. Gene flow and genetic drift can be easily incorporated into an EGT approach and enrich the existing proposals that usually rely on heuristic or genetic algorithms; EGT's main advantage over these two approaches is its sound mathematical basis that can facilitate system design and optimization.

Finally, merging different flavors of GT with EGT as is the case of iNet-EGT/C [34] is expected to provide more efficient and error-resilient schemes. A first approach toward the creation of a corresponding architecture is presented in [40], where a general framework modeling the creation of evolutionary coalitional applications for wireless networks is given. However, Khan et al. [40] do not deal with a number of actual wireless network challenges, ignoring several characteristics and imperfections of wireless networks. Other GT variants such as stochastic or hedonic games can be adjusted to work with EGT, in an attempt to multiply the gains from each game type considered.

5.5 Conclusion

This chapter presented the principles of EGT and how they can be applied to assist the operation of wireless ad hoc communication networks. EGT has been successfully used to model a variety of modern network problems. Its distributed nature may lead to feasible, scalable solutions suited for almost any kind of network. Enriching classical GT with the notions of reproduction, selection, and mutation, it is suitable for network environments whose characteristics change over time. Also, loosing some strict assumptions made by classical GT, network design is easier yet efficient. The existence of an ESS may assure that network behavior remains optimal under network environment changes and is not affected by random strategy changes in the population under consideration. Overall, EGT may facilitate network design in cases of changeable network environments, leading to scalable and stable solutions.

References

1. R. Landry, K. Grace, and A. Saidi, On the design and management of heterogeneous networks: A predictability-based perspective, *IEEE Communications Magazine*, 42(11), 80–87, November 2004.
2. S. Dobson, S. Denazis, A. Fernández, D. Gaïti, E. Gelenbe, F. Massacci, P. Nixon, F. Saffre, N. Schmidt, and F. Zambonelli, A survey of autonomic communications, *ACM Transactions on Autonomous and Adaptive Systems*, 1(2), 223–259, December 2006.
3. J. Webb, *Game Theory: Decisions, Interaction and Evolution*, Springer, New York, October 2006.
4. M. Nowak, *Evolutionary Dynamics: Exploring the Equations of Life*, Harvard University Press, Harvard, MA, September 2006.
5. J. Weibull, *Evolutionary Game Theory*, The MIT Press, Cambridge, U.K., August 1997.
6. I.F. Akyildiz, S. Weilian, Y. Sankarasubramaniam, and E. Cayirci, A survey on sensor networks, *IEEE Communications Magazine*, 40(8), 102–114, August 2002.
7. F. Garcin, M.H. Manshaei, and J.-P. Hubaux, Cooperation in underwater sensor networks, *International Conference on Game Theory for Networks, 2009. GameNets '09*, Istanbul, Turkey, pp. 540–548, May 2009.
8. J. Partan, J. Kurose, and B.N. Levine, A survey on practical issues in underwater networks, *WUWNet 06*, Los Angeles, CA, 2006.
9. G.V. Crosby and N. Pissinou, Evolution of cooperation in multi-class wireless sensor networks, *32nd IEEE Conference on Local Computer Networks (LCN 2007)*, Dublin, Ireland, pp. 489–495, 2007.
10. M.D. Perez-Guirao, R. Luebben, T. Kaiser, and K. Jobmann, Evolutionary game theoretical approach for IR-IWB sensor networks, *ICC Workshops '08*, Beijing, China, pp. 107–111, May 2008.
11. S. Durisi and G. Benedetto, Performance evaluation of TH-PPM UWB systems in the presence of multiuser interference, *IEEE Communication Letters*, 7(5), 224–226, May 2003.
12. M.D. Perez-Guirao, R. Lübben, Z. Zhao, T. Kaiser, and K. Jobmann, Cross-layer MAC design for IR-UWB networks, *International Workshop on Cross Layer Design 2007 (IWCLD '07)*, Jinan, China, pp. 113–116, September 2007.
13. V. Cerf et al., Delay tolerant network architecture, *Work-in-Progress as an IEFT 4838 Draft*, http://www.ieft.org/rfc/rfc4838.txt
14. K. Fall and S. Farrell, DTN: An architectural rertospective, *IEEE Journal on Selected Areas in Communications*, 26(5), June 2008.
15. K.H. Kabir, M. Sasabe, and T. Takine, Design and analysis of self-organized data aggregation using evolutionary game theory in delay tolerant networks, *IEEE Workshop on Autonomic and Opportunistic Communications (AOC 2009)*, Kos, Greece June 2009.
16. K. Fall and W. Hong, Custody transfer for reliable delivery in Delay Tolerant Networks, *Technical Report in Delay Tolerant Networks*, Intel Research, Berkeley-TR-03-030, 2003.
17. W. Zhao, M. Ammar, and E. Zegura, A message ferrying approach for data delivery in sparse mobile ad-hoc networks, *Proceedings of ACM Mobihoc*, Tokyo, Japan, 2004.

18. W. Lieberman, C. Hauert, and M. Nowak, Evolutionary dynamics on graphs, *Letters to Nature*, 433, 312–316, January 2005.
19. G. Szabó and G. Fáth, Evolutionary games on graphs, *Physics Reports*, 446(4–6), 97–216, July 2007.
20. H. Feng, S. Zhang, C. Liu, J. Yan, and M. Zhang, P2P incentive model on evolutionary game theory, *4th International Conference on Wireless Communications, Networking and Mobile Computing 2008 (WiCOM '08)*, Dalian, China, pp. 1–4, October 2008.
21. M. Sasabe, N. Wakamiya, and M. Murata, A caching algorithm using evolutionary game theory in a file-sharing system, *12th IEEE Symposium on Computers and Communications 2007 (ISCC 2007)*, Aveiro, Portugal, pp. 631–636, July 2007.
22. M. Sasabe, N. Wakamiya, and M. Murata, User selfishness vs. file availability in P2P file-sharing systems: Evolutionary game theoretic approach, *Peer-to-Peer Networking and Applications*, 3(1), 17–26, April 2009.
23. Y.F. Wang, A. Nakao, A.V. Vasilakos, and J.H. Ma, P2P soft security: On evolutionary dynamics of P2P incentive mechanism, *Computer Communications (COMCOM)*, 34(3), 241–249, 2011.
24. E. Altman, R. EIAzouzi, Y. Hayel, and H. Tembine, An Evolutionary Game Approach for the design of congestion control protocols in wireless networks, *6th International Symposium on Modeling and Optimization in Mobile, Ad Hoc, and Wireless Networks and Workshops, 2008 (WiOPT 2008)*, Berlin, Germany, April 2008.
25. H. Tembine, E. Altman, and R. EIAzouzi, *Asymmetric Delay in Evolutionary Games*, VALUETOOLS, Nantes, France, October 2007.
26. H. Tembine, E. Altman, and R. EIAzouzi, Delayed evolutionary game dynamics applied to the medium access control, in *Proceedings of Bionetworks*, Pisa, Italy, October 2007.
27. M.P. Anastasopoulos, P.-D. Arapolgou, R. Kannan, and P.G. Cottis, Adaptive routing strategies in IEEE 802.16 multi-hop wireless backhaul networks based on evolutionary game theory, *IEEE Journal on Selected Areas in Communications*, 26(7), September 2008.
28. A. Vasilakos and M. Anastasopoulos, Application of evolutionary game theory to wireless mesh networks, in *Advances in Evolutionary Computing for System Design*, 66, pp. 249–267, 2007.
29. C. Eklund, R.B. Marks, K.L. Stanwood, and S. Wang, IEEE Standard 802.16: A technical overview of the Wireless MAN air interface for broadband wireless access, *IEEE Communications Magazine*, 40(6), 98–107, June 2002.
30. S. Fischer and B. Vöcking, Adaptive routing with stale information, in *Proceedings of the 24th ACM Symposium on Principles of Distributed Computing*, Las Vegas, NV, pp. 276–283, July 2005.
31. M. Mitzenmancher, How useful is old information?, in *Proceedings of 16th Annual ACM SIGACT-SIGOPS Symposium on Principles of Distributed Computing (PODC)*, Santa Barbara, CA, pp. 83–91, 1997.
32. B. An, A. V. Vasilakos, and V. Lesser, *Evolutionary Stable Resource Pricing Strategies, ACM SIGCOMM 2009*, Barcelona, Spain, August 2009.
33. C. Lee, J. Suzuki, and A.V. Vasilakos, iNet-EGT: An evolutionarily stable adaptation framework for network applications, *4th International Conference on Bio-Inspired Models of Network, Information, and Computing Systems (BIONETICS 2009)*, Avignon, France, December 9–11, 2009.
34. C. Lee, J. Suzuki, and A.V. Vasilakos, An evolutionary game theoretic framework for adaptive, cooperative and stable network applications, in *Proceedings of the 5th International Conference on Bio-Inspired Models of Network, Information and Computing Systems (BIONETICS)*, Boston, MA, December 2010.
35. B. Wang, K.J.R. Liu, and C. Clancy, Evolutionary game framework for behavior dynamics in cooperative spectrum sensing, in *Proceedings of IEEE GLOBECOM 2008*, New Orleans, LA, 2008.
36. B.S. Manoj, R.R. Rao, and M. Zorzi, On the use of higher layer information for cognitive networking, in *Proceedings of IEEE GLOBECOM 2007*, Washington, DC, 2007.

37. M.P. Anastasopoulos, D.K. Petraki, R. Kannan, and A.V. Vasilakos, TCP throughput adaptation in WiMax networks using replicator dynamics, *IEEE Transactions on Systems, Man and Cybernetics— Part B: Cybernetics*, June 2010.

38. C. Barakat and E. Altman, Bandwidth tradeoff between TCP and link-level FEC, *Computer Network*, 39(2), 133–150, June 2002.

39. M.A. Khan, H. Tembine, and A.V. Vasilakos, Game dynamics and cost of learning in heterogeneous 4G networks, *IEEE Journal on Selected Areas in Communications* (to appear in 2012).

40. M.A. Khan, H. Tembine, and A.V. Vasilakos, Evolutionary coalitional games: Design and challenges in wireless networks, *IEEE Communications Magazine*, August 2011 (to appear).

Chapter 6

Game Theory for OFDM Systems with Incomplete Information

Gaoning He, Mérouane Debbah, and Samson Lasaulce

Contents

6.1 Introduction

In practical wireless communication networks, wireless devices have local information but can rarely access global information on the network status. The motivation of this chapter is exactly to study how game-theoretic tools [1] can be implemented in the mentioned situations where devices have

115

limited information. Recently, games of complete information have been introduced and studied in various types of communication networks, for example, multiple access channels (MACs) [2], independent parallel channels (e.g., OFDM) [3], multiple input and multiple output channels [4,5], Interference channels (ICs) [6] and combination of them. The game-theoretic models used in the previous works assume that the information/knowledge about other devices is available to all devices. However, this assumption is rarely true in practice.

For many practical reasons, it is meaningful to investigate communication scenarios where devices have *incomplete information* about their components. In game theory, the notion of incomplete information means that some players do not completely know the structure of the game, which may include player set (how many devices involved? who are they?), other player's action set (how can they behave?), and other players' payoff functions (what are their objectives?).

It is worth noting that the assumption of incomplete information is sometimes related to two other slightly different notions: *imperfect information* and *imperfect/partial channel state information* (*CSI*). Imperfect information is a specific notion used in game theory. It means that a player does not know exactly what actions other players take at that point in sequential games [1]. Imperfect/partial CSI is a specific notation used in wireless communications, which means that the CSI is not perfectly estimated/observed at the transmitter/receiver side. For example, we can imagine the following scenario with imperfect CSI as a game of incomplete information: a device has perfect CSI about its own channel, but it has imperfect CSI about any other device's channel. This is a common situation that usually happens in a real wireless network, since it may be too "expensive" for every device to keep track of the time-variant channels of all other devices. Although selective multiuser diversity algorithms (imperfect/partial feedback based on threshold) were introduced in [7] to reduce the feedback load, it is designed in the centralized framework where a central computing resource is needed, involving feedback and overhead communication whose load scales linearly with the total number of transmitters and receivers in the network. In order to satisfy the computational complexity and real-time QoS (quality of service) requirements of future wireless networks, it is necessary and urgent to consider game theory as a mathematical tool to understand the network resource conflicts in a distributed manner.

6.2 Bayesian Games

Strategic-form game is an appropriate model to interpret the interaction between decision makers (refer to players). The model captures interaction between the players by allowing each player to be affected by the actions of all players (not only the player's own action). The famous "Nash equilibrium" represents a common solution concept that is mutually steady, in the sense that no single player has any motivation to deviate from it. It perfectly answers the following question: what actions should the players choose in a noncooperative strategic-form game?

An important assumption behind the notion of Nash equilibrium is that players are clearly aware of the game they are playing. In particular, a player must know the preferences of all other players. However, in many situations, the participants are not perfectly informed about their opponents' characteristics. For example, companies may not know each others' production costs, chess players may not be familiar with each others' chess playing styles, and bargainers may not know each others' valuations of the object of negotiation.

What can be done beyond Nash equilibrium when players do not have complete information about the game? The early framework of "Bayesian game" was firstly defined and studied by Harsanyi in 1967 [8], which shows how to convert a game with incomplete information into one with complete yet imperfect information. In this section, we will describe the mathematical model of Bayesian games. This model allows us to analyze any situation in which players are imperfectly informed about the aspect of her environment that is relevant to her decision making.

The goal of this chapter is to introduce the model of Bayesian games and apply it to interactive situations, such as resource allocation conflicts in the multiuser communication networks. Note that game theory may not directly provide a magic formula that suddenly resolves a conflict situation. It does not specify about the resolution of conflicts; it is about modeling, predicting, and understanding conflicts. Once we understand the conflicts, perhaps we could use some of the insights to resolve them.

6.2.1 Description of Bayesian Games

Definition 6.1 (Bayesian Game) A Bayesian game $\mathcal{G}$ is a strategic-form game with incomplete information, which can be completely described as follows:

$$\mathcal{G} = \langle \mathcal{K}, \{\mathcal{T}_k, \mathcal{A}_k, \rho_k, u_k\}_{k \in \mathcal{K}} \rangle$$

which consists of

- A **player** set: $\mathcal{K} = \{1, \ldots, K\}$
  and for each player $k \in \mathcal{K}$
- A **type** set: $\mathcal{T}_k$ ($\mathcal{T} = \mathcal{T}_1 \times \mathcal{T}_2 \times \cdots \times \mathcal{T}_K$)
- An **action** set: $\mathcal{A}_k$ ($\mathcal{A} = \mathcal{A}_1 \times \mathcal{A}_2 \times \cdots \times \mathcal{A}_K$)
- A **probability** function set:

$$\rho_k : \mathcal{T}_k \to \Omega\left(\mathcal{T}_{-k}\right)$$

- A **payoff** function set:

$$u_k : \mathcal{A} \times \mathcal{T} \to \mathbb{R}$$

$u_k(a, \tau)$ is the payoff of player k when action profile is $a \in \mathcal{A}$ and type profile is $\tau \in \mathcal{T}$.

In order to give readers some insight into the definition of Bayesian games, we provide the following remarks that specify, one by one, the implications underlying the notions of type, strategy, probability, and payoff defined in $\mathcal{G}$.

Remark 6.1 Notions of "type," "strategy," "probability," and "payoff"

- Notion of type: player k's "type" $\tau_k \in \mathcal{T}_k$ represents any kind of private information that is relevant to her decision making. For example, her payoff function, her beliefs about other players' payoff functions, her beliefs about what other players believe her beliefs are, and so on.
- Notion of (pure) strategy: a (pure) strategy for player k is a function mapping her type set $\mathcal{T}_k$ into her action set $\mathcal{A}_k$

$$s_k : \mathcal{T}_k \to \mathcal{A}_k$$

We denote by $\mathcal{S}_k$ the set of (pure) strategies (maps from $\mathcal{T}_k$ to $\mathcal{A}_k$) for player k.

- Notion of probability: the probability function ρ_k represents what player k believes about the types of the other players, given her own type $\tau_k \in \mathcal{T}_k$. It is a conditional probability, that is, $\rho_k\left(\tau_{-k}|\tau_k\right)$, assigned to the type profile $\tau_{-k} \in \mathcal{T}_{-k} = \mathcal{T} \setminus \mathcal{T}_k$, given τ_k.
- Notion of payoff: player k's payoff function u_k is a function of the strategy profile $s(\cdot) \triangleq \{s_1(\cdot), \dots, s_K(\cdot)\}$ and the type profile $\tau \triangleq \{\tau_1, \dots, \tau_K\}$ of all players in the game $\mathcal{G}$,

$$u_k\left(s(\tau), \tau\right) = u_k\left(s_1(\tau_1), \dots, s_K(\tau_K), \tau_1, \dots, \tau_K\right).$$

Remark 6.2 $\mathcal{G}$ is a finite Bayesian game if the sets $\mathcal{K}, \mathcal{T},$ and $\mathcal{A}$ are all finite.

6.2.2 Bayesian Equilibrium in a Nutshell

What can we expect of the outcome of Bayesian games if every participant plays the game *selfishly* (only cares about her own payoff) and *rationally* (choosing the best response given her incomplete information)? Generally speaking, the process of such players' behaviors usually results in *Bayesian equilibrium* (BE), which represents a common solution concept for a Bayesian game. In many cases, BE represents a steady result of learning and evolution of all participants (who are not perfectly informed) in a distributed system. Therefore, it becomes important to define and characterize such equilibrium solutions.

Note that in a strategic-form game with complete information, each player chooses a concrete action. However, in a Bayesian game, each player k faces the problem of choosing a set or collection of actions (strategy $s_n(\cdot)$), one for each type she may encounter. It is also worth noting that the action set of each player is independent of the type set, that is, the actions available to her is the same for every type.

Now, letting $\{\widetilde{s}_k(\cdot), s_{-k}(\cdot)\}$ denote the strategy profile where all players play $s(\cdot)$ except player k who plays $\widetilde{s}_k(\cdot)$, we can then describe player n's payoff as

$$u_k\left(\widetilde{s}_k(\tau_k), s_{-k}(\tau_{-k}), \tau\right) = u_k\left(s_1(\tau_1), \dots, s_{k-1}(\tau_{k-1}), \widetilde{s}_k(\tau_k), s_{k+1}(\tau_{k+1}), \dots, s_K(\tau_K), \tau\right)$$

Definition 6.2 (Bayesian Equilibrium) The strategy profile $s^\star(\cdot)$ is a (pure strategy) **BE**, if for all $k \in \mathcal{K}$, and for all $s_k(\cdot) \in \mathcal{S}_k$ and $s_{-k}(\cdot) \in \mathcal{S}_{-k}$

$$\mathbb{E}_\tau\left[u_k\left(s_k^\star(\tau_k), s_{-k}^\star(\tau_{-k}), \tau\right)\right] \geq \mathbb{E}_\tau\left[u_k\left(s_k(\tau_k), s_{-k}^\star(\tau_{-k}), \tau\right)\right]$$

where we define

$$\mathbb{E}_\tau\left[u_k\left(x_k(\tau_k), x_{-k}(\tau_{-k}), \tau\right)\right] \triangleq \sum_{\tau_{-k} \in T_{-k}} \rho_k(\tau_{-k}|\tau_k) u_k\left(x_k(\tau_k), x_{-k}(\tau_{-k}), \tau\right)$$

as the expected payoff of player k, which is averaged over the joint distribution of all other players' types.

Remark 6.3 Note that the averaging effect on the payoff reflects the best effort a player can achieve (against other players) to the best of her knowledge, and this expected payoff does not depend on the actions of any other types of player n but only on the actions of the various types of the other players.

6.2.3 Methodology for Studying Bayesian Equilibria

Since a BE can be considered as a Nash equilibrium in the Bayesian game, we can adopt the analytical method of Nash equilibrium to study the BE set. Here, we briefly discuss this methodology in general situations [4,9], which involves the following three keywords in a row: "existence," "uniqueness," and "selection," as follows:

1. Existence: Does an equilibrium exist?
2. Uniqueness: Does there exist a unique equilibrium or multiple ones?
3. Selection: How to select a favoring equilibrium from the equilibrium set?

"Existence" is the very first question that naturally comes into our mind, since it is known that, in general, equilibrium point does not necessarily exist. Mathematically speaking, proving the existence of an equilibrium is equivalent to proving the existence of a solution to a fixed-point problem [10]. Since the existence of the fixed-point hints that there is some strategy set that is a best response to itself; therefore, no player could do any better by deviating, and so it is an equilibrium. Fortunately, there are many scenarios based on usual channel models and performance metrics (e.g., Shannon transmission rate and rate regions have desirable convexity properties [11] that are in favor of the existence of an equilibrium [12]) where existing theorems are sufficient.

"Uniqueness" is the second fundamental problem that we need to address when the existence is ensured. Ideally, we would prefer to have a unique equilibrium because it is the simplest solution form for general game-theoretic problems, and it is important not only for predicting the state of the network but also for convergence issues. Unfortunately, there are not so many general results on the topic of equilibrium uniqueness. One could find some useful results in the *concave N-person games* [12], where it is shown that there exists exactly a unique equilibrium if the payoff functions satisfy the condition of *diagonally strictly concave*.

However, there are many important scenarios where the equilibrium is not unique, for example, routing games [13], coordination games [14], noncooperative games with correlated constraints together with the concept of "generalized Nash equilibrium" [15], etc. Natural questions that arise concern the selection of an appropriate equilibrium, i.e.,

■ What can be done when one has to deal with a game having multiple equilibria?
■ Are there some equilibria "dominating" others?
■ Are there some equilibria more "fair" than others?
■ What is a "good" selection rule to follow?

As a matter of fact, "equilibrium selection" is a mature theory in itself [16]. But here, instead of paying attention to the general theory, we are more interested in the applications of concave games, for which Rosen has already introduced the notion of "normalized equilibria" [12] that shows a very neat way to tackle this selection problem. Obviously, the selection rule is strongly related to the fairness criteria being used, for example, max–min fairness [17], proportional fairness [18,19], Jain's fairness [20], global optimization and normalized equilibrium, etc. Specifically, the authors of [9] have shown that the max–min fairness, proportional fairness, and normalized equilibrium achieve the same rate allocation in the context of MACs with multiuser decoding applied at the receiver side. Up to now, equilibrium selection is still an open topic in many communication network models and applications, especially for the case of non-convex rate region, for example, achievable rate region of collision channels and ICs.

6.2.4 An Example

Example 6.1: (Battle of Standards)

Imagine two companies 1 and 2 negotiate for determining a wireless network standard. There are two choices: CDMA and GSM. Company 1 prefers CDMA due to its higher system performance; company 2 prefers GSM, since she does not want to change the network infrastructure. Both may prefer choosing the same standard; however, neither company knows exactly whether the other wants to cooperate. They must decide simultaneously (without communication) which standard to choose. This game can be described by

- A player set $\mathcal{K} = \{1, 2\}$.
- A type set $\mathcal{T}_k = \{X, Y\}$, $\forall n$, where X represents one player believes that the other player wants to cooperate with her and Y represents one player believes that the other player does not want to cooperate with her.
- An action set $\mathcal{A}_k = \{CDMA, GSM\}$, $\forall k$.
- A probability function set ρ_k. Suppose that the types X, Y are independently distributed. We let $\rho_k(X) = \Pr(\tau_{-k} = X)$ be the probability that player k believes the other player $-k$ wants to cooperate with her; $\rho_k(Y) = \Pr(\tau_{-k} = Y)$ be the probability that player k believes the other player $-k$ does not wants to cooperate with her.
- A payoff function set is decided by the following matrices:

$$\rho_2(X) = \frac{1}{2} \qquad\qquad \rho_2(Y) = \frac{1}{2}$$

$\rho_1(X) = \frac{2}{3}$

	CDMA	GSM
CDMA	(2, 1)	(0, 0)
GSM	(0, 0)	(1, 2)

	CDMA	GSM
CDMA	(2, 0)	(0, 2)
GSM	(0, 1)	(1, 0)

$\rho_1(Y) = \frac{1}{3}$

	CDMA	GSM
CDMA	(0, 1)	(2, 0)
GSM	(1, 0)	(0, 2)

	CDMA	GSM
CDMA	(0, 0)	(2, 2)
GSM	(1, 1)	(0, 0)

In each entry (x_1, x_2), the values x_1 and x_2 represent the payoff (for a certain type pair) of player 1 and 2, respectively. Based on the probability distribution of the types, one can also write the expected payoff matrix as

	CDMA	GSM
CDMA	$\left(\frac{4}{3}, \frac{1}{2}\right)$	$\left(\frac{2}{3}, 1\right)$
GSM	$\left(\frac{1}{3}, \frac{1}{2}\right)$	$\left(\frac{2}{3}, 1\right)$

Now the entry (x_1, x_2) represents the expected payoff of player 1 and 2, respectively. If this expected payoff matrix is considered as a payoff matrix in a strategic-form game with complete information, the Nash equilibrium would be $\{CDMA, CDMA\}$ and $\{GSM, GSM\}$, which are two cooperation strategies.

On the other hand, it is easy to verify that there are two Bayesian equilibria in this game, denoted as BE_1 and BE_2:

$$BE_1 : \begin{cases} s_1^\star(X) = CDMA, s_1^\star(Y) = GSM \\ s_2^\star(X) = CDMA, s_2^\star(Y) = GSM \end{cases}$$

$$BE_2 : \begin{cases} s_1^\star(X) = CDMA, s_1^\star(Y) = GSM \\ s_2^\star(X) = s_2^\star(Y) = GSM \end{cases}$$

The first equilibrium says that, at the BE_1, if a company believes the other company wants to cooperate with her, she would choose "CDMA"; if a company believes the other player doesn't want to cooperate with her, she would choose "GSM." The second equilibrium says that at the BE_2, company 1 follows the same strategy as in the BE_1; company would choose "GSM" regardless of her type.

6.3 Applications

In the previous decades, wireless communications research has mostly focused on systems built from point to point channels. In such systems, physical communication links are essentially interference free, and interference management is at most a peripheral issue. Whilst these approaches have obvious advantages in terms of simplicity of design and maintenance, they typically suffer from low spectral efficiencies. Interference management has become more and more a central problem that determines the system performance of several coexisting Tx–Rx links or coexisting systems. The "interference" is mainly due to the wide use of omnidirectional (radiates equally in all directions) antennas in current cellular systems.

In fact, every transmitter causes a degree of interference into every receiver, the key question being how much. It is worth noting that some techniques are capable of attenuating interference with some unnegligible cost (increasing complexity of signal processing, infrastructure change, etc.) at both Tx and Rx. The key idea of these techniques is to use the "degrees of freedom" to keep the transmission orthogonality between different Tx–Rx links.

- Time: Time division multiple access (TDMA)
- Frequency: Frequency division multiple access (FDMA)
- Space: Space division multiple access (SDMA)

Although these techniques are capable of interference avoidance, undesired mutual interference between different wireless links is inevitable for multicell networks. Thus, interference management becomes a central problem in wireless system design. In the following part, we will show different methods to analyze the wireless resource allocation problem as a Bayesian game. One example shows a Bayesian game with infinite states and finite action set, and another example shows finite states and infinite action set.

6.3.1 Multiple Access Channels

The first channel model that we study in detail is the MAC. In wireless communications, MAC is a basic channel model that allows several terminals connected to the same transmission medium to transmit through it and to share its capacity (Figure 6.1). In many previous game-theoretic studies, it is often assumed that the CSI is completely known at the transmitter side. However, this assumption of complete information is barely realistic in the wireless communication systems nowadays, due to the high cost of feedback signalings that provide CSI to all the transmitters.

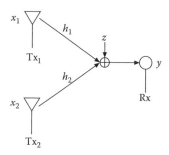

Figure 6.1 Two-user MAC.

Based on this realistic concern, it is important to shift the basic assumption on the knowledge level of CSI from "complete information" to "incomplete information." Therefore, we should consider this model as a Bayesian game, in which each transmitter has only "incomplete information" (e.g., the knowledge of her own channel gain). See an example in Figure 6.1, the "complete information" means that, at time slot t, the realization of channel gains $g_1(t)$ and $g_2(t)$ (which is defined as the magnitude squared of the fading channel coefficient, i.e., $g_k(t) = |h_k(t)|^2$) is known at both transmitters Tx_1 and Tx_2; "incomplete information" means that at time slot t, each transmitter Tx_k knows only her own channel realization $g_k(t)$, but not knowing the other user's channel gain $g_{-k}(t)$. Now, we can completely characterize the two-user MAC Bayesian game as follows:

DESCRIPTION OF TWO-PLAYER MAC BAYESIAN GAME

$$\mathcal{G}_{MAC-2} \triangleq \langle \mathcal{K}, \mathcal{T}, \mathcal{P}, \mathcal{Q}, \mathcal{U} \rangle$$

- Player set: $\mathcal{K} = \{1, 2\}$.
- Type set: $\mathcal{T} = \mathcal{T}_1 \times \mathcal{T}_2$,
  where $\mathcal{T}_k = \{g_-, g_+\}$. A player's type is defined as her channel gain, i.e., $g_k \in \mathcal{T}_k$. Note that each channel gain g_k is assumed to be independent of each other, and it has only two possible values g_- and g_+.
- Action set: $\mathcal{P} = \mathcal{P}_1 \times \mathcal{P}_2$,
  where $\mathcal{P}_k = \left[0, P_k^{\max}\right]$, and $P_k^{\max}$ is the power constraint of player k. A player's action is defined as its transmit power, i.e., $p_k \in \mathcal{P}_k$.
- Probability set: $\mathcal{Q} = \mathcal{Q}_1 \times \mathcal{Q}_2$,
  where $\mathcal{Q}_k = [0, 1]$. Let $\rho_- = \Pr(g_k = g_-)$ and $\rho_+ = \Pr(g_k = g_+)$ represent the probability that player k's channel gain g_k is g_- and g_+, respectively. We have $\rho_-, \rho_+ \in \mathcal{Q}_k$, and $\rho_- + \rho_+ = 1$.
- Payoff function set: $\mathcal{U} = \{u_1, u_2\}$,
  u_k is chosen as player k's average rate with single-user decoding*:

$$u_k(s_1, s_2) = \mathbb{E}_{g_1, g_2} \left[\log\left(1 + \frac{g_k s_k(g_k)}{\sigma^2 + g_{-k} s_{-k}(g_{-k})}\right)\right] \qquad (6.1)$$

---

* Single-user decoding means that a user decodes its own signal while treating other users' signals as noise.

Assumption 6.1   We assume that every channel gain g_k is independent and identically distributed (i.i.d.) between two values g_- and g_+ with probability ρ_- and ρ_+, respectively. Without loss of generality, we assume $g_- < g_+$.

With this assumption, we can rewrite the payoff (6.1) as

$$u_k(\mathbf{p}_1, \mathbf{p}_2) = \rho_-^2 \log\left(1 + \frac{g_- \bar{p}_k}{\sigma^2 + g_- \bar{p}_{-k}}\right) + \rho_- \rho_+ \log\left(1 + \frac{g_- \bar{p}_k}{\sigma^2 + g_+ p_{-k}^+}\right)$$

$$+ \rho_+ \rho_- \log\left(1 + \frac{g_+ p_k^+}{\sigma^2 + g_- \bar{p}_{-k}}\right) + \rho_+^2 \log\left(1 + \frac{g_+ p_k^+}{\sigma^2 + g_+ p_{-k}^+}\right) \tag{6.2}$$

where $\bar{p}_k$ and p_k^+ represent player k's transmit power when her channel gain is g_- and g_+, respectively. We write the power vector as $\mathbf{p}_k = \left[\bar{p}_k, p_k^+\right]$.

As mentioned in the previous part, the central question in such a game is whether a BE exists, and if so, whether the equilibrium is unique. In our game $\mathcal{G}_{MAC-2}$, the answer to the above two questions are fortunately "yes." It is easy to prove the existence part, since the strategy space s_k is convex, compact, and nonempty for each k; the payoff function u_k is continuous in both s_k and s_{-k}; and u_k is concave in s_k for any s_{-k} [1]. In order to prove the uniqueness part, we shall rely on a sufficient condition for the uniqueness of equilibrium, as introduced in [12]: the nonnegative weighted sum of the utility functions is *diagonally strictly concave*. We firstly give its definition:

Definition 6.3 (Diagonally Strictly Concave)   A weighted nonnegative sum function $f(\mathbf{x}, \mathbf{r}) = \sum_{k=1}^{K} r_i \varphi_k(\mathbf{x})$ is called diagonally strictly concave for vector $\mathbf{x} \in \mathbb{R}^{K \times 1}$ (vector length of K with real number element) and $\mathbf{r} \in \mathbb{R}_{++}^{K \times 1}$ (positive real number element), if for any two different vectors $\mathbf{x}^0, \mathbf{x}^1$, we have

$$\Omega(\mathbf{x}^0, \mathbf{x}^1, \mathbf{r}) \triangleq (\mathbf{x}^1 - \mathbf{x}^0)^{\mathrm{T}} \delta(\mathbf{x}^0, \mathbf{r}) + (\mathbf{x}^0 - \mathbf{x}^1)^{\mathrm{T}} \delta(\mathbf{x}^1, \mathbf{r}) > 0 \tag{6.3}$$

where $\delta(\mathbf{x}, \mathbf{r})$ is called pseudo-gradient of $f(\mathbf{x}, \mathbf{r})$, defined as

$$\delta(\mathbf{x}, \mathbf{r}) = \left[r_1 \frac{\partial \varphi_1}{\partial x_1} \; r_2 \frac{\partial \varphi_2}{\partial x_2} \; \cdots \; r_K \frac{\partial \varphi_K}{\partial x_K} \right]^{\mathrm{T}}$$

LEMMA 6.1   The weighted nonnegative sum of the payoffs in $\mathcal{G}_{MAC-2}$ is diagonally strictly concave.

Proof   Write the weighted nonnegative sum of the payoffs as

$$f^u(\mathbf{p}, \mathbf{r}) \triangleq r_1 u_1(\mathbf{p}) + r_2 u_2(\mathbf{p}) \tag{6.4}$$

where

u_k is player k's payoff defined in (6.1)
$\mathbf{p} = [p_1 \; p_2]^{\mathrm{T}}$ is a transmit power vector
$\mathbf{r} = [r_1 \; r_2]^{\mathrm{T}}$ is a nonnegative vector assigning weights r_1, r_2 to the payoffs u_1, u_2, respectively

Denote by $\delta^u(\mathbf{p}, \mathbf{r}) \triangleq [r_1 \partial u_1/\partial p_1 \; p_2 \partial u_2/\partial p_2]^T$ the pseudo-gradient of $f^u(\mathbf{p}, \mathbf{r})$. One can find that the payoffs u_1, u_2 are special cases of the following form (sum of th logarithm functions):

$$\begin{cases} u_1(\mathbf{p}) = \sum_i \log\left(1 + \dfrac{a_i + c_i p_1}{\sigma^2 + b_i + d_i p_2}\right) \\ u_2(\mathbf{p}) = \sum_i \log\left(1 + \dfrac{b_i + d_i p_2}{\sigma^2 + a_i + c_i p_1}\right) \end{cases}$$

where $a_i, b_i, c_i, d_i \in \mathbb{R}$ $(c_i \neq 0, d_i \neq 0)$, $a_i + c_i p_1 > 0$, $b_i + d_i p_2 > 0 \; \forall i, \sigma^2 > 0$. And we have

$$\delta^u(\mathbf{p}, \mathbf{r}) = \begin{bmatrix} r_1 \dfrac{\partial u_1}{\partial p_1} \\ r_2 \dfrac{\partial u_2}{\partial p_2} \end{bmatrix} = \begin{bmatrix} r_1 \sum_i c_i \phi_i^{-1}(\mathbf{p}) \\ r_2 \sum_i d_i \phi_i^{-1}(\mathbf{p}) \end{bmatrix} = \sum_i \begin{bmatrix} r_1 c_i \phi_i^{-1}(\mathbf{p}) \\ r_2 d_i \phi_i^{-1}(\mathbf{p}) \end{bmatrix}$$

where $\phi_i(\mathbf{p}) \triangleq \sigma^2 + a_i + b_i + c_i p_1 + d_i p_2$. Now, we check if our payoffs satisfy the diagonally strictly concave condition (6.3). Let $\mathbf{p}^0, \mathbf{p}^1$ be two different vectors satisfying the power constraint. We have

$$\begin{aligned} \Omega^u(\mathbf{p}^0, \mathbf{p}^1, \mathbf{r}) &\triangleq (\mathbf{p}^1 - \mathbf{p}^0)^T \delta^u(\mathbf{p}^0, \mathbf{r}) + (\mathbf{p}^0 - \mathbf{p}^1)^T \delta^u(\mathbf{p}^1, \mathbf{r}) \\ &= (\mathbf{p}^1 - \mathbf{p}^0)^T \left[\delta^u(\mathbf{p}^0, \mathbf{r}) - \delta^u(\mathbf{p}^1, \mathbf{r})\right] \\ &= \left[(p_1^1 - p_1^0) \; (p_2^1 - p_2^0)\right] \begin{bmatrix} r_1 \sum_i c_i \left(\phi_i^{-1}(\mathbf{p}^0) - \phi_i^{-1}(\mathbf{p}^1)\right) \\ r_2 \sum_i d_i \left(\phi_i^{-1}(\mathbf{p}^0) - \phi_i^{-1}(\mathbf{p}^1)\right) \end{bmatrix} \\ &= \sum_i \left\{ \left[r_1 c_i (p_1^1 - p_1^0) + r_2 d_i (p_2^1 - p_2^0)\right] \left(\phi_i^{-1}(\mathbf{p}^0) - \phi_i^{-1}(\mathbf{p}^1)\right) \right\} \\ &= \sum_i \left\{ \left[r_1 c_i (p_1^1 - p_1^0) + r_2 d_i (p_2^1 - p_2^0)\right]^2 \phi_i^{-1}(\mathbf{p}^0) \phi_i^{-1}(\mathbf{p}^1) \right\} \end{aligned}$$

It is easy to see that the term $\left[r_1 c_i (p_1^1 - p_1^0) + r_2 d_i (p_2^1 - p_2^0)\right]^2$ is nonnegative and must not be 0 simultaneously for each term with index i (since we have $c_i \neq 0, d_i \neq 0$ and there exists at least one c_i (d_i) that is different from any other c_j (d_j) $(j \neq i)$), and the term $\phi_i^{-1}(\mathbf{p}^0)\phi_i^{-1}(\mathbf{p}^1)$ is always positive. Therefore, we must have $\Omega^u(\mathbf{p}^0, \mathbf{p}^1, \mathbf{r}) > 0$, and from Definition 6.3, the sum-payoff function $f^u(\mathbf{p}, \mathbf{r})$ must be diagonally strictly concave. This completes the proof. □

THEOREM 6.1   There exists a unique BE in the two-user MAC game $\mathcal{G}_{MAC-2}$.

Proof   Refer to the proof of Theorem 2 in [12]. Since our sum-payoff function $f^u(\mathbf{p}, \mathbf{r})$ given in (6.4) is diagonally strictly concave, we have the uniqueness of Nash equilibrium in our game $\mathcal{G}_{MAC-2}$.   □

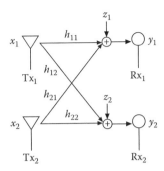

Figure 6.2   Illustration of 2 × 2 IC.

6.3.2 Interference Channels

Interference is a central phenomenon in wireless communication when multiple uncoordinated links share a common communication medium. An IC is a communication medium shared by multiple transmitter-to-receiver pairs. Signal transmission from each transmitter-to-receiver pair interferes with the communications between the other transmitter-to-receiver pairs.

An important information-theoretic model is the two-user IC (Figure 6.2). In this part, we consider a model of two-user IC with N sub-channels. These sub-channels can be interpreted as frequency subcarriers or clusters of subcarriers in a multi-carrier communication system, for example, OFDM. The input–output equations can be then written as

$$y_1^{(n)} = x_1^{(n)} h_{11}^{(n)} + x_2 h_{21}^{(n)} + z_1$$

$$y_2^{(n)} = x_2^{(n)} h_{22}^{(n)} + x_1 h_{12}^{(n)} + z_2$$

where
  n is the sub-channel index
  $x_k^{(n)}$ and $y_k^{(n)}$ stand for user k's input and output signals on sub-channel n, respectively
  $h_{kj}^{(n)}$ is the complex channel gain from transmitter Tx$_k$ to receiver Rx$_j$
  $z_k \sim \mathcal{N}(0, \sigma^2)$ is the zero mean Gaussian noise with variance σ^2
  We assume that z_1 and z_2 are independent

Assuming that each user treats other users' signals as noise and no interference cancelation techniques are applied, user k's Shannon transmission rate* is

$$R_k = \sum_{n=1}^{N} \log \left(1 + \frac{g_{kk}^{(n)} p_k^{(n)}}{\sigma^2 + g_{-kk}^{(n)} p_{-k}^{(n)}} \right) \qquad (6.5)$$

We make the simplifying assumption of flat fading, i.e., constant gains across sub-channels $g_{kj}^{(1)} = g_{kj}^{(2)} = \cdots = g_{kj}^{(N)}$. We also assume that users have "incomplete information" about the channel

---

* Note that the general capacity region of the two-user Gaussian IC is unknown and it has been an open issue for 30 years [21].

gain matrix $\mathbf{G}$, meaning that each user k is aware of the self channel gain g_{kk} and the incident channel gain g_{-kk}. For notational simplicity, we denote by $\mathbf{g}_k = \{g_{kk}, g_{-kk}\}$, whose realizations are completely known at user k. However, due to the difficulties involved in dissemination of CSI from other devices, we continue to assume that user k is unaware of the channel gains of all other users $\mathbf{g}_{-k}$, but she knows the distribution of $\mathbf{g}_{-k}$. Therefore, if each user selfishly maximizes its own rate R_k and if we consider the value of channel gain $\mathbf{g}_k$ as user k's "private information" or "type," this becomes a Bayesian game, in which a "strategy" is a family of functions $\mathbf{s}_k(\mathbf{g}_k) = \left\{ s_k^{(1)}(\mathbf{g}_k), \ldots, s_k^{(N)}(\mathbf{g}_k) \right\}$, where $s_k^{(n)}(\cdot)$ determines the power level that player k transmits over sub-channel n.

Thus, we can completely characterize the two-user IC Bayesian game by

---

DESCRIPTION OF TWO-PLAYER IC BAYESIAN GAME

$$\mathcal{G}_{IC-2} \triangleq \langle \mathcal{K}, \mathcal{T}, \mathcal{P}, \mathcal{Q}, \mathcal{U} \rangle$$

- Player set: $\mathcal{K} = \{1, 2\}$.
- Type set: $\mathcal{T} = \mathcal{T}_1 \times \mathcal{T}_2$,
  where $\mathcal{T}_k \equiv \mathbb{R}_{++}^2$. A player's type is defined as the channel gain vector, $\mathbf{g}_k \in \mathcal{T}_k$.
- Action set: $\mathcal{P} = \mathcal{P}_1 \times \mathcal{P}_2$,
  where $\mathcal{P}_k = \{P^{\max}\mathbf{e}_1, \ldots, P^{\max}\mathbf{e}_N, (P^{\max}/N)\mathbf{1}\}$, which is a finite action set (the meaning of this action set will be explained in Remark 6.4). A player's action is defined as its transmit power vector, i.e., $\mathbf{p}_k = \left\{ p_k^{(1)}, \ldots, p_k^{(N)} \right\} \in \mathcal{P}_k$.
- Probability set: $\mathcal{Q} = \mathcal{Q}_1 \times \mathcal{Q}_2$,
  it is assumed that the channel gains $\mathbf{g}$ are drawn from a certain distribution F, with continuous density f on a compact subset, and $\mathbf{g}_1$ is independent of $\mathbf{g}_2$.
- Payoff function set: $\mathcal{U} = \{u_1, u_2\}$,
  u_k is chosen as player k's average sum-capacity:

$$u_k(\mathbf{s}_1, \mathbf{s}_2) = \mathbb{E}_{\mathbf{g}_1, \mathbf{g}_2} \left[\sum_{n=1}^{N} \log\left(1 + \frac{g_{kk}^{(n)} s_k^{(n)}(\mathbf{g}_k)}{\sigma^2 + g_{-kk}^{(n)} s_{-k}^{(n)}(\mathbf{g}_{-k})} \right) \right] \tag{6.6}$$

Remark 6.4   [22] In this game, we investigate the *symmetric Bayesian equilibrium* (SBE), i.e., both players use the same strategy in the equilibrium. Since the functional strategic form of a player can be quite complex, we analyze the Bayesian games with the following finite actions:

$$\mathcal{P}_1 = \mathcal{P}_2 = \left\{ P^{\max}\mathbf{e}_1, \ldots, P^{\max}\mathbf{e}_n, \left(\frac{P^{\max}}{N} \right)\mathbf{1} \right\} \tag{6.7}$$

where

$\quad$ $\mathbf{e}_n$ represents a unit vector with all zero entries except a "1" in the nth position
$\quad$ $\mathbf{1}$ represents a vector with all entries being equal to 1
$\quad$ $P^{\max}$ represents the power constraint

We denote by $\mathbf{s}_k$ the power strategy vector (of length of N) chosen by player k, i.e., $\mathbf{s}_k = \left\{ s_k^{(1)}, \ldots, s_k^{(N)} \right\}$ where each element is $s_k^{(n)}(\mathbf{g}_k) = p_k^{(n)}$, a mapping from type $\mathbf{g}_k$ to action $\mathbf{p}_k$. The chosen actions in (6.7) have the following intuitive meanings:

- "$P^{\max}\mathbf{e}_n$" implies that player focuses the total power only on the nth sub-channel.
- "$(P^{\max}/N)\,\mathbf{1}$" implies that player spreads the total power uniformly across all the sub-channels.

So, this is a Bayesian game with finite (or discrete) actions. The next theorem shows that the IC two-player game $\mathcal{G}_{IC-2}$ has a unique SBE.

THEOREM 6.2 Assuming that $\mathbf{g_1}$ and $\mathbf{g_2}$ are *i.i.d.*, the unique pure strategy SBE in this game is that both players choose action $(P^{\max}/N)\,\mathbf{1}$.

Proof Define the probability that player k transmits with full power in sub-channel n as $\rho_n = \Pr\left(\mathbf{p}_k = P^{\max}\mathbf{e}_n\right)$, and the probability that player k transmits with equal power in all sub-channels as $\gamma = \Pr\left(\mathbf{p}_k = P^{\max}/N\right)$. Define $\bar{u}_k\left(\mathbf{p}_k; \mathbf{g}_k\right)$ as player k's expected payoff if she chooses action $\mathbf{p}_k$, given that the other player is choosing the equilibrium strategy set $\{s^{(1)}, \ldots, s^{(N)}\}$ and the channel gains are $\mathbf{g}_k$. $\square$

LEMMA 6.2 For any two sub-channels n and n', if $\rho_n < \rho_{n'}$, then $\bar{u}_k\left(P^{\max}\mathbf{e}_n; \mathbf{g}_k\right) > \bar{u}_k\left(P^{\max}\mathbf{e}_{n'}; \mathbf{g}_k\right)$ for all values of $\mathbf{g}_i$.

Proof From (6.6), we can write $\bar{u}_k\left(P^{\max}\mathbf{e}_n; \mathbf{g}_k\right)$ as

$$
\bar{u}_k\left(P^{\max}\mathbf{e}_n; \mathbf{g}_k\right) = \rho_n \log\left(1 + \frac{g_{kk}P^{\max}}{\sigma^2 + g_{-kk}P^{\max}}\right) + \gamma \log\left(1 + \frac{g_{kk}P^{\max}}{\sigma^2 + g_{-kk}P^{\max}/N}\right)
$$

$$
+ (1 - \rho_n - \gamma) \log\left(1 + \frac{g_{kk}P^{\max}}{\sigma^2}\right)
$$

Define Δ as

$$
\Delta = \log\left(1 + \frac{g_{kk}P^{\max}}{\sigma^2}\right) - \log\left(1 + \frac{g_{kk}P^{\max}}{\sigma^2 + g_{-kk}P^{\max}}\right)
$$

We have $\Delta > 0$, since we have assumed $g_{-kk} > 0$. It is easy to show that $\bar{u}_k\left(P^{\max}\mathbf{e}_n; \mathbf{g}_k\right) - \bar{u}_k\left(P^{\max}\mathbf{e}_{n'}; \mathbf{g}_k\right) = \Delta(\rho_{n'} - \rho_n)$. And since $\Delta > 0$ and $\rho_{n'} > \rho_n$, this lemma is proved. $\square$

Lemma 6.2 claims that if $\rho_n < \rho_{n'}$, player k strictly prefers to put full power into sub-channel n over putting full power into sub-channel n', which ensures that in a symmetric equilibrium, we cannot have $\rho_n < \rho_{n'}$ for any two sub-channels n and n'. Therefore, we must have $\rho_n = \rho_{n'}$, i.e., $\rho \triangleq \rho_n = (1 - \gamma)/N$ for all n. And since $N \log(1 + x/N) > \log(1 + x)$, it is easy to show that $\bar{u}_k\left(P^{\max}/N; \mathbf{g}_k\right) > \bar{u}_k\left(P^{\max}\mathbf{e}_n; \mathbf{g}_k\right)$ for all sub-channel n. Thus, in a symmetric equilibrium, we must have $\rho = 0$. This simply means that the unique symmetric equilibrium occurs where $\gamma = 1$, i.e., both players spread the total power uniformly across all sub-channels. $\square$

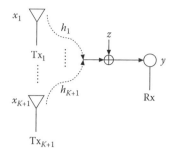

Figure 6.3 Illustration of RAC.

One may notice that we have a unique equilibrium in both Bayesian games $\mathcal{G}_{MAC-2}$ and $\mathcal{G}_{IC-2}$, which is a highly preferred property. However, we will show in the next example a Bayesian game model with unknown existence and uncertain number of equilibrium.

6.3.3 Random Access Channels

Random access channel (RAC) is a shared channel that is used by wireless terminals to access the access network (Figure 6.3), for example, TDMA/FDMA- and CDMA-based network. A key feature of RAC is that messages are not scheduled, i.e., there is no guarantee that a single device makes a connection attempt at one time, and collisions can happen at any time with uncertain number of devices. In the previous games, we always assumed that there were only two players in the game. However, here we consider the RAC as a time-slotted system with $K + 1$ players. At the current slot, each user has full knowledge about her own channel, but she does not know the channel condition of other users. In this situation, users can choose to transmit or keep silent. We formulate this distributed control scheme as a Bayesian game. We try to find the operating point for the transmission control problem, which is the BE of the game.

It is clear that a device will suffer from conflicts and result in a performance loss, if other devices transmit at the same time. In a certain time slot, when a device chooses to transmit, it can achieve a certain level of utility from the utility set $\{U_n\}$, where n represents the total number of other devices that choose to transmit. It is reasonable to assume that the utility sequence can generally be ordered as

$$U_0 > U_1 > \cdots > U_K \geq 0$$

since more devices bring more interference and therefore less utility. If device k chooses to keep silent, she receive "0" utility. Denote by C_k the cost of device k if she chooses to transmit (the cost usually refers to the energy consumed for a single transmission). The cost is "0" if she chooses to keep silent. It is therefore natural to consider a player's payoff as the difference between the utility U_n and the cost C_k (the payoff represents a player's net revenue). We illustrate the received payoff of device k in the following table:

Number of other players transmit	0	1	$\cdots$	K
Transmit	$U_0 - C_k$	$U_1 - C_k$	$\cdots$	$U_K - C_k$
Silent	0	0	$\cdots$	0

We consider a threshold-based transmission strategy: any player k will choose to transmit if her cost of transmission C_k is lower than or equal to a certain threshold c, i.e., $C_k \leq c$; she will choose to keep silent if her cost of transmission C_k is greater than the threshold c, i.e., $C_k > c$. Assume that the cost C_k is relevant to device k's transmit power. Due to the effect of power control, bad channel condition leads to more transmit power and therefore more cost. As it is well known, the wireless channel gain is usually assumed to be a random variable, which means that C_k (as a function of channel gain) is also a random variable.

Let $F_{C_k}(c)$ be the cumulative distribution function (CDF) of C_k when the threshold is set to c. Without loss of generality, we assume C_k to be i.i.d., and we have

$$F_c \triangleq F_{C_k}(c) = \int_{-\infty}^{c} f_C(x)\mathrm{d}x \ \ \forall k$$

where $f_C(c)$ is the probability density function of C_k. Note that under the above assumptions, F_c represents the transmission probability of each device as a function of the threshold c. We can completely characterize the RAC Bayesian game as follows:

DESCRIPTION OF RAC BAYESIAN GAME

$$\mathcal{G}_{RAC} \triangleq \langle \mathcal{K}, \mathcal{T}, \mathcal{A}, \mathcal{Q}, \mathcal{U} \rangle$$

- Player set: $\mathcal{K} = \{k\}, k \in \{1, \dots, K+1\}$.
- Type set: $\mathcal{T} = \{t\}, t \in \{1, \dots, K\}$, where t represents the total number of other players that choose to transmit.
- Action set: $\mathcal{A} = \{transmit, silent\}$.
- Probability set: $\mathcal{Q} = \{q_t\}, t \in \{1, \dots, K\}$, where q_t represents the probability of type t, and we have

$$q_t = \binom{K}{t} F_c^t (1 - F_c)^{K-t}$$

- Payoff function set: $\mathcal{U} = \{u_k\}, k \in \{1, \dots, K+1\}$, where u_k is the expected payoff of player k

$$u_k(c) = \mathbb{E}_{\mathcal{T}}[U_t - C_k]$$

$$= (U_0 - C_k)\binom{K}{0}(1 - F_c)^K + (U_1 - C_k)\binom{K}{1}F_c(1 - F_c)^{K-1}$$

$$+ \cdots + (U_K - C_k)\binom{K}{K}F_c^K$$

$$= \sum_{t=0}^{K} U_t \binom{K}{t} F_c^t (1 - F_c)^{K-t} - C_k$$

THEOREM 6.3 The solution of the following fixed-point equation, c^*, is the BE of the game $\mathcal{G}_{RMA}$.

$$\sum_{t=0}^{K} U_t \binom{K}{t} F_{c^*}^t (1 - F_{c^*})^{K-t} = c^* \tag{6.8}$$

Proof In short, the existence of a solution to a fixed-point problem is equivalent to the existence of an equilibrium. A detailed proof can be found in [10]. $\square$

However, in this game, it can be shown that the BE does not necessarily exist. To support this argument, we show a counter example as follows:

Consdier that an RAC with three players and the random variable C_k follows the uniform distribution with the following CDF:

$$F_c = \begin{cases} 0 & \text{if } c < 0 \\ c & \text{if } 0 \le c \le 1 \\ 1 & \text{if } c > 1 \end{cases}$$

We can write the expected payoff for player k ($k = 1, 2, 3$).

$$u_k(c) = \sum_{t=0}^{2} U_t \binom{2}{t} F_c^t (1 - F_c)^{2-t}$$

Now, let us check whether the fixed-point equation $u_k(c) = c$ can be satisfied. It is easy to verify that the BE exists under the following conditions:

$$\begin{cases} c_{(1)}^* = \dfrac{U_0 - U_1 - \sqrt{U_1^2 - U_0 U_2}}{U_0 - 2U_1 + U_2} & \text{if } U_1^2 - U_0 U_2 \ge 0 \text{ and } U_0 - 2U_1 + U_2 \ne 0 \\ c_{(2)}^* = U_2 & \text{if } U_2 \ge 1 \end{cases}$$

This result shows that in general the BE may not exist and may not be unique either.

6.4 Conclusion

This chapter presented the basic game-theoretic model considering games with incomplete information. Bayesian games together with the solution concept of BE have been introduced and specified in details. We emphasized the methodology for studying the equilibrium set, such as the problems of "existence," "uniqueness," and "equilibrium selection." We showed that Bayesian game is a powerful tool to understand the radio resource conflicts in a realistic multiuser communication network, where users have imperfect observation about other users' channel information. We demonstrated different ways to apply Bayesian games to the resource allocation and power control problems existing in three basic communication networks, i.e., MAC, IC, and RAC. This study is a step forward compared to the recent game-theoretic approach that considers users having complete information, which may not be realistic in many practical applications.

References

1. D. Fudenberg and J. Tirole, *Game Theory*, MIT Press, Cambridge, MA, 1991.
2. L. Lai and H. El Gamal, The water-filling game in fading multiple-access channels, *IEEE Transactions on Information Theory*, 54(5), 2110–2122, May 2008.
3. G. He, S. Beta, and M. Debbah, The waterfilling game-theoretical framework for distributed wireless network information flow, *EURASIT Journal on Wireless Communications and Networking*, 2010 (482975), p. 13, 2010.
4. S. Lasaulce, M. Debbah, and E. Altman, Methodologies for analyzing equilibria in wireless games, *IEEE Signal Processing Magazine*, 26(5), 41–52, 2009.
5. E. V. Belmega, S. Lasaulce, and M. Debbah, Power allocation games for MIMO multiple access channels with coordination, *IEEE Transactions on Wireless Communications*, 8(6), 3182–3192, June 2009.
6. E. Altman, M. Debbah, and A. Silva, Game theoretic approach for routing in dense ad-hoc networks, *Stochastic Networks Workshop*, Edinburg, U.K., July 2007.
7. D. Gesbert and M. S. Alouini, How much feedback is multi-user diversity really worth? *IEEE International Conference on Communications (ICC)*, Paris, France, vol. 1, 234–238, June 2004.
8. J. C. Harsanyi, Games with incomplete information played by "Bayesian" players, I–III. Part I. The basic model, *Management Science*, 14(3), 159–182, November 1967.
9. E. Altman, L. Cottatellucci, M. Debbah, G. He, and A. Suarez, Operating point selection in multiple access rate regions, *International Teletraffic Congress*, Paris, France, June 2009.
10. J. F. Nash, Equilibrium points in n-points games, *Proceedings of the National Academy of Science*, 36(1), 48–49, January 1950.
11. S. Boyd and L. Vandenberghe, *Convex Optimization*, Cambridge University Press, New York, 2004.
12. J. B. Rosen, Existence and uniqueness of equilibrium points for concave N-person games, *Econometrica*, 33, 520–534, July 1965.
13. T. Roughgarden and E. Tardos, Bounding the inefficiency of equilibria in nonatomic congestion games, *Games and Economic Behavior*, 47, 389–403, 2004.
14. R. Cooper, *Coordination Games*, Cambridge University Press, Cambridge, U.K. 1998.
15. E. Altman and A. Shwartz, Constrained markov games: Nash equilibria, *Annals of Dynamic Games*, 5, 213–221, 2000.
16. J. C. Harsanyi and R. Selten, *A General Theory of Equilibrium Selection in Games*, MIT Press, Cambridge, MA, March 2003.
17. Traffic management specification, version 4.0, Technical Report AF-TM-0056.000, ATM Forum Traffic Management Working Group, April 1996.
18. F.P. Kelly, Charging and rate control for elastic traffic, *European Transactions on Telecommunications*, 8, 33–37.
19. F.P. Kelly, A. Maullo, and D. Tan, Rate control for communication networks: Shadow prices, proportional fairness and stability, *Journal of the Operations Research Society*, 49, 237–252, 1998.
20. D. Chiu and R. Jain, Analysis of the increase and decrease algorithms for congestion avoidance in computer networks, *Computer Networks ISDM System*, 17, 1–14, 1989.
21. T. S. Han and K. Kobayashi, A new achievable rate region for the interference channel, *IEEE Transactions on Information Theory*, IT-27, 49–60, January 1981.
22. S. Adlakha, R. Johari, and A. Goldsmith, Competition in wireless systems via bayesian interference games, http://arxiv.org/abs/0709.0516, September 2007.

Chapter 7

Evolutionary Networking Games

Hamidou Tembine

Contents

We develop a framework for analyzing evolving games in large populations of players, in which each player interacts with the other random number of players, that is, the number of players that a given player can meet is unknown and varies in time. The actions of each player in an interaction together determine the instantaneous costs for all involved players. They also determine the rate of transition to use pure actions. We present various classes of evolutionary game-theoretic models (evolutionary stability and evolutionary game dynamics) and apply them to networking problems including intrusion detection, resource selection, and interference control in wireless networks. We will discuss some challenges and limitations in the application of game theory to the analysis of wireless communications and networks.

7.1 Introduction

Evolutionary games in a large population provide a simple framework for describing strategic interactions among a large number of players. Traditionally, predictions of behavior and outcome in game theory are based on some notion of equilibrium, typically Cournot equilibrium (Cournot, 1838), Bertrand equilibrium (Bertrand, 1883), Conjectural variation (Bowley, 1924), Stackelberg solution (Stackelberg, 1934), Nash equilibrium (NE) (Nash, 1951), Wardrop equilibrium (Wardrop, 1952), or some refinement and/or extensions thereof. Most of these notions require the assumption of equilibrium knowledge, which assume that each player correctly anticipates how the other players will react. The equilibrium knowledge assumption is too strong and is difficult to justify in particular in contexts with a large number of players in dense networks. As an alternative to the equilibrium approach, the evolutionary game approach proposes an explicitly dynamic updating choice, a model in which players myopically update their behavior in response to their current strategic environment. This dynamic procedure does not assume the automatic coordination of users' actions and beliefs, and it can derive many players' actions and transition rates. These procedures are specified formally by defining a revision of pure strategies called revision protocol (Sandholm, 2011). A revision protocol takes current costs and aggregate behavior as inputs. Its outputs are conditional switching rates which describe how frequently users in some class playing action who are considering switching strategies switch to another strategy, given the current cost vector and subpopulation state. Revision protocols are flexible enough to incorporate a wide variety of paradigms, including ones based on imitation, adaptation, learning, optimization, etc. The revision rules describe the procedures users follow in adapting their behavior to the dynamic evolving environment such as evolving networks.

7.1.1 Biology and Economics Meet Computer Science: The Birth of Evolving Games and Biologically Inspired Evolutionary Game Dynamics

Evolutionary game theory studies the evolution of populations of players interacting strategically. Applications were originally in biology and social sciences. Examples of situations that evolutionary game theory helps to understand include animal conflicts and communication behaviors (Hawk and Dove game, Chicken game). The interactions may be with players from the same population (a mobile with other mobiles, males fighting other males, etc.) or with players from other populations (males interacting with females or buyers with traders). The players are typically animals in biology, firms in economics, and mobiles in networks. The point of departure for an evolutionary game model is to believe that the players (animals, agents, mobiles, nodes, etc.) are not always rational. In the place of NE in classical game theory, game theory evolving uses the concept of evolutionary stability: the unbeatable state concept (Hamilton, 1967) and the evolutionarily stable state (ESS) concept (Smith and Price, 1973). A strategy is unbeatable if it is resilient to any fraction of deviants of any size (this notion generalizes the strong equilibrium concept or "acceptable point" to infinite population games). A strategy is evolutionarily stable if a whole population using this strategy cannot be invaded by a small fraction of deviants (mutants). An unbeatable strategy (when it exists) is then an ESS. Evolutionary stability concept is particularly interesting because it gives some information about the robustness and the dynamic properties of the population (refinement of equilibria). In situation where the game has multiple equilibria, the evolutionary stability notion can contribute to the selection of specific equilibria. A dynamic behavior of the populations is proposed using *evolutionary game dynamics* such as replicator dynamics (Taylor and Jonker, 1978), fictitious play (Brown, 1951; Gilboa and Matsui, 1991), Brown–von Neumann–Nash dynamics (Brown and von Neumann, 1950), Smith dynamics (Smith, 1984), projection dynamics, best-response dynamics (Fudenberg and Levine, 1991; Fudenberg and Tirole, 1991), logit dynamics (Fudenberg and Levine, 1998), etc. Much literature can be found in the extensive survey on evolutionary game dynamics in Weibull (1995), Hofbauer and Sigmund (2003), Vincent and Brown (2005), and in Sandholm (2011). Most of the applications using evolutionary game dynamics were interdisciplinary between economics and behavioral ecology. Recently, game theory evolving has found many applications in the field of wireless communications and networks. The standard formulations of evolutionary games consider only pairwise interactions such as pairwise random matching (Benaim and Weibull (2003), Smith (1982)). However, interactions that are not pairwise frequently arise in communication networks, such as the cases of code division multiple access (CDMA) system (Tembine, 2009; Tembine et al., 2008a), orthogonal frequency–division multiple access (OFDMA) in worldwide interoperability for microwave access (WiMAX) environment (Altman et al., 2008; Anastasopoulos et al., 2008) and next-generation mobile communications etc. In this work, we extend the basic pairwise interaction model to wireless communication systems with finitely many classes of users and arbitrary number of users corresponding to each local interaction (Tembine et al. (2008, 2009), Zhu et al. (2009)).

We consider large populations (multi-classes) of players in which small frequent interactions occur among a random number of players that are selected at random. Each player is thus involved in several interactions with the other random number of players. Each interaction in which a player is involved can be described as one stage of an evolving game. The set of players in each local interaction evolves in time. The actions of the players at each stage determine a cost (which can be delayed) for each player as well as the transition rates of actions. The transition rate is determined of the change of actions, the system state, and the fitnesses. This model extends the basic pairwise interaction model in evolutionary games by introducing a random number of interacting players and the rate of transition for several players that have opportunities to change its action. At each time slot, a one-shot game with an unknown number of players replaces the matrix game. Instead of a choice of an (possibly mixed) action, a player is now faced with the choice of decision rules and revision of theses strategies that determine what actions should be chosen at a given interaction for given present and past observations. In addition to the study of the equilibrium of the evolving game, we shall also consider some population dynamics and study its convergence properties. We shall assume that players revise their strategies and the strategies with lower cost are used with high probabilities. The system evolves under some evolutionary game dynamic processes (Cressman, 2003; Gintis, 2009; Hofbauer and Sigmund, 1998; Vincent and Brown, 2005; Weibull, 1995) that describes the change of strategies (incoming flow and outgoing flow) in the population.

The rest of this chapter is organized as follows: In the next section, we present evolving games with several classes of players. In Sections 7.3, 7.4, and 7.5, we then give examples of applications of this model in intrusion detection in ad hoc networks, interference control in distributed Aloha networks, and resource selection with an unknown number of interacting users. Finally, we briefly discuss some of the open problems and limitations that we identify in the field in Section 7.6.

7.2 Evolving Games

Consider several classes of players denoted by a finite set $\mathcal{E} = \{1, 2, \ldots, E\}$. Each class e has a large number of players. The class e has a mass m^e. The players of the class e select an action in the finite set $\mathcal{A}^e = \{a_1^e, \ldots, a_{n^e}^e\}$. Time is discrete or continuous. Let $x(t) = (x^e(t))_{e \in \mathcal{E}}$ denote the population profile at time t. The composition of class e is then described by $x^e(t) = [x_1^e(t), \ldots, x_{n^e}^e(t)]$ where $x_a^e(t)$ is fraction of players with the strategy a in class e at time t. For every time t,

$$\forall e, \quad \sum_{a \in \mathcal{A}^e} x_a^e(t) = m^e, \quad x_a^e(t) \geq 0.$$

We denote by $X^e = \Delta(\mathcal{A}^e)$ the set of probabilities over $\mathcal{A}^e$.

At each time t, there are several local interactions. Denote by $C_a^e(x(t))$ the expected cost of a player with a in class e when the population profile is $x(t)$ at time t. If K_t denotes the random variable that represents the number of players involved in a local interaction at time t, the expected cost can be expressed as

$$C_a^e(x(t)) = \sum_{k \geq 0} \mathbb{P}(K_t = k) C_a^{e,k}(x(t)) = \mathbb{E}_K \left(C_a^{e,K}(x(t)) \right)$$

where $C_a^{e,k}(x(t))$ is the expected cost of the strategy a obtained in a local interaction between k players when the population profile is $x(t)$. The number $k = \sum_{e \in \mathcal{E}} k_e$ where k_e is the number players in class e involved in the local interaction at time t.

7.2.1 *Homogeneous Population*

7.2.1.1 *Equilibrium and Stability*

- *Global optimum (social optimum)*: The mapping $x \longmapsto \sum_{a \in \mathcal{A}} x_a C_a(x)$ is minimized.
- *(Nash) equilibrium state*: A population profile x is an equilibrium if no player can unilaterally change actions to decrease its expected cost:

$$\forall y, \quad \sum_{a \in \mathcal{A}} (x_a - y_a) C_a(x) \leq 0 \tag{7.1}$$

This variational inequality is equivalent to the following characterization also known as Wardrop first principle: *At the equilibrium, the expected cost in all actions actually used are equal and less than those that would be experienced by a single player on any unused action*, that is, for any action a such that $x_a > 0$ one has $C_a(x) = \min_{b \in \mathcal{A}} C_b(x)$.

- *ESS*: A population profile x is an ESS if for any other population profile $y \neq x$, there exists a $\epsilon_y > 0$ such that $\forall \epsilon \in (0, \epsilon_y)$

$$\sum_{a \in \mathcal{A}} (x_a - y_a) C_a((1 - \epsilon)x + \epsilon y) < 0 \tag{7.2}$$

When the inequality (7.2) is non-strict, and $\epsilon = 0$, we obtain that the probability distribution x is a symmetric NE. When the inequality (7.2) is non-strict, the population profile x is neutrally stable state (NSS). An NSS is in particular an NE. Denote by Δ_{ESS} (resp. Δ_{NE}, Δ_{NSS}) the set of ESS (resp. NE, NSS). Thus, the following inclusions hold: $\Delta_{ESS} \subset \Delta_{NSS} \subset \Delta_{NE}$. For any evolving game with bilinear cost function (in particular in a bi-matrix game with cost (A, A^t)), x is an ESS if and only if for all mutant-strategy y,

$$\sum_a x_a C_a(x) \leq \sum_a y_a C_a(x) \quad \text{and}$$

$$\left[y \neq x, \sum_a x_a C_a(x) = \sum_a y_a C_a(x) \right] \Longrightarrow \sum_a (x_a - y_a) C_a(y) < 0$$

Note that an ESS may not exist. Consider the following family of constant-sum matrix game with strategies $\{1, 2, 3\}$ (Rock-Paper-Scissor games):

$$\begin{pmatrix} & 1 & 2 & 3 \\ \hline 1 & 0 & \mu_1 & -\mu_2 \\ 2 & -\mu_2 & 0 & \mu_1 \\ 3 & \mu_1 & -\mu_2 & 0 \end{pmatrix}, \quad \mu_i > 0, \quad i \in \{1, 2\}$$

It is easy to see that the unique mixed NE of this game is not an ESS.

Proposition 7.1 For any distribution of the discrete random variable K such that the cost functions are continuous, the evolutionary game with a random number of interacting players at each local interaction has at least one equilibrium.

Sketch of proof We show that the game has a symmetric NE. We first remark that the generating function of K is continuous in $(0, 1)$. Thus, the vector of cost functions C is lower semi-continuous in Δ_{n-1} (which is a non-empty, convex, and compact subset of Euclidean space $\mathbb{R}^n$). The existence of symmetric NE in mixed strategies follows from the existence of solutions of the following variational inequalities

$$\text{find } x^* \in \Delta_n \text{ s.t. } \langle (x^* - y), C(x^*) \rangle \leq 0, \ \forall y$$

where $\langle, \rangle$ is the inner product of $\mathbb{R}^n$. We prove that this is equivalent to the fixed point equation: $x^* = \Pi_{\Delta_n}(x^* - \varsigma C(x^*))$, $\varsigma > 0$ where $\Pi_{\Delta_{n-1}}$ denotes the projection into the simplex Δ_n. Multiplying the inequality $\langle (x^* - y), -C(x^*) \rangle \geq 0$ by $\varsigma > 0$ and adding $\langle x^*, y - x^* \rangle$ to both sides of the resulting inequality, one obtains $\langle y - x^*, x^* - [x^* - \varsigma C(x^*)] \rangle \geq 0$. Recall that the projection map on Δ_{n-1}, which is a convex and closed set, is characterized by

$$z \in \mathbb{R}^n, \ z' = \Pi_{\Delta_d} z \iff \langle z' - z, x - z' \rangle \geq 0, \quad \forall x \in \Delta_{n-1}$$

Thus,

$$x^* = \Pi_{\Delta_{n-1}} \left(x^* - \varsigma C(x^*) \right)$$

According to Brouwer–Schauder's fixed point theorem, given a map $\alpha: \Delta_{n-1} \longrightarrow \Delta_{n-1}$, with α continuous, there is at least one $x^* \in \Delta_{n-1}$, such that $x^* = \alpha(x^*)$. Since the projection $\Pi_{\Delta_{n-1}}$ and $(I - \varsigma C)$ are both continuous, we observe that $\Pi_{\Delta_{n-1}}(I - \varsigma C)$ is also continuous by composition where I stands for the identity operator. It follows from convexity and compactness of Δ_{n-1} and the continuity of $\Pi_{\Delta_{n-1}}(I - \varsigma C)$ that the Brouwer–Schauder fixed point theorem can be applied to the map $\Pi_{\Delta_{n-1}}(I - \varsigma C)$. We conclude that at least one stationary equilibrium state exists in the evolving game with a random number of players. This completes the proof. $\qquad\square$

7.2.2 Price of Anarchy in Population Games

One of the approaches used to measure how much the performance of decentralized decision-making systems is affected by the selfish behavior of its components is the so-called *price of anarchy* (PoA). We present a similar concept for the ESS. This notion of price of anarchy can be seen as an *efficiency metric* that measures the *cost of selfishness* or decentralization and has been extensively used in the context of congestion games or routing games (Roughgarden, 2005) where typically players have to minimize a cost function. If the evolutionary game has an ESS, we can define the analogue of the "price of anarchy" denoted by PoA_{ESS}, as the ratio between the payoff of the worst evolutionary equilibrium and the global optimum value. The "price of stability" (PoS_{ESS}) as the ratiobetween the payoff of the "best" evolutionary equilibrium and the social optimum (SO) value defined as the minimum of the expected global cost of $x \longmapsto \sum_{b \in A} x_b C_b(x)$. Without loss of generality, we assume that the global optimum is strictly positive.

$$PoA_{\text{ESS}} = \frac{\max_x \text{ ESS } \sum_{b \in A} x_b C_b(x)}{SO} \geq 1$$

$$PoS_{\text{ESS}} = \frac{\min_x \text{ ESS } \sum_{b \in A} x_b C_b(x)}{SO} \geq 1$$

Using the inclusions of Section 7.2.1, we obtain the following inequalities:

$$PoA_{\text{NE}} \geq PoA_{\text{NESS}} \geq PoA_{\text{ESS}} \geq PoS_{\text{ESS}} \geq PoS_{\text{NESS}} \geq PoS_{\text{NE}} \geq 1$$

7.2.3 Heterogeneous Population

Similar to homogeneous population, we define the following solution concepts for heterogeneous populations:

- *Global equilibrium:* $\forall y, \sum_e \sum_{a \in A^e} \left(x_a^e - y_a^e \right) C_a^e(x) \leq 0$. The existence of global optimum is guaranteed if C is continuous.
- *Global evolutionarily stable state:* $\forall y \neq x$ there exists a $\epsilon_y > 0$ such that $\forall \epsilon \in (0, \epsilon_y)$, $\sum_e \sum_{a \in A^e} \left(x_a^e - y_a^e \right) C_a^e((1 - \epsilon)x + \epsilon\, y) < 0$.

7.2.4 Evolutionary Game Dynamics

As an alternative to the equilibrium approach, evolutionary game dynamics propose an explicitly dynamic updating choice, a model in which players myopically update their behavior in response to their current strategic environment. This dynamic procedure does not assume the automatic coordination of users' actions and beliefs, and it can derive many players' choice procedures. These procedures are specified formally by defining a revision of pure strategies called *revision protocol* (Sandholm, 2011). The revision of strategies describes the procedures users follow in adapting their behavior in the dynamic evolving environment such as evolving networks.

Consider a finite set $\mathcal{E}$ of subpopulations. Players from subpopulation e select their actions in finite action set $\mathcal{A}^e$. Denote by $x(t) = \left(x_a^e(t) \right)_{a \in A^e,\ e \in \mathcal{E}}$ the population profile at time t and by $C_a^e(x(t))$ the expected cost of players from subpopulation e with a when the population profile $x(t)$. Let $\beta_{ba}^e(x(t), C)$ be the conditional switch rate from the strategy b to the strategy a in subpopulation e. The flow of the population is specified in terms of the functions $\beta_{ba}^e(x(t), C)$, which determine the rates at which a player who is considering a change in strategies opts to switch to his various alternatives.

The *incoming flow* of the action a is

$$\sum_{b \in A^e} x_b^e(t) \beta_{ba}^e$$

and the *outgoing flow* from a is

$$x_a^e(t) \sum_{b \in A^e} \beta_{ab}^e$$

where $x_a^e(t)$ represents the fraction of players of the population that use the action a at time t.

Let

$$V_{a,C}^e(x(t)) = \sum_{b \in A^e} x_b^e(t) \beta_{ba}^e - x_a^e(t) \sum_{b \in A^e} \beta_{ab}^e$$

The evolutionary game dynamics is given by

$$\dot{x}_a^e(t) = V_{a,C}^e(x(t)), \quad a \in A^e, \quad e \in \mathcal{E} \tag{7.3}$$

In the following, we briefly mention the standard evolutionary game dynamics.

- *Brown–von Neumann–Nash (BNN) dynamics:* Denote by g_a^e the positive part of the excess cost of subpopulation a of class e and by m^e the mass of subpopulation e. There is a close relation between BNN dynamics with Nash's original proofs of his equilibrium existence theorem (see Hofbauer and Sigmund, 2003). BNN dynamics can be interpreted as follows: Suppose there are large populations in which there is steady incoming and outgoing flow, New players joining the system use only strategies that are better than expected cost, and better strategies are more likely to be adopted. On the other hand, randomly chosen users leave the game. More precisely, the strategy x^e is adopted with probability proportional to the excess payoff $g_a^e(x)$. In continuous time, Brown–von Neumann–Nash Dynamics is given by

$$\dot{x}_a^e(t) = \mu^e \left[m^e g_a^e(x(t)) - x_a^e(t) \sum_{b \in \mathcal{A}^e} g_b^e(x(t)) \right] \tag{7.4}$$

where $\mu^e \in \mathbb{R}_+$

$$g_a^e(x(t)) = \max \left[0, -C_a^e(x(t)) + \frac{1}{m^e} \sum_{b \in \mathcal{A}^e} x_b^e(t) C_b^e(x(t)) \right]$$

The discrete time version is given by

$$x_a^e(t+1) = \frac{\kappa x_a^e(t) + g_a^e(x(t))}{m^e \kappa + \sum_{b \in \mathcal{A}^e} g_b^e(x(t))} \tag{7.5}$$

where κ is an appropriate constant. The Brown–von-Neumann–Nash dynamics has the nice property that every rest point of the dynamics is an NE state.

■ *Replicator dynamics*: One of the most studied evolutionary game dynamics is the *replicator dynamics*. It has been used for describing the evolution of road traffic congestion in which the fitness is determined by the strategies chosen by all drivers (Sandholm, 2011). It has also been studied in the context of the resource selection in hybrid systems and migration constraint problem in wireless networks in Tembine et al. (2008a). In the replicator dynamics, a fraction of members of a subpopulation of class e grows when its cost is less than the expected average cost of all the subclass of e in the population. In continuous time, the system of ordinary differential equations (ODE) is given by

$$\dot{x}_a^e(t) = \mu^e x_a^e(t) \left[-C_a^e(x(t)) + \frac{1}{m^e} \sum_{b \in \mathcal{A}^e} x_b^e(t) C_b^e(x(t)) \right] \tag{7.6}$$

$$x_a^e(t+1) = x_a^e(t) \frac{\kappa - C_a^e(x(t))}{\kappa m^e - \sum_{b \in \mathcal{A}^e}^{n^e} x_b^e(t) C_b^e(x(t))} \tag{7.7}$$

■ *Generalized Smith dynamics*: $\theta \geq 1$

$$\dot{x}_a^e(t) = \mu^e \sum_{b \in \mathcal{A}^e} x_b^e(t) \max(0, -C_a^e(x) + C_b^e(x))^\theta$$

$$- \mu^e x_a^e(t) \sum_{b \in \mathcal{A}^e} \max(0, -C_b^e(x) + C_a^e(x))^\theta \tag{7.8}$$

- *Best-response dynamics* (Gilboa and Matsui, 1991): Members of each subpopulation revise their strategy and choose the best replies $BR(x(t))$ at the current population state $x(t)$.

$$\dot{x}^e(t) \in m^e BR^e(x(t)) - x^e(t) \qquad (7.9)$$

where $BR^e(x^{-e}(t)) = \arg \min_{y^e \in \Delta(A^e)} \left\{ \sum_{b \in A^e} y_b^e C_b^e(x^{-e}(t)) \right\}$.

- *Fictitious play*: One widely used model of learning is the process of fictitious play and its variants. Fictitious play is a learning rule first introduced by Brown (1951). In this process, players behave as if they think they are facing a vector, but unknown, distribution of strategies of the others players. In it, each player presumes that her/his opponents are playing stationary (possibly mixed) strategies. A player is said to use fictitious play if, at every time instant t, he chooses an action that is a best response to the empirical distribution of the opponents' play up to time $t - 1$. At each round, each player thus best responds to the empirical frequency of play of his opponent. Such a method is of course adequate if the opponent indeed uses a stationary strategy, while it is flawed if the opponent's strategy is nonstationary. Fictitious play does not always converge. Shapley (1964) proved that in the game Rock, Paper, Scissors games, there is a cycle. Convergence of fictitious has been proved in the following configurations: Both players have only a finite number of strategies, the game is zero sum (Robinson, 1951), and the game is a potential game (Monderer and Shapley, 1996). The discrete time version of fictitious play is given by

$$x_t^e = \left(1 - \frac{1}{t}\right) x_{t-1}^e + \frac{1}{t} BR^e\left(x_{t-1}^{-e}\right)$$

As we can see, with an appropriate time normalization, the discrete-time fictitious play asymptotically is approximately the same as the continuous time best-response dynamic. More precisely, the set of limit points of discrete time fictitious play is an invariant subset for the continuous time best-response dynamics,

In continuous time, fictitious play is given by

$$\dot{x}^e(t) \in \frac{1}{t}(m^e BR^e(x^{-e}(t)) - x^e(t)) \qquad (7.10)$$

- Logit dynamics in games have been introduced by Fudenberg and Levine (1998). The logit dynamics is based on exponential weight of the cost function. This dynamics is sometimes referred to as Boltzman dynamics or smooth best-response dynamics.

$$\dot{x}_a^e(t) = \mu^e \left[m^e \frac{e^{-C_a^e(x(t))/\eta^e}}{\sum_{b \in A^e} e^{-C_b^e(x(t))/\eta^e}} - x_a^e(t) \right] \qquad (7.11)$$

Using an explicit representation in terms of the logit map, Hofbauer et al. (2009) have shown that the time average of the replicator dynamics is a perturbed solution of the best reply dynamics.

- Orthogonal projection dynamics is a myopic adaptive dynamic in which a subpopulation grows when its expected cost is greater than the arithmetic average cost of all the class.

$$\dot{x}_a^e(t) = \mu^e \left[-C_a^e(x(t)) + \frac{1}{n^e} \sum_{b \in A^e} C_b^e(x(t)) \right] \qquad (7.12)$$

Under the orthogonal projectional dynamics, every interior rest point is an NE.

■ Ray-projection dynamics is a myopic adaptive dynamic in which a subpopulation grows when its expected cost is greater than the ray-projection cost of all the class. In this formulation, the transition rate from a to b is independent of the next strategy b.

$$\dot{x}_a^e(t) = \mu^e \left[-m^e h_a^e(x(t)) + x_a^e(t) \sum_{b \in \mathcal{A}^e} h_b^e(x(t)) \right] \qquad (7.13)$$

It is easy to see that ray-projection dynamics is obtained from the revision protocol

$$\beta_b^{e,a} = h_a^e(C, x)$$

Notice that the replicator dynamics, best-response dynamics, and logit dynamics can be obtained as a particular case of the ray-projection dynamics.

■ *Target projection dynamics*

$$\dot{x}^e(t) = \mu^e \left[\Pi_{X^e} \left(x^e(t) - \varsigma C^e(x(t)) \right) - x^e(t) \right] \qquad (7.14)$$

From Proposition 7.1, x^* is an equilibrium is equivalent to

$$x^* = \Pi_{\prod_e X^e} \left(x^* - \varsigma C(x^*) \right)$$

Thus, the set of rest points of target projection dynamics coincide with the set of equilibria of the evolving game.

■ *Dynamic equilibrium*
  1. In many cases in evolving games, it is known that the trajectories of evolutionary game dynamics may not converge, can be chaotic, or may have cycle limits. Under several evolutionary game dynamics, the expected cost obtained in the cycle is not so far from candidate to be a "static" equilibrium cost. This suggests a *time dependent equilibrium* approach in evolving games.
  2. A trajectory is a *dynamic equilibrium* if the *time average cost under this trajectory leads to an equilibrium cost*.

The trajectory $x : [t_0, \infty[\longrightarrow \prod_e X^e$ is a dynamic equilibrium if for any other trajectory y, there exists t_1 sufficiently large such that

$$\forall e, \int_{t_1}^{t_1 + T} \langle x^e(t) - y^e(t), C^e(x(t)) \rangle \, dt \leq 0$$

7.2.5 Delayed Evolutionary Game Dynamics

We introduce in the following delays into classical models of population dynamics (Tembine, 2010). An action taken today will have its effect some time later. The delays can be symmetric or not. We then obtain delayed cost (or fitness) functions. The evolution of the system leads to delayed evolutionary game dynamics. Delayed evolutionary game dynamics are in general a system of first order non-regular, nonlinear differential equations or differential inclusions with time delays. We use the theory of delayed differential equation (DDE) to study stability, convergence, and non-convergence of the system under time delays. Evolutionary game dynamics with asymmetric time delays have been introduced in Tembine et al. (2007). See also Altman et al. (2008, 2009); Tembine (2009); Tembine et al. (2008b) for networking applications. Denote by τ_a^e the time associated with the strategy a of class e. The delayed evolutionary game dynamics is given by

$$\dot{x}_a^e(t) = \sum_{b \in \mathcal{A}^e} x_b^e(t) \beta_a^{e,b}\left(C, \left\{x\left(t - \tau_{a'}^{e'}\right)\right\}_{e',a'}\right) - x_a^e(t) \sum_{b \in \mathcal{A}^e} \beta_b^{e,a}\left(C, \left\{x\left(t - \tau_{a'}^{e'}\right)\right\}_{e',a'}\right) \quad (7.15)$$

7.2.5.1 Examples of Delayed Game Dynamics

■ *Delayed regular game dynamics*: In the delayed replicator dynamics, the share of a strategy in the population grows at a rate equal to the difference between the delayed payoff of that strategy and the average delayed payoff of the population. The fitness acquired at time t will impact the rate of growth τ_a^e time later. Thus, the delayed replicator dynamics of $x_a^e(t)$ is given by

$$\dot{x}_a^e(t) = \mu^e x_a^e(t)\left[-C_a^e\left(x\left(t - \tau_a^e\right)\right) + \frac{1}{m^e}\sum_b x_b^e(t) C_b^e\left(x\left(t - \tau_b^e\right)\right)\right]$$

The standard replicator dynamics is obtained when all the delays are zero.

■ *Delayed nonregular dynamics*: We suppose that users in the population review their strategy and imitate the better strategy at the time (imitation by dissatisfaction). The delayed imitate the better dynamics is given by

$$\dot{x}_a^e(t) = x_a^e(t)\left(\sum_{b \in \mathcal{A}^e} x_b^e(t)\left[\rho_{ab}^e\left(x\left(t - \tau_a^e\right), x\left(t - \tau_b^e\right)\right) - \rho_{ba}^e\left(x\left(t - \tau_b^e, x\left(t - \tau_a^e\right)\right)\right)\right]\right)$$

$$a \in \mathcal{A}^e \quad (7.16)$$

where $\rho_{ab}^e(x) = g\left(C_a^e(x), C_b^e(x)\right)$, $a, b \in \mathcal{A}^e$ and $g(a, b) = 0$ if $a > b$ and $g(a, b) = 1$ if $a < b$.

7.2.5.2 ESS Can Be Unstable

In contrast to the standard deterministic evolutionary game dynamics where the ESS is stable, the ESS can be unstable for large time delays under regular delayed game dynamics or can be completely unstable for any positive time delays (delayed imitate the better dynamics, delayed best-response dynamics, etc.).

7.2.5.3 Stability Bounds for ESS under Regular DDE

The population profile x^* is a rest point of the differential equation (7.15) if it is a critical point, that is, the right side of (7.15) is zero at x^*. Denote $\tau = \max_{e,a} \tau_a^e$. We use Lyapunov stability criteria: x^*

is stable if it is a stationary point with the property that for every neighborhood $\mathcal{V}$ of x^*, there exists a neighborhood $\mathcal{U} \subset \mathcal{V}$ with the property that if $x(t) \in \mathcal{U}$ for $t \in (-\tau, 0)$, then $x(t) \in \mathcal{V}$ for all $t > 0$. A population profile x^* is asymptotically stable if it is stable and there exists a neighborhood $\mathcal{W}$ of x^* such that $x(t) \in \mathcal{W}$ for all $t \in (-\tau, 0)$ implies $\lim_{t \to +\infty} x(t) = x^*$. x^* is exponentially stable if it is stable and there exists $t_0, L, \eta > 0$ such that $\forall\, t \geq t_0$, $|x(t) - x^*| \leq Le^{-\eta t}$.

From these definitions, it follows that exponential stability implies asymptotic stability implies stability which implies stationary point of system. For continuously differentiable evolutionary game dynamics, local stability and asymptotic stability areas of ESS (when its exists) can be established by linearizing the DDE at the rest point. If an interior point x^* is stable under the non-delayed dynamics, a necessary and sufficient condition of stability of x^* under delayed dynamics is that all roots of the *characteristic equation* is given

$$\det\left(\lambda I - K \sum_{e,b \in \mathcal{A}^e} B^{e,b} e^{-\tau_b^e \lambda} \right) = 0 \qquad (7.17)$$

negative real parts. I is the identity matrix with the same size as the matrix $B^{e,b}$, which is obtained from the Jacobian of the dynamics at the point x^*.

7.3 Intrusion Detection in Ad Hoc Networks

In recent years, there has been a growing interest in identifying and studying the behavior of potential intruders to networks or of malicious mobile nodes and in studying how to best detect these or to best protect the network from their actions. Defense strategies such as intrusion detection systems can be deployed at each mobile node; significant constraints are imposed in terms of the energy expenditure of such system. Depending on the capability of the intrusion detection system, the defending node can detect an attacking node in the neighborhood or any node in the network. We study in this paper security interactive behavior between the hackers (malicious mobile nodes, jammers) and the defenders (regular nodes). We consider a scenario where malicious mobile nodes attempt to jam the communications of their neighbor mobile or relay nodes to the destination. Regular nodes save the battery life for the communications; they do not intend to directly damage other nodes. We consider the distributed case and focus our study on large population of mobile nodes. Each regular mobile node seeks to maximize the difference between the rate of information that it transmits to some node destination and the cost of monitoring. Each malicious node seeks to minimize the total rate of transmission of the regular nodes that he can control. A regular node has two actions: *to monitor* (M) or not (nM). A malicious node has two actions: *to attack* (A) or not (nA).

7.3.1 Description of Model

Consider an ad hoc network consisting of a large number of mobile nodes with two types: malicious nodes and regular nodes. We shall consider two actions for each type (jammer or regular): *to attack* (A) or not (nA) for a jammer and *to monitor* (M) or not (nM) for regular node. We assume that any node will be nonregular with probability μ. The parameter μ can be estimated from the size of jammers in all the population (e.g., $\mu = m_{jam}/m_{re} + m_{jam}$ where m_{jam} [resp. m_{re}] is the mass of the malicious nodes [regular nodes] in the network), and with the probability $1 - \mu$ the node is regular.

We consider only interactions between different types of nodes, that is, attacker–defender and defender–attacker. We use evolutionary game with incomplete information framework to describe evolution and behavior of the network in long-run term.

7.3.1.1 Pure Actions of the Incomplete Information Game

The pure strategies of the incomplete information game are described as follows:

- a_1 plays *attack* (A) if malicious node and plays *monitor* (M) if regular node.
- a_2 plays *not attack* (nA) if malicious node and plays *monitor* if regular node.
- a_3 plays attack (A) if malicious node and plays *not monitor* (nM) if regular node.
- a_4 plays not attack (nA) if malicious node and plays *not monitor* (nM) if regular node.

7.3.1.2 Payoffs

The objective of each node depends on its type:

- The malicious node (attacker) would like to play a Bayesian strategy to minimize his chances of being detected.
- The defender would also want to play a Bayesian strategy in order to maximize his chance of detecting attacks without overspending his energy on monitoring.

We now describe an appropriate utility function of our model. The state of the population of nodes is given by a vector $s = (s_j)_j$, $j = 1,\ldots,4$ in the 3-simplex of the four-dimensional Euclidean space. To define the utility function, we introduce the following notions: c is the *cost of attack*, d is the *cost of monitoring*, and v is the value of security for a regular node. Typically the opposite of v is the cost of damage or the loss of security, and γ is a parameter that depends on the detection rate (true alarm rate).

The matrix payoff of the malicious node in an malicious-regular interaction is represented by the tabular

Malicious node:

	M	nM
A	u_1	u_2
nA	u_3	u_4

$u_1 = -\gamma v - c$ is the expected gain of detecting the attack when the regular node monitors. $u_2 = v - c$. When the malicious node does not attack, its payoff is zero: $u_3 = 0$ and $u_4 = 0$.

The matrix payoff of the defender in a regular malicious interaction is represented by the following tabular:

Regular node:

	A	nA
M	v_1	v_2
nM	v_3	v_4

where $v_1 = \gamma v - d, v_2 = v - d, v_3 = -v, v_4 = 0$.

The matrix payoff is $\mu U + (1 - \mu)V$ where

$$
U = \begin{array}{c|cccc}
 & a_1 & a_2 & a_3 & a_4 \\
\hline
a_1 & u_1 & u_1 & u_2 & u_2 \\
a_2 & u_3 & u_3 & u_4 & u_4 \\
a_3 & u_1 & u_1 & u_2 & u_2 \\
a_4 & u_3 & u_3 & u_4 & u_4
\end{array}
\,, \quad
V = \begin{array}{c|cccc}
 & a_1 & a_2 & a_3 & a_4 \\
\hline
a_1 & v_1 & v_2 & v_1 & v_2 \\
a_2 & v_1 & v_2 & v_1 & v_2 \\
a_3 & v_3 & v_4 & v_3 & v_4 \\
a_4 & v_3 & v_4 & v_3 & v_4
\end{array}
$$

A node with strategy a_1 is a malicious node with probability μ, and its strategy is then (A); it interacts with regular nodes of which s_j fractions are using the strategy a_j. The expected payoff of that node when he uses a_1 given that its type is malicious is

$$
s_1 u_1 + s_2 u_1 + s_3 u_2 + s_4 u_2 = (s_1 + s_2)u_1 + (s_3 + s_4)u_2
$$

With the probability $1 - \mu$, the node is regular, plays the strategy (M), and obtains in expectation the payoff

$$
(s_1 + s_3)v_1 + (s_2 + s_4)v_2
$$

In expectation on the type, this node receives

$$
F_{a_1}(s) = \mu(s_1 + s_2)u_1 + \mu(s_3 + s_4)u_2 + (1 - \mu)(s_1 + s_3)v_1 + (1 - \mu)(s_2 + s_4)v_2
$$

$$
= [\mu u_1 + (1 - \mu)v_1]s_1 + [\mu u_1 + (1 - \mu)v_2]s_2 + [\mu u_2 + (1 - \mu)v_1]s_3
$$

$$
+ [\mu u_2 + (1 - \mu)v_2]s_4
$$

which is exactly

$$
e_1[\mu U + (1 - \mu)V] \begin{pmatrix} s_1 \\ s_2 \\ s_3 \\ s_4 \end{pmatrix}
$$

where e_j is the jth element of canonical basis. Similarly, the expected payoff at a_j is

$$
F_{a_j}(s) = e_j[\mu U + (1 - \mu)V] \begin{pmatrix} s_1 \\ s_2 \\ s_3 \\ s_4 \end{pmatrix}
$$

We transform into cost function by taking $C_{a_j}(s) = -F_{a_j}(s)$. We can write the corresponding replicator dynamics as follows:

$$
\dot{s}_j(t) = s_j(t) \left[-C_{a_j}(s) + \sum_{j=1}^{4} s_j(t) C_{a_j}(s(t)) \right]
$$

The following result follows from the evolutionary version of the Folk theorem (Cressman, 2003): If s is asymptotically stable under the replicator dynamics, then s is an evolutionary stable strategy.

The replicator dynamics is not convergent but its trajectory remains at the relative interior of the simplex and then the time average trajectory defined by

$$\bar{s}_j^T = \frac{1}{T} \int\limits_0^T s_j(t) \, dt, \ j \in \{1, 2, 3, 4\}$$

converges to set of equilibria.

7.4 Interference Control in Distributed ALOHA Networks

Random medium access control (MAC) algorithms have played an increasingly important role in the development of wired and wireless networks and the performance and stability of these algorithms, such as slotted-Aloha, carrier sense multiple access (CSMA) is still an open problem. Distributed random MAC, starting from the first version of Abramson's Aloha to the most recent algorithms used in IEEE802.11, has enabled a rapid growth of both wired and wireless networks. It aims at efficiently and fairly sharing a resource among users even though each user must decide independently (eventually after receiving some messages or listening) when and how to attempt to use the resource. MAC algorithms have generated a lot of research interest, especially recently in attempts to use multi-hop wireless networks to provide high-speed access to the internet with low cost and low-energy consumption.

In this section, we consider a wireless communication network with distributed receivers in which some mobiles contend for access to a common, wireless communication channel. We characterize this distributed multiple access problem in terms of many random access games at each time. Random multiple access games introduce the problem of medium access. We assume that mobiles are randomly placed over an area and distributed receivers in the corresponding area, the channels are ideal for transmission, and all errors are due to collisions. A mobile decides to transmit a packet with some power level or not to transmit (null power level) to a receiver when they are within transmission range of each other. Interference occurs as in the Aloha protocol where the power control is introduced: if more than one neighbors of the receiver transmit a packet with a power level that is greater than the corresponding power of the mobile at the same time slot, there is a collision. The evolutionary random multiple access game is a nonzero-sum dynamic game; at each time, the mobiles in the same neighborhood have to share a common resource, the wireless medium. To cover non-saturated transmission case, the random selection of mobiles is done only on the set of mobiles that have a packet to transmit or mobiles that do not have a packet to sent at the current slot are automatically constrained to the state "0" (with null power).

We assume that for each packet, its source can choose the transmitted power among several power levels $\mathcal{A} = \{p_0, p_1, p_2, \ldots, p_n\}$ with $p_0 < p_1 < \cdots < p_n$, that is, a strategy for a mobile corresponds to the choice of a power level in $\mathcal{P}$. $p_0 = 0$ means that the mobile does not transmit; p_n is the maximum power level available to the mobiles. If mobile j transmits a packet using a power p_j, it incurs a transmission cost of $c(p_j) \geq 0$. The packet transmission is successful if the other users in the range of its receiver use some power levels strictly lower than p_j in that given time slot, otherwise there is a collision. If there is no collision, mobile j gets a reward of V from the successful packet transmission. If $c(a) > V$ for some power a, then a is dominated by 0 (not transmit). For the remainder, suppose that the reward V is greater than the cost of transmission $\max_{p_j} c(p_j) < V$. All packets of a lower power level involved in a collision are assumed to be lost and will have to be retransmitted later. In addition, if more than one packet of a higher

Table 7.1   Random Access Game with Two Mobiles and Three Actions: Mobile 1 Chooses a Row and Mobile 2 Chooses a Column $0 < V$

	p_2	p_1	p_0
p_2	$(0,0)^\star$	$(V,0)^{\sharp,\star}$	$(V,0)^{\sharp,\star}$
p_1	$(0,V)^{\sharp,\star}$	$(0,0)$	$(V,0)^{\sharp}$
p_0	$(0,V)^{\sharp,\star}$	$(0,V)^{\sharp}$	$(0,0)$

$\sharp$ Represents a Pareto optimal solution.
$\star$ Indicates a pure NE.

power level is involved in a collision, then all packets are lost. The power differentiation thus allows one packet of a higher power level to be successfully transmitted in collisions that do not involve other packets of the higher power level. Then, a transmission of mobile j is successful if its transmission power is strictly greater than the power levels used by the others' mobiles at the same slot. When the number of mobiles transmitting at the receiver is $k+1$, the payoff is given by $u^j_{k+1} : A^{k+1} \longrightarrow \mathbb{R}$

$$u^j_{k+1}(a^j, a^{-j}) = -c(a^j) + V \times \begin{cases} 1 & \text{if } a^j > \max_{l \neq j} a^l \\ 0 & \text{otherwise} \end{cases}$$

where
a^{-j} denotes $(a^1, \ldots, a^{j-1}, a^{j+1}, \ldots, a^{k+1})$
$c : \mathbb{R}_+ \longrightarrow \mathbb{R}_+$ is a pricing function

We assume that

$$c(p_0) = c(0) = 0 \le c(p_1) \le c(p_2) \cdots \le c(p_n)$$

Without pricing: Consider the case where there is no cost for transmission power, that is, $0 = c(p_0) = c(0) = c(p_1) = c(p_2) = \cdots = c(p_n) = 0$. Table 7.1 illustrates an example of this case.

With pricing: In this section, we study the case where

$$c(p_0) = c(0) = 0 < c(p_1) < c(p_2) \cdots < c(p_n)$$

We observe that if x is stochastically dominated by y, that is for all $i < n$, $\sum_{j=i+1}^{n} x_j \le \sum_{j=i+1}^{n} y_j$ (Table 7.2) then $C_b(x) \le C_b(y)$, $\forall b \in \mathcal{P}$. The function C is extended to a more general generating function (Vincent and Brown, 2005) G defined in a linear space that contains $\Delta_n \times \Delta_n$. The G-function satisfies $G(v, x)|_{v=p_j} = C_{p_j}(x)$. We then have the following results:

Proposition 7.2   The evolutionary random access game with $\sharp A \ge 3$ has no pure equilibrium and a unique strictly mixed equilibrium.

Table 7.2 Multiple Access Game with Two Mobiles and Three Actions: Mobile 1 Chooses a Row and Mobile 2 Chooses a Column. $0 < c(p_1) < c(p_2) < V$

	p_2	p_1	p_0
p_2	$(-c(p_2), -c(p_2))$	$(V - c(p_2), -c(p_1))$	$(V - c(p_2), 0)$
p_1	$(-c(p_1), V - c(p_2))$	$(-c(p_1), -c(p_1))$	$(V - c(p_1), 0)^\sharp$
p_0	$(0, V - c(p_2))$	$(0, V - c(p_1))^\sharp$	$(0,0)$

There is no pure equilibrium.

7.5 Resource Selection with an Unknown Number of Users

Consider a network with distributed Base Stations (BSs) and a large number of users. Let $B = \{1, 2, \ldots, n\}$ be the set of base stations ($n \geq 1$) (resources). Each BS j is associated with the signal to interference plus noise ratio

$$\text{SINR}_j^k = \frac{P_j h_j}{\sigma^2 + (k-1)P_j' h_j'}$$

if there are k players that choose the BS j. The term $P h_j$ represents the power level received at base station from an anonymous mobile using the BS j with the power level P; the SINR at BS j is the ratio between the power received at a base station j from mobile using the power P and the total power of noise and interference, and σ^2 is a constant that represents the noise power at the receiver. We assume that the wireless channels are considered to be the access for the mobiles to a global network, and this access can be done using one of the BSs. The action set of every player is the set of BS B. Mixed strategies are obtained by concerning the scenario where a mobile can transmit simultaneously to several BSs (Figure 7.1).

The instantaneous throughput of a user depends on the BS that he chooses and on the number of other players that choose this BS via the SINR functions. Let $x \in \Delta(B)$ be a mixed action of an arbitrary player. That is, $x = (x_1, \ldots, x_n)$, where x_j is the probability that a player who uses the mixed action x will select the BS j.

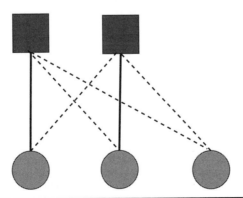

Figure 7.1 A resource selection game.

Base station choice with k players: Denote by $f_j(x)$ the expected payoff of a player that chooses BS j when each of the other $k-1$ players is using x. We denote the game with k users by Γ^k. Let $\sum_j x_j f_j(x)$ be the total expected payoff of every player when each of the k players use x. The expected payoff at the base station j is

$$f_j(x) = \sum_{i=0}^{k-1} g\left(\text{SINR}_j^{1+i}\right) \binom{k-1}{i} x_j^i (1-x_j)^{k-1-i}$$

where

$\quad g : \mathbb{R}_+ \longrightarrow \mathbb{R}_+$ is a nondecreasing and positive function

$\quad g\left(\text{SINR}_j^{1+i}\right)$ represents the instantaneous throughput (or instantaneous fitness, reward) of the player at the BS j

For example, g is the logarithm function log for the Shannon capacity, and $g(y) = (1 - exp(-\beta y))^M$, $\beta > 0$ for the bit error rate analysis.

Proposition 7.3 The BS choice game with k players possesses a unique symmetric mixed equilibrium.

Sketch of proof: *Step 1*: The function $x \longmapsto f_j(x)$ is strictly decreasing in x_j. Since

$$\forall j = 1, \ldots, n, \ \text{SINR}_j^{i+1} - \text{SINR}_j^i < 0, \ \forall i$$

one has

$$\frac{\partial}{\partial x_j} f_j(x) = \sum_i (k-1-i) \binom{k-1}{i} d_{i,j} x_j^i (1-x_j)^{k-1-i} < 0$$

where $d_{i,j} = \left[g\left(\text{SINR}_j^{i+1}\right) - g\left(\text{SINR}_j^i\right)\right]$.

Step 2: There is at most one mixed equilibrium: Suppose in negation that there is more than one mixed symmetric equilibrium of the game. Let x and x' be two symmetric equilibrium actions with $x \neq x'$. This means that $x_j > 0 \implies f_j(x) \geq f_{j'}(x)$. Since x and x' are different, there exists a base station j in B with $x_j \neq x_j'$. Without loss of generality, we can suppose that $x_j > x_j'$. There exists a base station j' such that $j' \neq j$ and $x_{j'} < x_{j'}'$ (because $\sum_l x_l = \sum_l x_l' = 1$). By applying step 1, we obtain that

$$f_j(x) < f_j(x') \leq f_{j'}(x') < f_{j'}(x) \leq f_j(x)$$

which is a contradiction.

Step 3: The existence of symmetric equilibrium is immediate by combining the symmetry of the game and Kakutani's fixed point theorem. This completes the proof. ◻

Resource selection as a population game: Altman et al. (2008) have analyzed evolutionary power and interference control games in wireless networks with two power levels. They have showed existence and uniqueness of evolutionary stable strategy under some sufficient conditions, and convergence to the ESS under replicator dynamics. In this subsection, we consider a network with many distributed base stations and a large number of users in an evolutionary game setting. Each user chooses one of the base stations available in its neighborhood. We say that a payoff f generates a *strict stable game* if for all $x \neq y$, $\sum_{j \in B}(x_j - y_j)(f_j(x) - f_j(y)) < 0$. We show that the corresponding population game with the local one-shot base station assignment game Γ^k is a nonzero-sum stable game. The function x_j characterizes aggressiveness of mobiles for the resource j. The result of Proposition 7.3 shows that the marginal throughput decrease with the aggressiveness of the mobiles: the less aggressiveness of mobiles, the bigger throughput.

Proposition 7.4 The game Γ^k is a nonzero-sum strict stable population game.

Sketch of proof By the monotonicity of f, one has $(x_j - y_j)(f_j(x) - f_j(y)) < 0$ all strategies $x \neq y$ such that $x_j \neq y_j$. Hence, $\sum_{j \in B}(x_j - y_j)(f_j(x) - f_j(y)) < 0$. □

Proposition 7.5 The base station selection game with a random number of players has a unique evolutionarily stable state.

Sketch of proof From the result of Proposition 7.3, the game possesses an equilibrium. Let x be the mixed equilibrium. Now we show that x is an ESS. Fix an arbitrary vector $y \neq x$. Since f is strict stable and x is an equilibrium, one has the system of inequalities:

$$\begin{cases} \sum_{j \in B}(y_j - x_j)(f_j(y) - f_j(x)) < 0 \\ \sum_{j \in B}(y_j - x_j)f_j(x) \leq 0 \end{cases}$$

Adding the two inequalities of the last system, we obtain that $\sum_{j \in B}(y_j - x_j)f_j(y) < 0$. Taking $y = (1 - \epsilon)x + \epsilon y$ for arbitrary $y \neq x$, we conclude that x is an ESS. The uniqueness is derived from 7.3. □

Remark Since the base selection game is a strict stable game with infinitely differentiable payoff function, it is not difficult to see that the game has also a unique correlated equilibrium in $\Delta(\Delta(A) \times \Delta(A))$. Also, it is easy to see that strict stable population game has a unique ESS. Thus, in this case, there is no benefit of correlation (correlated equilibrium approach does not improve the total equilibrium payoff).

7.5.1 Convergence to Equilibrium and Stability

We introduce here a class of generalized evolutionary game dynamics, which describes the evolution in the population of the various strategies. In this class of dynamics, the share of a strategy in the population grows at a rate that is a function of the difference between the payoff of that strategy and the payoff at the strategy he switches to. More precisely, let x^t be the state of the population at time t. Thus, we have $\sum_{j \in B} x_j^t = l$ and $x_j^t \geq 0$ where l is the mass of the population and x_j^t represents the fraction of players playing a strategy j in period t. This total mass can be normalized to one. We describe the evolution of the system by approximating from stochastic influence on the change in terms of frequency of actions. Suppose that in every period Δt, each player learns with probability $\alpha \Delta t > 0$ the expected payoff to the other opponent players and changes to the other's strategy if he perceives that the other's payoff is higher. However, the information about the difference in the expected payoffs of the strategies is imperfect, so the larger the difference in the payoffs, the more likely the player is to perceive it and change. Specially, we assume that the probability that a player using the BS j' will shift to the BS j is given by

$$x_{j' \to j}^{t+\Delta t} = \begin{cases} \mu^j [f_j(x^t) - f_{j'}(x^t)]^\theta & \text{if } f_j(x^t) > f_{j'}(x^t) \\ 0 & \text{otherwise} \end{cases}$$

where $\theta \geq 1$ and μ are sufficiently small such that $x_j^t \leq 1$ holds $\forall j, j'$. The expected fraction $\mathbb{E}x_j^{t+\Delta t}$ of the population using j in period $t + \Delta t$ is given by

$$x_j^t + \alpha \Delta t \sum_{j' \in \mathcal{I}} x_{j'}^t \mu^j [f_j(x^t) - f_{j'}(x^t)]^\theta - x_j^t \alpha \Delta t \sum_{j' \notin \mathcal{I}} \mu^{j'} [-f_j(x^t) + f_{j'}(x^t)]^\theta$$

where $\mathcal{I} = \{j', f_j(x^t) > f_{j'}(x^t)\}$. For large population, we can replace $\mathbb{E}x_j^{t+\Delta t}$ by $x_j^{t+\Delta t}$. Taking the limit of $(x_j^{t+\Delta t} - x_j^t)/\Delta t$ when Δt goes to zero, we then obtain

Continuous time: $\dot{x}_j^t = \alpha G_j(x^t), \ j \in B$

$$G_j(x^t) = \sum_{j' \in \mathcal{I}} x_{j'}^t \mu^j [f_j(x^t) - f_{j'}(x^t)]^\theta - x_j^t \sum_{j' \notin \mathcal{I}} \mu^{j'} [-f_j(x^t) + f_{j'}(x^t)]^\theta$$

The vector μ and constant α change the rate of adjustment to stationary point. The two parameters α and μ give us a framework for controlling game dynamics (changing or upgrading policy) through the choice of the gain parameters governing the dynamics. They have positive effect on the speed of convergence of the dynamics (see Figure 7.2).

Proposition 7.6   The dynamics (7.5) is positively correlated, that is,

$$\dot{x}^t \neq 0 \Rightarrow \sum_{j \in B} f_j(x^t) \left(\frac{d}{dt} x_j^t \right) > 0$$

Sketch of proof

$$\wp := \sum_{j \in B} f_j(x) \left(\frac{d}{dt} x_j(t) \right) = \alpha \sum_{j \in B} f_j(x) G_j(x)$$

$$= \alpha \sum_{j \in B} f_j(x) \sum_{j' \in B} x_{j'} \max\left(0, f_j(x) - f_{j'}(x)\right)^\theta - \alpha \sum_{j \in B} f_j(x) x_j \sum_{j' \in B} \max\left(0, -f_j(x) + f_{j'}(x)\right)^\theta$$

$$= \alpha \sum_{j,j' \in B} \max\left(0, f_j(x) - f_{j'}(x)\right)^\theta \left(f_j(x) - f_{j'}(x)\right) = \alpha \sum_{j,j' | f_j(x) > f_{j'}(x)} \left(f_j(x) - f_{j'}(x)\right)^{1+\theta} \geq 0$$

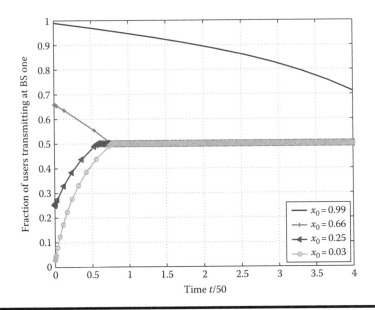

Figure 7.2 Convergence to ESS.

with equality if and only if $G(x) = 0$, that is, $(d/dt)x(t) = 0$. □

Proposition 7.7 Every equilibrium of the game is a rest point of the dynamics.

Sketch of proof Let x be an equilibrium point. Then there is no change rate from the BS j to the BS j'. This means that $G_j(x) = 0$, $\forall j \in B$. □

Proposition 7.8 Every rest point of the generalized Smith dynamics is an equilibrium of the evolutionary game.

Note that this result does not hold in general evolutionary game dynamics. It is known that the replicator dynamics may not lead to correlated equilibria.

Proposition 7.9 The ESS given in result 7.5 is globally asymptotically stable.

Sketch of proof We show that the following function $y \in \Delta(B) \longmapsto V(y)$ where

$$V(y) = \frac{1}{1+\theta} \sum_{j \in B} \sum_{j' \in B} y_j \mu^{j'} [\max(0, -f_j(y) + f_{j'}(y))]^{1+\theta}$$

is a Lyapunov function. One has $V(y) \geq 0$, $\forall y$ and $V(y) = 0$ if and only if y is an equilibrium point. Since f is stable, $(d/dt)V(y) \leq 0$. □

7.6 Open Issues

7.6.1 Dynamics of Multiobjective Evolutionary Networking Games

Most of the traditional analyses on the theories of decision making and games are based on the assumption of a single dimensional cost or payoff, which prevents their broader practical utilization. In reality, any alternative is likely to imply more than one utility function, for example, time, energy, quality, delay, loss/success probability, etc. These multiple objectives are usually in conflict with one another. This means that, in general, a multiple objective decision problem does not have a single solution that could optimize all objectives simultaneously. Because of this, multiple objective interactive decision making is not to search for optimal solutions but for efficient and stable solutions (best compromising solutions in some sense) that can best attain the prioritized multiple objectives as much as possible. Hence, it is interesting to study multicriteria problems in multi-population games with several (possibly coupled) constraints and extend evolutionary game dynamics for multiobjective evolutionary games and their application to networking and communication problems.

7.6.2 Evolutionary Network Formation Games

Network formation games and coalition structures have been studied in standard game theory. Coalitional formation game investigates players having interactions with their neighbors in a network, for example, people interacting with their friends or their colleague members of the network. Starting from von Neumann and Morgenstern seminal work on game theory, the question of the formation of coalitions has been the highly debated topic in the context of cooperative games. Noncooperative network formation game proposes an explicit representation of the coalition formation between players. In such games, players usually have some discretion to connect with each other; hence, the network structure both influences the result of the economic interactions and is shaped by the decisions of the players. We address the following questions: (i) How is the network structure important in determining the fitness or the outcome of socio-economic interaction? (ii) Is it possible to predict which networks are likely to form when individuals have the discretion to choose their connections or their preferential network coalition structure?

In addition to the economical and social aspects, this framework seems to be promising to investigate some important issues in computer networks. *Evolutionary network formation games* aim to describe the evolution of coalition and formation of new networks. The evolutionary game dynamics framework developed in this chapter can be extended to cover both individual players and group players in the same setting. The trajectories of the evolutionary game dynamics for both microscopic and macroscopic entities will indicate the survival of a coalition in the long term.

7.6.3 Multilevel Hierarchical Evolutionary Games

We extend our study to multilevel hierarchical evolutionary games. We keep the local leader roles and assume that there are many types of followers. Each local game becomes an extensive game with many rounds and additional information about the decisions of the players that move before. This is the case when there are some prioritized users depending on the demands (types of services). They play in second round after the local leader's decisions. Sequential evolutionary games with information rules of moving and perfect observation have been studied in the literature (Cressman, 2003). We can use this approach to solve each local sequential game by backward induction, but finding subgame perfect equilibria can be difficult in complexity. It has been shown that ESSs are also subgame perfect equilibria (Cressman, 2003; Gintis, 2009). This means that the constrained ESS concept (when it exists) is a well-adapted equilibrium concept for multilevel hierarchical population games with mobility constraints. We briefly describe the formulation of hierarchical solutions.

For the bi-level case, the problem can be seen as follows: a local leader minimizes the function $x^l \longmapsto C^l(x^l, x^f)$ such that the follower reacts as

$$x^f \in R^f(x^l) = \arg\min_{y^f} C^f(x^l, y^f)$$

where C^f, C^l denotes the cost of the follower (resp. leader). Note that $R^f(x^l)$ is a reaction set (it is not a singleton in general). This makes discontinuities of many maximization problems of the leaders and hierarchical solutions may not exist.

For the tri-level case with leader (l), primary follower (p), and secondary follower (s), the problem can be formulated as follows: the leader minimizes $x^l \longmapsto C^l(x^l, x^p, x^s)$ such that the primary follower reacts and minimizes $y^p \longmapsto C^p(x^l, y^p, x^s)$ such that $x^s \in R^s(x^l, x^p) = \arg\min_{y^s} C^s(x^l, x^p, y^s)$. One can define a tri-level hierarchical solution as follows:

$$\sup_{x^p \in R^p(x^l_*)} \sup_{x^s \in R^s(x^l_*, x^p)} C^l(x^l_*, x^p, x^s) = \inf_{y^l} \sup_{x^p \in R^p(x^l)} \sup_{x^s \in R^s(x^l, x^p)} C^l(x^l, x^p, x^s) \qquad (7.18)$$

$$R^p(x^l) = \left\{ x^p_* \,\middle|\, \sup_{x^s \in R^s(x^l, x^p_*)} C^p\left(x^l, x^p_*, x^s\right) = \inf_{x^p} \sup_{x^s \in R^s(x^l, x^p)} C^p(x^l, x^p, x^s) \right\} \qquad (7.19)$$

For multiple leaders–multiple followers, the problem can be formulated as follows: find $x^{\mathcal{L}}$ such that

$$\forall l \in \mathcal{L}, \ \forall y^l, \ C^l(y^l, x^{\mathcal{L}\setminus\{l\}}, x^{\mathcal{F}}) \geq C^l(x^l, x^{\mathcal{L}\setminus\{l\}}, x^{\mathcal{F}}), \ \text{such that}$$

$$x^{\mathcal{F}} \text{ is solution of}$$

$$\forall f \in \mathcal{F}, \ \inf_{y^f} C^f(x^{\mathcal{L}}, y^f, x^{\mathcal{F}\setminus\{f\}}) \geq C^f(x^{\mathcal{L}}, x^f, x^{\mathcal{F}\setminus\{f\}}).$$

The problem is that generally the vector $x^{\mathcal{F}}$ describes a set, and in indifference situation of followers the leaders will not be able to predict which strategy will be used by the followers. As an alternative, pessimistic and optimistic hierarchical game has been proposed. The pessimistic approach is given by the following problem: find $x^l_* \in X^l$ such that

$$\sup_{x^f \in R^f(x^l_*)} C^l\left(x^l_*, x^f\right) \leq \sup_{x^f \in R^f(x^l)} C^l(x^l, x^f), \ \forall x^l \in X^l \qquad (7.20)$$

where

$$R^f(x^l) = \arg\min_{x^f} C^f(x^l, x^f)$$

A pair $\left(x^l_*, x^f_*\right) \in X^l \times X^f$ with x^l_* satisfying (7.20) and $x^f_* \in R^f\left(x^l_*\right)$ is called a weak Stackelberg solution (so-called pessimistic Stackelberg solution). Equivalently, the pessimistic Stackelberg solution satisfies

$$\sup_{x^f \in R^f(x^l_*)} C^l\left(x^l_*, x^f\right) \leq \inf_{x^l \in X^l} \sup_{x^f \in R^f(x^l)} C^l(x^l, x^f) \qquad (7.21)$$

The optimistic Stackelberg solution can be defined as follows: find $x_*^l \in X^l$ such that

$$\inf_{x^f \in R^f(x_*^l)} C^l\left(x_*^l, x^f\right) \leq \inf_{x^f \in R^f(x^l)} C^l(x^l, x^f), \ \forall x^l \in X^l \tag{7.22}$$

or equivalently

$$\inf_{x^f \in R^f(x_*^l)} C^l\left(x_*^l, x^f\right) \leq \inf_{x^l \in X^l} \inf_{x^f \in R^f(x^l)} C^l(x^l, x^f) \tag{7.23}$$

In general there is no link between these two definitions.

Without considering the cost of leadership and cost of hierarchical systems, it is easy to see that the expected cost of the leader in the bi-level hierarchical population game at hierarchical solution (in mixed strategy) is less than the expected cost at NE. This result follows from the fact that

$$\sup_{x^f \in R^f(x_*^l)} C^l\left(x_*^l, x^f\right) \leq \inf_{x^l} \sup_{x^f \in R^f(x^l)} C^l(x^l, x^f) \leq C^l(x^{NE})$$

Nonexistence of Stackelberg Solutions: Given a two population game represented by the two matrix (A, B) with players from population one as leaders (with probability one). A population profile $y^* \in Y$ is a mixed Stackelberg equilibrium strategy for population one (leaders) if

$$\max_{z \in R^2(y^*)} \langle y^*, Az \rangle = \inf_{y \in Y} \left[\max_{z \in R^2(y^*)} \langle y, Az \rangle \right] = C^1$$

where $R^2(y) = \arg\min_z \langle y, Bz \rangle$. In Fact, the "inf" cannot always be replaced by "min." This is because the function

$$y \longmapsto \max_{z \in R^2(y)} \langle y, Az \rangle$$

is not necessarily continuous. Many formulations in the literature use the expression "$C^j(x^j, R^j(x^j))$," which is not well defined if R^j is a set.

7.6.4 Mean Field Game Dynamics for Networks

The aspect of mobility of players and their spatial repartition in the population play an important in the local interactions between players. Taking into account these parameters, we introduce a new class of game dynamics obtained from asymptotic of dynamic games with a variable number of interacting players. These dynamics cover the standard dynamics known in evolutionary games and population games. These dynamics contain also evolutionary game dynamics with migration developed in Tembine et al. (2008). In the case of multi-class of players, the dynamics describe both the intrapopulation interactions and interpopulation interactions. The *mean field game dynamics* are in general nonlinear (partial or not) differential equations or inclusions. This last class occurs when the mean process converges weakly to another process described by a drift plus a noise. The evolutionary stochastic stability properties and survival of weakly dominated strategies under the stochastic mean game dynamics remain open questions.

7.7 Conclusion

We have developed a framework for analyzing evolving games in large populations of players, in which each player interacts with the other random number of players. We presented various classes of evolutionary game-theoretic models (with and without hierarchy) and applied them to networking problems including intrusion detection, resource selection, and interference control in wireless networks.

Acknowledgment

The author would like to thank the two anonymous reviewers for their helpful comments.

References

Altman, E., Elazouzi, R., Hayel, Y., and Tembine, H. (2008), *Evolutionary Power Control Games in Wireless Networks*, Book chapter in A. Das et al. (Eds.): Networking 2008, Adhoc and sensor networks, wireless networks, next generation internet, Lecture Notes in Computer Science, 4982, pp. 930–942, DOI:10.1007/978-3-540-79549-0_82.

Altman, E., El-Azouzi, R., Hayel, Y., and Tembine, H. (2009), The evolution of transport protocols: An evolutionary game perspective, *Elsevier Computer Networks Journal*, 53(10), 1751–1759, Autonomic and Self-Organising Systems, July 2009.

Anastasopoulos, M.P., Arapoglou, P.-D.M., Kannan, R., and Cottis, P.G. (2008), Adaptive routing strategies in IEEE 802.16 multi-hop wireless backhaul networks based on evolutionary game theory, *IEEE Journal on Selected Areas in Communications*, 26(7), 1218–1225.

Benaim M. and Weibull J.W. (2003), Deterministic approximation of stochastic evolution in games, *Econometrica*, 71, 873–903.

Bertrand, J.L.F. (1883), Théorie des Richesses: Revue de Théories mathmatiques de la richesse sociale par Léon Walras et Recherches sur les principes mathématiques de la théorie des richesses par Augustin Cournot, *Journal des Savants*.

Bowley, A. (1924), *The Mathematical Groundwork of Economics*, Kelley: New York.

Brown, G.W. (1951), Iterative solutions of games by fictitious play, In *Activity Analysis of Production and Allocation*, T.C. Koopmans (Ed.), Wiley, New York.

Brown, G.W. and von Neumann, J. (1950), Solutions of games by differential equations. *Annals of Mathematics Studies*, 24, 73–79.

Cournot, A.A., *Recherches sur les principes mathématiques de la théorie des richesses*, 1838, Hachette, Paris. Researches into the Mathematical Principles of the Theory of Wealth, 1838 (1897, Engl. trans. by N.T. Bacon).

Cressman, R. (2003), *Evolutionary Dynamics and Extensive Form Games*, MIT Press, Cambridge, MA.

Fudenberg, D. and Tirole, J. *Game Theory*, MIT Press, Cambridge, Mass.: 1991, p. 579.

Fudenberg, D. and Levine, D.K. (1998), *The Theory of Learning in Games*, MIT Press, 292 pp. Cambridge, MA, ISBN 9780262061940.

Gilboa, I. and Matsui, A. (1991), Social stability and equilibrium. *Econometrica*, 59, 859–867.

Gintis H. (2009), *Game Theory Evolving: A Problem-Centered Introduction to Modeling Strategic Interaction*, 2nd edn., Princeton University Press, Princeton, NJ.

Hamilton, W.D. (1967), Extraordinary sex ratios. *Science*, 156, 477–488.

Hofbauer, J. and Sigmund, K. (1998), *Evolutionary Games and Population Dynamics*, Cambridge University Press, Cambridge, U.K.

Hofbauer, J. and Sigmund, K. (2003), Evolutionary game dynamics. *American Mathematical Society*, 40(4), 479–519.

Hofbauer, J., Sorin, S., and Viossat Y. (2009), Time average replicator and best reply dynamics, *Mathematics of Operation Research*, 34(2), 263–269.

Monderer, D. and Shapley, L.S. (1996), Potential games, *Games and Economic Behavior*, 14(1), 124–143.

Nash, J. (1951), Non-cooperative games, *The Annals of Mathematics*, 54(2), 286–295.

Robinson, J. (1951), An iterative method of solving a game, *Annals of Mathematics*, 54, 296–301.

Roughgarden, T. (2005), *Selfish Routing and the Price of Anarchy*, MIT Press, Cambridge, U.K.

Sandholm, W.H. (2011), *Population Games and Evolutionary Dynamics*, The MIT Press, ISBN 9780262195874, 560 pp. Cambridge, U.K.

Shapley, L.C. (1964), Some topics in two-person games. In: *Advances in Game Theory (Annals of Mathematical Studies 52)* M. Dresder, L.S. Shapley, and A.W. Tucker (Eds.), Princeton University Press, Princeton, 1–28.

Smith, J.M. (1982), *Evolution and the Theory of Games*, Cambridge University Press, Cambridge, U.K.

Smith, M.J. (1984), The stability of a dynamic model of traffic assignment: An application of a method of Lyapunov. *Transportation Science*, 18, 245–252.

Smith, J.M. and G.R. Price. (1973), The logic of animal conflict, *Nature* 246, 15–18.

Taylor, P. and Jonker, L. (1978), Evolutionary stable strategies and game dynamics. *Mathematical Biosciences*, 16, 76–83.

Tembine, H. (2009), Population games with networking applications, PhD disseration, University of Avignon, Avignon, France.

Tembine, H. (2010), Population games in large-scale networks: LAP, ISBN 3838363922.

Tembine, H., Altman, E., and El-Azouzi, R. (2007), Delayed evolutionary game dynamics applied to the medium access control, in *Proceedings of IEEE MASS*, Pisa, Italy.

Tembine, H., Altman, E., Elazouzi, R., and Sandholm, W.H. (2008a), Evolutionary game dynamics with migration for hybrid power control in wireless communications. In *Proceedings of 47th IEEE Conference on Decision and Control (CDC'2008)*, Cancun, Mexico.

Tembine, H., Altman, E., ElAzouzi, R., and Hayel, Y. (2008b), Evolutionary games with random number of interacting players applied to access control. In *Proceedings of IEEE/ACM WiOpt*, Berlin, Germany.

Tembine, H., Altman, E., Elazouzi, R., and Hayel, Y. (2008c), Stable networking games. In *Proceedings of 46th Annual Allerton Conference on Communication, Control, and Computing*, Monticello, IL.

Tembine, H., Altman, E., ElAzouzi, R., and Hayel, Y. (2009), Evolutionary games in wireless networks, *IEEE Transactions on Systems, Man, and Cybernetics, Part B: Cybernetics*, 40(3), 634–646.

Vincent, T.L. and Brown, J.S. (2005), *Evolutionary Game Theory, Natural Selection, and Darwinian Dynamics*, Cambridge University Press, Cambridge, U.K. 400 pp.

von Stackelberg, H.F. (1934), *Marktform und Gleichgewicht (Market Structure and Equilibrium)*, Springer, Vienna, Austria.

Wardrop, J.G. (1952), Some theoretical aspects of road traffic research. Proceedings, Institute of Civil Engineers, PART II, Vol. 1, pp. 325–378.

Weibull, J. (1995), *Evolutionary Game Theory*, The MIT Press, Cambridge, MA.

Zhu, Q., Tembine, H., and Basar, T. (2009), Evolutionary game for hybrid additive white Gaussian noise multiple access control. In *IEEE Proceedings of Globecom*, December 2009.

POWER CONTROL
GAMES

Chapter 8

Shannon Rate-Efficient Power Allocation Games

Elena Veronica Belmega, Samson Lasaulce,
and Mérouane Debbah

Contents

This chapter is dedicated to the study of distributed resource allocation problems in wireless communication networks. Noncooperative game theory proves to be a useful tool to investigate this type of problems. The players, the transmitter nodes, choose their power allocation (PA) policies in order to maximize their own information theoretical payoffs, namely, their individual achievable Shannon transmission rates. Our attention is mainly focused on two basic multiuser channel models: the multiple access channel (MAC) and the interference channel (IC). However, more complex channels such as the interference relay channel (IRC) and the cognitive radio channel (CRC) are also discussed. We provide an updated overview of the existing results with respect to the noncooperative solution of the game, the Nash equilibrium (NE): its existence, uniqueness, and convergence of distributed algorithms. Furthermore, we evaluate the performance gap between the NE point and the centralized solution where the network authority dictates the allocation policy of the overall network resources. We also discuss several methods that can improve the performance at the NE while introducing a supplementary cost in terms of signaling. We conclude with a critical discussion about the drawbacks and possible improvements of the game theoretical approach to solve resource allocation problems in general.

8.1 Introduction

Although game theory and information theory have both been extensively developed in the past sixty years, starting with the seminal contributions of von Neumann and Morgenstern [1], Nash [2], and Shannon [3], only recently the connections and interactions between the two theories started to be highlighted and exploited at a significant scale. However, the first application of game theoretical tools to reliable communications dates back to Mandelbrot's PhD dissertation [4] in 1952 and later in [5,6] where the communication between a transmitter and receiver was modeled as a two-player zero-sum game with a mutual information payoff function. The transmitter plays against a malicious nature that chooses the worse channel distribution minimizing the mutual information. It turns out that the noncooperative solution of this game is identical to the maximin worse case capacity assuming that the transmitter has no knowledge about the channel conditions (noise and channel gains statistics).

The recent surge of interest in applying game theoretical tools to communications was due to the development of wireless communications where multiple users share a common transmission environment. Thus, the competition for common resources (such as frequency bands, time, space, power, energy) naturally arises. These resources can be optimized by a central authority. However, this approach presents several inconveniences: (i) it is unrealistic in an environment where multiple concurrent service providers exist; (ii) the joint optimization problem over all the network parameters is generally a non-convex optimization problem involving a high computational cost; (iii) scalability issue; and (iv) it involves an important signaling cost to feed back the optimal allocation policies to each of the user in the network. For these reasons, a distributed solution in the sense of the decision may be desirable.

8.1.1 Noncooperative Game Theoretical Concepts

In the distributed approach, the competition for the resources gives rise to interactive situations among the network users. By definition, game theory is the mathematical framework dedicated to the study of interactive situations among decision makers or agents. In this chapter, we will consider the *rationality assumption* of the game players in the sense that a player chooses its best strategy to maximize its benefit [1]. The mathematical characterization of a strategic form game is given by the following definition.

Definition 8.1   A strategic form game is a triplet $\mathcal{G} = (\mathcal{K}, \{\mathcal{A}_k\}_{k \in \mathcal{K}}, \{f_k\}_{k \in \mathcal{K}})$, where $\mathcal{K} = \{1, \ldots, K\}$ represents the *set of players*, $\mathcal{A}_k$ represents the *set of strategies* or actions that player $i \in \mathcal{K}$ can take and $f_k : \times_{\ell \in \mathcal{K}} \mathcal{A}_\ell \to \mathbb{R}_+$ represents the benefit function of user k which is a measure of its satisfaction and can either represent a payoff function, $(f_k(\cdot) \equiv u_k(\cdot))$ to be maximized or a cost function, $(f_k(\cdot) \equiv c_k(\cdot))$ to be minimized.

In the case of noncooperative games, where the users act in a selfish and independent manner, the concept of NE [2] represents a solution concept of the game. The NE has been extensively studied in resource allocation problems because it represents a state of the system that is robust to unilateral deviations. This means that, once the system is operating in this state, no user has any incentive to deviate unilaterally because it will lose in terms of its own benefit. Assuming a payoff type of benefit function, the mathematical definition of the NE is as follows:

Definition 8.2   A strategy profile $\left(a_1^{\mathrm{NE}}, \ldots, a_K^{\mathrm{NE}} \right) \in \times_{\ell \in \mathcal{K}} \mathcal{A}_\ell$ is a Nash equilibrium if for all $k \in \mathcal{K}$ and for all $d'_{-k} \in \times_{\ell \neq k} \mathcal{A}_\ell$, $u_k \left(a_k^{\mathrm{NE}}, a_{-k}^{\mathrm{NE}} \right) \geq u_k \left(a_k^{\mathrm{NE}}, d'_{-k} \right)$, where $a_{-k} = (a_1, \ldots, a_{k-1}, a_{k+1}, \ldots, a_K)$ denotes the set of the other players' actions.*

The NE is a very important concept to network designers since it represents an operating state of the network that is both stable to a single deviation (which is realistic considering the fact that the players are assumed to be non cooperative and act in an isolated manner) and might be predictable. Based on the game structure, the topological properties of the strategy sets and the payoff functions, the main issues to be dealt with are: (i) first to establish the existence of NE; (ii) to establish whether the NE is unique or there are multiple NE states; (iii) to design distributed algorithms that allow the users to converge to an NE state using only local knowledge of the environment; and (iv) to determine the network performance of the NE. In general, the NE performance is suboptimal with respect to the overall network performance, which can be measured, for example, in terms of the sum of individual user payoffs $u(\underline{a}) = \sum_{k \in \mathcal{K}} u_k(a_k, a_{-k})$. Furthermore, the NE is not necessarily a fair state with respect to the users' payoffs. To cope with these issues, a different type of operating state such as the Pareto-optimal states can be investigated. A state of the system is Pareto-optimal if there is no other state that would be preferred by all the network users.

Definition 8.3   Let $\underline{a}$ and $\underline{a}'$ be two different strategy profiles in $\times_{\ell \in \mathcal{K}} \mathcal{A}_\ell$. Then, if

$$\forall k \in \mathcal{K}, u_k(a_k, a_{-k}) \geq u_k \left(a'_k, a'_{-k} \right) \tag{8.1}$$

with strict inequality for at least one player, the strategy $\underline{a}$ is Pareto-superior to $\underline{a}'$. If there exists no other strategy that is Pareto-superior to a strategy profile $\underline{a}^{\mathrm{PO}}$ then $\underline{a}^{\mathrm{PO}}$ is Pareto-optimal.

---

* Throughout this chapter we will use x_{-k} to denote the set of quantities x of all the players except player k.

However, a Pareto-optimal state is not necessarily a stable state in a selfish noncooperative user environment. Furthermore, the Pareto-optimal solution is rather a centralized type solution because global network information is needed to compute these states of the system. As we will discuss hereafter, there exist different techniques that can be used to improve the performance of the NE. Notice that there is always a trade-off between the performance obtained at the equilibrium state and the signaling cost involved. A more detailed study of the noncooperative game theoretical concepts is beyond the scope of this chapter and can be found in the specialized references [7,8]. Also, for a thorough analysis on the methodologies for analyzing NE in wireless games in general, the reader is referred to [9,10].

The purpose of this chapter is to review the most recent and relevant results with respect to noncooperative resource allocation games in wireless communication networks where Shannon transmission rates are considered as players' payoff functions. To be more precise, the *game players* are autonomous transmitter devices capable of sensing the environment and selfishly deciding *their strategies*, their own PA policies that maximize their *individual payoffs*, i.e., their Shannon transmission rates. Many reasons why this kind of payoffs is often considered can be found in the literature related to this problem. Here, we will mention only three of them. First, Shannon transmission rates allow one to characterize the performance limits of a communication system and study the behavior of (selfish) users in a network where good coding schemes are implemented. Second, because of the direct relationship between the achievable transmission rate of a user and his signal-to-interference plus noise ratio (SINR), they also allow one to optimize performance metrics like the SINR or related quantities of the same type (e.g., the carrier-to-interference ratio). Third, from the mathematical point of view, Shannon rates have many desirable properties (e.g., concavity properties), which allow one to conduct thorough performance analyses. Therefore, they provide useful insights and concepts that are exploitable for a practical design of decentralized networks. However, the literature on resource allocation games for wireless networks is very extensive. Different approaches where other types of payoff functions have also been considered, for example, in [11–14] the power control games with energy efficient payoffs (defined as the ratio between the received throughput and the consumed transmission power), have been studied. In these works, the battery life of the transmitters is considered a crucial resource whose consumption needs to be optimized. Recently, an evolutionary game theoretical approach has been applied to these problems [15,16]. For a more wide and detailed overview, the reader is referred to the following surveys: [17–23].

The network models we will investigate correspond to a given choice of the following degrees of freedom. We will address wireless networks that can be modeled by MAC [24,25], where multiple transmitters send their messages to a common receiver; IC [26], where several transmitter–receiver pairs coexist in a common environment creating mutual interference; IRC [27,28], which consists of an IC plus several relaying nodes that can be used by the transmitters to improve the performance of their transmissions; CRC [29,30], can be seen as an IC where the opportunistic users, the secondary users, are imposed supplementary constraints on their transmissions such that they create no interference or a low level of interference to the primary users, the legal holders of the spectrum. Another freedom degree is the number of antennas available at each node in the network: they can be equipped with either *single or multiple antennas*. As a function of the time varying properties, the channel can be *static* and considered fixed for the whole transmission duration or *fast fading* and considered as the realizations of random variables that change significantly over the transmission timescale. The transmitter may have perfect *channel state information* (CSIT), which is a reasonable assumption only in the case of static channel or *channel distribution information* (CDIT) assumed in the fast fading channel case. There is another case with respect to the transmitter knowledge of the channel which corresponds to the *compound channel* where the transmitter knows not the channel state at each moment but the finite set of possible values that the channel parameters may take. As far as the receiver knowledge is concerned, errorless decoding schemes are assumed, which imply perfect channel state information (CSIR). Furthermore, the decoding scheme can be *single user decoding* (SUD) where the interference coming from the other users is treated as noise or *successive interference cancelation* (SIC) where the users are decoded in a predetermined order such that only the first decoded user sees the interference coming from all the other users, while the last decoded user sees no interference whatsoever.

In the aforementioned scenarios, the important notion of network equilibrium will be addressed. We will show that in all PA games, an NE [7] exists, as a consequence of the concavity properties of Shannon type transmission rates. The multiplicity issue of such a stable states will also be tackled. From a more practical perspective, the efficiency of the equilibrium states and distributed algorithms will be discussed. Also, possible ways of improving the performance of the NE will be briefly reviewed (pricing mechanisms, introduction of hierarchy, coordination, cooperation, punishment/reward mechanisms, etc.).

The chapter is structured in three main parts. In the first part, we study the PA games in MAC for both the static and fast fading channels. Then we switch our attention to the PA games in IC and focus only on the static channel because the IC is a much more involving channel from an information theoretic point of view and no results are yet available with respect to the fast fading case. This is also the reason why, in the last part of the chapter, we consider only the static models when reviewing the PA games in IRC and CRC. The chapter will be concluded by a critical analysis on the game theoretical approach for solving distributed resource allocation problems. In particular, information and rationality assumptions at the user level will be discussed. Several rising and challenging problems will also be mentioned.

8.2 Power Allocation Games in Multiple Access Channels

In this section, we consider the MAC consisting of $K \geq 2$ transmitter nodes that send their private messages to a common receiver that has to decode reliably all the incoming messages. Unless otherwise specified, the players of the game are the K transmitters PA, which choose their own policies to maximize their individual Shannon achievable rates.

The received signal can be written in general as

$$\underline{y} = \sum_{k=1}^{K} \mathbf{H}_k \underline{x}_k + \underline{z} \tag{8.2}$$

where

$\underline{x}_k$ is the $n_{t,k}$-dimensional column vector of symbols transmitted by user k
$\mathbf{H}_k \in \mathbb{C}^{n_r \times n_{t,k}}$ is the channel matrix of user k
$\underline{z}$ is a n_r-dimensional complex white Gaussian noise distributed as $\mathcal{N}(\underline{0}, \sigma^2 \mathbf{I}_{n_r})$, assuming a number of $n_{t,k}$ antennas at the transmitter $k \in \{1, \ldots, K\}$ and n_r antennas at the common receiver

In what follows, we will briefly review the static channel case, where the channel matrices are deterministic and fixed for the whole duration of the transmission, before focusing on the more interesting and challenging case of fast fading channel, where the channel matrices will have random entries.

8.2.1 Static Multiple Access Channels

8.2.1.1 Single Input Single Output Links

Consider the simple case of single input single output (SISO) channel. Assuming the SUD scheme, the payoff of user k is $u_k(p_k, p_{-k}) = \log_2\left(1 + |h_k|^2 p_k \big/ \left(\sigma^2 + \sum_{\ell \neq k} |h_\ell|^2 p_\ell\right)\right)$, which is optimized with respect to $p_k \in [0, \overline{P}_k]$. The existence and uniqueness of the NE can be proved using Theorems 8.1 and 8.2 in [31] for concave games. The unique PA policy at the NE is trivial and consists in saturating the power constraint $p_k^{\text{NE}} = \overline{P}_k, \forall k \in \{1, \ldots, K\}$. The problem here is that by using all the available powers, the users create a high level of mutual interference, which results in a poor performance of the system in terms of achievable sum-rate (see Figure 8.1). One possibility to cope with multiuser interference is to consider a more sophisticated decoding technique such as SIC decoding. In [32], the authors considered a system composed of two mobile stations ($K = 2$) and a base station that broadcasts a coordination signal dictating the decoding order. To generate this signal, the base station flips an unfair coin (Bernoulli random variable of parameter $q \in [0, 1]$). The realization of the random signal indicates in which order the base station decodes the users with a perfect SIC. At each instant, the users are aware of the decoding order at the receiver and choose their strategies accordingly under an average power constraint. The existence and uniqueness of the NE were proven using the same theorems from [31]. It turns out that for any value of $q \in [0, 1]$, the performance of the system at the NE achieves the sum-capacity of the SISO MAC channel. Actually varying q from 0 to 1 allows one to move along the sum-rate segment of the Gaussian MAC capacity region illustrated in Figure 8.1. In conclusion, the Pareto-optimal points (i.e., points at which the achievable rate of one user cannot be increased without decreasing the achievable rate of another user) of the achievable rate region are achievable at the NE, and the base station can choose the operating point on the Pareto-optimal frontier by adjusting the parameter $q \in [0, 1]$ of the distribution of the coordination signal. Another way to improve the performance of the NE is to consider a power control pricing technique as proposed in [33]. The authors introduced a cost function defined as the difference between pricing and the payoff

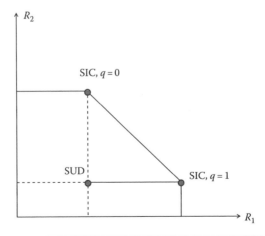

Figure 8.1 SISO static MAC achievable rate region. The NE of the PA game when SUD is used is suboptimal in terms of system sum-rate as opposed to SIC decoding, which achieves the sum-capacity of the MAC.

function $c_k(p_k, p_{-k}) = \lambda_k p_k - u_k(p_k, p_{-k})$. The intuition behind the modified game is that the system owner can manipulate the NE state by adjusting the price that each user is charged for the interference it creates in the system. Depending on the unit price $\lambda_k > 0$, transmitting with full power would not necessarily be the best strategy for user k and, thus, the overall interference level can be limited and the system performance improved. The values of the unit prices can be optimized in a centralized way by the base station or distributively via a market-based mechanism (usually considered directly proportional to the channel gain $|h_k|^2$).

Notice that, in both cases, the improvement of the network performance comes at a supplementary signaling cost (the coordination signal, the values of the prices have to be known by the users).

8.2.1.2 Multiple Output Multiple Input Links

In what the MIMO MAC static channel is concerned, the strategy of the user k consists of the input covariance matrices $\mathbf{Q}_k = \mathbb{E}\left[\underline{x}_k \underline{x}_k^H\right]$ in the set $\mathcal{A}_k = \{\mathbf{Q} \in \mathbb{C}^{n_{t,k} \times n_{t,k}} : \mathbf{Q} \succeq 0, \mathrm{Tr}(\mathbf{Q}) \leq \overline{P}_k\}$. The payoff of user k is its achievable rate:

$$u_k(\mathbf{Q}_k, \mathbf{Q}_{-k}) = \log_2 \left| \mathbf{I} + \rho \mathbf{H}_k \mathbf{Q}_k \mathbf{H}_k^H + \rho \sum_{\ell \neq i}^{K} \mathbf{H}_\ell \mathbf{Q}_\ell \mathbf{H}_\ell^H \right| - \log_2 \left| \mathbf{I} + \rho \sum_{\ell \neq k} \mathbf{H}_\ell \mathbf{Q}_\ell \mathbf{H}_\ell^H \right| \quad (8.3)$$

where $\rho = 1/\sigma^2$ and assuming SUD at the receiver. The existence of an NE can be proved by directly applying the results in [34]. As opposed to the SISO case, the uniqueness of the NE is no longer guaranteed in general. Sufficient ensuring the uniqueness of the NE have been given in [73]. The users converge to the unique NE by applying the *iterative algorithm based on the best response functions*:

$$\mathrm{BR}_k(\mathbf{Q}_{-k}) = \arg \max_{\mathbf{Q}_k \in \mathcal{Q}_k} u_k(\mathbf{Q}_k, \mathbf{Q}_{-k}) \quad (8.4)$$

The best response function of user k is given by the well-known water-filling solution [35] over the interference plus noise covariance matrix $\mathbf{\Sigma}_k = \mathbf{I} + \rho \sum_{\ell \neq k} \mathbf{H}_\ell \mathbf{Q}_\ell \mathbf{H}_\ell^H$ with the following spectral decomposition $\mathbf{\Sigma}_k = \mathbf{U}_{\mathbf{\Sigma}_k} \mathbf{\Lambda}_{\mathbf{\Sigma}_k} \mathbf{U}_{\mathbf{\Sigma}_k}$. Considering the spectral decomposition of the covariance matrix $\mathbf{Q}_k = \mathbf{U}_k \mathbf{P}_k \mathbf{U}_k^H$, the optimal eigenvector matrix $\mathbf{U}_k = \mathbf{U}_{\mathbf{\Sigma}_k}$ and the optimal eigenvalues $\mathbf{P}_k$ are the water-filling solutions over the eigenvalues of $\mathbf{\Sigma}_k$:

$$p_k(j) = \left[\frac{1}{\mu_k} - \frac{1}{\lambda_{\mathbf{\Sigma}_k}(j)} \right]^+ \quad (8.5)$$

where

$p_k(j)$ represents the diagonal elements of $\mathbf{P}_k$

$\lambda_{\mathbf{\Sigma}_k}(j)$ are the eigenvalues of $\mathbf{\Sigma}_k$

$\mu_k \geq 0$ is the water-filling level tuned such that the power constraint is satisfied $\sum_{j=1}^{n_{t,k}} p_k(j) = \overline{P}_k$

Using an iterative algorithm based on the best response functions, the users can converge to the unique NE point in a distributed manner. The algorithm is as follows:

■ Initialization: $\mathbf{Q}_k^{[1]}$, for all $k \in \mathcal{K}$

■ At every iteration step $t > 1$ the users update sequentially, in a round-Robin fashion, their covariance matrices, for all $k \in \mathcal{K}$:

$$\mathbf{Q}_k^{[t]} = \mathrm{BR}_k\left(\mathbf{Q}_1^{[t]}, \ldots, \mathbf{Q}_{k-1}^{[t]}, \mathbf{Q}_{k+1}^{[t-1]}, \ldots, \mathbf{Q}_K^{[t-1]}\right)$$

■ Repeat the previous step ($t = t + 1$) until convergence is reached

At every step of the algorithm, the players update sequentially their covariance matrices in a certain specified order (in our case, the user index represents the rank in the updating procedure). It turns out that the iterative algorithm based on best response functions coincides with the iterative water-filling algorithm studied in [35], which was proven to converge to the solution that maximizes the sum-capacity of the MIMO Gaussian MAC channel. However, because of the interference terms, the sum-capacity of the Gaussian MIMO MAC is not achieved at the NE, similarly to the SISO case.

If we consider the SIC decoding considering a similar coordination signal as described before, the existence and sufficient conditions ensuring the uniqueness of the NE are given in [73]. In order to determine the covariance matrices at the NE, we can again apply an iterative water-filling algorithm.

A fundamentally different approach was discussed in [6] for the MIMO MAC system. Assuming the compound channel where transmitter $k \in \mathcal{K}$ knows not the exact value of the channel matrix $\mathbf{H}_k$ but the fact that it takes one of the values in the set $\mathcal{H}_k$, the communication is modeled as a game against nature. Here, the set of players do not compete with each other, rather they are assumed to operate in a completely cooperative manner and to play jointly against a malicious nature that controls the channel matrices: $\mathbf{H}_k \in \mathcal{H}_k, \forall k \in \mathcal{H}_k$. The worst case capacity region is given by

$$C(\mathcal{H}) = \cup_{\mathbf{Q}_k \in \mathcal{A}_k} \cap_{\mathbf{H}_k \in \mathcal{H}_k} \mathcal{R}(\{\mathbf{Q}_k\}, \{H_k\}) \tag{8.6}$$

where $\mathcal{R}(\{\mathbf{Q}_k\}, \{H_k\}) = \left\{(R_1, \ldots, R_K) : 0 \leq \sum_{k \in \mathcal{S}} R_k \leq \log\left|\mathbf{I} + \sum_{k \in \mathcal{S}} \mathbf{H}_k \mathbf{Q}_k \mathbf{H}_k^H\right|, \forall \mathcal{S} \subseteq \mathcal{K}\right\}$. The capacity of the compound MIMO MAC (i.e., the transmitters know the finite discrete set of the possible values that the channel can take but don't know the actual channel state) was proven to be achieved when each user spreads its power uniformly over all the transmit antennas $\mathbf{Q}_k^* = (\overline{P}_k/n_{t,k})\mathbf{I}$ (see [6] for more details).

In the remaining of this section, we will focus on the case of the fast fading MAC channel. Without loss of generality and for clarity sake, we will focus on the two-user case ($K = 2$) which concentrates the concepts and ideas that remain valid for the general case of arbitrary number of users.

8.2.2 Fast Fading Single Input Single Output Links

We begin with the case of the fast fading SISO channel assuming both CSIT and CSIR, which was analyzed in [36]. If the receiver applies SUD, the payoff function of user $i \in \{1, 2\}$, its achievable Shannon rate is written as

$$u_k(p_k, p_{-k}) = \int_0^{+\infty} \int_0^{+\infty} \frac{1}{2} \log_2\left(1 + \frac{p_k(g_k, g_{-k})g_k}{p_{-k}(g_k, g_{-k})g_{-k} + \sigma^2}\right) f(g_1, g_2) \mathrm{d}g_k \mathrm{d}g_{-k} \tag{8.7}$$

where $g_k = |h_k|^2$ corresponds to the fading coefficients, which are exponentially distributed random variables. The NE PA policies, $p_k^{(\mathrm{NE})}(g_k, g_{-k})$, for all $k \in \{1, 2\}$, depend on the channel realizations and take values in the set:

$$\mathcal{A}_k = \left\{ p_k : [0, +\infty) \to [0, +\infty) \left| \int\limits_{0}^{+\infty} \int\limits_{0}^{+\infty} p_k(g_1, g_2) f(g_1, g_2) \mathrm{d}g_1 \mathrm{d}g_2 \leq \overline{P}_k \right. \right\}$$

The authors prove that the NE strategies are given by the simultaneous water-filling solutions, $\forall k \in \mathcal{K}$:

$$p_k(g_k, g_{-k}) = \left[\mu_k - \frac{\sigma^2}{g_k} - \frac{p_{-k}(g_k, g_{-k})g_{-k}}{g_k} \right]^+ \tag{8.8}$$

Using the fact that the fading coefficients have a continuous distribution, the authors proved the existence of only time-sharing Nash equilibria. The result is given in the following theorem.

THEOREM 8.1   The power allocation game described by the set of players $k \in \{1, 2\}$, the set of actions $\mathcal{A}_k$ and the payoff functions in (8.7) has at least an NE state. Furthermore the NE power allocation policies are given by the following water-filling type solutions:

$$p_k^{\mathrm{NE}}(g_k, g_{-k}) = \left| \begin{array}{ll} \left[\tilde{\mu}_k - \frac{\sigma^2}{g_k} \right]^+, & \text{if } \tilde{\mu}_k g_k \geq \tilde{\mu}_{-k} g_{-k} \\ 0, & \text{otherwise} \end{array} \right. \tag{8.9}$$

where the water-filling levels $(\tilde{\mu}_1, \tilde{\mu}_2)$ are the solution to the following system of equations (in μ_1, μ_2):

$$\begin{cases} \displaystyle\iint\limits_{\mu_1 g_1 \geq \mu_2 g_2} \left(\mu_1 - \frac{\sigma^2}{g_1} \right)^+ f(g_1, g_2) \mathrm{d}g_1 \mathrm{d}g_2 = \overline{P}_1 \\[2em] \displaystyle\iint\limits_{\mu_1 g_1 \leq \mu_2 g_2} \left(\mu_2 - \frac{\sigma^2}{g_2} \right)^+ f(g_1, g_2) \mathrm{d}g_1 \mathrm{d}g_2 = \overline{P}_2 \end{cases} \tag{8.10}$$

This theorem states that at every instant, only the user with the strongest channel will transmit (e.g., user k transmits whenever $\tilde{\mu}_k g_k \geq \tilde{\mu}_{-k} g_{-k}$, the rest of the time it remains silent) while the other one does not transmit anything. Knowing that the centralized policy corresponding to the sum-capacity of the SISO fading MAC channel is given by the same PA policy as (8.9) [37], the system sum-rate at the NE achieves the sum-capacity point of the achievable rate region, which is illustrated in Figure 8.2. We observe that, as opposed to the static case, the sum-capacity is achieved at the NE point when SUD is assumed at the receiver. In order to achieve the two corner points of the achievable rate region, the authors introduced the Stackelberg formulation, where the base station is the leader of the game and chooses its strategy to maximize a certain performance criteria. Here the receiver applies an SIC decoding such that the decoding order depends on the fading coefficients (g_1, g_2). The base station divides the whole possible space of the fading coefficients into two subsets: $\mathcal{D}_1$ and $\mathcal{D}_1^c$ such that if $(g_1, g_2) \in \mathcal{D}_1$ then user 1 is decoded first, user 2 is decoded second and sees no interference (conversely if $(g_1, g_2) \in \mathcal{D}_1^c$). The leader chooses $\mathcal{D}_1$ to maximize the weighted system sum-rate: $w_1 u_1(\mathcal{D}_1, p_1, p_2) + w_2 u_2(\mathcal{D}_1, p_1, p_2)$ where (w_1, w_2) are

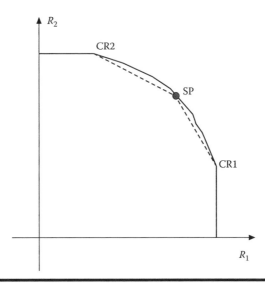

Figure 8.2 SISO fast fading MAC achievable rate region. As opposed to the static case, SUD achieves the sum-capacity point SR. The points of the segments CR1–SP and CR2–SP are achieved with SIC decoding. The other Pareto-optimal points are achieved using a repeated game formulation. (From Lai, L. and El Gamal, H., *IEEE Trans. Inf. Theory*, 54, 5, 2110, May 2008.)

positive constants representing the payment that users owe per unit rate. The authors proved the existence and uniqueness of admissible Nash equilibria and also the fact that the corner points of the achievable rate region can be achieved using either $\mathcal{D}_1 = \emptyset$ or $\mathcal{D}_1^c = \emptyset$. It turns out that, no other Pareto-optimal points are achieved using SIC but only suboptimal ones. In order to achieve these Pareto-optimal points (see Figure 8.2), the authors proposed a repeated game formulation. The players interact with each other many times which enables their cooperation and results in the improvement of their payoffs. They start at a cooperating point and if one player deviates, the other will punish it by choosing their strategies to induce a lower payoff to the deviator. Under these circumstances, no player has any incentive in deviating from the cooperative phase.

8.2.3 Fast Fading Multiple Input Multiple Output Channels

In this section, we study the more general fast fading MIMO MAC case, analyzed in [34]. The authors considered both SUD and SIC decoding with a coordination signal which is perfectly known to all the terminals. In order to take into account the antenna correlation effects at the transmitters and receiver, we will assume the different channel matrices to be structured according to the unitary-independent-unitary model introduced in [38]:

$$\forall k \in \{1, ..., K\}, \quad \mathbf{H}_k = \mathbf{V}\tilde{\mathbf{H}}_k\mathbf{W}_k \qquad (8.11)$$

where
   $\mathbf{V}$ and $\mathbf{W}_k$ are deterministic unitary matrices
   $\tilde{\mathbf{H}}_k$ is an $n_r \times n_t$ matrix (all the transmitters are assumed to have the same number of transmit antennas $n_{t,k} = n_t$) whose entries are zero-mean independent complex Gaussian random variables with an arbitrary profile of variances, such that $\mathbb{E}|\tilde{H}_k(i,j)|^2 = \sigma_k^2(i,j)/n_t$

The Kronecker propagation model with common receive correlation and for which the channel transfer matrices factorizes as $\mathbf{H}_k = \mathbf{R}^{1/2}\tilde{\mathbf{\Theta}}_k\mathbf{T}_k^{1/2}$ is a special case of the UIU model where the profile of variances is separable, that is, $\sigma_k^2(i,j) = d_k^{(T)}(j)d^{(R)}(i)/n_t$, $\mathbf{V} = \mathbf{U}_\mathbf{R}$ and $\mathbf{W}_k = \mathbf{U}_{\mathbf{T}_k}$ with $\mathbf{\Theta}_k$ being a random matrix with zero-mean i.i.d. entries, $\mathbf{T}_k$ being the transmit antenna correlation matrix, and $\mathbf{R}$ being the common receive antenna correlation matrix, such that $d_k^{(T)}(\ell)$ and $d^{(R)}(c)$ are their associated eigenvalues and $\mathbf{U}_{\mathbf{T}_k}$ and $\mathbf{U}_\mathbf{R}$ are their associated eigenvector matrices.

8.2.3.1 Single User Decoding

When the SUD is assumed at the receiver, each user has to choose the best precoding matrix $\mathbf{Q}_k = \mathbb{E}\left[x_k x_k^H\right]$, in the sense of his payoff function:

$$u_k^{\text{SUD}}(\mathbf{Q}_1, \mathbf{Q}_2) = \mathbb{E}\log\left|\mathbf{I} + \rho\mathbf{H}_1\mathbf{Q}_1\mathbf{H}_1^H + \rho\mathbf{H}_2\mathbf{Q}_2\mathbf{H}_2^H\right| - \mathbb{E}\log\left|\mathbf{I} + \rho\mathbf{H}_{-k}\mathbf{Q}_{-k}\mathbf{H}_{-k}^H\right| \tag{8.12}$$

The strategy set of user k becomes

$$\mathcal{A}_k^{\text{SUD}} = \left\{\mathbf{Q}_k \succeq 0, \mathbf{Q}_k = \mathbf{Q}_k^H, \text{Tr}(\mathbf{Q}_k) \leq n_t\overline{P}_k\right\} \tag{8.13}$$

where the average total power constraint is proportional to the number of transmit antennas. It turns out that the existence and uniqueness of the NE are guaranteed.

THEOREM 8.2   The space power allocation game described by the set of players $k \in \{1,2\}$; the sets of actions $\mathcal{A}_k^{\text{SUD}}$ and the payoff functions $u_k^{\text{SUD}}(\mathbf{Q}_k, \mathbf{Q}_{-k})$ given in (8.12), has a unique Nash equilibrium.

The existence of an NE can be proven using the results in [31]. For the uniqueness issue, the authors extended the second theorem in [31] to the case where the strategy sets are matrix spaces and obtained the desired result. In order to find the optimum covariance matrices, the authors proceeded in the same way as described in [39]. First, they determined the optimum eigenvectors and then they determined the optimum eigenvalues by approximating the payoff functions under the large system assumption. It turns out that there is no loss of optimality by choosing the covariance matrices $\mathbf{Q}_k = \mathbf{W}_k\mathbf{P}_k\mathbf{W}_k^H$, where $\mathbf{W}_k$ is the same unitary matrix as in (8.11) and $\mathbf{P}_k$ is the diagonal matrix containing the eigenvalues of $\mathbf{Q}_k$. Although this result is easy to obtain, it is instrumental in our context for two reasons. First, the search of the optimum precoding matrices boils down to the search of the eigenvalues of these matrices. Second, as the optimum eigenvectors are known, available results in random matrix theory can be applied to find an accurate approximation of these eigenvalues. Indeed, the eigenvalues are not easy to find in the finite setting. They might be found through extensive numerical techniques. Here, our approach consists in approximating the payoffs in order to obtain expressions not only easier to interpret but also easier to be optimized with respect to the eigenvalues of the precoding matrices. The key idea is to approximate the different transmission rates by their large system equivalent in the regime of large numbers of antennas ($n_r \to \infty$, $n_t \to \infty$, $n_r/n_t \to \beta$) and exploit results provided in [38] for single-user MIMO channels. The corresponding approximates can be found to be accurate even for relatively small numbers of antennas (see, e.g., [40,41] for more details). The NE PA policies will be given by water-filling type solutions [34].

As for the efficiency of the NE point, the SUD decoding technique is suboptimal in the centralized case (SUD does not allow the network to operate at an arbitrary point of the centralized MAC capacity region) and the sum-capacity is not reached at the NE similarly to the static MIMO MAC channel.

8.2.3.2 Successive Interference Cancelation

When SIC is assumed at the receiver, there is a coordination signal that dictates the decoding order and that has to be known by all terminals in the network. The coordination signal is represented by a random variable denoted by $S \in \mathcal{S}$. Since we study the 2-user MAC, $\mathcal{S} = \{1, 2\}$ is a binary alphabet. When the realization is $S = 1$ (resp. $S = 2$), user 1 (resp. user 2) is decoded in the second place and therefore sees no multiple access interference. The strategy of user $k \in \{1, 2\}$ consists in choosing the best pair of precoding matrices $\mathbf{Q}_k = \left(\mathbf{Q}_k^{(1)}, \mathbf{Q}_k^{(2)} \right)$ where $\mathbf{Q}_k^{(s)} = \mathbb{E}\left[\underline{x}_k^{(s)} \underline{x}_k^{(s),H} \right]$, for $s \in \{1, 2\}$, in the sense of his payoff function:

$$u_k^{\text{SIC}} \left(\mathbf{Q}_1^{(1)}, \mathbf{Q}_1^{(2)}, \mathbf{Q}_2^{(1)}, \mathbf{Q}_2^{(2)} \right) = p R_k^{(1)} \left(\mathbf{Q}_1^{(1)}, \mathbf{Q}_2^{(1)} \right) + (1-p) R_k^{(2)} \left(\mathbf{Q}_1^{(2)}, \mathbf{Q}_2^{(2)} \right) \qquad (8.14)$$

where

$$R_1^{(1)} \left(\mathbf{Q}_1^{(1)}, \mathbf{Q}_2^{(1)} \right) = \mathbb{E} \log \left| \mathbf{I} + \rho \mathbf{H}_1 \mathbf{Q}_1^{(1)} \mathbf{H}_1^H \right|$$

$$R_2^{(1)} \left(\mathbf{Q}_1^{(1)}, \mathbf{Q}_2^{(1)} \right) = \mathbb{E} \log \left| \mathbf{I} + \rho \mathbf{H}_1 \mathbf{Q}_1^{(1)} \mathbf{H}_1^H + \rho \mathbf{H}_2 \mathbf{Q}_2^{(1)} \mathbf{H}_2^H \right| - \mathbb{E} \log \left| \mathbf{I} + \rho \mathbf{H}_1 \mathbf{Q}_1^{(1)} \mathbf{H}_1^H \right|$$

$$R_1^{(2)} \left(\mathbf{Q}_1^{(2)}, \mathbf{Q}_2^{(2)} \right) = \mathbb{E} \log \left| \mathbf{I} + \rho \mathbf{H}_1 \mathbf{Q}_1^{(2)} \mathbf{H}_1^H + \rho \mathbf{H}_2 \mathbf{Q}_2^{(2)} \mathbf{H}_2^H \right| - \mathbb{E} \log \left| \mathbf{I} + \rho \mathbf{H}_2 \mathbf{Q}_2^{(2)} \mathbf{H}_2^H \right|$$

$$R_2^{(2)} \left(\mathbf{Q}_1^{(2)}, \mathbf{Q}_2^{(2)} \right) = \mathbb{E} \log \left| \mathbf{I} + \rho \mathbf{H}_2 \mathbf{Q}_2^{(2)} \mathbf{H}_2^H \right| \qquad (8.15)$$

with $\rho = 1/\sigma^2$. The main point to mention here is the power constraint under which the payoffs are maximized: $p \text{Tr} \left(\mathbf{Q}_k^{(1)} \right) + \bar{p} \text{Tr} \left(\mathbf{Q}_k^{(2)} \right) \leq n_t \bar{P}_k$. Thus, the strategy set of user k is

$$\mathcal{A}_k^{\text{SIC}} = \left\{ \mathbf{Q}_k = \left(\mathbf{Q}_k^{(1)}, \mathbf{Q}_k^{(2)} \right) \middle| \mathbf{Q}_k^{(1)} \succeq 0, \mathbf{Q}_k^{(1)} = \mathbf{Q}_k^{(1)H}, \mathbf{Q}_k^{(2)} \succeq 0, \mathbf{Q}_k^{(2)} = \mathbf{Q}_k^{(2)H} \right.$$
$$\left. p \text{Tr} \left(\mathbf{Q}_k^{(1)} \right) + \bar{p} \text{Tr} \left(\mathbf{Q}_k^{(2)} \right) \leq n_t \bar{P}_k \right\} \qquad (8.16)$$

The existence of the NE is similar to the previous sub-section. The uniqueness of the NE is more involving and requires the proof of a nontrivial matrix trace inequality [74].

THEOREM 8.3   The joint space-time power allocation game described by the set of players $k \in \{1, 2\}$; the sets of actions $\mathcal{A}_k^{\text{SIC}}$ and the payoff functions $u_k^{\text{SIC}}(\mathbf{Q}_k, \mathbf{Q}_{-k})$ given in Equation 8.14, has a unique Nash equilibrium.

Similarly, to the previous sub-section, there is no loss of optimality when choosing the structure of the covariance matrices $\mathbf{Q}_k^{(1)} = \mathbf{W}_k \mathbf{P}_k^{(1)} \mathbf{W}_k^H$ and $\mathbf{Q}_k^{(2)} = \mathbf{W}_k \mathbf{P}_k^{(2)} \mathbf{W}_k^H$ where the matrices

$\mathbf{W}_k$ are given in (8.11). The eigenvalues of the covariance matrices can be obtained assuming the asymptotic regime in terms of the number of antennas: $n_r \to \infty$, $n_t \to \infty$, and $n_r/n_t \to \beta$. The new approximated payoffs are as follows:

$$\tilde{R}_1^{(1)}\left(\mathbf{P}_1^{(1)}\right) = \frac{1}{n_r}\sum_{j=1}^{n_t}\log_2\left(1 + \rho P_1^{(1)}(j)\phi(j)\right) + \frac{1}{n_r}\sum_{i=1}^{n_r}\log_2\left(1 + \frac{1}{n_t}\sum_{j=1}^{n_t}\sigma_1(i,j)\psi(j)\right)$$

$$-\frac{1}{n_r}\sum_{j=1}^{n_r}\phi(j)\psi(j)\log_2 e$$

$$\tilde{R}_1^{(2)}\left(\mathbf{P}_1^{(2)},\mathbf{P}_2^{(2)}\right) = \frac{1}{n_r}\sum_{j=1}^{n_t}\left[\log_2\left(1 + 2\rho P_1^{(2)}(j)\gamma_1(j)\right) + \log_2\left(1 + 2\rho P_2^{(2)}(j)\gamma_2(j)\right)\right]$$

$$+\frac{1}{n_r}\log_2\left(1 + \frac{1}{2n_t}\sum_{j=1}^{n_t}(\sigma_1(i,j)\delta_1(j) + \sigma_2(i,j)\delta_2(j))\right)$$

$$-\frac{1}{n_r}\sum_{j=1}^{n_t}(\gamma_1(j)\delta_1(j) + \gamma_2(j)\delta_2(j))\log_2 e - \tilde{R}_2^{(2)} \tag{8.17}$$

where the parameters $\phi(j)$ and $\psi(j)$, $\forall j \in \{1,\ldots,n_t\}$ represent the unique solution to the system:

$$\begin{cases} \phi(j) = \dfrac{1}{n_t}\sum_{i=1}^{n_r}\dfrac{\sigma_1(i,j)}{1 + \frac{1}{n_t}\sum_{m=1}^{n_t}\sigma_1(i,m)\psi(m)} \\[4mm] \psi(j) = \dfrac{\rho P_1^{(1)}(j)}{1 + \rho P_1^{(1)}(j)\phi(j)} \end{cases} \tag{8.18}$$

and $\gamma_k(j)$ and $\delta_k(j)$ $\forall j \in \{1,\ldots,n_t\}$, $k \in \{1,2\}$ represent the unique solution to the system

$$\begin{cases} \gamma_k(j) = \dfrac{1}{2n_t}\sum_{i=1}^{n_r}\dfrac{\sigma_k(i,j)}{1 + \frac{1}{2n_t}\sum_{\ell=1}^{2}\sum_{m=1}^{n_t}\sigma_\ell(i,m)\delta_\ell(m)} \\[4mm] \delta_k(j) = \dfrac{2\rho P_k^{(2)}(j)}{1 + 2\rho P_k^{(2)}(j)\gamma_k(j)} \end{cases} \tag{8.19}$$

Also, the approximated functions $\tilde{R}_2^{(1)}(\cdot,\cdot)$ and $\tilde{R}_2^{(2)}(\cdot)$ can be obtained in a similar way and the approximated payoff of user $k \in \{1,2\}$ follows since

$$\tilde{u}_k^{\text{SIC}}(\mathbf{P}_1,\mathbf{P}_2) = p\tilde{R}_k^{(1)}(\mathbf{P}_1,\mathbf{P}_2) + (1-p)\tilde{R}_k^{(2)}(\mathbf{P}_1,\mathbf{P}_2) \tag{8.20}$$

where $\mathbf{P}_k \triangleq \left(\mathbf{P}_k^{(1)},\mathbf{P}_k^{(2)}\right)$.

Then, optimizing the approximated payoff $\tilde{u}_1(\cdot, \cdot)$ with respect to $P_1^{(1)}(j)$ and $P_1^{(2)}(j)$ leads to the following water-filling equations:

$$
P_1^{(1),\mathrm{NE}}(j) = \left[\frac{1}{\ln 2\lambda_1} - \frac{1}{\rho\phi(j)} \right]^+
$$

$$
P_1^{(2),\mathrm{NE}}(j) = \left[\frac{1}{\ln 2\lambda_1} - \frac{1}{2\rho\gamma_1(j)} \right]^+
\tag{8.21}
$$

where $\lambda_1 \geq 0$ is the Lagrangian multiplier tuned in order to meet the power constraint: $p\sum_{j=1}^{n_t} P_1^{(1),\mathrm{NE}}(j) + (1-p)\sum_{j=1}^{n_t} P_1^{(2),\mathrm{NE}}(j) = n_t\overline{P}_1$. Note that the same Lagrange multiplier (namely, λ_1) is used to adjust both the spatial and temporal PAs. Similar comments can be made when optimizing $\tilde{u}_2(\cdot, \cdot)$ with respect to $P_2^{(1)}(j)$ and $P_2^{(2)}(j)$. To solve the system of equations given above, we can use the same iterative PA algorithm as the one described in [39].

Notice that the existence and uniqueness of the NE issues have been analyzed in the finite setting (exact game) whereas the determination of the NE is performed in the asymptotic regime (approximated game). However, it is well-known that the large system approximates of ergodic transmission rates have the same properties as their exact counterparts (see, e.g., [42]) and therefore the optimality loss can be considered negligible.

Figure 8.3 illustrates the sum-rate at the NE assuming the Kronecker model for the channel matrices with exponential receive and transmit covariance matrices and also assuming a spatial PA

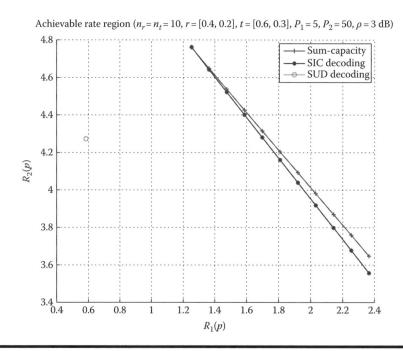

Figure 8.3 MIMO fast fading MAC. The achievable rate region at the NE versus the distribution of the coordination signal $p \in [0, 1]$ for $n_r = n_t = 10$, $r = [0.4, 0.2]$, $t = [0.6, 0.3]$, $\rho = 3\,\mathrm{dB}$, $P_1 = 5$, and $P_2 = 50$. Varying p allows moving along a segment close to the sum-capacity, similar to SISO MAC.

policy only when SIC decoding is used (the users choose an uniform PA policy over the time). We observe that the SIC decoding outperforms SUD. Furthermore, the gap between the achievable sum-rate at the NE with SIC and the sum-capacity upper bound is relatively small.

What is interesting is the fact that, when either the low or high signal to noise regime is assumed, the sum-capacity of the fast fading MAC is proved to be achieved at the NE regardless of the distribution of the coordination signal and in spite of being a suboptimal coordination mechanism (independent of the channel realizations).

We have seen that for the MAC system using a distributed PA policy, there are several spacial cases in which the centralized sum-capacity can be achieved. At the NE, water-filling type solutions are obtained for the MIMO cases. The SUD decoding technique is suboptimal in general (except for the SISO fading channel). By using a more complex decoding technique such as SIC, the performance of the system is improved and close or identical to the centralized case, at the cost of supplementary signaling cost (i.e., the decoding order has to be known by all users, and also global information is needed about the channel state or distribution of all users plus a complexity cost of the decoder). The next section will be dedicated to another basic multiuser channel model, the IC.

8.3 Power Allocation Games in Interference Channels

In this section, we consider the IC [26] composed of several uncoordinated transmitter–receiver pairs ($K \geq 2$) that share common resources and thus create mutual interference. In this case, the players of the game are again the transmitters that choose their own PA policies to maximize their Shannon achievable rates.

There are several fundamental differences between the IC and MAC. The greatest difficulty is the fact that the Shannon capacity region of the IC is unknown. Even in the two-user SISO static Gaussian case, only achievable rate regions are known. This is the reason why only the static channel will be considered here with perfect channel state information at the receiver (global channel state information, that is, information of all the links in the system) and transmitter (local or individual channel state information, that is, each transmitter knows only its own channel gain) nodes (CSIT,CSIR). Another common assumption, which is rather realistic in the proposed game framework of uncoordinated and noncooperative links, is not to allow multiuser coding/decoding schemes or SIC decoding scheme. Thus, in what follows, SUD will be implicitly assumed at the receiver nodes unless otherwise specified.

The received signal at receiver $k \in \mathcal{K}$ for the general MIMO case can be written as

$$\underline{y}_k = \mathbf{H}_{kk}\underline{x}_k + \sum_{\ell \neq k}\mathbf{H}_{k,\ell}\underline{x}_\ell + \underline{z}_k \tag{8.22}$$

where

  $\underline{x}_k$ is the $n_{t,k}$-dimensional column vector of symbols transmitted by user k
  $\mathbf{H}_{k,\ell} \in \mathbb{C}^{n_{r,k} \times n_{t,k}}$ is the channel matrix for the link between transmitter k and receiver ℓ user k
  $\underline{z}_k$ is an $n_{r,k}$-dimensional complex white Gaussian noise distributed as $\mathcal{N}(\underline{0}, \sigma^2 \mathbf{I}_{n_{r,k}})$, for all
    $(k, \ell) \in \mathcal{K}^2$

Let us consider the simple case of SISO IC. The payoff of user k writes as $u_k(p_k, p_{-k}) = \log_2\left(1 + |h_{kk}|^2 p_k / \left(\sigma_k^2 + \sum_{\ell \neq k} |h_{\ell k}|^2 p_\ell\right)\right)$. The NE solution is trivial, similarly to the SISO MAC with SUD, and consists in transmitting with full power $p_k^{\text{NE}} = \overline{P}_k$. We will focus on more challenging cases where the NE is no longer trivial and also review different ways to improve the efficiency of the NE.

8.3.1 Static Frequency Selective SISO Interference Channels

In [43], the authors studied the PA problem in a multiuser digital subscriber line system modeled as a frequency selective IC. The users allocate their powers over the available frequency spectrum under an average power constraint by tuning their power spectral densities $p_k(f) \in \mathcal{A}_k$, for all $k \in \mathcal{K}$:

$$\mathcal{A}_k = \left\{ p_k : [0, F_s] \rightarrow [0, +\infty) \,\middle|\, \int_0^{F_s} p_k(f) df \leq \bar{P}_k \right\}$$

where F_s is the sampling frequency. The payoffs for the players are their achievable rates:

$$u_k(p_k, p_{-k}) = \int_0^{F_s} \log_2 \left(1 + \frac{p_k(f) |h_{kk}(f)|^2}{\sigma_k(f)^2 + \sum_{\ell \neq k} |h_{\ell k}(f)|^2 p_\ell(f)} \right) df$$

The solution of the PA game is the simultaneous water-filling solution, for all $k \in \mathcal{K}$:

$$p_k(f) = \left[\frac{1}{\mu_k} - \frac{\sigma_k(f)^2 + \sum_{\ell \neq k} |h_{\ell k}|^2 p_\ell(f)}{|h_{kk}(f)|^2} \right]^+ \tag{8.23}$$

where the parameter $\mu_k \geq 0$ is the water level tuned such that the power constraint is saturated. Sufficient conditions that ensure the existence of such a solution and its uniqueness were provided. The authors proved the following result for the game under investigation assuming that there are only two users in the system ($K = 2$):

THEOREM 8.4   If $\frac{|h_{12}(f) h_{21}(f)|^2}{|h_{11}(f) h_{22}(f)|^2} \leq 1$ for all $f \in [0, F_s]$, then the existence of an NE is guaranteed. Furthermore, if $\varepsilon_0 < 1$, $\varepsilon_1 + \varepsilon_2 < \frac{1}{2}$ or $\varepsilon_1 + \varepsilon_3 < \frac{1}{2}$, where $\varepsilon_0 = \sup \left\{ \frac{|h_{21}(f)|^2}{|h_{11}(f)|^2} \right\} \sup \left\{ \frac{|h_{12}(f)|^2}{|h_{22}(f)|^2} \right\}$,

$\varepsilon_1 = \sup \left\{ \frac{|h_{21}(f) h_{12}(f)|^2}{|h_{11}(f) h_{22}(f)|^2} \right\}$, $\varepsilon_2 = \sup \left\{ \frac{|h_{21}(f)|^2}{|h_{11}(f)|^2} \right\} \frac{1}{F_s} \int_0^{F_s} \frac{|h_{12}(f)|^2}{|h_{22}(f)|^2} df$ and $\varepsilon_3 = \sup \left\{ \frac{|h_{12}(f)|^2}{|h_{22}(f)|^2} \right\} \frac{1}{F_s}$
$\int_0^{F_s} \frac{|h_{21}(f)|^2}{|h_{11}(f)|^2} df$, then the NE is unique.

A distributed iterative water-filling algorithm is shown to converge to the unique NE (under the sufficient conditions mentioned in Theorem 8.4) from any initial point. At every iteration, the users will successively update their own solution as follows. Each user, knowing the PA of the other users at the previous iteration, will apply the water-filling solution considering the other users' crosstalk as noise. The process is repeated until convergence is reached. Each user only needs to have the knowledge of its own channel transfer function (local information) and noise profile.

However, the NE point may be very inefficient and for this purpose several ways to improve the efficiency of the solution have been proposed. In [44], the authors study the flat fading frequency selective IC ($h_{k\ell}(f) = h_{k\ell} \forall f \in [0, F_s]$), assuming that the transmitter–receiver pairs coexist for a long period of time. In this situation, a repeated game is more appropriate to model the interaction among users. First, the authors of [44] studied the achievable rate region:

$$\mathcal{R} = \left\{ (R_1, \ldots, R_K) : R_k = \int_0^{F_S} \log_2 \left(1 + \frac{|h_{kk}|^2 p_k(f)}{\sigma_k^2 \sum_{k \neq \ell} |h_{\ell k}|^2 p_\ell(f)} \right) df \right.$$

$$\left. \int_0^{F_S} p_k(f) df \leq \overline{P}_k, p_k(f) \geq 0, \forall k \in \mathcal{K} \right\} \tag{8.24}$$

and showed that any point in the achievable rate region in (8.24) can be obtained with piece-wise constant PA. Also, in the high SNR regime, $|h_{k\ell}h_{\ell k}|^2 \leq |h_{kk}h_{\ell\ell}|^2$, the Pareto-efficient rates are achieved with orthogonal PAs such that the mutual interference is completely canceled: $p_k^*(f)p_\ell^*(f) = 0, \forall f \in [0, F_S]$, which is in accordance with the results in [45]. Second, the authors proved that the uniform power spreading over the spectrum is an NE in the noncooperative one shot game and also provide the sufficient condition that ensures the uniqueness of this solution $\sum_{\ell \neq k} |h_{\ell k}|^2 / |h_{kk}| < 1$. An infinite horizon repeated game is proposed that enables cooperation, and thus, the users can improve the performance of the NE. In the repeated games, the stage game or one-shot game is repeated over and over. At the end of each stage, the strategies and payoffs of the users are publicly revealed, such that the action of a certain user at a stage will be functions of the past observed history. The payoff of user k is defined by

$$U_k = (1 - \delta) \sum_{t=0}^{+\infty} \delta^t u_k(t) \tag{8.25}$$

where
  $u_k(t)$ is the payoff of user k in the one-shot game at time t
  $\delta \in (0, 1)$ is the discount factor that characterizes the player's degree of impatience (a payoff equal to one at stage t is equivalent to a payoff equal to δ at stage $t - 1$)

It turns out that any set of rates $(R_1, \ldots, R_M) \in \mathcal{R}$ that Pareto dominates the NE achievable rates, such that $R_k > R_k^{NE}, \forall k \in \mathcal{K}$ is supported or can be achieved by a sub-game perfect NE if δ is close to 1. The idea is that the users start the game at a cooperating point and keep on cooperating until there is one deviator. A punishment is then applied such that all the users play the NE strategy. If the players value sufficiently future payoffs compared to present ones (δ close to 1) and if past actions are observable then any state of the system that Pareto dominates the NE is a sustainable outcome. In order to achieve the cooperation state, the users are assumed to have global channel state information. In practice, these parameters have to be measured and exchanged by the users. In this context, the users may have an incentive to alter the parameter process to improve his payoff. Several methods to detect and punish such deviations are also discussed in [44].

A different way to improve the performance of the NE is to consider the Nash bargaining solution. This possibility has been exploited in [46] for the flat fading channel. The structure of a Nash bargaining problem contains a compact and convex set of outcomes, for example, an achievable rate region, $\mathcal{R}^{NB}$, and a disagreement outcome, that is, the achievable rates at the NE and $\underline{R}^{NE}$ (the problem is solved competitively if a disagreement arises). The Nash bargaining solution is a function $F : \mathcal{R}^{NB} \cup \{\underline{R}^{NE}\} \to \mathcal{R}^{NB} \cup \{\underline{R}^{NE}\}$ that satisfies four axioms introduced by Nash in [47]: (i) linearity, (ii) independence of irrelevant alternatives, (iii) symmetry, and (iv) Pareto optimality. The solution is obtained by maximizing this function given by:

$$F(\underline{\tau}) = \prod_{k=1}^{K} \left(R_k^{\mathrm{NB}}(\tau_k) - R_k^{\mathrm{NE}} \right) \tag{8.26}$$

Similarly to [44], the players at the cooperation state use their powers on orthogonal subspaces of the available spectrum such that the interference is canceled out. The users cooperate in the sense that they agree on a partition of the available spectrum, $\underline{\tau} = (\tau_1, \ldots, \tau_K)$, such that τ_k represents the fraction of the spectrum that is allocated to user k. The achievable rates are in this case: $R_k^{\mathrm{NB}}(\underline{\tau}) = \tau_k \frac{F_S}{2} \log_2 \left(1 + |h_{kk}|^2 \overline{P}_k / (\tau_k \frac{F_S}{2} \sigma_0^2) \right)$. The disagreement state is the NE of the SISO IC case, where the users spread their powers over the entire spectrum: $R_k^{\mathrm{NE}} = \frac{F_S}{2} \log_2 \left(1 + |h_{kk}|^2 \overline{P}_i / \left(\frac{F_S}{2} \sigma_0^2 + \sum_{\ell \neq k} |h_{\ell k}|^2 \overline{P}_\ell \right) \right)$. The existence of the bargaining solution is ensured if there is a partition of the band, $\underline{\tau}^* = (\tau_1^*, \ldots, \tau_K^*)$, such that each player gets more by cooperating than competing ($R_k^{\mathrm{NB}}(\underline{\tau}^*) \geq R_k^{\mathrm{NE}}, \forall k \in \mathcal{K}$, the cooperation point Pareto dominates the NE point). The authors also formulated the Nash bargaining solution as a convex optimization problem that can be easily solved.

8.3.2 Static Parallel Interference Channels

In this paragraph, we briefly review the parallel IC. In particular, this model allows one to study the noncooperative power allocation games in the OFDM systems. When the discrete multitone is assumed, the frequency selective channel can be modeled by a set of Q parallel flat fading Gaussian channels. The user payoff simply becomes

$$u_k(\underline{p}_k, \underline{p}_{-k}) = \sum_{q=1}^{Q} \log_2 \left(1 + \frac{\left| h_{kk}^{(q)} \right|^2 p_k^{(q)}}{(\sigma^{(q)})^2 + \sum_{\ell \neq k} \left| h_{\ell k}^{(q)} \right|^2 p_\ell^{(q)}} \right) \tag{8.27}$$

where the strategy of user k consists in the vector of powers it allocates to the available frequencies such that a total power constraint is satisfied $\sum_{q=1}^{Q} p_k^{(q)} \leq \overline{P}_k$. The existence of the NE in this case is always ensured [49]. The conditions that guarantee the uniqueness of the NE and convergence of the distributed water-filling algorithms have been provided in [49] and further refined in [48] using variational inequality theory. Several works have focused on different ways of improving the performance at the NE. In [51], a pricing technique is used to account for the interference introduced in the system similarly to [33]. In [50,52], the authors considered a certain hierarchy among the players and the Stackelberg formulation. Here, the leader or foresighted player chooses its own strategy that maximizes its payoff based on the perfect global knowledge of the system. The other players, the followers or myopic players, will react according to the leader's choice. It turns out that the Stackelberg equilibrium is unique in the two-user ($K = 2$) particular case [50]. Furthermore, a repeated game formulation was discussed in [53] for the case of two nonoverlapping frequency bands ($Q = 2$) and for discrete strategy spaces where the users choose either to transmit with full power on either of the available frequency bands or to spread uniformly their powers over the two bands. The space of possible strategies is explored using genetic algorithms, and the authors proved that it is possible to identify robust strategies with a high performance level.

8.3.3 Static MIMO Interference Channels

We start the analysis of the general MIMO IC, by first reviewing the PA game for the two-user $(K = 2)$ multiple input single output (MISO) IC $(n_r = 1)$ studied in [54,55]. The strategy spaces of the users are restricted only to the beamforming PA policies, that is, the transmitters concentrate all of their available power on a single direction such that the covariance matrix of user k is $\mathbf{Q}_k = \overline{P}_k \underline{w}_k \underline{w}_k^H$, where $\underline{w}_k$ is the beamforming vector. The strategy of user k is to select the optimal beamforming vector that maximizes the achievable rate:

$$u(\underline{w}_k, \underline{w}_{-k}) = \log_2 \left(1 + \frac{|\underline{w}_k^H \underline{h}_{kk}|^2}{\sigma^2 + |\underline{w}_\ell^H \underline{h}_{lk}|^2} \right) \tag{8.28}$$

For the two-user case, the NE beamforming vectors are aligned with the channel vectors: $\underline{w}_k^{\text{NE}} = \underline{h}_{kk}^* / \|\underline{h}_{kk}\|$. It turns out that any point on the Pareto optimal boundary of the achievable rates can be achieved by a linear combination between the NE point and the zero forcing point where the user's directions are chosen such that the interference terms are canceled $\left(\underline{w}_k^{\text{ZF}} = \pi_{\underline{h}_{k\ell}^*} \underline{h}_{kk}^* / \|\pi_{\underline{h}_{k\ell}^*} \underline{h}_{kk}^*\| \right.$ where $\pi_{\mathbf{X}} = \mathbf{I} - \mathbf{X}(\mathbf{X}^H \mathbf{X})^{-1} \mathbf{X}^H$ the projection onto the orthogonal complement of the column space of $\mathbf{X}$). Also in [54], the authors proposed a Nash bargaining approach such that any point on the Pareto optimal boundary $\left(R_k^{\text{NB}} \geq R_k^{\text{NE}} \right)$ of the achievable rate region can be achieved if the users agree to cooperate. The threat point or the disagreement point is given by the NE.

In what follows, we will focus on the more general MIMO IC case. The simultaneous water-filling problem was first studied in this case in [56] where the corresponding achievable rate-region was derived. However, the noncooperative PA game formulation was first described by the authors of [57] where the users choose their optimal covariance matrices to maximize their achievable rates:

$$u_k(\mathbf{Q}_k, \mathbf{Q}_{-k}) = \log \left| \mathbf{I}_{n_{r,k}} + \mathbf{H}_{kk}^H \mathbf{\Sigma}_{-k}^{-1}(\mathbf{Q}_{-k}) \mathbf{H}_{kk} \mathbf{Q}_k \right| \tag{8.29}$$

where $\mathbf{\Sigma}_{-k}(\mathbf{Q}_{-k}) = \mathbf{\Sigma}_{n_{r,k}} + \sum_{\ell \neq k} \mathbf{H}_{\ell k} \mathbf{Q}_\ell \mathbf{H}_{kk}^H$ is the multiuser interference plus the noise covariance matrix observed by user k. The existence of an NE has been proven using standard results on concave games [31] and it is given by the water-filling fixed point equation [58]:

$$\mathbf{Q}_k^{\text{NE}} = \text{WF}_k \left(\mathbf{Q}_{-k}^{\text{NE}} \right) \tag{8.30}$$

where the water-filling operator is defined as $\text{WF}(\mathbf{Q}_{-k}) = \mathbf{W}_k [\mu_k \mathbf{I} - \mathbf{\Lambda}_k]^+ \mathbf{W}_k^H$, with μ_k chosen such that the power constraint is satisfied. The matrices $\mathbf{W}_k$ and $\mathbf{\Lambda}_k$ are such that $\mathbf{H}_{kk}^H \mathbf{\Sigma}_{-k}^{-k}(\mathbf{Q}_{-k}) \mathbf{H}_{kk} = \mathbf{W}_k \mathbf{\Lambda}_k \mathbf{W}_k^H$. As far as the uniqueness of the NE is concerned, the authors showed that if the multiuser interference is sufficiently small, then the NE is unique and the distributed algorithm converges to this stable state. However, they did not quantify how low the interference level must be. The authors in [57] propose a two-level game approach to improve the efficiency of the equilibrium with respect to the achievable sum-rate of the system. At the first level, the users negotiate the limits on their number of independent transmit data streams, and, at the second level, given these limits, the users play the noncooperative game as before. A learning algorithm called *adaptive play* is proposed in order for the users to learn their maximum number of active modes while assuming a minimal information exchange between them.

In [58], the authors study the MIMO IC PA problem from a different perspective. They give a new interpretation of the water-filling operator as a projection onto a polyhedral set that can be written as

$$\mathrm{WF}_k(\mathbf{Q}_{-k}) = \left[-\left(\mathbf{H}_{kk}^H \mathbf{R}_{-k}^{-1}(\mathbf{Q}_{-k}) \mathbf{H}_{kk} \right)^{-1} \right]_{\mathcal{Q}_k}^+ \qquad (8.31)$$

where $\mathcal{Q}_k = \left\{ \mathbf{Q} \in \mathbb{C}^{n_{t,k} \times n_{t,k}} : \mathbf{Q} \succeq \mathbf{0}, \mathrm{Tr}(\mathbf{Q}) = \overline{P}_k \right\}$. Using tools from the fixed point theory and contraction theory, the authors provide sufficient conditions that ensure the uniqueness of the NE. This result is stated in the following theorem.

THEOREM 8.5   The power allocation game for MIMO IC described above admits a unique NE if $\rho(\mathbf{S}) < 1$ where $\rho(\mathbf{S})$ denotes the spectral radius of the matrix $\mathbf{S}$ defined as:

$$\mathbf{S}(k, \ell) = \begin{cases} \rho\left(\mathbf{H}_{\ell k}^H \mathbf{H}_{kk}^{-H} \mathbf{H}_{kk}^{-1} \mathbf{H}_{\ell k} \right), & \text{if } \ell \neq k \\ 0, & \text{otherwise} \end{cases} \qquad (8.32)$$

The following corollary quantifies how small the interference should be to insure a unique NE.

COROLLARY 8.1   A sufficient condition that implies $\rho(\mathbf{S}) \leq 1$ is given by one of the two following set of conditions:

$$\frac{1}{w_k} \sum_{\ell \neq k} \rho\left(\mathbf{H}_{\ell k}^H \mathbf{H}_{kk}^{-H} \mathbf{H}_{kk}^H, \mathbf{H}_{\ell k} \right) w_\ell < 1, \quad \forall k$$
$$\frac{1}{w_\ell} \sum_{k \neq \ell} \rho\left(\mathbf{H}_{\ell k}^H \mathbf{H}_{kk}^{-H} \mathbf{H}_{kk}^H, \mathbf{H}_{\ell k} \right) w_k < 1, \quad \forall \ell \qquad (8.33)$$

where $\underline{w} = [w_1, \ldots, w_Q]^T$ is a positive vector.

The first condition in (8.33) can be seen as a constraint on the maximum level of interference that each receiver can tolerate, while the second condition limits the maximum amount of interference that each transmitter is allowed to generate. The conditions that guarantee the uniqueness of the NE also ensure the convergence of the proposed distributed asynchronous algorithm. The most general case of noncooperative power allocation games in the MIMO IC is considered in [75] where the channel matrices are no longer assumed to be full-rank.

In [59], the authors study the noncooperative game for an MIMO wideband system where the strategies of the users are their optimal precoding and multiplexing matrices under constraints on the average transmit power plus a mask constraint. Two different games are defined in function of the users' payoff functions: i) the information theoretical achievable rates; ii) the achievable rates obtained using finite order constellations and considering a supplementary constraint on the average error probability. It turns out that the the optimal precoding strategies lead to the diagonalization of the channels such that the system is reduced to an equivalent parallel IC. Different distributed iterative algorithms (simultaneous water-filling, sequential water-filling, gradient-projection-based algorithms) that converge to the stable state of the system and their performance are discussed in detail in [60].

8.4 Power Allocation Games beyond the MAC and IC

In this section, we will consider more sophisticated channel models that can be seen as generalizations of the IC: the interference relay channel (IRC) and the cognitive radio channel (CRC).

8.4.1 Static Parallel Interference Relay Channels

A natural possibility of improving the transmission range and rate is to consider additional nodes in the system that can be exploited by some or all the devices operating in the same frequency band. This possibility is investigated in [61,63] where the PA game in a two-user parallel IRC is studied. We see that the parallel IC is a special case of the IRC where there is a supplementary relay node operating on each frequency band. The relays are assumed to operate in full duplex mode and the influence of three different protocols on the solution of the PA game were studied: (i) the zero delay scalar amplify-and-forward (ZDSAF) protocol, (ii) the decode-and-forward protocol (DF), and (iii) the estimate-and-forward protocol (see [62]). Because of the complexity of the expressions of the achievable rates derived in [28,63], perfect channel state information is needed both at the receivers and at the transmitters except for some particular cases (e.g., the case where the sources are located far away from their respective receivers and the direct and cross interference links are negligible, or the case of ZDSAF with constant amplification gain).

The baseband signals received in band q by the two destination nodes and the relay node are expressed as

$$
\begin{cases}
y_1^{(q)} = h_{11}^{(q)} x_1^{(q)} + h_{21}^{(q)} x_2^{(q)} + h_{r1}^{(q)} x_r^{(q)} + z_1^{(q)} \\
y_2^{(q)} = h_{12}^{(q)} x_1^{(q)} + h_{22}^{(q)} x_2^{(q)} + h_{r2}^{(q)} x_r^{(q)} + z_2^{(q)} \\
y_r^{(q)} = h_{1r}^{(q)} x_1^{(q)} + h_{2r}^{(q)} x_2^{(q)} + z_r^{(q)}
\end{cases}
\tag{8.34}
$$

where $Z_k^{(q)} \sim \mathcal{N}\left(0, N_k^{(q)}\right)$, $i \in \{1, 2, r\}$, represents the Gaussian complex noise on band q and, for all $k, \ell \in \{1, 2, r\}$, $h_{k\ell}^{(q)}$ is the channel gain between $\mathcal{S}_k$ and $\mathcal{D}_\ell$ in band q with the convention $h_{rr} = 0$. The expression of the signal sent by the relay, $x_r^{(q)}$, and also the achievable rates of the users in band q depend on the protocol used by the relay. The payoff function of user k is the sum of the achievable rates in all bands: $u_k(p_k, p_{-k}) = \sum_{q=1}^{(q)} R_k^{(q)}(p_k, p_{-k})$. If the ZDSAF protocol is used at the relaying nodes, then the achievable rate $R_k^{(q)}$ is written as follows:

$$
R_k^{(q),\mathrm{AF}} = C\left(\eta_k^{(q),\mathrm{AF}}\right)
\tag{8.35}
$$

where

$$
\eta_k^{(q),\mathrm{AF}} = \frac{\left| a_r^{(q)} h_{kr}^{(q)} h_{rk}^{(q)} + h_{kk}^{(q)} \right|^2 \rho_i^{(q)} p_k^{(q)}}{\left| a_r^{(q)} h_{\ell r}^{(q)} h_{rk}^{(q)} + h_{\ell k}^{(q)} \right|^2 \rho_\ell^{(q)} \dfrac{N_\ell^{(q)}}{N_k^{(q)}} p_\ell^{(q)} + \left(a_r^{(q)} \right)^2 \left| h_{rk}^{(q)} \right|^2 \dfrac{\left(\sigma_r^{(q)} \right)^2}{\left(\sigma_k^{(q)} \right)^2} + 1}
\tag{8.36}
$$

with $k \in \{1, 2\}$, $\ell = -k$ and $\rho_k^{(q)} = \dfrac{1}{\left(\sigma_k^{(q)} \right)^2}$.

As far as the achievable rates obtained with the other two relaying protocols (DF and EF) are concerned, the reader is referred to [61,63]. For all considered protocols, sufficient conditions that ensure the existence of the NE are provided. Assuming a time-sharing argument and at the cost of user coordination, the existence of the NE is always ensured. Similar to the parallel IC case, the NE is not unique in general. A thorough characterization of the number of NE as a function of the channel gains is intractable except for the special case of fixed relay amplification gain when ZDSAF protocol is assumed. The authors proposed iterative algorithms based on the best response functions that converge to an NE depending on the starting point. To further improve the overall performance of the NE, a Stackelberg formulation is investigated where the leader of the game is a central authority that can tune different parameters of the relay (i.e., spatial position or protocol-dependent parameters) such that a certain performance index is optimized. Let us consider the case where $Q = 2$ and assume that there is no relay operating in the second frequency band (an IRC in parallel with an IC). Figure 8.4 illustrates the powers that the two players dedicate to the transmission in the first frequency band at the NE as a function of the relay spatial position $(x_R, y_R) \in [-L, L]^2$. We can see that, at the equilibrium state, the regions where the users allocate a certain fraction of their power to the IRC are almost nonoverlapping. This means that, the

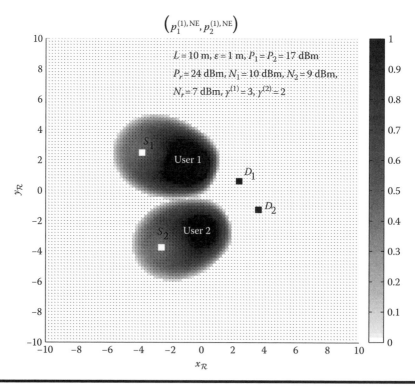

Figure 8.4 Static parallel IRC. PA policies at the NE $\left(p_1^{NE}, p_2^{NE}\right)$ as a function of $(x_R, y_R) \in [-L, L]^2$ for $L = 10\,\text{m}$, $\epsilon = 1\,\text{m}$, $P_1 = 20\,\text{dBm}$, $P_2 = 17\,\text{dBm}$, $P_r = 22\,\text{dBm}$, $N_1 = 10\,\text{dBm}$, $N_2 = 9\,\text{dBm}$, $N_r = 7\,\text{dBm}$, $\gamma^{(1)} = 2.5$, and $\gamma^{(2)} = 2$. The regions where the uses allocate their power to IRC are almost nonoverlapping. (From Belmega, E.V. et al., what happens when cognitive terminals compete for a relay mode? *IEEE International Conference on Acoustics, Speech and Signal Processing (ICASSP)*, Taipei, Taiwan, 2009.)

selfish users manage to coordinate their access to the supplementary frequency and to cancel out the mutual interference.

A different approach for PA games in IRCs was discussed in [64]. The system model is a two hop IRC; the first hop represents the IC between two sources and two relays and the second hop represents the IC between the two relays and the two transmitters. All the direct links between the transmitters and receivers are negligible and no transmission can occur without the relay nodes' intervention. This assumption is realistic when the transmitter nodes are situated far away from the destination nodes. The players are in this case the *source-relay-receiver pairs*. Each user splits its power between the two hops (the power allocates to its source and to its relay transmissions) rather than over two orthogonal frequency bands as in [61]. The relays are operating in half-duplex mode and assumed to apply the only DF relaying protocols. In this case, the authors prove the existence and uniqueness of the NE and they provide its closed form expression. Also, sufficient conditions that ensures that the NE is also Pareto-optimal is provided. However, as opposed to [61], the analysis presented here is greatly simplified by the assumption that the direct links between the two transmitters and the two receivers are negligible. Finally, a distributed asynchronous algorithm similar to the ones proposed in [58,60] that globally converge to the NE of the game is also provided.

8.4.2 Static MIMO Cognitive Radio Channel

The increasing number of wireless systems has made the radio spectrum a scarce resource. Furthermore, the fixed spectrum allocation that is used in most wireless systems is inefficient. This has led to the development of new technologies such as cognitive radios that allows the coexistence of primary users that are holders of the spectrum license and secondary users, opportunistic users, coexist on the same frequency band provided that the secondary users create a low level or no interference to the primary users. In [65], the authors study the PA game in a cognitive radio MIMO system where the secondary users chose their transmit covariance matrices to optimize their Shannon rates. The problem formulation is similar to the PA for the MIMO IC discussed in sub-section 8.3.3 and the system model is identical to (8.22). However, the analysis conducted there cannot be applied in this case because it does not provide a mechanism to limit the interference that the secondary users generate on the primary ones. The authors study the cases where the primary users can establish two types of constraints. The *null constraints* that prohibit secondary users to transmit over certain subspaces (portions of the spectrum or angular directions [66]) such that it creates zero interference to primary users:

$$\mathbf{V}_k^H \mathbf{Q}_k = 0 \tag{8.37}$$

where $\mathbf{V}_k$ is an $n_{t,k} \times n_{r,k}$ matrix, whose columns represent the directions along which user i is not allowed to transmit. The *soft shaping constraints* that allow the secondary users to transmit provided that the generated interference in the frequency bands or spatial locations be limited under a certain threshold (i.e., the interference-temperature limits):

$$\mathrm{Tr}\left(\mathbf{G}_k^H \mathbf{Q}_k \mathbf{G}_k\right) \leq \overline{P}_k^{\mathrm{ave}}$$
$$\lambda_{\max}\left(\mathbf{G}_k^H \mathbf{Q}_k \mathbf{G}_k\right) \leq \overline{P}_k^{\mathrm{peak}} \tag{8.38}$$

which limit the total average and peak power radiated along the spatial or frequency directions spanned by the matrix $\mathbf{G}_k \in \mathbb{C}^{n_{t,k} \times n_{g,k}}$. The authors showed that the solution of the game can still be computed efficiently via water-filling-like expressions, similar to (8.30) given in [58] for the MIMO IC, by introducing a proper projection matrix.

The NE was shown to exist and sufficient conditions very similar to the ones in (8.33) were derived to ensure its uniqueness and also the convergence of the distributed algorithms. Note that the intuition behind these conditions is that if the interference level is low enough, then these sufficient conditions are met. This actually is a very important problem in the cognitive radio scenario because the level of interference coming from the primary users can be large and uncontrollable by the secondary users. This means that the NE might not be unique and the proposed algorithms may not converge. To overcome this issue, the concept of *virtual interferers* is introduced, which leads to a new game with more relaxed conditions on the existence and uniqueness of the corresponding NE. What is interesting is that the new uniqueness conditions do not depend on the interference terms created by the primary users.

8.5 Conclusion and Open Issues

In this chapter, we have presented an overview of the most relevant literature on the noncooperative PA games in wireless networks where each transmitter node aims at maximizing its own achievable Shannon transmission rate. The basic multiuser network models such as the MAC, IC, IRC, and CRC have been investigated. In these cases, the concavity properties of the achievable Shannon rates guarantees the existence of stable solutions (i.e., NE) that can be reached from an initial state of the network, using distributed iterative algorithms based on the water-filling concept. We conclude this chapter by several critics of the game theoretical approach for resource allocation in wireless networks and by mentioning a few open issues.

A major drawback of the rate maximization games we have considered in this chapter is the fact that because of the inherent selfish nature of the players and the possible asymmetries in the system, they may lead to a very inefficient and more importantly unfair solution that associates high rates to the users with very good channel conditions and low rates to the users channel bad conditions. This formulation might not be suitable for communication systems with minimal quality of service constraints. For this purpose, there are several works that consider the *dual game* or the power minimization game [67,68] in the static parallel or MIMO IRCs, respectively. In these games, the user objective is to minimize their cost in terms of average consumed power under a minimal achievable rate constraint. This problem is more difficult to solve since the constraint of one user, the minimal rate constraint for user k, depends on the strategies used by the other players and the constraints induce coupling among the players' admissible strategies.

In the recent literature regarding the problem under investigation, only basic models of the multiuser networks have been investigated so far. One of the main difficulties in extending these results to a general multiuser network is the fact that the capacity region is still unknown. In these situations, achievable Shannon rates can be considered as payoff functions. Nevertheless, when considering a general network topology, the concavity property of the achievable rates may be intractable and different techniques have to be developed to prove the existence of the Nash equilibria.

Another important point is the convergence time of the distributed algorithms. We have seen that different iterative algorithms have to be implemented at the transmitter side to converge to the NE states. This is not a problem if channels are static, and thus, the coherence time (the interval of time in which the channel stays fixed) is larger than the convergence time of the algorithms. However, in a highly unstable environment (rapidly changing channel conditions or network topology), the NE might not be a realistic or practical operating network state and other solutions to the PA problem have to be envisioned.

A common assumption to reach the NE state is the complete knowledge of the game structure (i.e., set of players, actions, payoffs). Also, different assumptions with respect to the channel state knowledge at the transmitter side are considered (perfect CSIT for static channels or CDIT for fast fading ones). In wireless channels, these assumptions may turn out to be impractical (the time needed to acquire the necessary information may be too long and the equilibrium point not achieved before a change occurs in the channel) or even nonrealistic (errors may occur in the network parameter estimation process). Thus, one has to consider *games with incomplete information* and *imperfect channel state information*. A unifying framework that can be used to model incomplete or imperfect information is offered by *Bayesian games* [69–71].

Furthermore, we can question even the rationality which is also assumed common knowledge. For example, if at some point one or several autonomous devices break down and start misfunctioning, then achieving the stable state of the network may no longer be guaranteed. A possible solution would be to consider more general concepts such as *game situations* that take into account the interactive nature of games and *belief hierarchies* (i.e., beliefs about beliefs about beliefs, etc., about the other players' strategies) that one player takes into account when trying to maximize its *expected payoff* [72].

Abbreviations

SISO   Single input single output
MIMO   Multiple input multiple output
MAC   Multiple access channel
IC   Interference channel
SNR   Signal to noise ratio
IRC   Interference relay channel
CRC   Cognitive radio channel
NE   Nash equilibrium
PA   Power allocation
CSIT   Channel state information at the transmitter
CSIR   Channel state information at the receiver
CDIT   Channel distribution information at the transmitter
SUD   Single user decoding
SIC   Successive interference cancelation

References

1. J. von Neumann and O. Morgenstern, *Theory of Games and Economic Behavior*, Princeton University Press, Princeton, NJ, 1944.
2. J. Nash, Equilibrium points in N person games, *Proceedings of the National Academy of Sciences of the United States of America*, 36, 48–49, Jan. 1950.
3. C. E. Shannon, A mathematical theory of communications, *The Bell System Technical Journal*, 27, 379–423, 623–656, Jul., Oct. 1948.
4. B. Mandelbrot, Contribution à la théorie mathématique des jeux de communication, *PhD Dissertation*, Université de Paris, France, Dec. 1952.

5. N. M. Blachman, Communication as a game, *Proceedings of the WESCON Conference*, San Francisco, CA, Aug. 20–23, 1957, part 2, pp. 61–66.

6. D. P. Palomar, J. M. Cioffi, and M. A. Lagunas, Uniform power allocation in MIMO channels: A game theoretic approach, *IEEE Transactions on Information Theory*, 49, 7, 1707–1727, Jul. 2003.

7. D. Fudenberg and J. Tirole, *Game Theory*, The MIT Press, Cambridge, MA, 1991.

8. M. J. Osborne, *An Introduction to Game Theory*, Oxford University Press, Cary, NC, 2003.

9. S. Lasaulce, M. Debbah, and E. Altman, Methodologies for analyzing equilibria in wireless games, *IEEE Signal Processing Magazine, Special Issue on Game Theory for Signal Processing*, 26, 5, 41–52, Sep. 2009.

10. A. B. MacKenzie and L. A. DaSilva, *Game theory for Wireless Engineers, Morgan & Claypool Publishers' Series, Sinthesys Lectures on Communication*, San Rafael, CA, 2006.

11. V. Shah, N. B. Mandayam, and D. J. Goodman, Power control for wireless data based on utility and pricing, *IEEE Proceedings of the 9th International Symposium on Personal, Indoor, Mobile Radio Communications (PIMRC)*, Boston, MA, pp. 1427–1432, Sep. 1998.

12. D. J. Goodman and N. B. Mandayam, Power control for wireless data, *IEEE Personal Communications*, 7, 2, 48–54, 2000.

13. F. Meshkati, M. Chiang, H. V. Poor, and S. C. Schwartz, A game-theoretic approach to energy-efficient power control in multi-carrier CDMA systems, *IEEE Journal on Selected Areas in Communications*, 24, 6, 1115–1129, Jun. 2006.

14. F. Meshkati, H. V. Poor, and S. Schwartz, Energy-efficient resource allocation in wireless networks, *IEEE Signal Processing Magazine*, 58, 58–68, May 2007.

15. E. Altman, R. El-Azouzi, Y. Hayel, and H. Tembine, Evolutionary power control games in wireless networks, *Ad Hoc and Sensor Networks, Wireless Networks, Next Generation Internet Networking, Lecture Notes in Computer Science*, Springer, Berlin / Heidelberg, Germany, May 2008.

16. H. Tembine, E. Altman, R. El-Azouzi, and Y. Hayel, Evolutionary games with random number of interacting players applied to access control, *International Symposium on Modeling and Optimization in Mobile, Ad Hoc, and Wireless Networks (WIOPT)*, Berlin, Germany, Apr. 2008.

17. E. Altman, T. Boulogne, R. El-Azouzi, T. Jimenez, and L. Wynter, A survey on networking games in telecommunications, *Computers and Operations Research*, 33, 286–311, 2006.

18. S. Lasaulce and M. Debbah, Game theory for wireless communications, *Tutorial for the IEEE International Symposium on Personal, Indoor and Mobile Radio Communications (PIMRC)*, Cannes, France, Sep. 2008.

19. A. B. MacKenzie and S. B. Wicker, Game theory and the design of self-configuring adaptive wireless networks, *IEEE Communications Magazine*, 39, 11, 126–131, 2001.

20. *IEEE Journal on Selected Areas in Communications, Special Issue on Game Theory in Communication Systems*, 26, 7, Sep. 2009.

21. *Eurasip Journal on Advances in Signal Processing, Special Issue on Game Theory in Signal Processing and Communications*, 2009.

22. *Eurasip Journal on Advances in Signal Processing, Special Issue on Dynamic Spectrum Access for Wireless Networks*, 2009.

23. *IEEE Signal Processing Magazine, Special Issue on Game Theory in Signal Processing and Communications*, 26, 5, Sep. 2009.

24. A. D. Wyner, Recent results in Shannon theory, *IEEE Transactions on Information Theory*, 20, 2–10, Jan. 1974.

25. T. Cover, Some advances in broadcast channels, *Advances in Communication Systems*, vol. 4, Academic Press, New York, 1975.

26. A. B. Carleial, A case where interference does not reduce capacity, *IEEE Transactions on Information Theory*, 21, 5, 569–570, Sep. 1975.

27. O. Sahin and E. Erkip, Achievable rates for the Gaussian interference relay channel, *Proceedings of IEEE Global Communications Conference*, Washington, DC, pp. 786–787, Nov. 2007.
28. O. Sahin and E. Erkip, On achievable rates for interference relay channel with interference cancellation, *Proceedings of IEEE Asilomar Conference on Signal, Systems and Computers*, Pacific Grove, CA, Nov. 2007.
29. J. Mitola, Cognitive radio for flexible mobile multimedia communication, *IEEE International Workshop on Mobile Multimedia Communications*, San Diego, CA, pp. 3–10, Nov. 1999.
30. S. Haykin, Cognitive radio: Brain empowered wireless communications, *IEEE Journal on Selected Areas in Communications*, 23, 2, 201–220, Feb. 2002.
31. J. Rosen, Existence and uniqueness of equilibrium points for concave n-person games, *Econometrica*, 33, 520–534, 1965.
32. E. V. Belmega, S. Lasaulce, and M. Debbah, Power control in distributed multiple access channels with coordination, *International Workshop on Wireless Networks: Communication, Cooperation and Competition (WNC3)*, Berlin, Germany, Apr. 2008.
33. T. Alpcan, T. Basar, R. Srikant, and E. Altman, CDMA uplink power control as a noncooperative game, *Wireless Networks*, Springer, vol. 8, no. 6, pp. 659–670, Nov. 2002.
34. E. V. Belmega, S. Lasaulce, M. Debbah, M. Jungers, and J. Dumont, Power allocation games in wireless networks of multi-antenna terminals, *Springer Telecommunications Systems Journal, Special Issue on Game Theory in Communication Networks*, vol. 44, pp. 1–14, DOI:10.1007/s11235-010-9305-3, May–Jun. 2010.
35. W. Yu, W. Ree, S. Boyd, and J. M. Cioffi, Iterative water-filling for Gaussian vector multiple-access channels, *IEEE. Transactions on Information Theory*, 50, 1, 145–152, Jan. 2004.
36. L. Lai and H. El Gamal, The water-filling game in fading multiple-access channels, *IEEE Transactions on Information Theory*, 54, 5, 2110–2122, May 2008.
37. D.N.C. Tse and S. Hanly, Multi-access fast fading channels. Part I: Polymatroid structure, optimal resource allocation and throughput characterization, *IEEE Transactions on Information Theory*, 44, 7, 2796–2815, Nov. 1998.
38. A. Tulino and S. Verdu, Impact of antenna correlation on the capacity of multi-antenna channels, *IEEE Transactions on Information Theory*, 51, 7, 2491–2509, Jul. 2005.
39. S. Lasaulce, A. Suarez, M. Debbah, and L. Cottatellucci, Power allocation game for fading MIMO multiple access channels with antenna correlation, *The ICST/ACM Proceedings of the International Conference on Game Theory in Communication Networks (Gamecomm)*, Nantes, France, Oct. 2007.
40. E. Biglieri, G. Taricco, and A. Tulino, How far is infinity? Using asymptotic analyses in multiple-antenna systems, *Proceedings of International Symposium on Software Testing and Analysis*, Rome, Italy, vol. 1, pp. 1–6, 2002.
41. J. Dumont, P. Loubaton, and S. Lasaulce, On the capacity achieving transmit covariance matrices of MIMO correlated Rician channels: A large system approach, *IEEE Proceedings of Globecom Technical Conference*, San Francisco, CA, Nov./Dec. 2006.
42. J. Dumont, W. Hashem, S. Lasaulce, P. Loubaton, and J. Najim, Capacity of Rician MIMO Channels with antenna correlation: An asymptotic approach, *IEEE Transactions on Information Theory*, vol. 56, no. 3, pp. 1048–1069, Mar. 2010.
43. W. Yu, G. Ginis, and J. M. Cioffi, Distributed multiuser power control for digital subscriber lines, *IEEE Journal on Selected Areas in Communications*, 20, 5, 1105–1115, Jun. 2002.
44. R. Etkin, A. Parekh, and D. Tse, Spectrum sharing for unlicensed bands, *IEEE Journal on Selected Areas on Communications, Special Issue on Adaptive, Spectrum Agile and Cognitive Wireless Networks*, vol. 265, no. 3, pp. 517–528. Apr. 2007.
45. O. Popescu, C. Rose, and D.C. Popescu, Signal space partitioning versus simultaneous water filling for mutually interfering systems, *IEEE Globecom*, Dallas, TX, vol. 5, pp. 3128–3132, Nov. 2004.

46. A. Leshem and E. Zehavi, Bargaining over the interference channel, *IEEE Journal on Selected Areas in Communications*, 26, 7, Sep. 2008.

47. J. Nash, The bargaining problem, *Econometrica*, 18, 2, 155–162, 1950.

48. Z.-Q. Luo and J.-S. Pang, Analysis of iterative waterfilling algorithm for multiuser power control in digital subscriber lines, *Eurasip Journal on Applied Signal Processing*, 1–10, 2006, available at http://www.hindawi.com/journals/asp/2006/024012.abs.html.

49. S. T. Chung, S. J. Kim, J. Lee, and J. M. Cioffi, A game theoretic approach to power allocation in frequency-selective gaussian interference channels, *IEEE International Symposium on Information Theory (ISIT)*, Pacifico Yokohama, Kanagawa, Japan, pp. 316–316, Jun./Jul. 2003.

50. M. Bennis, M. Le Treust, S. Lasaulce, and M. Debbah, Spectrum sharing games on the interference channel, *IEEE International Conference on Game Theory for Networks (Gamenets)*, Istanbul, Turkey, May 2009.

51. J. Huang, R. A. Berry and M. L. Honig, Distributed interference compensation for wireless networks, *IEEE Journal of Selected Areas in Communications*, 24, 5, 1074–1084, May 2006.

52. Y. Su and M. Van der Schaar, A new perspective on multi-user power control games in interference channels, *IEEE Trans. on Wireless Communications*, vol. 8, no. 6, pp. 2910–2919, June 2009.

53. N. Clemens and C. Rose, Intelligent power allocation strategies in an unlicensed spectrum, *IEEE Symposium on New Frontiers in Dynamic Spectrum Access Networks*, Baltimore, MD, pp. 8–11, Nov. 2005.

54. E. G. Larsson and E. A. Jorswieck, Competition versus cooperation on the MISO interference channel, *IEEE Journal on Selected Areas in Communications*, 26, 7, 1059–1069, Sep. 2008.

55. E. A. Jorswieck and E. G. Larsson, The MISO interference channel from a game-theoretic perspective: A combination of selfishness and altruism achieves Pareto optimality, *IEEE International Conference on Acoustics, Speech and Signal Processing (ICASSP 2008)*, Las Vegas, NV, pp. 5364–5367, 31 Mar.–4 Apr. 2008.

56. O. Popescu, D. C. Popescu, and C. Rose, Simultaneous water filling in mutually interfering systems, *IEEE Transactions on Wireless Communications*, 6, 3, 1102–1113, Mar. 2007.

57. G. Arslan, M. F. Demirkol, and Y. Song, Equilibrium efficiency improvement in MIMO interference systems: A decentralized stream control approach, *IEEE Transactions on Wireless Communications*, 6, 8, 2984–2993, Aug. 2007.

58. G. Scutari, D. P. Palomar, and S. Barbarossa, Competitive design of multiuser MIMO systems based on game theory: A unified view, *IEEE Journal on Selected Areas in Communications*, 26, 7, 1089–1103, Sep. 2008.

59. G. Scutari, D. P. Palomar, and S. Barbarossa, Optimal linear precoding strategies for wideband non-cooperative systems based on game theory—Part I: Nash equilibria, *IEEE Transactions on Signal Processing*, 56, 3, 1230–1249, Mar. 2008.

60. G. Scutari, D. P. Palomar, and S. Barbarossa, Optimal linear precoding strategies for wideband non-cooperative systems based on game theory—Part II: Algorithms, *IEEE Transactions on Signal Processing*, 56, 3, 1250–1267, Mar. 2008.

61. E. V. Belmega, B. Djeumou, and S. Lasaulce, Resource allocation games in interference relay channels, *IEEE International Conference on Game Theory for Networks (Gamenets)*, Istanbul, Turkey, May 2009.

62. T. Cover, *Elements of Information Theory*, Wiley-Interscience, New York, 1991.

63. E. V. Belmega, B. Djeumou, and S. Lasaulce, Power allocation games in interference relay channels: Existence analysis of Nash equilibria, *Eurasip Journal on Wireless Communications and Networking (JWCN)*, to appear, 2011.

64. Y. Shi, J. Wang, K. Letaief, and R. Mallik, A game-theoretic approach for distributed power control in interference relay channels, *IEEE Transactions on Wireless Communications*, 8, 6, 3151–3161, Jun. 2009.

65. G. Scutari, D. P. Palomar, J.-S. Pang, and F. Facchinei, Flexible design of cognitive radio wireless systems, *IEEE Signal Processing Magazine, Special Issue on Game Theory for Signal Processing*, 26, 5, 107–123, Sep. 2009.

66. S. Medina Perlaza, M. Debbah, S. Lasaulce, and J.-M. Chaufray, Opportunistic spectrum sharing in MIMO Systems, *IEEE International Symposium on Personal, Indoor and Mobile Radio Communications (PIMRC)*, Cannes, France, 1–5 Sep. 2008.

67. J.-S. Pang, G. Scutari, F. Facchinei, and C. Wang, Distributed power allocation with rate constraints in Gaussian parallel interference channels, *IEEE Transactions on Information Theory*, 54, 8, 3471–3489, Aug. 2008.

68. G. Arslan, M. F. Demirkol and S. Yuksel, Power games in MIMO interference systems, *IEEE International Conference on Game Theory for Networks (Gamenets)*, Istanbul, Turkey, May 2009.

69. J. C. Harsanyi, Games with incomplete information played by Bayesian players, part I. The basic model, *Management Science*, 14, 3, 159–182, Nov. 1967.

70. J. C. Harsanyi, Games with incomplete information played by Bayesian players, part II. Bayesian equilibrium points, *Management Science*, 14, 5, 320–334, Jan. 1968.

71. J. C. Harsanyi, Games with incomplete information played by Bayesian players, part III. The basic probability distribution of the game, *Management Science*, 14, 7, 486–502, Mar. 1968.

72. R. J. Aumann and J. H. Dreze, Rational expectations in games, *The American Economic Review*, 98, 1, 72–86, 2008.

73. E. V. Belmega, On resource allocation problems in distributed MIMO wireless networks, Ph.D. Dissertation, Universit Paris Sud-11, 14 December 2010.

74. E. V. Belmega, S. Lasaulce, and M. Debbah, A trace inequality for positive definite matrices, *Journal of Inequalities in Pure and Applied Mathematics (JIPAM)*, vol. 10, no. 1, pp. 1–4, 2009.

75. G. Scutari, D. P. Palomar, and S. Barbarossa, The MIMO Iterative Waterfilling Algorithm, *IEEE Trans. on Signal Processing*, vol. 57, no. 5, pp. 1917–1935, May 2009.

Chapter 9

Noncooperative Power Control in CDMA Wireless Networks

Eirini-Eleni Tsiropoulou, Timotheos Kastrinogiannis, and Symeon Papavassiliou

Contents

191

With the growing demand for high data rates and support of multiple services with various quality of service (QoS) requirements, proficient power control mechanisms are essential toward efficient resource allocation and interference management in code division multiple access (CDMA) wireless networks. In this chapter, through a comprehensive analysis, concrete methodologies for treating various noncooperative power control games in CDMA wireless networks are presented, while the use and applicability of basic game theoretic tools in the design of power control approaches are illustrated. It is shown that game theory can be used as a unifying framework to study power control in CDMA wireless networks, as it provides a powerful theoretic tool that allows modeling and studying the interactions between self-interested users when competing for accessing scarce radio resources.

9.1 Introduction

The inner characteristics of wireless communications in code division multiple access (CDMA) cellular systems, in terms of network's scarce radio resources, mobile nodes' physical limitations, and users' time-varying channel conditions, have motivated the adoption of power control in both the uplink and downlink communication, toward proficient resource allocation and interference management. Power control aims at exploiting users' channel quality rapid variations toward improving system's and/or individual users' performance, as well as achieving fairness and/or meeting various services' Quality of Service (QoS) criteria. This is realized by opportunistically allocating system's resources in a time-slotted manner (i.e., transmission power and corresponding transmission rate) to the users with instantaneously "best channel" conditions, usually obtained through measurements and feedback.

Taking into account current and future trends in wireless networking, including the heterogeneity and diversity of the supported services and corresponding QoS requirements, utility (and pricing)-based theoretic frameworks have been applied to the QoS-driven power and rate control problem in CDMA wireless networks, triggering a tremendous amount of research efforts in that field. A utility function a concept adopted from the field of economics, reflects a user's degree of satisfaction with respect to his service performance, and therefore various user-centric or network-centric goals, as well as multiple services heterogeneous QoS prerequisites, can be modeled by forming appropriate utilities. As a result, network utility maximization (NUM) theory [1] became essential for studying power control either via analytical centralized solutions or distributed game theoretic approaches.

Despite the fact that rigorous analytical solutions have been derived toward studying the problem of power control in wireless networks within the framework of NUM theory, either concerning the downlink [2–8] or the uplink [9,10], in most cases it is assumed that users will update their powers aiming at overall social welfare, that is, overall network's utility maximization. Such an assumption implies that the goals of the network (operator) are in line with the goals of the users, which is not always true. As a result, game theory has been widely exploited as a powerful tool that allows studying interactions between self-interested users and, hence, modeling multiuser competitive nature via noncooperative games. Thus, game-theoretic approaches favor the analytical foundation and design of user-centric distributed power control algorithms, as an alternative to traditional approaches that have been based on decisions taken in a centralized or semi-centralized way at the base station, which tend to be complex and not easily scalable.

In a noncooperative game setting, each user individually selects his transmit power in order to maximize his degree of satisfaction (utility) under constraints imposed by physical limitations. In general, utilities assigned to users are functions of the consumed power and the achieved signal to noise and interference ratio, while their form and shape may vary significantly according to the goal they serve. Thus, the choice of the utility functions has a great impact on the nature of the game and how the users choose their respective actions.

Throughout our analysis in this chapter, we will see that a variety of utility functions, expressing multiple users' selfish behaviors, have been used in noncooperative power control formulations. For example, in Refs. [11,12] toward maximizing spectral efficiency, utilities proportional to the Shannon capacity for each user are adopted, thus treating all interference as white Gaussian noise Ref. [13]. In Refs. [14,15], the authors use a utility function that measures the number of reliable bits that are transmitted (i.e., goodput) per joule of energy consumed. This is further extended in Ref. [16] by introducing pricing to improve the efficiency of Nash equilibrium (NE). Joint energy-efficient power control and receiver design has been studied in Ref. [17], and a game theoretic approach to energy-efficient power allocation in multicarrier systems is introduced in Ref. [18]. Joint network-centric and user-centric power control is discussed in Ref. [19], while in Ref. [20], a QoS-driven power control supporting multiple services is considered.

This chapter discusses how game theory can be used as a unifying framework to study power control in CDMA wireless networks (Section 9.2). It does not only serve as extension to existing relevant survey works (e.g., Ref. [21]), but more importantly, it provides a comprehensive and concrete treatment of the noncooperative power control problem in CDMA wireless networks, by summarizing the foundations and basic theoretic components of noncooperative games and by illustrating their use and applicability in the design of power control approaches (Section 9.3). Therefore, in this chapter, an overview of current state-of-the-art approaches is provided and various power control problems are designed and formulated based on game theory. Furthermore, when a designed problem is formulated as a game, its goals are revealed; corresponding user's utilities design is discussed; and the analytical tools that game theory provides toward studying game's equilibrium properties in terms of existence, uniqueness, stability, and optimality are illustrated (Sections 9.4 through 9.13).

9.2 Power Control in Wireless Cellular Networks and Game Theory

Considering power control in wireless networks from a game-theoretic perspective, a noncooperative game can be modeled, where users act as individual players aiming at maximizing selfishly their degree of satisfaction, expressed by appropriately defined utility functions. Hence, considering a single cell of a CDMA wireless network with N users, a noncooperative power control game consists of three basic components:

1. *A set of players*, the set S of N users attached to the cell
2. *A set of actions*, the feasible strategy space A_i for each user $i \in S$, which is determined with respect to mobile terminal's physical limitations on the resources a_i that user i controls (e.g., transmission power $P_i \leq P_i^{\text{Max}}$ when $a_i = P_i$, or transmission power $P_i \leq P_i^{\text{Max}}$, and transmission rate $R_i \leq R_i^{\text{Max}}$ when $a_i = (P_i, R_i)$
3. *A set of preferences*, expressed via appropriately defined and assigned utilities U_i to the users, which in most cases are functions of their consumed power and the achieved bit energy to interference density ratio $E_b/I_o - (W/R_i)\gamma_i$ or signal to interference ratio (SIR) that is,

$$\gamma_i\left(P_i, \bar{P}_{-i}\right) = \frac{G_i P_i}{\theta \sum_{j=1}^{N} G_j P_j - \theta G_i P_i + I_0} = \frac{G_i P_i}{I_{-i}(\bar{P}_{-i})} \tag{9.1}$$

where

θ denotes the orthogonality factor

W denotes the system's spreading bandwidth

G_i denotes user's i channel gain

$\bar{P}_{-i}$ denotes the users' power allocation vector excluding user i

I_0 denotes the background noise and intercell interference

Thus, $I_{-i}(\bar{P}_{-i})$ actually denotes the network interference and background noise at the base station when receiving data from user i and is given by

$$I_{-i}(\bar{P}_{-i}) = \theta \sum_{j=1}^{N} G_j P_j - \theta G_i P_i + I_0 \tag{9.2}$$

while $I(P) = I_{-i}(\bar{P}_{-i}) - \theta G_i P_i$ represents overall intercell's interference plus background noise.

Based on the above definitions, let $G = [S, \{A_i\}, \{U_i\}]$ represent a noncooperative game, where each user aims at selecting its strategy (e.g., picks a transmission power) from his set of actions A_i in such a way as to maximize his own utility, that is,

$$\max_{a_i \in A_i} U_i(a_i, a_{-i}) \quad \text{for } i = 1, \dots, N \tag{9.3}$$

In other words, each player–user receives a payoff U_i relative to his selected action a_i given the selected actions of the rest of the users a_{-i}, in accordance with his best-response policy $BR_i = \max_{a_i \in A_i} U_i(a_i, a_{-i})$.

Toward studying the properties of game's G equilibrium (i.e., a stable outcome of game G), in the following we define the concept of *Nash equilibrium (NE)*, one of the most common approaches used in game theory.

Definition 9.1   NE is a set of strategies $\boldsymbol{a}^* = \left(a_1^*, \dots, a_N^*\right)$, such that no user can unilaterally improve his own utility, that is,

$$U_i\left(a_i^*, a_{-i}^*\right) \geq U_i\left(\alpha_i, a_{-i}^*\right) \quad \text{for all } \alpha_i \in A_i, \ \forall i \in S \tag{9.4}$$

where $a_{-i}^* = \left(a_{-1}^*, \dots, a_{-i-1}^*, a_{-i+1}^*, \dots, a_{-N}^*\right)$.

Specifically, in the case of a noncooperative power control game, where each user aims at $\max_{P_i} U_i(P_i, \bar{P}_{-i})$ s.t. $0 \leq P_i \leq P_i^{\text{Max}}$, NE is defined as the stable outcome of G, such that no user has the incentive to change his power level, since his satisfaction cannot be further improved by making any individual changes on his transmit power, given the transmit powers of other users, that is, the power allocation $\bar{P}^* = \left(P_1^*, \dots, P_N^*\right)$ for which $U_i\left(P_i^*, \bar{P}_{-i}^*\right) \geq U_i\left(P_i, \bar{P}_{-i}^*\right)$ for all $P_i^* \in A_i \forall i \in S$, where in this case $a_i = P_i$ and $A_i = \left[0, \ P_i^{\text{Max}}\right]$, in accordance with Definition 9.1.

In general, when studying power control in wireless networks as a noncooperative game, the game's formulation is based on (a) the under consideration system model that determines game's variables (i.e., users' actions sets) as well as users' strategy space limitations and (b) the designed users' or network's goals that determine users' utilities. Intuitively, the choice of the utility functions has a great impact on the nature of the game and how the users choose their respective actions. Furthermore, regardless of the designed problem's formulation, a noncooperative game's investigation (especially when NE is adopted) requires the consideration and analysis of the following key issues:

- Game's steady state via the existence of NE (**Existence of NE**)
- Equilibrium's properties via the characterization of NE (**Equilibrium's properties**)
- Optimality of NE (**Pareto optimality**)
- Convergence of NE via studying the convergence of users' corresponding best-response strategies toward selfishly maximizing their utilities, thus reaching game's equilibrium. (**Convergence of NE**)

Therefore, prior to concluding this section's analysis, we need to define three important concepts, namely *Pareto dominance*, *Pareto optimality*, and *supermodular games*.

As discussed earlier, NE can be regarded as a stable solution, at which none of the users has the incentive to change his strategy. Many games have several Nash equilibria. To compare the qualities of two different solutions, a commonly used concept is called Pareto dominance.

Definition 9.2   A set of strategies $\bar{\mathbf{a}} = (\bar{a}_1, \dots, \bar{a}_N)$ Pareto dominates another set of strategies $\mathbf{a} = (a_1, \dots, a_N)$ if for all $i \in S$, $U_i(\bar{\mathbf{a}}) \geq U_i(\mathbf{a})$ and for some $i \in S$, $U_i(\bar{\mathbf{a}}) > U_i(\mathbf{a})$.

Definition 9.3   A set of strategies $(\tilde{a}_1, \ldots, \tilde{a}_N)$ is Pareto optimal if there exists no other set of strategies for which one or more users can improve their utilities without reducing the utilities of other users.

In the case of power control noncooperative games, a power vector $\bar{P}$ Pareto dominates another vector P if for all $i \in S$, $U_i(\bar{P}) \geq U_i(P)$ and for some $i \in S$, $U_i(\bar{P}) > U_i(P)$, and thus, a power vector $\tilde{P}$ is Pareto optimal if there exists no vector that Pareto dominates $\tilde{P}$. Moreover, it should be noticed that analyzing a game (e.g., via adopting NE), system's performance is often not the optimal due to users/players selfish behavior. In order to improve the equilibrium utilities of a noncooperative power control game in Pareto sense, the concept of pricing has been introduced, which motivates users to adopt a more social behavior. Thus, overall system's performance can be increased (e.g., overall system utility) by implicitly inducing cooperation among the users and yet maintain the noncooperative nature of the resulting power control solution, as it will be shown later in this chapter.

Finally, the concept of supermodularity is introduced as it provides an effective game-theoretic tool toward treating noncooperative games [22].

Definition 9.4   A game is supermodular game if for all i

1. A_i is a compact subset (A_i is sub-lattice)
2. U_i is upper semi-continuous in (a_i, a_{-i}) (U_i is supermodular in A_i)
3. U_i has increasing differences in (a_i, a_{-i})

Concerning a single dimensional user strategy, as in the case of power control games, a formal definition of a supermodular game is simplified to the following.

Definition 9.5   A generic game $G = [S, \{A_i\}, \{U_i\}]$ with strategy space $a_i \subset \Re$ is supermodular, if for each user $i \in S$, the objective function $u_i(a_i, a_{-i})$ has nondecreasing differences in (a_i, a_{-i}).

Thus, the nondecreasing difference property of the objective function is defined as follows.

Definition 9.6   The objective function $u_i(a_i, a_{-i})$ has nondecreasing differences in (a_i, a_{-i}) if for all $a_{-i} \geq a'_{-i}$ the difference $u_i(a_i, a_{-i}) - u_i(a_i, a'_{-i})$ is nondecreasing in a_i. Equally, concerning continuous and twice differentiable objective functions, $u_i(a_i, a_{-i})$ has nondecreasing differences in (a_i, a_{-i}) if $(\partial^2 u_i(a_i, a_{-i})/\partial a_i \partial a_j) \geq 0$ for all $j \neq i$.

In Refs. [16,20,23,24], it has been shown that the best-response strategy of such (supermodular) games holds well-defined convergence properties; therefore, the concept of supermodularity has been widely used as one of the basic theoretic tools toward studying the convergence of NE.

9.3 Power Control Games, Goals, and Utility Functions

Aiming at performance optimality, a user's goal when participating in a power control game may vary from achieved throughput maximization to energy consumption minimization, and various service QoS prerequisites fulfilment to network's interference minimization. Since utility functions reflect in a normalized way users' degree of satisfaction with respect to the accomplishment of one or more of the previous goals, the selection of users' utilities is crucial in the nature of the noncooperative power control game and the choice of their corresponding actions when participating in the game. In addition, utilities provide enhanced flexibility on reflecting in a game theoretic problem formulation multiple objectives apart form power control, such as carrier selection, signature sequence control, and base station assignment. In general, users' utilities are normalized upper-bounded increasing functions of the network resources, while their form, shape, and goal may vary significantly.

Therefore, prior to the investigation and analysis of various power control games, in this section, we present a concrete introduction to the formulation and usage of different utility functions that have been proposed in the literature. Moreover, the presented utilities are related and classified in accordance with the user's or network's goals that they serve (summarized in Table 9.1). For each of the identified categories, throughout the rest of our analysis, corresponding noncooperative problems are formulated, methodologies for treating and investigated them via the tools of game theory are given, and corresponding power control algorithms are illustrated (summarized in Table 9.2).

9.3.1 Fixed SIR Constraints

Toward expressing users' degree of satisfaction with respect to their corresponding power consumption, when considering fixed predefined target SIR constraints, following the simplified utility function has been initially adopted in Ref. [25]:

$$U_i(P_i) = P_i \tag{9.5}$$

Thus, users aim to transmit with an optimal transmission power vector $\boldsymbol{P}$ such that their fixed predefined SIR constraints are satisfied (i.e., $\gamma_i = \gamma_i^{\text{fixed}}$, for all $i \in S$). The corresponding noncooperative power control game is extensively analyzed in Section 9.4.

9.3.2 Spectral Efficiency

Toward maximizing spectral efficiency, logarithmic, concave utility functions of users' SIR have been proposed in Ref. [12], which approximate the capacity via the classical Gaussian channel formula that corresponds in treating all interference as Gaussian noise, that is,

$$U_i(P_i, P_{-i}) = B \log_2(1 + \gamma_i) - c_i P_i \tag{9.6}$$

where
γ_i is the SIR for user i
B (Hz) is the system's unspread bandwidth
c_i is a positive scalar

The first term in (9.6) defines a proportional relationship between a user's degree of satisfaction and Shannon capacity (i.e., $U_i(P_i, P_{-i}) = \zeta_i \log(1 + \gamma_i)$ where ζ_i is a user-dependent constant) and was initially introduced in Refs. [11,12]. The second term aims at preventing users from transmitting with maximum power by introducing a linear cost on transmit power, toward improving the outcome of the power control game in Pareto sense via steering users to a more social behavior

Table 9.1 Noncooperative Power Control Games

Game	Utility Function	Game's Formulation	Game's Solution
Fixed SIR power control games	$U_i(P_i, \bar{P}_{-i}) = P_i$	$\min \sum_{i \in S} P_i$ s.t. $\gamma_i(P_i, \bar{P}_{-i}) \geq \gamma_i^t$	$P_i^* = (\gamma_i^t/G_i) L_{-i}(\bar{P}_{-i}), \forall i \in S$
Spectral efficiency via power control	$U_i(P_i, \bar{P}_{-i})$: quasi-concave of P_i	$\max_{P_i \in A_i} U_i(P_i, \bar{P}_{-i})$	$\mathbf{P}^* = F(\mathbf{P}) = F(\mathbf{MP + b})$
Energy-efficient power control	$U_i(P_i, \bar{P}_{-i}) = R_i f(\gamma_i(P_i, \bar{P}_{-i}))/P_i$	$\max_{P_i \in A_i} U_i(P_i, \bar{P}_{-i})$	$P_i^* = \min\left(\dfrac{\gamma_i^* R_i^{\text{Max}} L_{-i}(\bar{P}_{-i})}{WG_i}, P_i^{\text{Max}}\right)$
Energy-efficient power control and receiver design	$U_i(P_i, \bar{P}_{-i}, c_i) = f(\gamma_i(P_i, c_i))/P_i$	$\max_{P_i, c_i} \dfrac{f(\gamma_i(P_i, c_i))}{P_i}$	$P_i^{MF} = \dfrac{1}{G_i^2}\dfrac{\gamma^* \sigma^2}{1 - \alpha\gamma^*}$ for $\alpha < \dfrac{1}{\gamma^*}$ $P_i^{DE} = \dfrac{1}{G_i^2}\dfrac{\gamma^* \sigma^2}{1-\alpha}$ for $\alpha < 1$ $P_i^{MMSE} = \dfrac{1}{G_i^2}\dfrac{\gamma^*}{1-\alpha\frac{\gamma^*}{1+\gamma^*}}$ for $\alpha < 1 + \dfrac{1}{\gamma^*}$
Energy-efficient power control in multi-carrier CDMA wireless networks	$U_i^{MC} = \dfrac{\sum_{l=1}^D T_{il}}{\sum_{l=1}^D P_{il}}$	$\max_{P_{i,1},\dots,P_{i,D}} \dfrac{\sum_{l=1}^D T_{il}}{\sum_{l=1}^D P_{il}}$	$P_{il}^* = \Omega_i^* \, \Theta_{n(l)}$
Energy-efficient power and rate control with delay prerequisites	$U_i(P_i, \bar{P}_{-i}) = R_i f(\gamma_i(P_i, \bar{P}_{-i}))/P_i$	$\max_{P_i, R_i} U_i$ s.t. $\bar{W}_i \leq D_i$	$R_i = \left(\dfrac{M}{D_i}\right)^{\frac{1+D_i\lambda_i+\sqrt{1+D_i^2\lambda_i^2+2(1-f(\gamma^*))D_i\lambda_i}}{2f(\gamma^*)}}$, $\gamma_i = \gamma^*$
Multiservices QoS-aware power control	$U(R_{i^*}, P_i, \bar{P}_{-i}) = T_i(R_{i^*}, P_i, \bar{P}_{-i})/P_i$	$\max_{P_i \in P_{i^*}} U_i(P_i, \bar{P}_{-i})$ s.t. $0 \leq P_i \leq P_i^{\text{Max}}$	$P_i^* =$ $\min_{i \in S_{RT}}\left\{\dfrac{\gamma_i^*(R_{T,i}^* + MF_i)L_{-i}(\bar{P}_{-i})}{WG_i}, P_i^{\text{Max}}\right\}$, $\min_{i \in S_{NRT}}\left\{\dfrac{\gamma_i^* R_i^{\text{Max}} L_{-i}(\bar{P}_{-i})}{WG_i}, P_i^{\text{Max}}\right\}$
Power control via pricing	$U_i^c(P_i) = U_i(P_i) - c_i(P_i, \bar{P}_{-i})$	$\max_{P_i \in A_i}[U_i(P_i) - c_i(P_i, \bar{P}_{-i})]$	$P_i^* = \text{argmax}_{P_i \in A_i}[U_i(P_i) - c_i(P_i, \bar{P}_{-i})]$

Table 9.2 Noncooperative Power Control Games: Utility Function, Game's Formulation, and Game's Solution

Game	Utility Function	Existence of NE	Equilibrium's Properties	Convergence
Fixed SIR power control	$U_i(P_i, \bar{P}_{-i}) = P_i$	Standard interference function and supermodularity	Supermodular game $\Longrightarrow$ unique NE	Supermodular game $\Longrightarrow$ monotonic convergence to NE for any initial feasible power policy
Spectral efficiency via power control	$U_i(P_i, \bar{P}_{-i})$: quasi-concave of P_i	Quasi-concave utility function	Reductio ad absurdum, supposing the existence of two distinct equilibria	Quasi-concave utility function $\Longrightarrow$ monotonic convergence to NE for any initial feasible power policy
Energy-efficient power control	$U_i(P_i, P_{-i})$ $= R_i f(\gamma_i(P_i, \bar{P}_{-i}))/P_i$	Quasi-concave utility function	Unique NE $\gamma^* \longleftrightarrow P^*$, resulting from the unique positive solution of the equation $\gamma_i \partial f(\gamma_i)/\partial \gamma_i - f(\gamma_i) = 0$	Quasi-concave utility function $\Longrightarrow$ monotonic convergence to NE for any initial feasible power policy
Energy-efficient power control and receiver design	$U_i(P_i, P_{-i}, c_i)$ $= f(\gamma_i(P_i, c_i))/P_i$	Analytical mathematical method, the existence of NE, derives from the existence of a solution of the equation $\gamma_i \partial f(\gamma_i)/\partial \gamma_i - f(\gamma_i) = 0$	Unique NE for each receiver type (analytical mathematical method)	Best-response strategy via a distributed and iterative algorithm
Energy-efficient power control in multicarrier CDMA wireless networks	$U_i^{MC} = \dfrac{\sum_{l=1}^D T_{il}}{\sum_{l=1}^D P_{il}}$	Analytical mathematical method	No equilibrium, unique equilibrium, or more than one equilibriums depending on users' channel gains (analytical mathematical method)	Best-response strategy via a distributed and iterative algorithm
Energy-efficient power and rate control with delay prerequisites	$U_i(P_i, P_{-i})$ $= R_i f(\gamma_i(P_i, \bar{P}_{-i}))/P_i$	Analytical mathematical method, the existence of NE, derives from the achievement of γ^*, which is the unique positive solution of the equation $\gamma_i \partial f(\gamma_i)/\partial \gamma_i - f(\gamma_i) = 0$	Pareto-dominant NE, where each user receives the optimal transmission rate Ω_i^* for which $\gamma_i = \gamma^*$ (analytical mathematical method)	Best-response strategy via a distributed and iterative algorithm
Multiservices QoS-aware power control	$U(R_i^*, P_i, P_{-i})$ $= T_i(R_{iF}, P_i, \bar{P}_{-i})/P_i$	Quasi-concave utility function	Unique NE $\gamma^* \longleftrightarrow P^*$, resulting from the quasi-concave utility function	Supermodular game $\Longrightarrow$ monotonic convergence to NE for any initial feasible power policy
Power control via pricing	$U_i^c(P_i)$ $= U_i(P_i) - c_i(P_i, \bar{P}_{-i})$	Supermodular game	Supermodular game $\Longrightarrow$ Unique and Pareto dominant NE	Supermodular game $\Longrightarrow$ monotonic convergence to NE for any initial feasible power policy

(i.e., via the additional objective of interference minimization). The corresponding noncooperative power control game is extensively analyzed in Section 9.5.

Similar to the previous approach, instead of introducing a utility function for each user i, a cost function is defined in Refs. [11,26,27] as the difference between a linear pricing scheme proportional to transmitted power and a logarithmic, strictly concave utility function of user's achieved SIR.

$$J_i(P_i, P_{-i}) = c_i P_i - u_i \ln(1 + \gamma_i) \tag{9.7}$$

Extending the idea of utilities as cost functions in Ref. [28] a nonzero SIR level is considered necessary for accurate communication; therefore, users' cost is considered analogous to the difference between the actual SIR and the target SIR that depends on the desired service for each user i. Furthermore, a user has two conflicting objectives. On one hand, the higher the SIR, the better the service, and on the other hand, higher SIR is achieved at the cost of increased drain on the battery and higher interference to signals of other users, that is,

$$J_i(P_i, \gamma_i(P_i)) = c_i P_i + b_i \left(\gamma_i^{\text{target}} - \gamma_i\right)^2, \quad P_i \geq 0 \quad \forall i \tag{9.8}$$

where
$\quad c_i$ and b_i are constant nonnegative weighting factors
$\quad \gamma_i^{\text{target}}$ is the targeted SIR

Since the introduced cost function should be convex and nonnegative to allow existence of a nonnegative minimum and given that the transmission power is always positive, the SIR error may be either positive or negative. Thus, in order to ensure positivity and convexity of the cost function, the SIR error term is squared in the resulting cost function, as shown in (9.8).

9.3.3 Energy Efficiency and Transmission Awareness

When devising utilities from users' application point of view, the unique characteristics of the wireless transmission environment, in terms of transmission errors occurring at the receiver, which are strongly related with the achieved SIR, play a key role in the definition of the utility. On one hand, higher SIR level at the receiver will result in a lower bit-error rate and, hence, higher throughput, while on the other hand, achieving a high SIR level usually requires the user to transmit at a high power, which, in turn, results in low battery life and high interference. In Ref. [16], this trade-off is investigated and quantified by defining utility functions that measures energy efficiency and users' degree of satisfaction, as a function of SIR for fixed powers, and as a function of power for fixed SIR (Figure 9.1).

Toward that direction in Refs. [14–17,29], an alternative approach is adopted in which users' utilities express their degree of satisfaction with respect to the number of reliably transmitted bits per joule of energy consumed, that is,

$$U_i = \frac{T_i}{P_i} \tag{9.9}$$

where T_i is a function of user's i achieved *goodput*, that is, the number of bits that are transmitted successfully per unit time and is defined as

$$T_i = R_i f(\gamma_i) \tag{9.10}$$

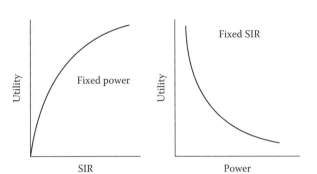

Figure 9.1 **Behavior of users' satisfaction as a function of SIR for fixed power and as a function of power for fixed SIR.**

where f_i denotes user's efficiency function. The latter represents the probability of a successful packet transmission for user i and is an increasing function of his SIR γ_i. A user's function for the probability of a successful packet transmission at fixed data rates depends on the transmission schemes (modulation and coding) being used, and is represented by a sigmoidal-like function of its power (and corresponding SIR) for various modulation schemes, as revealed in [3] and further depicted in Figure 9.2. Specifically, a user's i efficiency function f_i has the following properties [16]:

1. f_i is an increasing function of γ_i.
2. f_i is a continuous, twice differentiable sigmoidal function with respect to γ_i.
3. $f_i(0) = 0$ to ensure that $T_i = 0$ and $U_i = 0$ when $P_i = 0$.
4. $f_i(\infty = 1)$.

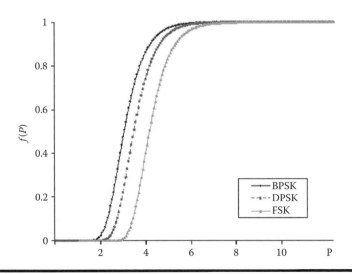

Figure 9.2 **Probabilities of packet transmission success for BPSK, DPSK, and FSK modulation schemes. (From Lee, J.-W. et al., *IEEE ACM Trans. Networking*, 13, 4, 854, 2005. With permission.)**

The utility function (9.9) and its variations of it have been adopted by a large number of noncooperative power control game-theoretic studies, aiming at different objectives, like energy efficiency (Section 9.6), energy efficiency and receiver design (Section 9.7), energy efficiency in multicarrier CDMA wireless networks (Section 9.8).

The above approach can be further extended by introducing and incorporating a pricing scheme to the definition of user's utilities defined in (9.9), toward improving overall system's utility-based performance efficacy by steering users to adopt a more social behavior. Hence, the authors in Refs. [14–16,29] have introduced a linear pricing function $C_i(P_i)$, which leads to the following definition of a user's i net utility function:

$$U_i^c = \frac{T_i}{P_i} - C_i = \frac{R_i f(\gamma_i)}{P_i} - c_i P_i \qquad (9.11)$$

where c_i is the pricing factor. Each cost coefficient is positive because any increase in the power of one user reduces the SIR, of every other user in the system and hence, should result in the decrease of his utility. The concept of charging a user according to the amount of harm he causes to the rest of the users results in encouraging users to transmit at a lower power level causing less interference to the others. The corresponding noncooperative pricing-driven power control game is extensively analyzed in Section 9.11.

9.3.4 Multiservice QoS-Aware Utilities

Even though utility (9.9) captures very well the trade-off between achieved goodput and battery life, which makes it in principle suitable for applications where energy efficiency is more important than just achieving high throughput, it cannot support services' differentiation. In other words, it can not be directly used for formulating a noncooperative game where users are competing for the available scarce wireless network's resources, while demanding the fulfilment of their individual services' diverse QoS prerequisites. Utility functions have been widely used in the literature [1] as enablers for expressing various services' QoS requirement in resource allocation problems, both in wired [30–32] and wireless networking environments [7,33,34]. In general, concerning non-real-time delay-tolerant data services, convex or concave utilities of the achieved throughput have been adopted [9,30,31] toward reflecting their high throughput expectations, while sigmoidal utilities are assigned to real-time delay-sensitive services, regarding either delay-adaptive (e.g., audio and video) or rate-adaptive real-time applications, as shown in Figure 9.3.

In Refs. [20,35,36], the pre-described utility-based framework for achieving service differentiation is introduced in noncooperative power control games over wireless networks, in order to simultaneously support multiple services' QoS prerequisites. Therefore, users' utilities are expressed as follows:

$$U\left(R_i^*, P_i, \bar{P}_{-i}\right) = \frac{T_i^{\text{act}}\left(R_i^*, P_i, \bar{P}_{-i}\right)}{P_i} = \frac{T_i^{\text{act}}\left(R_i \cdot f_i(\gamma_i), P_i, \bar{P}_{-i}\right)}{P_i} \qquad (9.12)$$

where a user's i *actual throughput utility* $T_i^{\text{act}}\left(R_i^*, P_i, \bar{P}_{-i}\right) \equiv T_i^{\text{act}}\left(R_i^*\right)$ is a function of his achieved goodput, properly selected to reflect within his overall utility U_i and respective power control strategy, his service performance requirements, and QoS prerequisites under the imposed physical limitations. Specifically, convex, concave, and sigmoidal actual throughput utilities are concerned and investigated for non-real-time data and real-time multimedia services, respectively. The corresponding noncooperative power control game is extensively analyzed in Section 9.10.

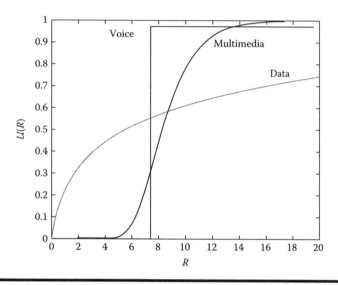

Figure 9.3 Utility function of different services (i.e., voice, data, multimedia).

In the following sections, a comprehensive and concrete analysis of applying game-theoretic approaches to the noncooperative power control problem in CDMA wireless networks is presented. Thus, toward increasing presentation's efficacy, in Table 9.2, we summarize the noncooperative power control games that will be extensively analyzed in the following sections. Specifically, Table 9.2 presents for each adopted utility function the corresponding noncooperative game formulation, and its concluded solution.

9.4 Fixed SIR Power Control

In the problem of uplink power control with fixed SIR, a fixed designed signal to interference constraint γ_i' is assigned to each user $i \in S$. Moreover, the corresponding problem of obtaining optimal users' transmission power vector $\boldsymbol{P}$ that satisfies fixed predefined target SIR constraints while minimizing total power, can be formally stated as follows:

$$\min \sum_{i \in S} P_i$$

$$\text{s.t.} \quad \gamma_i(P_i, \bar{P}_{-i}) \geq \gamma_i' \tag{9.13}$$

In Ref. [25], Yates introduced five different variations of problem (9.13), distinguished in accordance with the considered users to base stations assignment goal, namely fixed assignment, minimum power assignment, macro diversity, limited diversity, and multiple connection reception. Moreover, via expressing the SIR of each user by a vector inequality of interference constraints in the form $\boldsymbol{P} \geq \boldsymbol{I}(\boldsymbol{P})$, a general proof is provided for synchronous and totally asynchronous convergence of the iteration $p(t+1) = I(p(t))$ to a unique fixed point at which total transmitted power is minimized for all five problems, given that I satisfies the following properties that characterize it as *standard.*

Definition 9.7 Interference function $I(P)$ is standard if for $P \geq 0$ the following properties are satisfied.

1. Positivity: $I(P) > 0$
2. Monotonicity: If $P \geq P'$ then $I(P) \geq I(P')$
3. Scalability: For all $\alpha > 1$, $\alpha I(P) \geq I(\alpha P')$

From a game theoretic point of view, problem (9.13) can be formulated as a noncooperative game, at which each user aims at minimizing his transmission power while fulfilling the designed lower bounds of his achieved SIR, γ'_i, that depends on the transmission powers of all users, P, that is,

$$\min_{P_i \in A_i(\bar{P}_{-i})} P_i \quad \text{for } i = 1, \ldots, N \quad \text{where } A'_i(\bar{P}_{-i}) = \left\{ P_i \geq 0 : \gamma_i(P_i, \bar{P}_{-i}) \geq \gamma'_i \right\} \qquad (9.14)$$

9.4.1 Existence of Nash Equilibrium

In accordance with (9.14), each user i aims at achieving an SIR level above a designed threshold γ'_i with minimum transmission power. Intuitively, user i tries to maximize his SIR when the latter is lower than the predefined one, while aiming at minimizing his transmission power when his achieved SIR is higher than γ'_i, given other user's transmission power levels. In Ref. [37], it is proven that game (9.14) is supermodular via showing that users' strategy sets A_i are supermodular, that is,

$$\bar{P}_{-i} < \bar{P}'_{-i} \rightarrow A_i(\bar{P}_{-i}) \supset A_i\left(\bar{P}'_{-i}\right) \quad \forall i \in S$$

Thus, by using the tools from the supermodular game setting, the following properties can be justified:

■ A unique NE exists.
■ For any initial feasible power policy, the sequence of best responses defined as

$$\text{BR}_i = (\gamma'_i / G_i) I_{-i}(\bar{P}_{-i}) \, \forall i \in S \qquad (9.15)$$

converges monotonically to the game's equilibrium P^*_i.

In Ref. [38], a unified framework is proposed, based on the emergent potential games, to deal with a variety of network resource allocation problems. Generalizing and extending fixed SIR game theoretic power control setting, in the following sections, our interest is placed on variable SIR games where each user aims at jointly optimizing the achieved SIR and transmitted power. On one hand, such an enhanced freedom allows the formulation and investigation of various power control games by assigning to users more complex utility functions that can reflect multiple objectives, from throughput maximization to various QoS user's services prerequisites and power consumption minimization. On the other hand, this freedom is limited by user's terminal transmission power abilities, which are imposed as power constraints on the under-consideration games. Such limitations may raise feasibility issues, since not all SIR assignments are feasible due to interference-limited nature of the wireless network. In this chapter, we study various problems of jointly optimizing SIR assignment and transmit powers but only over the feasible SIR region. Thus, interested readers are referred to Ref. [39] for a detailed analysis on the characterization of feasible SIR regions in wireless networks.

9.5 Spectral Efficacy via Power Control

Aiming at spectral efficacy, a noncooperative power control game is proposed in Ref. [12], in which each user chooses his transmit power in such a way as to maximize his capacity (i.e., throughput). The objective is to formulate a game in terms of intrinsic properties of the channel, such as achieved SIR and transmission power, and, thus decouple it from lower layer decisions such as modulation and coding.

The corresponding noncooperative power control game can be written as

$$\max_{a_i \in A_i} U_i(P_i, \, P_{-i}) \quad \text{for } i = 1, 2, \ldots, N \tag{9.16}$$

where user i has the strategy set $A_i = [0, P_i^{\text{Max}}]$ and his utility function is given by $U_i = B \log_2(1 + \gamma_i) - c_i P_i$. In other words problem (9.16) implies that each user aims at maximizing his capacity by assuming the power of other users is given. As shown in Ref. [12], a NE in game (9.16) exists under the fulfillment of two sufficient conditions, detailed in the following theorem.

9.5.1 Existence of Nash Equilibrium

THEOREM 9.1 A NE exists in game (9.16) if for all $i \in S$

1. A_i is a nonempty, convex, and compact subset of some Euclidean space $\mathfrak{R}^n$.
2. $U_i(\boldsymbol{P})$ is continuous in $\boldsymbol{P}$ and quasi-concave in P_i.

A utility function is characterized as quasi-concave if the following attributes are satisfied.

Definition 9.8 A function U of a single variable P is quasi-concave if and only if either

- It is nondecreasing.
- It is nonincreasing.
- There exists P^* such that U is nondecreasing for $P < P^*$ and nonincreasing for $P > P^*$.

Hence, via showing the concavity of the adopted utility function and given that P_i is a compact, convex set, the existence of NE in the game (9.16) can be illustrated [12]. The corresponding best-response strategy gives

$$\boldsymbol{P}^* = F(\boldsymbol{P}) = F(\mathbf{M}\boldsymbol{P} + \mathbf{b}) \tag{9.17}$$

where $F(\cdot)$ is a diagonal mapping from $\mathfrak{R}^n$ to $\mathfrak{R}^n$ defined as

$$F_i(P_i) = P_i^{\text{Max}}, \quad \text{if } P_i \geq P_i^{\text{Max}} \tag{9.18}$$

$$F_i(P_i) = P_i \quad \text{if } 0 \leq P_i \leq P_i^{\text{Max}}$$

$$F_i(P_i) = 0 \quad \text{if } P_i \leq 0$$

the matrix $\mathbf{M} \in \mathbf{R}^{n \times n}$ is defined as

$$m_{ii} = 0 \quad \text{and} \quad m_{ij} = -\left(\frac{B}{W}\right)\frac{G_j}{G_i} \quad \text{for } i \neq j \tag{9.19}$$

and the ith component of the vector $b \in \mathfrak{R}^n$ is a constant given by

$$b_i = \frac{B}{\alpha_i \ln 2} - \frac{B\sigma^2}{WG_i} \tag{9.20}$$

9.5.2 Equilibrium's Properties

The uniqueness of the above NE is shown via reduction to the absurd (reductio ad absurdum), supposing the existence of two distinct equilibria P and P' and based on the following properties of mapping F for any $P, P' \in \mathfrak{R}^n$

1. $P \leq P'$ implies $F(P) \leq F(P')$
2. $F(P) - F(P') \leq F(P - P')$
3. $F(P + P') \leq F(P) + F(P')$
4. $F(P) + F(P') \leq |P|$ with equality if $-P_i^{\text{Max}} \leq P_i \leq P_i^{\text{Max}}$. Here $|P|$ denotes the vector obtained by replacing each component of P by its absolute value

Applying the previous methodology, it is concluded that one and only one NE exists [12].

9.5.3 Convergence of Nash Equilibrium

Following the previous analysis, a distributed and iterative (i.e., all users' responses are synchronously announced every $\tau_k \in T$, $k \in \aleph^+$) power control algorithm is proposed that converges to some power vector P^*, the NE of game (9.16).

9.5.3.1 Algorithm I (Spectral Efficiency Power Control Algorithm)

Consider the noncooperative game G given in (9.16) and suppose that terminals update their powers at time instances given by $T = \{\tau_1, \tau_2, \tau_3, \ldots\}$. Generate the sequence of powers as follows:

Step 1: Set the initial power vector $P(0) = P$, where P is any vector in the strategy space P. Set $k = 1$.
Step 2: For all k such that $\tau_k \in T$ and for all terminals, given $P(\tau_{k-1})$, compute $P_i^*(\tau_k) = \arg\max_{P_i} U_i(P_i, P_{-i}(\tau_{k-1}))$ for $i = 1, 2, \ldots, N$.

9.6 Energy-Efficient Power Control

With the objective of introducing users' energy expenditure, an inherent mobile nodes' vital restriction, in the noncooperative power control game proposed in Refs. [14–17,24,29], each user selects its transmission power in such a way as to maximize the achieved goodput while minimizing the corresponding energy consumption, that is, $U_i(P_i, \bar{P}_{-i}) = T_i(P_i, \bar{P}_{-i})/P_i = R_i f(\gamma_i(P_i, \bar{P}_{-i}))/P_i$ measured in bits/Joule. The corresponding optimization problem is formulated as follows:

$$\max_{P_i \in A_i} U_i(P_i, P_{-i}) \quad \forall i \in S \tag{9.21}$$

where $A_i = [0, P_i^{\text{Max}}]$. Toward investigating the maximization of a user's i utility, by requiring equality with zero of the partial derivative of U_i with respect to P_i (i.e., $\partial \mathbf{U_i}/\partial \mathbf{P_i} = 0$), the first order necessary condition demands that U_i is maximized when a specific SIR level is achieved γ_i^*, which corresponds to the unique (positive) solution of

$$\gamma_i \frac{\partial f(\gamma_i)}{\partial \gamma_i} - f(\gamma_i) = 0 \tag{9.22}$$

9.6.1 Existence of Nash Equilibrium

Equality (9.22) reveals that users' goal to maximize their utility-based performance and thus their degree of satisfaction in terms of throughput performance and corresponding energy consumption can be translated to a constant attempt of meeting specific SIRs γ_i^* (at the base station), which eventually leads to an signal to interference and noise ratio (SINR)-balanced network at game NE, if exists. Thus, γ_i^* depends only on the physical-layer characteristics of the communication such as modulation, coding, and packet size. Furthermore, if a user's maximum transmission power is not sufficient for reaching the targeted γ_i^*, due to his potentially bad channel conditions, then the best policy is to transmit with maximum power. As demonstrated in Ref. [40], the previous argument holds due to the quasi-concave form of user's i utility as a function of P_i, since when $P_i \leq P_i^*\left(\gamma_i^*\right)$ then $U_i(P_i)$ is an increasing function of P_i. Therefore, due to the one-to-one relationship between P_i and γ_i, based on (9.1), the best-response strategy of user i is given by

$$P_i^* = \arg\max_{P_i} U_i(\boldsymbol{P}) = \min\left\{\frac{\gamma_i^* R_i^{\text{Max}} I_{-i}(\bar{P}_{-i})}{W G_i}, P_i^{\text{Max}}\right\} \tag{9.23}$$

9.6.2 Equilibrium's Properties

In Ref. [15], it is shown that users' power control strategy $\boldsymbol{P}^* = \{P_1^*, \ldots, P_N^*\}$ leads to a game's (9.21) NE point, if there exists one. Moreover, the existence of the equilibrium follows from the fact that user's utilities are quasi-concave of their transit powers. Finally, equilibrium's uniqueness results from the uniqueness of γ_i^* and the one-to-one correspondence between P_i and γ_i.

THEOREM 9.2 [16] The power control game defined in (9.21) has a unique NE. At equilibrium point, a user either achieves the utility-maximizing power allocation $P_i^* = \gamma_i^* R_i^{\text{Max}} I_{-i}(\bar{P}_{-i})/WG_i$ or transmits at maximum power $P_i^* = P_i^{\text{Max}}$.

9.6.3 Convergence of Nash Equilibrium

Concluding this section's analysis, an iterative parallel (i.e., all users' responses are synchronously announced every $\tau_k \in T, k \in \aleph^+$) algorithm for obtaining game's (9.21) NE is presented [16].

9.6.3.1 Algorithm II (Energy-Efficient Power Control Algorithm)

Consider the noncooperative game G as given in (9.21). Generate a sequence of powers as follows:

> Step 1: Set the initial power vector at time $\tau_0 = 0 : \boldsymbol{P}(\tau_0) = \boldsymbol{P}$. Also let $k = 1$
> Step 2: For all k such that $\tau_k \in T$

1. For all users $i \in N$ such that $\tau_k \in T_i$
   (a) Given $\boldsymbol{P}(\tau_{k-1})$, compute $r_i(\tau_k) = \arg \max\limits_{P_i} U(P_i, P_{-i}(\tau_{k-1}))$
   (b) Assign the transmit power as $P_i(\tau_k) = \min(r_i(\tau_k))$, where $r_i(\tau_k)$ is the set of best transmit powers for user i at time instance k in response to the interference vector $P_{-i}(\tau_{k-1})$

9.7 Energy-Efficient Power Control and Receiver Design

With the intention of introducing receiver's design attributes in a game-theoretic power control setting, in Ref. [17] enhancing the previous game (Section 9.6), users are allowed to selfishly choose their uplink receivers as well as their transmit powers to maximize their own utilities. Emphasis is placed on linear receivers, where, apart from the simple matched filter (MF), the decorrelator (DE) [41] and the minimum mean-square error (MMSE) [42] receivers are studied as well. Therefore, even though a user's utility still measures the number of reliable bits transmitted per joule of energy consumed, as in (9.21), it is a function not only of his transmission power P_i but also of the receive filter coefficients c_i, for each user i. Thus, under the assumption of equal transmission rates, the corresponding noncooperative game can be formally stated as follows:

$$\max_{a_i \in A_i} U_i = \max_{P_i, c_i} U_i(P_i, c_i) = \max_{P_i, c_i} \frac{f(\gamma_i(P_i, c_i))}{P_i} \quad \forall i \in S \tag{9.24}$$

where
$$a_i = (P_i, c_i)$$
$$A_i = \left[0, P_i^{\text{Max}}\right] \times \Re^N$$

Via exploiting and then comparing the attributes of game's (9.24) equilibrium, in terms of resultant users' utilities values, it is shown that the MMSE receiver achieves higher utility than the other linear receivers.

9.7.1 Existence of Nash Equilibrium

THEOREM 9.3 The NE for the noncooperative game in (9.24) is given by $\left(P_i^*, c_i^*\right)$, where c_i^* is the vector of MMSE receiver coefficients, and $P_i^* = \min\left(P_i^{MMSE}, P_i^{\text{Max}}\right)$. Here, P_i^{MMSE} is the transmit power that results in an SIR equal to $\gamma_i \partial f(\gamma_i)/\partial \gamma_i - f(\gamma_i) = 0$, at the output of the MMSE receiver. Moreover, this equilibrium is unique (up to a scaling factor for the MMSE filter coefficients) [17].

9.7.2 Equilibrium's Properties

Furthermore, via using a large-system analysis, the best-response strategy of each user i is explicitly derived for all three power control games, under the assumption of fixed receiver types, that is,

$$P_i^{MF} = \frac{1}{G_i^2} \frac{\gamma^* \sigma^2}{1 - \alpha \gamma^*} \qquad \text{for } \alpha < \frac{1}{\gamma^*}$$

$$P_i^{DE} = \frac{1}{G_i^2} \frac{\gamma^* \sigma^2}{1 - \alpha} \qquad \text{for } \alpha < 1 \qquad\qquad (9.25)$$

$$P_i^{MMSE} = \frac{1}{G_i^2} \frac{\gamma^* \sigma^2}{1 - \alpha \dfrac{\gamma^*}{1 + \gamma^*}} \qquad \text{for } \alpha < 1 + \frac{1}{\gamma^*}$$

where
  α is the system load, which is defined as the ratio of the number of users to the processing gain
    (i.e., number of users per degree of freedom)
  σ^2 is the noise power, which includes other-cell interference

The previous results are also extended to multi-antenna systems, where it is observed that for the case of the MF and the MMSE detector, using more antennas at the receiver provides both power pooling and interference reduction. Specifically, the system behaves like a single-antenna system with processing gain mN, where m are the receiver antennas at the uplink receiver, and its received power is equal to the sum of the received powers at the individual antennas. On the other hand, the DE case benefits only from power pooling.

9.8 Energy-Efficient Power Control in Multicarrier CDMA Networks

Multicarrier CDMA combines the benefits of orthogonal frequency-division multiplexing (OFDM) with those of CDMA; therefore, it is considered as a potential candidate for next-generation high data-rate wireless systems [18]. Specifically, in such networks, the data stream for each user is divided into multiple parallel streams, and each stream is spread via a spreading sequence, which is then transmitted on a carrier. In a single-user scenario, assuming fixed total transmit power, the optimal power allocation strategy for maximizing the rate can be obtained via a waterfilling method over the frequency channels [43]. Considering multiuser scenario, several waterfilling methods have been investigated toward maximizing the overall throughput [44–46]. However, in many practical scenarios in wireless networking, where the main goal is to maximize the number of transmitted bits per joule of energy consumed, enhancing power efficiency is considered more important than maximizing throughput, which motivates a user-centric game-theoretic treatment of power control in multicarrier CDMA systems.

Focusing on multiple-access multicarrier direct-sequence CDMA wireless networks, a noncooperative game $G_D = [N, \{A_i^{MC}\}, \{U_i^{MC}\}]$ is formulated in Ref. [18], in which each user i determines his transmission power P_i over each frequency channel l (i.e., carrier) in order for his utility U_i^{MC} to be maximized. Moreover, the strategy set of the ith user is $A_i^{MC} = [0, P_i^{\text{Max}}]^D$, when considering D available carriers, which can be written as $a_i = (P_{i1}, \ldots, P_{iD})$. The utility of user i is defined as the ratio of the actual achieved throughput to the overall energy consumption, that is,

$$U_i^{MC} = \frac{\sum_{l=1}^{D} T_{il}}{\sum_{l=1}^{D} P_{il}} \tag{9.26}$$

where T_{il} is the actual achieved throughput by user i transmitting over the lth carrier defined as $T_{il} = (L/M)R_i f(\gamma_{il})$, L and M are the number of information bits and the total number of bits in a packet, respectively, and γ_{il} is the received SIR of user i over the lth carrier. In accordance with the previous system formulation, the corresponding noncooperative game can be expressed as the following maximization problem:

$$\max_{a_i \in A_i} U_i^{MC} = \max_{P_{i1},\ldots,P_{iD}} U_i^{MC} = \max_{P_{i1},\ldots,P_{i,D}} \frac{\sum_{l=1}^{D} T_{il}}{\sum_{l=1}^{D} P_{il}} \quad \forall i \in S \tag{9.27}$$

By characterizing the utility maximization of a singe user, the following theorem can be derived.

9.8.1 Existence of Nash Equilibrium

THEOREM 9.4   [18] For all linear receivers and with all other users' transmit power being fixed, user's i utility function is maximized when

$$P_{il} = \begin{cases} P_{iL_i}^*, & \text{for } l = L_i \\ 0, & \text{for } l \neq L_i \end{cases} \tag{9.28}$$

where $L_i = \arg\min_l P_{il}^*$ with P_{il}^* being the transmit power required by user i to achieve an output SIR equal to γ^* on the lth carrier, or P_i^{Max} if γ^* is not feasible. The parameter γ^* is the unique positive solution of equation $\gamma_i \partial f(\gamma_i)/\partial \gamma_i - f(\gamma_i) = 0$.

In other words, Theorem 9.4 implies that a user's i utility is maximized only over its "best" carrier and, thus, when his combined choice of power and rate is such that the output SIR equals γ^*. Intuitively, the "best" carrier is the one that requires the least amount of power to achieve γ^*. On one hand, when considering equal (or fixed) user's transmission rates, the uniqueness of the "best" carrier for each user reduces significantly the number of cases (i.e., potential power allocations) that need to be considered for possible NE. On the other hand, in the case of joint power and rate control, the satisfaction of γ^*, even over one carrier, implies that there are infinite power and rate combinations that maximize a user's utility, given the rates and powers of other users in the system fixed.

9.8.2 Equilibrium's Properties

Moreover, it is shown that for both cases, depending on the channel gains G_{il} of each user i transmitting over each lth carrier, the noncooperative game (9.27) may have no equilibrium, a unique equilibrium, or more than one equilibrium points. In contradiction with the previously examined case, the following theorem allows the identification of NE (if it exists) for a given set of channel gains [18].

THEOREM 9.5   [18] For a matched filter (MF) receiver, the necessary and sufficient condition for user i to transmit over the lth carrier at equilibrium is that

$$\frac{G_{il}}{G_{ik}} > \frac{\Theta_{n(l)}}{\Theta_{n(k)}}\Theta_0 \quad \text{for all } l \neq k \tag{9.29}$$

where

$n(l)$ is the number of users transmitting on the lth carrier

$\Theta_n = 1/(1 - (n-1)\frac{\gamma^*}{N})$, $n = 0, 1, \ldots, N$

The proof of Theorem 9.5 is based on the corresponding relations among a user's channel condition per carrier imposed by Theorem 9.4 (i.e., $G_{il} > G_{ik}$ in order for user i to transmit on carrier l at equilibrium). Thus, it is derived that the best-response strategy of user i transmitting over the lth carrier is given by

$$P_{il}^* = \frac{\gamma^*\sigma^2}{G_{il}}\Theta_{n(l)} \tag{9.30}$$

where σ^2 is the noise power, which includes other-cell interference. Consequently, the characterization of game's (9.27) equilibrium, in terms of existence and uniqueness, requires special treatment [18] for a given user's scenario (i.e., user's channel gains G_{il}).

9.8.3 Convergence of Nash Equilibrium

Further, an iterative and distributed algorithm for reaching the NE of the proposed multicarrier power control game, if it exists, is proposed in Ref. [16] applicable to all linear receivers including MF, decorrelating and MMSE detectors.

9.8.3.1 Algorithm III (Best-Response Multicarrier Power Control Algorithm)

Step 1: Initialize the transmit powers of all the users N over all the carriers D and let $k = 0$.

Step 2: Set $i = 1$ and $k = k + 1$.

Step 3: Given the transmit powers of other users, user i picks his "best" carrier and transmits only on this carrier at a power level that achieves an output SIR equal to γ^*. The "best" carrier is the one that requires the least amount of transmit power for achieving γ^*.

Step 4: $i = i + 1$.

Step 5: If $i \leq N$ then go back to Step 3.

Step 6: Stop if the powers have converged or if $k > MAX\_k$; otherwise go to Step 2.

The above algorithm is a best-response algorithm since at each stage, a user decides to transmit on the carrier that maximizes the user's utility given the current conditions of the system.

9.9 Energy-Efficient Power and Rate Control with Delay Prerequisites

The problem of energy-efficient joint power and rate control in CDMA wireless networks with delay-specific QoS constraints is addressed in Ref. [47], and in Ref. [48], users are organized in three basic classes and the transmission power and rate is allocated based on user's class. In the proposed game, each user aims at selfishly selecting his transmit power P_i and rate R_i (i.e., $\alpha_i = (P_i, R_i)$) in a distributed and iterative manner toward maximizing his own utility (i.e., $U_i = T_i(R_i f(\gamma_i))/P_i$) while satisfying his QoS requirements. The QoS prerequisites for each user i consist of an average source rate $R_{av,i}$ and an upper bound on the average delay D_i, which includes both transmission and queuing delays.

The incoming packets for each user are assumed to be stored in a queue and transmitted in a first-in-first-out (FIFO) fashion. Moreover, assuming that (a) the combination of user's i queue and wireless link is modeled as an $M/G/1$ queue and that (b) his incoming traffic has a Poisson distribution with parameter λ_i (in packets per second), which represents the average packet arrival rate with each packet consisting of M bits, the average source rate can be determined as $r_i = M\lambda_i$.

In accordance with the previous system model, the packet transmission time for each user i is estimated as $\tau_i = M/R_i$, while his service rate is defined as $\mu_i = f(\gamma_i)/\tau_i$, where $f(\gamma_i)$ is the efficiency function. Thus, the load factor is given by $\rho_i = \lambda_i/\mu_i = \lambda_i \tau_i/f(\gamma_i)$. In order to keep the queue of user i stable, the load factor must be $\rho_i < 1$, and thus, $f(\gamma_i) > \lambda_i \tau_i$ Furthermore, denoting $\bar{W}_i$ as the total packet delay for user i, the corresponding utility-maximizing strategy $\forall i \in S$, given the transmit powers and rates of the rest of the users, can be formally defined as

$$\begin{aligned} &\max_{P_i, R_i} U_i \\ &\text{s.t.} \quad \bar{W}_i \leq D_i \end{aligned} \tag{9.31}$$

where $\bar{W}_i = \tau_i(1 - \lambda_i \tau_i/2)/(f(\gamma_i) - \lambda_i \tau_i)$.

As shown in Ref. [47], the unconstraint utility maximization in (9.31) has an infinite number of solutions. Specifically, any combination of the transmit power P_i and rate R_i that achieves a level of SIR equal to γ^*, which is the unique positive solution of $\gamma_i \partial f(\gamma_i)/\partial \gamma_i - f(\gamma_i) = 0$, maximizes user's utility and achieves transmission rate $R_i \geq \Omega_i^*$, where Ω_i^* is the transmission rate for which $\gamma_i = \gamma^*$ defined as

$$\Omega_i^* = \left(\frac{M}{D_i}\right) \frac{1 + D_i\lambda_i + \sqrt{1 + D_i^2\lambda_i^2 + 2\left(1 - f\left(\gamma^*\right)\right) D_i\lambda_i}}{2f\left(\gamma^*\right)} \tag{9.32}$$

THEOREM 9.6 [47] The joint power and rate control game by (9.31) has more than one NE. In the Pareto-dominant equilibrium, each user achieves transmission rate $R_i = \Omega_i^*$ with $\gamma_i = \gamma^*$ (i.e., best-response strategy), and thus, $\bar{W}_i = D_i$.

However, if γ^* is not feasible due to the maximum transmit power limitation, the user adjusts his transmission rate and target SIR toward satisfying his QoS constraints. As shown in Ref. [47], this is achieved by adopting his transmission rate and target SIR, respectively, such that

$$\tilde{\Omega}_i = \left(\frac{M}{D_i}\right) \frac{1 + D_i\lambda_i + \sqrt{1 + D_i^2\lambda_i^2 + 2\left(1 - f\left(\tilde{\gamma}\right)\right) D_i\lambda_i}}{2f\left(\tilde{\gamma}\right)} \tag{9.33}$$

where $\tilde{\gamma} = BP_i^{\text{Max}}G_i/\tilde{\Omega}_i$.

Extending the previous analysis, the loss in energy efficiency and network capacity, due to the presence of real-time non-delay tolerant users, can be quantified by introducing the *"size"* of each user, which is defined as the indicator of the amount of system's resources that are occupied and thus consumed by each user. The user's "size" at the Pareto-dominant NE of joint power and rate control game (9.33) is defined as

$$\Phi_i^* = \frac{1}{1 + B/\Omega_i^* \gamma^*} \tag{9.34}$$

where B is the system's bandwidth. Moreover, the necessary and sufficient condition for this equilibrium to be feasible is given as

$$\sum_{i=1}^{N} \Phi_i^* < 1 \tag{9.35}$$

while the value of user's i corresponding utility is as follows

$$U_i = \left(\frac{B G_i f(\gamma^*)}{\sigma^2 \gamma^*} \right) \frac{1 - \sum_{i=1}^{N} \Phi_i^*}{1 - \Phi_i^*} \tag{9.36}$$

Concluding this section's analysis, it should be mentioned that based on (9.36) and considering the feasibility condition (9.35), the trade-offs among delay, energy efficiency, throughput, and network capacity can be analytically quantified.

9.10 Multiservices QoS-Aware Power Control

To address the need for supporting simultaneously various services with diverse QoS prerequisites over multiservice CDMA networks in Ref. [48], each user can be associated with a nested utility function, namely actual throughput utility T_i^{act}, which represents his degree of satisfaction in relation to the expected trade-offs between his QoS-aware actual uplink throughput (i.e., goodput) performance and the corresponding power consumption [20,35,36]. Concerning delay-tolerant high-throughput non-real-time services, users' actual throughput utilities are assumed to be convex or concave functions of their achieved goodput, while sigmoidal functions are adopted for delay-sensitive real-time applications. Founded on such a utility-based framework (detailed in Section 9.3), which favors services differentiations, the designed problem is formulated as a non-convex noncooperative multiservice uplink power control game, where users aim selfishly at maximizing their QoS-aware utility-based performance under the imposed physical limitation, that is,

$$\max_{P_i} U_i = \max_{P_i} U_i(P_i, \bar{P}_{-i}) \quad \text{for } i = 1, 2, \ldots, N$$
$$\text{s.t.} \quad 0 \le P_i \le P_i^{\text{Max}} \tag{9.37}$$

In other words, in the proposed optimization problem, each user i in the game (9.37) picks a transmission power from his strategy set A_i and receives a payoff $U_i = T_i^{\text{act}}/Pi$ (depending on his service performance requirements, QoS prerequisites fulfillment, and corresponding energy consumption) in accordance with his best-response policy $\text{BR}_i(\bar{P}_{-i}) = \max_{P_i \in A_i} U_i(P_i, \bar{P}_{-i})$.

The existence and uniqueness of NE in game (9.37) can be derived via studying the attributes of the corresponding users' nested utility function U. It is shown that, under specific conditions [20,36], the form of users' i actual throughout utility $T_i^{act}\left(R_i^*, P_i, \bar{P}_{-i}\right) \equiv T_i^{act}\left(\gamma_i\left(R_i^*, P_i, \bar{P}_{-i}\right), P_i\right)$, is a sigmoidal function of γ_i, for each one of the three considered user's actual throughput utilities (i.e., convex, concave, and sigmoidal). It is revealed that users' actual throughput utilities must comply with specific attributes regarding their form at the boundaries of their corresponding definition sets, in order (a) to assert overall utilities' soundness, in terms of expressing feasible users' QoS expectations given systems' and mobile nodes' physical limitations, and (b) for a unique NE of game (9.37) to exist. The absence of the NE not only means that the system is inherently unstable but also implies a strong correlation between users' QoS requirement expectations and systems' ability to fulfill them. Via the quantification of this correlation, specific QoS-aware call-admission-control (CAC) criteria can be derived [35].

9.10.1 Existence of Nash Equilibrium

In accordance with the previous analysis, a user's i overall utility function $U_i(P_i, \bar{P}_{-i})$ is a quasi-concave function of γ and, due to the one-to-one relationship between γ and P, is also a quasi-concave function of P. Thus, the quasi-concavity of user's utility function reveals the existence and uniqueness of Nash equilibrium P_i^*, considering both real-time and non-real-time services.

THEOREM 9.7   [20,36] The multiservice uplink power control game given in (9.37) has a unique NE. The best-response strategy of user i is given by

$$P_i^* = \begin{cases} \min\left\{\dfrac{\gamma_i^*\left(R_{T,i}^* + MF_i\right)I_{-i}(\bar{P}_{-i})}{WG_i}, P_i^{\text{Max}}\right\}, & i \in S_{RT} \\[4mm] \min\left\{\dfrac{\gamma_i^* R_i^{\text{Max}} I_{-i}(\bar{P}_{-i})}{WG_i}, P_i^{\text{Max}}\right\}, & i \in S_{NRT} \end{cases} \tag{9.38}$$

where

γ_i^* denotes the unique positive solution of the equation $\left(\partial T_i^{act}(\gamma_i)/\partial\gamma_i\right)\gamma_i - T_i^{act}(\gamma_i) = 0$

MF_i is the Margin Factor, which is determined in accordance with real-time services' QoS prerequisites [35]

S_{RT} and S_{NRT} are the sets of real-time and non-real-time users in the systems, respectively

9.10.2 Convergence of Nash Equilibrium

Furthermore, considering real-time users, sufficient conditions are given in Ref. [35], the fulfillment of which reassures not only users' overall utility-based performance maximization but also their service actual throughput expectation satisfaction. Moreover, it is demonstrated how this conditions can be used toward deriving appropriate CAC criteria (Proposition 2 in Ref. [35]). Finally, an iterative and distributed uplink power control algorithm is proposed, considering the above optimization problem, which converges to the unique NE beginning from any initial strategy of the strategy set A_i.

9.10.2.1 Algorithm IV (Multiservice Uplink Power Control Algorithm)

Step 1: At the beginning of time slot t, user i, $i \in S$ transmits with a randomly selected feasible power (i.e., $0 \le P_i^{*(0)} \le P_i^{\text{Max}}$). Set $k = 0$ and hence $P_i^{*(0)}$, $i \in S$.

Step 2: Given the uplink transmission powers of other users, which are implicitly reported by the base station when broadcasting its overall interference $I^{(k)}(\bar{P}^{(k)})$, the user computes $I_{-i}^{(k)}\left(\bar{P}_{-i}^{(k)}\right)$ and refines his power, that is, computes $P_i^{*(k+1)}$ in accordance with (9.39).

Step 3: If the powers converge (i.e., $\left|P_i^{*(k+1)} - P_i^{*(k)}\right| \le 10^{-5}$) then stop.

Step 4: Otherwise, set $k = k + 1$, go to step 2.

The previous multiservice uplink power control algorithm can be characterized as a single-valued low-complexity best-response algorithm for every user starting from a randomly selected feasible power of his non-empty orthogonal strategic space A_i (i.e., $P_i^{*(0)} \quad \forall i \in S$). Furthermore, under the assumption that P_i^{Max} is sufficiently large in order γ_i^* can be achieved by all users, the convergence of the proposed algorithm is always reached. In the general case, the convergence of the power control algorithm IV can be drawn based on game's supermodularity (i.e., Definitions 9.5 and 9.6) [20].

9.11 Power Control via Pricing

When participating in noncooperative power control game, the self-optimizing behavior of an individual user is said to create an externality when his actions correspond to the degradation of the performance of other users as well as of the overall system. Among the many ways to deal with externalities, pricing has been commonly used as an effective approach toward steering users/players to adopt a more social behavior, leading to the proposal and use of various pricing policies, such as flat rate, access based, usage based, priority based, etc. In general, the service provider determines both the pricing policy and the actual prices for the use of resources based on the system, the kind of resources it offers, and the type of demand for these services. Moreover, an efficient pricing method must reflect accurately the costs of usage of a resource and must take into account the nature of the demand for the offered service. One of the most commonly encountered pricing approaches in power control is usage-based pricing, in which the price a terminal pays for using the resources is proportional to the amount of resources consumed by the user.

As it has been demonstrated in Refs. [14–17,24,29,49], pricing constitutes an important design tool as it creates an incentive for the users to adjust their strategies, in this case power levels, in line with the goals of the network. In the special case already presented in Section 9.5 (i.e., $U_i = B \log_2(1 + \gamma_i) - c_i P_i$), the proposed pricing-based noncooperative power control game converges to a NE solution, without analytically computing the value of the pricing factor c_i (i.e., c_i is a predefined constant). Further, in Section 9.6, NE is adopted, considering utilities without pricing, although the achieved solution is inefficient, due to the selfish behavior of the users, who aim to maximize their own utility and fulfil their QoS requirements. In this section, the games presented in Section 9.6 are re-formulated appropriately by adding a pricing function to the initial user utilities and their solutions are derived via corresponding generic, iterative pricing-aware algorithms. Furthermore, complementing them by a power control algorithm (e.g., Algorithms I and II) the under consideration games coverage to the Nash equilibrium power vector, while simultaneously deriving the best pricing policy. Pricing is introduced as a linear function of transmit power P_i, alongside with the utilities (9.6) and (9.21).

The corresponding noncooperative power control game with pricing has also a set of users and a set of strategies and includes the net function $U_i^c(P_i)$, which is a linear combination of a utility and a pricing function. Let $G_c = [N, \{P_i\}, \{U_i^c\}]$ denote the N-users noncooperative power control game with pricing and user's net function [16], that is,

$$U_i^c(p) = U_i(p) - c_i(p_i, p_{-i}) \tag{9.39}$$

where $c_i : P \rightarrow \Re_+^1$ is the pricing function for user $i \in S$, which is a linear function of transmit power P_i, that is, $c_i(P_i, P_{-i}) = c_i P_i$ and c_i is a positive scalar.

Via the above formulation (9.39), efficiency in power control is promoted by a usage-based pricing strategy where each user pays a penalty proportional to his transmit power. Hence, the multi-objective optimization problem can be expressed as

$$\max_{P_i} \left[u_i(p) - c_i(p_i, p_{-i}) \right] \tag{9.40}$$

9.11.1 Existence of Nash Equilibrium and Pareto Optimality

Concerning the above noncooperative power control game with pricing, it is shown in Ref. [16] for positive values of the pricing factor, that there exist more than one Nash equilibria. Any NE of this set improves system's efficiency in Pareto sense in contrast to the NE achieved in the non-cooperative power control game without pricing. Moreover, the minimum power vector in the set of Nash equilibria achieves higher values of net utilities than any other equilibrium power vector. This equilibrium power vector is said to Pareto dominate all other power vectors.

9.11.2 Equilibrium's Properties

Furthermore, the theory of supermodular games is adopted to investigate Nash equilibria in the optimization problem illustrated in (9.40), due to their attribute of having a non-empty set of Nash equilibria. The definition of a generic supermodular game G given in Definition 9.6 (when $c = 0$) should be extended to the noncooperative power control game with pricing G_c, with the exogenous parameter of pricing c by imposing an additional nondecreasing difference condition regarding the pricing factor c. Therefore,

---

Definition 9.9 A noncooperative power control game with pricing, Gc, with an exogenous parameter c is said to be supermodular, it has nondecreasing differences in (a_i, a_{-i}) and in (a_i, c) for all i. Moreover, the Nash set has a largest element $a_L(c)$ and a smallest element $a_S(c)$, which are nondecreasing in c.

The proposed noncooperative power control game with pricing (9.40) can be shown that complies with the sufficient conditions of Definition 9.9; thus, it is proved that the game is supermodular.

9.11.3 Convergence of Nash Equilibrium

An iterative and distributed algorithm for noncooperative power control game with pricing is proposed and consists of two parts: the first one (i.e., Algorithm II), concerning the user, determines the transmit power vector and the second (i.e., Algorithm V), concerning the network, determines the optimal price. The network collaborates with the users in order for the noncooperative power control game with pricing to be solved, in terms of converging to the NE (i.e., user's goal) and the pricing factor (i.e., network's goal). Specifically, from user's perspective, Algorithm II converges to the NE of the proposed noncooperative power control game with pricing. In parallel, from network's perspective, Algorithm V, takes as an input the equilibrium with no pricing, which is already obtained and the noncooperative power control game with pricing is played again after incrementing the pricing factor, c, by a positive value, Δc. After, Algorithm II returns a set of powers at equilibrium with this value of the pricing factor. If the utilities at this new equilibrium with some positive price c improve with respect to the previous instance, the pricing factor is incremented and the procedure is repeated. The same procedure is continued until an increase in c results in utility levels worse than the previous equilibrium values for at least one user. The last value of c with Pareto improvement is declared to be the best pricing factor, c_{BEST}. The way c_{BEST} is determined by the network can be summarized in algorithmic format as follows:

9.11.3.1 Algorithm V (Pricing Factor Algorithm (Network))

Step 1: Set $c = 0$ and announce c to all users.
Step 2: Get U_i for all $i \in N$ at equilibrium, increment $c := c + \Delta c$ and announce to all users.
Step 3: If $u_i^c \leq u_i^{c+\Delta c}$ for all $i \in N$ then go to step 2, else stop and declare $c_{BEST} = c$.

As an alternative to the previously presented approaches, a distributed algorithm for the combined base station association and power control problem is proposed in Ref. [50], and subsequently, the corresponding problem is modeled as a player-specific congestion game. It is shown that the equilibrium states of such algorithms, which are Nash equilibria of the analogous games, may be far from the system optimal. Thus, a pricing mechanism is proposed to induce mobiles to behave in a way that optimizes system cost (i.e., network-centric goals). It is shown that such a mechanism can be employed in a distributed fashion. Toward this end, the network is modeled as having a continuum of (nonatomic) mobiles, each offering infinitesimal load, which leads to a population game formulation. Moreover, a marginal pricing mechanism is provided that motivates a pricing strategy for the discrete mobiles case.

9.12 Network-Centric Opposed to User-Centric Power Control and Base Station Assignment

Game models for power control can be further extended by allowing both network-centric and user-centric objectives to compete. In the power control problem considered in Ref. [19], the network is assumed as the leader, due to its ability to have access to more global information, and announces its decisions to the users, which are obliged to take this decision into account when selecting their reactions. The main goal via such a game is to adjust the transmission power of each user, the network pricing, and the base station assignment, while each user maximizes his net utility and the network maximizes its revenue. Specifically, the utility function is applied as the QoS for the user-centric problem, that is, $U_i = T_i/P_i$, $T_i = R_i f(\gamma_i)$, and for the network-centric problem; system's revenue is considered as the sum of payments from all the users, that is, $\rho = \sum_{i=1}^{N} \lambda_i T_i$, where λ_i is the unit price that each user is charged by the network in proportion to his throughput. Thus, user's i net function is defined as the difference between the utility and the payment for each user [51].

$$U_i^{\text{net}} \triangleq U_i - \lambda_i T_i \tag{9.41}$$

The joint user and network optimization problem is difficult to be solved and is not clear whether there exists NE. As a result, some heuristic algorithms that solve the problem have been proposed in Ref. [51], where the joint user and network optimization problem is decoupled into two different problems. From user's perspective, the objective is the maximization of his net utility, with two degrees of freedom: transmit power and base station assignment. Thus, given the network-decided unit price vector $\lambda = (\lambda_1, \lambda_2, \ldots, \lambda_N)$, the user objective is to maximize his net utility.

User Problem: $\quad \max_{P_i \in A_i} U_i^{\text{net}} = \max_{P_i \in A_i} U_i - \lambda_i T_i \quad \text{for } i = 1, \ldots, N \tag{9.42}$

where $A_i = \left\{ P_i | P_i^{\min} \le P_i \le P_i^{\max} \right\}$ is the strategy set of user i, with $P_i^{\min}$ and $P_i^{\max}$ denoting the minimum and the maximum transmitter power, respectively. In Ref. [51], it is shown that for the user-centric noncooperative game, there exists a unique NE.

THEOREM 9.8 [51] A unique NE exists for the noncooperative game defined in the user problem (9.42) if $BER(\gamma)$ decays exponentially in SIR γ, where BER is the standard transmission error rate, expressed as the ratio of error-bits received to the total bits sent, depending on γ, and is used as a measure of transmission quality.

The uniqueness of the NE derives from the properties of standard interference function of the mapping $X(\cdot)$ corresponding to the update rule $P(t + 1) = X(P(t))$, where $P(t) \in \Re^n$. Further, the output of the user optimization $P^*(\lambda)$ is applied as the input to the network optimization and it is explicitly a function of the unit prices. Therefore, payment and revenue are characterized by the unit price vector. Further, the network aims at finding its highest revenue by searching over a nonnegative price factor λ.

Network Problem: $\quad \max_{\lambda \ge 0} \rho(\lambda) = \max_{\lambda \ge 0} \lambda T_i(P^*(\lambda)) \tag{9.43}$

For the sake of simplicity, it is assumed that all users belong to the same priority class, and hence, the network applies a common unit price λ to all the users. However, this can be easily extended to cases with multiple classes of users. The existence of the optimal revenue derives from the finiteness property of the revenue.

The user problem defined by (9.42) can be further extended toward adjusting the base station assignment. However, searching over all possible base station assignments and performing net utility optimization over transmitter powers for every combination would be computationally intensive. Thus, the user problem can be simplified by noting that U_i^{net} is monotonically increasing in γ_i and therefore adjusts the base station assignment as follows [54]:

THEOREM 9.9 Given an interference vector P_{-i}, the base station assignment based on the net utility maximization is equivalent to the one based on maximizing SIR:

$$BS_i^* = \arg \max_{BS} U_i^{\text{net}} (BS, P_i, P_{-i}, \lambda) \equiv \arg \max_{BS} \gamma_i(BS, P). \tag{9.44}$$

The previous theorem states that the base station assignment based on net utility maximization is equivalent to that on SIR maximization. Further, user's i SIR maximization is independent of his power P_i. Therefore, the user problem can be solved by assigning base station first, followed by power control.

Concluding this section's analysis, two substitute game theoretic approaches for analyzing and studying the problem of distributed uplink power control are presented, namely Stackelberg [52] and potential [53] games. The foremost motivation behind both of illustrated approaches is the Pareto inefficiency of NE. Thus, alternative theoretic frameworks have been proposed and investigated toward studying the trade-off between cooperation and noncooperation in power control games.

In Ref. [47], a Stackelberg game is introduced into the power control game, which makes all users in system work in the best equal SIR. Specifically, opposed to the previously presented pricing-based studies, the following observations are made: (a) although all users' utility improves in the NE after the introduction of pricing, the algorithm is more complex than noncooperative game algorithm because of the uncertainty of the pricing factor; (b) in the NE of the SIR of all users is not equal, and, hence, the user whose distance is closer to base station has the maximum SIR and the farthest user has the minimum SIR. This results in unfairness between all users.

Therefore, Stackelberg game is introduced to the power control game to deal with these problems. Stackelberg game is a noncooperative two-stage leader–follower(s) game [54]. The two-stage leader–follower(s) game, categorized as a dynamic game, contains two types of players: a leader and follower(s), where one player, that is, the network, being the leader, announces its decision to the other players. The users, being the followers, then take this decision into account when designing their reactions.

Thus, a new distributed algorithm based on Stackelberg Power control Game (SPG) is proposed. Since in a realistic scenario, the network typically has access to more global information than the mobile users, it is more suitable for network to play the role of the leader instead of being another player in a common noncooperative game. In the first stage, the network, as the leader, attempts to find the best equal SIR for all users. In the second stage, after observing the leader's strategy, the users choose their transmitter power; this process is a noncooperative game. The existence and the uniqueness of the NE of the game are proved. In the NE, all users work in the best equal SIR, and thus, the utility of all users is maximal under the principle of the equal SIR for all users. It is shown that the algorithm can improve the capability of the system.

In Ref. [53], a potential game approach is exploited toward linking noncooperative and cooperative models of distributed uplink resource (power) allocation. The key motivation is that resource allocation games in congested networks are of strategic complements/substitutes with aggregation. In other words, the QoS of a mobile node only depends on its transmission power and an aggregate of others' strategies, measured as interference. Due to this aggregation, the players can be thought of as maximizing one common objective function. The common objective, namely potential function, formalizes the implicit joint target of the noncooperative players. Therefore, the potential game approach is a useful alternative approach in formalizing the system-level consequences of a given preference model (i.e., the potential function describing the system-level consequences of a given resource allocation game). With respect to the previous analysis and motivations, in Ref. [53], the authors formalize the potential function for distributed resource allocation games under congestion. Moreover, it is shown that via the proposed potential function approach, a distributed resource allocation framework can be derived that optimizes the trade-off between a sum of utilities and congestion costs.

9.13 Joint Power and Signature Sequence Control

Game models for power control can be further combined with signature sequence control under a unified game-theoretic framework. Specifically, each user is distinguished from the others by its signature sequence, which allows him to spread its information on the common channel through modulation using its signature sequence and to share the entire bandwidth with all the other users. In this section, a multi-objective joint power and signature control game formulation is outlined, and appropriate conditions that ensure the existence of NE are identified. The following noncooperative game setting has been proposed in Ref. [55], assuming that users are allowed to adjust both their transmit powers P_i and their signature sequences z_i, expressed as a unit-norm real-valued column vector in an L-dimensional vector space. Specifically,

The N-person power and sequence control game (G_{SP}) is considered as a tuple $(Z, \preceq_1, \ldots, \preceq_N)$, where Z is the strategic space defined as

$$Z = \left\{ (Z, P) \mid \ \|z_i\| = 1, 0 \le P_i \le P_i^{\text{Max}}, \forall i \in S \right\} \tag{9.45}$$

where $\mathbf{Z}$ is an $L \times N$ matrix whose columns consist of signature sequences and $\preceq_i$ is a preference relation for user i, represented by a properly assigned utility function U_i, as follows:

$$(Z', P') \preceq_i (Z, P) \quad \Leftrightarrow \quad U_i(Z', P') \le U_i(Z, P) \tag{9.46}$$

Moreover, in the context of such formulation, a pair $(\mathbf{Z}, \mathbf{P})$ is a NE of G_{SP} game, if for every user i the following inequality holds:

$$U_i(P_i, z_i, \mathbf{P}_{-i}, Z_{-i}) \ge U_i \left(P_i', z_i', \mathbf{P}_{-i}, Z_{-i} \right) \tag{9.47}$$

for every admissible P_i' and z_i'.

Toward studying the multi-objective G_{SP} game, users' utilities are assumed to posses the following two properties, the satisfaction of which will eventually lead to sufficient conditions for the existence of a NE point. Otherwise, no equilibrium exists and the systems in unstable.

1st Property. The utility of each user $i \in S$ can be transformed into the following form:

$$U_i(P_i, I_{-i}(S, \bar{P}_{-i})) \tag{9.48}$$

and for fixed power P_i the utility function $U_i(P_i, I_{-i})$ is a decreasing function of I_i.

2nd Property. Given any interference I, there is a power $P_i^*(I) \forall i \in S$ that results in

$$U_i \left(P_i^*(I), I \right) = \max_P U_i(P, I) \tag{9.49}$$

where function $P_i^*(I)$ is a continuous function of I and corresponds to the best responses of user i for fixed powers of other users.

The previous two properties allow the decomposition of game G_{SP} into two subgames namely, power control subgame $G_{SP.p}$ and sequence control subgame $G_{SP,s}$, due to the decomposability attributes of users' utilities as functions of P_i and $I_{-i}(Z, \bar{P}_{-i})$, respectively. Specifically, concerning the sequence control subgame $G_{SP,s}(P)$, the powers are fixed and the users are allowed to adjust their signature sequences only; thus, the strategic space is the Cartesian product of N unit spheres. The power control subgame $G_{SP.p}(Z)$ is the game obtained by restricting to a fixed set of signature sequences represented by Z, while the strategic space is the Cartesian product of intervals $[0, P_i^{\text{Max}}] \forall i \in S$. Moreover, in accordance with property 2 and (10.46), a power vector P^* is a NE for power control subgame $G_{SP,s}(P)$ if it satisfies

$$P_i = P_i^*(I_{-i}(Z, \bar{P}_{-i})) \qquad (9.50)$$

THEOREM 9.10 [54] Given a separable game G_{SP}, the pair (Z,P) is a NE if and only if P is a NE of subgame $G_{SP.p}(Z)$ and Z is a NE of subgame $G_{SP,s}(P)$.

Via the previous theorem, a generic methodology is provided and established for analyzing the overall multi-objective game. It is revealed that by independently studying the corresponding subgame's equilibria, the treatment and analysis of which can be performed in a more simple way, joint game's equilibrium properties can be derived. Specifically, if no equilibria exist for either of the two subgames, it can be concluded that the joint control game has no equilibrium. However, if equilibrium for the sequence control subgame exists and can be characterized by the following property (i.e., property 3), then the existence of NE for the joint control game can be verified as well (Theorem 2 in Ref. [55]).

3rd Property. For any P, there exists an equilibrium point $Z^*(P)$ in the subgame $G_{SP,s}(P)$ such that the function $I(Z^*(P), P_{-i})$ is continuous as a function of $P \forall i \in S$.

In Ref. [55], the previous methodology is applied for four typical cases in wireless networking environment, namely power control with fixed signature sequence, joint control for single-cell synchronous CDMA, joint control for multicell systems, and joint control for single-cell synchronous CDMA with finite sequence set. Thus, it is demonstrated that in the last two cases, no equilibrium point exists for the corresponding joint power and signature sequence control game.

9.14 Concluding Remarks and Open Issues

This chapter presents a comprehensive and concrete analysis of applying game-theoretic approaches to the noncooperative power control problem in CDMA wireless networks. Power control is envisioned as means of providing proficient resource allocation and interference management methodologies in wireless CDMA networks.

Specifically, in this chapter, foundations and basic theoretic components of noncooperative games, adopted for the power control problem, are initially presented and analyzed. Then, a number of noncooperative power control games are illustrated in which each user seeks to maximize his own utility while satisfying various user-centric or network-centric goals and/or QoS prerequisites. Through our discussion, it is revealed and demonstrated that game theory can be used as a unifying framework for studying radio resource management in wireless CDMA networks. As a concluding outcome of this chapter, Table 9.2 presents in a concrete way all the noncooperative power control games that have been presented, along with the used utility functions, and hence, for each one, appropriate game-theoretic tools are outlined for treating the corresponding game in terms of existence of NE and its properties and revealing the convergence.

It should be mentioned that in most of the presented games in this chapter, the network acts as a passive actor. Alternatively, we can consider the network as an active player who interacts with the users and, thus, tries to maximize its total profits. In such cases, a different class of joint user-network games can be formulated that will allow the investigation of operator's financial incentives and users' multiple degrees of freedoms via game theory. Such games, although of high research and practical importance, are yet to be studied.

Furthermore, despite the variety of the presented alternative game problem formulations, one common design attribute that characterizes them is the decentralized nature of the corresponding iterative power control algorithms toward obtaining equilibrium. This not only necessitates the participation of the users in system's resource allocation bargaining process but also implies the distribution of the decision-making procedures of the network among its components/users instead of traditional centralized approaches, which eventually increases their role. Such alternatives favor the development of mobile node self-optimization and self-manageability functionalities that are founded on theoretical frameworks toward enabling future networking vision of autonomicity. This still remains an unexploited research field.

References

1. S. Shenker, Fundamental design issues for the future Internet, *IEEE JSAC*, 13, 7, 1176–1188, Sept. 1995.
2. X. Liu, E. Chong, and N. Shroff, A framework for opportunistic scheduling in wireless networks, *Computer Networks*, 41, 451–474, Mar. 2003.
3. J. W. Lee, R. R. Mazumdar, and N. B. Shroff, Downlink power allocation for multi-class wireless systems, *IEEE/ACM Transactions on Networking*, 13, 4, 854–867, Aug. 2005.
4. T. Kastrinogiannis and S. Papavassiliou, Utility based short-term throughput driven scheduling approach for efficient resource allocation in CDMA wireless networks, *Wireless Personal Communications*, 52, 3, 517–535, 2010.
5. S. Shakkottai and A. Stolyar, Scheduling for multiple flows sharing a time-varying channel: The exponential rule, *American Mathematical Society Translations*, Series 2, 207, 2002.
6. M. Andrews, L. Qian, and A. Stolyar, Optimal utility based multi-user throughput allocation subject to throughput constraints, in *Proceedings of IEEE Infocom '05*, Volume 4, pp. 2415–2424, Miami, FL, Mar. 2005.
7. J.-W. Lee, R. R. Mazumdar, and N. B. Shroff, Downlink power allocation for multi-class wireless systems, *IEEE/ACM Transactions on Networking*, 13, 4, 854–867, Aug. 2005.
8. J. W. Lee, R. R. Mazumdar, and N. B. Shroff, Joint resource allocation and base station assignment for the downlink in CDMA networks, *IEEE/ACM Transactions on Networking*, 14, 1, Feb. 2006.
9. X. Duan, Z. Niu, and J. Zheng, A dynamic utility-based radio resource management scheme for mobile multimedia DS-CDMA systems, in *Proceedings of IEEE Global Telecommunications Conference 2002 (GLOBECOM'02)*, Taipei, Taiwan, Nov. 2002.
10. P. Hande, S. Rangan, M. Chiang, and X. Wu, Distributed uplink power control for optimal SIR assignment in cellular data networks, *IEEE/ACM Transactions on Networking*, 16, 6, 1430–1443, Nov. 2008.
11. T. Alpcan, T. Basar, R. Srikant, and E. Altman, CDMA uplink power control as a noncooperative game, *Wireless Networks*, 8, 659–669, Nov. 2002.

12. S. Gunturi and F. Paganini, Game theoretic approach to power control in cellular CDMA, in *IEEE Vehicular Technology Conference*, Volume 4, pp. 2362–2366, Orlando, FL, Oct. 2003.

13. K.-K. Leung and C. W. Sung, An opportunistic power control algorithm for cellular network, *IEEE Transactions on Networking*, 14, 3, 470–478, Jun. 2006.

14. V. Shah, N. B. Mandayam, and D. J. Goodman, Power control for wireless data based on utility and pricing, in *Proceedings of the 9th IEEE International Symposium on Personal, Indoor and Mobile Radio Communications*, Volume 3, pp. 1427–1432, Boston, MA, Sept. 1998.

15. D. J. Goodman and N. B. Mandayam, Power control for wireless data, *IEEE Personal Communications*, 7, 48–54, Apr. 2000.

16. C. U. Saraydar, N. B. Mandayam and D. J. Goodman, Efficient power control via pricing in wireless data Networks, *IEEE Transactions on Communications*, 50, 291–303, Feb. 2002.

17. F. Meshkati, H. V. Poor, S. C. Schwartz, and N. B. Mandayam, An energy-efficient approach to power control and receiver design in wireless data networks, *IEEE Transactions on Communications*, 53, 1885–1894, Nov. 2005.

18. F. Meshkati, M. Chiang, H. V. Poor, and S. C. Schwartz, A game-theoretic approach to energy-efficient power control in multicarrier CDMA systems, *IEEE Journal on Selected Areas in Communications (JSAC)*, 24, 1115–1129, Jun. 2006.

19. N. Feng, S. Mau, and N. Mandayam, Pricing and power control for joint network-centric and user-centric radio resource management, *IEEE Transactions on Communications*, 52, 9, 1547–1557, Sept. 2004.

20. E. E. Tsiropoulou, T. Kastrinogiannis, and S. Papavassiliou, QoS-driven uplink power control in multi-service CDMA wireless networks—A game theoretic framework, to appear in the *Journal of Communications*, 4, 9, 654–668, Academic Publisher, Oct. 2009.

21. F. Meshkati, H. V. Poor, and S. C. Schwartz, Energy-efficient resource allocation in wireless networks: An overview of game theoretic approaches, *IEEE Signal Processing Magazine: Special Issue on Resource-Constrained Signal Processing, Communications and Networking*, 24, 3, 58–68, May 2007.

22. D. M. Topkis, Equilibrium points in nonzero sum n-person submodular games, *SIAM Journal on Control and Optimization*, 17, 773–787, 1979.

23. C. Long, Q. Zhang, B. Li, H. Yang, and X. Guan, Non-cooperative power control for wireless ad hoc networks with repeated games, *IEEE Journal on Selected Areas in Communications*, 25, 1101–1112, Aug. 2007.

24. C. U. Saraydar, N. B. Mandayam, and D. J. Goodman, Pareto efficiency of pricing-based power control in wireless datanetworks, *IEEE WCNC*, 1, 231–235, 1999.

25. R. D. Yates, A framework for uplink power control in cellular radio systems, *IEEE Journal on Selected Areas in Communications*, 13, 1341–1347, Sept. 1995.

26. T. Alpcan and T. Basar, A hybrid systems model for power control in multicell wireless data networks, *Performance Evaluation Journal*, 57, 4, 477–495, Aug. 2004.

27. S. Koskie and Z. Gajic, A Nash game algorithm for SIR-based power control in 3G wireless CDMA networks, *IEEE/ACM Transactions on Networking*, 13, 1017–1026, Oct. 2005.

28. M. Xiao, N. B. Shroff, and E. K. P. Chong, A utility-based power-control scheme in wireless cellular systems, *IEEE/ACM Transactions on Networking*, 11, 210–221, Apr. 2003.

29. C. U. Saraydar, N. B. Mandayam and D. J. Goodman, Pricing and power control in a multicell wireless data network, *IEEE Journal on Selected Areas in Communications*, 19, 1883–1892, Oct. 2001.

30. F. P. Kelly, A. Maulloo, and D. Tan, Rate control for communication networks: Shadow prices, proportional fairness and stability, *Journal of Operations Research Society*, 49, 3, 237–252, Mar. 1998.

31. M. Chiang, S. Zhang, and P. Hande, Distributed rate allocation for inelastic flows: Optimization framework, optimality conditions, and optimal algorithms, in *IEEE Infocom'05*, Miami, FL, Mar. 2005.

32. D. Xu, Y. Li, M. Chiang, and A. Calderbank, Elastic service availability: Utility framework and optimal provisioning, *IEEE Journal on Selected Areas in Communications*, 26, 6, 55–65, Aug. 2008.

33. G. Bacci and M. Luise, A noncooperative approach to joint rate and power control for infrastructure wireless networks, in *Proceedings of the International Conference on Game Theory for Networks GameNets 2009*, Istanbul, Turkey, May 2009.

34. X. Lin, N. B. Shroff, and R. Srikant, A tutorial on cross-layer optimization in wireless networks, *IEEE Journal on Selected Areas in Communications* on "Non-Linear Optimization of Communication Systems", 24, 8, 1452–1463, June 2006.

35. T. Kastrinogiannis and S. Papavassiliou, Game theoretic distributed uplink power control in CDMA networks with real-time services, *Computer Communications Journal*, 32, 2, 376–385, Feb. 2009.

36. E. E. Tsiropoulou, T. Kastrinogiannis, and S. Papavassiliou, A utility-based power allocation non-cooperative game for the uplink in multi-service CDMA wireless networks, in *Proceedings of IEEE International Wireless Communications and Mobile Computing Conference*, pp. 365–370, Leipzig, Germany, Jun. 2009.

37. E. Altman and Z. Altman, S-modular games and power control in wireless networks, *IEEE Transactions on Automatic Control*, 48, 5, 839–842, May 2003.

38. G. Scutari, S. Barbarossa, and D. P. Palomar, Potential games: A framework for vector power control problems with coupled constraints, in *ICASSP 2006*, Volume 4, Toubuse, France, May 2006.

39. P. Hande, S. Rangan, M. Chiang, and X. Wu, Distributed uplink power control for optimal SIR assignment in cellular data networks, *IEEE/ACM Transactions on Networking*, 16, 6, 1430–1443, Nov. 2008.

40. V. Rodriguez, An analytical foundation for resource management in wireless communication, in *Proceedings of GLOBECOM'03*, Volume 2, No. 2, pp. 898–902, San Francisco, CA, 1–5 Dec. 2003.

41. R. Lupas and S. Verd, Linear multiuser detectors for synchronous code-division multiple access channels, *IEEE Transactions on Information Theory*, 35, 1, 123–136, Jan. 1989.

42. U. Madhow and M. L. Honig, MMSE interference suppression for direct-sequence spread-spectrum CDMA, *IEEE Transactions on Communications*, 42, 12, 3178–3188, Dec. 1994.

43. T. M. Cover and J. A. Thomas, *Elements of Information Theory*, Wiley, New York, 1991.

44. S. Vishwanath, S. A. Jafar, and A. Goldsmith, Adaptive resource allocation in composite fading channels, in *Proceedings of IEEE Global Telecommunications Conference*, pp. 1312–1316, San Antonio, TX, Nov. 2001.

45. G. Munz, S. Pfletschinger, and J. Speidel, An efficient waterfilling algorithm for multiple access OFDM, in *Proceedings of IEEE Global Telecommunications Conference*, pp. 681–685, Taipei, Taiwan, Nov. 2002.

46. Z. Shen, J. G. Andrews, and B. L. Evans, Optimal power allocation in multiuser OFDM systems, in *Proceedings of IEEE Global Telecommunications Conference*, pp. 337–341, San Francisco, CA, Dec. 2003.

47. F. Meshkati, H. V. Poor, S. C. Schwartz, and R. V. Balan, Energy-efficient resource allocation in wireless networks with quality-of service constraints, *IEEE Transactions on Communications*, 57, 11, 3406–3414, Nov. 2009.

48. Y. H. Lee and S. W. Kim, Generalized joint power and rate adaptation in DS-CDMA communications over fading channels, *IEEE Transactions on Vehicular Technology*, 57, 1, 603–608, Jan. 2008.

49. M. Rasti, A. R. Sharafat, and B. Seyfe, Pareto-efficient and goal-driven power control in wireless networks: A game-theoretic approach with a novel pricing scheme, *IEEE/ACM Transactions on Networking*, 17, 2, 556–569, Apr. 2009.

50. C. Singh, A. Kumar, and R. Sundaresan, Uplink power control and base station association in multichannel cellular networks, in *Proceedings of the International Conference on Game Theory for Networks, GameNets '09*, pp. 43–51, Istanbul, Turkey, May 2009.

51. N. Feng, S. C. Mau, and N. B. Mandayam, Joint network-centric and user-centric radio resource management in a multicell system, *IEEE Transactions on Communications*, 53, 7, 1114–1118, Jul. 2005.

52. H. Jiang and J. Ruan, The Stackelberg power control game in wireless data networks, in *Proceedings of the IEEE International Conference on Service operations and Logistics, and Informatics, IEEE/SOLI 2008*, Volume 1, pp. 556–558, Beijing, China, 12–15 Oct. 2008.

53. T. Heikkinen, A potential game approach to distributed power control and scheduling, *Computer Networks*, 50, 13, 2295–2311, Sept. 2006.

54. T. Basar and G. Olsder, *Dynamic Noncooperative Game Theory*, SIAM, New York, 1999.

55. C. W. Sung, K. W. Shum, and K. K. Leung, Stability of distributed power and signature sequence control for CDMA Systems—A game-theoretic framework, *IEEE Transactions on Information Theory*, 52, 4, 1775–1780, Apr. 2006.

Chapter 10

Hierarchical Power Allocation Games

Mehdi Bennis, Samson Lasaulce, and Mérouane Debbah

Contents

In this chapter, we look at the key issue of power allocation (PA) where the target is the competitive maximization of the information throughput sustained by each link over the network. More specifically, we focus on the concept of hierarchy that exists between different radios/systems sharing the same resources. This paradigm therefore requires a new design and framework aiming at distributed approaches. For this reason, game theory (GT) is used as a tool to model the interaction between several players and predict the outcome of the PA game. In particular, a special branch called *hierarchical* games is adopted wherein radios interact to maximize their respective payoffs following a leader–follower approach. The presented results corroborate the fact that the overall efficiency of the network is improved through hierarchy.

10.1 Introduction

Over the course of the last couple of years, we have witnessed a new *paradigm* wherein the radio spectrum allocation has gone from static to flexible. Due to the ever increasing needs for more stringent data rates and spectrum underutilization [1], spectrum sharing [2–6] has emerged as a new way to improve the spectral efficiency of radio systems. Two approaches basically exist under the radio resource sharing umbrella: the *underlaying* and the *overlaying* [7–9] scenario. In the first case, primary and secondary radio devices coexist as long as the interference temperature is satisfied. In the second case, secondary radios constantly sense the environment (i.e., spectrum holes) left vacant by primary radios where spectrum holes can be reused by secondary systems, provided that they might be required by the primary radios at any time.

In this chapter, we examine the problem of power allocation (PA) with a main emphasis on the concept of hierarchy existing between radios. This concept naturally arises in multiple practical situations: (a) when primary and secondary systems share the spectrum, (b) when users have access to the medium in an asynchronous manner, (c) when operators deploy their networks at different times, and (d) when some nodes have more power than others such as the base station. Within the realm of game theory (GT), the hierarchical spectrum sharing problem is naturally modeled using the Stackelberg [10,11] framework motivated by the fact that the noncooperative Nash equilibrium (NE) is generally inefficient and nonoptimal. As a result, Stackelberg approach provides (in general) better outcomes as compared to the noncooperative approach. Furthermore, it improves the overall network efficiency, hence bridging the gap between the selfish and fully centralized approach.

Clearly, radio devices are very likely to interact upon accessing spectral resources. This interaction happens in both short and long terms. In a noncooperative approach (i.e., competitive operators) [6], radios behave selfishly by maximizing their payoffs (throughput, revenues, etc.). Therefore, these interactions are modeled using GT [10], a branch of mathematics that has attracted considerable attention for analyzing wireless communication systems. In essence, GT models the strategic interaction between players (radios, transmitters, nodes, etc.) where an equilibrium point is found and analyzed. GT encompasses a rich plethora of disciplines such as noncooperative games [12], cooperative [13] games, coalitional games [14], static, and dynamic, with complete and incomplete [15] information. In this chapter, we specifically focus on hierarchical spectrum sharing games in wireless communication systems.

10.1.1 Two Important Multiuser Channel Models

Two elementary channel models exist in the literature, which are subject to interaction: the multiple access channel (MAC) and the interference channel (IFC) [16]. These are defined as follows:

■ **MAC**: The MAC consists of K transmitters aiming to communicate with a single receiver using a common channel. If $N > 1$ channels are available, then there exist N independent or parallel MACs, where transmitters in different MACs do not interfere with each other. For instance, this model corresponds to the uplink channel in a single-cell multi-carrier cellular system. Moreover, the channel gain from transmitter i to the receiver over channel

n is denoted by h_i^n. We assume a block flat-fading channel model such that the channel realizations remain constant during the transmission of M consecutive symbols. All the channel realizations, $\forall i = \{1, \ldots, K\}$ and $\forall n = \{1, \ldots, N\}$, are drawn from a Gaussian distribution with zero mean and unit variance. The power allocated by transmitter i to channel n is denoted by p_i^n. Each transmitter is power limited where for the ith transmitter, its transmit power cannot exceed $P_{i,\max}$, that is, $\forall i = \{1, \ldots, K\}$, $\sum_{n=1}^N p_i^n \leq P_{i,\max}$.

The symbol sent by transmitter i over channel n is represented by x_i^n. We consider that transmitted symbols $\forall i = \{1, \ldots, K\}$ and $\forall n = \{1, \ldots, N\}$ are random variables with zero mean and unit variance. The noise at the receiver is denoted by w_i^n and corresponds to an additive white Gaussian noise (AWGN) process with zero mean and variance σ^2. In vector notation, the channel realizations are written as $\mathbf{h}_i = \left(h_i^1, \ldots, h_i^N\right)$. Using a similar notation, the transmit powers, transmitted symbols, and noise are written as $\mathbf{p_i} = \left(p_i^1, \ldots, p_i^N\right)$, $\mathbf{x_i} = \left(x_i^1, \ldots, x_i^N\right)$, and $\mathbf{w_i} = \left(w_i^1, \ldots, w_i^N\right)$, respectively. Finally, the received signal can be expressed as

$$y^n = \sum_{i=1}^K h_i^n x_i^n + w_i^n \tag{10.1}$$

and the received signal to interference plus noise ratio (SINR) assuming single-user decoding (SUD) on channel n for transmitter i, denoted by $SINR_i^n$ for all $\forall i = \{1, \ldots, K\}$ and for all $\forall n = \{1, \ldots, N\}$, is given by

$$SINR_i^n = \frac{p_i^n \left|h_{ii}^n\right|^2}{\sum_{j \neq i}^K p_j^n \left|h_j^n\right|^2 + \sigma^2} \tag{10.2}$$

■ **IFC:** The IFC [16–18] consists of a set of K point-to-point links close enough to produce mutual interference due to the co-existence on the same channel. If $N \geq 1$ channels are available, there exists N independent or parallel IFCs, where transmitters in different IFCs do not interfere with each other. This topology typically appears in self-organized networks (SON) where nodes communicate in pairs over a set of sub-carriers. To describe the IFC model, we keep the same notation and assumptions presented in the MAC case. The only slight modification is introduced to denote the channel realization from transmitter i to receiver j on channel n, denoted by h_{ij}^n with $\forall n = \{1, \ldots, N\}$ and $(i, j) \in \{1, \ldots, K\}^2$. The noise at receiver i over channel n is denoted by w_i^n where $\mathbf{w}_i = \left(w_i^1, \ldots, w_i^N\right)$. The received signal at receiver i denoted by r_i^n is written as

$$r_i^n = \sum_{j=1}^K h_{ji}^n x_i^n + w_i^n \tag{10.3}$$

and the received SINR assuming SUD on channel n for transmitter i, denoted by $SINR_i^n$ for all $\forall i = \{1, \ldots, K\}$ and $\forall n = \{1, \ldots, N\}$, is given by

$$SINR_i^n = \frac{p_i^n \left|h_{ii}^n\right|^2}{\sum_{j \neq i}^K p_j^n \left|h_{ji}^n\right|^2 + \sigma^2} \tag{10.4}$$

Throughout this chapter and for both network topologies, unless otherwise stated, we assume that all transmitters have perfect channel state information (CSI), that is, each transmitter knows the channel realizations h_i^n for all $\forall i = \{1, \dots, K\}$ in the MAC case and h_{ij}^n for all $(i, j) \in \{1, \dots, K\}^2$ in the IFC case.

The chapter is organized as follows: In Section 10.2, we recall the fundamental and relevant concepts/notions of GT used throughout the chapter. First, we introduce some game-theoretical definitions such as the concept of NE for noncooperative games followed by the Stackelberg approach for hierarchical games. Section 10.3 looks at the applications of hierarchy in PA/control problems arising in wireless communication. Section 10.4 gives a summary of the hierarchical PA games. We finally conclude and discuss open problems in Section 10.5.

10.2 Review of Game-Theoretical Concepts

10.2.1 Normal Form Games

As pointed out earlier, GT provides a suitable mathematical framework to analyze the strategic interaction between different players. In general, a game is presented in a normal form as follows:

Definition 10.1 (Normal Form) [10]   A game in normal form is denoted by $\{\mathcal{K}, \mathcal{S}, \{u_k\}_{\forall k \in \mathcal{K}}\}$ and is composed of three elements:

- A set of players: $\mathcal{K} = \{1, \dots, K\}$
- A set of strategy* profiles: $\mathcal{S} = \mathcal{S}_1 \times \cdots \times \mathcal{S}_k$, where $\mathcal{S}_k$ is the strategy set of player k
- A set of utility functions: the kth player's utility function is $u_k: \mathcal{S} \to \mathcal{R}_+$ and is denoted by $u_k(s_k, s_{-k})$ where $s_k \in \mathcal{S}_k$ and $s_{-k} = (s_1, \dots, s_{k-1}, s_{k+1}, \dots, s_K) \in \mathcal{S}_1 \times \cdots \times \mathcal{S}_{k-1} \times \mathcal{S}_{k+1} \times \cdots \times \mathcal{S}_K$

In the PA game, the set of players includes transmitters, base stations, and mobile stations. A player's k strategy is to transmit with a certain power over a certain channel while the utility function is expressed in terms of achievable rates.

10.2.2 Noncooperative Games, Nash and Stackelberg Equilibria

In noncooperative games, players behave selfishly, not caring about other players' payoffs. Each player chooses its strategy to maximize its utility function. Players are moreover assumed to be rational and adopt the same selfish behavior. In contrast, in cooperative games, each player aims at maximizing a common and social benefit provided other players do the same. The information required to play games is of high importance. For noncooperative games, only local[†] information is needed whereas all CSI is generally needed for cooperative games.

---

* In the case of no coupling among players strategies, the set $\mathcal{S}$ can be written as the Cartesian product of the strategy set $\mathcal{S}_k$ of each player. In the more general case of coupled constraints, each player aims to restrict its strategy to a subset that depends also on the strategies chosen by the other players.

† By local, it is meant that a player needs only to know his own channel realizations.

The NE is an important concept in the field of GT where an NE corresponds to a profile of strategies $s^* = \left(s_1^*, \ldots, s_K^*\right)$ for which each player's strategy $s_k^* \in \mathcal{S}_k$ is an NE if it satisfies

$$\forall k \in \mathcal{K} \quad \text{and} \quad \forall s_k \in \mathcal{S}_K, u_k\left(s_k^*, s_{-k}^*\right) \geq u_k\left(s_k, s_{-k}^*\right) \tag{10.5}$$

That is, at the NE, any unilateral deviation from the strategy profile s_k of player k, $\forall k \in \mathcal{K}$ will not increase its utility function u_k. Hence, at the NE, there exists no motivation for a player to deviate from the NE strategy profile. As players are selfish and decide by themselves their strategy, one question arises: does an NE lead to an efficient game outcome? To answer this question, the notion of optimality comes into play where Pareto-optimality is a measure of optimality of a game defined as

---

Definition 10.2 (Pareto Optimality) [10]   Let $s = (s_1, \ldots, s_K)$ and $s' = \left(s_1', \ldots, s_K'\right)$ be two different strategy profiles in $\mathcal{S}$. Then, the strategy profile s is Pareto-superior to the strategy profile s' if

$$\forall k \in \mathcal{K} \quad u_k(s_k, s_{-k}) \geq u_k\left(s_k', s_{-k}'\right) \tag{10.6}$$

with strict inequality for at least one player. Moreover, if there exists no strategy that is Pareto-superior to s_i, then s_i is Pareto-optimal.

An interesting performance metric is the social welfare, which corresponds to the average utility of players:

Definition 10.3 (Social Welfare) [19]   The social welfare of a game is defined as the sum of the utilities of all players:

$$w = \sum_{i=1}^{K} u_i(s_i, s_{-i}) \tag{10.7}$$

The NE is generally inefficient and nonoptimal. However, it is a lower bound in the case of noncooperation between players. To measure the lack of cooperation, the notion of price of anarchy (PoA) is defined:

Definition 10.4 (PoA) [20]   The PoA of the game $\{\mathcal{K}, \mathcal{S}, \{u_k\}_{\forall k \in \mathcal{K}}\}$ is equal to the ratio of the highest value of the social welfare (joint optimization) to the worse NE of the game:

$$PoA = \frac{\max_{s \in \mathcal{S}} \sum_{k=1}^{K} u_k(s_k, s_{-k})}{\min_{s^* \in \mathcal{S}^{NE}} \sum_{k=1}^{K} u_k\left(s_k^*, s_{-k}^*\right)} \tag{10.8}$$

where $\mathcal{S}^{NE}$ is set of Nash equilibria of the game.

Power allocation games in which the concept of hierarchy is taken into account are referred to as *Stackelberg* games, initially introduced by Stackelberg in [11]. In these games, an implicit concept of hierarchy exists upon the set of players. Such hierarchy naturally occurs in a number of practical scenarios: (a) when primary and secondary systems share the spectrum, (b) when users have access to the medium in an asynchronous manner, (c) when operators deploy their networks at different times, and (d) when some nodes in the network have more power than others (such as the base station). For example, in a two-level Stackelberg game, the game leader moves first and the other players follow and play simultaneously. The game leader perfectly knows the set of strategies and the utilities of the followers who in turn can observe the actions of their leader(s).

A Stackelberg game is solved using the concept of *sub-game perfection* [10] NE where none of the players have an incentive to deviate in any sub-game. The inefficiency of the NE concept led Selten in [21] to devise the notion of sub-game perfect equilibria in extensive form games. These are defined as equilibria, which rely on threats that are really credible. In the first stage of a Stackelberg game, the leader who perfectly knows all the followers' set of actions and utility functions chooses the action that maximizes its benefits considering that each follower will react with the action that maximizes its own benefit as well. Thus, the game leader analyzes all the possible outcomes and picks up the action that maximizes its benefit considering the optimal moves for each player. A formal definition of a Stackelberg game and its equilibrium is given as follows:

Definition 10.5 (Stackelberg Game) [10]   A Stackelberg game is a two-stage game where one player (leader) moves in the first stage and all the other players (followers) react simultaneously in the second stage.

Definition 10.6 (Stackelberg Equilibrium) [20]   A vector $\underline{p}^{SE} = \left(p_i^{SE}, p_{-i}^{SE} \right)$ is called a Stackelberg Equilibrium (SE) if $p_{-i}^{SE} \in \mathcal{U}^*(p_i)$ where $\mathcal{U}^*(p_i)$ is the set of NE for the group of followers when the leader plays strategy p_i, and the power p_i^{SE} maximizes the utility function of user i (game leader). In other words, $p_i^{SE} = \arg\max_{p_i} u_i \left(p_1^{SE}(p_i), \ldots, p_{i-1}^{SE}(p_i), p_i, p_{i+1}^{SE}(p_i), \ldots, p_K^{SE}(p_i) \right)$, where $p_j^{SE}(p_i)$ is the power of the follower j at the NE for $j \neq i$.

10.3 Application to Wireless Communications

In this section, we investigate direct applications of hierarchical PA relevant in wireless communications problems. First, the problem of spectrum leasing between primary and secondary systems is investigated. Second, the concept of hierarchy is investigated in the context of energy-efficiency games. Finally, PA games in the context of multiuser channels are discussed.

10.3.1 Spectrum Leasing to Cooperative Secondary Systems

In [23], dynamic spectrum sharing is investigated in which primary systems lease the spectrum to secondary systems in exchange for cooperation (PA game). A primary transmitter PT wants to send information to its primary receiver PR either directly (with a rate of R_{dir}) or using cooperation from a subset $\mathcal{S} \subseteq \mathcal{S}_{\mathrm{tot}}$ of $|\mathcal{S}| = k \leq |\mathcal{S}_{\mathrm{tot}}| = K$ secondary nodes/transmitters. For transmission purposes, the primary system divides its data into two parts of αL bit durations and $(1-\alpha)L$ bit durations, with $0 \leq \alpha \leq 1$. The first $(1-\alpha)L$ bits are dedicated to a direct transmission from PT to PR whereas the second αL bits are again divided into two parts. One part, consisting of $\beta \alpha L$, with $0 \leq \beta \leq 1$, is dedicated to send information from PT to PR using the ad hoc (secondary) network by means of distributed space time coding (DSTC) [22], while the remaining $\alpha(1 - \beta)L$ bits are granted to the secondary network for its own data transmission. The leader maximizes its own utility function while deciding the portion of time-slots α, β, and $\mathcal{S} \subseteq \mathcal{S}_{\text{tot}}$ subset of secondary transmitters. In addition, two cases of CSI are investigated, namely, the full and partial CSI.

Given the set $\mathcal{S}$ and cooperation parameters α and β, the rate optimization problem for the leader (transmitter PT) is given by

$$\max_{\alpha,\beta,\mathcal{S}} R_P(\alpha, \beta, \mathcal{S})$$

$$\text{s.t.} \quad \mathcal{S} \subseteq \mathcal{S}_{\text{tot}}, \quad 0 \leq \alpha, \beta \leq 1 \tag{10.9}$$

where

$$R_P(\alpha, \beta, \mathcal{S}) = \begin{cases} \min\{(1 - \alpha)R_{PS}(\mathcal{S}), \alpha\beta R_{SP}(\mathcal{S})\}, & \alpha > 0 \\ R_{\text{dir}}, & \alpha = 0 \end{cases} \tag{10.10}$$

where

$R_{PS}(\mathcal{S})$ is the achievable rate between the primary transmitter and secondary receiver

$R_{SP}(\mathcal{S})$ is the achievable rate between secondary transmitter and primary receiver

R_{dir} is the achievable rate with no cooperation ($\alpha = 0$)

Once cooperation parameters α and β are calculated using (10.9), the secondary system exploits the $\alpha(1 - \beta)L$ bits within which it can transmit where transmitters play a noncooperative game, in which any active secondary terminal T_{S_i} from the subset $\mathcal{S}$ maximizes its utility function $u_i(p_i, p_{-i})$ toward its receiver R_{S_i} defined as the difference between the achievable transmission rate and the cost of transmitted energy c.

The optimization problem for the secondary transmitters T_{S_i}, $i = 1, \ldots, k$, is written as

$$\max_{p_i} u_i(p_i, p_{-i}) = \max_{p_i} \left((1 - \beta) \log_2 \left(1 + \frac{|h_{S,ii}|^2 p_i}{\sigma^2 + \sum_{j=1, j \neq i}^k |h_{S,ji}|^2 p_j} \right) - c p_i \right)$$

$$\text{s.t.} \quad 0 \leq p_i \leq P_{i,\max} \tag{10.11}$$

Note that in (10.11) depends only on k, α, and β and solving it yields the NE of the noncooperative game expressed as

$$\hat{p}_i = \left(\frac{1 - \beta}{c} - \frac{\sigma^2}{|h_{S,ii}|^2} - \sum_{j=1, j \neq i}^k \frac{|h_{S,ji}|^2}{|h_{S,ii}|^2} \hat{p}_j \right)^+ \tag{10.12}$$

where $x^+ = \max(0, x)$.

The existence of the NE is assured by the concavity of the utility function in (10.11) [20]. It is furthermore unique [23] if the following sufficient conditions are satisfied:

$$\sum_{j \in \mathcal{S}, j \neq i}^k \frac{|h_{S,ji}|^2}{|h_{S,ii}|^2} < 1 \tag{10.13}$$

The interaction between the primary and secondary networks is modeled as a *strategic* Stackelberg game. The game leader maximizes its own utility function knowing which in turn affects the secondary network. The Stackelberg equilibrium is solved using backward induction [10] where (10.12) is plugged back in (10.9) and the optimal α^*, β^*, and S^* are computed, maximizing thereby the leader's payoff. The Stackelberg equilibrium is obtained going back to (10.12) and solving the optimization problem. As a result, maximizing the revenue of the primary network results in several tradeoffs. First, if β increases, there is more cooperation time at the cost of lessening the cooperation from the secondary network. Moreover, a large value of subset S may limit the overall rate by reducing the term $(1 - \alpha)R_{SP}(S)$ while at the same time enhancing the term $\alpha\beta R_{SP}(S, \beta)$, thanks to cooperation.

In the second case of channel knowledge information (partial CSI), the primary system solves the optimization problem from a *probabilistic* standpoint (outage probability-based) as the primary system is unable to know the subset S of secondary ad hoc networks willing to cooperate. Moreover, the set of secondary transmitters able to decode the primary message is a random quantity, and thus the choice of a space–time code-book and the specific codeword to be transmitted by each node in S cannot be done by the primary system, hence randomized DSTC [24] is used. Similar to the full CSI case, PT wants to transmit at a given target rate R_P with a required bit error probability (BER), otherwise an outage is declared. The goal of the primary link is to minimize the outage probability $P_{out}(\alpha, \beta, C_i)$ for fixed R_P and BER with respect to α, β, and the space–time codebook $C_i \in \mathcal{C}$.

The outage probability $P_{out}(\alpha, \beta, C_i)$ is defined as the probability that the SNR on the cooperative link (denoted by $\gamma_{SP}(S, \beta, C_i)$) experiences an instantaneous SNR smaller than the desired threshold where

$$P_{out}(\alpha, \beta, C_i) = \sum_{S \subseteq S_{tot}} P_{PS}(S, \alpha) \cdot P_{out}(S, \alpha, \beta, C_i) \tag{10.14}$$

where $P_{PS}(S, \alpha)$ is the probability that the secondary nodes in S can decode the primary transmission in the first time-slot and

$$P_{out}(S, \alpha, \beta, C_i) = Pr\left[\gamma_{SP}(S, \beta, C_i) < \gamma_{th}\left(\frac{\bar{R}_p}{\alpha\beta R_{STC,i}}\right)\right] \tag{10.15}$$

is the outage probability of the randomized DSTC in the second time-slot when nodes in S are active and the orthogonal STC C_i is used. Note that the transmission rate in the second slot is $\bar{R}_p/R_{STC,i}.\alpha\beta$, which accounts for the duration of the second slot and the reduced rate $R_{STC,i} \leq 1$ of orthogonal STCs to attain an overall transmission rate $\bar{R}_P$.

In essence, the optimization problem for the primary link is written as

$$\min_{\alpha,\beta,C_i} P_{out}(\alpha, \beta, C_i)$$
$$\text{s.t.} \quad C_i \in \mathcal{C}, 0 \leq \alpha, \beta \leq 1,$$
$$\alpha N_S, \alpha\beta N_S, \alpha\beta N_S/q_i \in \mathcal{N} \tag{10.16}$$

where the outage probability of the primary link, $P_{out}(\alpha, \beta, C_i)$, is

$$P_{out}(\alpha, \beta, C_i) = P_{out,dir}, \quad \alpha = 0 \tag{10.17}$$

and $P_{out}(\alpha, \beta, C_i) = P_{out,SP}(S, \alpha, \beta, C_i)$ for $\alpha > 0$. $\mathcal{N}$ is the set of integers and $\beta\alpha N_S$ is the number of symbols in the second slot.

10.3.2 Hierarchy in Energy-Efficient Power Control Games

Unlike works in which the utility function to be optimized is often the Shannon rate, a hierarchy is introduced in [25], in the context of decentralized MAC with energy efficiency. In this approach, a user $1 \leq i \leq K$ aims at maximizing the ratio between the number of information transmitted bits without errors T_i and its transmit power level p_i. Moreover, the utility function of user i is given by

$$u_i(p_1, \ldots, p_K) = \frac{T_i}{p_i} = \frac{R_i f(SINR_i)}{p_i} \qquad (10.18)$$

where
  $SINR_i$ is defined in (10.2)
  f is an efficiency function* representing the packet success rate assumed identical for all users
  R_i is the transmission rate of user i

To cope with the inefficiency of the NE and improve the network equilibrium efficiency, a Stackelberg formulation of the power control game is proposed to bridge the gap between the fully centralized and the noncooperative approach. First, a Stackelberg approach is proposed with SUD at the receiver, followed by a more efficient receiver using successive interference cancelation (SIC).

In the first hierarchical game with a receiver-based SUD, one of the K users (denoted by user i) is chosen to be the leader whereas the others are the followers (users $j \neq i$). The solution of the Stackelberg game leads to an equilibrium whose existence and uniqueness are given in the following proposition.

Proposition 10.1 (Existence and Uniqueness of the stackelberg equilibrium [SE]) [25] There is a unique Stackelberg equilibrium $\underline{p}^{SE} = (p_i^{SE}, p_j^{SE})$ in the energy-efficient hierarchical game where user i is the leader:

$$p_i^{SE} = \frac{\sigma^2}{|h_i|^2} \frac{\gamma^*(1 + \Xi^*)}{1 - (K-1)\gamma^* \Xi^* - (K-2)\Xi^*} \qquad (10.19)$$

and for each follower $j \neq i$:

$$p_j^{SE} = \frac{\sigma^2}{|h_j|^2} \frac{\Xi^*(1 + \gamma^*)}{1 - (K-1)\gamma^* \Xi^* - (K-2)\Xi^*} \qquad (10.20)$$

if the following (sufficient) conditions hold:

$$\frac{f''(0)}{f'(0)} \geq 2 \frac{(K-1)\Xi^*}{1 - (K-2)\Xi^*} \quad \text{and} \quad \phi(x) = x\left[1 - \frac{(K-1)\Xi^*}{1 - (K-2)\Xi^*}\right] f'(x) - f(x)$$

has a single stationary point in $]0, \gamma^*[$, where Ξ^* is the positive solution of the equation $xf'(x) - f(x) = 0$ and γ^* is the positive solution of the equation $\phi(x) = 0$.

A number of questions regarding the hierarchical power control game are next looked at: (**1**) *Is it better to be a leader or follower in the proposed hierarchical game?* (**2**) *What is the gain provided by hierarchy?* (**3**) *Do all players benefit from it?*

---

* f is a sigmoidal function [26].

Interestingly, it turns out that it is better to be a follower rather than a leader. This is given in the following proposition:

Proposition 10.2 (Following is Better Than Leading) [25]   Every user has always a better utility being chosen as a follower instead of a leader.

Proof u_L^{*SE} (resp. u_F^{*SE}) the utility of user $i \in \{1, \ldots, K\}$ (resp. $j \neq i$) when he is chosen to be the leader (resp. a follower) of the game. First, we observe that at the Stackelberg equilibrium the SINR of the leader and follower are given by

$$SINR_L^{*SE} = \epsilon^* \quad \text{and} \quad SINR_F^{*SE} = \Xi^* \tag{10.21}$$

From [25], we have that

$$\forall x > 0 : x > \Xi^* \Leftrightarrow xf'(x) < f(x) \tag{10.22}$$

As for all $x > 0, x\left[1 - ((K-1)\Xi^*/(1-(K-2)\Xi^*))x\right]f'(x) < xf'(x)$, from a single geometrical argument, we see that $\epsilon^* < \Xi^*$. This means that the SINR of the follower Ξ^*, is higher than the SINR of the leader ϵ^*. □

It is also shown that both leaders and followers improve their payoffs expressed in terms of energy-efficiency and furthermore transmit with a lower power as compared to the non-cooperative approach.

In the second hierarchical power control game, the base station (chosen as a game leader) implements SIC where users are ranked and decoded successively. SIC is more complex than SUD where the decoding order has to be known to all users, nevertheless SIC partially removes multi-user interference. Similarly to the previous approach, the following proposition gives the existence and uniqueness of the NE:

Proposition 10.3 (Existence and Uniqueness of the NE) [25]   Let i denote the index of the decoded user with rank $K - i + 1$ in the successive decoding procedure at the receiver. In the noncooperative game where the utility is given by in (10.18) and the SINRs are those considered at the output of the SIC, there exists a unique (pure) NE $\left(p_1^{SIC}, \ldots, p_K^{SIC}\right)$ given by

$$\forall i \in \{1, \ldots, K\}, \quad p_i^{SIC} = \frac{\sigma^2}{|h_i|^2} \Xi^* \mu_i^{SIC} \tag{10.23}$$

---

where $\mu_i^{SIC} = (1 + \Xi^*)^{i-1}$ is a penalty term due to multiple access interference (MAI).

Next, a comparison is drawn between the SIC- and SUD-based receivers. It turns out that in contrast to the SUD case, the existence of a non-saturated NE is still insured when $\Xi^* > 1/(K-1)$. In addition, every user prefers to be in the game with a receiver implementing SIC instead of SUD. This is due to the fact that in the SIC-based receiver, less transmit power is used (more energy efficient) and less interference is generated in the network. Finally, the impact of the degrees of freedom on the overall network energy-efficiency is investigated based upon the social welfare and the energy-efficiency of the equivalent virtual multiple input multiple output (MIMO) system. In the social welfare case, the best choice for a leader is the one with lowest $R_i|h_i|^2$ and the best decoding order is to decode users in the increasing order of their energy weighted by the coding rate $R_i|h_i|^2$. In the equivalent virtual MIMO network with energy-efficiency and SUD, a sufficient condition is given for user i to be a game leader, while in the SIC case the best decoding order is to decode users in a decreasing order of their SNRs.

Having noted the above, let us now look at the case with K users and K hierarchy levels (one user per level) where the *super-leader* is assigned index $1, \ldots, K-1$ and the follower is assigned index K. User i has to tune its SINR at a value equal to the solution of the following equation (denoted by ζ_i):

$$x(1 - b_i x)f'(x) - f(x) = 0 \tag{10.24}$$

with

$$b_i = \prod_{j=i+1}^{K}\left[1 + \frac{\zeta_j(1 + b_j)}{1 - \zeta_j b_j}\right] - 1 = \prod_{j=i+1}^{K}\left(\frac{1 + \zeta_j}{1 - \zeta_j b_j}\right) - 1 \tag{10.25}$$

We see that all the SINRs can be recursively determined by initializing the recursion with $b_K = 0$.

Proof We do not provide the proof of this result here but it relies on [25] and reasoning by induction. □

The SINR for user i writes as

$$SINR_i = \frac{|h_i|^2 p_i}{(1 + b_i)\Psi_i^2 + b_i|h_i|^2 p_i} \tag{10.26}$$

with

$$\Psi_i^2 = \prod_{j=1}^{i-1}\left[1 + \frac{\zeta_j(1 + b_j)}{1 - \zeta_j b_j}\right] \tag{10.27}$$

Knowing that user i tunes its power in order to have $SINR_i = \zeta_i$, the generalized Stackelberg equilibrium (GSE) p_i^* is given by

$$p_i^* = \frac{\Psi_i^2}{|h_i|^2}\frac{\zeta_i(1 + b_i)}{1 - \zeta_i b_i}\prod_{j=1}^{i-1}\left[1 + \frac{\zeta_j(1 + b_j)}{1 - \zeta_j b_j}\right] \tag{10.28}$$

Proof Based on [25] plus induction. □

Proposition 10.4 (The Lower the Better) Any user has always a better utility being in a lower level in the hierarchy.

Proof Based on [25]. □

10.3.3 Hierarchical Power Allocation Games in Multiuser Channels

In [27], the authors tackle the problem of PA in the context of fading MACs. First, the problem is modeled as a static one-shot game [10] (without the intervention of the base station) where users selfishly compete between each other to maximize their respective payoffs using SUD subject to some power constraints given as follows:

$$\max_{p_i} \bar{R}_i(p_i, p_{-i}) \quad \text{s.t.} \quad p_i \in \mathcal{F}_i \tag{10.29}$$

where

$\mathcal{F}_i = \{p_i : \mathbb{E}_{\mathbf{h}}(p_i)\} \leq P_{i,\max}, p_i(\mathbf{h}) \geq 0\}$ is the set of all *feasible* power control policies of user i
p_{-i} represents the power control policy of user $j \neq i$

Finally, $\bar{R}_i = \mathbb{E}_{\mathbf{h}}(\mathcal{R}_i)$ is the average achievable rate with $\mathbf{h} = [h_1, h_2]$.
In the two-user case, the payoff of user i writes as

$$\bar{R}_i = \int \int \frac{1}{2} \log_2\left(1 + SINR_i\right) f(g_i, g_{-i}) dg_i dg_{-i} \tag{10.30}$$

where

$g_i = |h_i|^2$
$SINR_i$ is defined in (10.2) ($N = 1$ carrier)
$f(g_i, g_{-i})$ is the joint probability density function of the two fading coefficients

The solution to (10.30) is the well-known water-filling PA given by

$$p_i(g_i, g_{-i}) = \left(\lambda_i - \frac{\sigma^2}{g_i} - \frac{p_{-i}(g_i, g_{-i})g_{-i}}{g_i}\right)^+ \tag{10.31}$$

where λ_i is the power level satisfying the power constraint given in $\mathcal{F}_i$.
The maximum sum-rate of the capacity region is shown to be the unique NE of the water-filling game, in which users' selfish behavior leads them to *jointly* optimize the sum-rate of the channel. Stated otherwise, the user with the strongest channel perceives a relatively weak interference from the other user and, hence, decides to transmit with a higher power level. On the other hand, the other user sees a strong interference in addition to a weak channel and, hence, decides to conserve its power for a later usage. As a result, the users reach a distributely opportunistic time-sharing equilibria.
In the second part of their work and motivated by the fact that in the one-shot noncooperative approach, users treat each other as noise and hence are unable to reach all capacity region points other than the sum-rate, a hierarchy is introduced between players so that all boundary points of the capacity region are achieved. A Stackelberg game formulation is proposed with the base station posing as a *leader* announcing the decoding order in the high-level game, while the users are the followers in the low-level game. Moreover, the strategy space of user i is $\mathcal{F}_i$ and the payoff function of user 1 is defined as the supremum of the achievable rate given by

$$\bar{R}_1(D_1, p_1, p_2) = \int \int \frac{1}{2} \log_2\left(1 + \frac{p_1(g_1, g_2)g_1}{\sigma^2 + p_2(g_1, g_2)g_2 I_{(g_1, g_2) \in D_1}}\right) f(g_1, g_2) dg_1 dg_2 \tag{10.32}$$

where $I_{(g_1, g_2) \in D_1}$ is the indicator function. On the other hand, the payoff of the BS accounting for users' payments (μ_1 and μ_2) is written as

$$\mu_1 \bar{R}_1(D_1, p_1, p_2) + \mu_2 \bar{R}_2(D_1, p_1, p_2) \tag{10.33}$$

In the decoding order, the base station divides the whole possible space of (g_1, g_2) into two subsets D_1 and D_1^c. When $(g_1, g_2 \in D_1)$, the BS decodes user 1's information first whereas $(g_1, g_2 \in D_1^c)$ implies decoding user 2's signal first. Then, after the BS announces its strategy, D_1, the multiple access users play the low level using the NE concept.

The authors prove the existence and uniqueness of the NE using the concept of admissible* [10] NE. It is shown that the proposed Stackelberg game has a very desirable structure, in which for any given vector μ, the existence of a BS policy achieving a utility within an ϵ-difference from the optimal one is guaranteed. In addition, the optimal policy for every rational multiple access user in the low-level game is unique. Therefore, the users will have no difficulty in deciding the power and rate levels in a distributed way. As a result, the introduction of the BS as a game leader enlarges the achievable rate region as compared with the noncooperative game.

Next, the authors extend their work from one to N_r antennas at the base station. In the *one-shot* game, the authors prove the uniqueness of the NE (i.e., the one achieving the sum-rate point SP). However, the achievable rates are strictly smaller than the rates corresponding to SP. Moreover, and in contrast to the single antennas case in which the PA strategy was time-sharing, there are $\min(N, N_r)$ degrees of freedom, so more than one user is allowed to transmit at any fading state. In the Stackelberg game, a unique admissible NE for the low-level game is proven. It also turns out that the Stackelberg game achieves the two corner points of the capacity region but not the maximum sum-rate point. Additionally, unlike the static game, the users can achieve any point on the capacity region as a *sub-game* perfect equilibrium using the same strategies as in the single antenna case. Finally, to cope with the fact that not all the points of capacity regions are achieved (due to the structural difference between the successive decoding strategy and the optimal decoding strategy), the power control game is reformulated as a *dynamic and repeated game* [20]. It is proven that for infinitely repeated game, the boundary points of the capacity region are achievable with the base station as the game leader, and the corresponding equilibria are sub-game perfect.

In a frequency-selective interference channel setting, the problem of spectrum sharing has been well studied in [4,28–32] among others. Therein, the spectrum sharing problem is modeled as a *strategic* noncooperative (*one-shot*) game[†] in which transmitters selfishly maximize their utility function (Shannon rate) subject to their power constraint $P_{i,\max}, \forall i \in \{1, \ldots, K\}$:

$$\max_{p_i^1, \ldots, p_i^N} R_i = \max_{p_i^1, \ldots, p_i^N} \sum_{n=1}^{N} \log_2 \left(1 + SINR_i^n \right)$$

$$\text{s.t.} \quad \sum_{n=1}^{N} p_i^n \leq P_{i,\max}$$

$$p_i^n \geq 0 \tag{10.34}$$

where $SINR_i^n$ is defined in (10.4). The solutions to (10.34) are given by the water-filling power [33] allocation:

$$p_i^n = \left(\frac{1}{\mu_i} - \frac{\sigma_n^2 + \sum_{j \neq i} |h_{ji}^n|^2 p_j^n}{|h_{ii}^n|^2} \right)^+ \qquad i = 1, \ldots, K \qquad n = 1, \ldots, N \tag{10.35}$$

---

* This notion allows for eliminating Nash equilibria, which are Pareto-dominated by other equilibrium points.
[†] See [34] for a good survey on the application of GT for dynamic spectrum sharing.

where $(x)^+ = \max(x, 0)$ and $\mu_i > 0$ is the Lagrangian multiplier chosen to satisfy the power constraint: $\sum_{n=1}^{N} p_i^n = P_{i,\max}$.

It turns out that under certain channel realizations, the NE of the game is *unique* whereas other channel conditions exhibit several Nash equilibria. This renders the game unpredictable since players are distributed and hence there is no central entity enforcing a certain equilibrium. In the case of arbitrary number of transmitters K, the results for the two transmitters/operators and two-carrier case carry over where the sufficient conditions for the uniqueness are given by [30]:

$$\sum_{i=1,\,j\neq i}^{K} \frac{\left|h_{ji}^n\right|^2}{\left|h_{ii}^n\right|^2} < 1, \quad n = 1,\ldots,N \tag{10.36}$$

Stated otherwise, the physical meaning of (10.36) is that the uniqueness of the NE is ensured if the links are sufficiently far from each other.

A *Stackelberg* game is then proposed to model the spectrum sharing problem where the primary operator 1 is the leader and secondary operator 2 is the follower. The Stackelberg spectrum sharing game is formulated as follows. First, in the high-level problem, primary operator 1 maximizes his own utility function, and in the low-level problem secondary operator 2 maximizes his own utility taking into account the optimal PA of operator 1, p_1^{SE}. By denoting $\left(p_1^{SE}, p_2^{SE}\right)$ as the Stackelberg Equilibrium (Definition 10.6), the rate optimization problem for operator 1 (leader) is written as

$$\max_{p_1^1,\ldots,p_1^N} \sum_{n=1}^{N} \log_2\left(1 + \frac{\left|h_{11}^n\right|^2 p_1^n}{\sigma_n^2 + \left|h_{21}^n\right|^2 p_2^n\left(p_1^n\right)}\right)$$

$$\sum_{n=1}^{N} p_1^n \leq P_{1,\max}$$

$$p_1^n \geq 0 \tag{10.37}$$

The rate optimization problem for operator 2 is written as

$$\max_{p_2^1,\ldots,p_2^N} \sum_{n=1}^{N} \log_2\left(1 + \frac{\left|h_{22}^n\right|^2 p_2^n}{\sigma_n^2 + \left|h_{12}^n\right|^2 \left(p_1^n\right)^{SE}}\right)$$

$$\sum_{n=1}^{N} p_2^n \leq P_{2,\max}$$

$$p_2^n \geq 0 \tag{10.38}$$

where $p_2^{SE} = BR_2\left(p_1^{SE}\right)$ (i.e., the best response of player 2).

Using backward induction and given the best response of operator 2, (10.37) can be rewritten as

$$\max_{p_1^1,\dots,p_1^N} \sum_{n=1}^{N} \log_2 \left(1 + \frac{\left|b_{11}^n\right|^2 p_1^n}{\sigma_n^2 + \left|b_{21}^n\right|^2 \left(1/\mu_2 - \left(\sigma_n^2 + \left|b_{12}^n\right|^2 p_1^n\right)/\left|b_{22}^n\right|^2 \right)^+} \right)$$

$$\sum_{n=1}^{N} p_1^n \le P_{1,\max}$$

$$p_1^n \ge 0 \qquad\qquad (10.39)$$

The Stackelberg sharing game therefore boils down to solving (10.39) where several cases are considered, the details of which are given in [4].

The interoperator spectrum sharing in the context of two operators can be extended to the general case with arbitrary K operators sharing the spectrum. The problem is formulated in the same way where the leader's optimization problem is written as

$$\max_{p_1^1,\dots,p_1^N} \sum_{n=1}^{N} \log_2 \left(1 + \frac{\left|b_{11}^n\right|^2 p_1^n}{\sigma_n^2 + \sum_{j\ne1}^{K} \left|b_{j1}^n\right|^2 p_j^n \left(p_1^n\right)} \right)$$

$$\sum_{n=1}^{N} p_1^n \le P_{1,\max}$$

$$p_1^n \ge 0 \qquad\qquad (10.40)$$

and $p_j^{SE} = BR_j \left(p_1^{SE},\dots,p_{-j}^{SE} \right)$ is a function of p_1^n.

Solving (10.40) becomes much more involved in the general case in which the utility function of the primary operator is non-convex (p_j^n is function of p_1^n). Nevertheless, suboptimal and low-complexity methods provide neat solutions to solve the problem based on Lagrangian duality [35].

The Lagrangian of (10.40) is given by

$$g(\lambda) = \max_{p_1^1,\dots,p_1^N} \mathcal{L} \left(p_1^1,\dots,p_1^N,\lambda \right)$$

$$= \max_{p_1^1,\dots,p_1^N} \sum_{n=1}^{N} \log_2 \left(1 + \frac{\left|b_{11}^n\right|^2 p_1^n}{\sigma_n^2 + \sum_{j\ne1}^{K} \left|b_{j1}^n\right|^2 p_j^n} \right) + \lambda \left(P_{1,\max} - \sum_{n=1}^{N} p_1^n \right) \qquad (10.41)$$

where λ is the lagrangian dual variable associated with the power constraint.

Solving the Stackelberg problem is done by locally optimizing the Lagrangian function (10.41) via coordinate descent [26]. For each fixed set of λ, we find the optimal p_1^1 while keeping $p_1^2,\dots,p_1^N$ fixed and then find the optimal p_1^2 keeping the other p_1^n ($n \ne 2$) fixed and so on. Such process is guaranteed to converge because each iteration strictly increases the objective function. Finally, λ is found using sub-gradient [35] method (See Algorithm 1).

For the sake of illustration, Figure 10.1 depicts the best and worst NE where the best NE refers to the equilibrium maximizing the sum-rate of both operators whereas the worst NE case minimizes

Table 10.1   Algorithm 1

Algorithm 1
initialize λ, $\bar{P}_1$, $\bar{P}_2$
repeat
for $n = 1, \ldots, N$
set $p_1^n = \arg \max_{p_1^1, \ldots, p_1^N} \sum_{n=1}^{N} \log_2 \left(1 + \dfrac{\left\vert h_{11}^n \right\vert^2 p_1^n}{\sigma_n^2 + \sum_{j \neq 1}^{K} \left\vert h_{j1}^n \right\vert^2 p_j^n \left(p_1^n\right)}\right) + \lambda \left(P_{1,\max} - \sum_{n=1}^{N} p_1^n\right)$
by keeping $p_1^1, \ldots, p_1^{n-1}, p_1^{n+1}, p_1^N$ fixed.
end
until $\left(p_1^1, \ldots, p_1^N\right)$ converges
update λ using sub-gradient [35] method until it converges

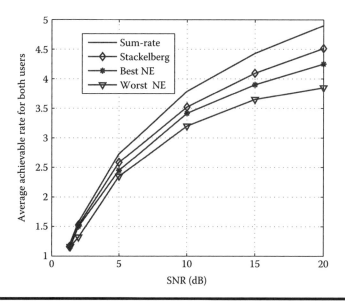

Figure 10.1   Average achievable rate for both users versus the signal-to-noise ratio for the centralized and Stackelberg approaches. Moreover, the best and worst Nash equilibria for the noncooperative game are illustrated.

it. It is also to be noted that the worst NE acts like a lower-bound for the NE. Furthermore, the Stackelberg approach is closer to the centralized approach as compared to the selfish case. This is due to the fact that in the Stackelberg approach, operators take into account other operators' strategies whereas in the selfish case, operators behave carelessly by using the water-filling technique. On the other hand, Figure 10.2 depicts the cumulative distribution function (CDF) of the ratio of both the achievable rates of the hierarchical and noncooperative approaches. In this scenario, we

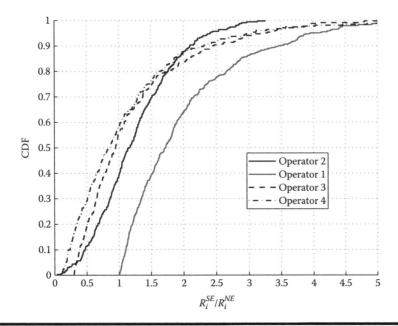

Figure 10.2 CDF of the ratio of the rates between the hierarchical (Stackelberg) and noncooperative (selfish) approaches.

assume $K = 4$ operators with one primary operator and three secondary wireless operators sharing the same spectrum composed of $N = 5$ carriers. As can be seen, the primary operator (operator 1) always improves his achievable rate compared to the selfish approach, while hierarchy provides good incentives for the other secondary operators.

10.4 Discussion

In what follows, the set of contributions dealing with hierarchical PA games (Section 10.3) is examined. First, all of these hierarchical games have the same leader choice (mobile station, base station, and operator). Second, the hierarchical PA game in multiuser channels exhibits Pareto-efficient outcomes as opposed to the energy-efficiency PA game. Third, the energy-efficiency game requires only individual CSI at each transmitter; the Shannon rate-based case requires full CSI, while the spectrum leasing problem with cooperative secondary nodes is investigated from a partial and full knowledge of channel information.

10.5 Concluding Remarks and Open Issues

In this chapter, the problem of hierarchical PA games is looked at where the concept of hierarchy is introduced. GT is a natural paradigm to study these network interactions in which nodes/terminals/transmitters compete with each other for the same resources. The existence and uniqueness of the Nash and Stackelberg equilibria are a nice feature as it predicts the outcome of the game and gives insights on its convergence.

We conclude this chapter by highlighting few open issues related to hierarchical PA problems. It has been shown in various applications that hierarchy is a relevant aspect in wireless communication. Since more and more terminals are foreseen to be deployed in unlicensed bands, more advanced sharing mechanisms have to be investigated to avoid the tragedy of the commons [36] due to the fact that device designers lack an incentive to conserve the shared spectrum resource. Other potential area of interest have to be sought after. For instance, due to the increasing interest for femtocell networks [37] in which the macro-user is given a higher priority than the femto-user, the Stackelberg framework investigated in this chapter can be readily applied to solve the radio resource allocation problem.

References

1. US FCC Spectrum Policy Task Force, Report of the spectrum efficiency working group, Technical Report, November 2002.
2. J. E. Suris, L. DaSilva, Z. Han, and A. MacKenzie, Cooperative game theory approach for distributed spectrum sharing, in *Proceedings of IEEE International Conference on Communications*, Glasgow, Scotland, June 2007.
3. J. Huang, R. Berry, and M. L. Honig, Auction-based spectrum sharing, *ACM Mobile Networks Appl. J.*, 11, 3, 405–418, June 2006.
4. M. Bennis, M. Letreust, S. Lasaulce, M. Debbah, and J. Lilleberg, Spectrum sharing games on the interference channel, *IEEE International Conference on Game Theory for Networks (Gamenets)*, Istanbul, Turkey, May 2009.
5. M. Bennis, S. Lasaulce, and M. Debbah, Inter-operator spectrum sharing from a game theoretical perspective, *EURASIP J. Adv. Signal Processing, Special Issue on Dynamic Spectrum Access*. DOI: 10.1155/2009/295739
6. M. Bennis, J. Lara, and A. Tolli, Non cooperative operators from a game theoretic perspective, *IEEE PIMRC*, Cannes, France, 2008.
7. M. Buddhikot, Understanding dynamic spectrum access: Models, taxonomy and challenges, *IEEE DySPAN*, Dublin, Ireland, April 2007.
8. Q. Zhao and B. M. Sadler, A survey of dynamic spectrum access, *IEEE Signal Processing Mag.*, 24, 3, 79–89, May 2007.
9. R. Menon, R. M. Buehrer, and J. H. Reed, On the impact of dynamic spectrum sharing techniques on legacy radio systems, *IEEE Trans. Wireless Comm.*, 7, 11, 4198–4207, November 2008.
10. D. Fudenberg and J. Tirole, *Game Theory*, Cambridge, MA: The MIT Press, 1991.
11. V. H. Stackelberg, *Marktform und Gleichgewicht*, Oxford University Press, Oxford, U.K., 1934.
12. J. Rosen, Existence and uniqueness of equilibrium points for concave n-person games, *Econometrica*, 33, 3, 520–534, July 1965.
13. A. Leshem and E. Zehavi, Bargaining over the interference channel, *IEEE J. Selected Areas Comm.*, 26, 7, September 2008.
14. S. Mathur, L. Sankaranarayanan, and N. B. Mandayam, Coalitional games in gaussian interference channels, *Proceedings of IEEE ISIT*, Seattle, WA, pp. 2210–2214, July 2006.
15. J. C. Harsanyi, Games with incomplete information played by Bayesian players, I–III, *Manage. Sci.*, 50, 12 Supplement, 1804–1817, 2004.
16. T. S. Han and K. Kobayashi, A new achievable rate region for the interference channel, *IEEE Trans. Inform. Theory*, 27, 1, 49–60, Jan. 1981.
17. R. Etkin, D. Tse, and H. Wang, Gaussian interference channel capacity to within one bit, *IEEE Trans. Inform. Theory*, 54, 12, 5534–5562, Dec. 2008.

18. J. Huang, R. Berry, and M. L. Honig, Distributed interference compensation in wireless networks, *IEEE J. Selected Areas Comm.*, 24, 5, 1074–1084, May 2006.
19. K. J. Arrow, *Social Choice and Individual Values*, Yales University Press, London, U.K. 1963.
20. S. Lasaulce, M. Debbah, and E. Altman, Methodologies for analyzing equilibria in wireless games, *IEEE Signal Processing Mag., Special Issue on Game Theory for Signal Processing*, 26, 5, 41–52, September 2009.
21. R. Selten, Reexamination of the perfectness concept for equilibrium points in extensive games, *Int. J. Game Theory (IJGT)*, 4, 25–55.
22. J. N. Laneman and G. W. Wornell, Distributed space-time coded protocols for exploiting cooperative diversity in wireless networks, *IEEE Trans. Inform. Theory*, 49, 10, 2415–2425, Oct. 2003.
23. O. Simeone, I. Stanojev, S. Savazzi, Y. Bar-Ness, U. Spagnolini, and R. Pickholtz, Spectrum leasing to cooperating secondary ad hoc networks, *IEEE J. Selected Areas Comm.*, 26, 1, 203–213, Jan. 2008.
24. A. Scaglione, Dl. L. Goeckel, and J. N. Laneman, Cooperative communications in mobile ad hoc networks, *IEEE Signal Processing Mag.*, 23, 5, 18–29, Sept. 2006.
25. S. Lasaulce, Y. Hayel, R. El Azouzi, and M. Debbah, Introducing hierarchy in energy games, *IEEE Trans. Wireless Comm.*, 8, 7, 3833–3843, July 2009.
26. S. Boyd and L. Vandenberghe, *Convex Optimization*. Cambridge University Press, Cambridge, U.K. 2004.
27. L. Lai and H. El Gamal, The water-filling game in fading multiple access channels, *IEEE Trans. Inform. Theory*, 54, 5, 2110–2122, May 2008.
28. R. Etkin, A. P. Parekh, and D. Tse, Spectrum sharing in unlicensed bands, *IEEE J. Selected Areas Comm.*, 25, 3, 517–528, April 2007.
29. Y. Su and M. van der Schaar, A new perspective on multi-user power control games in interference channels, *IEEE Trans. Wireless Comm.*, 8, 6, 2910–2919, 2009.
30. G. Scutari, D. P. Palomar, and S. Barbarossa, Optimal linear precoding strategies for wideband noncooperative systems based on game theory part I: Nash equilibria, *IEEE Trans. Signal Processing*, 56, 3, 1230–1249, March 2008.
31. G. Scutari, D. P. Palomar, and S. Barbarossa, Optimal linear precoding strategies for wideband noncooperative systems based on game theory part II: Algorithms, *IEEE Trans. Signal Processing*, 56, 3, 1250–1267, March 2008.
32. G. Scutari, D. P. Palomar, and S. Barbarossa, Competitive design of multiuser MIMO systems based on game theory: A unified view, *IEEE J. Selected Areas Comm.*, 26, 7, 1089–1103, Sept. 2008.
33. W. Yu, G. Ginis, and J. M. Cioffi, Distributed multiuser power control for digital subscriber lines, *IEEE J. Selected Areas Comm.*, 20, 5, 1105–1115, June 2002.
34. S. M. Perlaza, S. Lasaulce, M. Debbah, and J.-M. Chaufray, Game theory for dynamic spectrum sharing, in Y. Zhang, J. Zheng, and H. Chen (Eds.), *Cognitive Radio Networks: Architectures, Protocols and Standards*, Taylor & Francis Group, Auerbach Publications, Boca Raton, FL, 2010.
35. W. Yu and R. Lui, Dual methods for nonconvex spectrum optimization of multicarrier systems, *IEEE Trans. Comm.*, 54, 1310–1322, 2006.
36. Hardin G., The tragedy of the commons, *Science*, 162, 1243–1248, 1968.
37. V. Chandrasekhar, J. G. Andrews, and A. Gatherer, Femtocell networks: A survey, *IEEE Comm. Mag.*, 46, 9, 59–67, September 2008.

Chapter 11

Dynamical Transmission Control

Jane Wei Huang and Vikram Krishnamurthy

Contents

This chapter introduces a special type of game, namely, switching control Markovian dynamic game, and explores the structural result on its Nash equilibrium policy. The system it considers is an uplink time division multiple access (TDMA) cognitive radio network where multiple cognitive radios (secondary users) attempt to access a spectrum hole. It is assumed that each secondary user can access the channel according to a decentralized predefined access rule based on the channel qualities and the transmission delay of each secondary user. By modeling secondary user block-fading channel qualities as a finite state Markov chain, the transmission rate adaptation problem of each secondary user is formulated as a general-sum Markovian dynamic game with a delay constraint. It is shown that the Nash equilibrium transmission policy of each secondary user is a randomized mixture of pure threshold policies under certain conditions. A stochastic approximation algorithm can be applied based on the structural result on the Nash equilibrium policy. Such algorithm can adaptively estimate the Nash equilibrium policies and track such policies for nonstationary problems where the statistics of the channel and of the user parameters evolve with time.

11.1 Introduction

Traditional spectrum management polices are challenged with the increasing demand of spectrum resource. Federal Communications Commission (FCC) has reported that there are 15%–85% temporal and geographic variations in the allocated spectrum usage in Spectrum Policy Task Force 2002 [1]. Cognitive radio systems [2–4] have emerged as a potential technology to revolutionize spectrum utilization. Cognitive radios are defined as radio systems that continuously perform spectrum sensing, dynamically identify unused spectrum, and then operate in those spectrum holes where the primary users are idle. This new communication technology can dramatically enhance spectrum efficiency and is also referred to as the dynamic spectrum access (DSA) network.

Among the non-game-theoretic approaches to addressing resource management in cognitive radio systems, we discuss [5,6]. Shiang and van der Schaar [5] propose a distributed resource management algorithm that considers the delay and cost of exchanging network information for delay-sensitive applications over multi-hop cognitive radio networks. The algorithm learns the behaviors of interacting cognitive radio nodes using a simple interference graph to adjust and optimize the transmission strategies. Alternatively, Shao et al. [6] propose the use of decentralized cognitive media access control (MAC) protocols to address dynamic spectrum management for single-hop networks.

Applying game theory to address various problems in cognitive radio systems is currently a hot area. However, most games considered in wireless communication systems to date are static games. Static game-theoretic analyses have been applied in [7,8] to address the resource allocation problem in cognitive radio networks. Furthermore, multiuser power control problem in wireless networks are formulated under game-theoretic framework and various of iterative water-filling algorithms are proposed to obtain the Nash equilibrium policies [9,10].

Stochastic dynamic game theory is an essential tool for cognitive radio systems as it is able to exploit the correlated channels in the analysis of decentralized behaviors of cognitive radios. However, there is no existing algorithm to calculate the Nash equilibrium policy in a general stochastic game and the analysis of its Nash equilibrium remains a challenging open issue. Switching control game is a special type of dynamic games where the transition probability depends on only one player in each state [11–13]. Such type of game can be solved by a finite sequence of Markov decision processes (MDPs). MDP is a very well studied area and it can be generalized to constrained Markov decision processes (CMDPs) where more structural results are available. Specifically, this

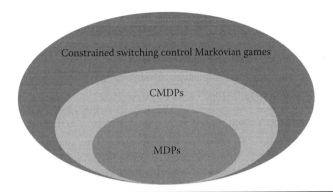

Figure 11.1 **The generality relationships among constrained switching control games, CMDP, and MDP.**

chapter formulates the rate adaptation problem in a cognitive radio system as a constrained switching control game and it can be solved by a finite sequence of CMDPs. Figure 11.1 shows the relationship among constrained switching control games, CMDPs, and MDPs.

This chapter considers the secondary user rate adaptation problem in cognitive radio networks where multiple secondary users attempt to access a spectrum hole. We assume a TDMA cognitive radio system model (as specified in the IEEE 802.16 standard [14]) that schedules one user per spectrum hole at each time slot according to a predefined decentralized scheduling policy. Therefore, the interaction among secondary users is characterized as a competition for the spectrum hole and can naturally be formulated as a dynamic game. By modeling transmission channels as correlated Markovian sources, the transmission rate adaptation problem for each user can be formulated as a general-sum switching control Markovian dynamic game with a latency constraint. The transmission policy of such a game takes into account the secondary user channel qualities, as well as the transmission delay of each secondary user.

11.1.1 Main Results

1. We formulate the secondary user rate adaptation problem in a cognitive radio network as a constrained general-sum switching control Markovian dynamic game. The TDMA cognitive radio system has a prespecified channel access rule that is typically a function of secondary user channel qualities and buffer occupancies. The Markovian transmission block-fading channels are formulated as a finite state Markov chain, and each secondary user aims to optimize its own utility under a transmission delay constraint. Since we are considering a TDMA system, the transmission rate control problem is formulated as a *constrained* switching control Markovian game [15].

2. Our main structural result is to show that the Nash equilibrium policy of a general-sum switching control game is monotone nondecreasing on the buffer occupancy state. Under the assumptions listed in Section 11.3.2, the Nash equilibrium transmission policy is a randomized mixture of two pure policies, each one is monotone nondecreasing on the buffer occupancy state. This result is illustrated in Figure 11.3, where the deterministic Nash equilibrium policy of a two-user cognitive radio system is shown with the Lagrangian multiplier fixed as a constant. From the figure, it can be seen that the optimal action policy of each user

under a certain channel state is monotone nondecreasing on the buffer state. This structured result can be exploited to derive efficient algorithms.

Lagrangian dynamic programming [16] has recently been applied to MDPs in transmission scheduling. Here, we extend its application to dynamic games. In our problem formulation, a Lagrangian multiplier is used to combine the latency constraint with the optimization objective. The original problem is then transformed into an unconstrained Markovian game. It is shown that the Nash equilibrium policy of such a game can be obtained by using a value iteration algorithm.

3. Based on the structural result on the Nash equilibrium policy, a stochastic approximation algorithm is proposed in Section 11.3.6 to learn the policy. The algorithm provides insight into the nature of the solution without brute force computation and can adapt the Nash equilibrium policy in real time to the nonstationary channel and user statistics. Numerical results of the stochastic approximation algorithm are provided in Section 11.4.

11.2 Rate Adaptation Problem Formulation

This section describes the system model (Figure 11.2). In a cognitive radio system [14], each secondary user tries to transmit its data to a specific receiver. We consider a TDMA system with K secondary users where only one user can access the channel at each time slot according to a predefined decentralized access rule. The access rule will be described later in this section. The correlated block-fading channel that each user accesses is modeled as a Markov chain. The rate control problem for each secondary user can then be formulated as a constrained Markovian dynamic game. More

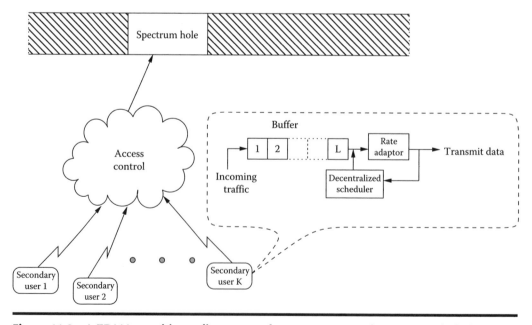

Figure 11.2 A TDMA cognitive radio system where users access the spectrum hole following a predefined access rule. Each user is equipped with a size L buffer, a decentralized scheduler, and a rate adaptor for transmission control.

specifically, under the predefined decentralized access rule, the problem presented is a special type of game, namely, a switching control Markovian dynamic game. Note that for simplicity, our problem formulation assumes that there is only one spectrum hole. However, our formulation also applies to systems with multiple spectrum holes. In such cases, the rate adaptation problem for each spectrum hole can be formulated independently as a switching control Markov game.

11.2.1 System Description and TDMA Access Rule

The channel quality of user k at time n is denoted as $h_k^{(n)}$. The channel model is assumed to consist of circularly symmetric complex Gaussian random variables that depend only on the previous time slot. After applying quantization on the channel quality, the channel state space is $h_k^{(n)} \in \{0, 1, 2, \dots\}$, where state 0 indicates that the primary user is using the channel. The composition of channel states of all the K users can be written as $\mathbf{h}^{(n)} = \left\{ h_1^{(n)}, \dots, h_K^{(n)} \right\}$. Assuming that the channel state $\mathbf{h}^{(n)} \in \mathcal{H}, n = 1, 2, \dots, N$ is block fading, the block length equals one time period. The channel states over time constitute a Markov chain. The transition probability of the channel states from time n to $(n + 1)$ can be denoted as $\mathbb{P}(\mathbf{h}^{(n+1)} | \mathbf{h}^{(n)})$.

Denoting the buffer size of each user as L, we can use $b_k^{(n)}$ to indicate the buffer occupancy state of user k at time n and $b_k^{(n)} \in \{0, 1, \dots, L\}$. The composition of the buffer states of all the K users can be denoted as $\mathbf{b}^{(n)} = \left\{ b_1^{(n)}, \dots, b_K^{(n)} \right\}$. $\mathbf{b}^{(n)}$ is an element of the secondary user buffer state space $\mathcal{B}$.

In the system model, there is new incoming traffic at the beginning of each time slot. We refer to the number of new incoming packets of the kth user at the nth time slot as $f_k^{(n)}$, $f_k^{(n)} \in \{0, 1, 2, \dots, \infty\}$. The composition of the incoming traffic of all the K users can be denoted as $\mathbf{f}^{(n)} = \left\{ f_1^{(n)}, \dots, f_K^{(n)} \right\}$, which is an element of the incoming traffic space $\mathcal{F}$. For simplicity, the incoming traffic is assumed to be independent and identically distributed (i.i.d.) in terms of time index n and user index k. The incoming traffic is not part of the system state but it affects the buffer state evolution.

The state of the kth secondary user at the nth time slot is denoted as $\mathbf{s}_k^{(n)}$ and it is specified as $\mathbf{s}_k^{(n)} = \left[h_k^{(n)}, b_k^{(n)} \right]$. The overall system state of the nth time slot is a composition of the states of all the K secondary users, which is denoted as $\mathbf{s}^{(n)} = \left\{ \mathbf{s}_1^{(n)}, \dots, \mathbf{s}_K^{(n)} \right\}$. We use $\mathcal{S}$ to denote the finite system state space, which comprises channel state $\mathcal{H}$ and secondary user buffer state $\mathcal{B}$. That is, $\mathcal{S} = \mathcal{H} \times \mathcal{B}$. Here, $\times$ denotes a Cartesian product. Furthermore, $\mathcal{S}_k$ is used to indicate the state space where user k is scheduled for transmission. $\mathcal{S}_1, \mathcal{S}_2, \dots, \mathcal{S}_K$ are disjoint subsets of $\mathcal{S}$ with the property of $\mathcal{S} = \mathcal{S}_1 \cup \mathcal{S}_2 \cup \dots \cup \mathcal{S}_K$.

This chapter adopts a TDMA cognitive radio system model (IEEE 802.16 [14]). A decentralized channel access algorithm is designed for the TDMA channel. The mechanism of the algorithm is described as follows: At the beginning of a time slot, each user tries to access the channel after a certain time delay t. The time delay of user k can be calculated using an opportunistic scheduling algorithm [17] and is specified as

$$t = \frac{\gamma}{b_k^{(n)} h_k^{(n)}} \tag{11.1}$$

where γ is a system parameter determined by the system. After the first user accesses the channel, all remaining users will detect the channel occupancy and stop accessing. However, when the spectrum is occupied by a primary user (assuming the system has perfect information concerning primary user activities), the channel states of all secondary users $h_k^{(n)}$ ($k = 1, 2, \ldots, K$) will be set to zero, which in turn will lead the value of t to infinity, meaning no user will access the channel during the nth period. We use t^* to denote the waiting time of the first secondary user who accesses the channel and k^* to denote the index of that user. The value of t^* and k^* can be specified as follows:

$$t^* = \min_{k \in \{1,2,\ldots,K\}} \frac{\gamma}{b_k^{(n)} h_k^{(n)}}$$

$$k^* = \arg \min_{k \in \{1,2,\ldots,K\}} \frac{\gamma}{b_k^{(n)} h_k^{(n)}}, \quad \text{if } h_k^{(n)} \neq 0 \tag{11.2}$$

$$k^* = 0, \quad \text{if } h_1^{(n)} =, \ldots, = h_k^{(n)} = 0$$

where
- $b_k^{(n)}$ and $h_k^{(n)}$ are two parameters affecting the scheduling algorithm
- $b_k^{(n)}$ evaluates the packet delay of each user at time n, and users with longer delay are preferred for scheduling
- $h_k^{(n)}$ denotes the channel quality of user k at the nth time slot, and the user with the better channel quality has a better chance of being scheduled for transmission

It is worth mentioning that the existing opportunistic scheduling approaches overlook the fact that the secondary users may be owned by different agents and they may work in competitive rather than cooperative manners. These selfish users can be so sophisticated that they lie about their states to optimize their own utility at the cost of reducing the overall system performance. It requires mechanism design theory in order to prevent this from happening. Huang and Krishnamurthy [18] justify this problem in the paper where the authors propose a pricing mechanism, which combines the mechanism design with the opportunistic scheduling algorithms, and ensures that each rational selfish user maximizes his own utility function, at the same time optimizing the overall system utility.

11.2.2 Action and Costs

If the kth user is scheduled for transmission at the nth time slot, its action $a_k^{(n)}$ represents the bits/symbol rate of the transmission. Different $a_k^{(n)}$ lead to different transmission rates. Assuming the system uses an uncoded M-ary quadrature amplitude modulation (QAM), different bits/symbol rates determine the modulation schemes as follows: $M = 2^{a_k^{(n)}}$.

11.2.2.1 Transmission Cost

Let $c_i\left(s^{(n)}, a_1^{(n)}, \ldots, a_K^{(n)}\right)$ denote the transmission cost of user i at time n. Specifically, $c_k\left(s^{(n)}, a_1^{(n)}, \ldots, a_K^{(n)}\right)$ is chosen to be the transmission bit error rate (BER) introduced by user k

during the transmission. When $\mathbf{s}^{(n)} \in \mathcal{S}_k$ and user k is scheduled for transmission at time instant n, the performance of that user depends only on itself, as all the other users are inactive. Thus, the costs of all the users in the system are

$$c_k \left(\mathbf{s}^{(n)}, a_1^{(n)}, \ldots, a_K^{(n)} \right) = c_k \left(\mathbf{s}^{(n)}, a_k^{(n)} \right) \geq 0$$

$$c_{i,i \neq k} \left(\mathbf{s}^{(n)}, a_1^{(n)}, \ldots, a_K^{(n)} \right) = 0 \tag{11.3}$$

For notational convenience, in the following sections, we will drop the subscript k, by defining

$$c \left(\mathbf{s}^{(n)}, a_k^{(n)} \right) := c_k \left(\mathbf{s}^{(n)}, a_k^{(n)} \right)$$

11.2.2.2 Holding Cost

Each user has an instantaneous Quality of Service (QoS) constraint denoted as $d_i \left(\mathbf{s}^{(n)}, a_1^{(n)}, \ldots, a_K^{(n)} \right)$, $i = 1, \ldots, K$. If the QoS is chosen to be the delay (latency), then $d_i \left(\mathbf{s}^{(n)}, a_1^{(n)}, \ldots, a_K^{(n)} \right)$ is a function of the buffer state $b_i^{(n)}$ of the current user i. The instantaneous holding costs will be subsequently included in an infinite horizon latency constraint. Since the transmission latency is independent of the actions of all the remaining users, it simplifies to

$$d_k \left(\mathbf{s}^{(n)}, a_1^{(n)}, \ldots, a_K^{(n)} \right) = d_k \left(\mathbf{s}^{(n)}, a_k^{(n)} \right) \geq 0, \quad \mathbf{s}^{(n)} \in \mathcal{S}_k \tag{11.4}$$

To allow the interaction among the secondary users, we couple the latency constraints of all the users. That is, we assume that the K holding costs among the users are zero-sum:

$$\sum_{i=1}^{K} d_i \left(\mathbf{s}^{(n)}, a_1^{(n)}, \ldots, a_K^{(n)} \right) = 0 \tag{11.5}$$

The zero-sum constraints can be interpreted as follows: User k, who is scheduled for transmission, pays a cost to other users, which is equivalent to the other users receiving a reward. Due to this "*budget balance property*" [19], the holding costs among all the users are zero-sum. In [19], a similar assumption has been used for resource allocation in a wireless multimedia system.

Since the reward is divided equally among the remaining $(K - 1)$ users, the holding cost of each can be written as

$$d_{i,i \neq k} \left(\mathbf{s}^{(n)}, a_k^{(n)} \right) = -\frac{1}{K-1} \cdot d_k \left(\mathbf{s}^{(n)}, a_k^{(n)} \right) \tag{11.6}$$

Omitting the subscript k of the user being scheduled for transmission, the holding cost of this user can be expressed as

$$d \left(\mathbf{s}^{(n)}, a_k^{(n)} \right) := d_k \left(\mathbf{s}^{(n)}, a_k^{(n)} \right)$$

Remark: The zero-sum holding cost system model is entirely equivalent to the model of constant-sum holding costs. (11.6) can also be generalized to $d_{i,i \neq k} \left(\mathbf{s}^{(n)}, a_k^{(n)} \right) = -\alpha_i \cdot d_k \left(\mathbf{s}^{(n)}, a_k^{(n)} \right)$ with $\sum_{i,i \neq k}^{K} \alpha_i = 1, \alpha_i \geq 0$.

11.2.3 Switching Control Game and Transition Probabilities

With the above setup, the decentralized transmission control problem in a Markovian block-fading channel cognitive radio system can now be formulated as a switching control game. In such a game [12], the transition probabilities depend only on the action of the kth user when the state $\mathbf{s} \in \mathcal{S}_k$. This feature enables us to solve such a game by a finite sequence of MDPs. The state transition probability of the switching control game is

$$
\mathbb{P}(\mathbf{s}^{(n+1)}|\mathbf{s}^{(n)}, a_1, a_2, \ldots, a_K) = \begin{cases} \mathbb{P}(\mathbf{s}^{(n+1)}|\mathbf{s}^{(n)}, a_1) & \text{if } \mathbf{s}^{(n)} \in \mathcal{S}_1 \\ \mathbb{P}(\mathbf{s}^{(n+1)}|\mathbf{s}^{(n)}, a_2) & \text{if } \mathbf{s}^{(n)} \in \mathcal{S}_2 \\ \cdots & \\ \mathbb{P}(\mathbf{s}^{(n+1)}|\mathbf{s}^{(n)}, a_K) & \text{if } \mathbf{s}^{(n)} \in \mathcal{S}_K \end{cases} \tag{11.7}
$$

According to the property of the switching control game, when the kth user is scheduled for transmission, the transition probability between the current composite state $\mathbf{s} = [\mathbf{h}, \mathbf{b}]$ and the next state $\mathbf{s}' = [\mathbf{h}', b']$ depends only on the action of the kth user a_k. The transition probability function of our system model can now be written mathematically as

$$
\mathbb{P}(s'|s, a_1, a_2, \ldots, a_K) = \mathbb{P}(s'|s, a_k) = \prod_{i=1}^{K} \mathbb{P}\left(b_i'|h_i\right) \cdot \prod_{i=1, i \neq k}^{K} \mathbb{P}\left(b_i'|b_i\right) \cdot \mathbb{P}\left(b_k'|b_k, a_k\right)
$$

Since each user is equipped with a size L buffer, the buffer occupancy of user k evolves according to Lindley's equation [20]

$$
b_k^{(n+1)} = \min\left(\left[b_k^{(n)} - a_k^{(n)}\right]^+ + f_k^{(n)}, L\right) \tag{11.8}
$$

The buffer state of user $i = 1, 2, \ldots, K, i \neq k$ evolves according to the following rule:

$$
b_i^{(n+1)} = \min\left(b_i^{(n)} + f_i^{(n)}, L\right)
$$

For user k, the buffer state transition probability depends on the distribution of its incoming traffic and its action. Its mathematical expression is

$$
\mathbb{P}\left(b_k^{(n+1)}\Big|b_k^{(n)}, a_k^{(n)}\right) = \begin{cases} \mathbb{P}\left(f_k^{(n)} = b_k^{(n+1)} - \left[b_k^{(n)} - a_k^{(n)}\right]^+\right) & b_k^{(n+1)} < L \\ \displaystyle\sum_{x=L-\left[b_k^{(n)}-a_k^{(n)}\right]^+}^{\infty} \mathbb{P}\left(f_k^{(n)} = x\right) & b_k^{(n+1)} = L \end{cases} \tag{11.9}
$$

For those users who are not scheduled for transmission, the buffer state transition probabilities depend only on the distribution of the incoming traffic, which can be written as

$$
\mathbb{P}\left(b_i^{(n+1)}\Big|b_i^{(n)}\right) = \begin{cases} \mathbb{P}\left(f_i^{(n)} = b_i^{(n+1)} - b_i^{(n)}\right) & b_i^{(n+1)} < L \\ \displaystyle\sum_{x=L-b_i^{(n)}}^{\infty} \mathbb{P}\left(f_i^{(n)} = x\right) & b_i^{(n+1)} = L \end{cases} \tag{11.10}
$$

11.2.4 Switching Control Markovian Game Formulation

We use π_i ($i = 1, 2, \ldots, K$) to denote the transmission policy vector of the ith user. With a slight abuse of notation, $\pi_i(\mathbf{s})$ is used to denote the transmission policy of user i in state $\mathbf{s}$ and is a component of π_i. $\pi_i(\mathbf{s})$ lives in the same space as the action of the ith user a_i. Assume at time instant n, user k is scheduled for transmission according to the system access rule as specified in (11.2). The infinite horizon expected total discounted cost* of the ith ($i = 1, 2, \ldots, K$) user under transmission policy π_i can be written as

$$C_i(\pi_i) = \mathbb{E}_{\pi_i} \left[\sum_{n=1}^{\infty} \beta^{n-1} \cdot c_i \left(\mathbf{s}^{(n)}, a_k^{(n)} \right) \right] \qquad (11.11)$$

where $0 \leq \beta < 1$ is the discount factor. The expectation of the above function is taken over the system state $\mathbf{s}^{(n)}$ evolution regarding time index n. If we denote the holding cost of user i at the nth time slot as $d_i \left(\mathbf{s}^{(n)}, a_k^{(n)} \right)$, the infinite horizon expected total discounted latency constraint can be written as

$$D_i(\pi_i) = \mathbb{E}_{\pi_i} \left[\sum_{n=1}^{\infty} \beta^{n-1} \cdot d_i \left(\mathbf{s}^{(n)}, a_k^{(n)} \right) \right] \leq \widetilde{D}_i \qquad (11.12)$$

where $\widetilde{D}_i$ is a system parameter depending on the system requirement. Note here that we assume the latency constraint is valid in our problem formulation. $\widetilde{D}_i$ is chosen so that the set of policies that satisfy such a constraint is nonempty. This assumption will be discussed more specifically in Section 11.3.2.

Equations 11.7,11.11, and 11.12 define a constrained switching control Markovian game. Our goal is to compute a Nash equilibrium[†] policy $\pi_i^*, i = 1, \ldots, K$ (which is not necessarily unique) that minimizes the discounted transmission cost (11.11) subject to the latency constraint (11.12). However, if the transmission cost (11.3) and holding costs (11.4,11.6) are both zero-sum among all users, then all the Nash equilibria have a unique value vector, which is a globally optimal solution of the game.

The following result shows that a Markovian switching control game can be solved using a sequence of MDPs. The reader is referred to [12,13] for the proof.

Result: [12, Chapter 3.2] A switching control Markovian game can be solved by a finite sequence of MDPs, as described in Algorithm 11.1. Thus, the constrained switching control Markovian game (11.11,11.12) can be solved by iteratively updating the transmission policy $\pi_i^{*(n)}$ of user i ($i = 1, \ldots, K$), with the policy of remaining users fixed. Here, n denotes the iteration index, which is equivalent to the time slot index. For each step, we aim to minimize the overall expected

---

[*] There are two criteria for evaluating the cost of an MDP: expected average cost criterion and expected total discounted reward criterion. We choose the discounted cost criterion because it is mathematically simpler than the average cost criterion, which also has several technicalities involved with obtaining the stationary optimal policies. In addition, if the stationary policy exists, when the discounted factor $\beta \to 1$, the policy obtained under the average cost criterion is the same as that from the discounted cost criterion, $C_{avg} = \lim_{\beta \to 1}(1 - \beta) \cdot C_{dis}$. Here, C_{avg} and C_{dis} denote the average cost and discounted cost, respectively.

[†] A Nash equilibrium [12] is a set of policies, one for each player, such that no player has incentive to unilaterally change its action. Payers are in equilibrium if a change in policy by any one of them would lead that player to earn less than if it keeps its current policy.

Algorithm 11.1 Value Iteration Algorithm

Step 1: Set $m = 0$; Initialize l;
Initialize $\mathbf{V}_1^0, \ldots, \mathbf{V}_K^0, \lambda_1, \lambda_2, \ldots, \lambda_K$.
Step 2: Inner Loop: Set $n = 1$;
Step 3: Update the transmission policy and value vector of each user:
for $k = 1 : K$ **do**
  for each $\mathbf{s} \in \mathcal{S}_k$,
  $$\pi_k^n(\mathbf{s}) = \arg\min_{a_k} \left\{ c(\mathbf{s}, a_k) + \lambda_k^m \cdot d_k(\mathbf{s}, a_k) + \beta \sum_{s'=1}^{|\mathcal{S}|} \mathbb{P}(s'|\mathbf{s}, a_k) v_k^n(s') \right\};$$
  $$v_k^{n+1}(\mathbf{s}) = c\left(\mathbf{s}, \pi_k^n(\mathbf{s})\right) + \lambda_k^m \cdot d_k\left(\mathbf{s}, \pi_k^n(\mathbf{s})\right) + \beta \sum_{s'=1}^{|\mathcal{S}|} \mathbb{P}\left(s'|s, \pi_k^n(\mathbf{s})\right) v_k^n(s');$$
  $$v_{i=1:K, i\neq k}^{n+1}(\mathbf{s}) = \lambda_i^m \cdot d_i(\mathbf{s}, a_k) + \beta \sum_{s'=1}^{|\mathcal{S}|} \mathbb{P}\left(s'|s, \pi_k^n(\mathbf{s})\right) v_i^n(s');$$
end for
Step 4: If $\mathbf{V}_k^{n+1} < \mathbf{V}_k^n, k = 1, \ldots, K$, set $n = n + 1$, and return to Step 3; otherwise, go to Step 5.
End of inner loop;
Step 5: Update the Lagrangian multipliers
for $k = 1 : K$ **do**
  $$\lambda_k^{m+1} = \lambda_k^m + \frac{1}{l}\left[D_k\left(\pi_1^n, \pi_2^n, \ldots, \pi_K^n\right) - \widetilde{D}_k\right]$$
end for
Step 6: The algorithm stops when $\lambda_k^m, k = 1, 2, \ldots, K$ converge, otherwise, set $m = m+1$ and return to Step 2.

discounted cost of the ith user subject to the latency constraint. The optimization problem can be written as follows:

$$\pi_i^{*(n)} = \left\{ \pi_i^{(n)} : \min_{\pi_i} C_i^{(n)}(\pi_i) \ \text{s.t.} \ D_i^{(n)}(\pi_i) \leq \widetilde{D}_i, i = 1, \ldots, K \right\} \tag{11.13}$$

11.3 Randomized Threshold Nash Equilibrium

This section first presents a value iteration algorithm, which is used to obtain the Nash equilibrium policy of a general-sum Markovian dynamic switching control game, as described earlier. We then present a structural result on the Nash equilibrium policy. Finally, we propose a computationally efficient stochastic approximation algorithm to search for the Nash equilibrium.

11.3.1 Value Iteration Algorithm

In [12], a value iteration algorithm is presented to compute the Nash equilibrium of an unconstrained general-sum Markovian dynamic switching control game. By using the Lagrangian relaxation method, we combine the optimization objective (11.11) with the latency constraint (11.12) via a Lagrangian multiplier λ_i, $i = 1, 2, \ldots, K$. We then compute the Nash equilibrium of the resulting constrained switching control Markovian dynamic game by the value iteration algorithm described in Algorithm 11.1.

The algorithm mainly consists of two parts: the outer loop and inner loop. The outer loop updates the Lagrangian multipliers of each user and the inner loop optimizes the transmission policy of each user under given Lagrangian multipliers. The outer loop index and inner loop index are m and n, respectively. We use $\mathbf{V}_{k=1,2,\ldots,K}^n$ and $\lambda_{k=1,2,\ldots,K}^m$ to represent the value vector and

Lagrangian multiplier, respectively, of each user at the nth and mth time slot. Note here that it could be seen from Algorithm 11.1 that the interaction among all the secondary users is through the update of value vectors since $v_{i=1:K,i\neq k}^{n+1}(s)$ is a function of $\pi_k^n(s)$.

In Step 1, we set the outer loop index m to be 0 and initialize the step size l, the value vector $\mathbf{V}_{k=1,2,\ldots,K}^0$ and Lagrangian multipliers $\lambda_{k=1,2,\ldots,K}^0$. Step 2 starts the inner loop where at each step, we solve kth user-controlled game and obtain the new optimal strategy for that user with the strategy of the remaining players fixed. Step 5 updates the Lagrangian multipliers based on the discounted delay value of each user given the transmission policies $\{\pi_1^n, \pi_1^n, \ldots, \pi_K^n\}$. $1/l$ is the step size, which satisfies the conditions for convergence of the Robbins-Monro algorithm. This sequence of Lagrangian multipliers $\{\lambda_1^m, \ldots, \lambda_K^m\}$ with $m = 0, 1, 2, \ldots$ converges in probability to $\{\lambda_1^*, \ldots, \lambda_K^*\}$, which satisfies the constrained problem defined in (11.13) [21,22]. The algorithm terminates when certain accuracy of $\lambda_{k=1,2,\ldots,K}^m$ is obtained, otherwise, go to Step 2.

Since this is a constrained optimization problem, the optimal transmission policy is a randomization of two deterministic polices [20]. Use $\lambda_{k=1,2,\ldots,K}^*$ to represent the Lagrangian multipliers obtained with the above algorithm. The randomization policy of each user can be written as

$$\pi_k^*(s) = q_k \pi_k^*(s, \lambda_{k,1}) + (1 - q_k)\pi_k^*(s, \lambda_{k,2}) \tag{11.14}$$

where

$0 \leq q_k \leq 1$ is the randomization factor

$\pi_k^*(s, \lambda_{k,1})$ and $\pi_k^*(s, \lambda_{k,2})$ are the unconstrained optimal policies with Lagrangian multipliers $\lambda_{k,1}$ and $\lambda_{k,2}$

Specifically, $\lambda_{k,1} = \lambda_k^* - \Delta$ and $\lambda_{k,2} = \lambda_k^* + \Delta$ for a perturbation parameter Δ. The randomization factor of the kth user q_k is calculated by

$$q_k = \frac{\widetilde{D}_k - D_k(\lambda_{1,2}, \ldots, \lambda_{K,2})}{D_k(\lambda_{1,1}, \ldots, \lambda_{K,1}) - D_k(\lambda_{1,2}, \ldots, \lambda_{K,2})} \tag{11.15}$$

THEOREM 11.1   The value iteration algorithm (Algorithm 11.1) converges to the Nash equilibrium of the constrained switching control general-sum Markovian game (with general-sum transmission cost (11.3) and zero-sum latency constraint (11.5)). □

Please refer to [12] for the proof of the convergence of the inner loop of Theorem 11.1. The intuition behind the proof can be shown as follows. The value vector $\mathbf{V}_k^{(n)}$ ($k = 1, 2, \ldots, K$) is nonincreasing on the iteration index n in the value iteration algorithm. There are only a finite number of strategies available for the optimal policy π_k^* for $k = 1, 2, \ldots, K$. It can be concluded that the algorithm converges in a finite number of iterations with given Lagrangian multipliers. The convergence of the outer loop, namely, Lagrangian multipliers, is proved in [21,22]. Thus, the Nash equilibrium of the switching control dynamic game under the discounted reward criterion can be obtained by the value iteration algorithm.

11.3.2 Structural Result on Randomized Threshold Policy

Since Algorithm 11.1 converges to a Nash equilibrium, we aim to characterize the structure of the Nash equilibrium policy. First, we list three assumptions. Based on these three assumptions, our main structural result is introduced in Theorem 11.2.

- *A1:* The delay constraint on the system (11.12) is valid and the set of policies that satisfies this constraint is nonempty.
- *A2:* Transmission cost $c(\mathbf{s}, a_k)$ and holding cost $d(\mathbf{s}, a_k)$ are submodular* functions of b_k, a_k given channel quality h_k of the current user and are independent of the incoming traffic f_k. $c(\mathbf{s}, a_k)$ and $d(\mathbf{s}, a_k)$ are also nondecreasing functions of b_k for any h_k and a_k.
- *A3:* $\sum_{b'_k = l}^{L} \mathbb{P}\left(b'_k | b_k, a_k\right)$ is a submodular function of b_k, a_k and is nondecreasing on b_k for any l and a_k.

The following theorem is our main result and the proof of the theorem is given in Section 11.3.3.

THEOREM 11.2 Consider the rate adaptation in a constrained Markovian dynamic switching control game system described in Section 11.2. Assume A1–A3 hold. Then the Nash equilibrium policy of the kth user (for $k = 1, 2, \dots, K$) π_k^* is a randomized mixture of two pure policies: π_k^1 and π_k^2. Each of these two pure policies is nondecreasing on the buffer occupancy state b_k. ◻

11.3.3 Proof of Theorem 11.2

Consider the case where the delay constraint parameter $\widetilde{D}_k$ in (11.12) is chosen so that a Nash equilibrium policy of the optimization problem exists. When the Nash equilibrium is obtained, the delay constraint (11.12) will hold with equality. This feature implies that by introducing the Lagrangian multiplier λ_k^m, we can get the Nash equilibrium policy when the objective function (11.11) is minimized and the equality of the delay constraint (11.12) is obtained.

According to Algorithm 11.1, the transmission policy and value matrix of the kth user are updated by the following steps:

$$\pi_k^*(\mathbf{s}, \mathbf{s} \in \mathcal{S}_k)^{(n)} = \arg \min_{\pi_k(\mathbf{s})^{(n)}} \left\{ c(\mathbf{s}, a_k) + \lambda_k^m \cdot d(\mathbf{s}, a_k) + \beta \sum_{s'=1}^{|\mathcal{S}|} \mathbb{P}(\mathbf{s}'|\mathbf{s}, a_k) v_k^{(n-1)}(\mathbf{s}') \right\} \quad (11.16)$$

$$v_k^{(n)}(\mathbf{s}, \mathbf{s} \in \mathcal{S}_k) = \min_{\pi_k^*(\mathbf{s})^{(n)}} \left\{ c(\mathbf{s}, a_k) + \lambda_k^m \cdot d(\mathbf{s}, a_k) + \beta \sum_{s'=1}^{|\mathcal{S}|} \mathbb{P}(\mathbf{s}'|\mathbf{s}, a_k) v_k^{(n-1)}(\mathbf{s}') \right\} \quad (11.17)$$

In order to prove that the optimal transmit action policy $\pi_k^*(\mathbf{s}, \mathbf{s} \in \mathcal{S}_k)^{(n)}$ is monotone nondecreasing on the buffer state b_k, we have to show that the right-hand side of (11.16) is a submodular function of (b_k, a_k). According to assumption A2, $c(\mathbf{s}, a_k) + \lambda_k^m \cdot d(\mathbf{s}, a_k)$ is a submodular function of (b_k, a_k); thus, we only need to demonstrate that $\sum_{s'=1}^{|\mathcal{S}|} \mathbb{P}(\mathbf{s}'|\mathbf{s}, a_k) v_k^{(n-1)}(\mathbf{s}')$ is also submodular in the pair (b_k, a_k).

---

* A function $f : \mathcal{A} \times \mathcal{B} \times \mathcal{C} \to \mathcal{R}$ is said to be submodular in (a, b) for any fixed $c \in \mathcal{C}$ if $f(a', b'; c) - f(a, b'; c) \leq f(a', b; c) - f(a, b; c)$ holds for all $a' \geq a$ and $b' \geq b$.

As mentioned before, the overall state space $\mathcal{S}$ is the union of all the K sub-spaces: $\mathcal{S} = \mathcal{S}_1 \cup \mathcal{S}_2 \cdots \cup \mathcal{S}_K$. Thus, we can write $\sum_{\mathbf{s}'=1}^{|\mathcal{S}|} \mathbb{P}(\mathbf{s}'|\mathbf{s}, a_k) v_k^{(n-1)}(\mathbf{s}')$ as a summation of K terms according to the partition of the state space. In order to show the submodularity of $\sum_{\mathbf{s}'=1}^{|\mathcal{S}|} \mathbb{P}(\mathbf{s}'|\mathbf{s}, a_k) v_k^{(n-1)}(\mathbf{s}')$, we only need to prove the submodularity property of each term. The decomposition is given as follows:

$$\sum_{\mathbf{s}'=1}^{|\mathcal{S}|} \mathbb{P}(\mathbf{s}'|\mathbf{s}, a_k) v_k^{(n-1)}(\mathbf{s}') = \sum_{\mathbf{s}', \mathbf{s}' \in \mathcal{S}_1} \mathbb{P}(\mathbf{s}'|\mathbf{s}, a_k) v_k^{(n-1)}(\mathbf{s}') + \cdots + \sum_{\mathbf{s}', \mathbf{s}' \in \mathcal{S}_K} \mathbb{P}(\mathbf{s}'|\mathbf{s}, a_k) v_k^{(n-1)}(\mathbf{s}')$$

(11.18)

If we consider the ith term of the right-hand side of (11.18), using the state transition probability expression from (11.8), we have

$$\sum_{\mathbf{s}', \mathbf{s}' \in \mathcal{S}_i} \mathbb{P}(\mathbf{s}'|\mathbf{s}, a) v_k^{(n-1)}(\mathbf{s}') = \prod_{l=1}^{K} \mathbb{P}\left(b_l'|h_l\right) \cdot \prod_{l=1, l \neq k}^{K} \mathbb{P}\left(b_l'|b_l\right) \cdot \mathbb{P}\left(b_k'|b_k, a_k\right) \cdot v_k^{(n-1)}(\mathbf{s}')$$

The proof of Theorem 11.2 requires the result from Lemma 11.1, whose proof will be provided following the proof of Theorem 11.2.

LEMMA 11.1   Under the assumptions of Theorem 11.2, $v_k(\mathbf{s})$ is nondecreasing on b_k when the state lives in state space $\mathcal{S}_i$, that is $\mathbf{s} \in \mathcal{S}_i$, $i = 1, 2, \ldots, K$.

Continuing with the proof of the theorem, the assumption says that $\sum_{b_k'=l}^{L} \mathbb{P}\left(b_k'|b_k, a_k\right)$ is submodular in (b_k, a_k) for any l. According to the definition of submodularity, this assumption can be mathematically written in the following way, for all the l and $\mathbf{s}' \in \mathcal{S}_i$:

$$\sum_{b_k'=l}^{L} \mathbb{P}\left(b_k'|b_k^-, a_k^-\right) + \sum_{b_k'=l}^{L} \mathbb{P}\left(b_k'|b_k^+, a_k^+\right) \leq \sum_{b_k'=l}^{L} \mathbb{P}\left(b_k'|b_k^-, a_k^+\right) + \sum_{b_k'=l}^{L} \mathbb{P}\left(b_k'|b_k^+, a_k^-\right) \quad (11.19)$$

Based on the result from Lemma 11.1, we can apply Lemma 4.7.2 from [23] to our model, which yields

$$\sum_{b_k'=0}^{L} \mathbb{P}\left(b_k'|b_k^-, a_k^-\right) \cdot v_k(\mathbf{s}') + \sum_{b_k'=0}^{L} \mathbb{P}\left(b_k'|b_k^+, a_k^+\right) \cdot v_k(\mathbf{s}')$$

$$\leq \sum_{b_k'=0}^{L} \mathbb{P}\left(b_k'|b_k^-, a_k^+\right) \cdot v_k(\mathbf{s}') + \sum_{b_k'=0}^{L} \mathbb{P}\left(b_k'|b_k^+, a_k^-\right) \cdot v_k(\mathbf{s}') \quad (11.20)$$

for $\mathbf{s}' \in \mathcal{S}_i$.

The summation of (11.20) over $\mathbf{h}'$ for $[\mathbf{h}', \mathbf{b}'] \in \mathcal{S}_i$ yields

$$\sum_{\mathbf{s}', \mathbf{s}' \in \mathcal{S}_i} \mathbb{P}\left(\mathbf{s}'|\mathbf{s}, b_k^-, a_k^-\right) v_k^{(n-1)}(\mathbf{s}') + \sum_{\mathbf{s}', \mathbf{s}' \in \mathcal{S}_i} \mathbb{P}\left(\mathbf{s}'|\mathbf{s}, b_k^+, a_k^+\right) v_k^{(n-1)}(\mathbf{s}')$$

$$\leq \sum_{\mathbf{s}', \mathbf{s}' \in \mathcal{S}_i} \mathbb{P}\left(\mathbf{s}'|\mathbf{s}, b_k^-, a_k^+\right) v_k^{(n-1)}(\mathbf{s}') + \sum_{\mathbf{s}', \mathbf{s}' \in \mathcal{S}_i} \mathbb{P}\left(\mathbf{s}'|\mathbf{s}, b_k^+, a_k^-\right) v_k^{(n-1)}(\mathbf{s}') \quad (11.21)$$

This is the definition of submodularity of $\sum_{s',s' \in S_i} \mathbb{P}(s'|s, a_k)v_k^{(n-1)}(s')$ in (b_k, a_k). Furthermore, the positive weighted sum of submodular functions is also submodular, which establishes the submodularity of the right-hand side of function (11.16). Thus, the optimal transmit action policy $\pi_k^*(s, s \in S_k)^{(n)}$ is monotone nondecreasing on the buffer state b_k with given Lagrangian multiplier λ_k^m.

According to [24], the constrained optimal transmission scheduling policy is a randomized mixture of two pure policies, which can be computed with two different Lagrangian multipliers. As discussed above, it has been shown that each of these two pure policies is nondecreasing on the buffer occupancy state. Thus, the mixed policy is also nondecreasing on the buffer state, which concludes the proof. ☐

11.3.4 Proof of Lemma 11.1

We use backward induction to prove that $v_k(s)$ is nondecreasing on b_k with state $s \in S_i$ for $i = 1, 2, \ldots, K$.

First, we assume that $v_k^{(n-1)}(s)$ is nondecreasing on b_k for $s \in S_i, i = 1, 2, \ldots, K$ at the time instant $(n-1)$. The value matrix at time instant n is updated according to (11.17). If we denote the optimal action policy to be π_k^*, the update of value matrix can be written more explicitly as follows:

$$
\begin{aligned}
v_k^{(n)}(s, s \in S_k) &= c\left(s, \pi_k^*(s)\right) + \lambda_k^m \cdot d\left(s, \pi_k^*(s)\right) + \beta \sum_{s'=1}^{|S|} \mathbb{P}\left(s'|s, \pi_k^*(s)\right) v_k^{(n-1)}(s') \\
&= c\left(s, \pi_k^*(s)\right) + \lambda_k^m \cdot d\left(s, \pi_k^*(s)\right) + \beta \sum_{i=1}^{K} \sum_{s',s' \in S_i} \mathbb{P}\left(s'|s, \pi_k^*(s)\right) v_k^{(n-1)}(s')
\end{aligned}
$$

(11.22)

Because of assumption A4, we know that $c\left(s, \pi_k^*(s)\right) + \lambda_k^m \cdot d\left(s, \pi_k^*(s)\right)$ is nondecreasing on b_k for any h, f and a^*. By applying the result from Lemma 4.7.2 in [23], we can prove that $\sum_{s',s' \in S_i} \mathbb{P}\left(s'|s, \pi_k^*(s)\right) v_k^{(n-1)}(s')$ is nondecreasing on b_k for any h, f and $\pi_k^*(s)$. Thus, $v_k^{(n)}(s)$ is nondecreasing on b_k for $s \in S_k, k = 1, 2, \ldots, K$ at time instant n.

For any $s \in S_i, i \neq k$, the value vector is updated according to (11.23). In this case, the instantaneous cost is independent of b_k; thus, the nondecreasing property of $v_k^{(n)}(s)$ on b_k when $s \in S_i$, $i \neq k$ is preserved from $v_k^{(n-1)}(s)$.

$$
v_k^{(n)}(s, s \in S_i, i \neq k) = c_k\left(s, \pi_i^*(s)\right) + \lambda_k^m \cdot d_k\left(s, \pi_i^*(s) + \beta \sum_{s'=1}^{|S|} \mathbb{P}\left(s'|s, \pi_i^*(s)\right) v_k^{(n-1)}(s')
$$

(11.23)

When the time horizon of the Markovian problem is very large, that is, $n \to \infty$, the initial value matrix $v_k^{(0)}(s)$ no longer affects $v_k^{(n)}(s)$. The chosen value of $v_k^{(0)}(s)$ is arbitrary. In our case, we can initialize $v_k^{(0)}(s)$ to satisfy the condition stated in Lemma 11.1, which concludes the proof. ☐

11.3.5 *Justification of Assumptions in Theorem 11.2*

This subsection justifies the assumptions listed in Section 11.3.2. In a TDMA cognitive radio system, the cost of a switching control dynamic game (11.3) is evaluated by the BER under unit transmission power and the constraint is defined to be the transmission delay (11.12). The channel model is assumed to consist of complex Gaussian random variables with a zero mean.

Assumption A1: First, we define the holding cost caused by user k when the state is $\mathbf{s} \in \mathcal{S}_k$:

$$d(\mathbf{s}, a_k) = \frac{b_k}{\bar{f}} \tag{11.24}$$

Here, $\bar{f}$ is the average number of incoming packets and is a parameter given by the system. Assumption A1 holds if there exists an action such that $a_k > \bar{f}$. In our problem setup, it is assumed that the system latency constraint is valid.

Assumption A2: In assumption A2, the system transmission cost (11.3) is chosen to be the transmission BER. Assume the channel states are quantized by quantization threshold parameters $\Gamma(h)_1, \Gamma(h)_2, \ldots$ which are set by the system. The system transmission cost $BER(\gamma, a_k)$ is a function of the random channel gain $\gamma \in [\Gamma(h)_i, \Gamma(h)_i)$. Therefore, the transmission cost is [25]

$$BER_k(\gamma, a_k) = 0.2 \times \exp\left[\frac{-1.6\gamma}{(2^{a_k} - 1)}\right] \tag{11.25}$$

$$BER_k^i(h_k, a_k) = \frac{\int_{\Gamma(h)_{i-1}}^{\Gamma(h)_i} BER(\gamma, a_k) g(\gamma) d\gamma}{\int_{\Gamma(h)_{i-1}}^{\Gamma(h)_i} g(\gamma) d\gamma} \tag{11.26}$$

where $g(\gamma)$ denotes the probability distribution of the signal-to-noise ratio (SNR) γ. For channel state h_k, which belongs to quantization region $[\Gamma(h)_{i-1}, \Gamma(h)_i)$, the expectation of $BER_k^i(h_k, a_k)$ is taken over γ for $\gamma \in [\Gamma(h)_{i-1}, \Gamma(h)_i)$. When the system uses uncoded M-ary QAM, $M = 2^{a_k}$.

From (11.25) and (11.26), it can be seen that the averaged transmission cost is independent of the buffer occupancy b_k. Furthermore, it is clear from (11.24) that the holding cost function $d(\mathbf{s}, a_k)$ is nondecreasing on b_k and independent of a_k. Thus, assumption A2 holds.

Assumption A3: As stated in (11.8), the buffer occupancy state evolves according to Lindley's recursion equation. Given the current state buffer occupancy and action, the transition probability depends on the probability of incoming traffic, which is shown in (11.9). Thus, the buffer state transition probability can be rewritten as $\mathbb{P}\left(b_k'|b_k, a_k\right) = \mathbb{P}\left(b_k'|(b_k - a_k)\right)$. Assume the incoming traffic is evenly distributed in $0, 1, \ldots, L$, which can be mathematically written as $\mathbb{P}(f_k < 0 \text{ or } f_k > L) = 0$ and $\mathbb{P}(0 \leq f_k \leq L) = 1/(L+1)$. Thus, the buffer state transition probability is

$$\mathbb{P}\left(b_k'|b_k, a_k\right) = \begin{cases} \frac{1}{L+1} & b_k' < L \\ \frac{1 + [b_k - a_k]^+}{L+1} & b_k' = L \end{cases} \tag{11.27}$$

Therefore, the buffer occupancy state transition probability is independent of b_k and a_k when $b_k' < L$ and is first order stochastically nondecreasing on $b_k - a_k$ when $b_k' = L$. This result verifies $\sum_{b'_k=l}^{L} \mathbb{P}\left(b'_k | b_k, a_k\right)$ is nondecreasing on b_k in A3. According to (11.27), we can see that $\mathbb{P}\left(b'_k = L | b_k, a_k\right)$ is submodular in b_k, a_k; thus, A3 holds.

11.3.6 Stochastic Approximation Algorithm

The structural result stated in Theorem 11.2 can be exploited to reduce the complexity of computing the Nash equilibrium policy. We now present a *stochastic approximation algorithm* that exploits this structure.

Consider the scenario where the action set of each user contains two actions, that is, $a_k = \{0, 1\}$. In a real system, the two actions can be chosen as *No Transmission* and *Transmission* with $a_k = 0$ denoting the action *No Transmission*, while $a_k = 1$ denoting the action *Transmission*. The two actions can also be chosen as two different transmission rates, namely, *High Transmission Rate* and *Low Transmission Rate*. Using the structural result on the Nash equilibrium policy from Theorem 11.2, the Nash equilibrium policy $\pi_k^*(\mathbf{s})$ is a randomized mixture with the following feature:

$$\pi_k^*(\mathbf{s}) = \begin{cases} 0 & 0 \le b_k < b_l(\mathbf{s}) \\ p & b_l(\mathbf{s}) \le b_k < b_h(\mathbf{s}) \\ 1 & b_h(\mathbf{s}) \le b_k \end{cases} \tag{11.28}$$

Here, $p \in [0, 1]$ is the randomization factor; $b_l(\mathbf{s})$ and $b_h(\mathbf{s})$ are the lower and higher buffer state thresholds, respectively. The search for the Nash equilibrium policy over the intractably large buffer space reduces to estimating just three parameters, namely, the parameters given here. This approach can be easily extended to apply to systems with bigger action sets, for example, when the action set for each secondary user is $a_k = \{1, 2, 3\}$, the Nash equilibrium policy can be parameterized by six parameters (four buffer state thresholds and two randomization factors). In a real system, it is most common that each secondary user has two modes, on and off. Thus, the stochastic approximation algorithm and the corresponding simulation result will focus on the scenario that each secondary user has two action choices. In the stochastic approximation algorithm, we first compute the continuous optimal values of $b_l(\mathbf{s}) \in [0, L]$ and $b_h(\mathbf{s}) \in [b_l(\mathbf{s}), L]$ $(b_h(\mathbf{s}) \ge b_l(\mathbf{s}))$ and then round off to the nearest discrete values. This is a relaxation of the original discrete stochastic optimization problem as the buffer states in the problem setup are discrete. For convenience, let θ_k denote the vector of all the parameters that will be estimated for user k, with $k = 1, 2, \ldots, K$. The composition of the parameter vectors of all the K users can be denoted as $\Theta = \{\theta_1, \theta_2, \ldots, \theta_K\}$.

We use the simultaneous perturbation stochastic approximation (SPSA) algorithm [21] to estimate the parameters. The essential feature of SPSA is the underlying gradient approximation, which requires only two objective function measurements per iteration, regardless of the dimension of the optimization problem. These two measurements are made by simultaneously varying, in a properly random fashion, all the variables in the problem. The SPSA algorithm is shown in Algorithm 11.2.

We denote $\Lambda^{(n)} = \{\lambda_1^{(n)}, \lambda_2^{(n)}, \ldots, \lambda_K^{(n)}\}$. β and α denote the constant perturbation step size and constant gradient step size, respectively. In the main part of the algorithm, the SPSA algorithm is applied to iteratively update the system parameters. When the kth user is scheduled for transmission at time slot n, parameters $\theta_k^{(n)}$ and the Lagrangian multiplier $\lambda_k^{(n)}$ can be updated after introducing a random perturbation vector $\Delta^{(n)}$. Meanwhile, the parameters of the other users remain unchanged.

Algorithm 11.2   Stochastic Approximation Algorithm

1: **Initialization:** $\Theta^{(0)}$, Λ^0; $n = 0$; $\rho = 4$;

2: Initialize constant perturbation step size β and gradient step size α;

3: **Main Iteration**

4: **if** $\mathbf{s}^{(n)} \in \mathcal{S}_k$ **then**

5:    $m_k = \left| \theta_k^{(n)} \right|$;

6:    Generate $\Delta^{(n)} = \left[\Delta_1^{(n)}, \Delta_2^{(n)}, \ldots, \Delta_{m_k}^{(n)} \right]^T$; $\Delta_i^{(n)}$ are Bernoulli random variables with $p = \frac{1}{2}$.

7:    $\theta_{k+}^{(n)} = \theta_k^{(n)} + \beta \times \Delta^{(n)}$;

8:    $\theta_{k-}^{(n)} = \theta_k^{(n)} - \beta \times \Delta^{(n)}$;

9:    $\Delta C_k^{(n)} = \dfrac{c\left(\mathbf{s}^{(n)}, \theta_{k+}^{(n)}\right) - c\left(\mathbf{s}^{(n)}, \theta_{k-}^{(n)}\right)}{2\beta} \left[\left(\Delta_1^{(n)}\right)^{-1}, \left(\Delta_2^{(n)}\right)^{-1}, \ldots, \left(\Delta_{m_k}^{(n)}\right)^{-1} \right]^T$;

10:    $\Delta D_k^{(n)} = \dfrac{d\left(\mathbf{s}^{(n)}, \theta_{k+}^{(n)}\right) - d\left(\mathbf{s}^{(n)}, \theta_{k-}^{(n)}\right)}{2\beta} \left[\left(\Delta_1^{(n)}\right)^{-1}, \left(\Delta_2^{(n)}\right)^{-1}, \ldots, \left(\Delta_{m_k}^{(n)}\right)^{-1} \right]^T$;

11:    $\theta_k^{(n+1)} = \theta_k^{(n)} - \alpha \times \left(\Delta C_k^{(n)} + \Delta D_k^{(n)} \cdot \max\left[0, \lambda_k^{(n)} + \rho \cdot \left(D\left(\mathbf{s}^{(n)}, \theta_k^{(n)}\right) - \widetilde{D}_k\right)\right] \right)$;

11:    $\lambda_k^{(n+1)} = \max\left[\left(1 - \frac{\alpha}{\rho} \cdot \lambda_k^{(n)}\right), \lambda_k^{(n)} + \alpha \cdot \left(D\left(\mathbf{s}^{(n)}, \theta_k^{(n)}\right) - \widetilde{D}_k\right) \right]$;

12: **end if**

13: The parameters of other users remain unchanged;

14: $n = n + 1$;

15: The iteration terminates when the values of the parameters $\Theta^{(n)}$ converge; else return to Step 3.

THEOREM 11.3 $\{\Theta^n(\alpha), \Lambda^n(\alpha)\}$ are system parameters generated by the stochastic approximation algorithm. Define the piecewise constant interpolated continuous-time processes of $\{\Theta^n(\alpha), \Lambda^n(\alpha)\}$ as $\{\Theta^t(\alpha), \Lambda^t(\alpha)\}$. For $t \in [n\alpha, (n+1)\alpha)$, the value of $\Theta^t(\alpha)$ is set to be $\Theta^n(\alpha)$ and $\Lambda^t(\alpha)$ is set to be $\Lambda^n(\alpha)$. The mathematical expressions are given as follows:

$$\Theta^t(\alpha) = \Theta^n(\alpha) \qquad t \in [n\alpha, (n+1)\alpha)$$

$$\Lambda^t(\alpha) = \Lambda^n(\alpha) \qquad t \in [n\alpha, (n+1)\alpha)$$

For sufficiently large ρ, as $\alpha \to 0$ and $t \to \infty$, $\{\Theta^t(\alpha), \Lambda^t(\alpha)\}$ converge in probability to the Kuhn Tucker (KT) pair of (11.13) [27], which is specified in (11.29).   □

The optimal policies of (11.13) satisfying the KT condition can be defined as follows: θ_i^* belongs to the KT set when

$$KT = \left\{ \theta_i^* : \exists \lambda_i > 0, \text{ such that } \nabla_{\theta_i} C_i + \nabla_{\theta_i} \lambda_i (D_i - \widetilde{D}_i) = 0, i = 1, \ldots, K \right\} \qquad (11.29)$$

where C_i and D_i are the optimization objective (11.11) and discounted time delay (11.12), respectively. Moreover, θ_i^* satisfies the second-order sufficiency conditions: $\nabla_{\theta_i}^2 C_i + \nabla_{\theta_i}^2 (D_i - \widetilde{D}_i) \geq 0$ is positive definite for all the i, and $(D_i - \widetilde{D}_i) = 0$, $\lambda_i > 0$, $i = 1, \ldots, K$.

Note that in the stochastic approximation algorithm, we first compute the continuous values of $b_l(\mathbf{s})$ and $b_h(\mathbf{s})$ and then round them off to the nearest discrete values. This relaxation leads to the continuous value of θ_i during calculation and, thus, (11.29) is differentiable on θ_i.

11.4 Numerical Examples

This section presents numerical examples of the structured Nash equilibrium transmission policy. For convenience, all the secondary users in the simulation setup have the same cost functions. The simulation results shown below are of the transmission policies of the first user. In the system model, each user has a size 10 buffer, and the channel quality measurements are quantized into two different states, namely, {1, 2}. In the system configuration, the transmission costs, the holding costs, and buffer transition probability matrices are chosen to ensure A2–A3, specified in Section 11.3.2. The channel transition probability matrices are generated randomly.

In the system models used in Figures 11.3 and 11.4, each user has three different action choices when it is scheduled for transmission. The system uses an uncoded M-QAM modulation scheme, and different actions are followed by different modulation modes, leading to different transmission rates. The three different actions are $a_k = 1, 2, 3$, and consequently, the modulation schemes are 2-QAM, 4-QAM, and 8-QAM. When a user has a Nash equilibrium policy equal to 0 for a certain state, that means the user is not scheduled for transmission in that state. The transmission policies shown in Figures 11.3 and 11.4 are obtained by first fixing the Lagrangian multiplier to be $\lambda_1 = \lambda_2 = 1.3$ and then applying the value iteration algorithm, which is specified in Algorithm 11.1.

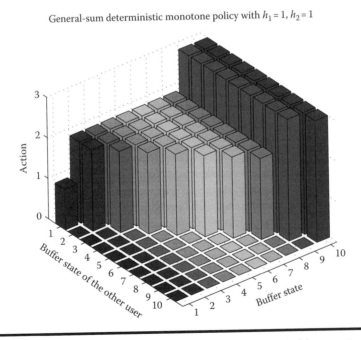

General-sum deterministic monotone policy with $h_1 = 1, h_2 = 1$

Figure 11.3 The transmission policy of a certain user in a switching control general-sum dynamic game system obtained by the value iteration algorithm. The result is obtained when the channel states of user 1 and 2 are $h_1 = 1$ and $h_2 = 1$, respectively.

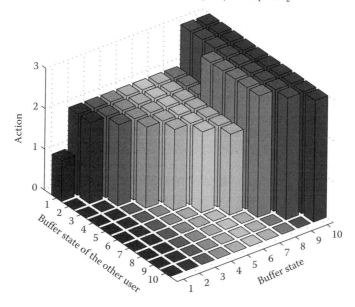

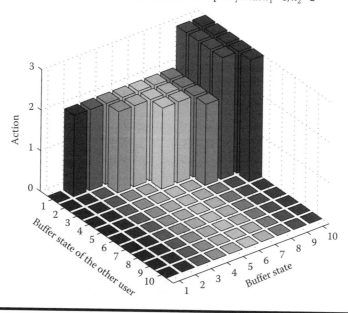

Figure 11.4 **The transmission policy of a certain user in a switching control zero-sum dynamic game system obtained by value iteration algorithm. The first subfigure shows the result when the channel states of both users are $h_1 = 1$ and $h_2 = 1$, while the second subfigure shows the result when the channel states of user 1 and 2 are $h_1 = 1$ and $h_2 = 2$, respectively.**

11.4.1 General-Sum Constrained Game: Structure of the Nash

Figure 11.3 shows the structured optimal transmission policy of a two-user general-sum system model with $h_1 = 1$ and $h_2 = 1$. The transmission costs are specified according to (11.3), and the holding costs are chosen to satisfy (11.5). The figure is the Nash equilibrium policy of a switching control game with general-sum transmission costs and zero-sum holding costs, and the transmission policy is monotone nondecreasing on the buffer occupancy state.

11.4.2 Zero-Sum Constrained Game: Structure of the Nash

Figure 11.4 considers a two-user system and shows the structural result on the optimal transmission policy. The first subfigure is of the zero-sum model with $h_1 = 1$ and $h_2 = 1$, and the second subfigure is of the zero-sum model with $h_1 = 1$ and $h_2 = 2$. It can be seen from the figure that the policy is deterministic and that under a certain channel state value, the optimal action policy is monotone nondecreasing on the buffer occupancy state. From the comparison of subfigures 1 and 2, we can see that when the channel state of the other user in the system becomes better, the current user has less chance to transmit and its transmission policy becomes more aggressive.

11.4.3 Stochastic Approximation Algorithm for Learning Nash Equilibrium Policy

Figures 11.5 and 11.6 consider a two-user system with each user having two different action choices, namely, $\{1, 2\}$. As it is a constrained switching control Markovian game, the Nash equilibrium policy is a randomized mixture of two pure policies. Each optimal transmit policy can be determined by three parameters, namely, lower threshold $b_l(\mathbf{s})$, upper threshold $b_h(\mathbf{s})$, and

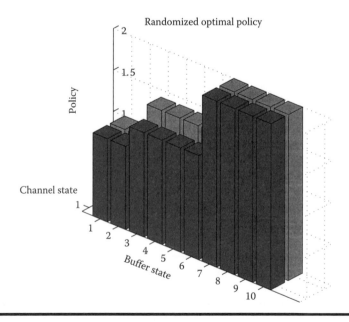

Figure 11.5 The Nash equilibrium transmission control policy obtained via stochastic approximation algorithm. A two-user system is considered, and each user has a size 10 buffer.

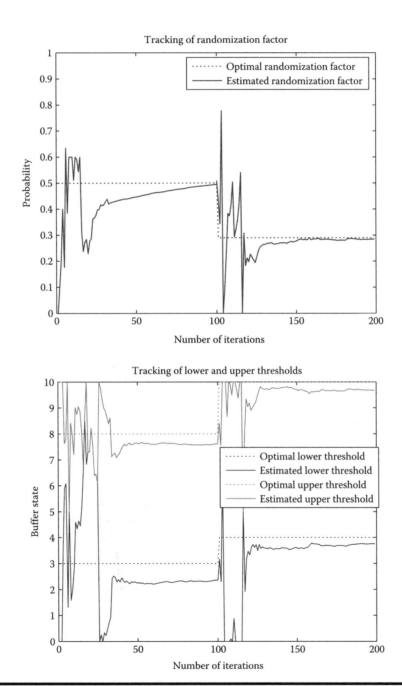

Figure 11.6 The tracking of Nash equilibrium policy according to the update of system parameters by stochastic approximation algorithm. System statics change at the 100th iteration, as specified in Section 11.3. These two figures compare the estimated randomization factor and buffer thresholds with the actual optimal values.

randomization factor p, as described in (11.28). The stochastic approximation algorithm is applied to find the Nash equilibrium policy. The simulation results of user 1 with $h_2 = 1$ and $b_2 = 1$ are shown in Figure 11.5. The figure shows that the optimal transmission policy is no longer deterministic but is a randomized mixture of two pure policies and that each Nash equilibrium policy is monotone nondecreasing on the buffer state.

The stochastic approximation algorithm proposed in Section 11.3.6 is able to adapt the Nash equilibrium policy according to the static of the system parameters. Figure 11.6 shows the time evolution of lower threshold $b_l(\mathbf{s})$, upper threshold $b_h(\mathbf{s})$, and randomization factor p for $h_1 = 1$, $h_2 = 1$, and $b_2 = 1$. The system configuration used from iteration 101 to 200 is different from that used from iteration 1 to 100. We can see from the simulation result that when the system parameters evolve with time, the stochastic approximation algorithm can update the transmission policy accordingly. The estimated three parameters of the policy at iteration 100 are $b_l(\mathbf{s}) = 2.4$, $b_h(\mathbf{s}) = 7.6$, and $p = 0.5$. After detecting the change of the system parameters, the new estimated parameters are updated to be $b_l(\mathbf{s}) = 3.8$, $b_h(\mathbf{s}) = 9.7$, and $p = 0.29$. The optimal parameters are shown with dashed lines in Figure 11.6. The actual optimal upper and lower thresholds of the system are discrete values, and the last step is to round the estimated parameters to the nearest discrete values.

11.5 Conclusions and Open Issues

The rate adaptation problem of secondary users in cognitive radio systems is formulated as a constrained general-sum switching control Markovian dynamic game, where the block-fading channels are modeled as a finite state Markov chain. We assume the system has perfect information of the behaviors of the primary users and each secondary user accesses the channel according to a decentralized access rule. It is shown that the Nash equilibria of the game are always available. Under reasonable assumptions, a Nash equilibrium policy is a randomized mixture of two pure policies, and each of them is monotone nondecreasing on the buffer occupancy state. This allows us to propose a more computationally efficient stochastic approximation algorithm. Numerical examples are provided to verify these results.

However, this chapter only focuses on a special type of game, namely, switching control Markovian game. The study of a more general type of Markovian game (e.g., prove the existence of the Nash equilibrium under certain conditions and propose algorithms to obtain the Nash equilibrium policy) can be one direction of the future work. A recent work of the authors [27] tries to use Q-learning to compute the correlated equilibrium of a general type dynamic Markovian game. Besides, the convergence speed of the algorithms proposed herein suffers from the "dimension curse" as the state space increases exponentially with the number of users. How to efficiently reduce the state space is yet another issue to solve.

References

1. Federal Communications Commission, Spectrum Policy Task Force Report, Rep. ET Docket No. 02-155, November 2002.
2. J. Mitola III, Cognitive radio for flexible mobile multimedia communications, in *Proceedings of IEEE Mobile Multimedia Conference*, San Diego, CA, November 1999, pp. 33–10.
3. S. Haykin, Cognitive radio: Brain-empowered wireless communications, *IEEE Journal on Selected Areas in Communications*, 23, 2, 201–220, Feburary 2005.

4. V. K. Bhargava and E. Hossain, *Cognitive Wireless Communication Networks*. New York: Springer-Verlag, 2007.

5. H. P. Shiang and M. van der Schaar, Distributed resource management in multihop cognitive radio networks for delay sensitive transmission, *IEEE Transactions on Vehicular Technology*, 58, 2, 941–953, Feburary 2009.

6. Q. Shao, L. Tong, A. Swami, and Y. Chen, Decentralized cognitive MAC for opportunistic spectrum access in ad hoc networks: A POMDP framework, *IEEE Journal on Selected Areas in Communications*, 25, 3, 589–600, April 2007.

7. N. Nie and C. Comaniciu, Adaptive channel allocation spectrum etiquette for cognitive radio networks, in *Proceedings of IEEE DySPAN*, Baltimore, MD, November 2005, pp. 269–278.

8. M. Felegyhazi, M. Cagalj, S. S. Bidokhti, and J. Hubaux, Non-cooperative multi-radio channel allocation in wireless networks, in *Proceedings of IEEE INFOCOM*, Anchorage, AK, May 2007, pp. 1442–1450.

9. W. Yu, G. Ginis, and J. M. Cioffi, Distributed multiuser power control for digital subscriber lines, *IEEE Journal on Selected Areas in Communications*, 20, 5, pp. 1105–1115, June 2002.

10. G. Scutari, D. Palomar, and S. Barbarossa, Asynchronous iterative water-filling for Gaussian frequency-selective interference channels, *IEEE Transactions on Information Theory*, 54, 7, 2868–2878, July 2008.

11. S. R. Mohan and T. E. S. Raghavan, An algorithm for discounted switching control stochastic games, *OR Spektrum*, 9, 1, 41–45, March 1987.

12. J. Filar and K. Vrieze, *Competitive Markov Decision Processes*. New York: Springer-Verlag, 1997.

13. O. J. Vrieze, S. H. Tijs, T. E. S. Raghavan, and J. A. Filar, A finite algorithm for the switching control stochastic game, *OR Spectrum*, 5, 1, 15–24, March 1983.

14. IEEE, "IEEE std 802.16-2004." IEEE Standard for Local and Metropolitan Area Networks, Part 16: Air Interface for Fixed Broadband Wireless Access Systems, Park Avenue, NY: IEEE Computer Society, 2004.

15. E. Altman, *Constrained Markov Decision Processes*. London, U.K.: Chapman and Hall, 1999.

16. K. Kar and L. Tassiulas, Layered multicast rate control based on Lagrangian relaxation and dynamic programming, *IEEE Journal on Selected Areas in Communications*, 24, 8, 1464–1474, August 2006.

17. A. Farrokh and V. Krishnamurthy, Opportunistic scheduling for streaming users in HSDPA multimedia systems, *IEEE Transactions on Multimedia Systems*, 8, 4, 844–855, August 2006.

18. J. W. Huang and V. Krishnamurthy, Truth reveling opportunistic scheduling in cognitive radio systems, in *Proceedings of IEEE SPAWC*, Perugia, Italy, June 2009.

19. F. Fu, T. M. Stoenescu, and M. van der Schaar, A pricing mechanism for resource allocation in wireless multimedia applications, *IEEE Journal of Selected Topics in Signal Processing*, 1, 2, 264–279, August 2007.

20. D. Djonin and V. Krishnamurthy, MIMO transmission control in fading channels—A constrained markov decision process formulation with monotone randomized policies, *IEEE Transactions on Signal Processing*, 55, 10, 5069–5083, October 2007.

21. J. C. Spall, *Introduction to Stochastic Search and Optimization*, Ser. Wiley-Interscience Series in Discrete Mathematics and Optimization. Wiley-Interscience, Hoboken, NJ, 2003.

22. V. Krishnamurthy and G. G. Yin, Recursive algorithms for estimation of hidden Markov models and autoregressive models with Markov regime, *IEEE Transactions on Information Theory*, 48, 2, 458–476, February 2002.

23. M. L. Puterman, *Markov Decision Processes: Discrete Stochastic Dynamic Programming*. New York: John Wiley & Sons, 1994.

24. F. J. Beutler and K. W. Ross, Optimal policies for controlled Markov chains with a constraint, *Journal of Mathematical Analysis and Applications*, 112, 236–252, 1985.

25. S. T. Chung and A. J. Goldsmith, Degrees of freedom in adaptive modulation: A unified view, *IEEE Transaction on Communication*, 49, 9, 1561–1571, September 2001.

26. H. J. Kushner and D. S. Clark, *Stochastic Approximation Methods for Constrained and Unconstrained Systems*, New York: Springer-Verlag, 1978.

27. J. W. Huang, Q. Zhu, V. Krishnamurthy, and T. Basar, Distributed correlated q-learning in dynamic transmission control of sensor networks, in *Proceedings of ICASSP*, 2010, pp. 1982–1985.

ECONOMIC APPROACHES

Chapter 12

Auction-Based Resource Management and Fairness Issues in Wireless Networks

Manos Dramitinos, Rémi Vannier, and Isabelle Guérin Lassous

Contents

Dynamic resource management in cellular 2.5+G networks is of prominent importance due to the scarcity of the radio spectrum and the increasing user demand for voice and data services. The fact that the timescales over which cellular networks are designed to allocate resources is much shorter than that of user services poses a significant challenge for mechanism design in this context. In multihop ad hoc networks (MANETs), a plethora of fair solutions have been proposed to improve the unfair way that the various flows are allocated bandwidth due to the inherent features of IEEE 802.11. So far, there has been limited research on the impact of the varying short-term allocations of these protocols, due to their inherent features and also nodes mobility, on the user-perceived quality of service (QoS) (and social welfare) for services of long duration. Thus, it remains questionable whether and to what extent these solutions are indeed beneficial. This chapter presents how auction and utility theory can be used as a means of (a) designing efficient auction-based resource management schemes for cellular networks and (b) quantifying the user-perceived QoS and assessing the performance of fair solutions in MANETs.

12.1 Introduction

This chapter presents how utility and game theory can be applied in the context of wireless cellular and ad hoc networks. In particular, we illustrate how history-dependent utility functions can be used as the basis of

- Auction schemes for the allocation of the resources of cellular 2.5+G networks to the competing users who bid against each other in a noncooperative game
- Quantifying the user-perceived quality of service (QoS) for the competing flows of mobile ad hoc networks (MANETs), thus providing a means of comparative assessment of the fairness schemes proposed in the literature for IEEE 802.11 networks

Multiunit auctions have recently received considerable attention as an economic mechanism for resource reservation and price discovery in networks. The case where users compete for reserving resources for *large timescales*, for example, minutes, is of particular interest for many practical cases. A prominent case is that of UMTS: the duration of services that users request, such as news downloading or video streaming, is significantly longer than the 10 ms "slot" of the UTRAN frame over which the resource manager allocates bandwidth on demand [1]. Furthermore, these services have QoS requirements, standardized by 3GPP report TR 23.107, which complicate the resource allocation. The same also applies to GPRS technology, including its enhanced version EDGE. The same issues are evident in the ongoing effort to evolve toward the Fourth Generation (4G). 4G prescribes that users can enjoy mobility, seamless access, and high QoS in an all-IP network on an "Anytime, Anywhere" basis. The 4G network access comprises multiple heterogeneous geographically coexisting wireless networks. The efficient allocation of the bandwidth of a 4G network whose access architecture is hierarchical in terms of radius and thus geographical coverage comprises a significant challenge for both game theory and traffic engineering.

Similar challenges for the handling of users' traffic and the resulting user-perceived QoS appear also for the users of the IEEE 802.11 MANETs. MANETs are self-configuring wireless networks of mobile nodes, the union of which forms arbitrary topology. The nodes that also serve as routers are free to move arbitrarily; thus, the network's topology may change rapidly and unpredictably. In addition to this, the small timescale of the network resource allocations, the degradation of the channel quality due to interference, and the absence of a central control/coordination point that regulates when each node transmits—as opposed to the UMTS case—further complicate this problem. For ad hoc networks, several fairness schemes and protocols have been proposed in order to improve the unfair way that the various flows are allocated bandwidth in the standard 802.11 protocols. In the remainder of this chapter, we elaborate on the shortcomings of the current performance evaluation metrics of fairness schemes* that cannot depict the impact of the sequence of the *varying over time* rate allocations on the user flows. This motivates the use of *utility functions*, which can serve as a common ground of comparison of the performance of various fairness schemes whose maximization goals are inherently different (e.g., max–min fairness, proportional fairness, etc.).

Furthermore, it is worth noting that auction schemes perform well in cases where demand is much larger than supply, as opposed to fairness schemes that are inapplicable in such cases but very attractive in cases of low demand. Therefore, auction and fairness schemes can be seen as complementary network management solutions under different network conditions. The challenging issue of investigating whether utility functions can be used in both these two worlds also motivates the material presented in this chapter. Finally, note that for brevity reasons we consider fairness or auction-based schemes proposed in the context of cognitive radios to be out of the scope of this chapter.

Concluding, this book chapter shows how a utility-based framework can be applied to GPRS, EDGE, UMTS, and WiFi networks as a means of (a) auction-based resource management in 2.5+G networks and (b) quantifying the user-perceived QoS of WiFi users. Overall, this chapter provides useful insight for both mechanism design issues and assessment of the user-perceived QoS in a wide range of wireless networks, both cellular and MANETs.

12.2 Auction-Based Resource Allocation in 2.5+G Networks

12.2.1 Auctions

We now present fundamental definitions, theorems, and results from auction theory; see [2,3]. An auction is a mechanism based on an *allocation rule* that defines which good is allocated to whom and a *payment rule* that defines the respective charges. A participant of an auction is called *bidder* while the entity conducting the auction *auctioneer*.

There are many ways to classify auctions. Auctions are classified as *simple* or *single-unit* if only one good is auctioned and *multiunit* if multiple units of a good (e.g., integral units of a link's bandwidth) are to be traded. Moreover, depending on whether bids are made in public or not, the auction is referred to as *open* or *sealed*, respectively. Auctions that maximize seller's revenue are referred to as *optimal* and those maximizing social welfare (the sum of agents' utilities) are referred to as *efficient*. If the auction is conducted in rounds, then it is called *progressive*. Depending on the payment rule, the auction can be either *uniform* if the same amount of money is paid by all for each unit of the good awarded and *discriminatory* (*pay-your-bid*) if each user pays his bid for each object he wins.

---

* The terms fair solution and fairness scheme or protocol are used interchangeably throughout this chapter.

12.2.1.1 Simple Auctions

A *bid* in the context of simple auctions is the amount of money offered by a bidder for the item auctioned. The best-known mechanism is by far the *English* auction, where the seller starts with a minimum price that is actually incremented until there is only one person claiming the item, whom the item is awarded to. The *Dutch* auction is exactly the opposite mechanism. The price is initially high and is gradually decremented until a bidder claims the object. The item is awarded to him for a charge equal to the current price. The sealed bid auctions (first *price* and second *price* or Vickrey) consist of two phases: (a) the first one where bidders submit sealed envelopes with their bids and (b) the second one where these envelopes are opened. The item is then awarded to the bidder who submitted the highest bid. The winner pays his bid at the first price auction and the highest losing bid,—that is, the second highest bid—at the Vickrey auction. It has been proved that under the Vickrey auction, it is best for each bidder to honestly bid his true value for the item being awarded. This property is referred to as *incentive compatibility*.

12.2.1.2 Multiunit Auctions

A *bid* in the context of multiunit auctions is defined to be the pair (p, q) of the per unit expressed willingness to pay p for a quantity q of units. All simple auctions can be generalized to multiunit auctions. Incentive compatibility holds only for the generalizations of the Vickrey auction. The rules of the Generalized Vickrey Auction (GVA) [5] prescribe that (a) each user reports his valuation for a subset or for all points of his demand function for the good auctioned, (b) units are allocated to the highest bids until demand exhausts supply, and (c) the VCG payment rule is applied, that is, each user is charged with the social opportunity cost that his presence entails. Formally, user's i charge equals $\mu_i(\theta) = SW_{-i}(0, \theta_{-i}) - SW_{-i}(\theta)$, where SW_{-i} is the social welfare of bidders other than i, θ is the set of the users' reported valuations, and $(0, \theta_{-i})$ is the efficient outcome if i's reported value is 0 while the reports by other users remain unchanged. Note that winners' charges are less than their respective bids.

12.2.2 Auction-Based Resource Management

The problem of auctioning network resources such as bandwidth in cellular networks initially seems to be identical to that of multiunit auctions. In particular, there is a quantity of resources, the capacity C of the access network, that is auctioned among the bidders—the users—who exhibit multiunit demand in order to accommodate the needs of their services. So why not apply straightforwardly any standard multiunit auction of the auction theory literature? The answer is that for a realistic mechanism, the *time dimension* must be taken into account in the model and the auction design. In particular, most of the complexity of the problem of resource management in UMTS lies in the fact that users demand sessions spanning partly overlapping intervals with different durations, which in general are much larger than the timescale t_a of a slot where network resources are allocated. This aspect of the problem has been neglected in the literature with the exception of the ATHENA auction [4].

The ATHENA approach [4] is to conduct a sequence of "mini-auctions," each concerning reservation of resources within one slot. Each mini-auction is a sealed-bid GVA [5], with atomic bids (i.e., bids that are either fully satisfied or rejected) of the type (p, q), where q is the quantity of resource units sought in the present slot and p is the price proposed for each such unit. Henceforth, unless otherwise specified, this mechanism restricts attention to services requiring constant bit rates. For UMTS, since resources are allocated in bits, if the service involves traffic of a specific bit rate m, then $q = m \cdot t_a$. (As explained later for each such service session, one bid is placed per mini-auction.) In a GVA, users with elastic utility functions have the incentive to bid truthfully their willingness to pay [5]. This is a very attractive property, motivating the selection of running a GVA in each mini-auction.

ATHENA consists of a series of GVAs, one per slot. However, in a realistic case of a UMTS network, these mini-auctions will need to be run so frequently that it would not be feasible for users to participate in all these mini-auctions. Since the user cannot bid on a per mini-auction basis, ATHENA defines utility functions, pertaining to the various services. These functions are provided by the network operator as *bidding functions* for the user to choose from; they are scaled by the user's total willingness to pay, which is to be given by the user himself (as part of his service request; see Figure 12.1). This approach is in line with 3GPP [6] regarding predefined QoS profiles. Then, the network runs all mini-auctions by bidding on behalf of each user, according to his respective selection of bidding function.

Furthermore, since users do not bid themselves, the details of the physical layer and the auction are hidden from them: A user demanding a service selects among the predefined bidding functions the one that better expresses his preferences and declares a total willingness to pay U_s. A session of time t_s is then created. Each user aims at achieving constantly the bit rate m by participating in a large number K_s of mini-auctions, where $K_s = t_s / t_a$. (Recall, however, that the network is bidding on each user's behalf.) For example, if the user wishes to watch his favorite music video clip lasting for 4 min, he simply declares the title of the video, a total willingness to pay U_s, the desired quality level, and the utility function type. Requests are sent from the user's terminal to the UMTS base station over the Random Access CHannel (RACH). The parameters t_s, K_s, and m are computed automatically by the network and are transparent to the user.

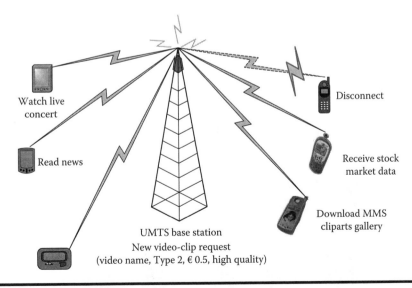

Watch live concert

Read news

Disconnect

Receive stock market data

Download MMS cliparts gallery

UMTS base station
New video-clip request
(video name, Type 2, € 0.5, high quality)

Figure 12.1 A user of a UMTS cell submits a video service request consisting of the video name, his selection of a utility function, and his total willingness to pay for that service of a certain quality (bit rate).

12.2.3 User Utility Functions

In this subsection, the ATHENA utility functions and their properties. Assume that the user's i value $u_{s,i}$ for obtaining the service s is the sum of the marginal "sub-utilities" $v_{s,i}^{(t)}$ attained due to each successful allocation (i.e., the allocated bit rate x_i equals the desired m_i); thus, $u_{s,i}(x_i^{(1)}, \ldots, x_i^{(K_{s,i})}) = \sum_{t=1}^{K_{s,i}} v_{s,i}^{(t)}(x_i^{(1)}, \ldots, x_i^{(t)})$. (Indices i and s denote the user and service, respectively; $K_{s,i}$ denotes the number of mini-auctions for user i and service session s; superscript t ranges from 1 to $K_{s,i}$, indicating the "current" mini-auction.) Next, utility functions, pertaining to the various services, are defined. These functions reflect the fact that, when satisfactory service cannot be provided, the position in time of the slots where user received a good rate makes considerable difference to the user-perceived QoS. Thus, by selecting one of the predefined ATHENA utility functions, each user declares his preferred form of allocation pattern for the cases where perfectly consistent resource reservation is not possible. The following utility functions are defined:

Type 1: *Indifferent to the Allocation Pattern.* This type pertains to throughput-sensitive users. Hence, the user's total value $U_{s,i}$ is equally apportioned among the various slots. Thus,

$$v_{s,i}^{(t)}\left(x_i^{(1)}, \ldots, x_i^{(t)}\right) = \mathbf{1}\left(x_i^{(t)} = m_i\right) \cdot \frac{U_{s,i}}{K_{s,i}},$$

where
  $U_{s,i}$ is the user's declared total willingness to pay
  $\mathbf{1}(\cdot)$ is the indicator function

An example of users belonging to this type is those accessing news or downloading files with FTP. This utility function is suitable for the UMTS Background Class services with m_i pertaining to the maximum bit-rate parameter of this class [6].

Type 2: *Sensitive to the Service Continuity.* This type pertains (among other cases) to users that prefer watching consistently half of a football match rather than watching multiple shorter periods. Thus, they prefer the allocation pattern of Figure 12.2a to that of Figure 12.2b. A sub-utility expressing this preference is

$$v_{s,i}^{(t)}\left(x_i^{(1)}, \ldots, x_i^{(t)}\right) = \mathbf{1}\left(x_i^{(t)} = m_i\right) \frac{U_{s,i}}{K_{s,i}} \cdot \alpha^{d_i(t)},$$

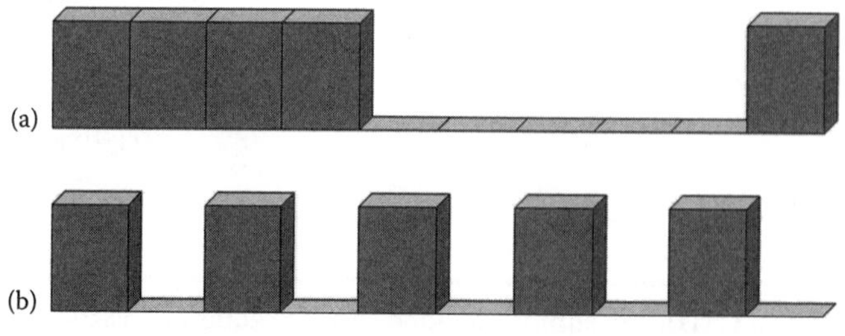

(a)

(b)

Figure 12.2   Inconsistent resource allocation patterns.

where

$\alpha \in (0, 1)$ is a discount factor

$d_i^{(t)}$ is the distance between the current and the previous slots during which user i achieved reservations

History of previous allocations influences $v_{s,i}^{(t)}$ through the value of d_i, which is kept track of. This utility function is suitable for the UMTS Streaming Class services [6].

Type 3: *Sensitive to the Regularity of the Resource Allocation Pattern.* Such users prefer the allocation pattern of Figure 12.2b to that of Figure 12.2a. The respective sub-utility is

$$v_{s,i}^{(t)}\left(x_i^{(1)},\ldots,x_i^{(t)}\right) = \mathbf{1}\left(x_i^{(t)} = m_i\right)\frac{U_{s,i}}{K_{s,i}}\cdot\alpha^{\max\{0,\Delta d_i\}},$$

where Δd_i is the difference of $d_i^{(t)}$ (as defined above) and the length of the previous gap. Note that the factor $\alpha^{\max\{0,\Delta d_i\}}$ equals 1 (resp. is less than 1) if Δd_i is non-positive, that is, when the received QoS improves (resp. deteriorates).

For all three types, if the user is constantly allocated resources, that is, served with the best QoS, then the attained utility is $U_{s,i}$. The ATHENA utility functions are not the only possible ones reflecting the user satisfaction with regard to the QoS attained. What is important is that the form of these utilities reflects rightly the preferences of each type of user.

12.2.4 User Incentives

ATHENA [4] also studies user incentives for bidding in a sequence of auctions. The proof sketch is as follows: incentive compatibility holds if user sub-utilities are independent among the various auctions due to the GVA incentive compatibility property. The interesting case where *complementarities* exist among subsequent auctions, for example, for Type 2 and 3 users, is also studied: placing a bid less than the user utility at any auction is not beneficial, since this only increases the probability of losing without affecting the payment. A bid greater than the corresponding sub-utility can be beneficial, due to the extra value that a present allocation (which becomes more likely by over-bidding) can bring to the overall service. However, this would not be the case for "uncertainty averse" (conservative) users, who—by definition—in cases of choice/behavior under uncertainty always opt to play the safest strategy. For such users, it is proven that truthful bidding comprises a subgame perfect equilibrium strategy. This "maximin" behavior was proposed by Wald [7] for situations of severe uncertainty, which is also encountered by the ATHENA bidders.

12.2.5 Assessment

The assessment of ATHENA, as well its variation HERA [8] for auction-based resource management in UMTS HSDPA networks, has revealed some very interesting properties.

In particular, simulations have revealed that the vast majority of the users either receive almost-perfect service or reserve negligible quantities of resources. Thus, there is a bimodal distribution of the resource allocation patterns with intermediate allocation patterns arising rarely in general; see Figure 12.3 where it is depicted that under medium congestion, the total percentage of users receiving mediocre service, that is, 25%–85% of the desired service are approximately 10%, as

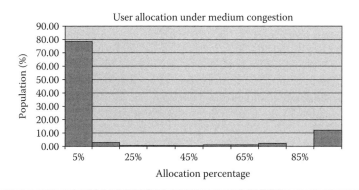

Figure 12.3 Distribution of resource allocation patterns with respect to the percentage of the total targeted resources that was attained under medium network congestion.

opposed to 90% of the users that receive perfect service or are not served at all. Overall, the percentage of users that experience frequent interrupts of service whose length in slots is significant is extremely limited; this is mostly due to the form of the utility functions and the fact that noncompetitive users are prohibited from entering the network due to the lower value of their bids compared to the auction cutoff price. This has also been analyzed theoretically by means of a random walk model for Type 2 users (see [9] for the analysis).

Moreover, a comparison of the auction with well-known scheduling policies, namely *First Come First Served* and *Round Robin*, indicate that (a) the auction is preferable to Round Robin, since the latter results in many intermediate resource allocation patterns, and (b) it is also better than FCFS, since the latter results in unacceptably high session setup delays.

Also, in terms of practical applicability, the auction is applicable to both UMTS and HSDPA 3G networks, where the 10 ms UTRAN frame and 2 ms HSDPA frame comprise the unit of resource allocation for these networks. The auction is also applicable to GPRS and its enhanced version EDGE. In GPRS, user data services compete for a number of time slots, with the unit of allocation being the radio block [10].

12.2.6 Generalizations and Extensions

The ATHENA mechanism can be extended so as to support multicast sessions as well; see [11]. This way, low-income users can form multicast groups and compete collectively in the auction. Simulations of [11] indicate that multicast always increases social welfare; it also improves revenue if there is demand in the market and wealth asymmetry among bidders.

So far in this chapter, we have addressed the issue of auction-based resource management in 2.5+G networks under the implicit assumption that users are served by means of accessing only one type of network interface. However, the current trend in wireless systems is evolving toward the 4G networks. 4G prescribes that users can enjoy mobility, seamless access, and high QoS in an all-IP network on an "Anytime, Anywhere" basis. The 4G network access comprises multiple heterogeneous geographically coexisting wireless networks [12]. This integrated network access approach has also been motivated by the 3G cellular mobile networks and their potential enhancement with WLAN radio access, and also WiMax, which can utilize the analog TV bands (700 MHz) that will be made available with the upcoming roll out of digital TV. Therefore, it

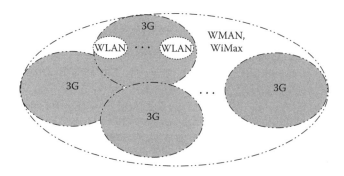

Figure 12.4 Coexisting wireless access networks.

is interesting to address the "generalized" problem of auction-based resource management in the context of 4G networks.

In this section, we describe a mechanism for allocating the downlink bandwidth of such a 4G network whose access architecture is hierarchical in terms of radius and thus geographical coverage; more details can be found by the interested reader in [13]. In particular, [13] addresses the problem of deciding on which user flows to admit and how the downlink bandwidth of the "integrated" access of a 4G network should be allocated to the competing user services so that efficiency is attained, that is, social welfare is maximized.

The proposed auction mechanism of [13] is applicable in 4G hierarchical networks, comprising multiple layers (tiers) of wide, middle, and local area access networks owned by one operator. Due to the hierarchical structure of the network, depicted as Figure 12.4, each user can be served by means of either a higher tier network of that area or some lower tier network (e.g., a WLAN). The standard 4G assumption that user terminals are capable of accessing multiple network interfaces is made. The model is motivated by the fact that wireless operators consider WLANs as a cheaper means of providing high-speed download services compared to other technologies such as HSDPA and opt to deploy WLANs [14].

A plethora of architectures, protocols, and resource management schemes for 4G have been proposed in the literature (see [12,15–19] and references therein). Those proposals generally lack economic merit, since they do not prioritize users in terms of their utility for the service, as opposed to [13]. In fact, most of them are complements to [13], since they provide the technological solutions by means of which [13] can be applied.

Hierarchical bandwidth allocation has been studied in [20], where in the top level a unique seller allocates bandwidth to intermediate providers (e.g., Internet Service Providers), who in turn resell it to their own customers in the lowest level. This model involves resale among the three tiers pertaining to different actors, as opposed to [13] where a single actor owns *multiple* tiers and aims to efficiently allocate their bandwidth. A utility-based load balancing scheme in WLAN/UMTS networks is proposed in [21]. For each network, a utility reflecting the current load is computed. The values of the network utilities, that is loads, are communicated to the clients who can switch to the less loaded network. Thus, this scheme cannot prioritize users in terms of their utilities. Closely related to [13] is the scheme of [22], where *fixed rate pipes* over *two* alternative paths are auctioned among synchronized users having different utilities for the two paths and thus submitting different bids. The latter contradicts the seamless access concept of 4G, as opposed to [13]. Also, the scheme of [22] cannot be generalized for multiple services, rates, and networks, as opposed to [13].

We now proceed to present the auction of [13].

Let us assume that each user $i \in \mathcal{I}$ has a certain utility u_i and declares a willingness to pay w_i for a service of rate m_i, by submitting w_i as part of his service request. Let p_i denote the per unit of bandwidth willingness to pay of user i, that is, $p_i = w_i/m_i$. For an l-tier network architecture, let $C_k^{(l)}$ denote the capacity of the l-tier access network k, with $l = 1, \ldots, L$; that is, $l = 1$ corresponds to the network technology having the greatest geographical coverage. k is the index of the l-tier network accessible by the user, for example, the index/ESSID of a WLAN inside the coverage of which the user is located. Users are both unaware of and indifferent to the internal routing of their traffic, as long as their service is provided.

When a service request is received, the operator* creates the bids $b_i^{(l)} = (p_i, m_i) \; \forall i \in \mathcal{I}$, $l = 1, \ldots, L$ and updates the respective active bid sets $\mathcal{B}_k^{(l)}$ for all networks k accessible by the user, one per tier. The basic idea is that winner determination is performed starting from the highest coverage network, where competition is most fierce. The users' bids are sorted by p_i and given the capacity constraint; the highest of them are declared as winning. The auction winners are propagated to the lower tier network auctions, from which their bids are deleted. Winner determination is then performed for the next layer, until the lowest tier is reached; this is done simultaneously, in a distributed fashion for same-tier networks. A sample auction execution for a two-tier network comprising a 3G network and three WLANs is provided as Figure 12.5. Thus, the proposed auction is defined as follows:

Step 0: Set $l = 1$. Sort($\mathcal{B}_k^{(l)}$)$\forall k, l$ // sort bids per p_i.

Step 1: Determine winning bids $\mathcal{W}_k^{(l)}$ of the l-tier network k to be the *largest* set of the *highest* bids of $\mathcal{B}_k^{(l)}$ that do not violate the capacity constraint $C_k^{(l)}$.

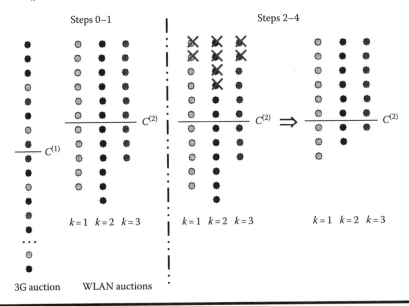

Figure 12.5 Sample algorithm execution for a UMTS/WLAN network. Different patterns and scales of gray denote the different WLANs that users can utilize.

---

* The terms network and operator are used to refer to the auctioneer.

Step 2: For every user i with $b_i \in \mathcal{W}_k^{(i)}$ delete user i's bids from $\mathcal{B}_k^{(j)}$, $\forall j > l$. Set $l = l + 1$.
Step 3: If $(l < L)$ go to *Step 1*.
Step 4: Compute payments.

In [13], a VCG payment rule is applied for *Step 4* and this mechanism is proven to be both incentive compatible, that is, it is best for users to bid truthfully ($p_i = w_i$), and of low computational complexity. Then, the auction is extended so that it can also accommodate multicast. Finally, it is shown how by replacing the user bids with operator weights that vary among services and over time, the auction is transformed to a cooperative bandwidth allocation mechanism capable of prioritizing certain classes of services and emulating DiffServ and time-of-day pricing schemes.

12.2.7 Concluding Remarks

We have presented auction-based resource management schemes for 2.5+G networks that deal successfully with data, audio, and video services in practical cases of networks with large numbers of competing users. The use of predefined utility and bidding functions enables the efficient utilization of the network and deals with the complex problem of efficiently serving long timescale services by means of reservations decided over short timescales. This approach also reduces the message and computational overhead of the resulting mechanism, thus facilitating its applicability. Indeed, due to these nice properties, this approach has also been adopted for wired network auctions by other researchers [45].

Overall, auctions are a promising tool of mechanism design in the context of dynamic resource management in 2.5+G networks. Though nowadays cellular networks are currently underutilized, the shift of the business models of the cellular operators toward serving business customers by means of providing VPNs over their networks combined with the exponentially increasing number of mobile users and the trend of evolving toward demanding 4G services may render such mechanisms applicable in the not-so-far future.

12.3 MANETs and Fairness Schemes

MANETs are infrastructureless self-configuring wireless networks of mobile nodes. They are ideal for a wide range of applications and can potentially become a common access network, as is the case now with the infrastructure-based wireless LANs (WLANs). MANETs are often based on the underlying wireless technology IEEE 802.11, due to the latter's simplicity and commercial availability. Some works have shown that this use raises issues in terms of efficiency and fairness [23]. Different protocols, henceforth referred to as fair solutions or schemes, have been proposed to improve the unfair way that the various flows are allocated bandwidth in the standard IEEE 802.11 protocol (see for instance [24–27]). This section presents these issues and solutions and introduces a utility-based performance evaluation framework for the assessment of the fair solutions.

12.3.1 Fairness: Issues, Protocols, and Evaluation

In order to prevent collisions, IEEE 802.11 DCF mode prescribes that before sending a frame, nodes must wait until they sense the channel free. Then, after a fixed time interval, the Distributed InterFrame Space (DIFS), during which the medium must be idle, a frame is emitted. If the medium is not idle, the sender randomly draws a *backoff* value from an interval, the *contention window*. Once the medium is idle again, the node waits for a DIFS and then keeps decrementing its backoff slot by slot. If the medium becomes busy, the process is stopped and later resumed after a new DIFS with the remaining number of backoff slots. When the backoff reaches 0, the frame is emitted. When a collision occurs, the contention window size of involved emitters is doubled, and the same process is repeated, until the maximum contention window value is reached, resulting in the frame to be dropped. Upon either the drop or the successful transmission of a frame, the contention window size is reset.

The IEEE 802.11 MAC layer raises issues in terms of efficiency and fairness in MANETs [23]. In particular, under certain configurations ([23]), some nodes may attain low network utilization, due to interfering node pairs and the presence of overlapping multihop flows. In order to deal with these issues, many schemes have been proposed in the literature: some solutions are based on the nodes exchanging information and/or knowledge of the topology, as in [28–33]. Others rely only on the data packets of the network or introduce a probabilistic behavior of the nodes, as in [24,34,35].

In [28], the authors propose a priority-based fair MAC protocol (P-MAC) by modifying the IEEE 802.11 DCF, and in particular the nodes contention window size, so that channel utilization is maximized subject to the weighted fairness among flows. The authors of [29] use knowledge of the network topology to compute the max–min fair share of the flows. In [30], a framework capable of mapping a given fairness model to its corresponding collision resolution backoff algorithm is proposed. The main merit of this approach is the ability to specify *arbitrary fairness models* and automatically creates corresponding contention resolution mechanisms. In [31], a two-phase scheduling scheme is proposed that computes a "fair" scheduling and attempts to maximize the aggregate channel reuse; it relies on both knowledge of the topology and exchange of flow information among nodes. In [32], $p_{i,j}$-persistent carrier sense multiple access-based algorithms are proposed in which a fair wireless access for each user is accomplished by modifying the backoff window using a pre-calculated link access probability and information exchanged by nodes; these algorithms are dynamic and sensitive to the changes in the network topology. The authors of [33] identify three main causes leading to the fairness problem, namely, the lack of synchronization, the double contention areas, and the lack of coordination problem. Then, based on this analysis, they propose a new P-MAC (EHATDMA), which strikes a good balance between performance and fairness and outperforms similar approaches such as [32].

In [24] DFWMAC, a different backoff scheme for IEEE 802.11 is proposed: nodes estimate dynamically all nodes throughput and then adjust their contention window according to a fairness index. This results in improved performance compared to 802.11 in terms of fairness, but not in terms of aggregate throughput. A problem of DFWMAC is that each node's fair share is computed by only considering the neighbor nodes instead of the nodes in carrier sensing range. Probabilistic NAV (PNAV) [34] attempts to improve fairness by preventing systematic successive transmissions of the same emitter through the probabilistic introduction of a waiting time, a virtual NAV, after each emission. The probability to set a NAV is adaptively computed. Simulations indicate that PNAV is fairer than IEEE 802.11 for some topologies, but PNAV global throughput is always smaller than the 802.11 aggregate throughput. In [35], the IdleSense access method is proposed: all hosts use similar values of the contention window to benefit from good short-term access fairness; this comprises a modification of the contention window based on idle slots perceived by each node, which however is not adapted to the multihop setting.

The MadMac protocol [25] deals with fairness and throughput in MANETs by resolving the unfairness of the MAC layer—by means of a non-probabilistic modification of 802.11 MAC—and subsequently maximizing aggregate throughput. It is only based on information provided by the 802.11 MAC layer. This scheme can avoid the hidden terminal problem without the use of RTS/CTS, and its collision avoidance scheme reduces the collision rate and thus increases the throughput.

The scheme of [26], henceforth referred to as Profiterole, addresses the fairness issues of the IEEE 802.11 standard by designing a rate regulation mechanism on top of 802.11. By exchanging information at the routing layer, this scheme uses a simple medium sharing model deduced by the network topology to compute the rates achievable by the MAC layer so that all flows are granted a part of the capacity according to a proportional fairness rationale.

The works that propose solutions to the fairness issues in ad hoc networks usually base their performance evaluation on flows throughput (see for instance [24,25]). Most of the tested scenarios are static: it means that the topology is given and the communicating source-destination nodes and respective flows are fixed and do not change during the simulation. For a general evaluation (and not restricted to one scenario), the flow throughputs are averaged over different runs. Even if a confidence interval is given, these averages give no indication on how the rates fluctuate *over time*.

The Jain index [36] is a metric often used in the literature to depict fairness. For N flows with throughputs x_i, $i \in N$, it is defined as

$$\frac{\left(\sum_i r_i\right)^2}{N \cdot \sum_i r_i^2}$$

$r_i = x_i/\phi_i$ is the normalized throughput, that is, the throughput attained x_i divided by the (theoretical) fair throughput ϕ_i of the flow i. The closer its value is to 1, the fairer the scheme. But in a long time period, traffic patterns and/or the network configuration may vary over time, partly due to the nodes' mobility and the traffic evolution. Thus, it is very likely that flows, penalized under a configuration during a given time period, do not encounter any fairness issues during other time periods. Moreover, average values (on throughput or on fairness index) give no clue on the volatility of flow throughput over time nor on the impact of the proposed schemes on different parameters, like delay for instance, that are also of importance on the users services.

The short-term fairness studies, like the one in [37], give an indication on the delays perceived by the stations for the channel access via the average Jain fairness index with a sliding window method or the distribution of the number of inter-transmissions. But once more, average values or even a distribution does not capture the realization of the rates fluctuation, which greatly affects user satisfaction for many services.

Finally, the Jain fairness index, often used in these studies, requires knowing the targeted fair allocation. This last point is tricky because the most appropriate fair allocation for ad hoc networks is not known a priori, and it is difficult to know if it is better to compare the rates achieved with the proposed solution with an allocation (that optimizes the overall throughput) or a max–min allocation or a proportional allocation or with an allocation in which all the flows have the same rate for instance. A similar metric relying on a precalculated fair share ϕ_i per flow i and its comparison with that truly attained W_i is the fairness index:

$$FI = \max \left\{ \forall i,j : \frac{\max\left\{\frac{W_i}{\phi_i}, \frac{W_j}{\phi_j}\right\}}{\min\left\{\frac{W_i}{\phi_i}, \frac{W_j}{\phi_j}\right\}} \right\}$$

Minimizing FI indicates that perfect fairness was attained [24]. Overall, these evaluations do not measure the user satisfaction.

Since the seminal work of von Neumann and Morgenstern [38], utility functions have been extensively used in economics, decision-making, game theory, grid, and computing systems (see [2,39,40] and references therein). There have been some research efforts to apply utility functions in order to measure user satisfaction in wireless networks [41–44]. The utility functions used in these works, depicted [42], quantify the benefit that users obtain from being allocated a certain bit rate at a given time. These benefit values are not correlated, and if the rate fluctuates, the utility values will also fluctuate according to the utility function regardless of the rates achieved in the past. Thus, this approach neglects the severe impact of important QoS parameters, such as delay, on the user-perceived quality for services of long duration.

12.3.2 A Utility-Based Framework for Assessing Fairness Schemes

Assessing the impact of the varying *over time* rate allocations of the ad hoc fair schemes upon the resulting user-perceived quality and the social welfare attained is both of high importance and an open research issue. *History-dependent utility functions* are an extension of standard utility functions for expressing the value attained over a long timescale from receiving various levels of quality at short timescales. This short timescale is henceforth referred to as *slot*. The slot is defined as the time interval over which the number of packets delivered per service flow is counted so that the average rate received by each flow in this interval is computed and subsequently inputted to the respective service utility function, so that the user-perceived quality of this flow is quantified. The value of the slot depends on the service type, due to the different tolerance of different services to quality degradation (e.g., high delay); hence, it is different per service type.

The main merit of history-dependent utility functions is that *multiple* quality parameters such as the vector of instantaneous bit rates, delay, and/or total quantity of resources allocated impact the values of the *correlated marginal utilities* and the overall expected level of users' satisfaction. The term "marginal utility" is used to denote the additional utility attained over each slot of the user's service session. Thus, these utility functions can accurately quantify the time-varying user-perceived quality. This kind of utility function was proposed in the ATHENA mechanism [4]—presented earlier in this chapter—for auction-based resource allocation in UMTS networks and subsequently used elsewhere [8,45].

In [46,47], a three-tier framework, depicted as Figure 12.6, is proposed. This framework uses history-dependent utility functions so as to quantify the satisfaction of the ad hoc users from the way their services are allocated bandwidth. In order to comparatively assess the performance of ad hoc fair protocols, the vector of the rates of the user flows computed over slots (either by real measurements or simulations—network plane) is inputted to the framework's utility functions (utility plane). The latter quantifies and outputs both the per-flow user-perceived quality and the social welfare (assessment plane), which are the performance values of the protocols under investigation. Social welfare is a widely used performance evaluation metric, also indicative of the acceptance of the proposed fairness schemes in practice: schemes resulting in low social welfare are expected to be of limited economic value and acceptance by the users. Note that this framework is a performance evaluation framework that has no impact on the protocols' running.

This approach is novel and can serve for the ad hoc research community as a performance evaluation tool of the fair protocols. This way it is possible to unambiguously compare the performance of various fair solutions whose maximization goals are inherently different (max–min fairness, proportional fairness, etc.) and rank their performance according to the social welfare attained, which is the framework's performance evaluation metric.

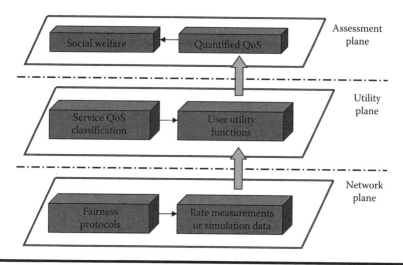

Figure 12.6   The proposed utility-based framework.

12.3.3 QoS Requirements and Utility Functions

We now focus on the various types of user services and their corresponding QoS requirements. There has been already substantial work in the context of UMTS networks by the 3GPP, in identifying the important QoS parameters of services and classifying them thereof. In particular, 3GPP report TS 23.107 [6] defines four QoS classes, namely *conversational, streaming, interactive,* and *background*. Conversational and streaming classes both pertain to delay-sensitive services such as voice and real-time audio/video streaming, respectively. The prominent QoS factors for both these classes are the *guaranteed bit rate* and the *delay*. In particular, the maximum delay tolerance of voice services is 100 ms, while it is 250 ms for streaming applications [6]. Interactive and background classes are well-suited to throughput-sensitive applications, such as web browsing and email (or downloading), respectively.

These QoS classes and their respective dominant QoS attributes are used for the definition of the framework's utility functions, presented in [46,47]. The idea is that there is one utility function definition per service class. This definition uses the maximum delay tolerance of the various services in order to partition the service duration to consecutive "slots," each size equal to this maximum delay tolerance. So for instance, for a video service of 60 s, the service duration is partitioned to 240 slots. For each of these slots, the vector of rates attained up to that slot are inputted to a utility function definition so that it quantifies the (dis)satisfaction attained due to the way the network serves the customer. This way, if at some slot, a user of video service is served with a rate less than the minimum demanded by the video application, the utility attained at this slot is taken to be 0. Interrupts and delays in serving the user also affect the value of this "marginal utility." Also, when more bandwidth than the minimum demanded is offered, the user's marginal utility also increases. Note that the fact that the slot duration equals the maximum delay tolerance of a service, along with the fact that the definition of each utility reflects the impact of various QoS parameters, ensures that these utilities are in line with the human perception of the QoS perceived and accurately quantify how well or poorly a user was served by the network.

The utility function definitions of this framework [46,47] are essentially not unique. This is not important per se for the assessment this framework performs, as long as (a) these definitions are rational, that is, reflecting the impact of the QoS parameters of importance on user utility, and (b) the *same utility definitions* are used for the comparative assessment of the protocols. Thus, the precise utility function definition and the resulting absolute values of the user-perceived QoS obtained by inputting the allocated rates under these protocols to the respective utility functions are not important; their respective ordering is important. This ordering is insensitive to the actual utility function definitions, as long as conditions (a) and (b) are met, and depicts which schemes perform better and for what kind of services, thus providing insight to the performance of the various ad hoc fairness schemes.

The utility functions of the streaming and conversational classes are affected by a multitude of parameters, as opposed to those of the background and interactive classes, which are only throughput sensitive. Note that for the former service classes, the entire history of rate allocations of each service session is taken into account by affecting the values of the marginal utilities. These values are monotonically increasing in terms of the rate allocated and monotonically decreasing with respect to delay. Furthermore, the relative reduction of the user utility due to service degradation is higher for the conversational class than for the streaming class, due to the latter's higher tolerance in delay. Thus, these utility functions provide an unambiguous ordering of users' preferences and subsequently of the value from being served under a certain fairness scheme, which is also in line with [6].

In [47], these utility functions are parametrized so that the social welfare attained over a slot if an additional user gets served even with the minimum acceptable rate exceeds the one attained if this bandwidth was given to a set of users already being served. This is clearly in line with the fairness rationale. Moreover, the proposed utility functions by construction take into account both the volume and the variance over time of the rates allocated to each service throughout its session, so as to express the impact of both channel utilization and variance of rate allocations on the user-perceived QoS. The value of the social welfare, that is, the sum of all users' utilities, under different traffic mixes and conditions indicates the sensitivity of the applications served to the underlying protocol behavior. The impact of admitting more flows or altering the nodes topology and mobility is depicted in the difference of values of the social welfare attained in these cases. Hence, this approach can provide better qualitative information compared to the average network metrics or the Jain fairness index.

12.3.4 Assessment

We demonstrate the usefulness of the framework by using it to comparatively assess the performance of Profiterole (scheme of [26]) and standard IEEE 802.11. Note however that it can be used for any scheme applying fairness in any network layer.

An extensive comparative assessment was conducted. The ns2 simulator, the protocol implementation code of [26], and a set of Perl scripts implementing the utility and assessment planes of the framework (see Figure 12.6) comprise the software used for this assessment. A full presentation and analysis of these results can be found in [47]. In this subsection, we refrain from presenting numerical results and instead present the main qualitative results and conclusions of this simulation analysis; refer to [46,47] for the data.

A sample simulation set regards the transmission of 10 video flows transmitting with 384 kb/s over a MANET of 25 nodes randomly deployed on a 500 m × 500 m terrain. The simulation set comprises 10 simulations (individual runs), each of which simulates such a MANET for 100 s. The average values of the social welfare for Profiterole are always higher than those attained under IEEE 802.11. IEEE 802.11, on the other hand, has slightly higher channel utilization. However, it exhibits higher variance in the rates allocated to the competing flows, thus more frequently failing to meet the minimum rate constraints of the video flows and limiting users' satisfaction; the rates allocated to the 10 competing flows for one simulation are depicted as Figure 12.7. This higher variance is evident both in the individual flow allocations of Figure 12.7 and somewhat surprisingly also on the aggregate channel utilization, depicted as Figure 12.8.

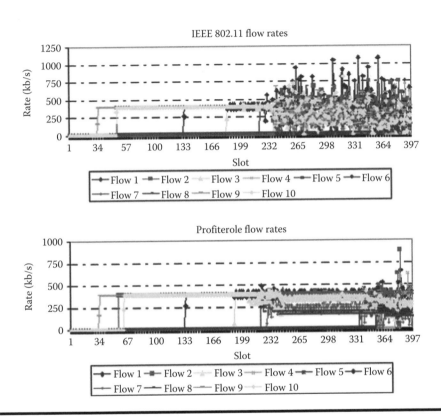

Figure 12.7   The flows rate allocations for a MANET of 30 nodes and 10 video flows.

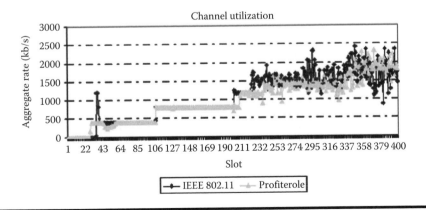

Figure 12.8   The aggregate throughput for a MANET of 30 nodes and 10 video flows.

The average value of the Jain index computed over all simulations of this set is better for Profiterole. Therefore, in this case, this also applies to the vast majority of simulations conducted; IEEE 802.11 has better performance than Profiterole with respect to the channel utilization. However, for the same simulations, IEEE 802.11 is generally worse than Profiterole in terms of fairness. Therefore, the ordering of the two metrics that are widely used in the literature to assess fair solutions is conflicting, thus failing to deduce to what extent the better fairness behavior of Profiterole is overall better or not and if it is indeed worth sacrificing some channel utilization to achieve it. On the contrary, the performance evaluation framework of [46,47] resolves this ambiguity by means of the social welfare metric, which quantifies the impact of the protocol performance on users' flows.

We now provide some indicative comparative plots of the rates and the respective marginal utilities attained by one flow under Profiterole—Figure 12.9a and b—and under IEEE 802.11—Figure 12.9c and d. The first flow is active between the 24th and 100th s. Note that since the attained rates of the flow at every 250 ms slot are in general higher than the minimum acceptable rate of video—assumed to be 128 kb/s for this simulation set—the marginal utility follows closely the trend of the rate allocations. As more flows emit over time, the fair rate of the flows is reduced, hence the different "steps" in the respective rates plot, and this is clearly depicted in the marginal utilities as well. Instantaneous spikes in the rates allocated also affect the marginal utility. Finally, when the minimum acceptable rate of 128 kb/s is not met—for slot 393—the marginal utility drops to zero. Overall, the marginal utility accurately reflects the QoS experienced by the user over time. Similar behavior is depicted for the second flow where the higher variance of 802.11 and its impact on the marginal utility are evident; see Figure 12.9c and d.

For lower loads of the network, both protocols have similar performance, with Profiterole being slightly better in terms of average values of social welfare. The higher the load, the better Profiterole performs in terms of fairness. The simulation assessment of [47] also contains simulation for traffic mixes comprising both UDP and TCP traffic with inherently different properties and sensitivity to QoS. For these cases, the social welfare value comprises a metric that can be used to comparatively assess the performance of the two schemes. On the contrary, it would not be possible to apply a metric such as the Jain index in this context, due to the heterogeneous traffic mix. These simulations

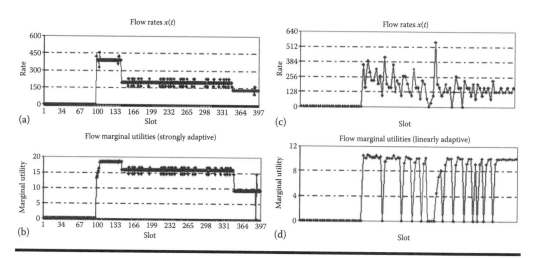

Figure 12.9   The rates and marginal utilities of two flows.

indicate that fairness schemes, such as Profiterole, allocate smoother rates over time due to rate regulation; this is highly desirable for multihop QoS-sensitive flows, thus resulting in higher values of the social welfare attained. Also, the proposed framework by means of the social welfare metric unambiguously ranks the performance of the schemes under evaluation.

12.3.5 Concluding Remarks

We presented the main issues, solutions, and evaluation methods of fairness studies in multihop ad hoc networks. We discussed their shortcomings and then presented a utility-based framework, based on *QoS-aware history-dependent* utility functions [46], which captures the inherent properties of the schemes under investigation. This framework unambiguously ranks the performance of different fair schemes by means of the social welfare metric. Its main merit is that it can answer if the adoption of a certain fair solution is indeed beneficial for (what kind of) services and to what extent it outperforms alternative approaches; the use of utility functions ensures that the social welfare metric reflects accurately how well users got served, which is clearly what really counts in networks.

12.4 Open Issues

This chapter has dealt with the issues of auction-based resource management and fairness in wireless networks. There are several open issues that are worth looking into in this field.

Regarding auction-based resource management, it is unclear how these schemes can be applied to networks where there is not a central coordination point, as is the base station in the cellular networks, that can enforce the auction outcome. Indeed, (malicious) users may not respect the auction outcome in networks such as MANETs and WLANs, thus limiting the applicability of these schemes in practice. Developing architectures and protocols that mitigate these threats is a very interesting topic for further research.

An additional open problem is how these schemes can efficiently handle mobility. Since auction-based schemes rank users with respect to the declared willingness to pay, they cannot prioritize handoff connections, which may be beneficial. Existing approaches where a certain part of the capacity is reserved for handoff traffic provide a simple but inefficient solution, since the valuable Radio Access Network (RAN) resources are not efficiently used [48].

A large number of open issues exist also for fairness schemes in wireless networks. Many schemes rely on unrealistic assumptions about what kind of information exists in user terminals and how this information is shared. Solutions typically assume cooperative nodes that respect the "suggestions" of the fair solution. Also, most of these schemes are only evaluated for homogeneous traffic. Assessing the applicability of these schemes is an open issue.

Also, many fair schemes rely on too detailed information regarding the network. Others rely on iterative convergence phase in order to attain the optimal fair solution. The impact of these issues on protocol performance must be further investigated. Also, it is worth investigating the trade-offs among applying a fair scheme on top of current protocols or modifying the unfair 802.11 MAC layer or adopting a vertical cross-layer optimization solution.

An additional issue of importance is that in order to apply any fair solution, the precise number (and topology) of flows crossing the network must be known. This implies that active flows of the network must be dynamically discovered and also a criterion for deciding whether they have terminated or not must be decided. Currently, most fair solutions implicitly assume that this number is always fixed and known to the protocol; their respective simulation assessment considers only aggressive flows: this means that these solutions if practically adopted over a network have no accurate way of knowing the actual flows over the network and thus may lead to misallocation of network resources. The design of fast and efficient protocols that could complement the fair solutions by providing reliable information regarding the set of active flows over time is an interesting open issue that is worth investigating.

The evaluations of fair solutions are somewhat problematic [49]; the results published are questionable to a certain extent, due to the lack of information that is required for a third party to repeat these experiments and deduce on the validity of the results.

Furthermore, as explained in this chapter, the metrics typically used for the evaluations of fair solution fail to depict the impact of these schemes on user QoS and network performance. To this end, we have presented a utility-based framework that attempts to mitigate these issues. Assessing this framework's usefulness combined with an in-depth study of the relationship of fairness and QoS in ad hoc networks comprises also open research topics.

Finally, it is also an open issue to study the performance degradation of fair solutions when the load of the network becomes substantially high. Combining solutions from auction mechanism design and fair solutions could lead to more robust and efficient solutions for resource management and QoS provisioning in wireless networks.

12.5 Conclusions

In this chapter, we have shown how history-dependent utility functions can be used as the basis of (a) auction-based resource management in 2.5+G networks and (b) a performance evaluation framework for the assessment of fair solutions in MANETs. We have discussed the pros and cons of various approaches regarding resource management and fairness in 2.5+G networks, presented some interesting solutions to the problems identified, and provided a set of unresolved issues that comprise interesting open research issues.

References

1. H. Holma and A. Toskala, *WCDMA for UMTS: Radio Access for Third Generation Mobile Communications*, John Wiley & Sons, Chichester, U.K., September 2000, ISBN 0-471-72051-8.
2. C. Courcoubetis and R. Weber, *Pricing Communication Networks: Economics, Technology and Modelling*, John Wiley & Sons, Chichester, U.K., 2003.
3. V. Krishna, *Auction Theory*, Academic Press, Burlington, MA, April 2002, ISBN 0-12426297-X.
4. M. Dramitinos, G.D. Stamoulis, and C. Courcoubetis, Auction-based resource reservation in 2.5/3G networks, *Kluwer/ACM Mobile Networks and Applications Special Issue on Mobile and Pervasive Commerce*, 9, 6, December 2004, 557–566.
5. T. Groves and J. Ledyard, Optimal allocation of public goods: A solution to the 'Free Rider' problem, *Econometrica*, 45, 4, May 1997, 85–96.
6. 3GPP, Specification 23.107 Quality of Service (QoS) concept and architecture, 2011, http://www.3gpp.org
7. A. Wald, *Statistical Decision Functions*, John Wiley & Sons, New York, 1950.
8. M. Dramitinos, G.D. Stamoulis, and C. Courcoubetis, Auction-based resource allocation in UMTS high speed downlink packet access, *First EuroNGI Conference on Traffic Engineering*, Rome, Italy, April 2005, pp. 434–441.
9. M. Dramitinos, G.D. Stamoulis, and C. Courcoubetis, A random walk model for studying allocation patterns in auction-based resource allocation, *Fifth International Workshop on Advanced Internet*

Charging and QoS Technologies (ICQT 2006), Springer LNCS 4033, St Malo, France, June 27, 2006, pp. 26–37.

10. S. Soursos, C. Courcoubetis, and G.C. Polyzos, Differentiated services in the GPRS wireless access environment, *IWDC 2001, Evolutionary Trends of the Internet*, Taormina, Italy, September 17–20, 2001.

11. M. Dramitinos, G.D. Stamoulis, and C. Courcoubetis, Auction-based resource reservation in 3G networks serving Multicast, *15th IST Mobile Summit*, Myconos, Greece, June 4–8, 2006.

12. J. Kristiansson, Creating always-best-connected multimedia applications for the 4th generation wireless systems, 2004. http://epubl.luth.se/1402-1757/2004/40/

13. M. Dramitinos and I. Guérin Lassous, A bandwidth allocation mechanism for 4G, *European Wireless Technology Conference (EuWiT) 2009*, Focused Session Infocity, Rome, Italy, September 2009.

14. J. Steven, Vaughan-Nichoks, Wi-Fi and 3G Together?, Wi-Fi planet article, http://www.wi-fiplanet.com/columns/article.php/2225411

15. V. Gazis, N. Alonistioti, and L. Merakos, Towards a generic always best connected capability in integrated WLAN/UMTS cellular mobile networks (and beyond), *IEEE Wireless Communications Magazine*, 12, 3, June 2009, 20–29.

16. Y. Xao, K.K. Leung, Y. Pan, and X. Du, Efficient 3G/WLAN interworking techniques for seamless roaming services with location-aware authentication, *Wireless Communications and Mobile Computing*, 5, 2005, 805–823.

17. H.-T. Lim, S.-J. Seok, and C.-H. Kang, A transport-layer based simultaneous access scheme in integrated WLAN/UMTS mobile networks, *EvoWorkshops 2007*, Valencia, Spain, 2007, pp. 137–144.

18. Y. Zhou, Y. Rong, H.-A. Choi, J.-H. Kim, J.K. Sohn, and H.I. Choi, A dual-mode mobile station modules for WLAN/UMTS internetworking systems, *OPNETWORK 2007*, Washington, DC, August 2007.

19. K.S. Munasinghe and A. Jamalipour, An analytical evaluation of mobility management in integrated WLAN-UMTS networks, *Elsevier Computers and Electrical Engineering*, 36, 4, 2008, 735–751.

20. M. Bitsaki, G.D. Stamoulis, and C. Courcoubetis, An efficient auction mechanism for hierarchically structured bandwidth markets, *Computer Communications*, 29, 7, 2006, 911–921.

21. Y. Zhou, Y. Rong, H.-A. Choi, J.-H. Kim, J.K. Sohn, and H.I. Choi, Utility-based load balancing in WLAN/UMTS internetworking systems, *IEEE Radio and Wireless Symposium*, Orlando, FL, January 2008, pp. 587–590.

22. S. Routzounis and G.D. Stamoulis, An efficient mechanism for auctioning alternative paths, *GameNets '06*, Pisa, Italy, 2006.

23. C. Chaudet, D. Dhoutaut, and I. Guérin Lassous, Performance issues with IEEE 802.11 in ad hoc networking, *IEEE Communication Magazine*, 43, 7, July 2005, 110–116.

24. B. Bensaou, Y. Wang, and C.C. Ko, Fair medium access in 802.11 based wireless ad-hoc networks, *ACM MobiHoc 2000*, Boston, MA, 2000, pp. 99–106.

25. T. Razafindralambo and I. Guérin Lassous, Increasing fairness and efficiency using the MadMac protocol in ad hoc networks, *Ad Hoc Networks Journal*, Elsevier Ed., 6, 3, May 2008, 408–423.

26. R. Vannier and I. Guérin Lassous, Towards a practical and fair rate allocation for multihop wireless networks based on a simple node model, *11th ACM/IEEE International Symposium on Modeling, Analysis and Simulation of Wireless and Mobile Systems (MSWiM)*, Vancouver, BC, Canada, October 2008, pp. 23–27.

27. Y. Xue, B. Li, and K. Nahrstedt, Optimal resource allocation in wireless ad hoc networks: A price-based approach, *IEEE Transactions on Mobile Computing*, 5, 4, April 2006, 347–364.

28. D. Qiao and K. Shin, Achieving efficient channel utilization and weighted fairness for data communications in IEEE WLAN under the DCF, *IEEE International Workshop on QoS*, Miami, 2002, pp. 227–236.

29. X.L. Huang and B. Bensaou, On max–min fairness and scheduling in wireless ad-hoc networks: Analytical framework and implementation, *MobiHoc'01*, ACM Press, New York, 2001, pp. 221–231.

30. T. Nandagopal, T. Kim, X. Gao, and V. Bharghavan, Achieving MAC layer fairness in wireless packet networks, *MobiCom'00*, ACM Press, New York, 2000, pp. 87–98.

31. H. Luo, S. Lu, and V. Bharghavan, A new model for packet scheduling in multihop wireless networks, *MobiCom'00*, ACM Press, New York, 2000, pp. 76–86.

32. T. Ozugur, M. Naghsineh, P. Kermani, and J.A. Copeland, Fair media access for wireless LANs, *GlobeCom*, Rio de Janeiro, Brazil, 1999.

33. J. He, and H. Pung, Fairness of medium access control protocols for multihop ad hoc wireless networks, *Computer Networks*, 48, 6, 2005, 867–890.

34. C. Chaudet, G. Chelius, H. Meunier, and D. Simplot-Ryl, Adaptive probabilistic NAV to increase fairness in ad hoc 802.11 MAC layer, *MedHoc NET*, Ile de Porquerolles, France, 2005.

35. M. Heusse, F. Rousseau, R. Guillier, and A. Duda, Idle sense: An optimal access method for high throughput and fairness in rate diverse wireless LANs, *SIGCOMM'05*, ACM Press, New York, 2005, pp. 121–132.

36. R. Jain, W. Hawe, and D. Chiu, A quantitative measure of fairness and discrimination for resource allocation in shared computer systems, DEC Research Report TR-301, September 1984.

37. G. Berger-Sabbatel, A. Duda, M. Heusse, and F. Rousseau, Short-term fairness of 802.11 networks with several hosts, *Sixth IFIP TC6/WG6.8 Conference on Mobile and Wireless Communication Networks (MWCN 2004)*, Paris, France, October 2004, pp. 263–274.

38. J. von Neumann and O. Morgenstern, *Theory of Games and Economic Behavior*, Princeton University Press, 2nd edn., Princeton, NJ, 1947.

39. S. Shenker, Fundamental design issues for the future internet, *IEEE Selected Areas in Communications*, 13, 7, September 1995, 1176–1188.

40. W.E. Walsh, G. Tesauro, J.O. Kephart, and R. Das, Utility functions in autonomic systems, *International Conference on Autonomic Computing*, New York, 2004, pp. 70–77.

41. K. Lee, Adaptive network support for mobile multimedia, *First Annual International Conference on Mobile Computing and Networking*, Berkeley, CA, 1995, pp. 62–74.

42. R.F. Liao and A.T. Campbell, A utility-based approach for quantitative adaptation in wireless packet networks, *Kluwer Wireless Networks*, 7, 2001, 541–557.

43. D. Miras, R.J. Jacobs, and V. Hardman, Content-aware quality adaptation for IP sessions with multiple streams, *Eighth International Workshop on Interactive Distributed Multimedia Systems*, London, U.K., 2001, pp. 168–180.

44. J. Sachs, M. Prytz, and J. Gebert, Multi-access management in heterogeneous networks, *Wireless Personal Communications*, 48, 1, January 2009, 7–32.

45. P. Maillé and B. Tuffin, An auction-based pricing scheme for bandwidth sharing with history-dependent utility functions, *First International Workshop on Incentive Based Computing (IBC'05)*, Compiegne, France, September 2005, pp. 30–49.

46. M. Dramitinos, R. Vannier, and I. Guérin Lassous, A utility-based framework for assessing fairness schemes in ad-hoc networks, INRIA Research Report RR-6843, 2009.

47. M. Dramitinos, R. Vannier, and I. Guérin Lassous, A performance evaluation framework for fair solutions in ad hoc networks, *12th ACM MSWiM*, Tenerife, Spain, October 2009, pp. 46–53.

48. S. Choi and K. Shin, Predictive and adaptive bandwidth reservation for hand-offs in QoS-sensitive cellular networks, *ACM SIGCOMM Computer Communication Review*, 28, 4, 1998, 155–166.

49. S. Kurkowski, T. Camp and M. Colagrosso, MANET simulation studies: The incredibles, *ACM SIGMOBILE Mobile Computing and Communications Review*, 9, 4, October 2005, 50–61.

Chapter 13

Cooperation Incentives in 4G Networks

Dimitris E. Charilas, Stavroula G. Vassaki,
Athanasios D. Panagopoulos, and Philip Constantinou

Contents

4G will introduce a user-centric system, which will personalize not only services but also terminal characteristics. Complex terminals should be able to function in multiple modes and access several radio technologies, so that the user receives a personally tailored service. Cooperation among users is of high importance, since resources may be "borrowed" from other users in exchange for certain "rewards"; however, powerful incentives need to be placed before full cooperation can be guaranteed and actually established. In this chapter, the state of the art in cooperation incentives is examined. The problem is motivated by considering cooperative schemes, through the creation of coalitions between players. More specifically, current approaches of both credit-based and reputation-based systems, followed by their strengths and weaknesses, are presented. This chapter is concluded by identifying some open research issues.

13.1 Introduction

In a 4G network environment, nodes will be able to connect to different access technologies at the same time. Such an availability of ubiquitous network access via several technologies is going to enable more options for network selection, trading off bandwidth, power, and cost, as well as create multiple opportunities for cooperation among network entities, that is, nodes and networks [1]. Apart from the direct interaction with the provider's infrastructure, nodes, in particular, may also be offered the possibility of cooperating with each other in order to take advantage of low-cost high-speed local connections to share content [2].

Nevertheless, cooperation cannot be taken for granted; even though, in most cases, players, meaning nodes and networks, do obtain the optimal result by sharing resources with others, in certain cases, it is not clear enough for them why they should not act selfishly or even not try to cheat. This assumption has created the need for orchestrating more and more complex incentive mechanisms, which will encourage players to act in a cooperative manner.

Incentives motivate a particular course of action and count as a reason for choosing one action over the alternatives and are in fact present in various application domains. Particularly, in wireless networks, incentive mechanisms may be applied to urge players to cooperate instead of pursuing their own interest as it is generally acknowledged that cooperation may help them perform better than if they were in competition [3].

In a cooperation incentive mechanism, cooperative behavior should self-evidently be more beneficial than uncooperative behavior. Reputation and pricing are the main concepts around which cooperation incentive mechanisms are built, providing, respectively, reputation-based and credit-based system mechanisms. Such systems are a valuable tool for making 4G networks work more efficiently, as the latter provide more opportunities for cooperation than traditional networks.

An important aspect of cooperation incentive mechanisms concerns their integration in game-theoretic frameworks developed in order to provide more efficient and fair equilibriums as far as resource allocation is concerned. Game theory may help one demonstrate that it is worth considering reputation (information) when analyzing the outcome of some competing situations with incomplete information. In a game-theoretical framework, players are continuously playing the same game. As a result, when an agent plays the game repeatedly in the same way, it can be assumed that he builds a reputation for playing certain kinds of actions that may prove useful to the rest of the players when forming their strategies. Similarly, pricing-based games have been also proven to provide more fair equilibriums compared to noncooperative approaches, mainly as far as the problem of routing and packet forwarding in ad hoc networks is concerned.

The discussion in this chapter proceeds as follows: Section 13.1 is an introduction to the issues addressed in this chapter. Section 13.2 introduces the concept of cooperation schemes in 4G networks and exposes two main strategies for establishing cooperative games, which are bargaining and coalition forming. Section 13.3 presents some basic characteristics of incentive mechanisms adopted in wireless networks. Sections 13.4 and 13.5 focus on the incentive mechanisms that can be deployed in order to both motivate players to cooperate with each other and discourage them from cheating. Two types of such mechanisms, credit-based and reputation-based systems, are distinguished and described, while the analysis performed includes as well a number of application approaches that have been proposed by the research community. Finally, emphasis is also put on the degree in which the proposed schemes succeed or fail in guaranteeing full cooperation as well as on the advantages and disadvantages of each cooperation incentive mechanism type. Lastly, the chapter concludes with a summary of ideas presented.

13.2 Cooperation Schemes in 4G

13.2.1 Cooperative Services

Cooperating terminals may share energy resources; as a result, energy efficiency can be achieved by sharing processing and spectrum resources among cooperating terminals. It is apparent that proper incentives, for example, in terms of energy efficiency, have to be in place before each terminal agrees to participate in a cooperating network and to spend energy resources to serve the needs from other cooperating terminals. It is expected that future mobile terminals will be less susceptible to channel variations and shadowing effects and will be able to transmit at lower power levels in order to achieve a certain throughput, thus increasing their battery life. Furthermore, cooperative transmission strategies may increase the end-to-end capacity and, hence, the spectral efficiency of the system. In a cooperation scenario, the terminals may communicate over a short-range communication in parallel to the cellular communication. Such architecture offers virtual high data rate, lower energy consumption, and new business models.

In order to successfully deploy cooperative schemes in wireless networks, new network architectures have to be developed and powerful incentives have to be placed. The decision about sharing resources may involve a reputation matrix and rewards according to user contribution. Rewards may involve monetary value, points, or even provision of higher QoS in future service requests.

13.2.2 Bargaining Games

13.2.2.1 Theory

A very promising element of cooperative game theory is bargaining. The Nash bargaining game is a simple two-player game used to model bargaining interactions. In the Nash bargaining game, two players demand a portion of some good. If the two proposals sum to no more than the total good, then both players get their demand. Otherwise, both get nothing. A Nash bargaining solution (NBS) is a Pareto efficient solution to a Nash bargaining game.

Bargaining approaches can be categorized in strategic and axiomatic bargaining. **Strategic bargaining**, such as the Rubinstein's model of bargaining, assumes that there is a bargaining process where the solution is achieved in a series of offers and counteroffers. The bargaining solution emerges as the equilibrium of a sequential game. The need for a bargaining process among the players is time-consuming and therefore unsuitable for wireless networks with many users. On the other hand, ***axiomatic bargaining*** ignores the bargaining process, assumes some desirable properties about the outcome, and then identifies process rules or axioms that guarantee this outcome. The operator may serve, for example, as the arbitrator in a cooperative resource bargaining game. The most used bargaining scheme is Nash bargaining in which the product of utilities u^* is maximized for player i, according to the following formula:

$$u^* = \underset{u \in U_0}{\text{argmax}} \prod_{i \in I} \left(u_i - u_i^0 \right) \tag{13.1}$$

where
   u is the utility space
   U_0 is the set of achievable utilities
   u_i is the achieved utility
   u_i^0 is the utility of player i at the disagreement point or disagreement utility
   I is the set of players that can achieve a utility strictly greater than the disagreement utility

13.2.2.2 Applications

The notion of axiomatic bargaining in cooperative game theory provides a good analytical framework to derive a desirable operative point that is fair and Pareto optimal. An allocation is Pareto optimal if there is no wasted utility, that is, it is impossible to make any one party better off without making any other worse off. Some very well known bargaining solutions are Nash, Raiffa–Kalai–Smorodinsky, utilitarian, and modified Thomson and each one corresponds to a unique point on the Pareto optimal boundary. Yaiche and Mazumdar [4] consider the NBS for bandwidth allocation for elastic traffic in broadband networks. Furthermore, Siew-Lee and White [5] proposed a cooperative resource bargaining framework among the users and the wireless network operator, dividing the problem in two subcategories: the *symmetric problem* (where all players have the same bargaining power) and the *asymmetric problem* (where players may submit bids to the arbitrator as a form of tactics to alter the final result and benefit from the resource distribution).

Moreover, in [6], the authors use Nash bargaining in order to solve the bandwidth allocation problem in modern satellite systems. Comparing this scheme to a simple noncooperative game (NG) and an NG with pricing, they prove that Nash bargaining results in a fairer outcome. As it can be seen in Figure 13.1, for a range of available bandwidth units, Jain's fairness index, defined as $\left(\sum_{i=1}^{N} x_i \right)^2 / \left(N \sum_{i=1}^{N} x_i^2 \right)$ (x_i being a bandwidth unit and N the number of all units), is higher for the NBS than for the NG and the NG with pricing (NGP).

13.2.3 Coalition Games

A cooperative game is a game in which the players have the option of planning as a group in advance their actions. By forming coalitions, players can strengthen their position in the game. Let $N = \{1, 2, \ldots, n\}$ be a set of players. A coalition S represents an agreement between the players to act as a single entity and is defined as a subset of N. The coalition form of an n-player game is defined by the pair (N, v), where v is a characteristic function of the game [7,8]. The quantity $v(S)$ is a real number for each coalition $S \subset N$, which may be considered as the value or worth of coalition S when its members act together as a unit.

The core is generally used to obtain stability region for the solution of an n-person cooperative game. It is defined as the set of all feasible outcomes that no player or coalition can improve upon

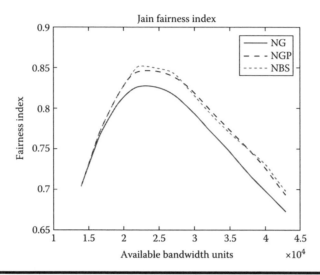

Figure 13.1 Bandwidth allocation problem using Nash bargaining.

by acting for themselves. In order to find each player's power in the coalition, several methods are proposed in the literature, as Shapley value, nucleolus, and *t*-value. The objective is to allocate the resources so that the total utility of the coalition is maximized. In wireless networks, the formation of coalitions involves the sharing of certain resources; however, as the costs of such resource sharing outweigh the benefits perceived by the nodes, users are less likely to participate, compromising overall network goals.

In [9], bandwidth allocation in 4G networks is modeled as a coalitional game corresponding to the well-known bankruptcy game. Specifically, a coalition among the different wireless access networks is formed to offer bandwidth to a new connection. The stability of the allocation is analyzed by using the concept of the core, and the amount of allocated bandwidth to a connection in each network is obtained by using the Shapley value.

13.3 Introduction to Cooperation Incentives

Incentive mechanisms intend to provide a framework that urges players to cooperate for the best interest of all participants. In other words, they provide a *motive* so that each individual prefers to work along with others, sometimes sacrificing their own resources and sometimes benefiting from the resources of others. Two types of incentive mechanisms are distinguished:

- Credit-based (or pricing-based) systems
- Reputation-based systems

We may distinguish two types of uncooperative nodes:

- *Faulty/malicious nodes*, which are either faulty and therefore cannot follow a protocol, or are intentionally malicious and try to attack the system.
- *Selfish nodes*, which are economically rational and their objective is to maximize their own welfare, which is defined as the benefit minus the cost of their actions.

For example, in some occasions, some nodes may refuse to forward packets in order to conserve their limited resources (e.g., energy), resulting in traffic disruption. Nodes exhibiting such behavior are termed selfish. The basic idea for node punishment is that nodes should be rewarded or penalized based on their behavior. Nodes that offer resources should be aided. On the other hand, selfish nodes should be gradually isolated from the network. This is bound to happen if nearby nodes refuse to forward packets deriving from the selfish ones. In addition, some systems may punish misbehaving nodes by isolating them from the network for a certain period of time in order to provide an incentive for users to cooperate.

A significant problem arising when dealing with node behavior issues is that sometimes, due to packet collisions and interference, cooperative nodes will be perceived as being selfish, which will trigger a retaliation situation. Such poor judgment may lead nodes to stop cooperating and thus degrade the overall network performance. In other words, disadvantaged nodes that are inherently selfish due to their precarious energy conditions should not be excluded from the network using the same basis as for malicious nodes. Furthermore, users on the outskirts of a network are found at a disadvantage unrelated to their willingness to participate. Those users will not have as much traffic routed through them due to their location. They will thus earn significantly less than a centralized node and be penalized for it resulting in low QoS.

In several cases, incentive mechanisms are required not only to motivate the user to cooperate but also to discourage him from cheating. For example, malicious users may refuse to pay. To this end, disincentives against cheating may also have to be considered.

13.4 Credit-Based Systems

13.4.1 General Description

A similar set of mechanisms called credit-based systems is used widely in ad hoc networks. The basic idea of these systems is to use notional credit, monetary or otherwise, to pay off users for forwarding packets coming from other users. This acts as a compensation for transmission and battery costs. These credits can then be used to forward their own packets through other users, resulting in an incentive to act as relay points, especially where there is the greatest excess demand for traffic since this is when they earn the most. Users who do not cooperate will not be able to use the network themselves, having not earned any credits.

Under the general token mechanism, a user's token counter is increased when it forwards and decreased proportionally to the number of hops it needs when it sends. This inevitably means that a user needs to forward more than he sends and also limit the amount of information that any user can send at any given time, dependent on their store of tokens.

13.4.2 Approaches

One of the most well-known credit-based schemes, called Sprite, is presented in [10]. The authors use the idea of credit to solve the problem of routing in ad hoc networks of self-interested nodes. The proposed system (Figure 13.2) consists of a credit clearance service (CCS) and a number of mobile nodes. In this system, the sender of the message is charging some credits, which will be used to cover the costs for packet forwarding by intermediate nodes. In order to earn credits, the nodes must transmit the CCS receipts of forwarded messages, so to cooperate with each other. To prevent cheating behavior, the CCS charges the sender with a higher amount than that due to relaying nodes. Moreover, the security architecture is based on public-key cryptography. Although

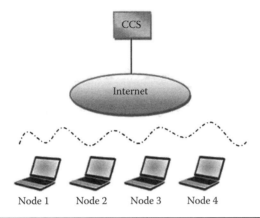

Figure 13.2   Architecture of Sprite.

Sprite focuses on combating cheating behavior, there are several potential drawbacks of this specific solution in terms of its overhead, security, and topology requirements such as that it does not prevent active attacks on the system (e.g., Denial of Service attacks).

Another well-known approach of credit-based schemes is presented in [11]. Buttyan and Hubaux introduce a virtual currency called nuglet, which is used in order to motivate the nodes of ad hoc networks to cooperate. Specifically, they propose two different models for charging for the packet forwarding service: the *Packet Purse Model* and the *Packet Trade Model* (Figure 13.3).

In the first model, the sender pays for the forwarding service and loads the packet with a number of nuglets sufficient to reach the destination. The intermediate nodes earn some nuglets from the packet when they forward it. If the packet does not have enough nuglets to be forwarded, then it is

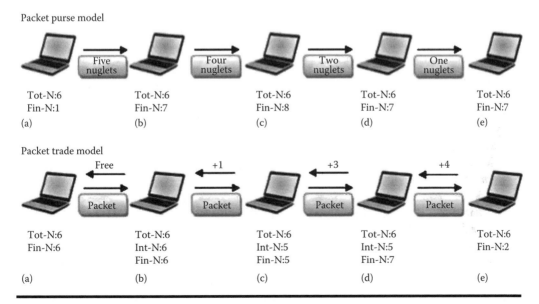

Figure 13.3   Charging with nuglets.

dropped. For example, in the upper part of Figure 13.3, node A wants to send a packet to node E. It is assumed that each node has a total number of six nuglets (Tot-N). In order node A to send the packet, he loads five nuglets in the packet and sends it to node B, which takes out one nuglet and forwards it to the next node. After node B forwards the packet, his final nuglets B increase by one nuglet and become seven nuglets (Fin-N). Similarly, node C acquires two nuglets from the packet and forwards it to node D, which takes out one nuglet and leaves one nuglet for the destination node E. Thus, all the nodes earn nuglets except from the sender of the packet, which results with a lower number of nuglets.

In packet trade model, it is the destination of a packet that pays for it and not the sender. In this model, each intermediate node buys the packet from the previous node for some nuglets and sells it to the next one in order to earn nuglets. In this way, the total cost of forwarding the packet is covered by the destination of the packet. In the lower part of Figure 13.3, node A wants to send a packet to node E. A sends the packet to node B for free, whereas node B sells the packet to node C for one nuglet. Thus, node C gives one nuglet to buy the packet and stays with five nuglets (Int-N). In order to forward the packet, node C sells it for more nuglets and results in a higher stock of eight nuglets. Similarly, node D sells the packet for even more nuglets, and thus, the destination node results with a lower number of nuglets.

In both models, the authors assume that each node has a tamper resistant hardware module so that their behavior cannot be modified by their users. The security infrastructure is based on public-key cryptography, with additional symmetric-key sessions between each communicating pair of neighbors.

Besides stimulating cooperation, the first model can also deter nodes from sending useless data and overloading the network. The basic problem of this model that is solved in the second one is that it might be difficult to predict the number of nuglets that are required to reach a given destination. Another advantage of the Packet Trade Model is that it can also be applied in case of multicast packets whereas a serious problem is that this approach for charging does not directly deter nodes from overloading the network.

As Buttyan and Hubaux refer in [12], in order to combine the advantages of the two proposed models, a *hybrid model* can be created. Specifically, in this hybrid scheme, the sender loads the packet with some nuglets before sending it and when the packet runs out of nuglets, the destination buys it. Finally, in [12], the reader can also find different extensions to the basic Packet Purse Model such as the Packet Purse Model with Fixed Per Hop Charges and the Packet Purse Model with Auctions.

In [13], the authors propose another incentive protocol against selfishness, called SIP. Specifically, they adopt a credit-based payment system, which charges or remunerates nodes for the service they receive or provide. Similar to the hybrid model proposed by Buttyan et al., in this system both the destination and the sender pay for packet forwarding. Moreover, the payment proportion between them is adjustable and can be negotiated during the session initialization phase.

In their next work, Buttyan and Hubaux [14] propose a scheme based on credit counter, called nuglet counter. This counter is decreased when the node sends its own packet and increased when the node forwards a packet. Thus, in order for a node to be able to send its own packet, it has to forward the packets of the other nodes. Moreover, the authors simulated four rules for a node to determine when to forward others' packets and when to send its own packets.

A different approach of credit-based mechanisms is presented in [15], where Ileri et al. analyze the network-geometric dependence of incentivized cooperation in wireless ad hoc networks with energy-efficient utility function. They design a pricing-based joint user-and-network-centric incentive mechanism that induces forwarding among selfish users by compensating the real and

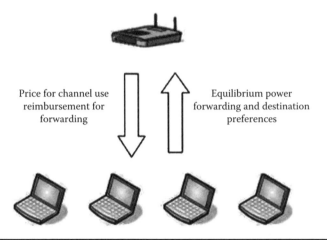

Figure 13.4 User-and-network-centric incentive mechanism.

opportunity costs of the forwarders. In their approach (Figure 13.4), the network announces prices for channel utilization, as well as reimbursements for forwarders, in order to maximize its net revenue. The users in response adjust their transmit powers as well as forwarding and destination preferences in order to maximize their net utilities. This interaction between network and users evolves until the network revenue is maximized.

A similar scheme is that of Shastry and Adve [16], in which they motivate the nodes of an ad hoc network to participate in cooperative diversity transmissions. More specifically, in their pricing scheme, the access point declares a set of prices, which are chosen such that the access point's revenue is maximized. Given the access point price and the source's reimbursement price for forwarding, the interaction between nodes is modeled as a strategic game that results in a Nash equilibrium. Thus, after the intermediate nodes have chosen their transmit power level and decided on whether to cooperate, the source chooses a reimbursement price that maximizes its utility. As illustrated in Figure 13.5, the basic difference from the proposed scheme in [15] is that here the source chooses the reimbursement price and not the access point as before.

In [17], Liu and Krishnamachari analyze a price-based reliable routing game in a wireless network of selfish users. In this game, each node is characterized by a probability of reliably forwarding a packet, and each link by a cost of transmission. In the specific pricing mechanism, the destination node pays some amount of virtual money to the source for each successfully delivered packet, and then the source offers some kind of payment to every node on its path in order to motivate them. Thus, given the payment from the source, each node has an incentive to participate in this routing game if it receives more payment than it pays for each transmission. The authors prove that for the specified routing problem, a polynomial-time solution exists to find efficient Nash equilibrium.

Moreover, Chen et al. in [18] focus on providing incentive for packet forwarding in a two-hop hot spot network. Specifically, they also adopt the "pay for service" incentive model where the mobile clients pay the relaying nodes to forward their packets. Although, this time, the objective is to determine a "fair" pricing for the packet forwarding service in the network. In order to investigate all the aspects of the game, the authors model the system as a market where the pricing for packet forwarding is determined by demand and supply. The relaying nodes compete for clients' traffic and clients can choose a relaying node who can offer a better price similar to a multiple-buyer multiple-seller market. The market structure in this network depends on the number of relaying

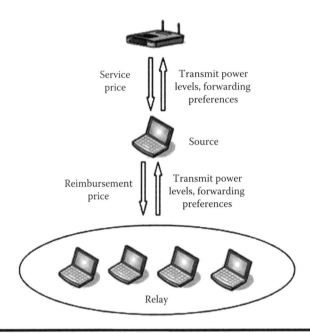

Figure 13.5 **Incentive mechanism for cooperative diversity transmissions.**

nodes, the communication among them, and the reachability of the clients to the relaying nodes. Therefore, the authors classify the network into four different scenarios (such as monopoly market and simple competitive market) and propose different pricing mechanisms for each one.

Another credit-based incentive mechanism is proposed in [19] for multi-hop cellular networks. Jakobsson et al. propose a micro-payment scheme that encourages collaboration in packet forwarding. Specifically, they use the concept of lottery tickets to reward the forwarding nodes in a probabilistic way, whereas at the same time they introduce appropriate mechanisms for detecting and punishing various forms of abuse. Similar to this work, in [20], the authors investigate a pricing game in multi-hop relay networks where nodes price their services and route their traffic selfishly and strategically. Although the analysis of price competition has been presented in prior work as it has been referred, this time the authors introduce a nonlinear pricing function to the game.

Zhang [21] proposes an incentive mechanism called bandwidth exchange, where a node can delegate a portion of its bandwidth to another node in exchange for relay cooperation. As it has been referred, Buttayan and Hubaux [14], as well as Zhong et al. [10], have developed pricing-based protocols where the amount charged per packet is determined exogenously and is the same for each node in the network. The drawback with the above approaches is the assumption of a simplified channel model—the energy required to forward a packet is assumed to be constant regardless of transmission distance. Stimulation mechanisms that take into account the fading channel have been developed during the last years. Results show that in this case cooperation is highly dependent on network geometry and configuration.

Finally, a different approach of credit-based schemes appears in [22], where Chen et al. propose an auction-based incentive scheme, called iPass, in order to enable cooperative packet forwarding in ad hoc networks. The basic idea of this scheme is that each intermediate node operates as a market; the users of the network put bids for their packets and are charged accordingly, when the packets are forwarded. The resource allocation mechanism in iPass is based on the generalized Vickrey auction with reserve pricing, which has several important properties such as truthful bidding and social welfare maximization in single market. Similar to this work, Denim and Comaniciu in [23] present another auction-based protocol based on incentivizing cooperation by balancing two different metrics: the residual energy and the current currency levels of the nodes in the network.

A hybrid system that combines principles of credit-based and reputation-based system appears in [24], called ARM. The basic idea of incentive scheme of ARM is to intelligently integrate reputation system with pricing-based model in order to avoid selfish nodes. Specifically, in this system, every node pays a price for the packet forwarding depending on his reputation. That means that a node with higher reputation value will pay less than a node with low reputation value for the packet forwarding. Thus, ARM can effectively prevent some selfish nodes from manipulating their reputation value just above some threshold value.

As it has been referred, pricing-based mechanisms can improve the outcome of the game by giving incentives for distributed users to cooperate for resource usage. Specifically, as the individual user has no incentive to cooperate with the other users in the system and may use inefficiently the other users' resources, the outcome of the game might not be optimal from the system point of view. An efficient pricing mechanism can make distributed decisions compatible with the system efficiency obtained by centralized control. A pricing policy is called incentive compatible if pricing enforces a Nash equilibrium that achieves the system optimum. In [25,26], a policy is proposed by usage-based pricing in which the price a user pays for using the resources is proportional to the amount of resources consumed by the user. Specifically, a pricing scheme is used in order to use more efficiently the network resources, such as power. In [27], a similar pricing scheme is used for power control in modern satellite systems. The authors conclude that the outcome of the pricing game is more optimal compared to the simple NG.

13.4.3 Advantages of Credit-Based Systems

The basic advantage of credit-based systems is that they succeed in a large scale to stimulate cooperation in networks with selfish nodes. Moreover, credits are useful when an action and its reward are not simultaneous. This is true for multi-hop wireless networks: the action is packet forwarding and the benefit is being able to send their own packet.

13.4.4 Disadvantages of Credit-Based Systems

Some systems propose using real money as credit, either directly or indirectly (to buy virtual credit). One problem that must be kept in mind is that the introduction of monetary rewards in the system may act not only as an incentive for collaboration but also as an incentive for cheating: a cheater may attempt to corrupt the routing tables (both his own and those of others) in order to gain rewards. Such malicious behavior can be avoided by using a set of protocols that allow the operator to collect information based on which it can decide which accounts should be charged and which accounts should be credited [28]. This leads to additional overhead making credit-based schemes complicated and difficult to implement.

Another problem with such systems is that it is very difficult to charge users fairly, without introducing additional complexity. The mechanisms used to implement these incentives take up resources themselves. If the number of cheating nodes is not high, then the benefit derived from the application of the incentive mechanisms may be outweighed by the resources they consume.

Credit-based systems suffer from the location privilege problem that means that nodes in different locations of the network will have different chances to earn credits. Specifically, nodes that are at the boundaries of the network will have less chance to be rewarded, something that results in unfair distribution of the credits.

Moreover, uneven distribution of wealth among a group of users may discourage rich users to forward for others when the rewards are of negligible value to the rich users. The difference between the rich and the poor nodes in the network may need a rebalancing mechanism to make the pricing scheme work properly.

Also, the most credit-based mechanisms require trusted third parties to administer remuneration of cooperative nodes. As it has been seen, tamperproof hardware like secure operating systems or smart cards has been suggested to enforce the fair exchange of the remuneration.

Concerning the micropayments, it is obvious that they are implemented as end-to-end schemes, requiring the exchange of information between all the nodes in the path from source to destination. Thus, for these schemes, the computation and communication overhead is a major concern.

When using tokens, there is the question of how the balance of tokens can be maintained for users [29]. The average token level within the system needs to be kept at a reasonable level in order for incentives to work properly. If it grows too high, everyone will possess lots of tokens and no longer have an incentive to cooperate; on the other hand, if there is not enough credit within the system, then hardly anyone will be able to transmit. However, if an individual's token level is regularly reset in order to maintain a certain token level, then there is no incentive to cooperate in the long term: nodes will stop cooperating once enough credit has been earned to complete their transmission, since excess credit will be lost anyway.

13.5 Reputation-Based Systems

13.5.1 General Description

Reputation is defined as the amount of trust inspired by a particular member (node) of a network community in a specific setting or domain of interest. Members that have good reputation, because they helpfully contribute to the community life, are able to use the network resources, while nodes with a bad reputation, because they usually refuse to cooperate, are gradually excluded from the community.

Reputation management systems can be categorized as centralized and decentralized and the reputation is estimated either in a central hub station or at each node individually [30]. The reputation-based mechanism consists of three main parts:

1. The first part of the reputation-based scheme is the *information gathering or collection of evidence*, where a reputation is created according to the observations by the nodes or recommended information by other nodes. The information may be specific or general based on the knowledge of its performance for a specific functionality or to all functionalities [30]. Moreover, the gathered information can be characterized as objective if the knowledge of a node about the reputation comes from the node's direct experience and subjective if the gathered information is indirect.

2. The second part of the reputation-based scheme is the *decision process*, where the nodes estimate the reputation of a node, which wish to communicate and make a decision whether to cooperate or not. In [31], a detailed survey of trust and reputation systems for online system provision is presented. For example, there are voting schemes, schemes that employ the average of all the rates by the nodes, probabilistic schemes [32], and flow models [33].

3. The last part of the reputation-based mechanism is the *cooperation evaluation*, where the node, after the interaction process, provides a score for the node that has been involved.

Reputation is a dynamic quantity that is formed and updated continuously through direct observations and through information provided by other members of the community. In [34], reputation is classified in three types similar to the above characterization of the three parts of reputation mechanism:

- **Subjective reputation** is calculated directly from a subject's observation. A subjective reputation is calculated using a weighted mean of the observations' rating factors, giving more relevance to past observations. The reason why more relevance is given to past observations is that a sporadic misbehavior in recent observations should have a minimal influence on the evaluation of the final reputation value: as a result, it is possible to avoid false detections due to link breaks and to take into account the possibility of a localized misbehavior caused by disadvantaged nodes.
- **Indirect reputation** adds the information provided by other members of the community to the final value given to the reputation of a subject.
- **Functional reputation** allows for the possibility to calculate a global value of a subject's reputation that takes into account different observation/evaluation criteria.

Each type of reputation is obtained as a combination of different observations made by a subject over another subject with respect to a predefined functionality. Furthermore, the above types of reputation information can be combined, as shown in Figure 13.6. When a node detects uncooperative behavior, it disseminates this observation to other nodes, which take action to avoid being affected by the node in question by changing traffic routes.

The following two components may allow a node to observe and evaluate neighboring nodes, reflecting their cooperative behavior [34]:

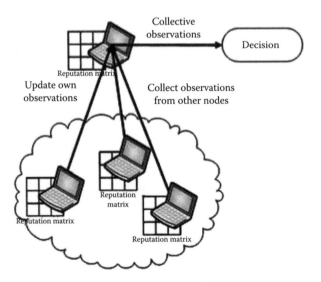

Figure 13.6 Collective observations in reputation-based systems.

- The *Reputation Table* is defined as a data structure stored in each network entity. Each row of the table includes the reputation data pertaining to a node.
- The *Watchdog* mechanism is used to detect misbehaving nodes.

Additional to the watchdog mechanism, a path-rater scheme is proposed in [35], which is in charge of defining the optimal route by avoiding the misbehaving nodes.

The reputation systems find application in self-organized networks, such as 4G, where there are various heterogeneous components [36]. Cooperation is the major issue in such self-organized communication systems because users/nodes are concerned mostly for their own profits. They usually show selfish behavior, which is catastrophic for the connectivity and the whole throughput of the wireless mesh network. On the other hand, it is not very clear how the altruistic behavior [37] can help the performance of the whole network. Generally, reputation in 4G networks is referred to the behavior of each node to be willing to forward the packets or not.

13.5.2 Approaches

The general idea of using collected information for reputation has also been extensively studied in economics and more specifically using game-theory approaches. The most accepted game-theoretic framework that is used to analyze reputation is that of repeated games [38]. In addition, some players are not certain for the payoff structures of their opponents. The key terms that are necessary to formulate a game-theoretic framework are the repeatability and the uncertainty [38]. It does not make any sense that a game for reputation is based on decisions without repetitions; moreover, the uncertainty of payoffs is necessary in order to model the behavior of nodes. The most important repeated games with limited information are Bayesian Games or games with incomplete information [39]. Each player is aware of his own type at the beginning of the game but the types of other players remain unknown to him throughout the game. Repeated games [38] differ greatly from one-shot games that encounter in the sense that they can allow for a whole range of equilibria, which are not normally found. There are two repeated interactions that can be used to model reputation-based mechanisms: the first one is with all the players participating in all the stages of the game (long-run) and the second one is one player playing long-run while all his opponents play only once (short-run). Prisoners' dilemma is a classic game in which the theorems of the repeated game can be applied. A great number of equilibria are possible in repeated games due to the possibility that the players can develop the ability to punish in the future stages any past misbehaviors of their opponents. Another way to model reputation-based systems is the inclusion of uncertainty. It is shown [38] that the long-run player is able to develop reputation for being a certain type and thus makes his intuition reliable that he may always pick the equilibrium he prefers the most.

Finally, a new game-theoretical model for reputation in online auctions is introduced [40]. It is shown that mixture models can be used as a proper framework for integrating information from various sources into the model, that the number of successful transactions increases significantly, and that proposed game is robust for changes in the behavior of the participants [40].

The most common approach to provide an incentive for cooperation is to develop a strategy such that the cooperation of a node is measured and if the fraction of packets it has dropped is above a threshold, it is considered selfish and is disconnected for a given amount of time. This approach is known as a ***Trigger Strategy*** [41]. We denote by $\tilde{p}_{i,S}^{(k)}$ the dropping probability player i should use at time slot k according to strategy S. For a pair of players i and $-i$, an n-step

Trigger Strategy for time slot k is defined as

$$\tilde{p}_{i,nT}^{(0)} = 0$$

$$\tilde{p}_{i,nT}^{(k)} = \begin{cases} 0 & \hat{p}_{-i}^{(j)} \leq T, \quad j = k-n, \ldots, k-1 \\ 1 & \text{otherwise} \end{cases} \tag{13.2}$$

where

   T is the threshold
   $\hat{p}_{-i}^{(k)}$ is the perceived dropping probability of player i for player $-i$
   n is the starting time slot

A second alternative is to use a ***Tit For Tat*** (**TFT**) strategy [42]. TFT is a highly effective strategy in game theory for the iterated prisoner's dilemma and is based on the English saying meaning "equivalent retaliation" (*tit for tat*). A player using this strategy will initially cooperate and then act regarding the opponent's previous action: if the opponent previously was cooperative then the player will be cooperative as well, otherwise he will not cooperate. As before, TFT can be modeled for wireless networks as

$$\tilde{p}_{i,\text{TFT}}^{(0)} = 0$$

$$\tilde{p}_{i,\text{TFT}}^{(k)} = \hat{p}_{-i}^{(k-1)}, \quad k \geq 1 \tag{13.3}$$

The problem with TFT is that it does not take into account the fact that it is not always possible to determine whether a packet was relayed or not due to network collisions. A proposed method to deal with this is the employment of a generosity factor g that allows cooperation to be restored. Such a strategy is known as ***Generous TFT*** (**GTFT**) [41]. According to this strategy, a player adopts an aspect of cooperation as long as the other side also does. If the opponent starts cheating, the player stops cooperating as well. If the cheater reverts, however, the first player might initiate renewed cooperation. GTFT can be modeled for wireless networks as

$$\tilde{p}_{i,\text{GTFT}}^{(0)} = 0$$

$$\tilde{p}_{i,\text{GTFT}}^{(k)} = \max\left[\hat{p}_{-i}^{(k-1)}, 0\right], \quad k \geq 1 \tag{13.4}$$

Another modification of TFT, known as ***Contrite Tit For Tat*** (**CTFT**) [43], has been proposed based on the idea of contriteness: a player that made a mistake and unintentionally defected should exercise contrition and try to correct the error instead of going into a retaliation situation. An extension of this game, called DARWIN (Distributed and Adaptive Reputation mechanism for Wireless ad hoc Networks), has been proposed in [43]. The authors propose a reputation strategy that does not depend on a perfect estimation of the possibility of rejected packets to achieve full cooperation and that is also more robust. More specifically, the dropping probability is in this case

$$\tilde{p}_{i,\text{DARWIN}}^{(k)} = \left[g \cdot \left(q_{-i}^{(k-1)} - q_i^{(k-1)}\right)\right]_0^1, \quad k \geq 0 \tag{13.5}$$

where $q_i^{(k)}$ measures the deviation between the estimated dropping probability and the dropping probability under DARWIN as

$$\hat{q}_{-i}^{(k)} = \begin{cases} \left[\hat{p}_{-i}^{(k)} - \tilde{p}_{i,\text{DARWIN}}^{(k)}\right]_0^1, & k \geq 0 \\ 0, & k = -1 \end{cases} \tag{13.6}$$

and g is equal to $(1 + 2p_e)/2p_e$, p_e being the probability that a packet that has been forwarded was not overheard by the originating node. The following function has also been utilized:

$$[x]_0^1 = \begin{cases} 1 & \text{if } x \geq 1 \\ x & \text{if } 0 < x < 1 \\ 1 & \text{if } x \leq 0 \end{cases} \tag{13.7}$$

Buchegger and Le Boudec proposed and evaluated in [44] their CONFIDANT (Cooperation of Nodes, Fairness In Dynamic Ad hoc NeTworks) protocol, which detects and isolates misbehaving nodes. In this approach, nodes have a monitor for observations, reputation records for first-hand and trusted second-hand observations, trust records to control trust given to received warnings, and a path manager for nodes to adapt their behavior according to reputation.

Another protocol for self-organized networks has been presented in [45], where users accept second-hand information only if this does not differ too much from the values of reputation that they already know. The authors analyze a simplified model for the special case of two users, one being liar and the other honest. In [45], it is shown that the reputation system displays a phase transition. That means that there is a threshold rate of misbehaving (saying lies) below which the value of reputation of the honest user remains unaffected. For greater values, the lying will have a detrimental effect and will corrupt the reputation-based system.

In [46], a Secure and Objective Reputation-based Incentive (SORI) framework for ad hoc networks is proposed, where the packet forwarding is encouraged and the selfish behavior is punished. The special features of SORI are that the reputation is quantified by objective measures, the propagation of reputation is computationally efficiently secured by one-way-hash-chain-based authentication scheme, and the reputation of each node is transmitted only to its neighbors so that the communication overhead is reduced.

While most previous efforts are based on exchanging reputation information about the other nodes, in [47], an Observation-based Cooperation Enforcement in Ad hoc Networks (OCEAN) scheme is proposed, where they use only the first-hand observations. It is shown that in many scenarios, OCEAN performs as well as or sometimes better than schemes that require second-hand reputation changes. OCEAN focuses on the packet forwarding and encounters misbehaving at the routing layer. The two types of routing misbehavior are considered the misleading and the selfish.

Finally, in [48], a novel dynamic reputation-based incentive for cooperative relays in heterogeneous networks is proposed. Taking into consideration reputation calculations for individual contribution of intermediate relay nodes in a cooperative relay service, within the proposed framework, incentive in terms of additional throughput is dynamically assigned. The dynamic incentive mechanism can be achieved [49] by adding a bonus parameter in the time slot allocation of the proportional fair scheduler. The incentive parameter consists of two components: a fixed-bonus variable b_{Fi} and a dynamic bonus variable b_{Di} that takes into account dynamically the user's contribution. The allocation of the time slots is expressed as

$$k = \max \left[\frac{b_i \times r_i(t)}{m(t)} \right] \quad \forall i \tag{13.8}$$

$$\text{where} \quad b_i = \begin{cases} b_{Fi} + b_{Di}, & \text{relaynodes} \\ 1, & \text{users} \end{cases} \tag{13.9}$$

where
  k is the time slot
  $r_i(t)$ is the maximum ratio of the instantaneous data rate
  $m(t)$ is the average of the former allocated rate

In this model, each end user dynamically calculates the reputation of every relay that has contributed to the traffic transmission.

13.5.3 Advantages of Reputation-Based Systems

The main idea of reputation-based systems is actually based on the essence of human society: one entity distributes a piece of information to the other members of the community, enabling this way a collective estimation of which members are worth helping and which ones are not. One of the main advantages of this kind of incentives is that it relies on observations from multiple sources, instead on the judgment of a single entity; it is therefore a rather subjective means of evaluation, relatively resistant to the diffusion of false information from a small number of lying nodes. Reputation-based systems may also be used for advanced protection to help organizations accurately detect and block all types of threats to their messaging, web, and network environments.

13.5.4 Disadvantages of Reputation-Based Systems

The major problems of reputation-based systems are the malicious active attacks such as denial of service, attacks to network functionalities, and a general attack to the reputation-based system (liar nodes) or random failures.

Reputation-based systems are also subject to some significant problems. The first problem is that they take up considerable resources due to the constant transmission of observation data, which serves no purpose other than to monitor node behavior. This spends valuable bandwidth and battery power that could be used to send actual data. Trust management systems also suffer from vulnerabilities due to exchange of inaccurate information: nodes may falsely accuse other nodes of misbehaving or collude with each other to cheat other users on the network. This forms one of the most important problems of reputation-based systems since colliding nodes may intensively disseminate false information in order to increase their benefits. This situation raises demands for authentication techniques of the accusing nodes, which further increases the complexity of the system.

Summing up, the state of the art of reputation-based systems is the fully distributed reputation-based systems. The distributed nature of the system leads to potentially very complex behavior [37]. All these systems are vulnerable to liars since some users might spread false information.

13.6 Conclusion

This chapter addressed the issue of cooperation incentives in 4G networks, featuring both reputation-based and pricing-based schemes. The main principles and key characteristics of both scheme types were presented. Furthermore, the authors have collected a variety of approaches that ensure both efficiency and establishment of cooperation in dynamic environments, featuring also benefits and drawbacks derived from the use of either type of incentives.

As it has been stated, both types of schemes suffer from certain weaknesses that may concern either the degree of cooperation reached or the fairness achieved. The development of more suitable and fair schemes that are also able to adapt to the network's status will offer significant value to the revolution of 4G. Hybrid schemes that combine both reputation and credit aspects are also of particular interest and, according to the authors' opinion, should be further investigated. Lastly, apart from cooperation incentives, disincentives against cheating also need to be considered.

Acknowledgments

This work is partially supported by the Irakleitos PhD programme in NTUA.

References

1. S. Frattasi, H. Fathi, and F.H.P. Fitzek, and R. Prasad, 4G: A user-centric system, *Mobile e-conference, Electronic Conference*, 2004.
2. S. Frattasi, B. Can, F. Fitzek, and R. Prasad, Cooperative services for 4G, *Proceedings of the 14th IST Mobile and Wireless Communications*, Dresden, Germany, 2005.
3. F.H.P. Fitzek and M. Katz, *Cooperation in Wireless Networks: Principles and Applications—Real Egoistic Behavior is to Cooperate!*, 2006, Springer, Dordrecht, the Netherlands.
4. H. Yaiche, R. Mazumdar, and C. Rosenberg, A game theoretic framework for bandwidth allocation and pricing in broadband networks, *IEEE/ACM Transactions on Networking*, 8, 5, 667–678, 2000.
5. H. Siew-Lee and L. White, Cooperative resource allocation games in shared networks: Symmetric and asymmetric fair bargaining models, *IEEE Transactions on Wireless Communications*, 7, 11, Part 1, 4166–4175, November 2008.
6. S. Vassaki, A. Panagopoulos, and P. Constantinou, Game-theoretic approach of fair bandwidth allocation in DVB-RCS networks, *International Workshop on Satellite and Space Communications (IWSSC 2009)*, Juscany, Italy, 321–325, 2009.
7. T. S. Ferguson, *Game Theory Text*, Electronic Text, Mathematics Department, UCLA. Available at: http://www.math.ucla.edu/~tom/Game_Theory/contents.html
8. W. Saad, Z. Han, M. Debbah, A. Hjrungnes, and T. Basar, Coalitional game theory for communication networks: A tutorial, *IEEE Signal Processing Magazine, Special Issue on Game Theory*, 26, 5, 77–97, 2009.
9. D. Niyato and E. Hossain, A cooperative game framework for bandwidth allocation in 4G heterogeneous wireless networks, *Proceedings of IEEE ICC'06*, Istanbul, Turkey, June 2006.
10. S. Zhong, J. Chen, and Y. R. Yang, Sprite: A simple, cheat-proof, credit-based system for mobile ad-hoc networks, *Proceeding of IEEE INFOCOM 2003*, San Francisco, CA, 2003.
11. L. Buttyan and J. P. Hubaux, Enforcing service availability in mobile ad-hoc WANs, *IEEE/ACM Workshop on Mobile Ad Hoc Networking and Computing (MobiHOC)*, Boston, MA, August 2000.

12. L. Buttyan and J.-P. Hubaux. Nuglets: A virtual currency to stimulate cooperation in selforganized ad hoc networks, Technical Report DSC/2001/001, Swiss Federal Institute of Technology, Lausanne, Switzerland, 2001.

13. Y. Zhang, W. Lou, and Y. Fang. SIP: A secure incentive protocol against selfishness in mobile ad hoc networks, *IEEE WCNC*, Atlanta, GA, March 2004.

14. L. Buttyan and J. P. Hubaux, Stimulating cooperation in self-organizing mobile ad hoc networks, *ACM/Kluwer MONET*, 8, 579–592, October 2003.

15. O. Ileri, M. Siun-Chuon, and N. B. Mandayam, Pricing for enabling forwarding in self-configuring ad hoc networks, *IEEE Journal on Selected Areas in Communications (JSAC)*, 23, 1, 151–162, January 2005.

16. N. Shastry and R.S. Adve, Stimulating cooperative diversity in wireless ad hoc networks through pricing, *IEEE International Conference on Communications (ICC '06)*, Istanbul, Turkey, 2006.

17. H. Liu and B. Krishnamachari, A price-based reliable routing game in wireless networks, *Proceedings of the First Workshop on Game Theory for Networks, GAMENETS-06*, Pisa, Italy, 2006.

18. K. Chen, Z. Yang, C. Wagener, and K. Nahrstedt, Market models and pricing mechanisms in a multihop wireless hotspot network, *Proceedings of ACM MobiQuitous Conference*, San Diego, CA, July 2005.

19. M. Jakobsson, J. P. Hubaux, and L. Buttyan, A micropayment scheme encouraging collaboration in multi-hop cellular networks, *Proceedings of Financial Crypto 2003*, La Guadeloupe, January 2003.

20. Y. Xi, E. Yeh, Pricing, competition and routing for selfish and strategic nodes in multi-hop relay networks, *INFOCOM 2008*, Phoenix, AZ, 2008.

21. D. Zhang, O. Ileri, and N. Mandayam, Bandwidth exchange as an incentive for relaying, *42nd Annual Conference on Information Sciences and Systems (CISS 2008)*, Princeton, NJ, 2008.

22. K. Chen and K Nahrstedt, iPass: An incentive compatible auction scheme to enable packet forwarding service in MANET, *Proceedings of IEEE International Conference on Distributed Computing Systems (ICDCS'04)*, Tokyo, Japan, pp. 534–542, March 23–26, 2004.

23. C. Demir and C. Comaniciu, An auction based AODV protocol for mobile ad hoc networks with selfish nodes, *IEEE International Conference on Communications (ICC '07)*, Glasgow, Scotland, June 24–28, 2007.

24. Z. Li and H. Shen, Analysis the cooperation strategies in mobile ad hoc networks, *5th IEEE International Conference on Mobile Ad Hoc and Sensor Systems 2008 (MASS 2008)*, Atlanta, GA, 2008.

25. T. Alpcan, T. Basar, R. Srikant, and E. Altman, CDMA uplink power control as a noncooperative game, *Wireless Networks*, 8, 659–670, 2002.

26. C.U. Saraydar, N.B. Mandayam, and D. Goodman, Efficient power control via pricing in wireless data networks, *IEEE Transactions on Communications*, 50, 2, 291–303, 2002.

27. S. Vassaki, A. Panagopoulos, and P. Constantinou, A game-theoretic approach of power control schemes in DVB-RCS networks, *15th Ka and Broadband Communications, Navigation and Earth Observation Conference*, Cagliari, Italy, September 2009.

28. N.B. Salem, L. Buttyan, J.-P. Hubaux, and M. Jakobsson, A charging and rewarding scheme for packet forwarding in multi-hop cellular networks, *ACM MobiHoc 2003*, Annapolis, MD, 2003.

29. E. Huang, J. Crowcroft, and I. Wassell, Rethinking incentives for mobile ad hoc networks, *Proceedings of the ACM SIGCOMM Workshop on Practice and Theory of Incentives in Networked Systems*, Portland, OR, 2004.

30. N. Oualha and Y. Roudier, Cooperation incentive schemes, *EURECOM*, Research Report, RR-06-0176, September 26, 2006.

31. A. Jsang, R. Ismail, and C. Boyd, A survey of trust and reputation systems for online service provision, *Decision Support Systems*, 43, 2, 618–644, 2007.

32. A. Jsang and R. Ismail, The beta reputation system, *Proceedings of the 15th Bled Electronic Commerce Conference*, Bled, Slovenia, June 2002.

33. L. Page, S. Brin, R. Motwan, and T. Winograd, The pagerank citation ranking: Bringing order to the web, Technical Report, Stanford Digital Library Technologies Project, 1998.

34. P. Michiardi and R. Molva, Core: A cooperative reputation mechanism to enforce node cooperation in mobile ad hoc networks, *Communications and Multimedia Security Conference (CMS)*, Portoroz, Slovenia, 2002.

35. S. Marti, T. J. Ciuli, K. Lai, and M. Baker, Mitigating routing misbehavior in mobile ad hoc networks, *International Conference on Mobile Computing and Networking*, 255–265, 2000.

36. S. Buchegger, J. Mundinger, J.-Y. Le Boudec, Reputation systems for self-organized networks: Lessons learned, *IEEE Technology and Society Magazine*, 27, 1, 41–47, Spring 2008.

37. J. Mundinger and J.-Y. Le Boudec, Reputation in self-organised communication systems and beyond, *ACM International Conference Proceeding Series*; Vol. 200 archive, *Proceedings from the 2006 Workshop on Interdisciplinary Systems Approach in Performance Evaluation and Design of Computer and Communications Systems*, Pisa, Italy, No. 3, 2006.

38. K. Aberer and Z. Despotovic, On reputation in game theory application on online settings, Working Paper, 2004.

39. J. Harsanyi, Games with incomplete information played by Bayesian Players, *Management Science*, Parts I-III, 4, 159–182, 320–334, 486–502, 1967–1968.

40. P. Nurmi, Bayesian game theory in practice: A framework for online reputation systems, pp. 121, *Advanced International Conference on Telecommunications and International Conference on Internet and Web Applications and Services (AICT-ICIW'06)*, Guadeloupe, 2006.

41. F. Milan, J. J. Jaramillo, and R. Srikant, Achieving cooperation in multihop wireless networks of selfish nodes, *Workshop on Game Theory for Networks (GameNets 2006)*, Pisa, Italy, October 14, 2006.

42. R. Axelrod, The emergence of cooperation among egoists, *The American Political Science Review*, 75, 2, 306–318, June 1981.

43. J. J. Jaramillo and R. Srikant, DARWIN: Distributed and adaptive reputation mechanism for wireless ad-hoc networks, *13th ACM International Conference on Mobile Computing and Networking*, Montreal, Canada, 2007.

44. S. Buchegger and J.-Y. L. Boudec, Performance analysis of the CONFIDANT protocol: Cooperation of nodes—Fairness in dynamic ad-hoc networks, *IEEE/ACM Workshop on Mobile Ad Hoc Networking and Computing (MobiHOC)*, Lausanne, Switzerland, June 2002.

45. J. Mundinger and J.-Y. Le Boudec, Analysis of a robust reputation system for self-organised networks, *European Transactions on Communication*, 16, 5, 375–384, 2005.

46. Q. He, D. Wu, and P. Khosla, SORI: A secure and objective reputation-based incentive scheme for ad-hoc networks, *IEEE Wireless Communications and Neworking Conference (WCNC 2004)*, Atlanta, GA, 2004.

47. S. Bansal and Mary Baker, Observation-based cooperation enforcement in ad hoc networks, Stanford Technical Report, 2003.

48. J. Hwang, A. Shin, and H. Yoon, Dynamic reputation-based incentive mechanism considering heterogeneous networks, *International Workshop on Modeling Analysis and Simulation of Wireless and Mobile Systems, Proceedings of the 3rd ACM Workshop on Performance Monitoring and Measurement of Heterogeneous Wireless and Wired Networks*, Vancouver, British Columbia, Canada, pp. 137–144, 2008.

49. D. Skraparlis, V. Sakarellos, A.D. Panagopoulos, and J. D. Kanellopoulos, Outage performance analysis of cooperative diversity with MRC and SC in correlated lognormal channels, *EURASIP Journal on Wireless Communications and Networking*, 2009, Article ID 707839, 7 pages, 2009.

Chapter 14

Dynamics of Coalition Games for Cooperation in Wireless Networks

Zaheer Khan, Savo Glisic, and Luiz A. DaSilva

Contents

14.1 Introduction

Recent advances in technology have led to the development of distributed and self-configuring wireless network architectures. Cooperation among network nodes has been proposed as a technique for the efficient use of network resources. However, several recent works have shown that when self-interested wireless nodes are allowed to cooperate, in certain scenarios, they may prefer to cooperate with a selected set of other nodes rather than with the entire set of nodes [1–3]. For example, when wireless receivers in an interference channel are allowed to cooperate, by jointly decoding their received signals, cooperation among all nodes in the network results in maximum gains. But when the receivers cooperate using linear multiuser detectors, the cooperation among all the network nodes depends on the SNR regime and the detector employed [1]. In such scenarios, it is important to analyze cooperative interactions between small groups of nodes, as well as cooperative interactions among all network nodes. To model such cooperative scenarios, we consider the cooperative network as a coalition game, using concepts from coalition game theory.

In the first part of this chapter, we recast the cooperative wireless network as a coalition game to analyze cooperative interactions among wireless nodes. Wireless nodes form coalitions because via coalitions they can achieve better performance than working alone. Most of the literature on coalition games for wireless networks ignores the presence of externalities [1,3,4]. In this chapter, we include an example where coalition formation in wireless networks can create externalities. Nodes make their decisions to form coalitions independently, but, due to the presence of externalities, their choice may impact all the nodes in the network. In the second part of this chapter, we model wireless nodes forming coalitions in a self-configuring distributed wireless network as a dynamic process using game theory. The central problem is to analyze whether the dynamics results in stable coalitions. Using dynamic models of coalition games, we address two important questions: (1) How are the coalitions formed? (2) How do players arrive at equilibrium?

14.2 A Coalition Game-Theoretic Framework

Using the standard framework of coalition game theory, let $\mathbf{N}$ denote the set of players playing the coalition game, $\mathbf{N} = \{1, 2, 3, \ldots, N\}$. A coalition, $\mathbf{S}$, is defined as a subset of $\mathbf{N}$, $\mathbf{S} \subseteq \mathbf{N}$. An individual player is called a singleton coalition and the set $\mathbf{N}$ is also a coalition, called the grand coalition (GC), where all players cooperate.

14.2.1 Coalition Forms of N-Person Games

To understand the recent contributions to the theory of coalition and network formation, it is useful to characterize three possible representations or forms of coalition games:

1. Characteristic function form (CFF)
2. Partition function form (PFF)
3. Graph function form (GFF)

The worth, or utility function, of a coalition in a game is called the coalition value and is denoted by v. The most common form of a coalition game is the characteristic function form [5].

14.2.1.1 Characteristic Function Form

In the CFF of coalition games, utilities achieved by the players in a coalition are unaffected by those outside it.

Definition 14.1   The CFF of an N-player coalition game is given by the pair $(\mathbf{N}, v)$, where $\mathbf{N}$ is the set of players and v is a real-valued function, called the value of the game, defined on the subsets of $\mathbf{N}$, with $v(\emptyset) = 0$.

When a coalition game is described using the CFF, it is assumed that the network environment has no externality. The characteristic function assigns to every coalition a value that is the aggregate payoff that a coalition can secure for its members, irrespective of the behavior of players outside this coalition. The usual way to assign a characteristic function is to define $v(\mathbf{S})$ for each $\mathbf{S} \subseteq \mathbf{N}$ in a way that assumes that the players within each coalition $\mathbf{S}$ cooperate to act together as a unit, or one player, and the nodes in the complementary coalition $\mathbf{S}^c$ act as the other player. We next present an example inspired by [6] and modified to illustrate the concept of CFF of coalition games.

Example 14.1

Consider a three-player buyer-seller game in which player 1 (a seller) has a car that is worthless to her (unless she can sell it). Players 2 and 3 (buyers) value the car at \$4000 and \$5000, respectively.
   If player 1 sells the car to player 2 at a price of x, she will effectively make a profit of x, while player 2's profit is $4000 - x$. The total value (profit) of the coalition $\{1, 2\}$ is $v(\{1, 2\}) = 4000$. Similarly, $v(\{1, 3\}) = 5000$. However, a single player or the two buyers together can obtain no profit. Thus, $v(\{i\}) = v(\{2, 3\}) = 0$. If some side payments that do not change the total amount of utility are allowed, the coalition of all three players, that is, the GC can do no better than $\{1, 3\}$. Thus, $v(\{1, 2, 3\}) = 5000$.

14.2.1.2 Partition Function Form

A coalition game with externalities is a game in which the value that a group of players can achieve through cooperation depends on what other coalitions form. Coalition games with externalities are described using the PFF. In partition function games, the real-valued function v is a function of a coalition $\mathbf{S}$ and a partition ρ. In other words, in partition function games, any coalition $\mathbf{S} \subset \mathbf{N}$ generates a value $v(\mathbf{S}; \rho)$ where ρ is a partition of $\mathbf{N}$, with $\mathbf{S} \in \rho$. We define $v(\emptyset; \rho) = 0$ for all partitions ρ of $\mathbf{N}$.
   To model all N-link coalitions, we define coalition structures as follows:

Definition 14.2   A *coalition structure* is a partition of $\mathbf{N}$ into exhaustive and disjoint subsets, where each subset is a coalition. The set of all possible coalition structures is denoted as $\mathbf{C}$, where $\mathbf{C} = \{\mathbf{C}_1, \mathbf{C}_2, \ldots, \mathbf{C}_{|\mathbf{C}|}\}$.

For example, **C** for $N = 3$ players is given as

$$\mathbf{C} = \left\{ \left\{ \{1\}, \{2\}, \{3\} \right\}, \left\{ \{1,2\}, \{3\} \right\}, \left\{ \{1,3\}, \{2\} \right\}, \left\{ \{1\}, \{2,3\} \right\}, \left\{ \{1,2,3\} \right\} \right\}.$$

In partition function games, the worth of a coalition depends on the entire coalition structure, that is, the partition of players inside and outside a coalition. One of the strengths of partition function games is that they take into account the impact of any positive or negative externality present in the network environment on coalition formation. For instance, positive externalities due to coalition formation may provide an incentive for wireless nodes to free ride, resulting in small stable coalitions. On the other hand, negative externalities may provide an incentive to cooperate, resulting in large stable coalitions. We next present an example inspired by [7] and modified to illustrate the concept of PFF of coalition games.

Example 14.2

Three players 1, 2, and 3 manufacture a product. Some toxic waste is generated as a by-product of this manufacturing process. If players remain on their own, they do not have enough money to buy the waste treatment plant and obtain singleton coalition values $v(\{i\}) = -25$ due to the ill effects of pollution. However, if any two of them form a coalition, they can buy a waste treatment plant that can partially clean up the waste and dump the remainder on the third player. In this scenario, with players i and j cooperating and player k outside the coalition, $v(\{i,j\}) = 0$ and $v(\{k\}) = -40$. If the three players form a GC, then together they can buy a better quality treatment plant and treat the entire waste safely, that is, $v\{1,2,3\} = 0$. It can be easily seen that the formation of coalition $\{i,j\}$ generates negative externality for the singleton coalition $\{k\}$ that is not participating in the coalition formation, that is,

$$\underbrace{v(\underbrace{\{k\}}_{S} ; \underbrace{\{\{i,j\}, \{k\}\}}_{\rho})} < \underbrace{v(\underbrace{\{k\}}_{S} ; \underbrace{\{\{i\}, \{j\}, \{k\}\}}_{\rho})},$$

and that player will be motivated to join the coalition.

Unfortunately, coalition games with externalities are difficult to analyze [1]. To overcome this problem, partition function coalition games are generally converted to characteristic function coalition games.

14.2.1.3 Graph Function Form

In the GFF, the value $v(g)$ of any graph g formed by the players is given. In general, graph function games are of two types: component additive graph value games and nonadditive graph value games. The component additive graph value games assume that the value of a component does not depend on the way other players are organized, that is, ignores externalities. In other words, it assumes that the value of a graph can be decomposed into the sum of values of its components. On the other hand, nonadditive value games allow for externalities across components.

We will provide examples for these three representations of coalition games for wireless networks in the next section of this chapter.

14.2.2 Solution Concepts for Coalition Games

Most solution concepts related to coalition games analyze stable coalition structures. Stable coalition structures correspond to the equilibrium state in which players do not have incentives to leave the existing coalitions. In analyzing stable coalition structures for networks, we essentially need to determine if the GC is stable, and this can be done using the concept of core.

In the study of coalition games, a payoff vector $\mathbf{x}$ is said to be in the core if it satisfies the following two properties: (1) $\mathbf{x}$ is an imputation; (2) imputation $\mathbf{x}$ is stable [8].

Definition 14.3 An *imputation* is a payoff vector that satisfies the following two conditions: (1) $\sum_{i \in \mathbf{N}} x_i = v(\mathbf{N})$; (2) $x_i \geq v(\{i\})$.

Definition 14.4 An imputation $\mathbf{x}$ is *unstable* if there exists a coalition $\mathbf{S} \subset \mathbf{N}$ such that $\sum_{i \in \mathbf{S}} x_i < v(\mathbf{S})$, otherwise $\mathbf{x}$ is said to be stable.

Definition 14.5 The set, γ, of stable imputations is called the *core*:

$$\gamma = \left\{ \mathbf{x} : \sum_{i \in \mathbf{N}} x_i = v(\mathbf{N}) \text{ and } \sum_{i \in \mathbf{S}} x_i \geq v(\mathbf{S}), \text{ for all } \mathbf{S} \subset \mathbf{N} \right\}. \tag{14.1}$$

A coalition game's core is exponentially hard to compute. However, if the size of coalitions is limited to a specified constant, then the kernel of the coalition game appears to be an attractive solution concept due to the existence of polynomial kernel-stable coalition configuration algorithms [9]. The kernel of a coalition game is based on two ideas: excess and surplus [6,10].

Definition 14.6 For the N-player coalition game $(\mathbf{N}, v)$, let $\mathbf{S}$ be a coalition and $\mathbf{x} = (x_1, \ldots, x_N)$ a payoff vector (not necessarily an imputation). Then the *excess* of $\mathbf{S}$ with respect to $\mathbf{x}$ is

$$e(\mathbf{S}, \mathbf{x}) = v(\mathbf{S}) - \sum_{i \in \mathbf{S}} x_i. \tag{14.2}$$

Definition 14.7 For the N-player coalition game $(\mathbf{N}, v)$, let $i \neq j$ be players, and $\mathbf{x} = (x_1, \ldots, x_N)$ a payoff vector. Then the *surplus* of i against j is

$$s_{ij}(\mathbf{x}) = \max_{\mathbf{S}: i \in \mathbf{S}, j \notin \mathbf{S}} e(\mathbf{S}, \mathbf{x}). \tag{14.3}$$

In other words, s_{ij} denotes the most that player i could hope to gain without the cooperation of j, under the best circumstances.

Objections and counter objections of the players against each other are defined as follows [10]:

Definition 14.8 A coalition **S** is an *objection* of i against j to **x** if **S** includes i but not j and $x_j > v(\{j\})$.

Definition 14.9 A coalition **R** is a *counter objection* to the objection **S** of i against j if **R** includes j but not i and $e(\mathbf{R}, \mathbf{x}) \geq e(\mathbf{S}, \mathbf{x})$.

Definition 14.10 The set **K** of all individually rational payoff vectors **x** with the property that for every objection **S** of any player i against any other player j to **x** there is a counter objection of j to **S** is called the *kernel* of a coalition game with transferable payoff.

In transferable payoff games, players are assumed to freely transfer among themselves a commodity, called money, such that any player's payoff increases one unit for every unit of money that he gets.

As an interpretation of the kernel we can envision one player presenting the following argument to another: "I could obtain greater payoff in a coalition that you do not belong to than you could in a coalition that I do not belong to. Therefore, I deserve greater payoff than you, while still preserving your individual rationality, that is, guaranteeing you a payoff at least as large as you could get from forming a singleton coalition."

Example 14.3

Consider the three-player coalition game in which $v(\{1\}) = 2$, $v(\{2\}) = 1.5$, $v(\{3\}) = 1$, $v(\{1, 2\}) = 6$, $v(\{1, 3\}) = 8$, $v(\{2, 3\}) = 7$, and $v(\{1, 2, 3\}) = 15$.

If maximum coalition size of $|\mathbf{S}| = 2$ is allowed, then using Equations 14.2 and 14.3, it is not too difficult to see that $(3.5, 2.5, 1) \in \mathbf{K}$, as, $e(\{1, 2\}, \mathbf{x}) = 0$, $s_{12}(\mathbf{x}) = v(\{1, 3\}) - (x_1 + x_3) = 3.5$, and $s_{21}(\mathbf{x}) = v(\{2, 3\}) - (x_2 + x_3) = 3.5$. Also, if maximum coalition size of $|\mathbf{N}|$ is allowed, then $(5, 4.25, 5.75) \in \mathbf{K}$.

As already stated, in certain scenarios, self-interested players may prefer to cooperate with only a selected subset of users to achieve maximum gains. In such scenarios, the GC will not be formed. We say that a coalition has reached individual stability or equilibrium if it is internally and externally stable (IES).

Definition 14.11 A coalition **U** is *internally stable* if no player has an incentive to leave its coalition to become a singleton, that is, $v_{i, i \in \mathbf{U}}(\mathbf{U}) \geq v(\{i\})$, $\forall i \in \mathbf{U}$, and a coalition **U** is *externally stable* if no other coalition has an incentive to join coalition **U**, that is, $v(\mathbf{V}) > v(\mathbf{U} \cup \mathbf{V}) - v(\mathbf{U})$, $\forall \mathbf{V} \subseteq \mathbf{U}^c$.

If each coalition in a coalition structure is IES, then a multi-coalition equilibrium results.

14.3 Coalition Forms of Games for Wireless Networks

In the previous section, we have presented a coalition game-theoretic framework that can be used to analyze key problems in wireless networks, such as radio resource management and cooperative communications. In this section, we present examples of different representations of coalition games for wireless networks.

14.3.1 A Packet-Forwarding Coalition Game in CFF

Numerous wireless networks, including mesh networks, ad hoc networks, and sensor networks, often require that nodes forward packets for one another. To improve network efficiency, a repeated game framework can be used to induce cooperation among selfish nodes to forward each others' packets. However, nodes at the network boundary cannot gain from this strategy, as the other nodes do not depend on them [11]. To overcome this problem, the authors in [11] have proposed coalition formation between boundary nodes and backbone nodes as shown in Figure 14.1. In this coalition formation game, boundary nodes provide certain transmission gains for the backbone nodes by relaying their received transmissions. As a compensation for this service, the backbone nodes forward the packets of the boundary nodes.

Assuming one backbone node and $N - 1$ nearby boundary nodes as shown in Figure 14.1, if no cooperative transmission is employed, the values of singleton coalitions for the backbone node s_1 and the boundary nodes are $v(\{s_1\}) = -P_d$ and $v(\{i\}) = -\infty$, respectively, where P_d represents the direct transmission power between s_1 and d_1. With cooperative transmission and a GC **N** that includes all the nodes, the payoffs for the backbone node and the boundary nodes are

$$v_{s_1, s_1 \in \mathbf{N}}(\mathbf{N}) = -P_0 - \sum_{i=1}^{N-1} \beta_i P_d \qquad (14.4)$$

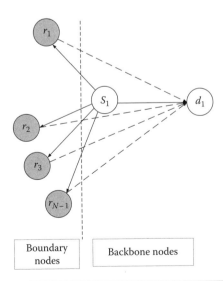

Figure 14.1 Coalition game with cooperative transmission for packet-forwarding wireless networks.

and

$$v_{i,i\in\mathbf{N}}(\mathbf{N}) = -\frac{P_i}{\beta_i}. \tag{14.5}$$

In Equations 14.4 and 14.5, P_0 is the transmitted power of the backbone node when cooperation is employed, P_i is relay i's transmit power, and β_i is the ratio of the number of packets that the backbone node is willing to forward for boundary node i, to the number of packets that the boundary node relays for the backbone node when a coalition is in place. The authors in [11] show that the core of the cooperative transmission game is not empty and the GC is formed if $\beta_i \geq 0$, $i = 1, \ldots, N - 1$, and β_i are such that $v_{s_1, s_1 \in \mathbf{N}}(\mathbf{N}) \geq v(\{s_1\})$, that is,

$$\sum_{i=1}^{N-1} \beta_i \leq \frac{P_d - P_0}{P_d}. \tag{14.6}$$

It is interesting to note that if several backbone nodes are included in the model, then any reduction in transmission power by a backbone node due to coalition formation (between a backbone node and the boundary nodes) may impact the received signal-to-noise ratios at the other backbone node destinations. In other words, the coalition formation between the boundary nodes and the backbone nodes may exhibit externalities. The presence of externalities may influence the stable solution of the coalition game proposed by the authors in [11].

14.3.2 Interference Channel with Externalities

To determine stable coalition structures in a coalition game, one must in general take into account whether the gain achieved by a coalition is also influenced by any externality. However, coalition game models for wireless networks are often analyzed with an assumption that the gain of any coalition is independent of coexisting coalitions in the network [1,3,11]. Thus, the possibility of interaction between coalitions is ruled out while analyzing the wireless network.

In what follows, we provide an example of wireless network environment in which there are externalities from coalition formation. We model a scenario in which different wireless systems operating in the same band (e.g., IEEE 802.11 and Bluetooth systems), each formed by a single transmitter–receiver pair, coexist in the same area. The received signal for any user i in a Gaussian interference channel is given by

$$Y_i = \sum_{j=1}^{N} h_{ji}X_j + z_i, \quad i \in \mathbf{N}, \tag{14.7}$$

where X_j and Y_i are input and output signals, respectively. The noise processes are independent and identically distributed (i.i.d.), zero mean, and unit variance Gaussian random variables, that is, $z_i \sim \mathcal{N}(0, N_0)$, with $N_0 = 1$. h_{ji} is the channel gain between the transmitter of user j and the receiver of user i. The channel from each transmitter to each receiver is assumed to be flat fading. The user i is assumed to have an average power constraint P_i. The transmission strategy for each user is the way that it allocates power in the given bandwidth. The user can either spread the power over the available bandwidth W or it can allocate the same power in a segment of W. We also assume that each wireless user treats multiuser interference as noise, and no interference cancelation techniques are employed. In many spectrum-sharing problems, the issue of fair and efficient solutions arises due to asymmetries between the systems. Figure 14.2 illustrates an example in which users 1 and 2 operate in symmetric situations while user 3 operates in an asymmetric situation. In Figure 14.2, users 1 and 2, that is, Tx1/Rx1 and Tx2/Rx2 with same power capabilities (e.g., two IEEE 802.11 systems) share the same spectrum band and, due to the locations of the transmitters and receivers, both 1 and 2 strongly interfere with each other [12]. There is also a third user with a low power capability (e.g., a Bluetooth system) sharing the same band with the two other users. Authors in [1] have shown that for arbitrary SNR values, the users in the stable coalitions benefit from the exclusion of the weak interferer. In Figure 14.2, all the channel gains are comparable, so intuitively if the three users decide to play a coalition game, then the users 1 and 2 will prefer to form a coalition, that is, share the spectrum with each other, rather than to form a coalition with the weak user. Assuming that the users 1 and 2, that is, Tx1/Rx1 and Tx2/Rx2 decide to form a coalition $\mathbf{S} = \{1, 2\}$ and share the spectrum bandwidth W, the value of $\mathbf{S}$ can be expressed as

$$v(\mathbf{S}) = \sum_{i \in \mathbf{S}} R_i = \sum_{i \in \mathbf{S}} \frac{W}{2} \eta_i \log_2 \left(1 + \frac{(P_i/\eta_i)}{1 + h_{3i}^2 P_3} \right), \quad \mathbf{S} \subset \mathbf{N}, \tag{14.8}$$

where

R_i is the data rate for user i, $h_{ii} = 1$, $\mathbf{N} = \{1, 2, 3\}$
h_{3i}^2 represents the interference channel power gain to the members of coalition $\mathbf{S}$
$0 \leq \eta_i \leq 1$ is the fraction of the band that user i uses, and $\sum_{i, i \in \mathbf{S}} \eta_i \leq 1$

To illustrate the above situation, we can construct an example, for the three users of Figure 14.2.

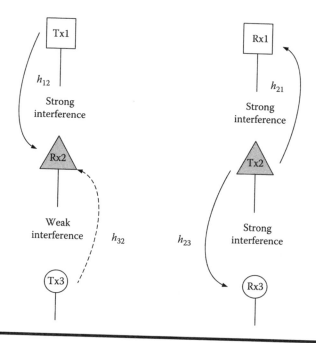

Figure 14.2 **Different wireless users coexisting in the same area and interfering with each other.**

Example 14.4

Consider, for example, a three-user scenario with bandwidth $W = 2$, $N_0 = 1$, link gain matrix $H = \begin{pmatrix} h_{11} & h_{21} & h_{31} \\ h_{12} & h_{22} & h_{32} \\ h_{13} & h_{23} & h_{33} \end{pmatrix}$, given as $H = \begin{pmatrix} 1 & .7 & .3 \\ .7 & 1 & .3 \\ .6 & .6 & 1 \end{pmatrix}$, $P_1 = P_2 = P$, with $P > N_0$, $P \le 20\,\text{dB}$, and $P_3 = P/5$. If all three users spread their powers through the entire band and prefer to form singleton coalitions, then they obtain coalition values

$$v(\{1\}) = v(\{2\}) = \log\left(1 + \frac{P}{(1 + 0.51P)}\right) \text{ and } v(\{3\}) = \log\left(1 + \frac{P}{5(1 + 0.72P)}\right).$$

However, if the users 1 and 2 form a coalition **S** and share the same spectrum band in a certain ratio $\sum_{i, i \in S} \eta_i \le 1$ (since $h_{12} = h_{21} = .7$, it is natural to assume that both users share the band in equal ratios, i.e., $\eta_1 = \eta_2 = 0.5$), the value of coalition **S** is

$$v(S) = \sum_{i \in S} R_i = \log\left(1 + \frac{2P}{1 + 0.02P}\right) \text{ and } v(\{3\}) = \log\left(1 + \frac{P}{5(1 + 0.72P)}\right).$$

In this scenario, the coalition value $v(\{3\})$ is not affected by the formation of new coalition **S**.

Consider now the scenario in which $H = \begin{pmatrix} 1 & .6 & .3 \\ .3 & 1 & .3 \\ .8 & .5 & 1 \end{pmatrix}$. If users 1 and 2 share the same spectrum band, then it is natural to assume that the two share the spectrum in a certain ratio $\eta_2 > \eta_1$, as $h_{21} > h_{12}$. The value of coalition **S** is

$$v(S) = \eta_1 \log\left(1 + \frac{P/\eta_1}{1 + 0.02P}\right) + \eta_2 \log\left(1 + \frac{P/\eta_2}{1 + 0.02P}\right) \text{ and}$$

$$v(\{3\}) = \eta_1 \log\left(1 + \frac{P}{5\left(1 + \frac{0.64}{\eta_1}P\right)}\right) + \eta_2 \log\left(1 + \frac{P}{5\left(1 + \frac{0.25}{\eta_2}P\right)}\right).$$

In this scenario, the coalition value $v(\{3\})$ is a function of η_i, that is, the coalition value of the third system depends on the way users 1 and 2 allocate the spectrum, although the third user is not participating in the coalition formation.

In other words, the coalition formation game process in an interference channel generates externalities, and the value of coalitions in an interference channel depends on a partition ρ of **N**.

Broadly speaking, the presence of positive externalities, that is, if players outside any coalition **S** do not either lose or gain by any player i joining **S**, may provide an incentive for wireless nodes to free ride, resulting in small stable coalitions. On the other hand, negative externalities, that is, if players outside any coalition **S** lose by any player i joining **S**, may provide an incentive to cooperate, resulting in large stable coalitions. Unfortunately, coalition games with externalities have PFF and such partition functions are difficult to analyze [1]. To overcome this problem, coalition games with PFF are generally converted into CFF.

14.3.3 Coalitions on Graph: Two Models

14.3.3.1 Myerson's Model

To study a cooperative self-managing wireless network using standard coalition game-theoretic framework, we need to describe coalition structures, that is, we need to describe who is cooperating

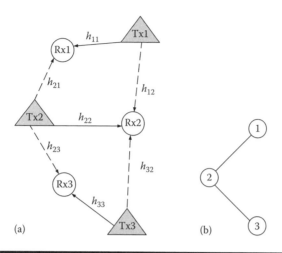

Figure 14.3 An asymmetric interference scenario and its undirected graph representation for bilateral cooperation links. (a) Different wireless nodes coexisting in the same area and interfering with each other asymmetrically. (b) Bilateral cooperation links represented as an undirected graph for the node cooperation preferences.

with whom among the network links. For instance, Figure 14.3a shows a scenario in which three wireless nodes with similar power capabilities coexist in the same area, but due to the locations of the transmitters and receivers, Rx2 experiences high interference from both the transmitters Tx1 and Tx2. On the other hand, receivers Rx1 and Rx3 experience high interference from transmitter Tx2, but weak interference from the other transmitters. In Figure 14.3a, all the channel gains are comparable, so intuitively in this three-player game, node 1 wants to cooperate with node 2 but not with node 3, node 3 wants to cooperate with node 2 but not with node 1, and node 2 wants to cooperate with both nodes 1 and 3. If we try to model this cooperation scenario by coalition structures $\big\{\{1,2\},\{3\}\big\}$ or $\big\{\{1\},\{2,3\}\big\}$ or $\big\{\{1,2,3\}\big\}$, each partition into coalitions seems to suppress some important aspects of this cooperative scenario.

To solve such cooperative dilemmas, Myerson introduces in [13] the concept of a cooperation graph in a coalition game, which incorporates the possible benefits from cooperation as modeled by the coalition game and the restrictions on communication reflected by the communication network. In such a graph, nodes represent the players and edges represent cooperation links. A cooperation structure is represented as a graph g. The graph models a list of players that are connected to each other, with the interpretation that if player 1 is connected to player 2 in the graph, then they can communicate and cooperate with each other. Using this graph representation model, for example, scenario of Figure 14.3a, there is a cooperation link between nodes 1 and 2, and between nodes 2 and 3, but not between nodes 1 and 3. The undirected graph representation illustrated in Figure 14.3b summarizes all significant aspects of cooperation for our example scenario. Under the graph g, the value of a coalition $\mathbf{S}$ is simply the sum of the values of the subsets of $\mathbf{S}$ across the partition of $\mathbf{S}$ imposed by g. For example, consider a coalition game where the value of coalition $\{1,2\}$ is 2, the value of coalition $\{3,4\}$ is 2, and the value of coalition $\{1,2,3,4\}$ is 5. If under the graph g only player 1 is connected to 2 and player 3 is connected to 4, then the value of coalition $\{1,2,3,4\}$ is $2+2=4$, rather than 5, as this is the only way in which the coalitions can form.

While Myerson's model provides an important enrichment of a coalition game, it falls short of providing a general model where value is network dependent. For example, in the case of a three-player coalition game, the value of a coalition $\{1, 2, 3\}$ is the same when the underlying network is represented by a graph that only connects 1 to 2 and 2 to 3, or whether it is a complete network whose graph also connects 1 to 3.

14.3.3.2 Jackson and Wolinsky Model

Jackson and Wolinsky in [14] specified the value function in a way that is defined on networks directly, rather than on coalitions. Thus, the authors in [14] start with a value function v that maps each network into a worth or value. Different networks connecting the same players can result in different values, allowing the value of a coalition to depend not only on who is connected but also on how they are connected. This allows for both direct and indirect costs and gains for the connections to be analyzed. In the Jackson–Wolinsky model, an allocation rule specifies a distribution of payoffs for each pair of network and value function. One of the central issues examined by Jackson and Wolinsky is whether efficient (value maximizing) networks will be formed when self-interested individuals can choose to form links and break links. They define a network to be pairwise stable if no pair of players wants to form a link by adding a pair of links to the existing graph and no player gains by severing a pair of links in the existing graph. The work analyzes the question of the compatibility of efficiency and stability. The main drawback of the model is that for high and low costs the efficient networks are pairwise stable, but this is not always the case for somewhere between low and high costs.

14.4 Coalition Formation Process

Traditionally, the study of coalition games in wireless networks has focused on cohesive games [1,15], that is, games where the value of the GC formed by the set of all players **N** is at least as large as the sum of the values of any partition of **N**. The authors in [1,15] also assume that there is no cost to the coalition formation process. In such coalition games, the coalition structure generation is trivial because the wireless nodes always benefit by forming the GC.

However, many coalition game models of wireless node cooperation are not cohesive (see, e.g., [2]), because in wireless networks there is some cost to the coalition formation process itself. For instance, coalition formation in wireless networks may require wireless links to activate some coalition message exchange link, which may be costly. Some coalitions, in such scenarios, may gain by merging while the others may not. In such scenarios, it is important to analyze cooperative interactions between small groups of nodes, as well as cooperative interactions between all the network nodes. In other words, we need to analyze the entire coalition formation process until we find the social welfare maximizing coalition structure that satisfies the equilibrium conditions. The coalition formation process is generally analyzed using either static or dynamic coalition formation game models. In static coalition formation models, the objective is to analyze a certain coalition structure that is imposed by an exogenous factor.

Stable coalition structures in coalition games correspond to the equilibrium state in which users do not have incentives to leave the already-established coalitions. However, the models proposed for the analysis of stable coalition structures are often static, in the sense that they fail to specify *how* the players arrive at equilibrium. In dynamic coalition formation models, the objective is to address this issue and analyze the coalition formation process in which at any time,

1. Players may enter or leave coalitions.
2. A time-evolving sequence of steps is used by players to reach stable coalition structures. At each time step, players interact with each other and through this interaction form stable coalition structures.
3. A player's environment may vary and players are required to adapt to their environment.

14.4.1 Dynamic Models of Coalition Formation Process for Wireless Networks

Establishing cooperation in a wireless network is a dynamic process and two important questions must be addressed: (1) How are coalitions formed? (2) How do players arrive at equilibrium? In this part of the chapter, we will focus on dynamic models of coalition formation games for cooperation in wireless networks.

14.4.1.1 A Merge-and-Split Approach

In cognitive radio networks, the effects of fading or shadowing may result in unreliable detection of the primary user by the cognitive radio user. To overcome the problem of unreliable detection, cognitive radios may perform cooperative spectrum sensing. However, in cooperative spectrum sensing, the gain in terms of detection of the primary user may increase the costs in terms of false alarm probability. Using a coalition formation game model, the authors in [16] study the impact of this trade-off on the topology and the dynamics of a cognitive radio network of secondary users. The authors devise distributed cooperative sensing strategies for secondary users using simple merge-and-split rules for coalition formation. In merge-and-split rules, users autonomously interact, and based on this interaction, a group of coalitions (or users) decides to merge if it is able to improve its total utility through the merge. Similarly, an existing coalition splits into smaller coalitions if it is able to improve the total utility through the split [17]. After the termination of merge-and-split operations, the users in a particular coalition send their individual sensing decisions to a cognitive user selected as coalition head. The coalition head combines these individual sensing decisions to make a final sensing decision. The authors in [16] show that the proposed distributed coalition formation model based on merge-and-split reduces the average missing probability per cognitive user up to 86.6% compared to the noncooperative spectrum-sensing case. The drawback of the model in [16] is that the proposed utility function fails to take into account the fact that before the distributed cognitive radios decide to transmit in the primary user band, they are required to satisfy the sensing reliability with at least a certain target probability of detection [18].

14.4.1.2 A Constrained Coalition Game Approach

Self-managing wireless networks depend on cooperation between their nodes in the sense that they can achieve performance gains that are not possible to accomplish without such cooperation. For instance, in wireless packet-forwarding, network nodes need to forward one another's packets. However, cooperation involves some cost on the part of the cooperating nodes. This cost of cooperation may require some constraints to be imposed over the way coalitions can be formed. Recently, the authors in [3] have analyzed constrained coalition games in self-managing wireless networks. The authors in this work study coalitions with the fundamental view that nodes in wireless networks are dynamic entities that cooperate, because via cooperation they improve individual and network performance. However, such cooperation imposes some costs, for instance, due to communications or energy overhead. The authors in [3] analyze the fundamental trade off between gain and cost in the context of user cooperation in self-managing networks.

The coalition formation model proposed in [3] is formulated as a two-step process. In the first step, nodes decide whether to join a coalition and in the second step nodes in a coalition negotiate the payoff allocation based on the total value of the coalition. The authors point out that, due to the dynamic nature of node interaction, the central problem in dynamic coalition games is not only to analyze the convergence of the two-step game but also to study whether the dynamics result in a stable solution. For a comprehensive study of the model, we refer the reader to [3], which presents the conditions under which the formed coalitions are unstable. One limitation of [3] is that the authors derive coalition gains based on a saturated model rather than using specific network traffic patterns. For instance, the authors assume that users have information to transmit and the transmission failures are modeled by an abstract parameter called the depreciation rate.

14.4.2 Dynamic Modeling of Coalition Formation with Markov Chains

In the previous subsection, we have discussed two different dynamic models of coalition formation games and their applications to self-managing wireless networks. In this subsection, we discuss the dynamic modeling of coalition formation problem with Markov chains.

To allow nodes in a self-managing wireless network to form coalitions, it is essential for the nodes to be self-aware. The wireless nodes require knowledge of their current status, as well as knowledge of their interactions with the other nodes in the network. To analyze cooperative interactions in such networks with coalition games, we need a dynamic model that incorporates the time-evolving sequence of steps that is followed by the participating nodes to reach equilibrium. In this context, the authors in [19–21] present coalition game models for economic behavior in which, at each time step, a player observes its environment and decides which of the existing coalitions to join. In [22], the authors model the dynamic coalition formation decision process as an absorbing Markov chain, and using Markov chain theory the authors analyze mean μ and variance σ^2 of the time for the coalition game to reach a stable coalition structure, that is, to reach an absorbing state. In [22], the authors show that by analyzing the mean μ and variance σ^2 of the time to absorption, we can optimize the time interval between two coalition formation decisions in the system. The model also shows that depending on the coalition values, the coalition formation process converges either to the absorbing state of the GC or to another coalition structure satisfying the condition of internal and external stability.

Some of the main characteristics of the dynamic model of the coalition game are [22] as follows:

■ The model is dynamic in the sense that a time-evolving sequence of steps is followed by a player to reach stable coalition structures.
■ Each state of the Markov chain represents a coalition structure. A finite set $\mathbf{C}$ of all possible coalition structures for N players forms the state space of the coalition formation game. Let

$$\mathbf{C} = \{\mathbf{C}_1, \mathbf{C}_2, \ldots, \mathbf{C}_{|\mathbf{C}|}\}, \tag{14.9}$$

where each element of $\mathbf{C}$ is a state representing a coalition structure. We define the value of a coalition structure as

$$V(\mathbf{C}_j) = \sum_{\mathbf{S}_i \in \mathbf{C}_j} v(\mathbf{S}_i), \quad \text{where } j = 1, 2, \ldots, |\mathbf{C}|.$$

■ The cardinality of the set of all possible coalitions structures, that is, **C** is given by the Bell number $B(N)$

$$B(N) = \sum_{k=0}^{N-1} \binom{N-1}{k} B_k, \qquad (14.10)$$

with

$$B(0) = 1.$$

■ The dimension of the transition matrix is $B(N) \times B(N)$, where N is the number of players playing the coalition game and $B(N)$ is the Bell number function. The dimension of the matrix can be quite large; however, the transition matrix is sparse.
■ A group of players or a coalition can deviate only if all players within the coalition are at least as well off as a result of the proposed deviation. In other words, once players decide to form a coalition, they enter a binding agreement and cannot unilaterally deviate.
■ The dynamic coalition formation game involves two steps:
  1. In the first step, each of n coalitions proposes a coalition structure change with some probability of success p to any of the existing coalitions. In the case of singleton coalitions, each player individually proposes a coalition structure change. When two or more players form a coalition $\mathbf{S_1}$, then any player within $\mathbf{S_1}$ is selected to propose the coalition structure change on behalf of that coalition. Any player i residing in any existing coalition $\mathbf{S_1}$ may propose to form a new coalition $\mathbf{S}$ to another coalition $\mathbf{S_2}$, that is, $\mathbf{S_1} \cup \mathbf{S_2} = \mathbf{S}$. Each player participating in the formation of the proposed coalition $\mathbf{S}$ will put forward its rational demand $d_i(\mathbf{S})$, where $i \in \mathbf{S}$. A demand is said to be individually rational if a demand payoff gives each player at least as much payoff as that link receives without joining the new coalition. Individual demands by the participating links are calculated in the presence of any expected cost incurred due to coalition formation. If demands are feasible for the proposed coalition, that is, $\sum_{i,i\in\mathbf{S}} d_i(\mathbf{S}) \le v(\mathbf{S})$, then each player is promised at least the payoff $v_{i,i\in\mathbf{S}}(\mathbf{S}) \ge d_{i,i\in\mathbf{S}}(\mathbf{S})$ and the proposed coalition is formed. Since any proposed coalition $\mathbf{S}$ is formed only if all players within $\mathbf{S}$ are at least as well off as without $\mathbf{S}$, whenever players agree to form $\mathbf{S}$, the new coalition is internally stable, that is, no player has an incentive to become a singleton.
  2. In the second step, the total value $v(\mathbf{S})$ of the coalition can be arbitrarily partitioned among the players, subject to the minimum rational payoff condition. We define a minimum rational payoff as a payoff partition that gives each player at least as much value as that player receives without joining the new coalition.

The transition probabilities for the general N link coalition game with $B(N)$ possible CSs as state space **C** are given as [22]

$$P_{\mathbf{C}_k\mathbf{C}_l} = \frac{2p(1-p)^{(|\mathbf{C}_k|-1)}}{|\mathbf{C}_k|-1} 1(V(\mathbf{C}_l), V(\mathbf{C}_k)), \qquad P_{\mathbf{C}_k\mathbf{C}_k} = 1 - \sum_{\mathbf{C}_l \in \Lambda_l} P_{\mathbf{C}_k\mathbf{C}_l}, \qquad (14.11)$$

where

$$1(V(\mathbf{C}_l), V(\mathbf{C}_k)) = \begin{cases} 1 & \text{when } V(\mathbf{C}_l) \geq V(\mathbf{C}_k), \\ 0 & \text{otherwise.} \end{cases}$$

$V(\mathbf{C}_k)$ and $V(\mathbf{C}_l)$ are the values of CS states $\mathbf{C}_k$ and $\mathbf{C}_l$, respectively
$|\mathbf{C}_k|$ represents the number of coalitions in the present CS state $\mathbf{C}_k$
$\mathbf{C}_l$ represents any one of the new possible CS states to which coalitions can transit from $\mathbf{C}_k$
Λ represents the set of all new possible CS states to which coalitions can transit from $\mathbf{C}_k$

The set Λ is given as $\Lambda = \{\{\mathbf{S}_1 \cup \mathbf{S}_2, \mathbf{S}_3, \ldots, \mathbf{S}_{(|\mathbf{C}_k|)}\}, \{\mathbf{S}_1, \mathbf{S}_2 \cup \mathbf{S}_3, \ldots, \mathbf{S}_{(|\mathbf{C}_k|)}\}, \ldots, \{\mathbf{S}_1, \mathbf{S}_2, \ldots,$
$\mathbf{S}_{(|\mathbf{C}_k|-1)} \cup \mathbf{S}_{(|\mathbf{C}_k|)}\}\}$.
Detailed descriptions of the absorbing Markov chain model of the dynamic coalition game for the N players can be found in [22].

Using standard theory of absorbing Markov chains, one can calculate the mean time μ and its variance σ^2 for the dynamic coalition game starting from the initial state of all singleton coalitions to reach a stable coalition structure. As a first step, the state transition probability matrix should be arranged in a canonical form. If there are r absorbing states and t transient states, the transition matrix will have the following canonical form [23,24]:

$$\bar{P} = \begin{pmatrix} I & O \\ R & Q \end{pmatrix},$$

where
   I is an r-by-r identity matrix
   R is a nonzero t-by-r matrix of transition probabilities from transient to absorbing states
   O is an r-by-t matrix with all zero entries
   Q is a t-by-t matrix of transition probabilities between the transient states

The matrix $F = (I - Q)^{-1}$ is called the *fundamental matrix* for $\bar{P}$. Using F, one can calculate the mean time μ and its variance σ^2 for the coalition game process to converge to the absorbing state:

$$\mu = F\tau, \tag{14.12}$$

$$\sigma^2 = (2F - I)\mu - \mu_{sq}, \tag{14.13}$$

In Equations 14.12 and 14.13, μ is a column vector whose ith entry, μ_i, is the expected number of steps before the process reaches a stable coalition structures, given that the process starts in state $\mathbf{C}_i$, μ_{sq} is a column vector whose entries are the square of the entries of μ, and τ is a column vector whose components are the respective state dwell times.

Example 14.5

We present an example of calculating the mean time μ and its variance σ^2 for a 4 person coalition game to reach the stable coalition structure, whose characteristic function is given in the form

$$v(\{1\}) = v(\{2\}) = v(\{3\}) = 1, v(\{4\}) = 4,$$

$$v(\{1,2\}) = v(\{2,3\}) = v(\{1,3\}) = 2.5,$$

$$v(\{1,4\}) = v(\{2,4\}) = v(\{3,4\}) = 4.5, \tag{14.14}$$

$$v(\{1,2,3\}) = 4, v(\{1,2,4\}) = v(\{2,3,4\}) = v(\{1,3,4\}) = 5,$$

$$v(\{1,2,3,4\}) = 6.$$

There are 15 coalition structures for this four-player game. Hence, we may associate 15 states with the 15 coalition structures as shown in Figure 14.4. Any coalition with some probability of success p proposes a coalition structure change to any other coalition in the current state $\mathbf{C}_k$, and if the proposed coalition satisfies individual rational demands $d_i(\mathbf{S})$ of each participating link, then the game transitions to $\mathbf{C}_l$.

It is easy to verify that this game has a unique stable coalition structure $\{\{1,2,3\},\{4\}\}$ that satisfies the internal and external stability of the coalition structures.

Using Figure 14.4 and Equation 14.11, the canonical form matrix $\bar{P}$ of our example game is given as

$$\bar{P} = \begin{array}{c} \\ \mathbf{C}_8 \\ \mathbf{C}_1 \\ \mathbf{C}_2 \\ \mathbf{C}_3 \\ \mathbf{C}_5 \end{array} \begin{array}{ccccc} \mathbf{C}_8 & \mathbf{C}_1 & \mathbf{C}_2 & \mathbf{C}_3 & \mathbf{C}_5 \\ \left(\begin{array}{ccccc} 1 & 0 & 0 & 0 & 0 \\ 0 & 1-2p(1-p)^3 & \frac{2p(1-p)^3}{3} & \frac{2p(1-p)^3}{3} & \frac{2p(1-p)^3}{3} \\ p(1-p)^2 & 0 & 1-p(1-p)^2 & 0 & 0 \\ p(1-p)^2 & 0 & 0 & 1-p(1-p)^2 & 0 \\ p(1-p)^2 & 0 & 0 & 0 & 1-p(1-p)^2 \end{array} \right) \end{array}$$

In Figure 14.5a and b, we provide results for the mean time and its variance to reach the IES (absorbing state) stable coalition structure $\{\{1,2,3\},\{4\}\}$. We can observe from the figures, when the probability of success $p = 0.3$, the mean time and its variance to absorption to the IES coalition structure state achieve the minimum value. Figure 14.5a shows that for small p, μ is high because the mean time between the two coalition structure change proposals is long. If p is high, then the mean time between the two coalition structure change proposals is shorter, but the number of coalitions attempting to play the coalition game at the same is higher, resulting in a longer time between successful decisions to initiate new coalition structure changes. This suggests that depending on the number of players in the game there should be an optimum value for p.

14.5 Computational Complexity

In coalition games that are not cohesive, some coalitions gain by merging while the others do not. In such network scenarios, the network welfare maximizing coalition structure varies, and this variation results in coalition structure generation that is computationally intensive. The aim in such scenarios is to maximize the social welfare of $\mathbf{N}$ by finding a coalition structure [25]

$$\mathbf{C}_j^* = \arg \max_{\mathbf{C}_j \in \mathbf{C}} V(\mathbf{C}_j), \tag{14.15}$$

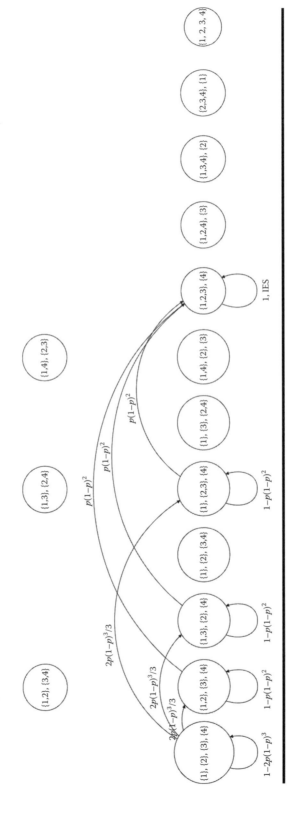

Figure 14.4 Transition probability graph for the four-player coalition game (Example 14.5). Coalition structure state $\{\{1, 2, 3\}, \{4\}\}$ is an IES (absorbing) state as demands of players 1, 2, and 3 are satisfied, that is, $1 (V(C_I), V(C_k)) = 1$ and they form the coalition $\{1, 2, 3\}$, while the demands of player 4 are not satisfied and it prefers to stay as a singleton coalition.

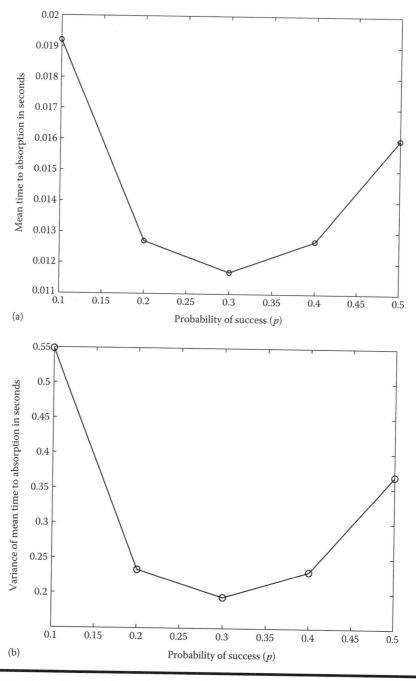

Figure 14.5 **Mean time and its variance for the coalition game (starting from the singleton coalitions state $\{\{1\}, \{2\}, \{3\}, \{4\}\}$) to form the IES stable coalition structure $\{\{1, 2, 3\}, \{4\}\}$. (a) Mean time to absorption to form the IES stable coalition structure. The state dwell time is set to 1 ms. (b) Variance of time to absorption to form the IES stable coalition structure. The state dwell time is set to 1 ms.**

where

$$V(\mathbf{C}_j) = \sum_{S_i \in \mathbf{C}_j} v(S_i), \; j = 1, 2, \ldots |\mathbf{C}|,$$

and **C** denotes the set of all possible coalition structures. The problem is that the number of coalition structures follows the Bell number function (Equation 14.10), which grows super-exponentially. In case of a large number of players, it may be computationally more efficient to search through a subset ($\mathcal{C} \subset \mathbf{C}$) of coalition structures and select the best coalition structure seen so far:

$$\mathbf{C}_{j,\mathcal{C}}^* = \arg \max_{\mathbf{C}_j \in \mathcal{C}} V(\mathbf{C}_j). \tag{14.16}$$

The coalition structure generation process can be considered as search in a coalition structure graph. Such a graph for $N = 4$ nodes is presented in Figure 14.6. The nodes in the coalition structure graph represent coalition structures. When followed downward, the arrows represent the merger of coalitions and when followed upward, the arrows represent the splitting of coalitions. If there are too many nodes in the graph of Figure 14.6, then it is high that the selected coalition structure be within a worst case bound from the optimal structures, that is, that $k \geq V(\mathbf{C}_j^*)/V(\mathbf{C}_{j,\mathcal{C}}^*)$ be as small as possible [25]. Different coalition structure search algorithms have been proposed in the literature to select the best coalition structure with the minimal amount of search [26,27]. In [25], it is shown that to bound k, it is sufficient to search the lowest two levels of the coalition structure graph of Figure 14.6. In other words, the number of nodes to be searched is $n = 2^{|\mathbf{C}|-1}$. If this result is interpreted positively, then it means that a worst case bound from the optimum can

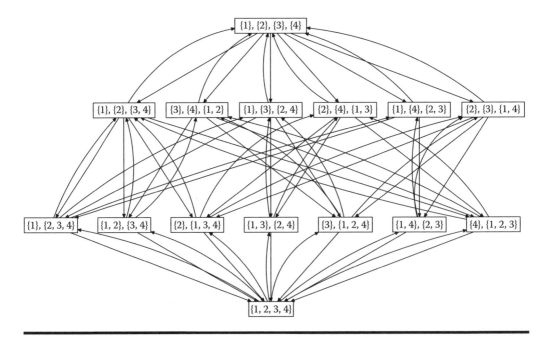

Figure 14.6 Coalition structure graph. Arrows in the graph represent coalition structure search directions. When followed downward, the arrows represent the merger of coalitions, and, when followed upward, the arrows represent the splitting of coalitions.

be guaranteed without seeing all coalition structures. If it is interpreted negatively, then it means that, if the number of players is large, then still exponentially many coalition structures have to be searched before a bound can be established.

14.5.1 Computational Complexity for Dynamic Wireless Networks

Let us consider a self-configuring wireless network of N nodes randomly placed with a uniform, independent distribution in a square area. We assume that any pair of wireless nodes within a radius r can communicate/interfere with each other. For a particular configuration of wireless nodes, as we increase the radius r, generally the network graph becomes denser. The network is generally quite sparse when r is small and exhibits low level of communication/interference. We can get a complete network graph when r is sufficiently high. In this complete graph, each node can communicate/interfere with every other node. However, it can be seen from Equation 14.10 that $B(N)$ grows super-exponentially. For large N, coalition interaction becomes an issue, leading to a combinatorial explosion. To find stable coalition structures, imposing some hierarchy, through clustering or limiting communication/interference radius r, becomes the practical solution to this problem in large scale wireless networks, with links interacting with a small subset of peers. This can be modeled as a network of wireless nodes only interacting with a limited subset of links known to them to form stable coalition structures. If the size of coalitions is limited to a specified constant, then we can find kernel-stable coalitions for the network with reduced computational complexity as explained in Section 14.2.2.

14.6 Conclusions and Open Challenges

The next generation of wireless networks are evolving toward distributed and self-configuring network architectures. The deployment of such networks raises new research challenges related to the adaptation of the participating nodes to the changing network environment. Cooperation among distributed and self-configuring wireless network nodes may lead to the efficient use of network resources. It is important to analyze such interactions between small groups of nodes, as well as between all network nodes.

As we discussed in this chapter, the idea of wireless nodes cooperating with each other to accomplish performance gains can be captured by coalition game theory. In self-configuring wireless networks, nodes make their decisions to form coalitions independently, but due to the presence of externalities, their choice has an impact on all the nodes in the network. There are many open problems unresolved regarding the presence of externalities in coalition-forming wireless networks. For instance, one issue of concern is the design of incentives for nodes to form coalitions in games with positive externalities. Furthermore, cooperative interactions among nodes in a self-configuring wireless network are a dynamic process. Protocols within the coalition game framework must be designed to incorporate the dynamic interactions of self-configuring nodes for resource allocation.

References

1. S. Mathur, L. Sankar, and N. B. Mandayam, Coalitions in cooperative wireless networks, *IEEE Journal on Selected Areas of Communications*, 26, 7, 1104–1115, Sept. 2008.
2. W. Saad, Z. Han, M. Debbah, and A. Hjorungnes, A distributed merge and split algorithm for fair cooperation in wireless networks, in *Proceedings of the ICC08*, Beijing, China, pp. 311–315, May 2008.

3. J. S. Baras, T. Jiang, and P. Purkayastha, Constrained coalitional games and networks of autonomous agents, in *Proceedings of the ISCCSP*, St. Julians, Malta, pp. 972–979, Mar. 2008.

4. T. Jiang and J. S. Baras, Fundamental tradeoffs and constrained coalitional games in autonomic wireless networks, in *Proceedings of the WiOpt07*, Limassol, Cyprus, Apr. 2007.

5. Thomas S. Ferguson, Games in Coalitional form, University lecture, University of California, Los Angeles, 2005.

6. G. Owen, *Game Theory*, New York: Academic Press, 1982.

7. E. Maskin, Bargaining, coalitions, and externalities, in *Proceedings of 13th WZB Conference on Markets and Political Economy*, Berlin, Germany, Oct., 2004.

8. R. B. Myerson, *Game Theory: Analysis of Conflict*, Cambridge, MA: Harvard University Press, 1990.

9. M. Klusch and A. Gerber, Dynamic coalition formation among rational agents, *IEEE Intelligent Systems*, 17, 3, 42–47, 2002.

10. M. J. Osbourne, *An Introduction to Game Theory*, New York: Oxford University Press, 2004.

11. Z. Han and H. V. Poor, Coalition games with cooperative transmission: A cure for the curse of boundary nodes in selfish packet-forwarding wireless networks, *IEEE Transactions on Communications*, 57, 1, 203–213, Jan. 2009.

12. R. Etkin, A. Parekh, and D. Tse, Spectrum sharing for unlicensed bands, *IEEE Journal on Selected Areas of Communications*, 25, 3, 517–528, Apr. 2007.

13. R. B. Myerson, Graphs and cooperation in games, *Journal on Mathematics of Operations Research*, 2, 3, 225–229, Aug. 1977.

14. M. O. Jackson and A. Wolinsky, A strategic model of social and economic networks, *Journal of Economic Theory*, 71, 44–74, 1996.

15. S. Mathur, L. Sankaranarayanan, and N. B. Mandayam, Coalitional games in receiver cooperation for spectrum sharing, in *Proceedings of CISS*, Princeton, NJ, 2006.

16. W. Saad, Z. Han, M. Debbah, A. Hjørungnes, and T. Basar, Coalitional games for distributed collaborative spectrum sensing in cognitive radio networks, in *Proceedings of IEEE INFOCOM*, Rio de Janeiro, Brazil, Apr. 2009.

17. K. Apt and T. Radzik, Stable partitions in coalitional games, arXiv:cs/0605132v1 [cs.GT], May 2006.

18. C. Cordeiro, E. Sofer, and G. Chouinard, Functional requirements for the 802.22 WRAN Standard r47, *IEEE Transactions on Aerospace and Electronic Systems*, AES-22, 1, pp. 98–101, Jan. 1986.

19. T. Dieckmann and U. Schwalbe, Dynamic coalition formation and the core, *Journal of Economic Behavior and Organization*, 49, 3, 363–380, Nov. 2002.

20. H. Konishi and D. Ray, Coalition formation as a dynamic process, *Journal of Economic Theory*, 110, 1–41, Apr. 2002.

21. S. Airiau and S. Sen, A fair payoff distribution for myopic rational agents, in *Proceedings of the Eighth International Conference on Autonomous Agents and Multi Agent Systems*, Budapest, Hungary, May 2009.

22. Z. Khan, S. Glisic, L. DaSilva and J. Lehtomaki, Modeling the dynamics of coalition formation games for cooperative spectrum sharing in an interference channel, to appear in IEEE Transactions on Computational Intelligence and AI in Games.

23. S. Glisic and B. Vucetic, *Spread Spectrum CDMA Systems for Wireless Communications*, Boston, MA: Artech House, 1997.

24. J. G. Kemeney and J. L. Snell, *Finite Markov Chains*, New York: Springer-Verlag, 1976.

25. T. Sandholm, K. Larson, M. Anderson, O. Shehory, and F. Tohme, Anytime coalition structure generation with worst case guarantees, *Journal of Artifical Intelligence*, 10, 209–238, Jul. 1999.

26. T. Sandholm, K. Larson, M. Anderson, O. Shehory, and F. Tohme, Coalition structure generation with worst case guarantees, *Journal of Artificial Intelligence*, 10, 209–238, Jul. 1999.

27. T. Sandholm, Distributed rational decision making, in: G. Weiss (Ed.), *Multi-Agent Systems*, MIT Press, Cambridge, MA, pp. 201–258, 1999.

Chapter 15

Auction Algorithms for Dynamic Spectrum Access

Bin Chen, Anh Tuan Hoang, and Ying-Chang Liang

Contents

In realization of cognitive radio technology, spectrum utility potential is expected to be fully released, along with the market power. In the future agile spectrum access market, the bargaining between service providers and users will become an inevitable topic as the lines between competition, cooperation, and deviation turn out to be fainter due to the complexity of the network. To deal with dynamic spectrum access from heterogeneous users, the network has to be self-maintainable and the role of the user has to be clearly defined in order for the "game" to play on. This chapter introduces a second-price auction algorithm to address the channel allocation and spectrum access problems for cognitive radio. The Nash equilibrium convergence of the budget-constraint auction and the strategic interactions between auctioneer and secondary users are examined. To cater for the demand of both parties, the implementation issues such as profit maximization, resource utilization maximization, and user performance maximization are also studied. The chapter is concluded by unveiling some open research questions in the application of game-theoretic algorithms.

15.1 Introduction

In many countries, almost all frequency bands have been allocated. As the wireless devices become more popular and new wireless systems are being introduced, spectrum scarcity has become a serious problem for the future development of wireless communications technologies. The FCC (Federal Communications Commission) report prepared by Spectrum Policy Task Force in 2002 points out that the spectrum shortage is caused by the current inefficiency of spectrum usage, rather than the physical spectrum scarcity. The report indicates that while some bands are heavily used—such as those bands used by cellular base stations (BS)—many other bands are not in use or are used only part of the time. For example, the report discloses some usage data regarding a particular police dispatch channel in New York State. The data show that "for the measurement period, typical channel occupancy was less than 15%, while the peak usage was close to 85% [1]." Recent measurements by FCC show that 70% of the allocated spectrum in the United States is not utilized. However, due to the existing legacy command-and-control regulation, it is hard to reallocate frequency bands.

15.1.1 Cognitive Radio

The seemingly spectrum shortage induces the notion of frequency reuse, which grants the secondary users/network to access the radio spectrum licensed to the primary users/network when the spectrum is temporarily unused. The very promising technology behind that allows this idea to come true is cognitive radio. It characterizes a communication system that does not need to depend on the heavily used frequency band and is capable of detecting the unused spectrum licensed to the primary users, thus gaining temporary access right to the spectrum. In December 2003, FCC issued a Notice of Proposed Rule Making that identifies cognitive radio as the candidate for implementing negotiated/opportunistic spectrum sharing [2]. In response to this, the IEEE (Institute of Electrical and Electronics Engineers) has formed the 802.22 Working Group to develop a standard for WRAN (wireless regional area networks) based on cognitive radio technology [3]. The primary users of a WRAN system include TV users and wireless microphones, and the secondary users include both WRAN BS (base station) and WRAN CPE (customer premise equipment).

15.1.2 Auction

Auction is a fundamental application of game theory, which is the formal study of conflict and cooperation. After William Vickrey's pioneer work in 1961, auction theory underwent a main development in 1980s and 1990s. New auction models were developed to address practical issues. Today, auctions have become very popular in the internet sales. Online sales magnates like eBay and GoIndustry all employ auctions to sell millions of goods and help the auctioneers maximize their revenues. Many economists regard auction theory as the best application of game theory to economics [4]. Many types of auctions have been developed to date; however, most of them are various forms of traditionally four standard types:

- Open ascending-bid auction (English auction)—An open-bid auction where the price is raised until one bidder remains.
- Open descending-bid auction (Dutch auction)—An open-bid auction where the price is successively lowered from a very high starting price until one bidder bids.
- First-price sealed-bid auction—A sealed-bid auction where the highest bidder wins and pays its own bid price.
- Second-price sealed-bid auction—A sealed-bid auction where the highest bidder wins, but pays the amount bid by the second-highest bidder. It is also called the Vickrey Auction.

Out of the four standard types of auctions, this chapter has placed particular interest on the second-price sealed-bid auction. Open auctions are not suitable for communication as they involve too much signaling. For sealed-bid auctions, one of the reasons that we favor second-price auction over first-price auction is that the former has a simple optimal strategy of bidding the true value, whatever other players do. In other words, "truth telling" is a dominant strategy equilibrium (and so also a Nash equilibrium), so the person with the highest value will win at a price equal to the value of the second-highest bidder.

To confirm the above statement, consider bidding $v - j$ when your true value is v [5]. If the highest bid other than yours is w, then if $v - j > w$, you win the auction and pay w, just as if you bid v. If $w > v$, you lose the auction and get nothing, just as if you bid v. But if $v > w > v - j$, bidding $v - j$ causes you to lose the auction and get nothing, whereas if you had bid v, you would have won the auction and paid w for a net surplus of $v - w$. So you never gain, and might lose, if you bid $v - j$. Now consider bidding $v + j$ when your true value is v. If the highest bid other than yours is w, then if $v > w$ you win and pay w, just as if you bid v. If $w > v + j$ you lose and pay nothing, just as if you bid v. But if $v + j > w > v$, having bid $v + j$ causes you to "win" an auction; you otherwise would have lost, and you have to pay $w > v$ so you get negative surplus. So bidding $v + j$ may hurt you compared with bidding v, but it never helps you.

In contrast, a bidder has to lower its true value to win the bid in a first-price auction. Masking of bid is essential as the payment is equal to the bid if the auction is won. That makes the formulation much more complicated as compared with second-price auction. Also, this computational complexity will create unnecessary communication overhead. Furthermore, these two types of auctions are equally effective in allocating goods to the bidder who values it most, while maximizing the seller's revenue, according to Revenue Equivalence Theorem [5]:

Assume each of n risk-neutral potential buyers has a privately known value independently drawn from a common distribution $F(v)$ that is strictly increasing and atomless. Suppose that no buyer wants more than one of the k available identical indivisible objects. Then any auction mechanism in which (a) the objects always go to the k buyers with the highest values, and (b) any bidder with the lowest possible value expects zero surplus, yields the same expected revenue and results in a buyer with value v making the same expected payment.

15.1.3 Modeling

There are three essential components for cognitive radio as being defined in [6]:

- *Spectrum Sensing*—The secondary users are required to sense the radio spectrum environment within their operating range to detect the frequency bands that are not occupied by primary users.
- *Dynamic Spectrum Management*—Cognitive radio networks are required to dynamically select the best available bands for communications and monitor the radio environment in order to protect primary users.
- *Adaptive Communications*—A cognitive radio device can configure its transmission parameters (carrier frequency, bandwidth, transmission power, etc.) to opportunistically make best use of the ever-changing available spectrum.

Spectrum sensing can be divided into two categories: in-band and out-band sensing. In a cognitive radio network, secondary party needs to periodically monitor the spectrum in which it is operating on; this is called in-band sensing. All users have to be quiet in order for in-band sensing to be carried out. This guarantees that the possibility that any transmitting secondary users being mistaken as the active primary user is zero. The methodology of sensing could be in a cooperative or opportunistic way to keep track of the active list of channels. In case it detects the incumbent user, BS can switch to back up channels that are maintained by out-band sensing. Out-band sensing can be carried out by BS or can be assigned to secondary users to free up resources of BS. The later is recommended by this chapter, as it would impose cooperative sensing in an incentive-driven manner and allow BS to focus its role as an auctioneer. During out-band sensing, secondary users search for the new channels, which become available for the cognitive radio network, also known as spectrum hole search. Spectrum hole is defined as a band of frequencies assigned to a primary user, but, at a particular time and specific geographic location, the band is not being utilized by that user [7]. Users do not have to be quiet during this spectrum hole "searching" time. While in-band sensing has to impose on all secondary users to avoid interference with the primary users, allocation of the out-band sensing time requires a standard to achieve fairness. For dynamic spectrum management, the spectrum allocation problem comes into play if there are several users competing for the same available channel. The cognitive radio network is required to allocate the channel to the user who can best utilize it. Due to the increased wireless network complexity nowadays, a centralized allocation scheme is too rigid to handle the unpredictability of mobility and traffic models, coupled with the dynamic topology and the uncertainty of the link quality. Distributed decision making is therefore a better alternative, which allows independent users to make rational decisions in the ever-changing environment and play the games among themselves. Auction suits well in this context, where each decision maker has to bear the consequence of its own decision. As the name "cognitive radio" suggests, secondary users should possess the ability to analyze the environment and make the best decision to maximize their gain.

We apply auction algorithm in this dynamic spectrum access game for the cognitive radio network while modeling out-band sensing time as the price to pay upon winning the transmission time slot. There are some universal assumptions for this chapter:

- The algorithm is built upon the basic operating mode of IEEE 802.22 WRAN, that is, the non-hopping mode [8], in which sensing and data transmission are not done concurrently as each user is regarded as an isolated and independent one. Users are limited by the single antenna of the CPE, and they have to manage sensing and data transmission on the same time axis.

- There are less than enough vacant channels, say during the peak hour, thereby leading to competitions among users. The limited resources result in myopic and selfish behavior, so some of the users may break away from the cooperative rules designated by the protocol, turning it into a noncooperative game. By "myopic," it means that these users only care about short-term benefits. They are ready to tamper with their wireless interface in order to increase their own share of the common transmission resource while neglecting out-band sensing task initially assigned to them assumed in the cooperative mode. By "selfish," it means that these users are rational, but not malicious. They are profit-driven, capable of calculating and maximizing their outcome. In fact, even if there are enough resources, selfish behavior still cannot be ruled out especially when users are aware of the increasingly programmable features of the network adapters nowadays [9].

15.1.4 Organization

The rest of the chapter is organized as follows. In Section 15.2, we propose our model of cognitive radio and formulate the second-price auction. In Section 15.3, we discuss the strategic interactions among users and implementation issues of the auction model. Several optimization approaches are evaluated to upgrade the auction algorithm. In Section 15.4, we conclude our findings.

15.2 Game Formulation

15.2.1 The Proposed Operating Model of Cognitive Radio

In [6], the operating time of the cognitive radio network is divided into frames where each frame consists of one in-band sensing slot and one transmission slot. The typical frame duration is about 100 ms while the optimal in-band sensing time duration is about 2.55 ms to achieve the targeted 90% detection probability. In this long duration, the channel condition can change drastically. Hence, it is suggested to further divide a frame into more subframes, with a much shorter duration, say 10 ms. Those subframes will be allocated to users according to the preferred allocation scheme. When the time frame is sufficiently short, the assumption of constant channel state coefficient over the transmission time slot can be held. The proposed operating model is shown in Figure 15.1, where the in-band sensing time slot is kept at the beginning of every frame, and the remaining part of the frame is divided into m subframes. During the in-band sensing time, all users will remain silent. After that, subframe allocation will be carried out using second-price sealed-bid auction. The winner of each subframe will be granted the transmission right for a duration of $(1 - w) T$, and then it will have to carry out out-band sensing for a duration of wT, which is the second-highest bid.

15.2.2 Second-Price Auction Formulation

For a given power level, it is assumed that the throughput is a linear function of the channel state. This is justified by the Shannon capacity at low signal-to-noise ratio [10]. X_i is used to represent a random variable of the channel state for the channel between the transmitter and receiver i, for

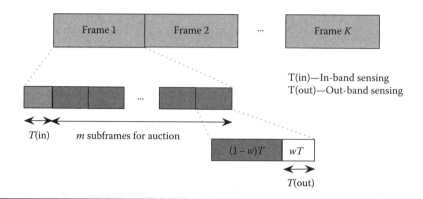

Figure 15.1   The proposed operating model of cognitive radio in non-hopping mode.

$i = 1, \ldots, N$. A general throughput function will then be $P \cdot X_i$, where P is the transmit power of the base station. Without the loss of generality, P is assumed to be 1 and constant for all users. Moreover, the distribution of X_i is assumed to be known by all users for all i.

Assume that the exact value of the channel state X_i is revealed to user i only at the beginning of each subframe. Let α_i denote the budget constraint of user i. It is also assumed that users know the average amount of budgets available to each user. The assumption is valid in the case that each user has the knowledge of the historical play of other users. Thus, they can estimate the budgets of other users according to their historical bids. This is a typical *complete information game* in game theory. The assumption of complete information will be released in the next section when the implementation issues are raised. During the auction, the following actions will take place:

■ Each user submits a bid according to the channel condition at the particular moment of submission.
■ The auctioneer (the transmitter or the BS) chooses the one with the highest bid to transmit.
■ The winner pays the price of the second-highest bid. In case of a tie, the winner will be chosen among the equal bidders with equal probability.

Note that there is a significant communication overhead in terms of time to submit a bid for every resource. In the auction implementation, it is proposed to update bids and corresponding resource iteratively. This is similar to the distributed power control case in which signal-to-noise ratio and power are obtained iteratively. It will slow down the equilibrium converging speed but overall signaling can be significantly reduced. The convergence might not be a priority of a cognitive radio network in which the amount of users and channels are subjected to dynamical change. The optimal strategy can also be recommended by BS.

The main objective of the formulation is to design a bidding strategy, so that a user will act accordingly in every possible condition to maximize its expected throughput subject to its average budget constraint. The expectation might depend on its Quality of Service (QoS) requirement, say the packet arrival rate. When the strategic function of user i is chosen, say f_i, it will bid an amount equal to $f_i(x')$, where its channel condition is $X_i = x'$. The analysis of the second-price auction begins by looking at a simple two-user case, $N = 2$ and then further extends to N-user games. It is similar to the derivation in [11].

For a general N-player game, it can be represented as $\Gamma = [N, \{S_i\}, \{g_i(\cdot)\}]$, which specifies for each player i a set of strategies, or bidding functions S_i (with $s_i \in S_i$), and a payoff function

$g_i(s_1, \ldots, s_N)$ giving the throughput associated with outcome of the auction arising from strategies $(s_1, \ldots, s_N)$.

Consider the case in which two users choose their strategies from the set F_1 and F_2, respectively. Here it is assumed that the two players are of the same type. Each user's strategy is a function of its own channel state X_i. Thus, F_i is defined to be the set of continuous real-valued, and square integrable increasing functions over the range of X_i, which is continuously distributed over a finite interval of $[l_i, u_i]$. l_i and u_i are both nonnegative real number with $l_i < u_i$.

Given user 1's strategy $f_1 \in F_1$, with its range from $f_1(l_1) = a$ to $f_1(u_1) = b$, user 2 aims to maximize its own expected throughput, while satisfying its expected budget constraint α_2. Denoting user 2's throughput as g_2, it is a function of both user 1 and user 2's strategies f_1 and f_2. It is written as

$$g_2(f_1, f_2) = E_{X_1, X_2}[X_2 \cdot (1 - f_1(X_1))_{f_2(X_2) \geq f_1(X_1)}],$$

where

$$(1 - f_1(X_1))_{f_2(X_2) \geq f_1(X_1)} = \begin{cases} 1 - f_1(X_1) & \text{if } f_2(X_2) \geq f_1(X_1) \\ 0 & \text{otherwise} \end{cases} \tag{15.1}$$

As the price that the winner has to pay follows second-price rule, user 2's strategy set depends on f_1 and has to meet its budget constraint α_2. Therefore, user 2's set of feasible bidding functions $S_2(f_1)$ is given by

$$S_2(f_1) = \{f_2 \in F_2 | E_{X_1, X_2}[f_1(X_1) \cdot 1_{f_2(X_2) \geq f_1(X_1)}] \leq \alpha_2\} \tag{15.2}$$

To evaluate the opponent's strategy, an inverse function $h(y)$ is introduced, where y is the user 2's bid at a particular subframe. For a strictly increasing f_1, $h(y)$ is defined as

$$h(y) = \begin{cases} l_1 & \text{if } y \leq a \\ f_1^{-1}(y) & \text{if } a < y < b \\ u_1 & \text{if } y \geq b \end{cases} \tag{15.3}$$

$h(y)$ serves to inverse user 2's bid to reflect user 2's position in user 1's range of channel state condition $[l_1, u_1]$ using user 1's strategy function, such that user 2's winning probability given its bid y can be determined:

$$P_2^{win}(y) = P(f_1(X_1) \leq y) = P(X_1 \leq h(y)) = \int_{l_1}^{h(y)} p_{X_1}(x_1) \, dx_1 \tag{15.4}$$

where

p_{X_1} is the probability density function of user 1's channel state condition X_1

x_1 is a possible channel state realization of X_1 ranging from l_1 to u_1

Similarly, p_{X_2} and x_2 are the probability density function and the channel state realization of X_2, respectively, as they will appear in the later part of the formulation.

User 2 now wants to find a strategy f_2 to maximize its expected throughput subject to its budget constraint α_2. The optimization problem is written as

$$\max_{f_2} \int_{l_2}^{u_2} \int_{l_1}^{h(y)} x_2(1-f_1(x_1))p_{X_2}(x_2)P_2^{win}(y)dx_1dx_2 = \max_{f_2} \int_{l_2}^{u_2} x_2p_{X_2}(x_2) \int_{l_1}^{h(f_2(x_2))} (1-f_1(x_1))p_{X_1}(x_1)dx_1dx_2$$

$$\text{subject to } \int_{l_2}^{u_2} \int_{l_1}^{h(f_2(x_2))} f_1(x_1)p_{X_1}(x_1)p_{X_2}(x_2)dx_1dx_2 \leq \alpha_2 \tag{15.5}$$

Solving the above problem gives the user 2's *best response* as

$$\begin{aligned} f_2(x_2) &\leq a & \text{for } x_2 \in [l_2, \Theta_1] \\ f_2(x_2) &= \frac{x_2}{x_2 + \lambda_2} & \text{for } x_2 \in [\Theta_1, \Theta_2] \\ f_2(x_2) &\geq b & \text{for } x_2 \in [\Theta_2, u_2] \end{aligned} \tag{15.6}$$

where

$$\Theta_1, \Theta_2 \in [l_2, u_2]$$
$$\Theta_1 = \frac{a\lambda_2}{1-a}, \quad \Theta_2 = \frac{b\lambda_2}{1-b}$$

A strategy f_2 is said to be the best response for user 2 given its opponent's strategy if $g_2(f_1, f_2) \geq g_2(f_1, f_2')$ for all $f_2' \in S_2(f_1)$. Applying the same analysis method to user 1 given user 2's strategy, the best response of user 1 gives the similar form. A strategy pair (f_1^*, f_2^*) is said to be in Nash equilibrium if f_1^* and f_2^* are both the best response.

Assuming the channel state random variable X_i is uniformly distributed over $[0, 1]$, and user 2's budget is higher than that of user 1's (i.e., $\alpha_2 > \alpha_1$), it gives the following equations:

$$\int_0^1 \int_0^{f_2^{-1}(f_1(x_1))} f_2(x_2)\, dx_2dx_1 = \alpha_1 \tag{15.7}$$

$$\int_0^{\frac{\lambda_2}{\lambda_1}} \int_0^{f_1^{-1}(f_2(x_2))} f_1(x_1)\, dx_1dx_2 + \int_{\frac{\lambda_2}{\lambda_1}}^1 \int_0^1 f_1(x_1)\, dx_1dx_2 = \alpha_2 \tag{15.8}$$

Solving the above equations simultaneously, the solution is given by

$$\alpha_1 = \frac{\lambda_2}{2\lambda_1} + \lambda_2 + (\lambda_2 + \lambda_2\lambda_1)\ln\left(\frac{\lambda_1}{1+\lambda_1}\right) \tag{15.9}$$

$$\alpha_2 = 1 - \frac{\lambda_2}{2\lambda_1} + \lambda_2 + \lambda_1(1+\lambda_2)\ln\left(\frac{\lambda_1}{1+\lambda_1}\right) \tag{15.10}$$

15.3 Numerical Results and Discussion

15.3.1 Simulation Models

15.3.1.1 Data Traffic Model

In this simulation model, it is assumed that data packets arrive at the buffers randomly, following a Poisson distribution. Denoting A_i as the packet arrival event, it gives $A_i \sim P_o(\mu_i)$, for $i = 1, \ldots, N$, where μ_i is the mean packet arrival rate for user i. It should be noted that secondary users might not always have packets to transmit in this traffic system. Furthermore, it is assumed that all data packets of user i have the same size and deadline constraint. If a packet is not transmitted by its deadline, it is dropped and considered lost. The transmitter then has to retransmit the lost packets. The deadline constraint characterizes the timeout constraints existing in many network services like file downloading. Buffer overflow is assumed to be negligible.

15.3.1.2 Fading Channel Model

Consider a simple downlink communication scenario with one sub-channel, N receivers, and one transmitter. Assuming that h_i is the block-fading process of receiver i on the sub-channel, P_i is the transmit power of the BS when receiver i is granted the transmission opportunity, n_i is the additive Gaussian noise at receiver i. h_i is assumed to stay constant over each block-fading length (i.e., coherent communication). The assumption of coherent reception is reasonable if the fading is slow in the sense that the receiver is able to track the channel variations. The additive Gaussian noise n_i at the receiver is i.i.d. (independent, identically distributed) circularly symmetric with zero mean and variance $E[|n_i|] = N_o$, for $i = 1, \ldots, N$. The transmit power is assumed to be constant over time. The *instantaneous capacity* is given by

$$C_i = \log_2 \left(1 + \frac{P_i |h_i|^2}{N_o} \right) \quad \text{for } i = 1, \ldots, N \tag{15.11}$$

For convenience, $|h_i|^2$ is equalized to X_i in order to match the notation of the channel state coefficient used in the theoretical model. As such, the achievable channel capacity is solely dependent on the channel state condition, which is time varying.

15.3.2 Nash Equilibrium Establishment

The author in [12] states that a noncooperative game is confined to an equilibrium condition, and the Nash equilibrium does not tell about the underlying dynamics involved in establishing that equilibrium. We show that a simple learning strategy is enough to establish a Nash equilibrium state as illustrated by Figures 15.2 and 15.3. Since the auction model here is a typical second-price auction with budget constraint, users can adopt the derived equilibrium bidding strategy and play the game, while adjusting the bidding coefficient λ_i in a designated step size to meet their budget constraints.

Given budget constraints $\alpha_2 > \alpha_1$, and the channel state coefficient X_i uniformly distributed over $[0, 1]$ for all i, user 2's best response strategy is given by

$$f_2(x_2) = \frac{x_2}{x_2 + \lambda_2} \quad \text{for } x_2 \in \left[0, \frac{\lambda_2}{\lambda_1} \right]$$

$$f_2(x_2) = \frac{1}{1 + \lambda_1} \quad \text{for } x_2 \in \left[\frac{\lambda_2}{\lambda_1}, 1 \right] \tag{15.12}$$

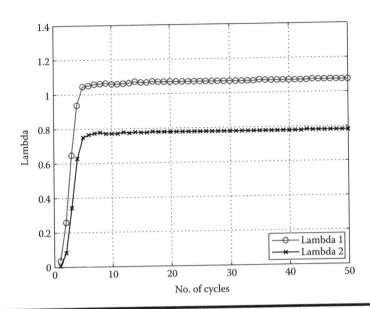

Figure 15.2   Convergence of 2-user game.

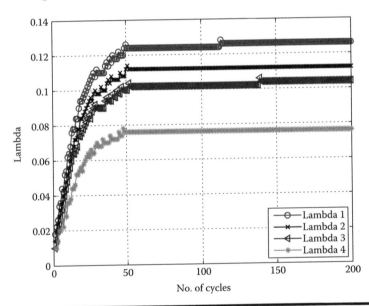

Figure 15.3   Convergence of 4-user game.

User 1's best response strategy is given by

$$f_1(x_1) = \frac{x_1}{x_1 + \lambda_1} \quad \text{for } x_1 \in [0, 1] \tag{15.13}$$

Figures 15.2 and 15.3 show that a unique pure strategy Nash equilibrium exists for both 2-user and 4-user games. The implication is valid for any multiuser games. The convergence is also valid when the traffic is taken into account as shown in Figure 15.4.

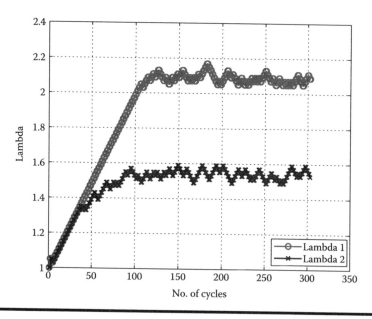

Figure 15.4   Convergence of 2-user game in the traffic system.

If one of the users (user 2) deviates from the Nash equilibrium strategy and bids more aggressively, Figure 15.5 shows that a Nash equilibrium can still be established. In this case, user 2's bidding strategy is $f_2(x_2) = \lambda_2 \sqrt{x_2}$, and both users have the same budget constraint. As shown in Figure 15.6, aggressive bidding does earn user 2 some benefits during the equilibrium-establishing phase, reflected by a higher throughput for user 2. However, the throughput for both users diminishes to 0 after equilibrium is established. The prophecy of Prisoner's Dilemma has been fulfilled

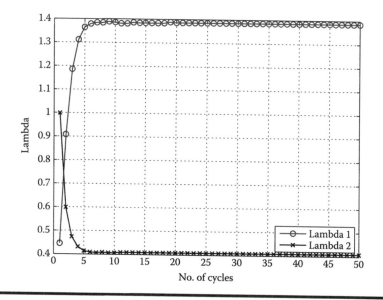

Figure 15.5   Convergence of 2-user game with a deviant user.

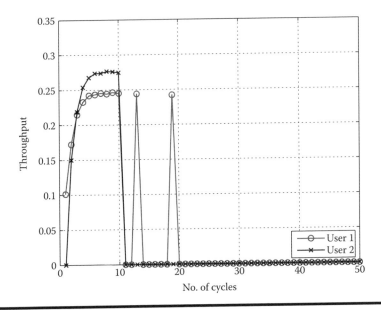

Figure 15.6 Throughput of 2-user game with a deviant users.

here as the defective equilibrium drains both users. In fact, the greedy user may face a real dilemma to deviate from recommended Nash equilibrium strategy, as it may or may not benefit from it depending on its current strategic interaction state with other users. Nevertheless, it can still gain something by trial and error and revert to the original strategy as a disguise. As illustrated by the simulation, this kind of opportunistic behavior will need to be contained as the disastrous outcome brought by the undesirable equilibrium is sudden.

15.3.3 Bidder's Best Budget

If the system does not impose any constraints on the budget, each user will have the freedom to choose the budget based on its QoS requirement. However, strictly obeying the QoS criteria may not always be a wise choice especially in the case when a new user enters a system in which several other users have already established an equilibrium state. We use a 2-user game to illustrate the optimal budget selection decision making as shown in Figure 15.7. In this simulation of 2-user game, the throughput of user 1 is simply $X_1 \cdot (1-f_2)$ if it wins the game, according to the second-price rule. We are able to find the optimal budgets of user 2 indicated by its optimal throughput when user 1 varies its budget. In this case, user 2 is considered a new user and user 1 is an existing one.

The game outcome is summarized in Table 15.1. We can see that there is always an optimal budget for user 2 to help it achieve the optimal throughput. When the system cannot satisfy the needs of every user, it will rely on users themselves to make the best bargain of it. If users are rational, the undesirable equilibrium in the classic example of Prisoner's Dilemma should not emerge.

15.3.4 Optimal Bidding Slot Length

The bidding slot duration has been presumably defined to be 10 ms earlier in this chapter without any justification. In a real traffic system, the length of the bidding slot will have an impact on the system efficiency. If the slot is too long, the winner will have plenty of time for transmission,

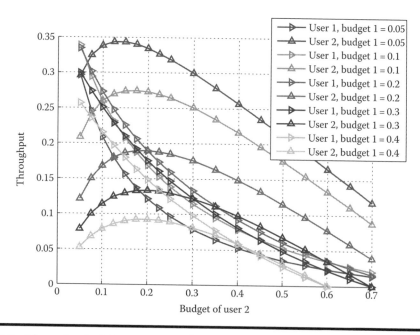

Figure 15.7 **Throughput of 2-user game when user 2's budget varies. With budget, fixed at 0.05, 0.1, 0.2, 0.3, and 0.4, respectively.**

Table 15.1   Game Outcome When User 2 Plays against User 1 of Different Budgets

Game	1	2	3	4
User 1's budget	0.05	0.1	0.2	0.3
User 2's best budget	0.14	0.17	0.2	0.2
User 2's best throughput	0.35	0.275	0.18	0.13
User 1's corresponding throughput	0.17	0.2	0.18	0.175
Total throughput	0.52	0.475	0.36	0.305

but other players will have to wait if there are insufficient sub-channels to share around. The long waiting time simply deprives their opportunities to commit to the packet deadline constraints. In contrast, if the slot duration is too short, there will be insufficient time to clear up the packets with an imminent deadline violation, and it will also increase the packet loss rate.

Figure 15.8 is plotted to find the existence of the optimal slot duration. It is assumed that users do not change their strategies. The simulation also takes different packet deadlines and traffic loads into consideration. It is found that the user with higher packet arrival rate (user 2) prefers longer slot length while the user with lower packet arrival rate (user 1) prefers a shorter one. The system has to make a choice in between. When the deadline constraint becomes more stringent (Short D), even user 2 prefers shorter slot duration because its packet loss rate increases and losing the bid creates a more negative impact on its throughput.

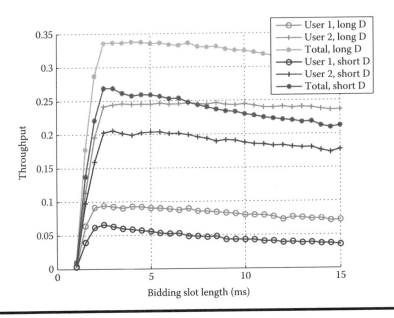

Figure 15.8 **Throughput vs. slot length for two different packet deadlines.**

15.3.5 Supplementary Bidding Rules

In the game formulation, traffic is not taken into consideration. It implicitly assumes that a user always submits a nonzero bid in each subframe, as it always has packets to transmit. In reality, if the achievable channel capacity is sufficiently large to support the traffic load, the user will encounter a situation in which there is no packet to transmit when it wins the bid, thus wasting the system resources and others' opportunities. From a selfish point of view, it would also be better to give up the bidding of the current round and save the budget for the future.

Considering the fact that bidders may not be able to fully utilize the resource, the auctioneer may also want to impose an additional bidding rule that a user only bids when its channel utility can exceed a certain limit at the instant of bidding. The channel utility here refers to the estimated percentage of time the user will be spending on transmission rather than waiting for the incoming packets if it wins the slot. The utility is dependent on its packet arrival rate and the buffer length. A user with a higher traffic load and longer buffer length will be able to utilize the transmission time more efficiently. The utility ratio r is expressed as follows:

$r_i = (L_i + \mu_i \cdot T)/(c \cdot T)$, for $i = 1, \ldots, N$. L_i is user i's instantaneous buffer length, and c is the average transmission rate. The denominator $(c \cdot T)$ is the normalizer. Note that r_i can exceed 1.

The buffer length alone may not truly reflect the waiting time. For two equal lengths of buffers, the queuing time may be very different. The longer a packet waits in the buffer, the higher the chance it will violate the deadline constraint. Therefore, another rule is required to take the packet waiting time into account. Here, we introduce a weighting factor w_i, which is expressed as follows:

$w_i = Q_{max}/Q_i \cdot f_i$, for $i = 1, \ldots, N$. Q_{max} denotes the longest packet waiting time in user i's buffer, Q_i is the average packet waiting time in the buffer, and f_i is user i's bid.

"Wait-time rule" deals primarily with the waiting time of the data packets. If there is no packet waiting in the buffer, normal second-price auction rule will be applied. This rule becomes effective only when the buffer contains at least one packet. In that case, the weighting factor w_i will be used

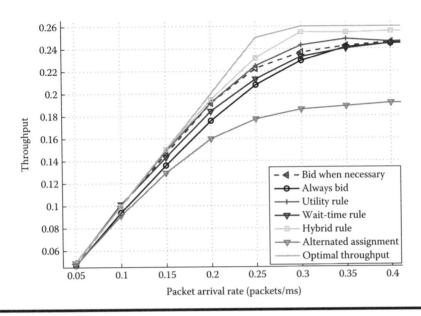

Figure 15.9 **Throughput vs. packet arrival rate for different bidding rules.**

to decide the winner instead. A user with a higher w_i wins the auction and pays an amount equal to the second-highest weighting factor.

To compare the above-mentioned supplementary bidding rules, throughput is plotted against different packet arrival rates for a 2-user game with the same budget, as shown in Figure 15.9. "Hybrid rule" refers to the combination of "Utility rule," "Wait-time rule," as well as the strategy of bidding only when necessary. It is observed that a simple "bid when necessary" strategy has significantly improved the system efficiency. The other supplementary rules all outperform the original rule in the light-traffic condition. In the heavy-traffic condition, the effectiveness of these supplementary rules diminishes as the throughput of the network caps at the maximum.

15.3.6 Auctioneer's Revenue

The revenue here refers to the total out-band sensing time that the auctioneer can collect from the users. The objective of the auctioneer is substantially different from that of the secondary users. The auctioneer aims gather more out-band sensing commitment from users to maximize its profit. When there are sufficient available channels detected through out-band sensing, the auctioneer can sell them to more users and maintain a good service at the same time.

To maintain a good service, the auctioneer will have to satisfy users' QoS requirements. Figure 15.10 is plotted to examine the revenue that the auctioneer can earn while meeting a minimum throughput requirement claimed by the users, based on a simple setup where two users have the same throughput requirement. The auctioneer will in turn impose a minimum average budget that users must set. The system would therefore be able to operate in a more regulated manner.

In a system with traffic load, instead of proposing a minimum throughput requirement, the user may request a minimum packet loss, to address its deadline and energy constraints, considering

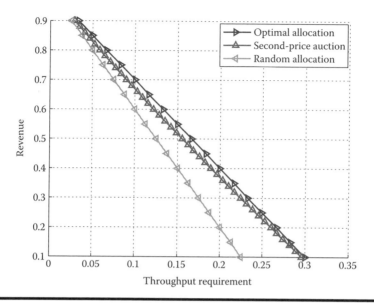

Figure 15.10 Auctioneer's revenue vs. users' throughput requirements.

the possibility that a user with a high throughput or traffic load may have a large packet loss at the same time. To study the auctioneer's revenue under different packet loss requirements, Figure 15.11 is plotted. In Figure 15.11, the revenue decreases drastically when the user demands the packet loss rate to be lower than 20%. The auction scheme earns the auctioneer significantly more revenue when users' packet loss requirement is high and performs equivalently as the optimal scheme. These three schemes do not differ much when the user has a low demand of the packet

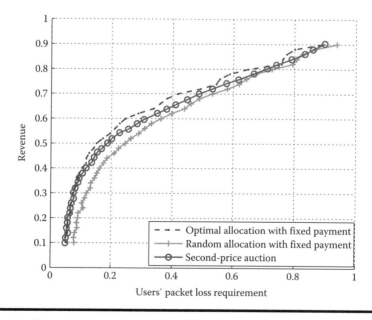

Figure 15.11 Auctioneer's revenue vs. users' packet loss requirements.

loss rate, which implies a low necessity for the efficient allocation. As a result, auction becomes redundant.

In a market full of competition and price war, the auctioneer may want to set a reserve price so as to safeguard a certain amount of profit. If the reserve price is too low, there is a high chance that the object is sold at a low price, because it would not be able to exclude users who do not value it much and conveniently drop a low bid for it due to the second-price rule. Due to the market mechanism (demand and supply), price war can also cause the value of the object to drop. If the reserve price is too high, it will scare away the bidders, and the object will not be sold. By setting an optimal reserve price, the auctioneer essentially places a bid against every other potential bidder, thus preserving the auctioneer's private value about the object.

Unlike the average budget constraint, a reserve price is a hard constraint, which means that the bidder has to submit a bid higher or equal to the reserve price to make its bid a valid one. Consider a situation where two users carry out the bidding for a single sub-channel, and each of them has an equal throughput requirement to meet. For simplicity, we assume that the equilibrium state has been established. Due to the presence of other concurrent auctions, the user can choose not to bid if the reserve price is too high. The auction model is exactly the same as before except the existence of a reserve price.

Figure 15.12 is plotted to illustrate the auctioneer's revenue in the auction mentioned earlier. We observe that the optimal reserve price exists for both pairs of users with different throughput requirements. The optimal reserve price and revenue gained by the auctioneer are higher for the pair of user with low traffic requirement, because their average budgets are high, and so they can afford to bid higher. The auctioneer is then able to sell the channel at a higher price. If the user has a higher throughput demand, it would be very sensitive to the raise of the reserve price and would quickly drop out after it reaches a threshold price.

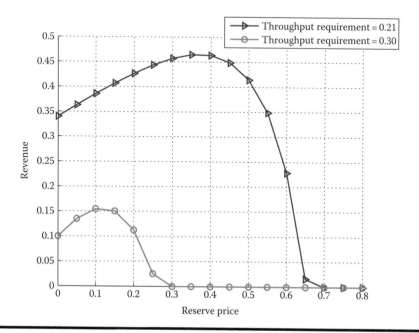

Figure 15.12   Auctioneer's revenue vs. different reserve prices.

15.4 Future Work and Conclusions

In this chapter, we design a second-price auction algorithm to address the channel allocation and spectrum access problem for cognitive radio. We model the out-band sensing time as the price to pay if one wins the auction and introduce a budget constraint to help establish the Nash equilibrium. We focus on the implementation issues that might draw the policy maker's attention, such as the strategic interactions among users, spectrum utilization efficiency, user's QoS requirement, auctioneer's profit, and so on.

We have proved the convergence toward Nash equilibrium state when users play to fulfill their budget constraints. Whether the equilibrium is constructive or destructive will depend on the obedience of user. The auctioneer may recommend a calculated initial state and strategy to accelerate the convergence in a dynamic network where the user may not possess sufficient knowledge about others' historical play.

From the system point of view, the efficiency of the auction can be improved by optimizing the bidding slot length, imposing supplementary bidding rules, etc. In the ideal model, it is assumed that users always bid and commit to the auction. This assumption is relaxed when we take the traffic condition and market mechanisms into account. A selfish user can choose bid or not to bid when it tries to fulfill its own QoS requirement, especially when there are more options available in the market. To safeguard the profit as well as to select appropriate customers, the auctioneer can impose a minimum payout on users, which can be in the form of a reserve price while still satisfying their minimum QoS requirements. In this way, both parties can benefit from the auction. To cater for different QoS requirements, the setting of reserve price should also vary.

The analysis and formulation in this chapter rely much on two-player scenario, as well as on Nash equilibrium strategy. A multiple-player game in which users are succumbed to incomplete information certainly is a strong direction for the future work. The users will have to maximize their gain with a budget constraint in a more uncertain manner, probably based on historical play, and without even establishing an equilibrium. In this case, BS will have to reveal certain network information for users to find a gain-maximizing solution, probably the network size besides historical winning bids, which have to be published as a standard auction procedure. To reduce communication overhead, the idealistic bidding per subframe as presented in this chapter can be replaced by a multiple-unit auction model. The advantage of multiple-unit auction against iterative bid updating can be weighted in the perspective of network maintainability, utility, and user satisfaction level. At the mean time, the private value model due to fading can be combined with a common value model, as the importance of future value will diminish in a multiple-unit auction.

References

1. Federal Communications Commission, Spectrum Policy Task Force Report, FCC 02-155. Nov. 2002.
2. Federal Communications Commission, Facilitating Opportunities for Flexible, Efficient, and Reliable Spectrum Use Employing Cognitive Radio Technologies, Notice of Proposed Rule Making and Order, FCC 03-322. Dec. 2003.
3. IEEE 802.22 Wireless RAN, Functional requirements for the 802.22 WRAN standard, IEEE 802.22-05/0007r46, Oct. 2005.
4. E. Maasland and S. Onderstal, Auction theory, *Medium Econometrische Toepassingen*, 13, 4, 4–8, 2005.
5. P. Klemperer, *Auctions: Theory and Practice*. Princeton University Press, Princeton, NJ, 2004.
6. Y.-C. Liang, Y. Zeng, E. Peh, and A. T. Hoang, Sensing-throughput tradeoff for cognitive radio networks, *IEEE Transactions on Wireless Communications*, 7, 4, Apr. 2008.

7. P. Kolodzy et al., Next generation communications: Kickoff meeting, in *Proceedings of DARPA*, Oct. 2001.

8. IEEE P802.22/D0.1 Draft Standard for Wireless Regional Area Networks Part 22: Cognitive wireless RAN medium access control (MAC) and physical layer (PHY) specifications: Policies and procedures for operation in the TV Bands, 2008.

9. M. Raya, J.-P. Hubaux, and I. Aad, DOMINO: A system to detect greedy behavior in IEEE 802.11 hotspots, in *Proceedings of ACM MobiSys*, Boston, MA, pp. 84–97, 2004.

10. A. Fu, E. Modiano, and J. Tsitsiklis, Optimal energy allocation for delay-constrained data transmission over a time-varying channel, in *IEEE INFOCOM*, San Francisco, CA, vol. 2, pp. 1095–1105, Apr. 2003.

11. J. Sun, E. Modiano, and L. Zheng, Wireless channel allocation using an auction algorithm, *IEEE Journal on Selected Areas in Communications*, 24, 5, 1085–1096, May 2006.

12. S. Haykin, Cognitive radio: Brain-empowered wireless communications, *IEEE Journal on Selected Areas in Communications*, 23, 2, 201–220, 2005.

Chapter 16

Bargaining Strategies for Camera Selection in a Video Network

Bir Bhanu and Yiming Li

Contents

Due to the broad coverage of an environment and the possibility of coordination among different cameras, video sensor networks have attracted much interest in recent years. Although the field of view (FOV) of a single camera is limited and cameras may have overlapping or nonoverlapping FOVs, seamless tracking of moving objects can be achieved by exploiting the handoff capability of multiple cameras. In this chapter, we will provide a new perspective to the camera selection and handoff problem that is based on game theory. In our work, game theory is used for multi-camera multi-person seamless tracking based on a set of user-supplied criteria in a network of video cameras for surveillance and monitoring. The bargaining mechanism is considered for collaborations as well as for resolving conflicts among the available cameras. Camera utilities and person utilities are computed based on a set of criteria. They are used in the process of developing the bargaining mechanisms. The merit of our approach is that it is independent of the topology of how the cameras are placed in the network. When multiple cameras are used for tracking and where multiple cameras can "see" the same object, we are able to choose the "best" camera based on multiple criteria that are selected a priori. The algorithm can automatically provide an optimal as well as stable solution of the camera assignment quickly. The detailed camera calibration or 3D scene understanding is not needed in our approach. Experiments for multi-camera multi-person tracking are provided to corroborate the proposed approach. We also provide a comprehensive comparison of our work and some non-game-theoretic approaches, both theoretically and experimentally.

16.1 Introduction

The growing demand for security in airports, banks, shopping malls, homes, etc. leads to an increasing need for video surveillance, where camera networks play an important role. Significant applications of video network include object tracking, object recognition, and object activities from multiple cameras. The cameras in a network can cooperate with each other and perform various tasks in a collaborative manner. Multiple cameras enable us to have different views of the same object at the same time, such that we can choose one or some of them to monitor a given environment. This can help to solve the occlusion problem to some extent, as long as the FOVs of the cameras have some overlaps. However, since multiple cameras may be involved over long physical distances, we have to deal with the handoff problem as well. *Camera hand-off* is the process of finding the next best camera to see the target object when it is leaving the FOV of the current camera, which is being used to track it [17]. This has been an active area of research and many approaches have been proposed. Some camera networks require switches (video matrix) to help monitor the scenes in different cameras [1]. The control can be designed to switch among cameras intelligently. Both distributed and centralized systems are proposed. Some researchers provide hardware architecture design, some of which involve embedded smart cameras, while others focus on the software design for camera assignment and algorithm development. This chapter first gives a comprehensive review for the existing related works and then focuses on an introduction to the game-theoretic approach to do camera selection and hand-off, followed by a systematic comparison of this game-theoretic technique with some other non-game-theoretic approaches. Detailed experimental comparisons are provided for four selected techniques.

The rest of this chapter is organized as follows: Section 16.2 gives a comprehensive background of the current and emerging approaches for camera selection and handoff. Comparison tables are provided to help the readers to have a macroscopic view of the existing techniques. Section 16.3 focuses on the theoretical approach description and the comparison with two other non-game-theoretic approaches. Experimental results are provided in Section 16.4. Finally, the conclusions are drawn in Section 16.5.

16.2 Related Work and Our Contributions

The research work in camera selection and handoff for a video network consisting of multiple cameras can be classified according to many different aspects, such as whether it is embedded/PC based; distributed/centralized; calibration needed/calibration free; topology based or topology free; statistics based/statistics free, etc.

16.2.1 Comparison for Existing Works

Some researchers work on the design for embedded smart cameras, which, usually, consist of a video sensor, a DSP or an embedded chip, and a communication module. In these systems, such as [2–6], since all the processing can be done locally, the design work is done in a distributed manner. There are also some PC-based approaches that consider the system in a distributed manner, such as [7–10]. Meanwhile, a lot of centralized systems are proposed as well, such as [11–15]. Some works, such as [15], require the topology of the camera network, while some are image based and do not have requirements for any *a priori* knowledge of the topology. As a result, calibration is needed for some systems, while some systems, such as [16–19] are calibration free. Active cameras (pan/tilt/zoom cameras) are used in some systems, such as [14,15,17], to obtain a better view of objects. However, to our knowledge, only a small amount of work has been done to propose a large-scale active camera network for video surveillance. More large-scale camera networks generally consist of static cameras. Images in 3D are generated in some systems, such as [6]. However, in most approaches proposed for the camera selection and handoff, only 2D images are deployed. There are also other considerations, such as resource allocation [20], fusion of different types of sensors [21], etc. In Table 16.1, we compare the advantages and disadvantages for some of the important issues discussed earlier.

Table 16.2 lists sample approaches from the literature and their properties. It is to be noticed that, not all the distributed systems are realized in an embedded fashion. For instance, a distributed camera node can consist of a camera and a PC as well, although the trend is to realize distributed systems via embedded chips. That is why we treat distributed systems and embedded systems separately in Table 16.1. In Table 16.2, some approaches are tested using real data, while some provide only the simulation results. There is no guarantee that the systems, which are experimented using synthetic data, can still work satisfactorily and realize real-time processing when using real data. So, the real-time property is left blank for those approaches whose experiments use simulated data. Similarly, most of the experiments are done for a small-scale camera network. The performance of the same systems for a large-scale camera network still needs to be evaluated.

Table 16.1 Merits of Various Characteristics Encountered in Distributed Video Sensor Networks

Properties	Advantages	Disadvantages
Distributed	Low bandwidth requirement; no time requirement for image decoding; easy to increase the number of nodes; the system is hard to die fully	Lack of global cooperation
Centralized	Easy for cooperation among cameras; hardware architecture is relatively simple compared with distributed systems	Require more bandwidth; high computational requirements for the central server; may cause severe problem once the central server is down
Embedded	Easy to realize distributed system; low bandwidth	Limited resources, such as memory, computing performance, and power; only simple algorithms have been used
PC based	Computation can be fast; no specific hardware design requirements, like for embedded chips or DSPs	A bulky solution for many cameras
Calibrated	Can help to know the topology of the camera network; a must for PTZ cameras, if a precise zoom is required	Preprocessing is required; calibration process may be time-consuming
Uncalibrated	No off-line camera calibration is required	Exact topology of cameras difficult
Active cameras	Provide better view of objects; can save the number of cameras by pa/tilt to cover larger monitoring range	Camera calibration may be required, especially when zooming; complex algorithms to account camera motions
Static/mobile cameras	Low cost, high for mobile; easy to determine topology of the camera network; relatively simpler algorithms as compared with those for active (and mobile) cameras	More (statitc) cameras are needed to have a full coverage; have no close-up if the object is not close to any cameras

16.2.2 Our Contributions

The contributions of our work are as follows:

■ Game-theoretic approach to do camera selection and handoff is provided. Bargaining mechanism is applied to get to the stable solution.

Table 16.2 A Comparison of Some Properties for Selected Approaches

Approaches	HW		Algorithm/SW				Experiment Details			
	E	A	D	C	RT	RD	N_C	N_P	T	O
Quaritsch et al. [3]	Yes	No	Yes	No	Yes	Yes	2	1	Camshift	No
Flech and Straßer [5]	Yes	No	Yes	No	Yes	Yes	1	1	Particle filter	Yes
Park et al. [7]	No	No	Yes	No	N/A	No	20	N/A	N/A	Yes
Morioka et al. [8]	No	No	Yes	No	N/A	No	6	1	N/A	Yes+
Micheloni [9]	No	Yes	Yes	Yes	Yes	Yes	3	3	Kalman filter	Yes
Qureshi and Terzopoulos [10]	No	Yes	Yes	Yes	No	No	16	100	N/A	Yes+
Kattnaker and Zabih [12]	No	No	No	No	Yes	Yes	4	2	Bayesian	No
Everts et al. [14]	No	Yes	No	Yes	Yes	Yes	1	1	Histogram-based	No
Javed et al. [16]	No	No	No	No	Yes	Yes	2	2	N/A	Yes
Jo and Han [19]	No	No	No	No	Yes	Yes	2	N/A	Manual	Yes
Gupta et al. [22]	No	No	No	No	Yes	Yes	15	5	M2Tracker	Yes
Song et al. [23]	No	No	Yes	Yes	No	Yes	7	9	Particle filter	No
Song et al. [24]	No	Yes	Yes	Yes	No	No	14	N/A	N/A	Yes
Our approach	No	No	No	No	Yes	Yes	3	2	Particle filter	Yes+

Legends for the table: E—Embedded; A—Active camera; D—Distributed; C—Calibration needed; RT—Real-time; RD—Real data; N_C—Number of cameras; N_P—Number of objects; T—Tracking algorithm used; O—Overlapping FOVs, Yes+—Yes but not necessary.

■ A comprehensive comparison of recent work for camera selection and handoff is provided. Two non-game-theoretic approaches are compared with our work both theoretically and experimentally.

■ Results with real data and simulations in various scenarios are provided for an in-depth understanding of the advantages and weaknesses of the key approaches. The focus of comparison is solely on multi-object tracking using non-active multi-cameras in an uncalibrated system. The comparison considers software- and algorithm-related issues. Resource allocation, communication errors, and hardware considerations are not considered.

16.3 Technical Approach

16.3.1 Motivation and Problem Formulation

Game theory can be used for analyzing the interactions as well as conflicts among multiple agents [25,26]. Analogously, in a video sensor network, communications as well as competitions among cameras exist simultaneously. The cooperation lies in the fact that all the available cameras, those which can "see" the target person, have to work together to track the person so that he can be followed as long as possible. On the other hand, the available cameras also compete with each other for the rights of tracking this person, so that a camera can maximize its own *utility*. This enlightens us to view the camera assignment problem in a game-theoretic manner. The interactive process is called a game [27], while all the participants of the game are called players, who strive to maximize their *utilities*. The utility of a player refers to the welfare that he can get in the game. In our problem, for each person to be tracked, there exists a multiplayer game, with the available cameras being the players. If there are multiple persons in the system, this becomes a multiple of multiplayer game being played simultaneously.

Vehicle–target assignment [28] is a multiplayer game that aims to allocate a set of vehicles to a group of targets and achieve an optimal assignment. Viewing the persons being tracked as "vehicles" while the cameras as "targets," we can adopt the vehicle–target assignment model to choose the "best" camera for each person. In the following, we propose a game theory–based approach that is well suited to the task at hand.

16.3.2 Game-Theoretic Framework

Game theory involves *utility*, the amount of "welfare" an agent derives in a game. We are concerned with three different utilities:

1. *Global utility*: the overall degree of satisfaction for tracking performance.
2. *Camera utility*: how well a camera is tracking persons assigned to it based on the user-supplied criteria.
3. *Person utility*: how well a person is satisfied while being tracked by some camera.

Our objective is to maximize the global utility as well as to make sure that each person is tracked by the "best" camera. During the course of competition among available cameras, they *bargain* with each other, and finally a decision is made for the best camera assignment based on a set of probabilities.

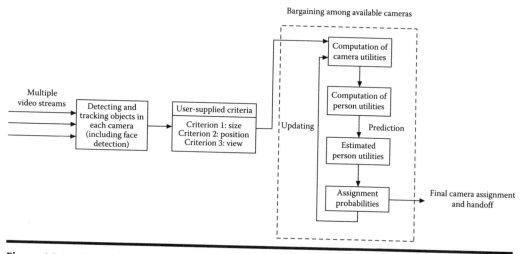

Figure 16.1 Game-theoretic framework for camera assignment and handoff.

An overview of the approach is illustrated in Figure 16.1. Moving objects are detected in multiple video streams. Their properties, such as the size of the minimum bounding rectangle and other region properties (color, shape, location within FOV, etc.), are computed. Various utilities (camera utility, person utility, and global utility) are computed based on the user-supplied criteria, and bargaining processes among available cameras are executed based on the prediction of person utilities at each step. The results obtained from the strategy execution are in turn used for updating the camera utilities and the person utilities until the strategies converge. Finally, those cameras with the highest converged probabilities are used for tracking. This assignment of persons to the "best" cameras leads to the solution of the handoff problem in multiple video streams.

A set of symbols are used in the discussion of our approach and their descriptions are given in Table 16.3.

16.3.2.1 Computation of Utilities

We first define the following properties of our system:

1. A person P_i can be in the FOV of more than one camera. The available cameras for P_i belong to the set A_i. C_0 is assumed to be a virtual (null) camera.
2. A person can only be assigned to one camera. The assigned camera for P_i is named as a_i.
3. Each camera can be used for tracking multiple persons.

For some person P_i, when we change its camera assignment from a' to a'' while assignments for other persons remain the same, if

$$U_{P_i}\left(a'_i, a_{-i}\right) < U_{P_i}\left(a''_i, a_{-i}\right) \Leftrightarrow U_g\left(a'_i, a_{-i}\right) < U_g\left(a''_i, a_{-i}\right) \tag{16.1}$$

Table 16.3 Notations of Symbols Used in the Paper

Symbols	Notations
P_i	Person i
C_j	Camera j
N_p	Total number of persons in the entire network at a given time
N_c	Total number of cameras in the entire network at a given time
A_i	The set of cameras that can see person i, $A_i = \{a_1, a_2, \ldots, a_{n_C}\}$
n_c	Number of cameras that can *see object i, number of elements in* A_i
n_p	Number of persons currently assigned to camera C_j
a_i	The assigned "best" camera for person i
a_{-i}	The assignment of cameras for the persons excluding person i
a	Assignment of cameras for all persons, $a = (a_i, a_{-i})$
$U_{C_j}(a)$	Camera utility for camera j
$U_{P_i}(a)$	Person utility for person i
$U_g(a)$	Global utility
$\bar{U}_{P_i}(k)$	Predicted person utility for person i at step k, $\bar{U}_{P_i}(k) = [\bar{U}^1_{P_i}(k), \ldots, \bar{U}^l_{P_i}(k), \ldots, \bar{U}^{n_c}_{P_i}(k)]$, where $\bar{U}^l_{P_i}(k)$ is the predicted person utility for P_i if camera a_i is used
$p_i(k)$	Probability of person i's assignment at step k, $p_i(k) = [p^1_i(k), \ldots, p^l_i(k), \ldots, p^{N_c}_i(k)]$, where $p^l_i(k)$ is the probability for camera a_i to track person P_i

the person utility U_{P_i} is said to be aligned with the global utility U_g, where a_{-i} stands for the assignments for persons other than P_i, i.e., $a_{-i} = (a_1, \ldots, a_{i-1}, a_{i+1}, \ldots, a_{N_p})$. We define the global utility as

$$U_g(a) = \sum_{C_j \in C} U_{C_j}(a) \tag{16.2}$$

where $U_{C_j}(a)$ is the camera utility and defined to be the utility generated by all the engagements of persons with a particular camera C_j. Now, we define the person utility as

$$U_{P_i}(a) = U_g(a_i, a_{-i}) - U_g(C_0, a_{-i}) = U_{C_j}(a_i, a_{-i}) - U_{C_j}(C_0, a_{-i}) \tag{16.3}$$

The person utility $U_{P_i}(a)$ can be viewed as a marginal contribution of P_i to the global utility. To calculate (16.3), we have to construct a scheme to calculate the camera utility $U_{C_j}(a)$. We assume that there are N_{Crt} criteria to evaluate the quality of a camera used for tracking an object. Thus, the camera utility can be built as

$$U_{C_j}(a_i, a_{-i}) = \sum_{s=1}^{n_p} \sum_{l=1}^{N_{Crt}} Crt_{sl} \tag{16.4}$$

where n_p is the number of persons that are currently assigned to camera C_j for tracking. Plugging (16.4) into (16.3) we can obtain

$$U_{P_j}(a_i, a_{-i}) = \sum_{s=1}^{np} \sum_{l=1}^{N_{Crt}} Crt_{sl} - \sum_{\substack{s=1 \\ s=P_i}}^{np} \sum_{l=1}^{N_{Crt}} Crt_{sl} \tag{16.5}$$

where $s \neq P_i$ means that we exclude person P_i from those who are being tracked by camera C_j. One thing to be noticed here is that when designing the criteria, we have to normalize them.

16.3.2.2 Bargaining among Cameras

As stated previously, our goal is to optimize each person utility as well as the global utility. Competition among cameras finally leads to the Nash equilibrium. Unfortunately, this Nash equilibrium may not be unique. Some of them are not stable solutions, which are not desired. To solve this problem, a bargaining mechanism among cameras is introduced, to make them finally come to a compromise and generate a stable solution.

When bargaining, the assignment in the kth step is made according to a set of probabilities

$$p_i(k) = \left[p_i^1(k), \ldots, p_i^l(k), \ldots, p_i^{n_C}(k) \right]$$

where n_c is the number of cameras that can "see" the person P_i and $\sum_1^{n_C} p_i^l(k) = 1$, with each $0 \leq p_i^l(k) \leq 1$, $l = 1, \ldots, n_C$. We can generalize $p_i(k)$ to be

$$p_i(k) = \left[p_i^1(k), \ldots, p_i^l(k), \ldots, p_i^{N_C}(k) \right]$$

by assigning a zero probability for those cameras that cannot "see" the person P_i, meaning that those cameras will not be assigned according to their probability. Thus, we can construct an $N_p \times N_C$ probability matrix

$$\begin{bmatrix} p_1^1(k) & \cdots & p_1^{N_C}(k) \\ \vdots & \ddots & \vdots \\ p_{N_p}^1(k) & \cdots & p_{N_p}^{N_C}(k) \end{bmatrix}$$

At each bargaining step, we will assign a person to the camera that has the highest probability. Since in most cases, a person has no information of the assignment before it is made, we introduce the concept of predicted person utility $\bar{U}_{P_i}(k)$: Before we decide the final assignment profile, we predict the person utility using the previous person's utility information in the bargaining steps. As shown in (16.5), person utility depends on the camera utility, so we predict the person utility for every possible camera that may be assigned to track it. Each element in $\bar{U}_{P_i}(k)$ is calculated by the following equation:

$$\bar{U}_{P_i}^l(k+1) = \begin{cases} \bar{U}_{P_i}^l(k) + \dfrac{1}{p_i^l(k)}(U_{P_i}(a(k)) - \bar{U}_{P_i}^l(k), \; a_i(k) = A_i^l \\ \bar{U}_{P_i}^l(k), \quad otherwise \end{cases} \tag{16.6}$$

with the initial state $\bar{U}_{P_i}^l(1)$ to be assigned arbitrarily as long as it is within the reasonable range for $\bar{U}_{P_i}(k)$, for $l = 1, \ldots, n_C$. Once these predicted person utilities are calculated, it can be proved that the equilibrium for the strategies lies in the probability distribution that maximizes its perturbed predicted utility [10],

$$P_i(k)' \bar{U}_{P_i}(k) + \tau H(p_i(k)) \tag{16.7}$$

where

$$H(p_i^l(k)) = -p_i^l(k)' \log(p_i^l(k)) \tag{16.8}$$

is the entropy function and τ is a positive parameter belonging to $[0,1]$ that controls the extent of randomization. The larger the τ is, the faster the bargaining process converges; the smaller the τ is, the more accurate result we can get. So, there is a trade-off when selecting the value of τ, and we select τ to be 0.5 in our experiments. The solution of (16.7) is proved [28] to be

$$p_i^l(k) = \frac{e^{((1/\tau)\bar{U}_{P_i}^l(k))}}{e^{((1/\tau)\bar{U}_{P_i}^l(k))} + \cdots + e^{((1/\tau)\bar{U}_{P_i}^{rC}(k))}} \tag{16.9}$$

After several steps of calculation, the result of $p_i(k)$ tends to converge. Thus, we finally get the stable solution, which is proved to be at least suboptimal [28].

16.3.2.3 Criteria for Camera Assignment and Handoff

A number of criteria, including human biometrics, can be used for camera assignment and handoff. For easier comparison between the computed results and the intuitive judgment, four criteria are used for a camera selection:

1. The **size** of the tracked person. It is measured by the ratio of the number of pixels inside the bounding box of the person to that of the size of the image plane. Here, we assume that neither a too-large nor a too-small object is convenient for observation. Assume that λ is the threshold for best observation, i.e., when $r = \lambda$ this criterion reaches its peak value, where

$$r = \frac{\text{\# of pixels inside the bounding box}}{\text{\# of pixels in the image plane}}.$$

$$Crt_{i1} = \begin{cases} \dfrac{1}{\lambda}r, & \text{when } r < \lambda \\[2mm] \dfrac{1-r}{1-\lambda}, & \text{when } r \geq \lambda \end{cases} \tag{16.10}$$

2. The **position** of the person in the FOV of a camera. It is measured by the Euclidean distance that a person is away from the center of the image plane

$$Crt_{i2} = \frac{\sqrt{(x - x_C)^2 + (y - y_C)^2}}{\left(\frac{1}{2}\right)\sqrt{x_C^2 + y_C^2}} \tag{16.11}$$

where

(x, y) is the current position of the person

(x_c, y_c) is the center of the image plane

3. The **view** of the person, as measured by the ratio of the number of pixels on the detected face to that of the whole bounding box, which is similar to Criterion 1. We assume that the threshold for best frontal view is R, i.e., when $R = \xi$ the view of the person is the best, where

$$R = \frac{\text{\# of pixels on the face}}{\text{\# of pixels on the entire body}}$$

$$Crt_{i3} = \begin{cases} \dfrac{1}{\xi} r, & \text{when } R < \xi \\ \dfrac{1-R}{1-\xi}, & \text{when } R \geq \xi \end{cases} \qquad (16.12)$$

4. **Combination** of criteria (1) through (3), which is called the *combined criterion*, is given by the following equation:

$$Crt_{i4} = \sum_{m=1}^{3} w_m Crt_{im} \qquad (16.13)$$

where w_m is the weight for different criteria.

It is to be noticed that all these criteria are normalized for calculating the corresponding camera utilities.

16.3.3 Theoretical Comparison with Two Non-Game-Theoretic Approaches

We selected four approaches [8,19] for comparison. They are chosen as typical approaches because these approaches cover both distributed system [8] and centralized system [19]. Although none of these approaches needs camera calibration, some of them do a geometry correspondence [19], while some do not [8]. This section focuses on the comparison of theoretical ideas while experimental comparison is provided in the next section.

In this section, we first describe the key ideas of these approaches. Analysis of the advantages and disadvantages are provided in Table 16.4.

16.3.3.1 Descriptions of the Key Ideas of Selected Approaches

16.3.3.1.1 Approach 1: The Co-Ocurrence to Occurrence Ratio Approach

This approach decides whether two points are in correspondence with each other by calculating the co-occurrence to occurrence ratio (COR). If the COR is higher than some predefined threshold, then the two points are decided to be in correspondence with each other. When one point is getting close to the edge of the FOV of one camera, the system will hand off to another camera that has its corresponding point.

Table 16.4 Relative Merits of the Selected Approaches

Approaches	Pros	Cons
COR approach [19]	Intuitive efficient approach; acceptable results when there are few occlusions and few cameras and objects	Time-consuming correspondence of point pairs; when correspondence fails or occlusion happens, there is handoff ambiguity and the error rate increases; computing structure becomes complicated with the increase of the number of camera nodes/objects; FOVs have to be overlapped
Fuzzy-based approach [8]	Distributed approach; camera state transition and handoff rules are both intuitive; no requirement for overlapping FOVs	Only simulation results are provided; tracking has to be accurate; not robust when occlusion happens; no guarantee for convergence in a large-scale network
Our approach	Provides a mathematical framework; can deal with the cooperation and competition among cameras; can perform camera selection based on user-supplied criteria; no need for overlapping FOVs	Communication among cameras is not involved; can be extended for distributed computation

The COR is defined as

$$R(x, x\prime) = \frac{p(x, x\prime)}{p(x)} \tag{16.14}$$

where

$$p(x) = \frac{1}{T} \sum_{t=1}^{T} \sum_{i=1}^{N_t} K_2 \left(x - x_t^i \right) \tag{16.15}$$

is the mean probability that a moving object appears at x, i.e., the occurrence at x. K_2 is claimed to be circular Gaussian kernel. Similarly,

$$p(x, x\prime) = \frac{1}{T} \sum_{t=1}^{T} \sum_{i=1}^{N_t} K_2 \left(x - x_t^i \right) \sum_{i=1}^{N_t'} K_2 \left(x\prime - x\prime_t^i \right) \tag{16.16}$$

is the co-occurrence at x in one camera and x' in another camera.

It is intuitive that if two points x and x' are in correspondence, i.e., the same point in the views of different cameras, then the calculated COR should be 1 ideally. On the contrary, if the x and

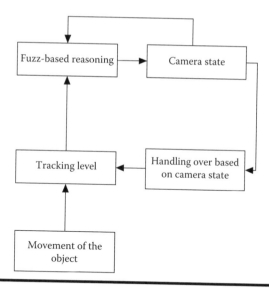

Figure 16.2   Diagram for camera state transition.

x' are completely independent of each other, i.e. two distinctive points, then $p(x, x) = p(x)p(x')$, which leads the COR $R(x, x')$ to be $p(x')$. These are the two extreme cases. If we chose some threshold θ_r such that $p(x') < \theta_r < 1$, then by comparing with θ_r, the correspondence of two points in two camera views can be determined. Another threshold θ_0 is needed to be compared with $p(x)$ to decide whether a point is detected in a camera. Thus, camera handoff can be taken care of by calculating the correspondence of pairs of points in the views of different cameras and performed when necessary.

16.3.3.1.2 Approach 2: Fuzzy-Based Approach

This is another decentralized approach. Each candidate camera has two states for the object that is in its FOV: the **nonselected** state and the **selected** state for tracking. Then, camera handoff is done based on the camera's previous state S_i and the tracking level state SS_i, which is defined by estimating the position measurement error in the monitoring area. The two states for the tracking level are **unacceptable**, meaning that the object is too far away, and **acceptable**, meaning that the object is within the FOV and the quality is acceptable.

The block diagram for camera state transition and the fuzzy rule for camera handoff are given in Figures 16.2 [8] and 16.3 [8], respectively.

16.3.3.2 Pros and Cons Comparison of the Selected Approaches

We list the Pros and Cons for these approaches in Table 16.4.

16.4 Experimental Results

In this section, we perform experiments in different cases for the proposed approach and compare it with the other two non-game-theoretic approaches introduced previously in Section 16.3.3.

(1) If S_i = **Selected** And SS_i = **Acceptable** Then S_i = **Selected**
(2) If S_i = **Non-selected** And SS_i = **Unacceptable** Then S_i = **Non-selected**
(3) If S_i = **Selected** And SS_i = **Non-selected** And SS_k = **Unacceptable** Then S_i = **Selected**, $\forall k \in [1, N]$, $k \neq i$, where N is the number of camera candidates
(4) If S_i = **Non-selected** And SS_i = **Acceptable** And S_k = **Non-selected** And SS_k = **Unacceptable** Then S_i = **Selected**, $\forall k \in [1, N]$, $k \neq i$
(5) If S_i = **Non-selected** And SS_i = **Acceptable** And S_k = **Selected** And SS_k = **Acceptable** Then S_i = **Non-selected**, $\equiv k \in [1, N]$, $k \neq i$
(6) If S_i = **Selected** And SS_i = **Unacceptable** And S_k = **Non-selected** And SS_k = **Acceptable** Then S_i = **Non-selected**, $\equiv k \in [1, N]$, $k \neq i$
(7) If S_i = **Non-selected** And SS_i = **Acceptable** And S_k = **Selected** And SS_k = **Unacceptable** Then S_i = **Selected**, $\equiv k \in [1, N]$, $k \neq i$

Figure 16.3   Fuzzy-based reasoning rules.

Although some of the approaches [8] do not have results with real data, in this chapter, both indoor and outdoor experiments with real data are carried out for all the approaches. For convenience of comparison among different approaches, no cameras are actively controlled.

16.4.1 Data

The experiments are done using commercially available AXIS 215 cameras. Three experiments are carried out with an increase in complexity. *Case 1:* two cameras and three persons, indoor. *Case 2:* three cameras and five persons, indoor. *Case 3:* four cameras and six persons, outdoor. The frames are dropped whenever the image information is lost during the transmission. The indoor experiments use cable-connected cameras, with a frame rate of 30 fps. However, for the outdoor experiment, the network is wireless. Due to the low quality of the images, the frame rate is only 10–15 fps on average. The images are 60% compressed for the outdoor experiment to save bandwidth. Image quality is 4CIF, which is 704×480. They are overlapped randomly in our experiments, which is not required by some of the approaches but required by some others.

16.4.2 Tracking

None of the approaches discussed here depends on any particular tracker. Basically, ideal tracking can be assumed for comparing the camera selection and handoff mechanisms. It should be noted that tracking is not the focus of our work.

Trackings in all the experiments are initialized by a human observer manually at the very beginning and then done with color-based particle filter [29] automatically. The dynamic model used is random walk. Measurement space is two dimensional: hue and saturation values of a pixel. The sample number used for each object to be tracked is 200 for indoor experiments and 500 for outdoor experiments. Tracking can be done in real time by implementing the OpenCV structure CvConDensation and the corresponding OpenCV functions. Matches for objects are done by calculating the correlation of the hue values using cvComparehist. (We compare the hue values of the upper bodies first. If there is ambiguity, then lower body is considered.) Minor occlusion is recoverable within a very short time. Tracking may fail when severe occlusion takes place or in the case where an object is not in the scene for too long and then reenters. Theoretically, this can be solved by spreading more particles. However, more particles may be very computationally expensive. Thus, we just re-initialize the tracking process manually to avoid non-real-time processing.

16.4.3 Parameters

We first define the following properties of our system:

- A person P_i can be in the FOV of more than one camera. The available cameras for P_i belong to the set A_i.
- A person can only be assigned to one camera. The assigned camera for P_i is named as a_i.
- Each camera can be used for tracking multiple persons.
  1. *Our Approach:* In our experiments, we use the combined criterion to perform the bargaining mechanism. This is because it can comprehensively consider all the criteria provided by the user. We give value to the parameters empirically. $\lambda = \frac{1}{15}$, $\xi = \frac{1}{6}$, $w_1 = 0.2$, $w_2 = 0.1$, $w_3 = 0.7$. For instance, the criterion for P_i is calculated as

$$Crt_i = 0.2 Crt_{i1} + 0.1 Crt_{i2} + 0.7 Crt_{i3} \qquad (16.17)$$

  The weights are like this because we want to have the frontal view of a person, which contains much information, whenever it is available. The utility functions are kept exactly the same as stated in Section 16.3.2.3.
  2. *The COR Approach:* The COR approach in [19] has been applied to two cameras only. We generalize this approach to the cases with more cameras by comparing the accumulated COR in the FOVs of multiple cameras. We randomly select 100 points on the detected person, train the system for 10 frames to construct the correspondence for these 100 points, calculate the cumulative CORs in the FOVs of different cameras, and select the one with the highest value for handoff.
  3. *Fuzzy-Based Approach:* We apply the same fuzzy reasoning rule as the one in Figure 16.2, which is given in [8]. The tracking level state is decided by Criterion 2, i.e., Crt_{i2}, which is used for the utility-based game-theoretic approach.

16.4.4 Experimental Results and Analysis

Due to limited space, only those frames with camera handoffs are shown (actually, only some typical handoffs, since the video is long and there are too many handoffs.). These camera handoffs for Cases 1–3 are shown in Figures 16.4 through 16.6, respectively. Since no topology of the camera network is given, tracking is actually performed by every camera all the time. However, for easy observation, we only draw the bounding box for an object in the image of the camera that is selected to track this object. Cases 1 and 2 are simple in the sense that there are fewer cameras and objects, and the frame rate is high enough to make the objects trajectories continuous. So, we only show some typical frames for these cases and give more handoff examples in Case 3, which is more complicated. We show some typical handoffs for Cases 1 and 3, while for Case 2, we show the same frames for the four approaches to see the differences caused by performing handoffs by different approaches.

Figure 16.4   Selective camera handoff frames for the four approaches (Case 1).

Figure 16.5   Selective camera handoff frames for the four approaches (Case 2).

It is clear that the proposed game-theoretic approach considers more criteria when performing the camera selection. Camera handoffs take place whenever a better camera is found based on the user-supplied criterion in this case. So, cameras that can see persons' frontal views, which have the highest weight in Crt_i, are more preferred most of the time. The other two approaches have similar results in the sense that they all consider handoff based on the position of the objects. In this sense, the game-theoretic approach is more flexible to perform camera handoffs based on different criteria. The modification of a criterion will have no influence on the decision-making mechanism.

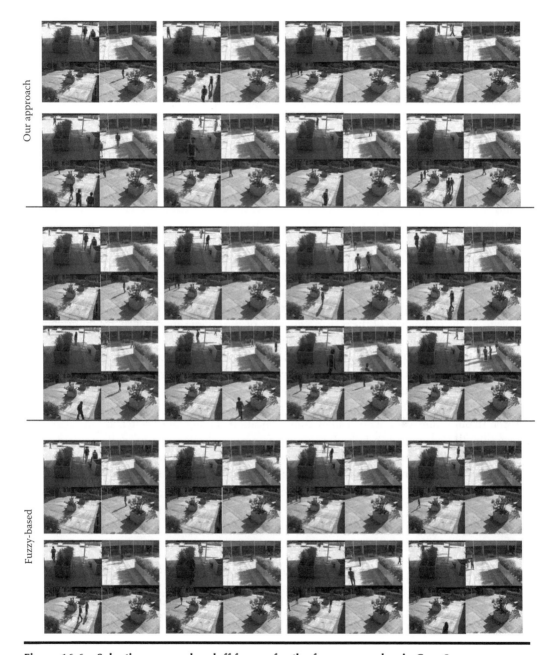

Figure 16.6 Selective camera handoff frames for the four approaches in Case 3.

Figure 16.4 shows the camera handoff results for a very simple case. All the three approaches achieve similar results, although the game-theoretic approach prefers frontal view.

CAs the scenario being more complex, i.e., more objects and more cameras are involved and occlusions happen frequently, the COR approach and the fuzzy-based approach have less satisfactory results. Error rates for different approaches in each case are given in Table 16.5.

Table 16.5 Error Rates of the Selected Approaches

	Our Approach (%)	COR (%)	Fuzzy Based (%)
Case 1	3.86	4.23	4.64
Case 2	4.98	10.01	7.11
Case 3	7.89	45.67	21.33

16.5 Conclusions and Future Work

In this work, we propose the novel idea of doing the camera selection and handoff problem in a game-theoretic manner. Some intuitive criteria are designed for easy observation. The bargaining mechanism in game theory is applied to obtain a stable as well as optimal solution to the problem. We also compare our work with two selected non-game-theoretic approaches, which are discussed in detail. Experimental results are provided to show the merits of the proposed game-theoretic approach. It is obvious, from the shown results, that our approach is flexible to deal with multiple predefined criteria. Meanwhile, this provides a systematic method to solve the camera selection problem. As the complexity of the scenario goes up, or the criteria are changed, we do not bother to modify the algorithm and can still have acceptable results.

We also analyzed existing and emerging techniques for the camera selection and handoff problem in the related work part. Advantages and disadvantages of some properties, such as distributed or centralized systems, are discussed.

There is the trend to have a hierarchical structure, which hybrids the distributed and centralized control. In our future work, we will allow communications among cameras to make the algorithm decentralized. Also, there is a lack of research on camera selection and handoff in a large-scale network of active cameras. Current research is short on experimental results with real data processed in real time. We also want to extend our work in a large-scale camera network and realize real-time control of the cameras.

References

1. M. Valera and S.A. Velastin, Intelligent distributed surveillance systems: A review, *IEEE Proc. Vis. Image Signal Process.*, 152, 2, April 2005, 192–204.
2. H. Kim, C. Nam, K. Ha, O. Ayurzana, and J. Kwon, An algorithm of real time image tracking system using a camera with pan/tilt motors on an embedded system, ICMIT, *Proceedings of SPIE*, Vol. 6041 604112–1, 2005.
3. M. Quaritsch, M. Kreuzthaler, B. Rinner, H. Bischof, and B. Strobl, Autonomous multicamera tracking on embedded smart cameras, *EURASIP J. Embedded Systems*, Vol. 2007, 1, January 2007.
4. B. Rinner, M. Jovanovic, and M. Quaritsch, Embedded middleware on distributed smart cameras, *ICASSP 2007*, Honolulu, HI, pp. IV-1381–1384, 2007.
5. S. Flech and W. Straβer, Adaptive probabilistic tracking embedded in a smart camera, *CVPR 2005*, San Diego, CA, 2005.
6. S. Flech, F. Busch, P. Biber, and W. Straβer, 3D surveillance—A distributed network of smart cameras for real-time tracking and its visualization in 3D, *CVPRW 2006*, New York.
7. J. Park, P.C. Bhat, and A.C. Kak, A look-up table based approach for solving the camera selection problem in large camera networks, *Workshop on DSC 2006*, Boulder, CO, 2006.

8. K. Morioka, S. Kovacs, J. Lee, P. Korondi, and H. Hashimoto, Fuzzy-based camera selection for object tracking in a multi-camera system, *IEEE Conference on Human System Interactions*, Krakow, Poland, pp. 767–772, 2008.

9. C. Micheloni, G.L. Foresti, and L. Snidaro, A network of co-operative cameras for visual surveillance, *IEE Proc. Vis. Image Signal Process.*, 152, 2, April 2005, 205–212.

10. F. Bashir and F. Porikli, Multi-camera control through constraint satisfaction for persistent surveillance, *AVSS*, Santa Fe, NM, pp. 211–218, 2008.

11. S. Lim, L.S. Davis, and A. Elgannal, A scalable image-based multi-camera visual surveillance system, *AVSS*, Miami, FL, 2003.

12. V. Kattnaker and R. Zabih, Bayesian multi-camera surveillance, *CVPR*, Vol. 2, pp. 253–259, 1999.

13. Y. Lu and S. Payandeh, Cooperative hybrid multi-camera tracking for people surveillance, *Can. J. Elect. Comput. Eng.*, 33, 3/3, Summer/Fall, 145–152, 2008.

14. I. Everts, N. Sebe, and G. Jones, Cooperative object tracking with multiple PTZ cameras, *ICIAP*, Modena, Italy, pp. 323–330, 2007.

15. S. Kang, J. Paik, A. Koschan, B. Abidi, and M.A. Abidi, Real-time video tracking using PTZ cameras, *Proceedings of SPIE*, Getlinburg, Tennessee, Vol. 5132, pp. 103–112.

16. O. Javed, S. Khan, Z. Rasheed, and M. Shah, Camera handoff: Tracking in multiple uncalibrated stationary cameras, *IEEE Workshop on Human Motion*, Austin, TX, pp. 113–118, 2000.

17. U.M. Erdem and S. Sclaroff, Look there! Predicting where to look for motion in an active camera network, *AVSS*, Como, Italy, pp. 105–110, 2005.

18. B. Moller, T. Plotz, and G.A. Fink, Calibration-free camera hand-over for fast and reliable person tracking in multi-camera setup, *ICPR 2008*, Tampa, FL, 2008.

19. Y. Jo and J. Han, A new approach to camera hand-off without camera calibration for the general scene with non-planar ground, *VSSN'06*, Santa Barbara, CA.

20. C. Chen, Y. Yao, D. Page, B. Abidi, A. Koschan, and M. Abidi, Camera handoff with adaptive resource management for multi-Camera multi-target surveillance, *AVSS 2008*, Santa Fe, NM, 2008.

21. J. Nayak, L. Gonzalez-Argueta, B. Song, A. Roy-Chowdhury, and E. Tuncel, Multi-target tracking through opportunistic camera control in a resource constrained multimodal sensor network, *ICDSC 2008*, Stanford, CA, 2008.

22. A. Gupta, A. Mittal, and L.S. Davis, COST: An approach for camera selection and multi-object inference ordering in dynamic scenes, *ICCV 2007*, Rio de Janeiro, Brazil, pp. 1–8.

23. B. Song and A.K. Roy-Chowdhury, Robust tracking in a camera network: A multi-objective optimization framework, *IEEE J. Selected Topics Signal Process.*, 2, 4, Aug. 2008.

24. B. Song, C. Soto, A.K. Roy-Chowdhury, and J.A. Farrell, Decentralized camera network control using game theory, *ICDSC Workshop 2008*, Stanford, CA, 2008.

25. M. Isard and A. Blake, CONDENSATION—Conditional density propagation for visual tracking, *IJCV 1998*, 29(1), pp. 5–28.

26. R.B. Myerson, *Game Theory—Analysis of Conflict*. Harvard University Press, Cambridge, MA, 1991.

27. M.J. Osborne, *An Introduction to Game Theory*. Oxford University Press, New York, 2003.

28. Stanford Encyclopedia of Philosophy, http://plato.stanford.edu/entries/game-theory/#Uti

29. G. Arslan, J.R. Marden, and J.S. Shamma. Autonomous vehicle-target assignment: A game-theoretical formulation. *ASME J. Dynamic Systems Measurement Control*, Special Issue on *Analysis and Control of Multi-Agent Dynamic Systems, September 2007*, pp. 584–596.

30. F. Qureshi and D. Terzopoulos, Distributed coalition formation in visual sensor networks: A virtual vision approach, *3rd IEEE International Conference on Distributed Computing in Sensor Systems*, 2007.

RESOURCE MANAGEMENT

Chapter 17

Game-Theoretic Radio Resource Management in OFDMA-Based Cognitive Radio

Hanna Bogucka

Contents

The orthogonal frequency division multiple access (OFDMA) is considered as one of the most promising techniques for multiuser high data-rate wireless communication. One can envision future OFDMA-based cognitive radio (CR) systems, in which the unlicensed users can access the spectrum in an opportunistic manner while applying OFDMA as a flexible air interface. Noticeable worldwide interest in OFDMA-based CR has motivated research on efficient and opportunistic spectrum access and dynamic resource allocation (DRA) for this technology. This research focuses on finding spectrally efficient DRA algorithms, which also take fairness of the resource distribution into account. Noteworthily, game theory provides good tools to approach the problems of efficient, yet fair DRA. The main purpose of using them in the CR framework is to model strategic interactions among the network users considered as players who access and make the best use of the available spectrum resources in a cognitive and opportunistic manner.

17.1 Introduction

The demand for the access to electromagnetic spectrum and for the efficient usage of radio resources has been noticeably growing in the last decades together with the development of the mobile telecommunication industry and business. Intelligent and flexible spectrum access procedures and resource allocation methods are needed to increase the efficiency of the utilization of the scarce spectrum resources. Apart from the major objective to maximize the spectral efficiency, the ambition of the modern radio network design is to rationalize the distribution of radio resources and the cost of their usage. This means that some rationalizing mechanisms should be applied in the DRA procedures, resulting in a balanced network and users' perceived throughput, high Quality of Service (QoS), and high Quality of Experience (QoE). For the CR concept, where the nodes are expected to be intelligent, and possibly autonomous entities, these mechanisms are of particular importance.

The concept of the CR has been introduced by Joseph Mitola III in [1,2]. Through the development of the appropriate software control, it is envisioned that the CR-network node could orient itself in the radio environment, create plans, decide, and take actions [3]. The cognitive decision making should reflect the *rationality* paradigm, trading off between optimality (revenue) and its cost. Some cooperation between the CRs can also be envisioned, which would result in globally and socially desirable outcome. Moreover, learning algorithms that impact the decision making should reckon with the behavior and the character of other players, namely, other CRs. We can thus identify high relevance of the considered scenery of CRs accessing a network and its resources to game-theoretic models.

17.1.1 Why Focus on OFDM-Based Cognitive Radio?

The orthogonal frequency division multiplexing (OFDM) is a well-known multicarrier modulation scheme, which makes use of orthogonal SubCarriers (SCs) for the transmission of parallel data streams. Both the OFDM modulation and demodulation can be implemented with relatively low complexity by the means of the inverse fast Fourier transform (IFFT) and the fast Fourier transform (FFT), respectively. The major drawbacks of the OFDM technique are the sensitivity to the phase noise and frequency offsets resulting in the intercarrier interference (ICI), as well as high peak-to-average power ratio (PAPR) resulting in the transmit signal liability to nonlinear distortions. The OFDM is described in a number of books and papers, and an interested reader should refer to [4–7] for details of this technology and its applications.

It is considerable that OFDM has been applied in a number of wireless communication standards and has been recommended by relevant working groups for use in diverse environments. Wireless personal area networks (WPANs) based on IEEE 802.15.3.a, wireless local area networks (WLANs) based on IEEE 802.11.a/g/n/e, wireless metropolitan area networks (WMANs) based on IEEE 802.16, wireless regional area networks (WRANs) based on IEEE 802.22 standard, or the future mobile communication system based on the 3GPP long-term evolution (LTE) recommendation—all use OFDM for their diverse radio access technologies (RATs), mobility, and range. Apart from the above-mentioned standards, OFDM is used by the broadcasting systems, such as Digital Audio Broadcasting (DAB), Digital Video Broadcasting Terrestial (DVB-T), and Handheld (DVB-H) standards. It seems that the application of the OFDM and related OFDMA answers to the demand of interoperability, one of the key paradigms for the concept of CR.

OFDM has an inherent flexibility since one can adapt the error-correcting coding scheme, symbol constellation order and power individually at distinct the SCs depending on the quality of the channel observed for these SCs [8–10]. Other parameters, such as the number of SCs, their frequency separation, the actual pattern of used SCs, interleaving pattern, or the duration of the CP, can also be adjusted in a flexible manner. Furthermore, using OFDM has the advantage that fragmented bands of available spectrum can be relatively easily aggregated to convey the secondary user's (SU's) (unlicensed user's) traffic, and that the spectrum utilization increases significantly. The spectrum shaping in available spectrum holes can be done by assigning data traffic to available SCs and feeding prohibited SCs occupied by the primary users (PUs) (licensed users) with zeros. This type of noncontiguous OFDM transmission has been described in [11]. As shown in [12], the subband spectrum shaping can be done in a number of ways, taking the fact into account that the interference generated by the SUs is dominated by the SCs neighboring the PU's spectrum. Moreover, the complexity of the FFT and IFFT algorithms in a CR noncontiguous OFDM transceiver can be lowered by removing the operations on deactivated SCs [11]. Thus, OFDM-based technologies seem to be capable of easy shaping of the signal into the spectrum mask and of taking the advantage of the fragmented spectrum opportunities.

We may thus summarize that OFDM and OFDMA are suitable for the application in CR due to their inherent interoperability, flexibility in the transmission parameters adjustment, efficient spectrum utilization, aggregation, and spectrum shaping capabilities, as well as to localization and positioning proficiency (resulting from the usual application of pilot sequences and cyclic prefix).

17.1.2 What Is Specific about DRA in OFDMA?

Introduction of the dynamic allocation of resources should aim at the maximization of the multiuser system performance. In the case of OFDMA, dynamic subcarrier assignments (DSA), also called scheduling, can be employed to increase the network spectral efficiency and users' perceived throughput. The problem of optimal DSA is known as *multiuser water-filling* and is considered as quite complex in terms of finding the absolute optimum. In practical applications, it can be done in two steps: first, the DSA algorithm assigns the SCs to users, and then, a water-filling algorithm for each single user determines the power allocated to these SCs.

The DSA and adaptive bit and power loading (BPL) in OFDMA have attracted considerable research interest, mainly to find some more practical algorithms increasing QoE for an average user. For the users' high QoE, it is desirable that the DSA algorithms incorporated some fairness criteria on individual users' performance. Some practical approaches to this problem have been proposed, which aim at the maximization of the spectral efficiency while respecting transmit power constraints [13,14] or at the minimization of the total transmit power while guaranteeing a minimum bit rate for each user [15]. A practical suboptimal solution is also proposed in [16].

The key issue of the DRA in the framework of CR is to assure autonomous, possibly decentralized, yet efficient and fair distribution of radio resources. This major challenge in the design of CRs reflects the dilemma of optimality versus complexity, efficiency versus fairness, and centralized DRA with extensive control-traffic overhead versus localized sub-optimum decisions. Apart from the optimality and fairness problems, the DRA algorithms in OFDMA-based CR should take other distinctive issues into account, such as dynamically changing availability of radio resources, the necessity of interference management, decision on when to use the single-band and multiband OFDM schemes. These decisions depend on various parameters: required throughput, the interference temperature, hardware limitations, the computational complexity, the number of spectrum holes, etc.

17.1.3 Why Game Theory?

The game theory is considered to be a good tool to study conflicts and cooperation among decision makers as well as to find stable solutions of these conflics or Preto-optimal outcomes of the players' cooperation. The players are usually assumed to be intelligent and rational; however, irrationality of one player can also be considered, as in the concept of *the game against the nature* [17]. The main purpose of using it in radio resource acquisition is to model competitive behavior and strategic interactions among network nodes (the players) with a need and a potential to access the spectrum resources. The application of the economic concepts, such as competition, cooperation, or the mixture thereof, allows analyzing the problem of flexible usage of limited radio resources in a competitive environment [18,19]. In the scenery of the CR network, our player is an intelligent decisive unit (DU) of a CR node, an entity with a key role in the so-called cognition cycle discussed in [3]. Here, we refer to her decision-making functionality, which matches the DRA problems in changing radio environment. (Note that female pronouns are established in the game-theoretic convention.)

17.2 Overview of Game-Theoretic OFDMA Resource Allocation

Similarly, as in the case of more general spectrum allocation, game-theoretic scheduling for OFDM has also been considered in the literature in two major directions: The first one is centralized SC allocation, which allows for more efficient and fair spectrum utilization and applies cooperative game theory, Nash-bargaining, and arbitrary Pareto-optimal solutions. The second one is distributed decision making, which, on the contrary, deploys noncooperative games and results in usually inefficient Nash equilibrium as a game solution. The motivation behind the choice of the game model shall reflect desired properties of the system, availability of the channel state information (CSI), and various criteria of the system performance.

17.2.1 Cooperative Solutions

As stated in the previous sections, our focus is on cognitive and rational decision making by the individual CRs. However, it does not mean that the CRs must take the decisions totally isolated from each other. In fact, some cooperation between them is also possible.

In [20–23], centralized and fair schemes of the OFDMA SC, rate, and power allocation are proposed. In [20] cooperative solutions based on Nash bargaining and finding, the Pareto-optimal solutions are proposed. There, the overall system rate is maximized under the constraints of each user's minimal rate requirement and maximal transmitted power of the base station. In order to reduce the complexity, the DSA (implemented by the base station) is separated from the power allocation procedure (performed locally by the mobile station). The fairness in [22,23] is considered as proportional fairness based on the Nash bargaining solution (NBS) (in [22]) or the Raiffa–Kalai–Smorodinsky bargaining solution (RBS) (in [23]) within the coalitions of two users. In fact, the problem of finding the NBS or RBS for more than two users becomes extremely complex. Therefore, a multiuser bargaining algorithm is based on grouping of pairs of users into coalitions either by a random choice or by using the Hungarian method [24].

An interesting study on fairness versus efficiency is presented in [25]. There, some most popular game-theoretic solutions for cooperative bargaining applied in OFDMA schedulers are compared with respect to sum throughput, per-user throughput, frequency-band sharing, and scaling capabilities. It is shown that scheduling according to the RBS turns out to be an alternative to proportional fairness obtained with NBS. Both solutions offer a compromise between efficiency and fairness.

Finally, let us note that we shall not concentrate on these strictly centralized solutions, as they are not so much fit in the general framework of the CR. Although optimal and applicable in the downlink transmission in the networks with infrastructure and centralized management, they require a lot of control traffic and neglect the smartness of individual CRs.

17.2.2 Noncooperative Solutions

A number of noncooperative, decentralized algorithms for the DRA in OFDMA have been considered in the literature [26–28]. Usually, the problem is narrowed to centralized SC allocation and distributed power allocation based on noncooperative game models. These models often apply the power-pricing function, which induces fairness among users and enhances the rate per joule of consumed energy. Alternatively, power allocation problem in OFDM or OFDMA is defined in such a way that it can be solved independently by every user irrespective of other users' behavior, and thus, it narrows to regular optimization problem.

Some well-known microeconomic oligopoly models can also be applied to address the problems of OFDM spectrum sharing and pricing, that is, Cournot, Bertrand, and Stackelberg game models, which present different nature and degree of competition [29]. In these models however, the knowledge of the CSI for all users is assumed (the game model is full-information model), which is not the case in practical systems with distributed decision making. In [30,31], an iterative English auction process is proposed for the resource-control algorithm in OFDMA wireless systems. The proposed solution for the efficient and fair DRA in OFDMA is considered as distributed, although it requires a number of iterations and some control traffic between the BS and the users. An interesting study on the game formulation for this problem and the Nash equilibrium existence is presented in [32]. There, two game models are considered for the distributed DSA and BPL. One accounts for the partial knowledge of the CSI (i.e., every user has a perfect CSI of her own channel and only statistical data on the other users channels) while the other considers only statistical knowledge of the CSI.

Some papers consider distributed DRA in a multicell environment, where players are the base stations [33–35]. They consider either iterative algorithms for distributed DRA, or the power-pricing [34], or the spectrum-usage pricing [35]. The power and spectrum-usage pricing concepts developed for the multicell scenario, where the base stations act as players, cannot be considered straightforward for the single-cell distributed resource allocation, because a base station has the information of all links in the cell area, while a mobile user may only have the CSI of its own link.

The drawbacks of the existing solutions for the DRA problem in OFDMA networks are either lack of optimality (NBS is usually not the optimal solution), lack of rationality (the utilities are defined so as to maximize the total sum-throughput or neglect QoE, the power economy, or computational complexity), lack of generality (Pareto-optimal solutions are found for a very limited number of users), or lack of suitability for the CR networks (some optimal solutions can be found in a cooperative manner, what requires centralized management). Moreover, distributed solutions are found in the noncooperative games with full information and thus assume perfect CSI knowledge of all links (each player knows other players' channel states and their associated strategic options), which is not the situation met in practical mobile scenarios. Rather, each player knows her own channel conditions at most. Some promising decentralized iterative algorithms require some time for the convergence (which must be shorter than the shortest coherence time of all involved links' channels) and in most cases some control traffic to improve or adjust the players' strategies.

17.3 Cognitive DRA in OFDMA Based on Resource Taxation

As stated in the previous section, many existing algorithms of the DRA in OFDMA networks lack the suitability for the CR. Below, we will focus on the distributed allocation of SCs for multiple CRs within a cell in the OFDMA-based network, which aims at rational and efficient spectrum utilization. Rationality in our case means that apart from maximizing the spectral efficiency (the number of bits per second per Hertz), the network and each individual CR aim at lowering the cost of this efficiency (in terms of the power consumption and the total transmitted power) as well as at increasing the QoE (resulting from the number of served users). This can be done based on the noncooperative game model defined below, which includes the concept of *coercive taxation* [36]. This concept is related to the idea of pricing; however, the pricing function is introduced for the spectrum acquisition mechanisms and is calculated based on the local CSI only (for more details, see [37,40]).

17.3.1 Efficiency versus Rationality: The Tragedy of Commons

The DV of a CR node is a player in our game and makes the decision as to which SCs out of those available ones it is going to use. Naturally, the selfish player would occupy the whole available spectrum, because such an action would maximize her throughput. However, from the network perspective, this behavior would decrease its capacity in terms of the number of users served. The problem is related to the classical common resource utilization dilemma known as the *Tragedy of Commons* and described in [36], which shows that selfish behavior of the common-resource users leads to inefficient (or even the poorest possible) utilization of these resources. The difference in our case is that no user can transmit on exactly the same SC as another user, unlike in the example given in [36].

17.3.2 *The Scenery of a Cognitive Radios Network: The Players*

Let us consider the scenery of multiple mobile CRs appearing in the OFDMA-based network area. These nodes, which can be viewed as secondary (unlicensed) users of the network, are able to sense the radio environment and detect available spectrum resources. Each node transmits and receives data streams of various nature and is not necessarily the source or the destination of the transmission. Relying is also one of its functionalities. Therefore, the first goal of the CR in the considered situation is to acquire radio resources and make the best use of them, that is, maximize the throughput. Our considerations are application-blind, that is, we do not care how the achieved data rates are mapped to data streams. Although in some particular standardized applications of the OFDMA-based networks, an access channel is a quantum set of SCs (collocated or spread in a given bandwidth), we consider the possibility of acquiring any subset of SCs (of any cardinality) by a CR node as a proof of concept.

A CR terminal senses the spectrum for the available SCs. This information obtained from sensors, measurements, a so-called cognitive pilot channel, or a dedicated database is passed to the DU. In our CR-network scenery, there exists a collision-avoidance mechanism, for example, the one that makes use of the cognitive pilot channel, when two nodes try to access the spectrum at the same time instant. The CR senses a number of available SCs and makes decision concerning how many of them it is going to use for the transmission. Below, we consider the utility-based resource acquisition procedure as a game. The players in this game are the CRs, to be exact the DUs of the CRs in the considered OFDMA-based network area. In order to prevent a single CR from occupying all resources, there may be a limit set on the maximum number of SCs a player can acquire at a time. Moreover, some "social consciousness" mechanisms (such as network capacity factor) in the utility function should be employed. One such mechanism is the resource taxation.

17.3.3 *Considerations on the Strategies of the CRs*

After sensing the available spectrum resources, each player can occupy a number of available SCs. If we assume that there are N available SCs and K players, and each player can take any subset of these SCs but no more than a maximum number of I SCs, the problem of finding the game (either cooperative or noncooperative) solution becomes extremely complex. This is because the number of each user strategies equals $\sum_{i=0}^{I} N!/i!(N-i)!$, and the game is K-dimensional. Analyzing the existence of the Nash equilibrium becomes very complicated particularly for high I, N, and K values. It is therefore crucially important to narrow the space of this analysis. For this purpose, we let each CR take decisions independently while treating the rest of the players as a whole (the CRs community), and by eliminating its strategic choices that are disadvantageous.

In the considered case, when the player wants to use a number of i SCs, for sure, these should be her i strongest SCs in terms of the highest carrier-to-interference and noise ratio (CINR), because from both the individual CR perspective and the whole network perspective, making use of the strongest SCs results in higher spectral efficiency. No other combination of i SCs would be more beneficial from the spectral efficiency point of view. Moreover, from a single player standpoint, it does not matter how efficient the other players are in utilizing potentially acquired resources, that is, what are their CINRs at various SCs and the resulting throughput, but rather how many of these SCs the other players are going to use. The number of occupied SCs affects the network ability to serve the incoming users including the users who are already being served, but would like to use more resources when they decide so. It also affects the interference level between the users. In such a case, the game for each player becomes two dimensional, what simplifies the problem grossly.

Let us define the strategies of a CR as possible numbers of SCs (form 0 to I) to be occupied. The ith strategy relates to i strongest SCs detected. The strategies of the rest of the CR-nodes community are defined as the numbers of SCs this community may occupy. The collision-avoidance mechanism allows to consider this game as dynamic, and it is impossible for different nodes to occupy the same SCs.

17.3.4 Considerations on Payoff Functions for the DSA Game in OFDMA

After the definition of the strategies, let us define possible payoffs. For the sake of clarity, let us number the players, which make their decisions at subsequent game stages, and use an index k, where $k \in [0, K-1]$. The value of the utility (payoff) function $p_{i,j}^{(k)}$ for the kth player should exhibit the game outcome, when the player k chooses strategy i (i strongest SCs) while the rest of the community occupies j SCs. In our game, we may consider a number of payoff functions reflecting the player's benefit, for example, the player's contribution to the overall spectral efficiency (her throughput divided by the available OFDM bandwidth):

$$p_{i,j}^{(k)} = \frac{1}{N} \sum_{s \in \mathbf{S}_i^{(k)}} \log_2 \left[1 + \alpha^{(k)} P^{(k)}(f_s) \gamma^{(k)}(f_s) \right], \tag{17.1}$$

where

$\gamma^{(k)}(f_s) = \left| H^{(k)}(f_s) \right|^2 / (\mathcal{N}_0 \Delta f + \mathcal{I}_0)$ is the CINR measured at the SC frequency f_s, whose index s belongs to the set $\mathbf{S}_i^{(k)}$ of indices of i strongest SCs (the cardinality of $\mathbf{S}_i^{(k)}$ is i)

Δf is the SC spacing

$H^{(k)}(f_s)$ and $P^{(k)}(f_s)$ are the kth user's channel characteristic and the power level allocated to this SC, respectively

$\mathcal{N}_0$ is the noise power spectral density

$\mathcal{I}_0$ is the interference power

$\alpha^{(k)}$ is the factor depending on the assumed player's BEP $P_e^{(k)}$

(In case of M-QAM, $\alpha^{(k)} = -1, 5 / \ln \left(0, 5 P_e^{(k)} \right)$, while for $M \geq 4$ and signal-to-noise-and-interference ratio (SINR) in the range of 0–30 dB, it can be set more precisely as $\alpha^{(k)} = -1, 5 / \ln \left(5 P_e^{(k)} \right)$ [38].) Moreover, $i \leq I$, since we restrict each player from occupying more than I SCs at a time. Let us note that before a decision is taken by the player, the CINR is measured, and the player does not have the knowledge as to what will be the interference level after the last player takes her decision. In a perfectly synchronized system, the interference level would be zero, since the SCs used by the players are orthogonal. In a more realistic approach, the interference level would depend on the number of SCs occupied by other players, their relative velocity and Doppler frequency shifts, and the power levels assigned to their occupied SCs. To an average OFDMA user, this interference occurs as noise whose power at a particular SC depends on the number of occupied neighboring SCs. Assuming that at the end of the game all available SCs will be occupied, a single player may assume a stable level of expected interference, which will also be an assumption in our game. Moreover, the optimal power allocated to SCs should be calculated according to the water-filling principle [38], assuming that the channel characteristic and the noise power spectral density are perfectly known by the DU of a CR, that is, the CSI is perfect, although in reality it may contain an error. The player uses estimated CINR for each chosen SC frequency f_s. When the noise or interference level changes, the user has to adjust the power levels at her SCs adequately. Alternatively, if the users' channels change dramatically, we may start a new game and redistribute the resources.

Let us note that the right-hand side of Equation 17.1 does not contain an index j, which means that if we define our payoff function like that, the whole problem narrows to the optimization interrogation, that is, to find a set of SCs $\mathbf{S}_i^{(k)}$ out of those available ones that would maximize (17.1). The player does not factor other players' behavior and the network's overall performance. Let us modify (17.1), so that it also reflects the network potential to serve other users [40]:

$$p_{i,j}^{(k)} = \left\{ \frac{1}{N} \sum_{s \in S_i^{(k)}} \log_2 \left[1 + \alpha^{(k)} P^{(k)}(f_s) \gamma^{(k)}(f_s) \right] \right\} \cdot \left\{ N^{(k)} - i - j \right\}, \qquad (17.2)$$

where $N^{(k)}$ is the number of SCs available at the kth game stage. (Note that at the beginning of the game, $N^{(1)} = N$.) Thus, the first factor in (17.2) represents the throughput of the considered CR divided by the available bandwidth (contribution of this player to the network spectral efficiency) and its potential benefit, while the second one represents the network potential to serve other nodes. One may interpret (17.2) as the total normalized throughput (throughput divided by Δf), which could be obtained by the new incoming users in case they occupied the remaining SCs and had the same average spectral efficiency as the considered player. This way, in the decision making on how many SCs to occupy, the players factor the social aspect of the network (to serve multiple users) and not just their own benefit.

Let us note that for every player k and for any value of the strategy of the rest of the CRs community j, it is always that $p_{I,j}^{(k)} \geq p_{i,j}^{(k)}$ $(i = 0, ..., I)$, and thus, $p_{I,j}^{(k)}$ represents the highest payoff regardless of the strategy j of the rest of the CRs community. Thus, the dominating strategy for every considered player is to use the maximum allowable number of SCs I. As a result, the common radio resources would not be utilized efficiently and rationally. Again it is worth mentioning that this is a typical example illustrating the *Tragedy of Commons*. Therefore, the utility defined as in formula (17.2) has to be modified so as to force the players to behave more "prudently" with respect to the scarce resources. The modification to (17.2) is to introduce a tax (a pricing function), which reduces the possible payoff, so that the utility is now defined as [37]

$$\tilde{p}_{i,j}^{(k)} = p_{i,j}^{(k)} - \tau_i \cdot i, \qquad (17.3)$$

where τ_i is the tax-rate parameter determined by the base station based on the values of the players' average SINRs $(\overline{\Upsilon}^{(k)})$ in the network area. Note that the tax that each player has to pay depends on the amount of resources used by a player for achieving the *revenue* defined by $p_{i,j}^{(k)}$. Thus, the tax rate τ_i is measured in revenue units per spectrum-resource units, that is, in bit per second per Hertz per Hertz ([bit/s/Hz/Hz]). If a single player faces poor channel quality and therefore can potentially transmit with low data rates, and yet occupies a lot of SCs, its benefit should be small (or there should be even a loss reflected in negative payoff) due to poor spectral utilization and relatively high tax "to pay."

It is crucial to understand that the payoff (benefit associated with the player's revenue and its cost) is not the same as the actual data rate that can be achieved by a player, when she acquires and uses her resources. It is a utility value calculated to assist her decision making. As a result of choosing the ith strategy (i SCs), the player still obtains the average throughput (averaged over N SCs) as defined by (17.1); however, she also "pays" a tax for this data rate equal to $\tau_i \cdot i$. Additionally, a fairness factor incorporated in the utility function as in (17.2) allows the player to take the future network capacity into account (resources remaining after her decision making, which may be used by other network nodes in the future, including herself). Such a definition of the payoff (as by (17.2) or (17.3)) is in accordance with the utility theory.

Similarly as in common life, properly established taxes aim at extorting the desirable players' behavior and in some cases at the Pareto-optimal solution of the game [17]. However, calculation of such an optimal tax would require the knowledge of the CINRs at all SCs for all links, complex analysis to find Pareto-optimal vectors of payoffs, and the calculation of the tax (perhaps not common for all nodes). These complex calculations and extensive analysis can be done only centrally, for example, at an access point or at a base station, where the required CSI for all potential links may be available, and from where the tax values (and their updates every time the CINRs change) could be broadcasted. However, in the considered CR network scenery, the analysis of the game and all decisions are expected from the intelligent DU of a CR node. Therefore, we look at a simplified case of common and invariant (with respect to CINR changes) way of taxing the resource utilization.

With a relation to common-life economic models, the tax rate can be fixed (linear tax) and not dependent on the chosen strategy, or contrary, it may depend on the number of chosen SCs (progressive tax). For example, the progressive tax can be arithmetically progressive:

$$\tau_i = \tau_0 + i \cdot \Delta_\tau, \tag{17.4}$$

or the geometrically progressive:

$$\tau_i = \tau_0 \cdot (\Gamma_\tau)^i, \tag{17.5}$$

or defined in some other manner. Naturally, progressive taxation can be defined in a number of ways, for example, with the use of increasing tax-rate thresholds between which taxation is linear. Nevertheless, as long as our taxation function $\tau_i \cdot i$ is a nondecreasing function of i (which is always a common-sense assumption for taxation), our considerations presented below are general for any type of taxation.

Let us note that if $\Delta_\tau = 0$ in (17.4) or $\Gamma_\tau = 1$ in (17.5), the taxation becomes linear. The tax-rate parameters τ_0, Δ_τ, or Γ_τ should be carefully adopted to the number of resources, the number of players, and their anticipated behavior. As mentioned above, these parameters although common for all CRs should depend on the set of players' average SINRs $\{\overline{\Upsilon}^{(k)}\}$ in the system. Too high taxes established in the CR network area would result in poor usage of available resources, if the average SINRs experienced by all CRs were low. (The resulting payoffs would encourage every player not to use any SC, and thus, the resources would not be used at all.)

Let us also consider the utilities for the other player, here referred to as "the rest of the CR-nodes community." It seems natural that this player would care for resources that can be potentially occupied, and that the utility of the community at the kth stage of the game should be defined as $q_{i,j}^{(k)} = j$.

After we have introduced taxation of the occupied number of SCs, there may not be one dominating strategy of the single player, whose turn is to take actions (acquire resources). Although the values of payoffs $p_{i,j}^{(k)}$ decrease with increasing j, it is not the case for increasing i [40]:

$$\forall j \in [0, N^{(k)} - I] \; \exists! \; i^\star : \begin{cases} \tilde{p}_{i-1,j}^{(k)} \le \tilde{p}_{i,j}^{(k)} & \text{for } 0 < i \le i^\star, \\ \tilde{p}_{i,j}^{(k)} \ge \tilde{p}_{i+1,j}^{(k)} & \text{for } i^\star \le i < I. \end{cases} \tag{17.6}$$

Moreover, there may still exist some dominated strategies after the tax has been introduced. If, for instance, the ith strongest SC is a very bad one (has low CINR), an associated tax exceeds the benefit of using it for every considered strategy j of the rest of the community. In such a case, this dominated strategy i should be removed from the considerations of the rational player.

The above considerations lead to the conclusion that there exist the Nash equilibrium for the earlier defined game for $i = i^\star$ and $j = N^{(k)} - I$. This equilibrium strategy for a considered CR would be strategy $i^\star$ defined by (17.6) for $j = N - I$. However, it is well known that a Nash equilibrium may be far from Pareto-optimality. This is because it is based on the assumption of the rationality of both players aiming at achieving the payoff no lower than necessary. In our game, the player named as "the rest of the CR-node community" does not strictly rationalize its behavior this way. Although the behavior of each individual player is hard to anticipate (because it depends on the channel quality they face), the behavior of the whole community of the CR nodes can be assessed. This is because her strategy j is the aggregated number of SCs this community occupies, and thus, the probability of this strategy approximates the Gaussian distribution. (The higher the number of the competing nodes K, the more accurate this approximation is.) Thus, an optional choice of the player's strategy could be to apply mixed strategy as a response for a given probability of the community strategies.

17.3.5 Experimental Results

In the following, example experimental results are presented for the game model described earlier. The following simulation setup has been considered: the number of available SCs at the beginning of the game $N = 256$, the maximum number of SCs each player can take I, and the number of competing nodes K satisfy $KI = N$ (potentially there is enough SCs to serve all nodes). Each node is allowed to transmit with the same power limit. For a given number of competing nodes K, the total power in the CR-node community (being the sum of the transmit powers of the CR nodes) has also been fixed, so that if some nodes do not use available SCs, the other nodes are allowed to increase their power. Thus, the interference level is constant. Moreover, two scenarios have been considered: equal average-SINR case (the "E" case) when the average SINR $\overline{\Upsilon}^{(k)} = 30$ dB is the same for every CR due to the power-control (PC) mechanism ($\overline{\Upsilon}^{(k)} = \overline{\Upsilon}$), and diverse average-SINR case ("D" case) when the PC mechanism has a tolerance of 2.5 dB, so that random deviation not exceeding 2.5 dB from the average SINR is possible for any node. The channel model considered is the two-path Rayleigh-fading channel with the delay spread ranging from zero to one-fourth of the OFDM symbol, and the average power of the second path being -3 dB relative to the average power of the first path. The assumed BEP for uncoded QAM modulation is $P_e^{(k)} = P_e = 10^{-3}$.

In Figures 17.1 through 17.4, simulation results are presented for the case of linear taxation of resources, that is, for $\Delta_\tau = 0$ and $\Gamma_\tau = 1$, and for various K and I values. In Figure 17.1, the spectral efficiency (sum throughput averaged over the number of used SCs) is presented, while in Figure 17.2, the throughput divided by the whole available bandwidth (available SCs N) is

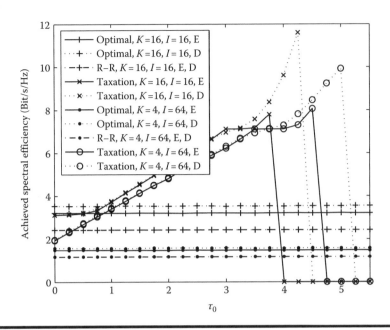

Figure 17.1 Spectral efficiency vs. the linear tax-rate parameter value τ_0 ("E"—equal average SINR for all nodes $\overline{\Upsilon}^{(k)} = \overline{\Upsilon} = 30$ dB, "D"—diverse average SINRs for nodes $\overline{\Upsilon}^{(k)} \in [30 - 2.5$ dB, $30 + 2.5$ dB$]$).

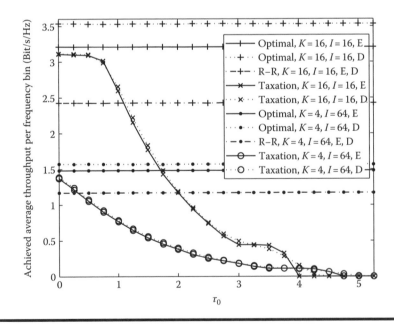

Figure 17.2 The overall throughput averaged over N available SCs vs. the linear tax-rate parameter value τ_0 ("E"—equal average SINR for all nodes $\overline{\Upsilon}^{(k)} = \overline{\Upsilon} = 30$ dB, "D"—diverse average SINRs for nodes $\overline{\Upsilon}^{(k)} \in [30 - 2.5$ dB, $30 + 2.5$ dB$]$).

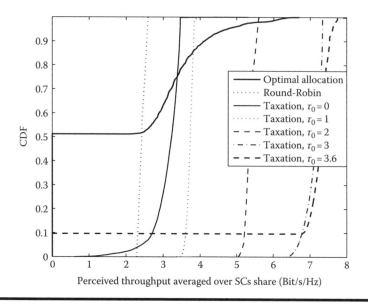

Figure 17.3 **CDF of the perceived throughput per frequency unit in the game with linear taxation for the "E" case, $\overline{\Upsilon}^{(k)} = \overline{\Upsilon} = 30$ dB.**

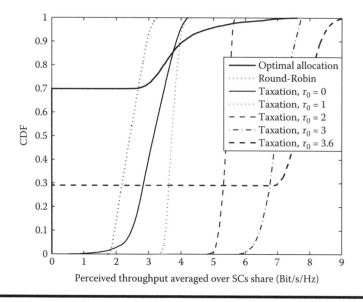

Figure 17.4 **CDF of the perceived throughput per frequency unit in the game with linear taxation for the "D" case, $\overline{\Upsilon}^{(k)} \in [30 - 2.5$ dB, $30 + 2.5$ dB].**

plotted vs. the tax-rate parameter value τ_0. Moreover, these results are compared with the Round-Robin (R-R) SC allocation and with the *optimal* (greedy) SC allocation. Greedy algorithm assigns subsequent SCs to CRs, which experience the highest CINRs at these SCs.

Let us note that for the appropriately high tax values, the spectral efficiency is higher in our game model than even this optimal allocation of SCs, although the average throughput per SC decreases with an increase of the relevant tax-rate parameter τ_0. Low average throughput per SC for high tax values occurs due to poor SC utilization, which become too "costly" (in terms of low payoff). Moreover, as expected, the spectral efficiency for the "D" case is slightly higher than for the "E" case for the appropriately high tax-rate values. This is because in the diverse-average-SINR scenario, the nodes with higher average SINRs acquire more SCs than the nodes with lower average SINRs. In case of linear taxation of resources, the average throughput per SC is approximately the same for both cases "E" and "D."

In Figures 17.3 and 17.4, the cumulative distribution function (CDF) of the nodes' perceived throughput averaged over their SC share (perceived spectral efficiency) is presented for the greedy SC allocation discussed earlier, for the R-R SC allocation and for our linear resource-taxation model. Let us note that for the optimal SC allocation, there is a high number of nodes, which are not served at all (the probability that the perceived spectral efficiency is lower than even very low but positive value is quite high). In the R-R algorithm, every CR receives approximately equal share of SCs, regardless of the channel quality. Therefore, their perceived throughput does not drop below a certain level; however, this level is rather low.

The CDF curves obtained for our game model show that depending on the τ_0 value, high perceived spectral efficiency can be approached with adequately high taxes. Too high taxes however result in low percentage of the utilized SCs and served nodes. More fair SC allocation, better SC utilization (resulting in a decent percentage of used SCs), higher average throughput, lower perceived, and network spectral efficiency can be obtained with lower taxes, for example, with $\tau_0 = 1$ for $K = 16$ and $I = 16$ for linear taxation. An interesting observation can also be made that even with no taxes, that is, $\tau_0 = 0$, the perceived-spectral-efficiency CDF for the taxation-based models can be found below the respective CDF for the optimal SC allocation for adequately low CDF argument values. This is because in our game model, we have defined the payoff, which reflects not only the potential throughput, but also the fairness, that is, network potential to serve multiple nodes. Some more results for other scenarios, other simulation parameters, and arithmetically progressive taxation of resources can be found in [40].

17.4 OFDM Adaptive Transmission as a Game

After acquiring the OFDM SCs, a question arises on how the CR we can further increase the efficiency of radio resource utilization. To maximize the throughput, DU of the CR should allocate appropriate power levels at SCs. This problem has been solved based on the water-filling principle [38], which however assumes the perfect knowledge of the CSI. Since we consider the autonomous CR, and very limited control traffic, the CSI may be erroneous due to the channel estimation or prediction errors, or obsolete due to the feedback delay. In our scenario, the CINR values at SCs are measured or estimated for the purpose of the rational spectrum sharing. However, if we consider wireless mobile channel and its dynamic changes, then in the course of time, the DU has to assume increasing uncertainty concerning these measured values or its fixed or variable error, which may depend on the actual SINRs at distinct SCs. (In the case of a static or stationary channel, having more observations of the CSI leads to more accurate estimations; however, we do not consider this trivial case, which is rarely met in practical wireless scenarios even with low users mobility.) Thus, the BPL procedure appears as a game, which a CR plays against the wireless environment, represented by the CINRs at the OFDM SCs. The values of CINRs are not exactly known by the rational player (the CR), and thus, the problem of assigning a number of bits and a power level to each SC cannot be approached as a typical optimization interrogation.

17.4.1 The Idea of a Game against the Nature

Let us consider a game, which the CRs play independently against their radio environment. In order to distinguish our game model from the typical optimization task, let us stress that the CSI representing the environmental conditions is not exactly known by this player (the CR has some environmental awareness based on a limited feedback). The impact of the imperfect CSI on the adaptive OFDM performance has been discussed by a number of papers, but no one tries to account for this imperfection. Here, the environment is viewed as a player, whose behavior is not rational, that is, she does not try to maximize her payoff nor is reactive to the behavior of the rational player. Such a model is referred to as *the game against the nature* in [17]. Because in our considerations, rationality is a key factor, apart from throughput maximization, one should pay a special attention to the power economy, which should be reflected in the payoff function of this game.

Some initial results in modeling the erroneous CSI and the BPL problem as a game have been presented in the previous work [39,40]. There, the CSI uncertainty has been modeled in a simplified way. As a result, the so-called water level has been calculated based on the erroneous values of CINRs and therefore has been containing a relatively large error. Here below, we present a better model of the CSI-error probability, resulting from the application of the channel prediction.

17.4.1.1 Strategies

In our game against the channel, the strategies of the kth CR node are related to the choice of the constellation sizes and power levels at the acquired SCs f_s (where $s \in \mathbf{S}_i^{(k)}$) under the total transmit power constraint and for the assumed target BEP. The DU is supposed to choose a certain (mth) strategy (the power level at each SC) as a response to the one (lth) of L possible channel strategies and maximize the throughput on the chosen SCs. Channel strategies are related to possible values of the CINRs at distinct SCs (L values appearing in a given range defined for each SC). The actual value of CINR for every SC is in the considered range; however, it is unknown.

By adopting an mth strategy at SC f_s as a response to the lth strategy of the channel, the DU considers the following throughput that could be perceived at this SC:

$$R_{m,l}^{(k)}(f_s) = \Delta f \cdot \log_2 \left[1 + \alpha^{(k)} P_m^{(k)}(f_s) \hat{\gamma}_l^{(k)}(f_s) \right] \text{ [bit/s]}, \tag{17.7}$$

where

$P_m^{(k)}(f_s)$ is the mth possible value of the considered power levels allocated to SC frequency f_s by the kth player

$\hat{\gamma}_l^{(k)}(f_s)$ is the lth possible value of the CINR at this SC considered by this player

Note that (17.7) reflects the bit rate that would be achieved if $\hat{\gamma}_l^{(k)}(f_s)$ was the actual CINR value. In practical systems, it is common that the estimates of $\gamma^{(k)}(f_s)$ for each f_s are used for various purposes (channel estimation, correction, adaptive transmission) and that $\hat{\gamma}_l^{(k)}(f_s)$ contains an error; thus, if the DU applies the modulation scheme reflecting exactly the rate of $R_{m,l}^{(k)}(f_s)$, the target BEP might not be met. In the approach proposed, the DU considers a number of $\hat{\gamma}_l^{(k)}(f_s)$ values at each SC and makes her choice based on some preferences, as defined in the next section.

Since every CR plays the same type of a game against her channel independently, and based on the same principles, for the simplicity of mathematical description, we will discard the upper index

(k) without the loss of generality. Allocation of the power level of $P_m(f_s)$ to the SC frequency f_s allows transmitting data of the following M-QAM constellation size $M_{m,l}(f_s)$ at this SC:

$$M_{m,l}(f_s) = 1 + \alpha P_m(f_s)\hat{\gamma}_l(f_s). \qquad (17.8)$$

The strategies of the CR are chosen as a response to a set of considered $\hat{\gamma}_l(f_s)$ values (for $l \in [1, L]$) and follow the optimal BPL principle with the assumed total power constraint $\overline{P}$. Let us note that the actual CINR values are not known by the DU, which impacts the water level calculations.

17.4.1.2 Payoffs

In various approaches aspiring to the spectral efficiency optimization, the utility function reflects the throughput as the objective to be maximized. In our game, however, rationality is a key factor, which means that apart from the throughput maximization, we pay special attention to the power economy. If we define our utility function by (17.7) for each SC, the DU would choose to transmit with the highest power (out of all considered strategies) for a given SC. This is because in face of any strategy of a channel $\hat{\gamma}_l(f_s)$, allocation of higher $P_m(f_s)$ level to a SC results in higher $R_{m,l}(f_s)$. Such a strategic choice at each SC would not be favorable for the transmission with limited power budget. Therefore, it is proposed to define utilities $U_{m,l}(f_s)$ for each SC, which reflect not just the throughput but also the *power economy*. The amount of power saved per frequency unit $S_{m,l}(f_s)$, when the mth DU's strategy ($P_m(f_s)$) is chosen as a response for the lth channel's strategy, is defined as

$$S_{m,l}(f_s) = P_l^{\mathrm{opt}}(f_s) - P_m(f_s), \qquad (17.9)$$

where $P_l^{\mathrm{opt}}(f_s)$ is the optimum (based on water-filling) best-response strategy of the DU to the lth channel's strategy. Now, let us define the utilities $U_{m,l}(f_s)$ to reflect the benefit of BPL (achieved throughput and power savings):

$$U_{m,l}(f_s) = x\varphi\left\{S_{m,l}(f_s)\right\} + (1-x)\varphi\left\{R_{m,l}(f_s)\right\}, \qquad (17.10)$$

where
  x is the weighting factor
  $\varphi\left\{Y_{m,l}(f_s)\right\}$ is linear nondecreasing and normalizing function of $Y_{m,l}(f_s)$

$$\varphi\left\{Y_{m,l}(f_s)\right\} = \frac{Y_{m,l}(f_s) - \min\left\{Y_{m,l}(f_s)\right\}}{\max\left\{Y_{m,l}(f_s)\right\} - \min\left\{Y_{m,l}(f_s)\right\}}. \qquad (17.11)$$

Such a linear nondecreasing function does not affect the game solution [17]. The weighting factor x reflects the player's preferences, that is, to what degree she cares for maximizing the throughput and to what degree for economizing the power consumption.

When choosing the mth strategy, the DU expects the payoff $B_m(f_s)$, which can be calculated taking the probabilities $\Pr\left\{\hat{\gamma}_l(f_s)\right\}$ of the channel's strategies into account:

$$B_m(f_s) = \sum_{l=1}^{L} U_{m,l}(f_s) \cdot \Pr\left\{\hat{\gamma}_l(f_s)\right\}. \qquad (17.12)$$

Let us note that the probabilities $\Pr\{\hat{\gamma}_l(f_s)\}$ have to be known by the DU. Thus, modeling of the CSI uncertainty is a key issue in this game. One of the typical radio-transceiver functionalities is to estimate the channel characteristic (or the channel impulse response). If the estimation errors of the real and imaginary parts of the channel characteristic $H(f_s)$ are Gaussian, the CINR error has the Rayleigh distribution. In the CR case, one can assume that the CINR-error variance is known by the DU. It can be assumed as constant for a given transmission scheme (if it results from the feedback delay) or as dependent on the average SINR (if it results from the estimation error) of the considered CR link. Thus, $\Pr\{\hat{\gamma}_l(f_s)\}$ follows the shifted CINR-error distribution (with maximum probability related to the recently available value of CINR at the SC frequency f_s). The standard deviation of the above-mentioned distribution assumed by the DU should reflect the level of uncertainty concerning the CINR [39]. Moreover, the actual CINR values change with time, and so does the actual water level W. Application of the channel prediction at the CR transceiver is advantageous in order to calculate W and the CINR values more accurately.

Finally, the DU adopts the strategy m^+ and the resulting $P_{m^+}(f_s)$, which maximizes (17.12):

$$m^+ : B_{m^+}(f_s) = \max B_m(f_s). \tag{17.13}$$

Apart from the adopted power spectral density, the DU also adopts the M-QAM constellation size at respective SCs according to (17.8) for $m = m^+$.

17.4.2 Experimental Results

In our experiments, we have assumed the same channel statistics as in the computer simulation of the game for resources based on resource taxation. A fixed number ($L = 7$) of the channel's strategies have been considered for every accessed SC at every OFDM symbol period. In Figure 17.5, the results of our adaptive transmission game in terms of the achieved average throughput per SC

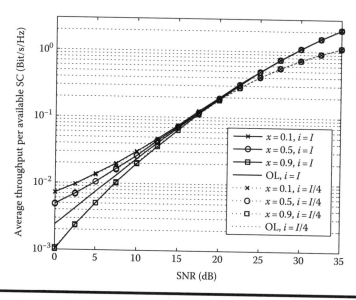

Figure 17.5 Maximum achieved throughput for various values of the utility contribution weighting factor x and $\sigma_\varepsilon = 1/\overline{\Upsilon}^{(k)}$ ($\overline{\Upsilon}^{(k)} = \overline{\Upsilon}$); OL—optimal loading.

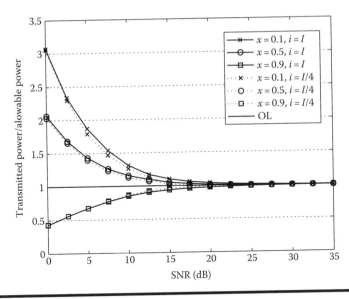

Figure 17.6 Total transmitted power relative to the assumed power limit for various values of the utility contribution weighting factor x and $\sigma_\varepsilon = 1/\overline{\Upsilon}^{(k)}$ ($\overline{\Upsilon}^{(k)} = \overline{\Upsilon}$); OL—optimal loading.

(achieved player's throughput divided by the total available bandwidth: $N\Delta f$) are presented for various values of the $U_{m,l}(f_s)$ utility-component weighting factor x ($x \in [0.1, 0.5, 0.9]$), for the assumed BEP ($P_e = 10^{-3}$) and for the standard deviation of the CINR error dependent on the average SINR value: $\sigma_\varepsilon = 1/\overline{\Upsilon}$ (note that $\overline{\Upsilon} = \overline{\Upsilon}^{(k)}$ and $P_e = P_e^{(k)}$ since we have discarded the player's index (k)). In Figure 17.6, the transmitted power relative to the assumed power limit is presented. The throughput has been calculated using expression (17.7) for the strategies adopted at all acquired SCs for an example player, who admits the network and chooses a number of SCs (i out of $N^{(k)}$ available SCs). The results in Figures 17.5 and 17.6 are presented for $N^{(k)} = I = 16$ and $i = 16$ or $i = 4$ (an example player takes the maximum or one-fourth of the available SCs). The *optimal-loading* (OL) curves presented in Figures 17.5 and 17.6 have been obtained when the perfect CSI has been available at the CR transmitter. Similar (but not the same) results for other scenarios and only one considered total number of available SCs have been presented in [39,40].

Let us note that for small SINR values and for $x = 0.1$, the transmitted power significantly exceeds the power limit. The reason for this effect is that σ_ε is high for small SINR values, which has an impact on the transmit power level, particularly when the power-economizing utility component $x\varphi\{S_{m,l}(f_s)\}$ is small (due to low value of x). In such a case, throughput-maximizing utility component $(x - 1)\varphi\{R_{m,l}(f_s)\}$ is increased, and higher throughput is achieved than in the case of the BPL algorithm with perfect CSI. This happens at the expense of the total transmitted power exceeding the assumed limit. If $x = 0.9$, the opposite results are obtained because the DU is forced to be more power efficient. This leads us to the conclusion that by choosing the appropriate value of the utility-component weighting factor x, the player can approach the ideal bit-and-power-loading algorithm results. Figures 17.5 and 17.6 show that by assuming $x = 0.5$ in the considered game model, we obtain results close to the optimal BPL curves of the achieved throughput and the total transmitted power.

17.5 Concluding Remarks

Multicarrier technologies, which include OFDM and OFDMA, are considered well suited for the application in the CR due to their inherent features: interoperability, flexibility in the transmission parameters adjustment, efficient spectrum utilization, aggregation, and spectrum shaping capabilities, as well as localization and positioning proficiency. The challenges posed by the OFDM-based CR include the development of the intelligent, decentralized, efficient, and rational DRA procedures of reasonable complexity under multiple limitations posed by the network and limited environmental awareness.

The game-theoretic approach to these challenging problems based on resource taxation allows for the efficient dynamic spectrum access and for the rational spectrum utilization in the OFDMA-based CR network scenery. Moreover, the adaptive BPL game presented above can handle the impact of the imperfect CSI. The considered games are noncooperative and with limited information, which means that each player makes decisions alone based on limited (or even erroneous) information available. For the sake of efficiency, the utility functions in these games have been defined so as to reflect the throughput achieved by a CR node and the network potential to serve the CR-node community. For the sake of rationality, taxation of the SCs has been introduced in the DSA game, as well as the power-economizing component in the utility function in the BPL game. Moreover, by treating the CR-node community as a whole and by introducing the above-mentioned efficiency and rationality measures, we can approach optimal usage of the available spectrum with decent computational complexity.

17.5.1 Open Issues

Resource taxation can be the right approach to efficient and rational resource utilization. However, further research is needed to come out with the proper optimization routines that should be executed at the base station to calculate and beacon the optimal tax-rate parameter. The definition of the objective function that should be optimized for this purpose is an open issue. For sure, it should reflect the network general objectives, which however may vary with the network load, users' channel conditions, users' requirements concerning the QoS and QoE, etc.

The game for optimality modeling the adaptive OFDM with imperfect CSI seems to approach the perfect water-filling performance when the utility parameters are properly chosen. An open issue is the extension of this game for OFDM adaptive modulation and coding or for other popular OFDM transmission schemes, for example, with spatial diversity. In these cases, the formula for the player's throughput would be much more complex than (17.7) and would require numerical approximation.

References

1. J. Mitola, Cognitive radio: An integrated agent architecture for software defined radio, Dissertation, Doctor of Technology, Royal Institute of Technology (KTH), Stockholm, Sweden, 2000.
2. J. Mitola, *Cognitive Radio Architecture: The Engineering Foundation of Radio XML*, John Wiley & Sons, Hoboken, NJ, 2006.
3. S. Haykin, Cognitive radio: Brain-empowered wireless communications, *IEEE Journal Selected Areas in Communications*, 23, 2, February 2005, 201–220.
4. L. Hanzo, M. Munster, B.J. Choi, T. Keller, *OFDM and MC-CDMA for Broadband Multi-User Communications, WLANs and Broadcasting*, John Wiley & Sons, Chichester, U.K., July 2003.

5. R. Van Nee, R. Prasad, *OFDM for Wireless Multimedia Communications*, Artech House Publishers, Boston, MA, 2000.

6. T. Hwang, C. Yang, G. Wu, S. Li, G.Y. Li, OFDM and its wireless applications: A survey, *IEEE Transactions on Vehicular Technology*, 58, 4, May 2009, 1673–1694.

7. Y. (G.) Li, G. Stber, *Orthogonal Frequency Division Multiplexing for Wireless Communications*, Springer-Verlag, Boston, MA, January 2006.

8. T. Keller, L. Hanzo, Adaptive modulation techniques for duplex OFDM transmission, *IEEE Transactions on Vehicular Technology*, 49, 5, September 2000, 1893–1906.

9. L. Hanzo, C.H. Wong, M.S. Yee, *Adaptive Wireless Transceivers: Turbo-Coded, Turbo-Equalised and Space-Time Coded TDMA, CDMA and OFDM Systems*, John Wiley, Chichester, West Sussex, England, March 2002.

10. A. Goldsmith, *Wireless Communications*, Cambridge University Press, New York, 2005.

11. R. Rajbanshi, A.M. Wyglinski, G.J. Minden, OFDM-based cognitive radios for dynamic spectrum access networks, Chapter 6 in E. Hossain, V.K. Bhargava, eds., *Cognitive Wireless Communication Networks*, Springer Science and Business Media, New York, 2007.

12. H.A. Mahmoud, T. Yucek, H. Arslan, OFDM for cognitive radio: Merits and challenges, *IEEE Wireless Communications*, 16, 2, April 2009, 6–14.

13. C. Zeng, M. Luise, C. Hoo, J.M. Cioffi, Efficient water-filling algorithms for a Gaussian multiaccess channel with ISI, *Vechicular Technology Conference Fall 2000*, Boston, MA, September 2000.

14. G. Munz, S. Pfletschinger, J. Speidel, An efficient waterfilling algorithm for multiple access OFDM, *IEEE Global Telecommunications Conference, 2002, GLOBECOM'02*, Taipei, Taiwan, Vol. 1, November 17–21, 2002, pp. 681–685.

15. C.Y. Wong, R.S. Cheng, K.B. Letaief, D. Mursh, Multiuser OFDM with adaptive subcarrier, bit, and power allocation, *IEEE Journal on Selected Areas in Communications*, 17, 10, October 1999, 1747–1757.

16. S.H. Ali, K.-D. Lee, V.C.M. Leung, Dynamic resource allocation in OFDMA wireless metropolitan area networks, *IEEE Wireless Communications*, 14, 1, February 2007, pp. 6–13.

17. P.D. Straffin, *Game Theory and Strategy*, The Mathematical Association of America, Washington, DC, 2002, Chapter 10.

18. A. MacKenzie, L. DaSilva, *Game Theory for Wireless Engineers*, Morgan and Claypool Publishers, San Rafael, CA, 2006.

19. V. Srivastava et al., Using game theory to analyze wireless ad hoc networks, *IEEE Communication Surveys and Tutorials*, 7, 4, 2005, pp. 46–56.

20. G. Zhang, H. Zhang, Adaptive resource allocation for downlink OFDMA networks using cooperative game theory, *IEEE Singapore International Conference on Communication Systems, ICCS 2008*, Guangzhou, China, November 19–21, 2008, pp. 98–103.

21. T. Zhang, Z. Zeng, C. Feng, J. Zheng, D. Ma, Utility fair resource allocation based on game theory in OFDM systems. *16th International Conference Computer Communications and Networks, ICCCN 2007*, August 13–16, 2007, pp. 414–418.

22. Z. Han, Z.J. Ji, K.J. Ray Liu, Fair multiuser channel allocation for OFDMA networks using nash bargaining solutions and coalitions, *IEEE Transactions on Communications*, 53, 8, August 2005, 1366–1376.

23. T.K. Chee, C.-C. Lim, L.J. Choi, A cooperative game theoretic framework for resource allocation in OFDMA systems, *IEEE Singapore International Conference on Communication Systems, ICCS 2006*, Singapore, October 2006, pp. 1–5.

24. H.W. Kuhn, The Hungarian method for the assignment problem, *Naval Research Logistics*, 2, 1955, 83–97.

25. A. Ibing, H. Boche, Fairness vs. efficiency: Comparison of game theoretic criteria for OFDMA scheduling, *Asilomar Conference on Signals, Systems and Computers, ACSSC 2007*, Pacific Grove, CA, November 4–7, 2007, pp. 275–279.

26. D. Wu, D. Yu, Y. Cai, Subcarrier and power allocation in uplink OFDMA systems based on game theory, *International Conference on Neural Networks and Signal Processing, 2008*, Zhenjiang, China, June 7–11, 2008 pp. 522–526.

27. F. Chen, L. Xu, S. Mei, T. Zhenhui, L. Huan, OFDM bit and power allocation based on game theory, *International Symposium on Microwave, Antenna, Propagation and EMC Technologies for Wireless Communications, 2007*, Hangzhou, China, August 16–17, 2007, pp. 1147–1150.

28. D. Yu, D. Wu, Y. Cai, W. Zhong, Power allocation based on power efficiency in uplink OFDMA systems: A game theoretic approach, *11th IEEE Singapore International Conference on Communication Systems, ICCS 2008*, Guangzhou, China, November 19–21, 2008, pp. 92–97.

29. D. Niyato, E. Hossain, Microeconomic models for dynamic spectrum management in cognitive radio networks, Chapter 14 in E. Hossain, V.K. Bhargava, eds., *Cognitive Wireless Communication Networks*, Springer Science and Business Media, New York, 2007.

30. W. Noh, A distributed resource control for fairness in OFDMA systems: English-auction game with imperfect information, *IEEE Global Telecommunications Conference, GLOBECOM 2008*, New Orleans, LA, November 30–December 4, 2008, pp. 1–6.

31. J. Sun, E. Modiano, L. Zheng, Wireless channel allocation using an auction algorithm, *IEEE Journal on Selected Areas in Communications*, 24, 5, May 2006, 1085–1096.

32. G. He, S. Gault, M. Debbah, and E. Altman, Distributed power allocation game for uplink OFDM systems, *6th International Symposium on Modeling and Optimization in Mobile, Ad Hoc, and Wireless Networks, 2008*, Berlin, Germany, April 1–3, 2008, pp. 515–521.

33. H. Kwon, B.G. Lee, Distributed resource allocation through noncooperative game approach in multi-cell OFDMA Systems, *IEEE ICC'06*, Istanbul, Turkey, pp. 4345–4350, 2006.

34. L. Wang, Y. Xue, E. Schulz, Resource allocation in multicell OFDM systems based on noncooperative game, *IEEE 17th International Symposium on Personal, Indoor and Mobile Radio Communications*, Helsinki, Finland, 2006, September 11–14, 2006, pp. 1–5.

35. Z. Liang, Y. Huat Chew, C. C. Ko, Decentralized bit, subcarrier and power allocation with interference avoidance in multicell OFDMA systems using game theoretic approach, *IEEE Military Communications Conference, MILCOM 2008*, San Diego, CA, DOI:10.1126/science.162.3859.1253, November 16–19, 2008, pp. 1–7.

36. G. Hardin, The tragedy of the commons, *Science*, 162(3859), December 1968, pp. 1243–1248.

37. H. Bogucka, Linear and arithmetically-progressive taxation of spectrum resources in cognitive OFDMA, *3rd ICST/ACM International Workshop on Game Theory in Communication Networks, GameComm 2009*, October 23, 2009, Pisa, Italy.

38. G.J. Foschini, J. Salz, Digital communications over fading radio channels, *Bell Systems Technical Journal*, 62, 2, February 1983, 429–456.

39. H. Bogucka, Bit and power loading game of the flexible OFDM transceiver in a wireless environment, *10th IEEE International Symposium on Spread Spectrum Techniques and Applications, ISSSTA'08*, August 25–28, 2008, Bologna, Italy.

40. H. Bogucka, Efficient and Rational Spectrum Utilization in Opportunistic OFDMA Network with Imperfect CSI: a Utility-Based Top Down Approach, *Wireless Communications and Mobile Computing*, 2010. Published online in Wiley Interscience, DOI:10.1002/WCM.973, www.interscience.wiley.com

Chapter 18

Noncooperative Resource Management in Wireless Systems

Zhen Kong and Yu-Kwong Kwok

Contents

Resource allocation and management are important in a wireless communication system, in which the scarce bandwidth spectral resources are shared by multiple users. In this chapter, we consider resource allocation problems in *noncooperative* environments, where users are assumed to be selfish and each user's resource is obtained by competition. Specifically, in a noncooperative environment, wireless users exhibit noncooperative behaviors due to self-interests, compete for the resource with each other, try to overuse the radio resource, and maximize their own advantages without regard to the performance of other users and the overall system.

Noncooperative radio resource allocation has emerged significantly in many practical situations, where the mobile users are assumed to be autonomous and selfish. The noncooperative behavior often leads to detrimental competition among users. For instance, in a public wireless local area network (WLAN) hot spot with one access point (AP), only one mobile user can transmit at a time. As the radio link is shared between all stations, a selfish user obviously has the temptation to overuse the radio resource and increase its bandwidth by deviating from the protocol, at the expense of a poor performance in the other stations [3]. Furthermore, when multiple service providers or APs coexist in this hot spot, the competition for subscribers among wireless service providers will inevitably occur so as to increase their respective revenues. Such phenomenon makes the research of noncooperative radio resource allocation more and more crucial in reality.

In view of the selfish and autonomous nature of users, development of decentralized control mechanisms has become a major issue in wireless network research. Specifically, game theory [22] has been used as an efficient tool to gain a deeper understanding of these complex problems in noncooperative systems and has been widely used in the area of wireless communications. For example, in terms of selfish behaviors at the physical layer, Meshkati et al. [20] have modeled power control for multicarrier CDMA systems as a noncooperative game in which each user tries to selfishly maximize its overall utility. They have also proposed an iterative and distributed algorithm for reaching an equilibrium with a significant improvement in the total network utility. As to routing at the network layer, Stephan et al. [29] have focused on designing a routing algorithm with the feature of individual rationality, truthfulness, and energy efficiency for selfish devices in ad hoc networks based on an auction model. While for noncooperative behavior at the MAC layer, Tan and Guttag [31] have examined a kind of selfish misbehavior under 802.11 distributed coordination function (DCF) mode, where the selfish nodes intentionally transmit at a lower data rate so as to achieve a higher channel share and a higher data rate. With the observation that the wireless channel is inefficiently utilized in Nash equilibrium (NE) state, they have also shown that, by guaranteeing the allocation of long-term shares of channel time to competing nodes, the MAC protocol can enforce rational nodes to efficiently use the shared medium, thereby improving the achieved data rates of all competing nodes.

Kong and Kwok have recently studied the noncooperative resource management in wireless systems by using game-theoretic methods [14,15,17]. In this chapter, we aim to analyze and solve the problem of how to efficiently allocate radio resource in noncooperative wireless environments based on this research. Specifically, we first study the noncooperative resource scheduling in a single-carrier TDMA-based wireless network in Section 18.1, then extend it to a multicarrier OFDM system, and propose an auction-based scheduling method in Section 18.2. Besides the competition for radio resources among wireless users, the revenue competition between wireless service providers is also important in a noncooperative environment. Consequently, the issue of pricing competition among WLAN providers is also discussed in Section 18.3. In Section 18.4, we suggest some future research avenues. We conclude this chapter in Section 18.5.

18.1 Noncooperative Scheduling in TDMA Wireless Networks

18.1.1 Introduction

In a centralized infrastructure-based wireless network, packet scheduling is a very important issue to manage the precious radio resource while satisfying users' Quality-of-Service (QoS) requirements. Specifically, in a traditional downlink packet scheduling protocol, wireless users are required to report their channel conditions, such as signal-to-noise ratio (SNR) or maximal achievable data rate, to the scheduler located at the base station (BS) or AP. Then the scheduler can select some users' packets and allocate radio resources, such as power and frequency bandwidth, to these users for transmission according to some scheduling policies, such as maximum rate (MR) [32] and proportional fairness (PF) [11].

Usually, these scheduling algorithms are based on the assumption that the wireless users cooperate with each other, comply with the predefined scheduling algorithm, and honestly report their real channel conditions to the scheduler. Then the wireless user will accept the scheduling results passively, and it will not affect the scheduling policy employed by the scheduler at all. However, in many practical scenarios, the wireless users are autonomous and, thus, may exhibit noncooperative behaviors due to self-interests. For instance, in a public WLAN hot spot, individual users may attempt to deviate from the standard protocols or algorithms and behave in a rational but selfish manner so as to gain advantages in radio resource allocation, without regard to the overall system performance [31].

In a typical wireless packet scheduling process, the scheduling policy is one of the major factors governing a user's data rate because the scheduler determines which user can be selected for transmission. Therefore, in a noncooperative environment, a wireless user experiencing a bad channel condition might find out that if it honestly reports its channel condition to the scheduler, it may not be scheduled or just be assigned with a low data rate. Consequently, with a rationally selfish motivation, such user might report a bogus channel condition so as to get a higher probability to be scheduled for transmission or get a higher data rate. Though this noncooperative behavior could increase the data rate for this selfish user, it may lead to inefficient resource utilization for the whole system. Due to the proliferation of open-source software technologies and software-defined radios, such noncooperative behaviors become more and more practicable because wireless devices could be easily programmed by users to behave in a selfish manner [3,27]. Thus, whether the traditional packet scheduling algorithms are still effective in allocating resource in a noncooperative environment is in doubt and needs to be scrutinized carefully. In the following, we focus on the issue of downlink packet scheduling for selfish and rational wireless users in a noncooperative TDMA wireless network, in which each user attempts to locally and selfishly choose its transmission rate so as to maximize its utility. Specifically, we will first investigate the impact the selfish users, then analyze, and solve the problem by means of game-theoretic methods.

18.1.2 Impact of Selfish Users

To study the impact of selfish users, we consider a time-slotted system with one BS serving N wireless users, where there are S selfish users deliberately deviating from the packet scheduling algorithm. The BS transmits in slots with fixed duration, and only one user can be scheduled in one time slot. All users are assumed to be either static or moving slowly and within the same communication range (i.e., each user can overhear any other users). We assume users always have packets to transmit. At the beginning of time slot t, each user i measures the downlink channel condition, and returns, via a feedback channel, a measured data rate $r_i(t)$ to the BS. Based on this information, the BS determines which user to transmit its packet in this time slot.

In our analysis, we assume that adaptive modulation and ideal phase detection are used in a Rayleigh fading channel with bandwidth W, and no retransmission is considered. Typically, at time slot t, the maximal achievable symbol rate $c_i^a(t)$ (bit/symbol) for user i can be decided by its current channel SNR $\gamma_i(t)$ and the required BER P_{ber}, then $c_i^a(t)$ can be expressed as [26]

$$c_i^a(t) = \log_2 \left(1 + \frac{-1.5}{\ln(5 \cdot P_{ber})} \cdot \gamma_i(t) \right). \tag{18.1}$$

We assume that the symbol rate belongs to the symbol rate set $C = \{c_i : 0 \leq c_i \leq M\}$, that is, $c_i^a(t) \in C$, where M can be interpreted as the maximal modulation mode. Then the corresponding maximal achievable data rate is given by $r_i^a(t) = c_i^a(t) \cdot W$. Correspondingly, we define the set of data rate as $R = \{r_i : 0 \leq r_i \leq r^M\}$, where $r^M = M \cdot W$ and $r_i^a(t) \in R$.

We assume that each $r_i(t)$ is an independent and stationary random variable, and let the reported $r_i(t)$ also belong to the set R, that is, $r_i(t) \in R$. Within this framework, in the literature, there are several well-known packet scheduling algorithms, such as the MR algorithm [32], which is designed to maximize the data rate at each slot, as well as the system aggregate data rate, by scheduling the user with the largest $r_i(t)$ for transmission.

These algorithms are optimal under their respective objectives with the assumption that every wireless user i will comply with the algorithms and report its maximum feasible rate $r_i^a(t)$ to the BS honestly. However, they do not have any consideration of the impact of selfish behavior on their predefined performance and objectives. For example, if a user knows that it will not be scheduled for transmission if it reports its real channel condition, it could tell a bogus $r_i(t) > r_i^a(t)$ to the BS so as to increase its chance for transmission. Then the corresponding scheduling results may be totally different with what will be realized in cooperative environments. Of course, a higher assigned data rate $r_i(t)$ may result in a higher BER or lower PTSR under the same SNR $\gamma_i(t)$, making its actual achievable rate lower than the intended value. Nevertheless, a rationally selfish consideration is that the smaller realized rate may be compensated by the rate improvement induced by the increased transmission probability. Consequently, a noncooperative device could still have the incentive to report a different data rate to the BS so as to increase its own potential payoff.

Let BER $\left(r_i, r_i^a \right)$ and $\alpha \left(r_i, r_i^a \right)$ be the BER and PTSR, respectively, for user i when its maximum feasible rate is r_i^a and feedback data rate is r_i, where $r_i^a \in R$ and $r_i \in R$. When there are L bits in one packet, we have $\alpha \left(r_i, r_i^a \right) = \left(1 - \text{BER} \left(r_i, r_i^a \right) \right)^L$.

Since the probability that it can transmit at this time slot is $Pr(r_i(t) > r_j(t), \forall j \neq i)$ and we assume all the users are within the same communication range, the selfish user i can overhear or estimate the feedbacks of other users and report a bogus rate $r_i(t) > r_j(t), \forall j \neq i$. For example, one of the simplest way to report a bogus rate is to just report the highest possible rate r^M as $r_i(t)$ to the BS. Thus, its expected rate can be expressed as

$$R_i(r_i(t)) = \begin{cases} r_i^a \cdot \alpha \left(r_i^a, r_i^a \right) \cdot Pr(r_i(t) = r_i^a > r_j(t), \forall j \neq i) & \text{cooperative,} \\ r_i \cdot \alpha \left(r_i, r_i^a \right) \cdot Pr(r_i(t) = r_i > r_j(t), \forall j \neq i) & \text{noncooperative.} \end{cases} \tag{18.2}$$

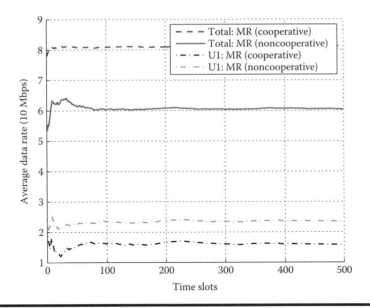

Figure 18.1 **The impact of selfish behavior on the average data rate of packet scheduling algorithm.**

To empirically quantify the performance impact of this kind of selfish behavior, we conduct preliminary simulations based on the above model. The entire system bandwidth is 10 MHz. We set the time slot duration as 100 ms, and let $c_i \in [0, 10]$. We also assume that the packet length is 100 bits/packet and the maximum BER requirement is 10^{-5}. There are eight wireless users. User 1 (U1) deliberately fails to adhere to the algorithm and tries to misbehave, following the selfish model presented above. In a cooperative environment, all users will report their maximum feasible data rate to the BS, whereas in noncooperative environments, U1 will report a bogus data rate to the BS and all other users still report their maximum feasible rates honestly. In this simulation, we can see in Figure 18.1 that U1 can achieve about 100% increase in average data rate by reporting a higher data rate and behaving selfish. However, as a result of U1 using this selfish strategy, the system average data rate decreases by 30%. Therefore, the existence of selfish behavior in noncooperative wireless networks could significantly degrade the data rate performance of packet scheduling algorithm.

18.1.3 Static Game Analysis

The above preliminary analysis clearly shows that a selfish user has an incentive to act in a noncooperative way so as to improve its own data rate. Such selfish behavior can be modeled as a noncooperative game in which each player has a continuum of actions [22]. In this game, we denote wireless users $\Omega = \{1, \dots, N\}$ as the set of players. When player i's maximal feasible rate is r_i^a, its action is its reported rate r_i and the set of its possible actions is the interval from r_i^a to r^M, that is, $A_i = \{r_i \mid r_i^a \leq r_i \leq r^M\}$. The action combination is denoted as $r = (r_1, r_2, \dots, r_N) \in A$, where $A = \times_{i \in \Omega} A_i$ is the Cartesian product of the N players' action profile. We identify each player's mixed strategy with a cumulative distribution function (CDF) $F_i(r_i)$ on this interval, for which $0 \leq F_i(r_i) \leq 1$ for every action r_i, and thus, the number $F_i(r_i)$ is the probability that player i's data rate is at most r_i. Each player i's preference is represented by the expected value of the data rate:

$$u_i(r) = \begin{cases} r_i \cdot \alpha\left(r_i, r_i^a\right) & r_i > r_j, \\ \dfrac{1}{N} \cdot r_i \cdot \alpha\left(r_i, r_i^a\right) & r_i = r_j, \\ 0 & r_i < r_j. \end{cases} \tag{18.3}$$

where the parameter $1/N$ captures the situation that when these players have the same rate, they will be chosen with the equal probability. We assume that all user rates r_is are independent and identically distributed with the same CDF, and then the expected payoff is

$$U_i(r, F) = r_i \cdot \alpha\left(r_i, r_i^a\right) \cdot Pr(r_i > r_j) + \frac{1}{N} \cdot r_i \cdot \alpha\left(r_i, r_i^a\right) \cdot Pr(r_i = r_j) + 0 \cdot Pr(r_i < r_j)$$

$$= r_i \cdot \alpha\left(r_i, r_i^a\right) \cdot Pr(r_i > r_j), \forall j \neq i) = r_i \cdot (F_j(r_i))^{N-1}. \tag{18.4}$$

Definition 18.1 An action combination $r^* \in A$ and the corresponding mixed strategy $F^* \in [0, 1]$ are said to achieve the state of NE if for every player $i \in \Omega$, we have

$$U_i\left(r_i^*, r_{-i}^*; F_i^*, F_{-i}^*\right) \geq U_i\left(r_i', r_{-i}^*; F_i^*, F_{-i}^*\right) \quad \forall r_i' \in A_i, F_i' \in [0, 1]. \tag{18.5}$$

where

  r_{-i} and F_{-i} denote the actions and the corresponding mixed strategies chosen by everyone else other than i respectively

  r_i' is a reported rate other than NE rate r_i^*

  $U_i(r_i, r_{-i}; F_i, F_{-i})$ is user i's utility under rate r_i, r_{-i} and CDF F_i, F_{-i}

It is known that for a game in which each player has finitely many actions, when a mixed strategy profile is a mixed strategy NE, the expected payoff to every action assigned with positive probability is the same. Correspondingly, as described in the Proposition 142.2 in [22], for the game in which each user has infinitely many actions $A_i = \left\{r_i \mid r_i^a \leq r_i \leq r^M\right\}$, the mixed strategy is determined by the probabilities assigned to sets of actions; and the expected payoff should be constant from r_i^a to r^M in NE. Moreover, because $F_j(r_i = r^M) = Pr\left(r_i(t) \leq r_i = r^M\right) = 1$, we have

$$U_i(r, F) = \begin{cases} r_i \cdot \alpha\left(r_i, r_i^a\right) \cdot (F_j(r_i))^{N-1} = C & r_i^a \leq r_i \leq r^M, \\ r^M \cdot \alpha\left(r^M, r_i^a\right) \cdot (F_j(r_i))^{N-1} = C & r_i = r^M. \end{cases} \tag{18.6}$$

where C is a constant. Thus, for $0 \leq r_i^a \leq r_i \leq r^M$, we get $F_j(r_i) = \left(r^M \cdot \alpha\left(r^M, r_i^a\right)/r_i \cdot \alpha\left(r_i, r_i^a\right)\right)^{1/N-1}$.

When the maximum achievable rate for player i is r_i^a, it will choose its feedback rate r_i according to the above NE strategy. Because in practice r_i is a discrete random variable, if $r_i^a \leq r^A \leq r_i \leq r^B \leq r^M$, the probability that it reports r_i to BS is expressed as $Pr(r_i) = \frac{1}{2} \cdot (F_i(r^A) + F_i(r^B))$, where $r^A = \lfloor r_i \rfloor$ and $r^B = \lceil r_i \rceil$, $\lfloor x \rfloor$ express the largest integer smaller than x, $\lceil x \rceil$ express the least integer larger than x.

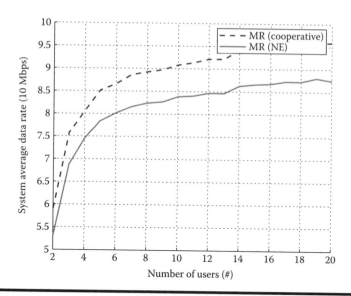

Figure 18.2   System average data rate comparison with different users.

In a cooperative environment, the average data rate of the system is maximized by the MR algorithm. Furthermore, it is also *Pareto efficient* [22] because it is impossible to make one user get a higher rate without adversely affecting other users. While in a noncooperative environment, selfish behavior breaks this property. With the same simulation environments in Section 18.1.2, the system average data rate is plotted for a network in Figure 18.2, where the network size varies from 2 users to 20 users. We can see that the system average rate in NE state is much smaller than that in a cooperative environment. Thus, the achieved NE rate is *Pareto inefficient* [22], which is a common characteristic for a noncooperative game.

18.1.4 Repeated Game and Striker Strategy

The main reason for the data rate decrease in a noncooperative environment is that a selfish user intends to report a higher data rate. Thus, if the user in a bad channel condition gives up the competition and lets the others to transmit, the average data rate may be increased. However, in a noncooperative environment, a rational user has little incentive to give up its channel if there is no mechanism to enforce cooperation. In this section, we propose a repeated game to enforce cooperation. We assume that the users do not know the end of the game; hence, we study the problem in an infinite repeated game model with discounting [22].

We extend the NPS game as follows: we assume that the game is split up steps denoted by h. In each step, user $i \in \Omega$ adjusts the rate according to its strategy. Furthermore, let us define the discounted average utility in $H_i < +\infty$ time steps as

$$\overline{U}_i(H_i) = (1 - \omega) \cdot \sum_{h=0}^{H_i} U_i(h) \cdot \omega^h, \tag{18.7}$$

where $0 < \omega < 1$ is the discounting factor, which can be interpreted as the probability that the game ends in the next step. We assume ω is identical for all users in our model.

We have found that the selfish users are in an inefficient equilibrium when they all play NE strategy, whereas the maximal data rate can be achieved by using a cooperative strategy. From Folk Theorem [22], we know that in an infinitely repeated game, any feasible outcome that gives each player better payoff than the NE can be obtained. We can now determine the conditions that enable the users to enforce cooperation and prove that they can do better by applying a strategy called *Striker*, as detailed in the following.

Definition 18.2 A wireless user i is said to employ the *Striker* strategy if it plays r_i^A in the first time step, and for any subsequent time steps, it plays:

- r_i^A in the next time step if the other player j played r_j^A in the previous time step
- r^M for the next H_i time steps, if the other played anything else

The punishment interval H_i defines the number of time steps for which a player punishes the selfish player [5,8]. To simplify our analysis, we assume that the overall channel conditions remain relatively unchanged. Then, r_i^A over each step in the repeated game is similar. However, our simulation results show that our analysis still holds in wireless fading situations. Consequently, cooperation can be enforced using the *Striker* strategy as formalized in the following proposition.

Proposition 18.1 An efficient NE can be enforced by the *Striker* strategy.

Proof We consider the *Striker* strategy, and suppose user i adheres to it and chooses r_i^A. If user $j \neq i$ uses the same strategy, then the outcome is (U_i^{COP}, U_j^{COP}) in every step, so that it obtains the stream of payoffs, which gives a discounted average of $(1-\omega) \cdot \sum_{h=0}^{H_i} U_i(h) \cdot \omega^h = (1 - \omega^{H_i+1}) \cdot U_i^{COP}$.

If user j adopts a rate r_j^X so as to get a larger utility $U_j^{NCOP} > U_j^{COP}$ in all subsequent steps, user i will choose r^M since user j's choice of r_j^X triggers the punishment. Since all other users choose to report the maximal data rate r^M, the selfish user j will have to choose r^M in every subsequent step with payoff U_j^M. Consequently, it obtains the stream of payoffs with discounted average utility:

$$(1-\omega) \cdot \left(U_j^{NCOP} + U_j^M + \omega \cdot \left(U_j^M \right) + \cdots + \omega^{H_i} \cdot \left(U_j^M \right) \right)$$
$$= (1-\omega) \cdot U_j^{NCOP} + \left(1 - \omega^{H_i+1} \cdot U_j^M \right). \tag{18.8}$$

Thus, user j cannot increase its utility by deviating if and only if

$$(1-\omega) \cdot U_j^{NCOP} + \left(1 - \omega^{H_i+1} \cdot U_j^M \right) < \left(1 - \omega^{H_i+1} \cdot U_j^{COP} \right). \tag{18.9}$$

$$\omega^{H_i+1} < 1 - (1-\omega) \cdot \frac{U_j^{NCOP}}{U_j^{COP} - U_j^M}. \tag{18.10}$$

The inequality cannot be fulfilled if the right side is negative, and therefore,

$$(1-\omega) \cdot \frac{U_j^{NCOP}}{U_j^{COP} - U_j^M} < 1. \tag{18.11}$$

When this condition holds, since $\omega < 1$, we have

$$H_i \geq \log_\omega \left((1 - \omega) \cdot \frac{U_j^{NCOP}}{U_j^{COP} - U_j^M} \right) - 1. \qquad (18.12)$$

Thus, when the discounting factor is chosen as in (18.11) and punishment interval is set according to (18.12), the selfish user j will be forced to cooperate with user i such as to get higher payoff by *Striker* strategy. Correspondingly, an efficient NE is achieved. □

To demonstrate the effect of Striker strategy, we perform simulations with 20 users. And the performance result shown is averaged over 20 channel and location realizations. The punishment interval is 10 steps. U1 deviated from the cooperative action at time slot 100. In the beginning, all wireless users perform honestly, and the individual average data rate is about 8 Mbps, which is also used as a threshold to detect deviation. At time slot 100, U1 begins to deviate from the cooperative action. As shown in Figure 18.3, its average rate increases up to 13 Mbps dramatically. This defection is retaliated by other users in the system soon, which then make the throughput of this selfish user decrease dramatically. After that, the throughput returns to the cooperative state via *Striker* strategy. Thus, a more efficient equilibrium is achieved as well as the scheduling performance is optimized in noncooperative environments. We further show the average individual throughput over the whole 500 time slots for four users to investigate the long-term fairness in Figure 18.4, where only U1 is selfish. The results illustrate that the fairness performance is much worse in a noncooperative situation because selfish U1 consumes large part of radio resources by lying to the AP. With the introduction of *Striker* Strategy, the impact of selfish behavior is restricted, and then the users are more likely to report their channel conditions honestly. Consequently, the users are scheduled in a fairer manner.

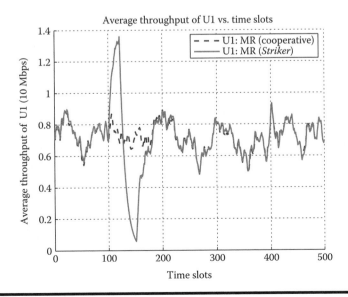

Figure 18.3 Average throughput of U1 under different strategies.

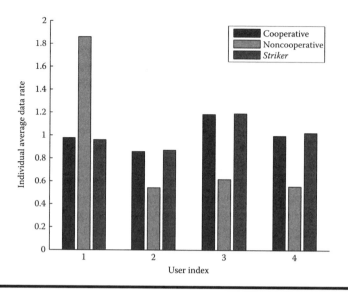

Figure 18.4 Long term fairness comparison.

18.1.5 Summary

In this section, we investigate the impact of selfish behavior on the performance of MR packet scheduling algorithm in noncooperative wireless networks. We first set up a novel mixed strategy game model to analyze this problem and deduce the corresponding NE, in which the system throughput is found to be significantly reduced. By using repeated game to enforce cooperation, we further propose a novel game-theoretic approach that can lead to an efficient equilibrium.

18.2 Auction-Based Scheduling in Noncooperative OFDM Systems

18.2.1 Introduction

In Section 18.1, we have studied the impact of selfish behaviors on efficiency performance for MR scheduling algorithm in a noncooperative single-carrier TDMA system, and proposed a repeated game-theoretic approach to thwart selfish behaviors and enforce users to report channel conditions truthfully to the scheduler. In many practical situations, besides throughput efficiency, user fairness is also an important factor to consider in wireless resource allocation. It is well known that PF [11] has been introduced as a good solution to guarantee an attractive trade-off between the system throughput and user fairness. On the other hand, some more advanced multicarrier techniques, such as orthogonal frequency-division multiplexing (OFDM) [12], have also been utilized in current high-speed wireless networks. In this section, we intend to extend our previous work from single-carrier systems to multicarrier systems while achieving PF resource allocation.

There are several difficult challenges to be tackled. Firstly, in this noncooperative situation, wireless users may be selfish and report their own channel conditions dishonestly so as to maximize their own benefits. As we have shown in Section 18.1, such a selfish behavior can dramatically reduce the system throughput. Thus, one of the most important objectives is to design an efficient mechanism to enforce reporting true channel conditions. Secondly, in OFDM systems, multiple subcarriers need to be allocated to different users. This greatly increases the complexity of the scheduling process. Thirdly, the quality of every subcarrier may be different for different users. Indeed, wireless users in OFDM systems compete for multiple units of radio resource instead of single object in the single channel TDMA system. Thus, the resource allocation algorithm must be capable of resolving competition for multiple objects among selfish users. Finally, to be of practical use, the proposed method must be suitable for practical implementation with a low complexity.

Auctions [18] have recently been introduced into several areas of wireless communications to handle the problem of resource competition among selfish users. Sun et al. [30] proposed a channel allocation algorithm based on the second-price auction mechanism to allow wireless users to fairly compete for a wireless fading channel, but it did not consider multicarrier situation. Huang et al. [10] studied auction methods for cooperative communication systems. But whether truthful revealing is a dominant strategy was not discussed. Fu and van der Schaar [6] proposed a resource management algorithm based on Vickrey–Clarke–Groves (VCG) mechanism [18] to ensure that wireless users truthfully declare their resource requirements. However, in their models, they just assumed that the qualities of different spectrum bands are identical. In OFDM systems, the quality of every subcarrier may be different for different users. Thus, the resource allocation problem needs to be formulated as a multiunit auction with nonidentical value, which is usually referred to as combinatorial auction [33]. Pal et al. [24] have proposed a combinatorial reverse auction-based scheduling algorithm to maximize the number of satisfied users in wireless systems. But, the bidding language and computational process are highly complex. The closest work to ours is [7], where an auction algorithm is applied to the OFDM subcarrier allocation problem. Though it can achieve efficient resource allocation, it cannot prevent selfish users from reporting their valuations for subcarriers dishonestly.

In this section, we consider the design of an auction-based resource allocation algorithm with a low complexity to realize PF in a noncooperative multiuser OFDM system. In this system, wireless users compete for subcarriers and evaluate the valuations of different subcarriers according to their channel conditions and QoS, and submit bids to the scheduler, which also acts as the auctioneer. The auctioneer uses a VCG-based auction method to calculate payment paid by different users so as to enforce selfish users to truthfully report their valuation of subcarriers, and utilizes the greedy MC PF scheduling method we proposed in [16] to achieve low-complexity resource allocation with PF in a multiuser OFDM system.

18.2.2 Model

Here we consider an OFDM system with K users and M equi-width orthogonal frequency subcarriers. The total spectrum bandwidth is W. At the physical layer, we use an L-path Rayleigh fading channel model. By using adaptive modulation, at time slot t, the kth user's data rate (bit per second) on the mth subcarrier $r^a_{k,m}(t)$ can be usually decided by the current channel SNR and the required BER [26], that is,

$$r^a_{k,m}(t) = \left\lfloor \log_2 \left(1 + \frac{-1.5}{\ln(5 \cdot P_{ber})} \cdot \gamma_{k,m}(t) \right) \cdot \nabla f \right\rfloor. \tag{18.13}$$

where

$\gamma_{k,m}(t)$ is the SNR for the kth user's mth subcarrier signal at time instant t

P_{ber} is the required BER for this transmission

∇f is the frequency spacing between two adjacent subcarriers

Let $x_{k,m}(t)$ be the channel assignment status taking value 1 and 0 when the mth subcarrier is and is not occupied by the kth user, and then the kth user's instantaneous data rate is $r_k^a(t) = \sum_{m=1}^{M} r_{k,m}^a(t) \cdot x_{k,m}(t)$.

In a cooperative situation, the objective of the scheduler is to achieve PF in multiuser OFDM system. In [16], it has been formulated as a global optimization problem based on the PF definition for MC systems in [13]. Here we redefine it as follows:

$$\max_{\{x_{k,m}\}} \sum_{k=1}^{K} \log \left(1 + \frac{\sum_{m=1}^{M} x_{k,m}(t) \cdot r_{k,m}^a(t)}{(t_c - 1)\overline{R_k(t)}} \right) \qquad (18.14)$$

subject to

$$(I) \sum_{k=1}^{K} x_{k,m}(t) = 1, \quad x_{k,m}(t) = \{0, 1\},$$

where the average throughput $R_k(t)$ can be approximated by a moving average value with average window size t_c time slots, that is, $\overline{R_k(t + 1)} = (1 - 1/t_c) \cdot \overline{R_k(t)} + 1/t_c \cdot r_k^a(t)$.

But in a noncooperative environment, a selfish user may report a bogus channel rate $r_{k,m}(t)$ other than $r_{k,m}^a(t)$ or lie about its valuation $v_{k,m}(t)$ on the radio channel. Without reliable inputs, the resource allocation result is then totally different from that in a cooperative environment. Consequently, the radio resource may be utilized inefficiently and the fair allocation of resource for wireless users is not realized.

18.2.3 Auction-Based Algorithm

Motivated by the above considerations, we set out to solve the problem of resource competition among selfish users by applying the concept of auction from economics. Specifically, an auction is a decentralized market mechanism for allocating resources, consisting of three key elements: (1) the good to be allocated; (2) an auctioneer, who determines the allocation of the good and the payment paid by bidders; and (3) a group of bidders, who want to obtain the good from the auctioneer. To formulate the scheduling problem in noncooperative OFDM systems as an auction, the good to be allocated are M subcarriers, the auctioneer is just the scheduler in the BS, and the bidders are K users.

In a noncooperative environment, the bidder k can untruthfully declare (exaggerate) its resource requirement $r_{k,m}(t)$ or its valuation on subcarrier $v_{k,m}(t)$ in order to obtain a higher throughput. With inaccurate bidding information, the auctioneer cannot allocate resource among users efficiently and fairly. Thus, we need a mechanism to prevent the bidder from lying to the auctioneer about its valuation for the subcarriers. VCG auction [18] has been regarded as one of the most effective mechanism to induce truth-revealing strategies. Specifically, in a VCG auction, by charging each bidder a payment corresponding to the inconvenience it causes to other bidders, the auctioneer can encourage the bidder to report truthfully, and thus resource is allocated in an efficient manner. Thus, we design an auction-based algorithm using the VCG auction model so as to enforce bidders to tell their truthful valuations of the subcarriers.

As described in Section 18.2.1, another important requirement is to allocate subcarriers among users and achieve PF through a multiunit auction. Most of the existing solutions for multiunit auctions apply combinatorial auctions as the most general framework [24,33]. Unfortunately, these auctions usually have complex bid expression that grows exponentially with the size of goods, and require solving NP-hard optimization problems. However, in a practical scheduling scenario, we should avoid the complex combinatorial auctions while achieving PF. In our previous study [16], we have proposed a low-complexity greedy MC PF method in multiuser OFDM systems. With this method, PF can be achieved in real time by assigning subcarriers to the user with the highest system PF value one by one. Specifically, in [16], the system PF value if user k occupies subcarrier m is defined as

$$\text{PF}(t, k) = \log\left(1 + \frac{r_k'(t) + r_{k,m}^a(t)}{(t_c - 1)\overline{R_k(t)}}\right) + \sum_{j \neq k} \log\left(1 + \frac{r_j'(t)}{(t_c - 1)\overline{R_k(t)}}\right)$$

$$= \text{PF}(t, k, m) + \sum_{j \neq k} \log\left(1 + \frac{r_j'(t)}{(t_c - 1)\overline{R_k(t)}}\right), \tag{18.15}$$

where $r_k'(t)$ is user k's rate before assigning subcarrier m. Because the subcarrier is assigned to the user so as to achieve the highest system PF value, a rationally selfish user k may just evaluate its valuation $v_{k,m}^a(t)$ of subcarriers based on the PF value and report a higher one $v_{k,m}(t)$ so as to increase its opportunity to be scheduled. Specifically, when letting user k's valuation for sending traffic on subcarrier m with data rate $r_{k,m}^a(t)$ be $v_{k,m}^a(t)$, it can be expressed by user k's PF value on subcarrier m, that is,

$$v_{k,m}^a(t) = \text{PF}(t, k, m) = \log\left(1 + \frac{r_k'(t) + r_{k,m}^a(t)}{(t_c - 1)\overline{R_k(t)}}\right). \tag{18.16}$$

With the above analysis, we can propose an auction-based scheduling algorithm combining the VCG auction, which is employed to calculate payment and realize incentive compatibility, as well as the greedy MC PF algorithm, which is used to achieve low-complexity PF scheduling for a multiuser OFDM system. Our algorithm can be summarized as follows:

- Channel condition measurement and valuation
  At the beginning of time slot t, every user k measures $r_{k,m}^a(t)$ on each subcarrier m and calculates its valuation $v_{k,m}^a(t)$ on each subcarrier m
- Bidding
  Because there are M subcarriers, every user k submits a set of bid $b_{k,m}(t)$ according to its valuation $v_{k,m}^a(t)$ on each subcarrier m. $b_{k,m}(t)$ can also be expressed as

$$b_{k,m}(t) = f_b\left(v_{k,m}^a(t)\right) = v_{k,m}(t) = \log\left(1 + \frac{r_k'(t) + r_{k,m}^a(t)}{(t_c - 1)\overline{R_k(t)}}\right) \tag{18.17}$$

- Allocation and payment

After collecting all bids from these K users, for every subcarrier m, the auctioneer determines the allocation results $x^*_{k,m}(t)$ according to the greedy MC PF scheme and calculates the payment $p_{k,m}(t)$ to be paid by user k according to VCG auction, that is,

$$x^*_{k,m}(t) = \arg\max_{\{x_{k,m}(t)\}} \sum_{k=1}^{K} \left(b_{k,m}(t) + \sum_{j \neq k} \log\left(1 + \frac{r'_j(t)}{(t_c - 1) \cdot \overline{R_j(t)}} \right) \right) \cdot x_{k,m}(t) \quad (18.18)$$

$$p_{k,m}(t) = \max_{\{x_{j,m}(t)\}} \left(\sum_{j \neq k} b_{j,m}(t) \cdot x_{j,m}(t) \right) - \sum_{j \neq k} b_{j,m}(t) \cdot x^*_{j,m}(t) \quad (18.19)$$

The first term of (18.19) is the maximum aggregated valuation that other users can derive if user k does not participate in the auction, and the second term is the sum of aggregated valuation of the other users except user k under optimal resource allocation results $x^*_{k,m}(t)$ in the presence of user k. Thus, the payment paid by each user corresponds to the loss of declared valuation it imposes the others through its presence.

It should be noticed that one subcarrier can only be occupied by just one user. Thus, the sealed-bid second-price auction [18], which is a special case of VCG auction, can be used to simplify the auction process. Let $k^*_m(t)$ be the user chosen by the auctioneer to occupy subcarrier m; (18.18) and (18.19) can be replaced by the following expressions:

$$k^*_m(t) = \arg\max_k \left(b_{k,m}(t) + \sum_{j \neq k} \log\left(1 + \frac{r'_j(t)}{(t_c - 1)\overline{R_k(t)}} \right) \right). \quad (18.20)$$

$$p_{k,m}(t) = \begin{cases} \max\limits_{j \neq k^*_m(t)} b_{j,m}(t) & k = k^*_m(t), \\ 0 & k \neq k^*_m(t). \end{cases} \quad (18.21)$$

In terms of computational complexity of the algorithm, when the auctioneer determines the allocation results according to (18.20) and calculates the payment according to (18.21), for a system with K users and M subcarriers, the computational complexity is KM.

18.2.4 Simulations

In this section, we present our simulation results. The entire system bandwidth is 20 MHz, which is divided into 256 subcarriers. The wireless channel is also modeled as a six-path frequency-selective Rayleigh fading channel. Each path is simulated by Clark's fading model and suffers from different Rayleigh fading with the maximum Doppler frequency of 30 Hz, which corresponds to the average speed of 4.5 km/h for the carrier frequency of 5 GHz. The maximum BER requirement is 10^{-5}. In this simulation, we compare the throughput as well as the payments paid by users under two scenarios: (1) no user is lying about its valuation, and (2) only User 5 is lying about its valuation, but the others are telling the truth. Table 18.1 shows the efficiency and fairness performance, as well as the payments, for the two cases.

From this table, we can see that when all users reveal the true valuations of the subcarriers, the resources are allocated among users efficiently and in a fair manner. However, when User 5 exaggerates its valuation, its throughput is improved by 94.39%, but the payment to be paid is also significantly increased by 1525.88%. Furthermore, the throughput of any other user is decreased

Table 18.1 Comparisons of Throughput, Fairness, and Payments in the Two Cases

User	(1) No Users Lying		(2) Only User 5 Lying		∇ Throughput (%)	∇ Payment (%)
	Throughput (Kbps)	Payment	Throughput (Kbps)	Payment		
1	6.202	5.777	4.696	80.07	−24.28	1386.01
2	6.536	6.050	4.835	82.02	−26.03	1355.70
3	5.477	5.527	3.952	72.60	−27.84	1313.33
4	5.879	4.870	4.781	65.80	−18.68	1351.75
5	6.953	7.013	13.516	107.01	+94.39	1525.88

dramatically. Thus, the fairness performance also deteriorates due to the selfish behavior of User 5. Now, it becomes clear that by using the VCG-based auction scheduling, the lying of selfish users is penalized through a significantly increased payment. Thus, a rational user will not lie about its valuation of subcarriers because the increased cost cannot be compensated by the improved throughput. Consequently, with the proposed auction-based scheduling scheme, selfish users will be enticed to report their channel conditions and corresponding valuation of channels honestly. Moreover, its bid is just its true valuation of the channels, that is, $b_{k,m}(t) = v_{k,m}^A(t)$. Thus, PF can be achieved in noncooperative multiuser OFDM system as that in a cooperative environment.

18.2.5 Summary

In this section, we extend our previous research in Section 18.1 about noncooperative scheduling in TDMA system in two aspects. Firstly, we focus on the problem of how to allocate resource with PF, while Section 18.1 mainly studies throughput performance. Secondly, we study this noncooperative resource allocation in a multiuser OFDM environment instead of a single-carrier TDMA system. We propose an auction-based scheduling algorithm, which combines the merits of the VCG auction and the greedy MC PF algorithm, to ensure that wireless users truthfully declare their resource requirements and achieve a low-complexity PF scheduling in a noncooperative multiuser OFDM system. Through simulations, we find that users lying about their resource requirements are severely penalized by very high payments so that they should rather declare true valuation of subcarriers to the scheduler. We also observe that a conventional PF scheduling scheme heavily depends on the truthful report from wireless users, then it will result in significantly worse performance when users dishonestly report their requirements, because it does not provide a mechanism to penalize selfish users for misusing resources. In summary, the proposed auction-based scheduling can be used efficiently in a noncooperative situation to realize PF (Table 18.1).

18.3 Price Competition between WLAN Providers

18.3.1 Introduction

Besides the competition for radio resources among wireless users, the competition for revenue between wireless service providers is also an important problem to be tackled in a noncooperative environment. In particular, in the current highly competitive market environments for WLAN access service providers, they have to compete with each other and attract users in order to generate profits. In a typical competitive hotspot environment (e.g., a public cyber-café), several WLAN providers may coexist to provide wireless access services for the same group of users. In attempting to attract users and optimize their revenues, WLAN providers need to price their services by taking into account a wide range of factors, including preferences of end users, the QoS limitations of used technology, and the potential competition from other providers. Furthermore, when the QoS and prices are observed, wireless users will allocate their demands across the providers. This process will in turn affect the revenue of providers. Thus, with the capability to affect users' demand and providers' revenue, pricing in wireless communication networks has gained much attention recently [4].

Studying the impact of price competition is receiving increasing attention in the networking community. For general ideas on pricing, the reader is referred to [4]. The pricing game among wireless providers with fixed capacity is analyzed in [19], and it is shown that the price of anarchy (PoA) [25] is 1. Acemoglu and Ozdaglar [1] study a pricing problem in which the users are sensitive to the price and a convex congestion delay, and give tight bounds on PoA. In [9], Hayrapetyan et al. give an improved bound for the special case in which packet delay is a pure negative effect, which is linear or convex. Niyato et al. [21] investigate two levels of competitions in cognitive wireless mesh network and propose game-theoretic solutions to choose price and transmission rate for primary users and secondary users, but the demand distribution of secondary users is not analyzed. In [28] a cost-price mechanism based on users' channel occupancy is proposed in WLAN to maximize the system throughput, but the competition between different WLAN providers is not analyzed. Furthermore, because WLAN standard employs a contention-based random multiple access method based on CSMA/CA, that is, DCF, as the fundamental MAC technique, there exists packet collision, which will result in packet loss. On the other hand, wireless transmission error may lead to packet loss too. Thus, in this section, we consider packet loss rate (PLR) instead of delay or rate as negative externality [4] and find that it is indeed concave other than convex in WLAN environments. Thus, the previous finding cannot be used in WLAN situations directly. To the best of our knowledge, the problem of pricing competition for WLAN providers with the consideration of price and PLR is relatively unexplored.

Here, we focus on a *duopoly* competition, in which two WLAN providers compete for a same group of wireless users by adjusting the price charged. The provider's aim is to maximize its own profit, while the users, driven by self-interests, rationally choose the service provider offering the best combination of QoS and price. We assume that users are sensitive to the experienced PLR as well as charged prices. The PLR at each AP occurs because of packet collision or packet transmission error. Generally, the users are more likely to connect to the provider with good channel condition and/or low price, but when more users connect to a provider, the probability of collision will increase too. This feature is also known as negative externality [4], since the decision made by a user will have a negative effect on the payoff of others connected to the same provider. Thus, it can affect user's willingness to accept a provider's service.

Consequently, each user has to calculate the experienced cost to choose a provider according to the charged price and estimated PLR. We assume that each user likes to choose the provider with the minimum experienced cost and the number of users (i.e., demand) decreases with the cost. This leads us to research some important questions: How is the users' demand split among providers when their price strategies are given and the PLRs are estimated? Does there exist NE price vector, under which no provider can unilaterally improve its own revenue by changing its price? Further, whether there exists the efficiency loss, that is, PoA, as what is shown by many prior works, for example, [1,9]?

To answer these questions, in this section, we first set up a game model to analyze negative externalities induced by PLR, give its approximation, and find it is concave indeed. Then we characterize the Wardrop equilibrium (WE) [34] for the distribution of users' demands. After that, we prove the existence of NE on providers' charged prices. Then through numerical analysis, we find that PoA can be close to 1.1, which indicates that the social welfare (SW) does not suffer too much in this competitive environment. Furthermore, we find that the competition does not lead to revenue loss when the users are very sensitive to charged prices.

18.3.2 Model

We consider an IEEE 802.11 hotspot area covered by two APs, and let $\Omega = 1, 2$ be the set of APs. Each AP $i \in \Omega$ is controlled by a distinct provider, which charges a price p_i per user for using its service. Thus, here the terms of AP and provider are interchangeable. We assume that APs share the same frequency bandwidth in the network, whereas different APs are being operated on different frequency channels and using different PHY mode. Thus, there is no interference among different users. DCF is used as MAC level multiuser access method. We also ignore hidden node problem, so that every user can sense all others' transmission as in [2]. Furthermore, we operate in saturation conditions, that is, the transmission queue of each user is assumed to be always nonempty. And there is only one user that can transmit its packets successfully to the belonged AP at one time slot.

Let x_i be the number of users connected to AP i. Note that users are assumed to be infinitesimal, each of them having a negligible impact on others. We assume that when the x_ith user connects to the AP i, it will get a utility of $U_i(x_i)$, which is its willingness to pay to get service. Similar to what is done in [1], we use the assumption that when a user decides to receive the WLAN access service, it will get a reservation preference of R, that is, $U_i(x_i) = R$. But a cost $c_i(x_i)$ will also be experienced when there are x_i users, which is directly related to charged price p_i and packet loss rate $\text{PLR}_i(x_i)$. Furthermore, this user will not connect to AP i if the experienced cost exceeds its reservation preference. Then the aggregate user's surplus in AP i is $S_i = (R - c_i(x_i)) \cdot x_i$.

The cost $c_i(x_i)$ can be expressed as $c_i(x_i) = p_i + f_i(\text{PLR}(x_i)) = p_i + f_i(x_i)$, where $f_i(\cdot)$ is used to express the experienced cost resulting from PLR. For the sake of simplicity, we assume that users are homogeneous and let $f_i(x_i) = \alpha \cdot \text{PLR}(x_i)$, where $\alpha \geq 0$. The experienced cost is an important parameter since users' demand and providers' revenue are highly related to it. In this work, we define the revenue of provider i as $\Pi_i(p_i) = p_i \cdot x_i$. While the SW can be expressed by adding the utilities of all users and providers, that is, $\text{SW} = \Sigma_{i \in \Omega}(R - f_i(x_i)) \cdot x_i$.

In 802.11 systems, PLR can be expressed as

$$\text{PLR}_i(x_i) = P_i^l = 1 - \left(1 - P_i^c\right)\left(1 - P_i^t\right) = P_i^c + P_i^t - P_i^c \cdot P_i^t, \qquad (18.22)$$

where
P_i^l and P_i^c are PLR and packet collision rate experienced at AP i with x_i users, respectively
P_i^t is the minimal packet transmission error rate supported by AP i

In fact, due to location-dependent characteristics, the transmission error rate will be different for different users. To focus on the price setting for different providers and ease the analysis, the exogenous information used in this game is the minimal transmission error rate P_i^t that AP i

can support under its geography situation and provisioned transmission techniques, for example, coding and modulation. Furthermore, a user may also automatically move to a satisfactory place for an acceptable transmission condition. Then because there is no interference between providers in this model, we can assume that P_i^t at a particular provider is maintained as a constant value and different providers will support different P_i^t.

In terms of packet collision rate P_i^c, here we refer to it as the packet collision observed at each individual user [2], that is, it is the probability that one user encounters collisions when it transmits packets. Define P_i^0 as the transmission probability for each user. Then if x_i users are connected to the provider i, we have $P_i^c = 1 - (1 - P_i^0)^{x_i - 1}$. Since the providers want to maximize their profits, the assumption of maximum saturation throughput is reasonable in this research. In [2], an approximation for P_i^0 under maximum achievable saturation throughput is given by

$$P_i^0 = \frac{\sqrt{(x_i + 2(x_i - 1)(T_c^* - 1)/x_i) - 1}}{(x_i - 1)(T_c^* - 1)} \approx \frac{1}{x_i\sqrt{T_c^*/2}} = \frac{1}{x_i K}, \qquad (18.23)$$

where
$K = T_c^*/2$
T_c^* is the average time when the channel is sensed busy by each station during a collision, which is determined as a constant by given PHY and MAC mechanism

Thus, we have

$$\mathrm{PLR}_i(x_i) = 1 - \left(1 - \frac{1}{x_i K}\right)^{x_i - 1} \left(1 - P_i^t\right) = 1 - \left(1 - P_i^t\right) \cdot e^{(x_i - 1) \cdot \ln(1 - 1/x_i K)}$$

$$\approx 1 - \left(1 - P_i^t\right) \cdot e^{(x_i - 1)[(-1/x_i K) + 1/(2(x_i K)^2)]} \approx 1 - e^{\frac{-1}{K}} \cdot \left(1 - P_i^t\right) \cdot e^{1/x_i((1/K) + 1/2K^2)}. \qquad (18.24)$$

We can further yield

$$x_i = \mathrm{PLR}_i^{-1}\left(P_i^l\right) = \frac{\nu}{\ln\left(1 - P_i^l\right) + \omega_i}. \qquad (18.25)$$

where
$\omega_i = 1/K - \ln\left(1 - P_i^t\right)$
$\nu = 1/K + 1/2K^2$

Remark that $\mathrm{PLR}_i(x_i)$ is strictly increasing with regard to x_i and P_i^t. Also, since the second derivative $\mathrm{PLR}_i'' < 0$, it is concave.

18.3.3 Distribution of Users' Demand

When a user wants to access a WLAN, it will face several competitive providers with different charged prices and provisioned QoS and therefore will feel different experienced costs $c_i(x_i)$. Naturally, it prefers to choose the provider with the minimum cost $\underline{C} = \min(c_i(x_i))$. Furthermore, if a provider's cost is too high, there will be no additional user willing to choose it, and even already connected ones will be expected to switch. Then at equilibrium state, the experienced cost will be identical at all providers having a positive demand, or the number of users connected to one provider will be zero because of a too high access price. This is also known as WE, which can be mathematically defined as follows:

Definition 18.3 The users' demand $X^{\mathrm{WE}} = (x_1, x_2)$ is said to achieve WE if $x_i \cdot (c_i(x_i) - \underline{C}) = 0$ $\forall i \in \Omega$, and $\underline{C} = \min(c_i(x_i))$, $x_i > 0$. Also the total level of demand verifies $\sum_{i \in \Omega}(x_i) = X(\underline{C})$.

This equation means that the providers with positive demand will have the same experienced cost $\underline{C}$; otherwise, the provider will have zero demand due to the high cost. Since $c_i(x_i) - \underline{C} \geq 0$ and $\sum_{i \in \Omega}(x_i) = X(\underline{C})$, the existence of WE in our model can be verified as that in [1]. Next, we will derive the uniqueness of WE in this research.

We assume that the total number of users in the system decreases with the minimum experienced cost. Then we can express the aggregate demand function $X(\underline{C})$ as

$$X(\underline{C}) = X_0 - d \cdot \underline{C}, \quad d \geq 0, \tag{18.26}$$

where d is a demand parameter to express user's sensitivity to the experienced cost. Then according to the definition of WE, we can directly characterize the WE vector X^{WE} as follows:

Proposition 18.2 With demand function $X(\underline{C})$ and price vector $P = (p_1, p_2)$, WE vector $X^{\mathrm{WE}} = (x_1, x_2)$ can be characterized as the smallest solution of the inequation $X(\underline{C}) \leq \sum_{i \in \Omega} x_i(\underline{C})$ with $\underline{C} \in [0, X^{-1}(X_0)]$, and

$$
x_i = \begin{cases}
X(\underline{C}) & \dfrac{\underline{C} - p_i}{\alpha} \geq \mathrm{PLR}_i(X_0), \\[2ex]
\mathrm{PLR}_i^{-1}\left(\dfrac{\underline{C} - p_i}{\alpha}\right) & \mathrm{PLR}_i(X_0) > \dfrac{\underline{C} - p_i}{\alpha} > \mathrm{PLR}_i(0), \\[2ex]
0 & \dfrac{\underline{C} - p_i}{\alpha} \leq \mathrm{PLR}_i(0).
\end{cases} \tag{18.27}
$$

From Proposition 18.2, we can see that after one user chooses a provider, the minimum experienced cost $\underline{C} = \min(c_i(x_i))$ increases and $X(\underline{C})$ is decreased. If $X(\underline{C}) > \sum_{i \in \Omega} x_i(\underline{C})$, new user can enter the system to choose a provider; otherwise, the demand limit is reached, and the corresponding WE is achieved too.

18.3.4 Price Competition Analysis

In this duopoly environment, each provider seeks to set its price in order to maximize its own revenue. Thus, we set up a price game and the set of providers Ω represents the players. For provider $i \in \Omega = \{1, 2\}$, its action is defined as the price choice p_i. Then the action profile is denoted as the price vector $p = (p_1, p_2)$. Its utility is expressed by the revenue $\Pi_i(p_i, p_{-i})$, where p_{-i} denotes the prices chosen by providers else other than i. Then we define NE as

Definition 18.4 A price vector $p^{\text{NE}} = \left(p_1^{\text{NE}}, p_2^{\text{NE}}\right)$ is said to be NE if for all $i \in \Omega$, $\Pi_i\left(p_i^{\text{NE}}, p_{-i}^{\text{NE}}\right) \geq \Pi_i\left(p_i, p_{-i}^{\text{NE}}\right) \quad \forall p_i > 0$.

We first give a lemma that establishes some results with a price configuration p.

LEMMA 18.1 Assume that both providers have positive demand $x = (x_1, x_2)$ for price $p = (p_1, p_2)$. If provider i decreases its price to $p_i^n = p_i - \varepsilon < p_i$, and p_j for $\forall j \neq i$ is unchanged, where superscript "n" is used to refer the values corresponding to a new situation; then

1. The PLR increases to $P_i^{l,n} > P_i^l$.
2. The least experienced cost decreases to $\underline{C}^n = \underline{C} - \psi < \underline{C}$, and $\psi < \varepsilon$, where $\psi < 0$, $\varepsilon < 0$.

With this observation, we can get the existence of NE.

Proposition 18.3 When PLR is much smaller than 1, there exists a price vector $p^{\text{NE}} = \left(p_1^{\text{NE}}, p_2^{\text{NE}}\right)$ to achieve NE in this price competition.

The detailed proofs of Lemma 18.1 and Proposition 18.3 are given in [17] and omitted here due to space limitations.

18.3.5 Efficiency Analysis

We investigate the efficiency of this game in terms of SW and providers' revenue. Here we use PoA as a measure of the worst case difference between the SW of a cooperatively optimal solution (social optimum) and that in a noncooperative NE state. Similar to that in [9], we define PoA as follows:

Definition 18.5 If $P^S = \left(p_1^S, p_2^S\right)$ is the price vector that maximizes SW, $P^{\text{NE}} = \left(p_1^{\text{NE}}, p_2^{\text{NE}}\right)$ is the price vector in NE state; then PoA is defined as PoA $= \text{SW}(P^S)/\text{SW}(P^{\text{NE}})$.

Since $\text{SW}(P^S) \geq \text{SW}(P^{\text{NE}})$, we get PoA ≥ 1. When PoA is close to 1, it means the SW arrived at competitive environments is nearly as good as that reached through cooperative optimization. Whereas a large PoA means that the competition is less efficient.

Based on above analysis, we can investigate the PoA through numerical methods with two APs that either compete or cooperate with each other, in which the total number of users X_0 is 20 and R is 1. We also set $K = 9.334$ as that in [2].

Figure 18.5 shows the best response under different packet transmission error rates and demand parameter combinations. We find in Figure 18.5a that there exists a unique NE at prices (0.4, 0.4) when $P_t^1 = 0.01$, $P_t^2 = 0.01$, and $d = 10$. A similar result with NE at prices (0.32, 0.22) also exists in Figure 18.5b, where $P_1^t = 0.01$, $P_2^t = 0.03$, and $d = 3$. These verify the existence of NE under different situations.

From Figure 18.6, we can see that SW increases with demand parameter, while in NE state, it is close to that in cooperative situations. We also find that PoA is near 1.1 from Figure 18.7. Then letting the providers compete with each other will yield almost the same SW as a global market regulator would have given. Thus, this competition is not as inefficient as what we usually look in a

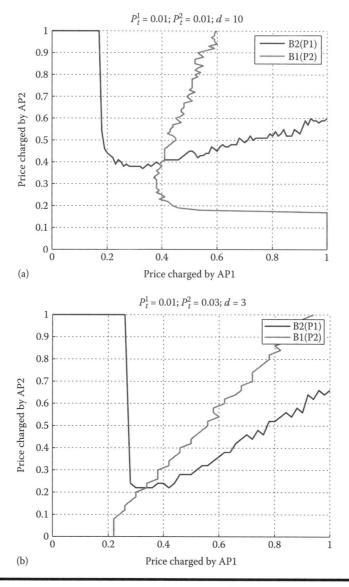

Figure 18.5 Best response functions: (a) $P_t^1 = 0.01$, $P_t^2 = 0.01$, $d = 10$; (b) $P_t^1 = 0.01$, $P_t^2 = 0.03$, $d = 3$.

noncooperative situation. Though the efficiency of SW is not decreased much under competition, the providers' revenue in NE state is much smaller than that in cooperative situations when the demand parameter is small as shown in Figure 18.8. But when the demand parameter increases, the aggregate revenues in NE state will eventually approach the cooperative maximal revenue. This is because the larger the demand parameter, the more sensitively that the user demand responds to the change of prices. If the user is not sensitive about the price, the providers would like to cooperate or even collude with each other to gain high revenue by setting high price. But when the user has higher price sensitivity, it will not receive any providers' service at all if the charged prices are too high. Then the providers will be more rational and prefer to set relatively lower prices so as to attract

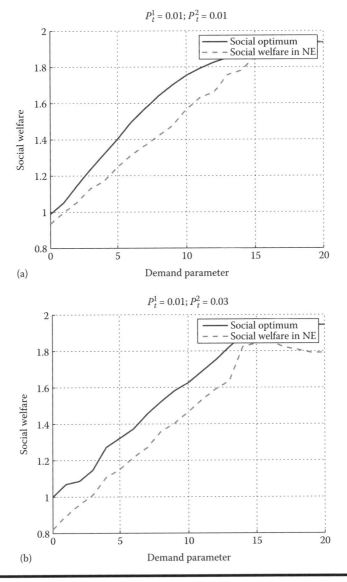

Figure 18.6 SW vs. demand parameter: (a) $P_t^1 = 0.01$, $P_t^2 = 0.01$; (b) $P_t^1 = 0.01$, $P_t^2 = 0.03$.

users. Though the revenue is decreased with demand parameter, it will consequently be same as cooperative maximal revenue. Furthermore, as shown in Figure 18.6, the SW also improves with the increase of user's price sensitivity. Usually the users are very sensitive to the price of wireless services; thus, all entities including providers and users will not suffer from this competition.

18.3.6 Summary

In this section, the impact of charged price and the provisioned PLR by different WLAN providers on users' demand, providers' revenue, and SW in a duopoly environment is investigated. Based

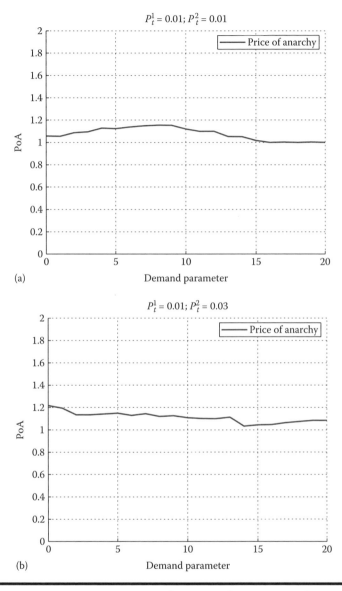

Figure 18.7 PoA vs. demand parameter: (a) $P_t^1 = 0.01$, $P_t^2 = 0.01$; (b) $P_t^1 = 0.01$, $P_t^2 = 0.03$.

on a game-theoretic model, we first analyzed the negative externalities associated with PLR. Then we found the existence and uniqueness of a WE for users' demand distribution between providers and determined the existence of the NE of price competition. Furthermore, through numerical analysis, we showed that the SW will not suffer too much under price competition. Thus, even without administrative enforcement, the competitive market itself can still determine the right price without degrading SW or even providers' revenues, especially when the users are very sensitive to the price. (In fact, this is just the normal case for users' attitude on price for wireless services.)

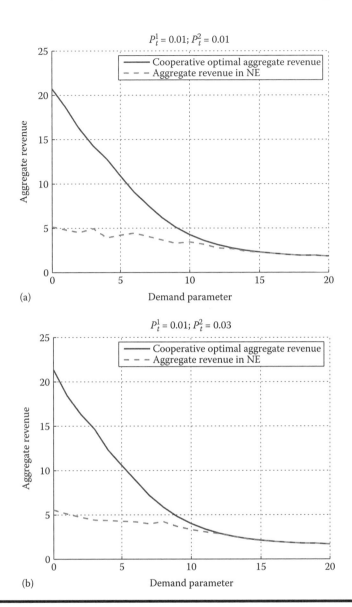

Figure 18.8 **Aggregate revenue vs. demand parameter: (a) $P_t^1 = 0.01$, $P_t^2 = 0.01$; (b) $P_t^1 = 0.01$, $P_t^2 = 0.03$.**

18.4 Future Work

This chapter has studied the problems of wireless resource management in noncooperative environments. Some of the future research avenues are suggested as follows:

Our current study on noncooperative packet scheduling is based on infrastructure cellular wireless networks. In the next step, we consider studying the impact of selfish behavior for resource management in an ad hoc environment. Furthermore, the proposed *Striker* protocol does not support dynamic situation. Because when a new user enters the network and competes for the radio resource, the data rate of an old user may reduce and then lead to the misjudgment of a selfish behavior. We plan to solve this problem in the future work. In addition, current analyzed performance metrics are throughput efficiency and fairness. In the future, we also want to analyze other performances, such as packet delay and power consumption, in a noncooperative environment.

For auction-based OFDM scheduling in a noncooperative environment, the type of algorithm considered in the thesis is suboptimal. In the future research, we will seek to design a pricing mechanism to achieve global optimum.

For price competition, firstly, we want to extend this research to a more general oligopoly environment, where there exist more than two competitive providers. Secondly, our current work does not consider elastic traffic [23], and thus, we consider using a demand-related utility function similar to that used in [19] to express the utility of user with elastic demand. Thirdly, because providers' revenue in an NE state is much smaller than that in a cooperative situation when the demand parameter is small, we consider designing an efficient mechanism, such as a strategy based on the idea of repeated game, to increase their revenues.

18.5 Conclusions

In this chapter, we analyze the problem of how to efficiently allocate radio resource in noncooperative wireless environments.

Firstly, the impact of selfish behavior on wireless packet scheduling algorithm in a noncooperative TDMA wireless network is investigated. We find that the existence of selfish behavior degrades significantly the system data rate performance of MR packet scheduling algorithm. Based on this observation, we further set up a novel mixed strategy game model to analyze this problem and deduce the corresponding NE, in which the system average data rate is found to be significantly reduced. We then further propose a distributed *Striker* strategy based on a repeated game model to enforce cooperation among users and achieve a more desirable NE, in which the data rate performance can be increased significantly.

This chapter: Firstly, we focus on the problem of how to achieve PF-based resource allocation, while the fairness performance is not considered in the above part. Secondly, we study the noncooperative resource allocation in a multiuser OFDM environment instead of a single-carrier TDMA system. Then based on VCG auction, we propose an auction-based scheduling algorithm to ensure that wireless users truthfully declare their resource requirements and achieve a low-complexity proportional fair scheduling in a noncooperative multiuser OFDM system.

Thirdly, a duopoly price competition game among WLAN providers is analyzed. We first show that the users' demands are distributed between providers according to a WE and then prove the existence of an NE on providers' charged prices. Through an extensive numerical analysis, we further find that in an NE state, the SW is very close to its maximal value in a cooperative situation. Furthermore, the providers' aggregate revenues also do not decrease when the users have high sensitivity about the charged prices. Thus, the competitive duopoly WLAN market can still run in an efficient manner even in the absence of complex regulation schemes.

References

1. D. Acemoglu and A. Ozdaglar, Competition and efficiency in congested markets, *Math. Operations Res.*, 32, 1, 1–31, Feb. 2007.

2. G. Bianchi, Performance analysis of the IEEE 802.11 distributed coordination function, *IEEE J. Select. Areas Commun.*, 18, 3, 535–547, Mar. 2000.

3. L. Buttyán and J. P. Hubaux, *Security and Cooperation in Wireless Networks*, Cambridge University Press, New York, 2007.

4. C. Courcoubetis and R. Weber, *Pricing Communication Networks: Economics, Technology and Modelling*, John Wiley & Sons, Chichester, U.K., 2003.

5. M. Felegyhazi and J. P. Hubaux, Wireless operators in shared spectrum, in *Proceedings of the 25th Annual IEEE Conference on Computer Communications (INFOCOM 2006)*, pp. 1–11, Barcelona, Catalunya, Spain, Apr. 2006.

6. F. Fu and M. van der Schaar, Noncollaborative resource management for wireless multimedia applications using mechanism design, *IEEE Trans. Multimedia*, 9, 4, pp. 851–868, June 2007.

7. S. Han and Y. Han, A competitive fair subchannel allocation for OFDMA system using an auction algorithm, in *Proceedings of IEEE the 66th Vehicular Technology Conference (VTC'07-Fall)*, pp. 1787–1791, Baltimore, MD, Sept.–Oct. 2007.

8. Z. Han, Z. Ji, and K. J. R. Liu, A cartel maintenance framework to enforce cooperation in wireless networks with selfish users, *IEEE Trans. Wireless Commun.*, 7, 5, Part 2, 1889–1899, May 2008.

9. A. Hayrapetyan, E. Tardos, and T. Wexler, A network pricing game for selfish traffic, in *Proceedings of the 24th Annual ACM Symposium on Principles of Distributed Computing (PODC 2005)*, pp. 284–291, Las Vegas, NV, July 2005.

10. J. Huang, Z. Han, M. Chiang, and H. V. Poor, Auction-based distributed resource allocation for cooperative transmission in wireless networks, in *Proceedings of the 50th Annual IEEE Global Communications Conference (Globecom 2007)*, pp. 4807–4812, Washington, DC, Nov. 2007.

11. A. Jalali, R. Padovani, and R. Pankai, Data throughput of CDMA-HDR a high efficiency-high data rate personal communication wireless system, in *Proceedings of the 51th IEEE Vehicular Technology Conference (VTC2000-Spring)*, vol. 3, pp. 1854–1858, Tokyo, Japan, Jan. 2001.

12. J. Jang and K. B. Lee, Transmit power adaptation for multiuser OFDM systems, *IEEE J. Select. Areas Commun.*, 21, 2, 171–178, Feb. 2003.

13. H. Kim and Y. Han, A proportional fair scheduling for multicarrier transmission systems, *IEEE Commun. Lett.*, 9, 210–212, Mar. 2005.

14. Z. Kong, Y. K. Kwok, and J. Wang, On the impact of selfish behaviors in wireless packet scheduling, in *Proceedings of the 2008 IEEE International Conference on Communications (ICC 2008)*, Beijing, China, May 2008.

15. Z. Kong, Y. K. Kwok, and J. Wang, Auction-based scheduling in non-cooperative multiuser OFDM systems, in *Proceedings of the 69th Vehicular Technology Conference (VTC2009-Spring)*, Barcelona, Spain, Apr. 2009.

16. Z. Kong, Y. K. Kwok, and J. Wang, A low complexity QoS-aware proportional fair multicarrier scheduling algorithm for OFDM systems, *IEEE Trans. Vehicular Technol.*, 58, 5, 2225–2235, June 2009.

17. Z. Kong, B. Tuffin, Y. K. Kwok, and J. Wang, Analysis of duopoly price competition between WLAN Providers, in *Proceedings of the 2009 IEEE International Conference on Communications (ICC 2009)*, Dresden, Germany, June 2009.

18. V. Krishna, *Auction Theory*, Academic Press, San Diego, CA, 2002.

19. P. Maille and B. Tuffin, Analysis of price competition in a slotted resource allocation game, in *Proceedings of the 27th Annual IEEE Conference on Computer Communications (INFOCOM 2008)*, Phoenix, AZ, Apr. 2008.

20. F. Meshkati, M. Chiang, H. V. Poor, and S. C. Schwartz, A game-theoretic approach to energy-efficient power control in multicarrier CDMA systems, *IEEE J. Select. Areas Commun.*, 24, 1115–1129, June 2006.

21. D. Niyato, E. Hossain, and L. Le, Competitive spectrum sharing and pricing in cognitive wireless mesh networks, in *Proceedings of IEEE 2008 Wireless Communications and Networking Conference (WCNC 2008)*, pp. 1431–1435, Las Vegas, NV, Mar.–Apr. 2008.
22. M. J. Osborne, *An Introduction to Game Theory*, Oxford University Press, New York, 2004.
23. A. Ozdaglar, Price competition with elastic traffic, *Networks*, 52, 141–155, Oct. 2008.
24. S. Pal, S. R. Kundu, M. Chatterjee, and S. K. Das, Combinatorial reverse auction-based scheduling in multirate wireless systems, *IEEE Trans. Computers*, 56, 10, 1329–1341, Oct. 2007.
25. C. Papadimitriou, Algorithms, games, and the Internet, in *Proceedings of 33rd Annual ACM Symposium on Theory of Computing (ACM STOC 2001)*, pp. 749–753, Hersonissos, Greece, July 2001.
26. X. Qiu and K. Chawla, On the performance of adaptive modulation in cellular systems, *IEEE Trans. Commun.*, 47, 6, pp. 884–895, June 1999.
27. S. Radosavac, J. S. Baras, and I. Koutsopoulos, A framework for MAC protocol misbehavior detection in wireless networks, in *Proceedings of the 4th ACM Workshop on Wireless Security (ACM WiSe'05)*, pp. 33–42, Cologne, Germany, Sept. 2005.
28. S. Shakkottai, E. Altman, and A. Kumar, Multihoming of users to access points in WLANs: A population game perspective, *IEEE J. Select. Areas Commun.*, 25, 6, pp.1207–1215, Aug. 2007.
29. E. Stephan, R. Giovanni, and S. Paolo, The COMMIT protocol for truthful and cost-efficient routing in ad hoc networks with selfish nodes, *IEEE Trans. Mobile Computing*, 8, pp. 19–33, Jan. 2008.
30. J. Sun, E. Modiano, and L. Zheng, Wireless channel allocation using an auction algorithm, *IEEE J. Select. Areas Commun.*, 24, 5, 1085–1096, May 2006.
31. G. Tan and J. Guttag, The 802.11 MAC protocol leads to inefficient equilibria, in *Proceedings of the 24th Annual IEEE Conference on Computer Communications (INFOCOM 2005)*, vol. 1, pp. 1–11, Miami, FL, March 2005.
32. B. S. Tsybakov, File transmission over wireless fast fading downlink, *IEEE Trans. Information Theory*, 48, pp. 2323–2337, Aug. 2002.
33. S. de Vries and R. Vohra, Combinatorial auctions: A survey, *INFORMS J. Computing*, vol. 15, pp. 284–309, Mar. 2003.
34. J. G. Wardrop, Some theoretical aspects of road traffic research, in *Proceedings of the Institute of Civil Engineers*, Part II, vol. 1, pp. 325–378, London, U.K., 1952.

Chapter 19

Multistage Congestion Games for Wireless Real-Time Streaming

Giovanni Rossi, Gabriele D'Angelo, and Stefano Ferretti

Contents

This work mainly proposes to model real-time streaming in congestion game form: at each stage, given the prevailing content distribution, peers (or players) choose who to ask for an additional content unit. In this way, (stage- and peer-wise) congestion is measured precisely by counting how many requests to forward are addressed to each peer (and to the broadcaster). While in the wired setting, all nodes (or peers, together with the broadcaster) are assumed to directly communicate with each other, in the wireless framework, an ad hoc network formalizes transmission constraints due to spatial positions: the broadcaster and each peer can send content only to those at unitary hop distance from them in such a network. For the wired (or fully connected) setting, we develop a strategy restriction mechanism whose equilibrium outcomes minimize both streaming length (or number of stages needed to spread the whole content over the whole population) and congestion, while also fulfilling a fairness condition. As this mechanism heavily relies upon connectivity among nodes, in general it does not apply to the wireless (or non-fully connected) case. Accordingly, the focus next turns on equilibrium outcomes with no strategy restriction apart from communication constraints due to the ad hoc network. In the wired setting, simulations allow to compare equilibrium outcomes with and without the proposed strategy restriction mechanism, while in the wireless one, they are used for evaluating how performance deteriorates, at equilibrium, as the size of the given ad hoc network decreases.

19.1 Introduction and Background

In real-time streaming, media content is produced and distributed in time: a broadcaster periodically generates one *new* content unit and makes it available to interested users. Timely dissemination of content is a main issue, commonly addressed with the creation of overlay networks where nodes or peers are interested users themselves, who collaborate toward data dissemination by forwarding the received data. Such a peer-to-peer (P2P) functioning can be designed in different ways, and three main concerns are as follows: (i) the overall load due to forwarding activity should be evenly shared among the participants; (ii) the time needed for full dissemination should be minimized; and (iii) the expected time needed to receive the whole content should be the same across peers.

Recently, game theory proves very useful for modeling communication systems and dynamic distributed environments. Here, we model live streaming scenarios in multistage congestion game form. Roughly speaking, in congestion games, there is a set of facilities and players have to choose what facilities to use. Each facility provides some common utility level to all its users, which depends on the number of users, and finally each player gets the sum over used facilities of the utility levels that these latter provide. In particular, the larger the number of players who use a facility, the lower the common utility level that they attain from that facility (monotonicity). Congestion games are potential games [19], thereby possessing pure-strategy (possibly strong [11,27]) equilibrium [25]. When using congestion games for modeling real-time streaming, facilities are players themselves: at any stage of the game, some content distribution over peers prevails, and therefore, each peer faces a facility set containing all other participants with some additional content units (i.e., possible data forwarders).

Wireless P2P communication may be framed in these terms by assuming that, given the broadcaster's and peers' spatial positions (in a given region), each of them can forward content units only to neighbors, which constitute a proper subset of the remaining nodes. In particular, given the exogenous network with peers and the broadcaster as nodes, communication may be assumed to be possible only between nodes that are *directly* linked in the network. This choice seems the more reasonable one, while also enabling to tightly cope with the wired setting. In fact, if wireless live streaming is dealt with as depicted thus far, then a whole range of alternative, more specific choices become available, thereby allowing for a rich set of open issues and possible developments. For example, in *mobile* wireless systems, the focus would be on a MANET (mobile ad hoc network) changing in time as peers move around, which in turn would require monitoring network changes during the streaming and next reconfiguring (optimally) the exchange of content units in incoming stages.

Within this wide spectrum of scenarios that become conceivable, our analysis next focuses on nonmobile settings and, together with constraining communication only between directly linked nodes, also adds further rules governing stage-wise P2P exchange. The reason for this, apart from the need to start somewhere, is that for the wired (or fully connected) case, our modeling of real-time streaming in multistage congestion game form allows neatly distinguishing between streaming length (or number of stages through which the whole content gets distributed over the whole population) on the one side and stage- and peer-wise congestion on the other. Most importantly, it enables to identify a strategy restriction mechanism RM, which at each stage prevents peers from asking certain content units, given the prevailing content distribution over the population. With such restrictions, at equilibrium, both streaming length and stage-wise congestion are minimized. This is detailed by a distributed algorithm that we develop for implementing the proposed equilibrium selection method.

Our RM is, in fact, a P2P exchange protocol, and thereby in the wireless setting, it can only be used when the given ad hoc network displays certain (connectivity) features. In fact, for generic network, determining whether the RM is implementable or not seem hard. Accordingly, when concerned with wireless scenarios, we rather concentrate on streaming performance, at equilibrium, with no strategy restriction at all apart from communication constraints as formalized precisely by the ad hoc network itself. More specifically, if at each stage, peers can choose any neighbor when asking for additional content, then the fully connected case becomes simply one (although special) among others, and thereby neatly comparable. Along this line, our simulation results are divided into slot 1 and slot 2. The first deals exclusively with the fully connected case, comparing equilibrium outcomes with and without the RM in terms of both average latency (or delay) in receiving content units and its standard deviation. As predicted by former analytical results, the RM is shown to perform very well. In particular, the aim is at figuring out the gap separating such an equilibrium selection method from average equilibrium outcomes. In the second slot, the concern is only with situations where no superimposed coordination method (such as the RM) is applied, but with communication constraints due to an hoc network. Here, equilibrium streaming performance is shown to deteriorate when the number of edges in the network decreases.

As multimedia streaming is mostly approached by means of a P2P (adaptive) overlay network through which to distribute content across peers [7], these latter are assumed to behave cooperatively, at equilibrium, as long as interaction is suitably designed. That is to say, so to avoid free riding [20], which here means entering the system with the aim to receive while not (or little) re-forwarding the content. This yields fair protocols, enabling to rapidly distribute content units produced by broadcasters. In one direction, BitTorrent philosophy led to design *tit-for-tat*-based policies, thereby guaranteeing that every participant contributes to satisfying others' needs [6]. Along another route, P2P streaming may resort to a pseudo-random partner selection in a gossip scheme where peers have no incentive to deviate from the protocol [16]. The general issue is to model multicast protocols in terms of noncooperative games [13] whose equilibrium outcomes display (socially) desirable features.

In our analysis, participants are assumed to fully cooperate, in that they all satisfy one request whenever they receive some valid one (see below). In fact, a main tool for limiting free riding is the capability to tag those who do not cooperate, so that everybody can recognize them. This is especially true in dynamic environments, where each participant has to interact one to one with several others, several times, as in our case. In this respect, the proposed method constructs a new P2P network for each content unit to be distributed, and therefore, each peer is expected to interact with all other peers several times during the whole course, making incentive-based schemes applicable [22]. In fact, current approaches to wireless live streaming ignore free riding while concentrating on congestion in routing problems as well as on issues dealing with aggregation, transport errors, and buffer schedule [12,21,33,34].

Among noncooperative approaches, potential games [19] include congestion ones as a subclass. Routing problems probably constitute the main networking issue, which is fruitfully framed in terms of congestion (or, more generally, potential) games. Broadly speaking, in basic routing problems modeled as games, there is a given network one of whose nodes is the common destination. Then, each player is assigned to some origin node and has to reach the destination through a route (or origin–destination path) provided by the network. Each arc induces a cost, which grows with the number of players who choose a route using that arc, so that players selfishly aim at minimizing their own route cost. This game can be enriched in different ways; for example, it may be assumed to be repeated [1], so to exploit best response dynamics [4,5]. In addition, for certain cases, even strong equilibrium has been analyzed in detail [8,11,28]. Finally, from another perspective, important results have also been derived through cooperative game-theoretic approaches [24].

In this work, a main focus is how to minimize, through peers' selfish behavior, the number of stages needed to completely distribute the content over peers. To this end, together with some general assumptions such as perfect information and cooperative behavior on the forwarding side, we also propose strict rules for content dissemination. For example, we do not consider multisource environments (e.g., video conferences). Still, we provide a general modeling of real-time streaming in terms of congestion games that might be extended for studying alternative scenarios.

The remainder of this work is organized as follows: Section 19.2 introduces the needed basic concepts and notation. In Section 19.3, the problem of multimedia live streaming is presented as a multistage congestion game. In Section 19.4, we focus on the fully connected case and, in particular, on how to distribute the whole content over the whole population in a minimum number of stages. In Section 19.5, we concentrate on equilibrium conditions, distinguishing between unconstrained (Section 19.5.1) and constrained (Section 19.5.2) communication. In Section 19.6, we simulate equilibrium evolutions, again distinguishing between unconstrained (Section 19.6.1) and constrained (Section 19.6.2) communication. We conclude discussing open issues and possible developments (Section 19.7).

19.2 Preliminaries

Congestion game modeling of live streaming crucially has to take into account that the game intrinsically must be multistage: for given numbers of content units and peers, any course of the streaming is fully characterized by a time sequence of directed trees, each spanning all nodes, and specifying how each content unit reaches each peer from the broadcaster. Note that what game is actually played at each stage depends on what content distribution over peers does prevail at that stage. In fact, a main concern in this dynamic setting is streaming length. More specifically, each player aims at receiving all the content as fast as possible, but how long the whole dissemination will actually take depends on how peers distribute their requests to receive at each stage. This is a typical coordination issue, where selective and timely dissemination of content units deeply affects the overall performance of the system. For the wired setting, we provide a strategy restriction mechanisms enabling selfish agents to reach overall coordination through optimizing behavior. Accordingly, costs are expressed in terms of (expected) streaming length or duration, and thereby, congestion seems most suitably measured by counting, at each stage, how many requests for additional content reach each peer and the source or broadcaster.

Stream production is assumed to occur over time $t = 0, 1, \ldots, T$. Starting at $t = 0$, a *source* provides one new unit c^t of content at the beginning of each time period (or stage/round) $t \to t+1$, until the end T is reached, while n peers aim at receiving all the content as soon as possible. This may be modeled as a *multistage game* [17] by means of *congestion game forms* [11], with player set $N = \{1, \ldots, n\}$ and naturals $1, \ldots, n$ being peers' identifiers. Also denote the source by 0 and let $N_0 = \{0, 1, \ldots, n\}$. A multistage game is identified by a *tree*, each of whose nodes corresponds to a moment at which at least one player has to take action, and thereby *rooted* at the start, which here is when the source has just finished producing the very first content unit c^0 and peers begin to line up in order to receive it. All paths from the start to some node referring to a fixed time t bijectively correspond to all distinct courses the game may take up to t. What nodes are actually reached, during any course, obviously depends on what actions players actually take. In *stochastic* games, for given per-node actions, what successive node is actually reached depends on some random event. We assume *perfect information*: when asked to take action, at any time t, all players know exactly what node has been reached at t. The *leaves* of the tree correspond to outcomes, over which players have preferences. A *strategy*, for a player, specifies an (admissible) action to take at each node. The *rules* are as follows:

Ru1: The source can send each unit $c^t, 0 \le t \le T$ over time period $t \to t+1$ only (just after that unit is produced) and, additionally, to only one (some) peer

Ru2: If a peer at the end t of any time period $t-1 \to t$ has already received units $c^{t_1}, \ldots c^{t_h}$, with $0 \le t_1, \ldots, t_h < t$, then in the following time period $t \to t+1$, or equivalently in the following *stage* or *round* t, this peer can send only one of such units and to one (i.e., some) other peer only

Ru3: In any round, each peer may ask to receive some unit either from one other peer or from the source

With these rules, spreading the whole content (or time sequence of units) $CON = \{c^0, c^1, \ldots, c^t, \ldots, c^T\}$ over the whole peer set surely requires much P2P exchange, as the source sends each unit precisely once, to only one peer, and next the unit has to be P2P-exchanged exactly $n-1$ times. Looking at the first rounds, at $t = 0$ all peers will surely ask for c^0 from the source, but only one, possibly picked at random, and denoted $i_1^0 \in N$, will actually get it over time period $0 \to 1$. Hence, at $t = 1$ some peers will ask for c^1 from the source (with i_1^0 surely among them), while the remaining ones will ask for c^0 again, but now from i_1^0. Picking at random from each of these two peer subsets (which at equilibrium are complements of each other) will yield someone, denoted i_1^1, receiving c^1 from the source (in round 1), and someone else, denoted i_2^0, receiving c^0 from i_1^0 (in round 1). This is where the cost of coordination failure may begin to be paid, because if $i_1^0 = i_1^1$, in next round 2 only two content units will be sent: c^2 from the source to some peer (hopefully not $i_1^0 = i_1^1$ again), and *either* c^0 *or else* c^1 from $i_1^0 = i_1^1$ to some other peer. Such a scheduling is clearly suboptimal, because although three units are already available, only two of them get actually distributed.

For turning this into a game, the first step is determining the players, which clearly are the peers. Next, the nodes in the game tree have to be identified: a time-indexed node C^t, referring to time (or beginning of stage/round) t, is an $n+1$ set $C^t = \{C_0^t, C_1^t, \ldots, C_n^t\}$ where each $C_i^t \subseteq \{c^0, c^1, \ldots, c^t\}$ specifies what content units up to t peer $i \in N_0$ has received. Hence, a game tree node C^t may be looked at as a $0-1$ matrix with i, t' entry $C_i^t(t') \in \{0, 1\}$ defined by

$C_i^t(t') = 1$ if $c^{t'} \in C_i^t$ (i.e., if in node C^t peer/source $i \in N_0$ has unit $c^{t'}$) and 0 otherwise. Notice that, given Ru1, the source is constrained to provide at each time only the current unit: $C_0^t(t') = 1$ if $t' = t$ and 0 otherwise for all $0 \le t \le T$, while $C_0^t(t') = 0$ for all $t > T, t' \ge 0$. That is, at any time, the source only provides the unit that it has *just* finished producing and therefore no unit at all after T. For each $i \in N$, a strategy specifies, for each game tree node C^t, some $j \in N_0$ from whom to ask, in round t, a content unit $c^{t'}, 0 \le t' \le t$. These strategies are finite sequences as long as some upper bound $T_* < \infty$ on duration (in stages) of the game, over all conceivable courses, exists.* In the sequel, this worst-case (or longest) streaming length at equilibrium T_* is observed in terms of the *streaming tree*, whose vertex set N_0 contains all peers together with the source, and through which each content unit spreads over the whole peer population. In fact, the streaming tree is a nested time sequence of *sub-streaming trees*, one for each content unit. Specifically, the whole streaming, from an ex post perspective, is identified by the $T + 1$ trees, with peers as vertexes, describing how each of the $T + 1$ content units actually reaches the whole population. In each of these trees, edges are directed and, in particular, the in-degree is 1 for each vertex (apart from the source). That is, while in general, peers may well forward a content unit several times (to several different other peers, in different rounds), they receive each content unit only once.

A MANET is here most suitably modeled as a time sequence $G_t = (V_t, E_t)$ of (simple) graphs, with V_t and E_t denoting, respectively, the set of vertexes (or nodes or peers) and the set of edges (or links or hops) at time $t = 0, 1, 2, \dots$. Nodes are commonly assumed to move around in a spatial region, causing variations in network topology, that is, both the vertex set V_t and the edge set $E_t \subsetneq V_t^{(2)}$ frequently change over time, where $V_t^{(2)} = \{A \subset V_t : |A| = 2\}$ denotes the set of all pairs $\{v, v'\} \subset V_t$ and with $|\cdot|$ being the cardinality of sets. As already outlined, this work is mainly concerned with nonmobile settings, which formalizes into the following: $V_t = N_0$ as well as $E_t = E \subsetneq N_0^{(2)}$ for all $0 \le t \le T_*$. In words, all peers and the broadcaster remain online during the whole streaming length, and an exogenously given and fixed ad hoc network E constrains communication throughout the streaming. In particular, E is conceived to result from users' access terminal positions and to enable per-round transmission only over unitary distance. That is to say,

Ru4: Over each round, the source and each peer can send content only to peers who are at unitary hop distance from them in the given ad hoc network.

Given edge set E, for each $i \in N_0$ set $E_i = i \cup \{j \in N_0 : \{i, j\} \in E\}$. In words, $E_i \subseteq N_0$ is the set containing i together with those at unitary hop distance from i in the ad hoc network. Also let $\mathcal{N}^t$ denote the set of all game tree nodes referring to time $t \ge 0$. That is, $\mathcal{N}^t$ is the family of all $(0 - 1$ matrices specifying) content distributions over peers that may be reached along some game course up to t, when the above main constraints Ru1–4 apply: in any round, each peer can forward at most one unit to some neighbor and can also submit at most one request for receiving from a neighbor. In our model, a strategy A^i for peer $i \in N$ has a form $A^i : \mathcal{N} \to E_i$, where $\mathcal{N} = \cup_{0 \le t \le T_*} \mathcal{N}^t$, and with $A^i(C) = j$ denoting the one (i.e., peer or source) $j \in E_i$ from whom i asks to receive at game tree node $C \in \mathcal{N}$. Hence, in valid requests $j \in \{0, \dots, i-1, i+1, \dots n\} \cap E_i$, although we interpret $A^i(C) = i$ as one possible way in which peer i at node C does not ask to receive any content, from

---

* Intuitively, as peers are concerned with receiving some content while this latter is produced, sometime after content production is finished nobody will any longer be interested in receiving any unit at all. More formally, rational strategy profiles, under perfect information (and without free riding), surely provide a finite game course, as shown below.

anybody (see below). Note that all game tree nodes are unified into a single set $\mathcal{N}$, independently from what different game courses lead to them and when. The resulting analysis provides behavioral rules according to which players respond to any realized content distribution over the population, at any time. The main concern is what game course length (or number of rounds needed to distribute the whole content over all peers) may be the outcome of selfish, stage-wise optimization by players.

19.3 Congestion in P2P Streaming

In a congestion game, there is a set N of players and a set M of facilities, and each player $i \in N$ has a set $\Sigma^i \subseteq 2^M$ of strategies, where 2^M is the (power) set of all subsets of M. Usually, M is the edge set of a graph, and each player $i \in N$ has to reach a destination v_i^d from an origin v_i^o. Then, the set Σ^i of strategies for i contains all (edges of) $v_i^o - v_i^d$ paths. A *congestion game form* $F = (N, M, \Sigma^1 \times \cdots \times \Sigma^n)$ identifies a whole class of congestion games, each obtained by specifying the payoffs $\pi^i : \Sigma \to \mathbb{R}_+$ of players $i \in N$, where $\Sigma = \Sigma^1 \times \cdots \times \Sigma^n$. Profile $A = \{A^1, \ldots, A^n\} \in \Sigma$ of strategies identifies *congestion vector* $\sigma(A) = \{\sigma_a(A) : a \in M\}$ specifying how many players have each facility $a \in M$ in their strategy A^i. That is, $\sigma_a(A) = |\{i \in N : a \in A^i\}|$. The game is *monotone* when each $a \in M$ has an associated utility function $u_a : \mathbb{Z}_+ \to \mathbb{R}_+$ satisfying $u_a(k) < u_a(k')$ whenever $k > k'$, and each $i \in N$ gets a payoff given by the sum over all the chosen facilities $a \in A^i$ of the corresponding utility: $\pi^i(A) = \sum_{a \in A^i} u_a(\sigma_a(A))$. Finally, F (and any game derived from it) is *symmetric* when all players share the same strategy set: $\Sigma^1 = \cdots = \Sigma^n$ [11].

Wireless P2P streaming systems may be approached in congestion game form with facilities being players themselves (together with the source), and with every strategy profile $A = (A^1, \ldots, A^n)$ corresponding to a *congestion matrix* $\sigma(A) = \{\sigma_C^i(A) : C \in \mathcal{N}, i \in N_0\}$, whose generic entry

$$\sigma_C^i(A) = |\{j \in N : A^j(C) = i \in E_j, C_i \not\subseteq C_j\}|$$

is the number of peers $j \in N$ who are at unitary hop distance from $i \in N_0$ and who ask to receive from this latter some content unit in $C_i \backslash C_j$, where $\cdot \backslash \cdot$ denotes the difference between sets (of content units). In particular (and for notational convenience), at any reached game tree node $C \in \mathcal{N}$, if a peer asks to receive from someone who has no additional content, then such a request is simply ignored by the system, causing null congestion. A request is *valid* if it contributes to congestion.

Denote by $\kappa = |\mathcal{N}|$ the whole number of game tree nodes. A strategy A^i for a peer $i \in N$ can be regarded as a point $A^i \in E_i^\kappa$, as it specifies somebody (at hop distance ≤ 1, but possibly with no additional content) to ask from at each node $C \in \mathcal{N}$ that may be reached. Hence, the corresponding congestion game form is $F = (N, N_0^\kappa, E_1^\kappa \times \cdots \times E_n^\kappa)$. Let $\mathcal{E} = E_1^\kappa \times \cdots \times E_n^\kappa$. Players' payoffs $\pi^i : \mathcal{E} \to \mathbb{R}_+$ ($i \in N$) are assumed to consist of a sum over nodes of some (possibly 0) utility or *per-node* payoff received at each $C \in \mathcal{N}$. This utility depends on the prevailing content distribution (which is precisely what the game tree node C specifies), and on the profile $A^1(C), \ldots, A^n(C)$ of *per-node* strategies that players choose at node C. Hence, per-node payoffs received by peers $i \in N$ may depend on congestion, which here is the number of other peers $i' \in N$ with the same (valid) per-node strategy $A^{i'}(C) = A^i(C)$. This models real-time streaming in congestion game form with facilities being pairs (j, C), where $j \in N_0$ is either a peer or the source and $C = C^t$ is a game tree node or content distribution that may prevail at some time t.

In a simplest form* for $\pi^i : \mathcal{E} \to \mathbb{R}_+$ ($i \in N$), the payoff $\pi^i(A)$ of any given strategy profile $A \in \mathcal{E}$ to a peer $i \in N$ is the sum, over all conceivable game tree nodes $C \in \mathcal{N}$, of the values taken by utility u_{C_j}, which in turn depends only on congestion $\sigma_C^j(A)$. That is, for any strategy profile A, each peer $i \in N$ gets a utility at each node $C \in \mathcal{N}$, which depends exclusively on the number $\sigma_C^j(A)$ of those with the same (valid) per-node strategy:

$$\pi^i(A) = \sum_{\substack{C \in \mathcal{N} \\ A^i(C)=j}} \sum_{j \in E_i : C_j \not\subseteq C_i} u_{C_j}\left(\sigma_C^j(A)\right). \tag{19.1}$$

Profile $A = (A^1, \ldots, A^n)$ is *Pareto-optimal* if no $B = (B^1, \ldots, B^n) \in \mathcal{E}$ satisfies $\pi^i(B) \geq \pi^i(A)$ for all $i \in N$, with strict inequality for at least one i. Hence, from an aggregate perspective, these profiles are efficient: there is no way of improving someone's payoff without deteriorating someone else's one. Congestion games allow for neat conditions under which desirable properties, such as Pareto-optimality and strength of equilibrium, attain. In fact, in such monotone (possibly symmetric) games, these properties depend on the structure of the union $\Sigma^U = \cup_{i \in N} \Sigma^i$ of strategy spaces, in particular on whether a *bad configuration* appears or not. Formally, Σ^U displays a bad configuration when there are three strategies $X, Y, Z \in \Sigma^U$ and two facilities $x, y \in M$ such that $x \in X \not\ni y$ and $x \notin Y \ni y$ but $x \in Z \ni y$. "Thus, two facilities give rise to a bad configuration if there are strategies in Σ^U which use one of them but not the other, and there is also a strategy in Σ^U which uses both of them. The latter never occurs if Σ^U consists of singletons" [11, pp. 87–88]. As the name suggests, it is desirable that no bad configuration exists. Here facilities are players themselves, although any fixed player corresponds to two distinct facilities when referring to two distinct game tree nodes. By Ru3, strategies are time sequences of singletons, and hence, the safe case applies.

An equilibrium profile $A = (A^1, \ldots, A^n)$ satisfies $\pi^i(A^{-i}, A^i) \geq \pi^i(A^{-i}, B^i)$ for all $i \in N$ and $B^i \in E_i^\kappa$, where $A^{-i} \in \underset{j \in N \setminus i}{\times} E_j^\kappa$ is an $n-1$ profile for peers $j \in N \setminus i$ as well as $A^i \in E_i^\kappa$ is a strategy for peer i. In particular, A is a *strong* equilibrium if for no coalition $\emptyset \neq S \subseteq N$ is there a choice of $B^i \in E_i^\kappa$ for coalition members $i \in S$ such that $\pi^i(B^S, A^S) > \pi^i(A)$ for all coalition members $i \in S$, where (B^S, A^S) denotes the profile in which each $i \in S$ chooses B^i and each $j \in S^c = N \setminus S$ chooses A^j. In words, no nonempty coalition can deviate from strong equilibrium profiles and thereby strictly increase the payoffs of *all* its members. When considering the implications of strong equilibrium for $S = N$, one gets rather similar conditions as those identifying Pareto-optimal profiles. In fact, as strategies are time sequences of singletons, the model provided thus far yields a monotone congestion game with no bad configuration, where therefore the set of strong equilibria is nonempty, coincides with the set of equilibria, and, generically, is (weakly) included in the set of Pareto-optimal profiles [11].

Given payoffs $\pi^i : \mathcal{E} \to \mathbb{R}_+$ for peers $i \in N$, any function $P : \mathcal{E} \to \mathbb{R}$ is a potential if $[P(A^{-i}, A^i) - P(A^{-i}, B^i)][\pi^i(A^{-i}, A^i) - \pi^i(A^{-i}, B^i)] \geq 0$ for all $i \in N$, all $n-1$ profiles $A^{-i} \in \underset{j \in N \setminus i}{\times} E_j^\kappa$ and all pairs $A^i, B^i \in E_i^\kappa$. A potential is *exact* when the two differences within square parentheses are equal. Hence, potentials take values on strategy profiles, and for any such a profile and unilateral deviation from it, the deviating player's payoff and the potential itself change in the same direction.

---

* Further payoff functions, possibly also depending on the difference between the sender's and the receiver's contents (hence *P2P-specific* or *player-specific* [14]), are considered in [26].

Claim 19.1 (see [25]): With payoffs (19.1), an exact potential is (for $A \in \mathcal{E}$)

$$P(A) = \sum_{C \in \mathcal{N}} \sum_{j \in N_0} \sum_{k=1}^{\sigma_C^j(A)} u_{C_j}(k). \tag{19.2}$$

Proof Fix $i \in N$ and $A^i, B^i \in E_i^\kappa$ as well as an $n-1$ profile $A^{-i} \in \underset{j \in N \backslash i}{\times} E_j^\kappa$ of strategies for peers $j \in N \backslash i$. Consider congestion matrices $\sigma(A)$ and $\sigma(B)$ associated, respectively, with $A = (A^{-i}, A^i)$ and $B = (A^{-i}, B^i)$. Their elements $\sigma_C^j(A)$ and $\sigma_C^j(B)$ $(j \in N_0)$ differ only if $C_j \not\subseteq C_i$ and either $A^i(C) = j \neq B^i(C)$ or $A^i(C) \neq j = B^i(C)$. In the former case, $\sigma_C^j(A) = \sigma_C^j(B) + 1$, while in the latter, $\sigma_C^j(A) = \sigma_C^j(B) - 1$. Then, $P(A) - P(B) =$

$$\sum_{C \in \mathcal{N}} \left(\sum_{\substack{j \in E_i : C_j \not\subseteq C_i \\ A^i(C) = j \neq B^i(C)}} u_{C_j}\left(\sigma_C^j(A)\right) - \sum_{\substack{j \in E_i : C_j \not\subseteq C_i \\ A^i(C) \neq j = B^i(C)}} u_{C_j}\left(\sigma_C^j(B)\right) \right)$$

$$= \pi^i(A) - \pi^i(B). \qquad \square$$

For monotone congestion games with no bad configuration, (19.2) is a strong potential, any of whose maximizers is a strong equilibrium [11, Theorem 5.2].

19.4 Optimal P2P Content Exchange

This section deals with the fully connected case: $E_i = N_0$ for all $i \in N_0$ and thus $\mathcal{E} = N_0^{\kappa n}$ (making Ru4 redundant). The focus is on optimal scheduling of P2P exchange of content units. In this view, an indicator of streaming efficiency is simply the number of rounds needed to spread the whole content CON over the whole peer set N. Assume that the number of peers is a power of 2, that is, $n = 2^m$ for some natural m. Under our assumptions Ru1–3, any content unit can spread over the whole population no faster than through $m + 1$ (consecutive) rounds. For example, in round 0, the very first content unit c^0 will go from the source to some peer i_1^0. In round 1, it will go from i_1^0 to some other peer $i_2^0 \in N \backslash i_1^0$. In round 2, it will go from i_1^0, i_2^0 to two distinct other peers $i_3^0, i_4^0 \in N \backslash \{i_1^0, i_2^0\}$, and so on, doubling the forwards in each round, until in round m (which is the $m + 1$th round since unit c^0 started circulating), exactly half of the population sends the content unit to the other half through one-to-one matching. A crucial fact from now on is that *all* the $T + 1$ content units, for any T, can spread over the whole peer set in exactly $m + 1$ rounds. This can be observed in terms of the different possibilities for building the whole streaming tree (see above). In particular, if a generic content unit c^t reaches everybody in $m + 1$ rounds, then the number of peers who send (and therefore also the number of those who receive) this content unit c^t in round $t + k$ is 2^{k-1} for $k = 1, \ldots, m$.

If *all* content units must reach everybody in $m + 1$ rounds, then whenever a peer receives a unit c^t in round $t + k$, it must forward c^t for the remaining rounds $t + k + k'$, where $k' = 1, \ldots, m - k$. Hence, this peer for these latter rounds cannot receive units to be further forwarded. That is to say, in any round $t + k + k'$, with $k' = 1, \ldots, m - k$, if this peer receives some unit, then such a unit must be $c^{t+k+k'-m}$, in which case this round $t + k + k'$ is precisely the $m + 1$th (i.e., last) one in which this unit $c^{t+k+k'-m}$ circulates. In other words, the peer must be among those 2^{m-1} who are the last ones to receive that unit. This not only is feasible but also can be obtained through many different streaming trees.[*]

Definition: Profile $A \in \mathcal{N}_0^{\kappa n}$ (or $A \in \mathcal{E}$) is *deterministic* if both the following

(I-a) $\left| \{ C \in \mathcal{N} : A^i(C) = j, C_j \nsubseteq C_i \} \right| = T + 1$
(I-b) $A^i(C) = j, C_j \nsubseteq C_i \Rightarrow |C_j \backslash C_i| = 1$ for all $C \in \mathcal{N}$

hold for all $i \in N$.

Hence, each peer makes exactly $T + 1$ valid requests to receive (which under Ru1–3 clearly is the minimum number of such valid requests needed to receive all the $T + 1$ content units $c^0, c^1, \ldots c^T$) and therefore receives some (distinct) content unit every time a valid request is made. Also, the $T + 1$ valid requests made by any peer i are all addressed, each at a different game tree node, to someone who at that node has precisely one additional content unit.

The name *deterministic* is due to the assumption that transitions from one game tree t node $C^t \in \mathcal{N}^t, t \geq 0$ to $t + 1$ nodes $C^{t+1} \in \mathcal{N}^{t+1}$ are stochastic: a generic strategy profile A does not yield a unique game course, but a probability distribution over game courses. Whatever its form, an underlying probabilistic model essentially decides who gets what when multiple peers $i_1, \ldots, i_k$ ask to receive from a common $j \in N_0$ such that $C_j^t \nsubseteq C_{i_{k'}}^t, 1 \leq k' \leq k$. As peers (and the source) can forward at most one unit per round, such a model has to select precisely one peer $i_{k'} \in \{i_1, \ldots, i_k\}$ and one unit $c^{t'} \in C_j^t \backslash C_{i_{k'}}^t$ to be received by the former. Deterministic profiles actually allow to ignore the underlying probabilistic model (which is intended to be complex and mostly unknown), because the transition from any node to successive ones becomes deterministic.

Deterministic profiles put probability 1 on one game course and probability 0 on all other courses. Accordingly, consider the unique content distribution over peers (or game tree node) reached at t by deterministic profile $A \in N_0^{\kappa n}$. In particular, denote it by $C^t(A) = \{ C_0^t(A), C_1^t(A), \ldots, C_n^t(A) \}$.

Definition: A deterministic profile $A \in N_0^{\kappa n}$ is *fastest streaming* if

(II) $\left| \{ i \in N : C_i^t(A) \ni c^{t-k} \} \right| = \min\{2^m, 2^{k-1}\}$ for all $0 < k \leq t \leq T_*$.

These profiles spread each content unit c^t over $2^0 = 1$ peer in round t, over (new) $2^0 = 1$ peer in round $t + 1$, over (new) $2^1 = 2$ peers in round $t + 2$, and so on, until (new and final) 2^{m-1} peers receive unit c^t in round $t + m$, which is the $m + 1$th (i.e., final) round where this unit circulates. Let $\mathcal{A}^*$ be the set of fastest streaming profiles. As $\sum_{0 \leq h < k} 2^h = 2^k - 1$, summing newly reached peers across these $m + 1$ rounds yields that each unit c^t reaches the whole population in $m + 1$ rounds. That is, the whole peer set is covered at the end of round $t + m$, as $2^0 + 2^0 + 2^1 + \cdots + 2^{m-1} = 2^0 + 2^m - 1 = 2^m$. In addition, in these profiles $A \in \mathcal{A}^*$ peers $i \in N$ receive each and every time t they ask something from someone (i.e., some $j = A^i(C^t)$

---

[*] Although in all of them the sub-streaming tree for each unit must result from a suitable permutation of vertexes/peers, thereby always reproducing the same essential condition that for each unit c^t the number of senders doubles after each round $t + h$ for $h = 1, \ldots, m$, vanishing afterward.

with $C_j^t \nsubseteq C_i^t$). Given Ru1–3, these conditions just listed are rather demanding, and one may well wonder whether $\mathcal{A}^* \neq \emptyset$ at all. Accordingly, $|\mathcal{A}^*|$ is now determined.

Consider a generic t such that $m \leq t \leq T$. For any $A \in \mathcal{A}^*$, in round t, there are exactly $m + 1$ content units $c^t, c^{t-1}, \ldots, c^{t-m}$ being distributed across the whole population, out of which precisely m (i.e., $c^{t-1}, c^{t-2}, \ldots, c^{t-m}$) are sent by some peers to some other peers, while one unit (i.e., c^t) is sent from the source to some suitably chosen peer. Hence, in this round t, each peer is a receiver (of some unit $c^{t-k}, 0 \leq k \leq m$). Conversely, only $2^m - 1$ peers also send (units $c^{t-k}, 1 \leq k \leq m$), as the source forwards c^t.

Conditions (I-a,b) and (II) may be turned into a useful recursive method for establishing, for any course of the game reached up to any time $t \geq 0$ (and thus applying since the very beginning), how to proceed in round t in order to have an A-induced streaming for some $A \in \mathcal{A}^*$. In fact, all scheduling priorities can be captured by the following main constraint: for any $t \geq 0$, if in the previous round, a peer has received and/or forwarded some unit that will have to be forwarded in round $t + 1$ as well, then in this round t, this peer cannot receive any unit that will also have to be forwarded in round $t + 1$. Let S_t^k be the subset of peers who *send* unit c^{t-k} in round t ($1 \leq k \leq m$) and R_t^k be the subset of peers who *receive* unit c^{t-k} in round t ($0 \leq k \leq m$).

Recurrence

Re1: If $i \in S_{t-1}^k$ or $i \in R_{t-1}^k$ for some $k \leq m - 2$

Re2: Or $t \geq m$ and $i \notin S_{t-1}^k$ for all $k \geq 1$

Re3: Then $i \notin R_t^k$ for all $k \leq m - 1$

Consider that any unit $c^{t'}$ has to be (still) forwarded in round $t + 1$ (i.e., over $t + 1 \to t + 2$) when $t' + m \geq t + 1$. Now focus on a generic unit c^{t-1-k} that a peer is either sending or else forwarding in round $t - 1$. Letting $t' = t - 1 - k$, we have that this unit will have to be forwarded in round $t + 1$ if $t - 1 - k + m \geq t + 1$, that is, if $k \leq m - 2$. If this is the case, then the peer cannot now (i.e., in round t) receive any unit c^{t-k} such that $t - k + m \geq t + 1$ or $k \leq m - 1$. On the other hand, Re2 entails precisely that when we reach any round $m \leq t \leq T + m$, all peers who still have not received unit c^{t-m} are matched with the other half of the population, so to ultimately receive such a unit that will no longer be distributed throughout the whole streaming.

We now proceed to counting all streaming trees that satisfy this recurrence Re1–3. Recall that for naturals $a \geq b$ product $[a]_b = a(a-1)(a-2) \cdots (a-b+1)$ is the *falling factorial*. Our enumerative concern is with situations where $T > m$, and the whole streaming evolves through the following *three phases*:

Ph1: It comprehends all initial rounds $t = 0, 1, \ldots, T_1$ where at least one peer does not receive any content unit

Ph2: It comprehends all rounds $t = T_1 + 1, T_1 + 2, \ldots, T_1 + T_2$ where each peer receives a unit

Ph3: It comprehends all rounds $t > T_1 + T_2$ where some peer does not receive any unit but some other peer receives one

---

Claim The number of fastest streaming profiles is

$$|\mathcal{A}^*| = \left(\prod_{t=0}^{m-1} [2^m - 2^t + 1]_{2^t} \right) (2^{m-1}!)^{2^{T+2-m}} \left(\prod_{t=1}^{m} [2^{m-1}]_{2^{m-1}-2^{t-1}} \right).$$

Proof The needed counting procedure (detailed in [26]) may be sketched in three phases, out of which the first and final ones display an excess demand of content units, while demand equals supply in the central phase.

Ph1: (II) entails that in rounds $t = 0, 1, 2, \ldots, m - 1$, all needed receivers must be chosen among those who still have not received anything, yielding 2^t new peers involved in the streaming at each t. As $\sum_{0 \leq h < k} 2^h = 2^k - 1$, at time m (i.e., at the end of round $m - 1$ and at the beginning of round m), all peers apart from one (i.e., $2^m - 1$) have received precisely one unit. The number of distinct ways to achieve this, or number of alternative streaming tree evolutions induced by profiles $A \in \mathcal{A}^*$ in rounds 0 to $m - 1$ inclusive, is

$$\prod_{t=0}^{m-1} \binom{2^m - 2^t + 1}{2^t} 2^t! = \prod_{t=0}^{m-1} [2^m - 2^t + 1]_{2^t}.$$

Ph2: In the starting round m, a unique peer $i^* \in N$ received nothing in previous (initial) rounds. By Re1–3, from now on until the end, in any round t peer i^* must be in the half of the population who receives from the other half precisely the unit c^{t-m-1} whose distribution terminates in that round, thereby never forwarding any unit. In all successive rounds $t = m, m + 1, \ldots, T$, the set of senders contains the source together with $2^m - 1 = |N \setminus i^*|$ peers, and these latter are partitioned into m blocks with cardinalities $2^0, 2^1, \ldots, 2^{m-1}$ whose members send, respectively, units $c^{t-1}, c^{t-2}, \ldots, c^{t-m-1}$. Counting the number of distinct ways to match demand and supply in these central rounds $t = m, m+1, \ldots, T$ (where they both involve 2^{m-1} participants, but twice) gives $\left(2^{m-1}!\right)^{2(T+1-m)}$ different evolutions available between round m and round T inclusive.

Ph3: In the first of final rounds $t = T + 1, T + 2, \ldots, T + m$, a unique peer $i_{T+1}^* \neq i^*$ does not receive any unit. Hence, at time $T + 2$ (i.e., at the end of round $T + 1$), an additional peer i_{T+1}^*, like i^*, still misses units $c^{T+2-m}, c^{T+3-m}, \ldots, c^T$. In view of Re2, in round $T + 2$, this peer i_{T+1}^*, as a receiver, matches some c^{T+2-m} sender. Till the end of the streaming, this peer will never be among those who receive units to be further forwarded. The number of peers such as i_{T+1}^* (or i^*) increases by 2^{t-T-1} in each remaining round t until the end. In fact, in the last round $T + m$, half of the peer send (without receiving anything) what the other half receives. Repeating the same argument used earlier for phase Ph1, but in an opposite manner, between rounds $T + 1$ and $T + m$ inclusive, the streaming tree may be checked to evolve in

$$\prod_{t=1}^{m} 2^{m-1}! \binom{2^{m-1}}{2^{m-1} - 2^{t-1}} (2^{m-1} - 2^{t-1})! = (2^{m-1}!)^m \prod_{t=1}^{m} [2^{m-1}]_{2^{m-1} - 2^{t-1}}$$

different possible ways. ☐

Hence, many different streaming trees distribute the whole content over all peers in a way such that each unit c^t reaches *new* 2^{k-1} peers in each round $t + k$ for $k = 1, \ldots, m$, so that the whole streaming completes in exactly $T + m + 1$ stages. Note that all of them satisfy condition (I-a), as at any node C and for any two peers $i, j \in N$, we have $|C_i \setminus C_j| \in \{0, 1\}$. Still, with payoffs given by (19.1), fastest streaming is not sustainable at equilibrium, because condition (I-b) is too demanding: selfish (and myopic) peers try to receive some unit in any round until they get the whole content, as detailed hereafter.

19.5 Equilibrium P2P Content Exchange

Given the P2P setting, where peers always satisfy precisely one (randomly selected) valid request among those received, in each round, the number of distributed units equals the number of those who are asked to forward through some valid request. At equilibrium, this equals the minimum between (1) the number of those who (at the beginning of the round) have some neighbors who miss a unit that they have and (2) the number of those who miss some unit that some of their neighbors already have. Accordingly, the analysis now distinguishes between the fully connected case and the non-fully connected one.

19.5.1 Unconstrained Communication

Without (or with loose) communication constraints, at an equilibrium strategy profile A, it may well be that each content unit $c^t, 0 \le t \le T$ reaches 2^{k-1} *new* peers in each round $t + k, k = 1, \ldots, m$ and thus the whole population in $m + 1$ rounds, which is optimal in terms of streaming length. Still, such an A cannot be deterministic, as any equilibrium profile A must result in a congestion $\sigma_C^j(A) > 1$ for some (j, C) entries of the associated matrix $\sigma(A)$. In fact, each (greedy) peer $i \in N$ makes a valid request whenever possible.

Claim In the fully connected case, the upper bound for equilibrium streaming length is $T_* = T + 2^m$.

Proof Let $N_0^t \subseteq N_0, t \ge 0$ be the set $N_0^t = \{j \in N_0 : C_j^t \nsubseteq C_i^t \text{ for some } i \in N\}$ containing all those who at t have some content that someone else still misses. Also, let $N^t \subseteq N, t \ge 0$ be the set $N^t = \{i \in N : C_i^t \nsupseteq C_j^t \text{ for some } j \in N_0\}$ containing all those who at t still miss some unit. Note that $N = N^t, 0 \le t \le T$, because at t no peer has unit c^t yet, but the source $0 \in N_0$ does. Recall that players always prefer to make a valid request from a node where congestion is 1 (i.e., from which they are the only one asking to receive) rather than from one where congestion is > 1 (i.e., independently from what nonempty additional content may be obtained from these nodes). On the other hand, at equilibrium, each peer makes a valid request in each round as long as the whole content CON is not fully received. Hence,

$$\left| \left\{ j \in N_0^t : \sigma_{C^t}^j(A) > 0 \right\} \right| = \min \left\{ |N_0^t|, |N^t| \right\} \tag{19.3}$$

and, more specifically, $j \in N_0^t \Rightarrow \sigma_{C^t}^j(A) \in \left\{ \left\lfloor \dfrac{N^t}{N_0^t} \right\rfloor, \left\lceil \dfrac{N^t}{N_0^t} \right\rceil \right\}$ $\tag{19.4}$

for any node C^t that may be reached at any time $0 \le t \le T_*$ along some game course. This states precisely that in any round t, the number of forwarded (and received) units equals the minimum between the number of those who at t have a unit that at least one peer is still missing and the number of those who at t still miss some unit (see above). Accordingly, streaming length is maximized when $|N_0^t|$ is kept to its minimum at each t. Any of the units minimally takes $m + 1$ rounds to be distributed over the whole population. While the source puts units in circulation, the number of those who have one unit that someone else misses minimally increases by 1 in each round, till either all peers have a unit that someone else is missing or the content production ends, whatever first: $|N_0^t| \ge |N_0^{t-1}| + 1$ for $0 < t < \min\{2^m - 1, T\}$, with $N_0^0 = \{0\}$. Maximum duration occurs when this last inequality is an equality, requiring in turn that each unit, up to round

$t = \min\{2^m - 1, T\}$ inclusive, is streamed along a same *chain*: each unit is received by peers in a fixed order, such as the natural one $1 < 2 < \cdots < n$. If $T \le 2^m - 1$, then such a linearly ordered streaming induces, at the end of round T, the game tree node $C^{T+1} = (C_1^{T+1}, \ldots, C_n^{T+1})$ where $C_1^{T+1} = CON$ and $C_i^{T+1} = C_{i-1}^{T+1} \backslash c^{T-i+1}$ for $i = 2, \ldots, n$. That is, maintaining the example given by the natural order for the sake of simplicity, at $T + 1$, peer 1 has the whole content CON, while each peer $i = 2, \ldots, n$ misses the last $i - 1$ units $c^T, c^{T-1}, \ldots, c^{T-i+1}$. Now consider the whole streaming tree: it must (eventually) consist of $(T+1)n$ (directed) edges, out of which $T+1$ have the source as one end vertex, while $(T+1)(n-1)$ are P2P links. At game tree node C^{T+1}, the source has exhausted its role, and $\sum_{2 \le i \le n}(i - 1) = \binom{n}{2}$ P2P links are still missing. In particular, peer n, at time $T + 1$, still misses units $C^T, C^{T-1}, \ldots, C^{T-n+1}$. Accordingly, for getting the whole content, this peer needs further rounds $T + 1, T + 2, \ldots, T + n$, because in each round only one unit can be received. Then, the whole streaming length (in stages) is $T + n = T + 2^m$. The case $T > 2^m - 1$ is handled in the same fashion (see [26] for details). $\square$

Although worst-case equilibrium streaming length is linear in both the whole number $T + 1$ of produced units and the whole number $n = 2^m$ of peers, it can be rather greater than the socially optimal streaming length. This gap may be closed by means of a dynamic mechanism for constraining per-node strategies.

The general issue addressed through strategy restriction, which is also sometimes dealt with by introducing *mediators* in the routing system [28], arises most interestingly when players can choose more facilities [11]. Still, as our setting is intrinsically dynamic, it seems worth being addressed even when strategies are time sequences of singletons, as described earlier. In particular, the focus is on simple rules of the form: certain peers $i \in N$ at certain nodes $C \in \mathcal{N}$ cannot ask to receive from certain $j \in N_0$. Hence, restrictions apply to per-node strategies, which in any case (already) are singletons. In fact, the proposed model provides a monotone congestion game with no bad configuration, where equilibria are (generically) strong and Pareto-optimal. In this setting, adding per-node strategy restrictions has to deal with *symmetry* [11], but here such restrictions result anyhow: peers' per-node payoffs depend only on valid requests; hence, at any $C \in \mathcal{N}$ and for any $i, i' \in N, j, j' \in N_0$, it shall be $C_j \not\subseteq C_i \supseteq C_{j'}$ and $C_{j'} \not\subseteq C_{i'} \supseteq C_j$. That is, i can make a valid request to j but not to j', and the converse for i'. From a social planner's perspective, there are two priorities when designing restrictions: at equilibrium (with restrictions) streaming length should be $T + m + 1$, the same as with fastest streaming profiles, and congestion should be minimized. In this respect, a key fact is that those restrictions yielding minimal streaming length also minimize congestion. Indeed, in fastest streaming profiles, there is a whole central phase Ph2 where streaming occurs through one-to-one matching involving all peers both as senders and as receivers (apart from one peer, who never forwards, and the source, who never receives). Therefore, minimal (equilibrium) streaming duration also quantitatively translates into null congestion over phase Ph2.

Consider a strategy restriction mechanism that specifies from what $j \in N_0$ each peer $i \in N$ can ask for content at each node $C = C^t = (C_1^t, \ldots, C_n^t)$ the game may reach up to any time t. In other terms, the mechanism specifies for any node $C = (C_0, C_1, \ldots, C_n)$ and for any $i \in N, j \in N_0$ such that $C_j \not\subseteq C_i$, whether it may be $j = A^i(C)$ or not. In view of the above recurrence Re1–3, consider the following per-node restriction mechanism: for all $i \in N$ and $C^t \in \mathcal{N}$

Rm1: if $C_i^t \ni c^{t-k}$ for some $k \le m - 1$, then $A^i(C^t) \ne j$ for all $j \in N_0$ such that $C_j^t \ni c^{t-k'}$ for some $k' < m$

Rm2: if $C_i^t \not\ni c^{t-m}$, then $A^i(C^t) = j$ for some $j \in N_0$ such that $C_j^t \ni c^{t-m}$

In this way, for any reachable game tree node C^t, strategies for round t are constrained precisely in the manner established by constraint Re1–3 applying to streaming tree evolution. If, given previous history, a peer in t has some content unit c^{t-k} that must be forwarded in round $t + 1$ (i.e., such that $t - k + m \geq t + 1$), and then in this round t the peer cannot ask to receive from those $j \in N_0$ who in t have units $c^{t-k'}$ to be also forwarded in round $t + 1$ (i.e., such that $t - k' + m > t$). Still, note that while Re1–3 are stated from the perspective of an overall coordinator (i.e., specifying forwarders and receivers), this Rm1–2 is more broadly stated in terms of contents. Simple though it is, this mechanism enables the demand for new units to get synchronized (i.e., timely coordinated) and thereby to reach social optimality by means of selfish behavior.

Any profile $A \in N_0^{\kappa n}$ yields a probability distribution $p_{C^t}^A$ over $\mathcal{N}^{t+1}$ for each game tree node C^t that may be reached at any time $0 \leq t \leq T_*$. That is, $p_{C^t}^A(C^{t+1})$ is the probability of reaching node C^{t+1} from node C^t when chosen strategies are $A^i(C^t), i \in N$. Accordingly, $\sum_{C^{t+1} \in \mathcal{N}^{t+1}} p_{C^t}^A(C^{t+1}) = 1$ for all $t \geq 0, C^t \in \mathcal{N}^t$ and $A \in N_0^{\kappa n}$. For given underlying probabilistic model (anyhow handling multiple valid requests whenever there are, see above), any strategy profile $A \in N_0^{\kappa n}$ puts a probability p^A on each game course $\{C^0, C^1, \ldots, C^{T_*}\}$ (or sequence of content distributions) obtained as the following product

$$p^A(\{C^0, C^1, \ldots, C^{T_*}\}) = \prod_{t=0}^{T_*-1} p_{C^t}^A(C^{t+1}) \tag{19.5}$$

of conditional probabilities, where any tth round* starts at t and ends at $t + 1$. Let $\mathcal{C}^A$ denote the set of all game courses that may prevail with strictly positive probability through profile A. In other terms, $\mathcal{C}^A$ contains all $T_* + 1$ sequences $\{C^0, \ldots, C^{T_*}\}$ or game courses on which A puts strictly positive probability, that is, such that $p^A(\{C^0, \ldots, C^{T_*}\}) > 0$ as defined by (19.5).

Claim   If $A \in N_0^{\kappa n}$ is an equilibrium under Rm1–2, then the following two conditions hold for all $\{C^0, \ldots, C^{T_*}\} \in \mathcal{C}^A$:

(a) $\left|\{i \in N : c^t \in C_i^{t+k}\}\right| = 2^{k-1}$ for $k = 1, \ldots, m + 1$ and all $0 \leq t \leq T$,

(b) $\displaystyle\sum_{j \in N_0} \max\{0, \sigma_{C^t}^j(A) - 1\} = \begin{cases} 2^m - 2^{t+1} + 1 & \text{for } 0 \leq t < m \\ 0 & \text{for } m \leq t \leq T, t = T + m \\ 2^{t-T-1} & \text{for } T < t < T + m \end{cases}$

Proof   The demand for content units (to be further forwarded) is always provided by peers' utility maximization, at any node $C^t \in \{C^0, \ldots, C^{T_*}\} \in \mathcal{C}^A$. Under constraint Rm1–2, content demand at each node is convoyed toward valid requests, which, whenever satisfied, allow for minimal streaming length. On the supply side, peers (and the source) always satisfy precisely one (random) valid request among those received, and, thus, equilibrium conditions under Rm1–2 yields that any resulting game course distributes each content unit $c^t, 0 \leq t \leq T$ in the desired manner.

---

* If T_* is the maximum conceivable streaming length, expressed in rounds and starting with 0, and then a game course ends at time T_*, when round $T_* - 1$ ends, although the whole content shall generically be completely distributed much in advance.

Concerning (b), $\sigma^j_C(A) - 1$ is the number of *excess* valid requests for any pair (j, C), that is, the number of non-satisfied valid requests that $j \in N_0$ receives at node C. Accordingly, $\sum_{j \in N_0} \max\{0, \sigma^j_{C^t}(A) - 1\}$ measures the whole (i.e., aggregate) number of excess valid requests at any reached node C^t. Given (a), any game course in $\mathcal{C}^A$ provides a streaming with minimal duration and thus evolves through the same phases Ph1–3 as fastest streaming, but nondeterministically. In other words, there is congestion, but only in the initial and final phases (Ph1,3), and in order to measure it, we have to check how many times the above reasoning on fastest streaming profiles uses one-to-one matchings between sets of different cardinalities. More specifically, if cardinality is the same, then the problem of finding some one-to-one matching is solved by equilibrium condition (19.3).* Conversely, if cardinality is different, then some of those in excess shall be left out. The number of these latter, at node C^t, is precisely $\sum_{j \in N_0} \max\{0, \sigma^j_{C^t}(A) - 1\}$. In fact, in any initial round $t = 0, 1, \ldots, m-1$, if a peer receives a unit, then this peer will make no valid requests until round $t = m$ (i.e., the initial of Ph2).† Summing up, in initial rounds $t = 0, 1, \ldots, m-1$, the numbers of (restricted) excess valid requests, respectively, are $2^m - 2^0, 2^m - 2^0 - 2^1, \ldots, 1$, that is, $2^m - 2^{t+1} + 1$. Similarly, in rounds $t = T+1, T+2, \ldots, T+m-1$, the numbers of (restricted) excess valid requests, respectively, are $2^0, 2^1, \ldots, 2^{m-2}$, that is, 2^{t-T-1}. Finally, in round $T + m$, clearly there are no excess valid requests. $\qquad\qquad\qquad\qquad\qquad\qquad\qquad\qquad\qquad\qquad\qquad\qquad\qquad\quad\square$

Restriction mechanism Rm1–2 is useful for exploiting selfish behavior toward socially desirable outcomes: it is simple and, most importantly, specifies conditions only in terms of the generic node C^t that may be reached at some time t during game course. The pattern through which the system reaches this node is irrelevant; all that matters for strategy restriction is content distribution over peers, which is precisely captured by the node itself. Any outcome or game course constrained through the mechanism provides minimal streaming length and (consequently, given Ph2) also minimizes congestion.

19.5.2 Constrained Communication

Attention now turns to a (generic) graph or network $G = (N_0, E)$, with the broadcaster and peers as vertexes, and edges $\{i, j\} \in E \subsetneq N_0^{(2)}$ providing the available communication links for per-stage data transmission (see above). Notice immediately that, as a (exact) potential function exists anyhow (see (19.2)), the whole game-theoretic framework still provides existence of (pure strategy) equilibrium. Yet, when the given graph does not fulfill certain basic requirements, the whole situation becomes totally trivial: rules Ru1–4 evidently require much P2P exchange for distributing *each* content unit, and, hence, if the graph does not allow, in practice, for such an exchange, then there is no upper bound for game course length.

Hereafter, an edge or link or unordered pair $\{i, j\} \in E$ is denoted, more simply, by $ij = ji$. A $G' = (V, E')$ such that $V \subseteq N_0$ and $E' \subseteq E$ is a subgraph $G' \subseteq G$. A vertex subset $V \subseteq N_0$ spans subgraph $G(V) = (V, E(V))$, whose edge set $E(V) = E \cap V^{(2)}$ contains all (existing) edges both of whose end vertexes are in V. For $V, V' \subseteq N_0$, subgraph $G(V \setminus V')$ obtains by deleting from $G(V)$ all vertexes in V' together with all edges one or both of whose end vertexes are in V'. In view of the notation $E_i, i \in N_0$ introduced in Section 19.2, the degree $d_G(i)$ of a vertex i is simply

---

* When a set of senders and a set of receivers, both of same cardinality, have to match, at equilibrium receivers make their valid request each to a different sender.

† Otherwise, if a valid request was made and satisfied, then this peer would have two units to forward.

the number of its links: $d_G(i) = |E_i| - 1$. For $i, i' \in N_0$, a $i - i'$ path $P_{ii'} \subseteq G$ is a subgraph with vertex set $\{j_0, j_1, \ldots, j_k\}$ such that $j_0 = i, j_k = i'$, and edge set $\{j_0 j_1, j_1 j_2, \ldots, j_{k-1} j_k\}$. For $V \subseteq N_0$, spanned subgraph $G(V)$ is connected if there is an $i - j$ path $P_{ij} \subseteq G(V)$ for all pairs $i, j \in V^{(2)}$. More generally, the connectivity of $G(V)$ is the minimum cardinality $|V'|$ of a vertex subset $V' \subset V$ such that $G(V \setminus V')$ either is disconnected or has a singleton as vertex set [3, p. 3]. A spanning tree, in G, is a connected subgraph whose vertex set is N_0 and with size or number of edges n. If Θ denotes the set of all spanning trees available in G, then $\Theta \neq \emptyset \Leftrightarrow G$ is connected.

It seems rather evident that, apart from any equilibrium condition, Ru1–4 allows bounding streaming length only if the whole given graph $G = G(N_0)$ is connected. Otherwise, there is no chance of distributing *any single unit* over the whole peer set. In fact, for any two peers not connected by any path, if one receives a unit, then definitely the other one may not (through P2P exchange) receive that unit. Now, invoking the same argument, assume that $G(N_0 \setminus 0)$ is disconnected. Then, there is a partition* $\{S_1, \ldots, S_h\}$ of the peer set $N_0 \setminus 0 = N$ such that $h \geq 2$ and blocks bijectively correspond to $G(N)$ components (or maximal connected subgraphs). As the broadcaster sends each unit precisely once, to only one peer, this means that each unit may be sent to only one block $\hat{S} \in \{S_1, \ldots, S_h\}$, and next constrained P2P exchange will prevent this unit from reaching any other block. Hence, connectedness of both G and $G(N)$ is a necessary condition for finiteness of streaming length.

When considering sufficiency, equilibrium conditions must also be crucially taken into account, requiring further tools. A bipartite graph $\Gamma = (V_1 \dot\cup V_2, W)$ has vertex set given by the union $\dot\cup$ of two disjoint subsets V_1, V_2, and edge set $W \subseteq V_1 \times V_2$ such that each edge has one end vertex in V_1 and the other one in V_2. In particular, for our purposes, edges may well be directed (or oriented) or *ordered* pairs $(v_1, v_2) \in V_1 \times V_2$, that is, going from one vertex in V_1 to one vertex in V_2. Now let $N_0^t(G)$ denote the set of those $j \in N_0$ who, at the beginning of any stage $t \geq 0$, have at least one neighbor missing some unit that they already have, that is, $N_0^t(G) = \{j \in N_0 : C_j^t \not\subseteq C_i^t \text{ for some } i \in E_j \setminus j\}$. Analogously, let $N^t(G)$ denote the set of those (peers only) $i \in N$ who, at the beginning of the stage, have at least one neighbor with some unit that they still miss, that is, $N^t(G) = \{i \in N : C_i^t \not\supseteq C_j^t \text{ for some } j \in E_i \setminus i\}$. Here again, recall that players always prefer to make a valid request from a neighbor where, given the other peers' requests to receive, congestion is lower. Hence, in particular, they prefer to make a valid request from a neighbor where congestion is 1 (i.e., from which they are the only one asking to receive) rather than from a neighbor where congestion is > 1 (i.e., independently from what nonempty additional content may be obtained from such neighbors). Also, at equilibrium each peer makes a valid request in each stage where the given graph $G = (N_0, E)$ allows for, until the whole content CON is not fully received.

Equilibrium P2P content exchange, at a generic stage t, can be investigated by means of a bipartite graph $\Gamma = \Gamma_{G,C^t} = (N_0^t(G) \dot\cup N^t(G), W)$ with edge set $W = \{(i, j) \in N_0^t \times N^t : ij \in E, C_i^t \not\supseteq C_j^t\}$, where $ij = ji$ (see above). Hence, in Γ a peer $i \in N$ has associated two distinct vertexes $v_1^i \in N_0^t, v_2^i \in N_t$ whenever $i \in N_0^t, N^t$. In fact, peers are players when deciding from what neighbor to ask for additional content, while they are facilities when receiving requests to forward. On the other hand, edge set W contains all one-to-one P2P content exchange possibilities that graph G and game tree node C^t allow for. A vertex cover in a graph is a set of vertexes that includes at least one endpoint of each edge, and a vertex cover is minimum if no other vertex cover has fewer vertexes. A matching in a graph is a set of edges no two of which share an endpoint, and a matching is maximum if no other matching has more edges [3, ch. 2]. In a bipartite graph,

---

* A *partition* of a set X is a collection of nonempty and pair-wise disjoint subsets of X whose union yields X; its elements are called *blocks* [2].

the number of edges in a maximum matching equals the number of vertexes in a minimum vertex cover. Let $\#\Gamma_{G,C^t}$ denote this number for bipartite graph Γ_{G,C^t}. Then, at any equilibrium strategy profile A,

$$\left| \left\{ j \in N_0^t(G) : \sigma_{C^t}^j(A) > 0 \right\} \right| = \#\Gamma_{G,C^t}. \tag{19.6}$$

This is the analog of expression (19.3), but when communication constraints are taken into account. In words, at any equilibrium, the number of vertexes (either peers or the broadcaster) who receive a valid request to forward (which equals the number of peers who receive a unit), over any stage $t \geq 0$, equals the cardinality $\#\Gamma_{G,C^t}$ of a maximum matching in Γ_{G,C^t}. If both G and $G(N)$ are connected, then $\#\Gamma_{G,C^t} \geq 1$ until the whole content is not fully distributed (i.e., until $N_0^t \neq \emptyset \neq N^t$), which guarantees that such a (necessary) condition is also sufficient for finiteness of equilibrium streaming length. Still, for generic connected G and $G(N)$, providing a tight upper bound is far more demanding.

Conversely, the longest conceivable equilibrium streaming length, over *all* possible connected graphs $G = (N_0, E)$ such that $G(N)$ is also connected, may be checked to attain when $d_G(i_*) = n$ for one peer $i_* \in N$ while $d_G(j) = 1$ for all $j \in N_0 \backslash i_*$. This means that peer i_* is directly linked to the broadcaster as well as to every other peer, and there is no other edge apart from these n ones. In fact, in this case, at any equilibrium with $T + 1$ units to distribute, streaming length is precisely $1 + (n - 1)(T + 1)$. To see this, firstly note that the given graph allows for a unique spanning tree through which a (i.e., any) unit may reach everybody. Specifically, the unit has to firstly go from 0 to i_*, and next all remaining links needed for P2P exchange have i_* for sender and some $j \in N \backslash i_*$ for receiver. Hence, these links must take place each over a different stage, entailing that we need an initial stage (when the first unit reaches i_* from 0) and next additional $(n - 1)(T + 1)$ stages.

Concerning minimal equilibrium streaming length, our previous results lead to ask whether any given connected G (and $G(N)$) allows for fastest streaming or not. Specifically, the issue is whether G includes at least one subgraph displaying certain features or not. Again, let $n = 2^m$ for some natural m and consider a generic fastest streaming tree evolution up to stage $2m - 1 \rightarrow 2m$ included, that is, only over the initial phase Ph1 and the first m stages of the central phase Ph2 (see above). If the given graph allows for any of the $\prod_{t=0}^{m-1} [2^m - 2^t + 1]_{2^t} \left(2^{m-1}! \right)^{2m}$ different fastest streaming tree evolutions over rounds from 0 to $2m - 1$ included, then in the starting round $2m \rightarrow 2m + 1$ we can make again repeat the same scheme initiated from the beginning and thereby continue indefinitely. More precisely, for any connected $G = (N_0, E)$ and $i, j \in N_0^{(2)}$, let $d_G(i, j)$ denote the distance or length of a shortest path between i and j. For $h \in N_0, k \in \mathbb{N}$, let $V_h^k = \{i \in N_0 : d_G(h, i) = k\}$ denote the set of vertexes at distance k from vertex h. Fastest streaming tree evolution, for the first $2m$ rounds, requires a set of *discrete isoperimetric inequalities* [15] to be satisfied. Perhaps, the most basic such an inequality states the following: given a graph G, for any vertex subset $V \subset N_0$ such that $|V| = h$, the number of vertexes at distance no more than k from some vertex in V is no smaller than $g(h, k) = g_G(h, k) \in \mathbb{N}$. Along this line, here we need to ask for more specific conditions. In particular, looking at the requirements that fastest streaming casts on G in the very first stages, the broadcaster must be able, minimally, to send the first m units each to a different peer.* That is, there must be an m subset $S_0 \subseteq V_0^1, S_0 = \{i_1^0, \ldots, i_m^0\}$ of the set

---

* In fact, the source can use again these m distinct peers to send units $c^m, c^{m+1}, \ldots, c^{2m-1}$, but still these stages $m, m+1, \ldots, 2m-1$ require many P2P links whose existence is not needed in initial stages $0, 1, \ldots, m-1$. This is precisely why existence or nonexistence of the sought subgraph structure can only be checked by considering (minimally) the first $2m$ stages.

V_0^1 of vertexes at distance 1 from the source (i.e., 0). Each of them must be able, in turn, to send the unit received from the broadcaster to $m-1$ different *new* peers. Hence, for each $i_h^0 \in S_0$, there must be an $m-1$ subset $S_{i_h^0} \subseteq V_{i_h^0}^1, S_{i_h^0} = \{i_1^{i_h^0}, \ldots, i_{m-1}^{i_h^0}\}$ of i_h^0's neighbors. In particular, these neighbor subsets must be pair-wise disjoint, that is, $S_{i_h^0} \cap S_{i_{h'}^0} = \emptyset, 1 \leq h < h' \leq m$, while also having empty intersection with S_0, that is, $S_{i_h^0} \cap S_0 = \emptyset, 1 \leq h \leq m$. And so on, until all one-to-one links required by fastest streaming up to the $2m$th stage inclusive are checked to be available in the given G. Although simple, at least in principle, such a search is clearly computationally rather demanding.

An example may further clarify the issue: let $N_0 = \{0, 1, 2, 3, 4\}$, so that $n = 2^2$, and next let edge set $E \subset N_0^{(2)}$ be $E = \{01, 02, 12, 13, 23, 34\}$. Although $G(N)$ (and thus, *a fortiori*, G) is non-fully connected, still it admits fastest streaming. To see this, let $i \xrightarrow[k \to k+1]{c^t} j$ indicate that in stage $k \to k+1$ unit c^t goes from $i \in N_0$ to $j \in N$. Now consider the following evolution for $t = 0, 1, 2, 3, 4$

Stage 0 : $0 \xrightarrow[0 \to 1]{c^0} 1$

Stage 1 : $0 \xrightarrow[1 \to 2]{c^1} 2$ and $1 \xrightarrow[1 \to 2]{c^0} 3$

Stage 2 : $0 \xrightarrow[2 \to 3]{c^2} 1$ and $2 \xrightarrow[2 \to 3]{c^1} 3$ as well as $1 \xrightarrow[2 \to 3]{c^0} 2$ and $3 \xrightarrow[2 \to 3]{c^0} 4$

Stage 3 : $0 \xrightarrow[3 \to 4]{c^3} 2$ and $1 \xrightarrow[3 \to 4]{c^2} 3$ as well as $2 \xrightarrow[3 \to 4]{c^1} 1$ and $3 \xrightarrow[3 \to 4]{c^1} 4$

Afterwards, we can repeat this same scheme indefinitely through time, for any number of content units. The only available further fastest streaming tree evolution is simply the same but starts with the source 0 sending unit c^0 to 2 rather than to 1, and then again repeating the same scheme for all (pairs of) units. In both cases, 3 receives each unit c^t with unitary delay (i.e., over stage $t + 1 \to t + 2$), while 4 receives each unit with delay 2 (i.e., 4 is the peer who never forwards any unit, which is in fact needed in any fastest streaming evolution; see above). Finally, both 1 and 2 receive each unit either with no delay (i.e., directly from the source), or with delay 2, alternating the two delays one stage after the other. If the number of units is large enough (and, in particular, even), then these two fastest streaming equilibria are (payoff) equivalent for all the four players. Hence, from the perspective of a social planner (possibly with lexicographic preferences [17] giving priority to streaming length minimization, and next considering fairness issues), a fair protocol makes the source (randomly) choose who is the first (either 1 or 2) to receive c^0 and next proceeds precisely in the manner just detailed.

It seems worth emphasizing that in this example if we add restriction mechanism rules Rm1–2 to communication constraints as given by G, then equilibrium outcomes not only provide minimal streaming length but also yield no congestion at all in any stage apart from the first one, when both 1 and 2 ask to receive from the broadcaster. In all remaining stages, equilibrium with both communication and Rm1–2 restrictions provides no congestion, as any $i \in N_0$ with some unit that some neighbor still misses receives precisely one request to forward. Such a situation is clearly a very lucky one,* in that fastest streaming is achievable and, in addition, the conjunction of both Rm1–2

---

* It is not difficult to construct examples where the conjunction of Rm1–2 and communication constraints may well yield, in the worst case, very bad outcomes in terms of equilibrium streaming length.

and communication constraints avoids the congestion characterizing equilibrium fastest streaming in phases Ph1 and Ph3 for the fully connected case (see above), thereby most efficiently resolving the crucial coordination issue raised by P2P live streaming. Thus, detecting whether the given graph allows or not to exploit the conjunction of both Rm1–2 and communication constraints seems quite important. Evidently, adding further edges in the proposed example, say 03 and 24, provides many more possible fastest streaming tree evolutions, making again nontrivial how to optimally schedule P2P content exchange.

This leads to conclude that, even when there was available a rapid method for finding (if any) a subgraph structure allowing for the first $2m$ stages of fastest streaming evolution, still what is optimal with constrained communication would remain an issue. In fact, if one such a subgraph structure is found, then repeating it indefinitely throughout the whole streaming, as proposed above, is not as fair, in general, as the fully connected case, where senders always have large sets of potential receivers to randomly choose from. The reason for this, of course, is that here the choice is only among neighbors. Accordingly, further improvements, toward more fairness, could be searched for. Technically, this means checking whether the given graph includes further (i.e., richer) subgraph structures, allowing to depart from repeating always the same scheme found available for the first $2m$ stages (see the above example). Conversely, if G is found *not* to admit any sought subgraph structure, then one must determine the actual minimal streaming length, and then check how fairly (i.e., in how many different ways) it can be achieved, which seems rather hard.

19.6 Simulation Design and Findings

Our simulations address different issues depending on whether the given network $G = (N_0, E)$ is fully connected or not. In the former case, the target is quantifying the gain from Rm1–2 with respect to generic equilibrium outcomes. In the latter case, the aim is to measure how equilibrium outcomes deteriorate as the size (or number $|E|$ of edges) of G decreases. For the fully connected situation, the overall idea is to quantify the inefficiency costs due to coordination failure. Put it differently, the general aim is at comparing socially optimal outcomes with equilibrium ones. Of course, such a comparison can be made from different perspectives, and quantified in alternative ways. In a game, where players' utilities (or payoffs) are fully specified, the *social utility* associated with any strategy profile is simply the sum over players of the values taken by their utility on that profile. Two common measures of equilibrium inefficiency are the *price of anarchy* and that *of stability*, which are commonly defined for typical routing games [4,5] (see above), where players selfishly strive to *minimize* a cost function [23, ch. 17–18]. The former compares the *best* value taken by social utility over all strategy profiles, with its *worst* value over all equilibrium profiles. The latter compares the best value taken by social utility over all strategy profiles, with its *best* value over all equilibrium profile. These concepts also apply to games with positive payoffs [18, footnote 2] and allow for mixed strategies [17,24]. Albeit it is customary to "quantify the inefficiency of the worst or the best equilibrium of a game, a third interesting approach is to analyze a *typical* equilibrium. Such *average-case analyses* are notoriously difficult to define in a meaningful and analytically tractable way, however, and this approach has not yet been used successfully to study the inefficiency of equilibria" [23, p. 446].

Now, rather than on a specific congestion game, the focus here on congestion game forms and thus on the whole class of monotone games derived from them [11]. Thus, the improvement induced by Rm1–2 over a generic equilibrium varies as players' payoff specification varies, but in any case the sooner peers receive all content units, the better. In particular, in our framework, socially optimal strategy profiles are fastest streaming ones, which display no congestion at all (being deterministic), and thus are not sustainable at equilibrium. On the other hand, among equilibrium strategy profiles, the best ones are those that manage to distribute the whole content over the whole population in a minimum number of stages, although naturally displaying some congestion in phases Ph1 and Ph3. Hence, strictly speaking, the price of stability here is only due to congestion (which characterizes best equilibrium outcomes but not fastest streaming ones), and thereby quite low. As for the price of anarchy, the worst-case equilibrium length has been determined analytically as well.

Accordingly, what becomes more interesting, in our view, is quantifying the gain, at equilibrium, obtained by introducing Rm1–2. More precisely, this latter restriction mechanism is a powerful and simple instrument for coordination or, equivalently, an *equilibrium selection method*: although possible, it is very unlikely that an equilibrium strategy profile picked at random provides minimal streaming length. Conversely, with Rm1–2, *any* equilibrium provides minimal streaming length. The purpose of our first slot of simulations is precisely the comparison between average equilibrium outcomes with constraints Rm1–2 and average equilibrium outcomes without any constraint. Still, as our analytical results already show that Rm1–2 always provide minimal streaming length, the sought comparison is most fruitfully made in terms of *average latency*. This is the positive real quantity $\Lambda \in \mathbb{R}_+$ such that, by picking at random both a peer and a unit c^t, the former receives the latter, on average, over time period $t + \Lambda \to t + \Lambda + 1$. In order to have information about fairness as well, we also plot average latency standard deviation as a measure of dispersion.

In the non-fully connected case, our concern is different. Given that determining whether Rm1–2 may be successfully superimposed or not (and, if yes, how fairly) is hard, we only consider generic equilibrium outcomes. In particular, we check how equilibrium performance (in terms of average latency and its standard deviation) deteriorates in random graphs with decreasing size.

19.6.1 Unconstrained Communication

In order to simulate a generic equilibrium outcome, one basically has to make sure that condition (19.3) is fulfilled at each stage, while also randomizing insofar as possible when matching senders and receivers. Specifically, at the beginning $t \geq 0$ of each round $t \to t + 1$, the simulation model inherits a content distribution or game tree node $C^t \in \mathcal{N}$ from the past. In turn, this identifies the sets N_0^t (of those who have some content that someone else still misses) and N^t (of those peers who miss some content that someone else already has) appearing in (19.3). The elements in these two sets are iteratively considered in a random order, that is, $N_0^t = \{i_1^t, \ldots, i_K^t\}$ as well as $N^t = \{j_1^t, \ldots, j_H^t\}$. In particular, for $k = 1, \ldots, K$, we select the smallest $h \in \{1, \ldots, H\}$ such that $C_{j_h}^t \not\supseteq C_{i_k}^t$, and set $C_{j_h}^{t+1} = C_{j_h}^t \cup \{c^{t'} \in C_{i_k}^t \setminus C_{j_h}^t : t' < t'' \text{ for all } c^{t''} \in C_{i_k}^t \setminus C_{j_h}^t\}$. In words, i_k sends to j_h the oldest unit in $C_{i_k}^t \setminus C_{j_h}^t$. Before doing the same for $k + 1$, we remove i_k from N_0^t and j_h from N^t. We stop at the $\min\{K, H\}$th iteration, that is, either at the Kth one, if there are available more receivers than senders, or at the Hth one, if there are available more senders than receivers. For all peers $j \in N$ not involved as receivers in any of the $\min\{K, H\}$ iterations, we simply set $C_j^{t+1} = C_j^t$. In this way, we define the novel game tree node C^{t+1} where to start the whole procedure again, ensuring a random streaming tree evolution with (19.3) being satisfied at each stage.

In order to implement a random equilibrium game course under Rm1–2 or, equivalently, in order to achieve minimal streaming length, when the number of peers is a power of 2, we have developed an algorithm, hereinafter referred to as ConGaS (*Con*gestion *Ga*mes for *S*treaming). If peers must be able to coordinate themselves toward fastest streaming, then they must be constantly endowed with perfect information. That is, they must be able to "see," during each round, who sends ongoing content units to who. Technically, peers share a seed to randomly generate same sequences of pseudo-random numbers. Such shared seed serves as the needed coordination mean among nodes (i.e., it is employed to randomly select those peers that receive any given content unit). During the initialization, the broadcaster (i.e., node 0) sends to all peers a generated seed value. The distribution loop consists of an iterative behavior: each iteration t corresponds to the production, at the broadcaster, of a novel content unit c^t to be distributed. Meanwhile, ongoing content units $c^{t-k}, k = 1, \ldots, m$ not yet delivered to all peers are disseminated according to our method. Correspondingly, a per-stage distribution procedure, executed by all nodes, defines who sends what content unit to who. Differences in peers' *actions* are simply determined according to their *ids* $1, \ldots, n$.

At each iteration t, each peer is selected to receive a (single, new) content unit. This is achieved by picking nodes from an auxiliary list, which is initialized to N, and then progressively emptied through different calls of distribution procedure. Specifically, given a content unit c^k being distributed, $t - m \leq k \leq t$, we determine a bijection between those 2^{k-t+m} who have the unit and a corresponding random 2^{k-t+m}-cardinal subset of those who do not and are available according to Rm1–2. Thus, for any content unit, at each hth stage of distribution, 2^h nodes have the content unit and are selected to be the receivers. Once the distribution for the ongoing content units is specified, a new content unit is produced at the broadcaster and the distribution procedure is called for the novel content unit.

It seems worth emphasizing that in this whole reasoning applying to the fully connected case, what appear to be "bad" outcomes are nevertheless equilibrium ones, thereby realizing a great deal of coordination anyhow. More precisely, although ConGaS provides outcomes outperforming average equilibrium ones (in a monotone fashion as the number of content units and/or peers increases), still this latter term of comparison already performs rather good because of equilibrium conditions. In practice, assuming condition (19.3) to be satisfied over all stages (until content distribution is not complete) entails much coordination among peers. In fact, these average equilibrium outcomes outperform, in turn, typical gossiping protocols and other unstructured mechanisms.

Figure 19.1 shows (on the vertical axis) the average latency, measured in stages, experienced by peers in receiving content units. In all our experiments reported below, we maintain a fixed number $n = 2^8$ of peers. The horizontal axis measures the stream size or number $|CON| = T + 1$ of content units to be distributed. For each value of $|CON| = 1, 2, \ldots, 10^4$, we run both ConGaS (straight line below) and a generic equilibrium outcome (dots above), making a total of 2×10^4 simulation runs. As predicted by our former analysis, ConGaS obtains steadily very good results, much better than average equilibrium ones, and the greater the gap, the larger the size $|CON|$.*

Another index we consider has to deal with fairness, in that we compute the variance across peers of the average latency (over all units) they experience during the whole streaming. In fact, Figure 19.2 displays the (average) standard deviation of latency, which is much more limited with ConGaS rather than in average equilibrium outcomes. This is an important result for the implementation of viable live streaming systems, as it corresponds to a jitter reduction, which is a main requirement in most multimedia systems.

---

* Other experiments, not reported here for reasons of space, show that the gain of coordination provided by ConGaS with respect to average equilibrium outcomes also monotonically (and rapidly) increases as the number n of peers gets larger.

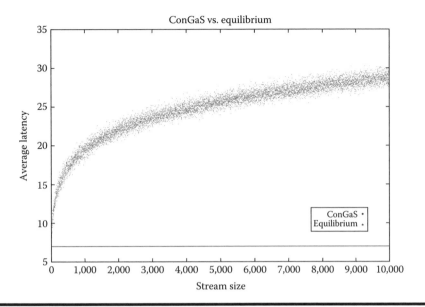

Figure 19.1 ConGaS vs. unconstrained equilibrium: average latency.

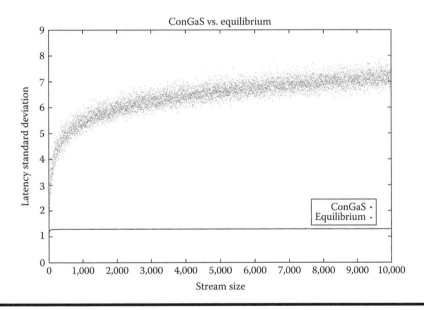

Figure 19.2 ConGaS vs. unconstrained equilibrium: latency standard deviation.

19.6.2 Constrained Communication

Commonly, $G(n, p)$ denotes the random graph [31, p. 13] with n vertexes and where for each of the $\binom{n}{2}$ unordered pairs the probability that the pair is an existing edge is precisely $p \in [0, 1]$. In order to simulate equilibrium outcomes in the non-fully connected case, for each value of $p = 1 - k/10^2, k = 0, 1, 2, \ldots, 99$, we generate 30 random graphs $G(n + 1, p)$ and use them as the given ad hoc network $G = (N_0, E)$ constraining per-stage data transmission. For each of

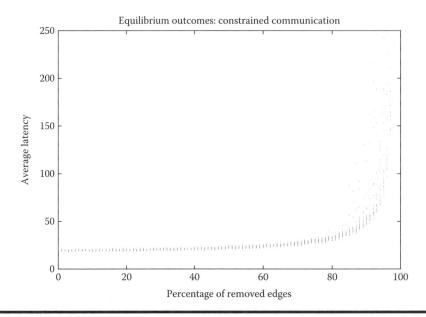

Figure 19.3 Graph-restricted equilibrium: average latency.

the 3000 networks thus generated (and checked to be connected with $G(N)$ also connected), we consider one single run with condition (19.6) (randomly) satisfied at each stage, until content distribution is not complete. In particular, the search for maximum matchings in bipartite graphs has a long tradition, with alternative efficient algorithms available [4,9,30]. Here again, like in the fully connected case, at each stage, for each randomly found maximal matching, each matched sender forwards to the corresponding receiver the oldest unit that the former has while the latter has not. In all 3000 runs, stream size $|CON| = 2^{10}$ is maintained fixed.

Figure 19.3 shows the average latency, measured in stages, experienced by peers in receiving content units, while Figure 19.4 displays latency (average) standard deviation. Clearly enough, they both increase as p decreases, but this occurs in a seemingly peculiar fashion, more or less steadily and almost negligibly until p reaches approximately 0.2 (i.e., 80% of removed edges), where the raise suddenly becomes huge. The only plausible explanation seems that when p approaches such values the resulting graphs most likely display, locally, the worst-case bottleneck situation already described. That is, the randomly generated G is such that for some peers $i \in N$ we have $d_G(i) > k$ or, equivalently, $|E_i| > k + 1$, but $d_G(j_1) = \cdots = d_G(j_k) = 1$ for a k subset $\{j_1, \ldots, j_k\} \subset E_i \backslash i$ of i's neighbors.

19.7 Open Issues and Developments

The main novelty of this work is modeling P2P real-time streaming in multistage congestion game form: in each stage, peers aim at receiving some additional content unit, until all units get received. As each peer receives only one unit, at most, per stage, and also forwards precisely one (randomly selected) unit whenever receiving a valid request, our setting is crucially constrained in terms of streaming length (or duration), intended as the number of stages needed to spread the whole content over the whole population. Content distribution over peers changes in time, and for any

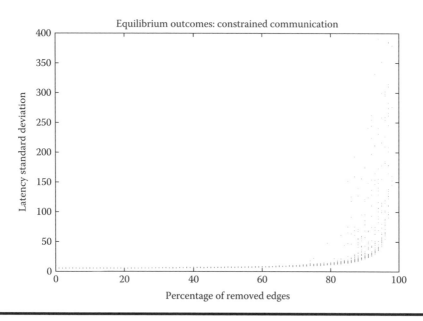

Figure 19.4   Graph-restricted equilibrium: latency standard deviation.

such a distribution (or game tree node) prevailing at any stage, the associated peers' per-stage payoffs depend on how many requests to forward are addressed to each peer and to the source. The idea is that an underlying and unknown probabilistic model handles all those cases where two or more peers make a valid request to receive from a common other peer or from the source: however, these cases are handled, the greater the number of peers who all (validly) ask to receive from a common other peer or from the source (i.e., the higher congestion), the less likely it becomes, for each of them, to be precisely the one who actually receives. For a simple (monotone) payoff specification, an exact potential is provided, any of whose maximizers is a pure-strategy equilibrium. In particular, strategies become sequences of singletons: for each peer a strategy specifies, for any (per stage) content distribution over the whole population, who to send a request for additional content. With these singleton strategies, the potential is strong: any of its maximizers is a strong (and Pareto-optimal) equilibrium. Wireless communication may be framed into this game-theoretic model by means of a (fixed) network, intended to result from peer's and the broadcaster's different positions in a spatial region. Then, per-stage data transmission is assumed to be possible only between nodes at unitary hop distance in the network. It may be worth mentioning, once again, that if peers and the source satisfy, at each stage, precisely one of the received valid requests to forward, then of course free riding (intended as non-forwarding content) is not a feasible strategy here. In fact, the proposed P2P exchange protocols build different substreaming trees specifying how different units get routed through the whole population, and, thus, any reputation-based mechanism that makes free riding a dominated strategy in tree-based multicast systems is applicable [23, ch. 23]. In particular, in our setting in order to detect deviations, at the end of each stage, all peers must know who received at least one valid request but did not satisfy any. Then this information enables to apply tit-for-tat policies in the following stages.

For the fully connected case, here conceived for wired settings, we show that if the number $n = 2^m$ of peers is a power of 2, then P2P content exchange can be scheduled so that *each unit* reaches everybody in $m + 1$ (consecutive) stages, with corresponding whole duration $T + m + 1$.

Thus, for any (monotone) payoff specification, this scheduling is socially optimal, allowing to spread each unit in the minimum conceivable number of stages. Next, we firstly investigate worst-case equilibrium outcomes and secondly provide a strategy restriction mechanism which, for any (per stage) content distribution over peers, at equilibrium selects one *feasible* forwarder–receiver matching. Although simple, this mechanism is very useful: it impedes the satisfaction of all those requests submitted by selfish peers at equilibrium which have to be ignored in order to achieve minimal duration. In fact, with these strategy restrictions, any equilibrium provides minimum streaming length, while also (and consequently, see above) minimizing congestion. These selected equilibria are compared with average equilibria through simulations.

For the non-fully connected case, here conceived for wireless settings, we show that the strategy restriction mechanism cannot be successfully applied, in general, and, thereby, we concentrate on average equilibrium outcomes, which are those performing a maximum matching, at each stage, between those with some neighbor missing some content that they have on one side and those with some neighbor having some content that they still miss on the other. Here simulations allow to observe how average equilibrium performance deteriorates as the size of the ad hoc network decreases.

In this novel setting, there are many open issues and possible developments. Firstly, for the fully connected case, given the realized gain of coordination due to Rm1–2, it seems worth trying to extend this mechanism so to allow for (i) a generic number of peers and (ii) a varying number of peers. As for (i), if the number n of peers is not a power of 2, then one may distribute the content by partitioning the population into blocks, each of which with cardinality equal to some power of 2. Specifically, blocks' cardinalities are $n_1 = 2^{\lfloor \log(n) \rfloor}$, $n_2 = 2^{\lfloor \log(n-n_1) \rfloor}, \ldots,$ $n_k = 2^{\lfloor \log(n-n_1-\cdots-n_{k-1}) \rfloor}$. Then, under rules Ru1–3, the source may keep on serving the n_1-cardinal block, and the peer who, in this block, is never required to forward (but receives each unit with latency $\lfloor \log(n) \rfloor$) acts as the source for the (next) block with cardinality n_2, and so on, so that latency adds through blocks of (decreasing) cardinalities $n_1 \leq n_2 \leq \cdots \leq n_k$. Turning to (ii), allowing peers to leave the system as well as new ones to join it, at any time, clearly has great impact on the functioning of Rm1–2. From a system design perspective, the issue might be tackled, again, by keeping the peer population constantly partitioned into groups. Then, additional restriction rules should govern exchanges between blocks as well as allocation of *new entries*, to be performed at the beginning of each stage, depending on how many peers left each block in the previous stage.

Perhaps, the most challenging issue, applying to both the fully and non-fully connected cases, is how to efficiently implement equilibrium (possibly restricted through Rm1–2 whenever possible). More precisely, as already outlined, equilibrium conditions require a great deal of coordination among peers. The issue is not how to endow players with perfect information, but more effectively how to make them actually choose the right equilibrium strategy. In other terms, there exist equilibria from which deviations are not profitable, neither individually nor at a group level. Clearly enough, players will stay with such strategy profiles whenever suitably directed. That is to say, they would most likely delegate their strategy choice to some mediator who performed a random maximum matching at each stage if everybody did so [27,28]. Hence, the issue is how to efficiently perform perfect matching (possibly satisfying Rm1–2 whenever possible) in applicative (mobile) scenarios. In such settings, there is typically a trade-off, which seems mostly harsh here, between figuring out what is optimal on the one side and concretely implementing near-optimal outcomes on the other. More precisely, in a relatively large MANET where nodes change their positions and may also leave and enter the network at any time, what P2P content exchange is actually optimal seems hard to say. Even when all players cooperate thereby allowing mediators to fully act as coordinators, still one cannot centralize optimization but rather has to locally provide near-optimal P2P content exchange scheduling. This leads naturally to envision typical super-peer-based architectures. Along another line of development, the hypothesis of perfect information maybe relaxed by assuming that peers only have "local" or partial information, thus leading to the issue of learning equilibrium [29,32].

From another perspective, there also appears a whole set of other issues more strictly related to exploring the ad hoc network topology. In particular, as already observed, given the gain of coordination realizable through Rm1–2, it seems worth investigating how to efficiently determine whether the network allows for (at least one) fastest streaming tree evolution or not. If not, it then becomes interesting to determine the maximal streaming length, at equilibrium (that is, with a maximum matching between senders and receivers at each stage), for a generic given network.

Another issue concerning the non-fully connected case is fairness. In the fully connected case (assuming, at a higher level of abstraction, that all links are identical in terms of transmission cost), a fair protocol may be seen as one that evenly distributes average latency across peers. Conversely, in the non-fully connected situation, fairness crucially has to deal with peers' different positions in a spatial region, in that those with more central positions and closer to the source are naturally privileged. Therefore, a formal definition of fairness in this contest seems desirable.

Finally, concerning the problem of determining whether a given network allows for some (at least one) fastest streaming tree evolution, it might be interesting to tentatively tackle it in terms of multiprocessor scheduling problems.

References

1. M. Afergan and R. Sami, Using repeated games to design incentive-based routing systems, In *Proceedings of INFOCOM06*, Barcelona, Spain, 2006.
2. M. Aigner, *Combinatorial Theory*, Springer, New York, 1979.
3. B. Bollobás, *Extremal Graph Theory*, Academic Press, New York, 1978.
4. M. Charikar, H. Karloff, C. Mathieu, J. S. Naor, and M. Saks, Online multicast with egalitarian cost sharing, In *Proceedings of 20th Symposium on Parallelism in Algorithms and Architectures*, Munich, Germany, pp. 70–76, 2008.
5. C. Chekuri, J. Chuzhoy, L. Lewin-Eytan, J. S. Naor, and A. Orda, Non-cooperative multicast and facility location games, *IEEE Journal on Selected Areas in Communications*, 25(6):1193–1206, 2007.
6. Y. R. Choe, D. L. Schuff, J. M. Dyaberi, and V. S. Pai, Improving VoD server efficiency with bittorrent, In *Proceedings of 15th Conference on Multimedia*, Augusburg, Germany, 2007.
7. Y. Chu, S. G. Rao, and H. Zhang, A case for end system multicast, In *Proceedings of SIGMETRICS* Santa Clara, CA, 2000.
8. A. Epstein, M. Feldman, and Y. Mansour, Strong equilibrium in cost sharing connection games, *Games and Economic Behavior*, 67(1):51–68, 2008.
9. Z. Galil, Efficient algorithms for finding maximum matching in graphs, *ACM Computing Surveys*, 18(1):23–38, 1986.
10. M. Hańćkowiak, M. Karoński, and A. Panconesi, On the distributed complexity of computing maximal matchings, In *Proceedings of SODA 98, the Ninth Annual ACM–SIAM Symposium on Discrete Algorithms*, pp. 219–225, 1997.
11. R. Holzman and N. Law-Yone, Strong equilibrium in congestion games, *Games and Economic Behavior*, 21:85–101, 1997.
12. W. Jiang, X. Liao, H. Jin, and Z. Yuan, WiMA: A novel wireless multicast agent mechanism for live streaming system, In *Proceedings of International Conference on Convergence on Information Technology*, Gyeongju, South Korea, pp. 2467–2472, 2007.

13. I. Keidar, R. Melamed, and A. Orda, Equicast: Scalable multicast with selfish users, In *Proceedings of PODC06*, Denver, CO, pp. 63–71, ACM, 2006.

14. H. Konishi, M. Le Breton, and S. Weber, Equilibrium in a model with partial rivalry, *Journal of Economic Theory*, 72:225–237, 1997.

15. I. Leader, Discrete isoperimetric inequalities, *Proceedings of the Symposia in Applied Mathematics*, 44:57–80, 1991.

16. H. C. Li, A. Clement, E. L. Wong, J. Napper, I. Roy, L. Alvisi, and M. Dahlin, Bar gossip, In *Proceedings of OSDI06*, Seattle, WA, pp. 191–204, 2006.

17. A. Mas-Colell, M. D. Whinston, and J. R. Green, *Microeconomic Theory*, Oxford University Press, Oxford, U.K., 1995.

18. D. Monderer, Solution-based congestion games, *Advances in Mathematical Economics*, 8:397–409, 2006.

19. D. Monderer and L. S. Shapley, Potential games, *Games and Economic Behavior*, 14:124–143, 1996.

20. T. Moscibroda, S. Schmid, and R. Wattenhofer, On the topologies formed by selfish peers, In *Proceedings of PODC06*, Denver, CO, pp. 133–142, 2006.

21. V. Navda, A. Kashyap, S. Ganguly, and R. Izmailov, Real-time video stream aggregation in wireless mesh network, In *Proceedings of IEEE 17th International Symposium on Personal, Indoor and Mobile Radio Communications*, Helsinki, Finland, pp. 1–7, 2006.

22. T. Ngan, D. S. Wallach, and P. Druschel, Incentives-compatible peer-to-peer multicast, In *Proceedings of 2nd Workshop on Peer-to-Peer Systems*, Cambridge, MA, 2004.

23. N. Nisan, T. Roughgarden, É. Tardos, and V. Vazirani, *Algorithmic Game Theory*, Cambridge University Press, New York, 2007.

24. M. Quant, P. Borm, and H. Reijnierse, Congestion network problems and related games, *European Journal of Operational Research*, 172:919–930, 2006.

25. R. W. Rosenthal, A class of games possessing pure-strategy Nash equilibria, *International Journal of Game Theory*, 2:65–67, 1973.

26. G. Rossi, S. Ferretti, and G. D'Angelo, Equilibrium selection via strategy restriction in multi-stage congestion games for real-time streaming, Technical Report UBLCS-2009-11, University of Bologna, 2009, www.cs.unibo.it/pub/TR/UBLCS/2009/2009-11.pdf

27. O. Rozenfeld, Strong equilibrium in congestion games, Research Thesis, Computer Science, Techion-Haifa, Israel, 2007.

28. O. Rozenfeld and M. Tennenholtz, Routing mediators, In *Proceedings of IJCAI07*, Hyderabad, India, 2007.

29. M. Scarsini and T. Tomala, Repeated congestion games with local information, In *Proceedings of the Workshop on Optimization, Transport and Equilibrium in Economics (OTAE)*, Paris, France, 2009.

30. A. Shapira, An exact performance bound for an $O(n + m)$ time greedy matching procedure, *The Electronic Journal of Combinatorics*, 4:1–14, 1997.

31. J. Spencer, *The Strange Logic of Random Graphs*, Springer, Berlin, Germany, 2001.

32. M. Tennenholtz and A. Zohar, Learning equilibria in repeated congestion games, In *Proceedings of 8th International Conference on Autonomous Agents and Multiagent Systems*, Budapest, Hungary, 2009.

33. G. Yang, Real-time streaming over wireless links: A comparative study, In *Proceedings of ISCC '05*, Cartagena, Spain, pp. 249–254, 2005.

34. G. Yang, T. Sun, M. Gerla, M. Y. Sanadidi, and L.-J. Chen, Smooth and efficient real-time video transport in the presence of wireless errors, *ACM TOMCCAP 2006*, 2(2):109–126, 2006.

Chapter 20

Friends or Foes for OFDM Interference Channel

Rajatha Raghavendra and Zhu Han

Contents

Modern wireless communication requires broadband and high-speed connections to meet the increasing demands of the users for applications such as multimedia. Frequency selective fading makes broadband channels prone to high errors. An efficient way to overcome the frequency fading effects is by employing orthogonal frequency division multiplexing (OFDM). Also, the increasing number of users gives rise to co-channel interference among the users. Therefore, resource allocation over OFDM interference channel becomes crucial to cater to the needs of the users while minimizing the co-channel interference. A distributive approach to resource allocation is essential to ensure simplicity of the system. A distributed solution to the resource allocation problem can be implemented in either a cooperative or a noncooperative way. In this chapter, we concentrate on the friends and foes (among distributed users) formed during the resource allocation for interference channel using game theory. Three specific case studies are discussed in detail.

20.1 Introduction

Communication systems have undergone major changes in the recent years with the advent of the wireless communication systems. Cellular communications and broadband internet connections form the core of wireless applications that have gained immense popularity because of the reliable, high-speed, and low-cost services offered to the users. This huge increase in the users led to the concept of frequency reuse which allows the limited number of frequency bands to be used in different cellular areas separated by a minimum distance to prevent interference. To accommodate a large number of users in crowded areas like metropolitan cities, denser networks are formed, which leads to increase in the co-channel interference among the cells. Co-channel interference can also occur within a cell when the number of users or the bandwidth requirement becomes higher. This is the motivation to the recent rise in popularity of the study of the interference channel.

The future of wireless communication depends on reliable broadband connections. OFDM transmission technique was introduced to accommodate the need of users to access broadband applications. OFDM is a modulation technique that enables transmission of high bandwidth applications by splitting the data and sending it through multiple low data rate sub-carriers. These sub-carriers can be individually modulated adaptively according to the channel conditions. In this case, the co-channel interference exists within the cell where the users share the same sub-carriers. Resource allocation for an OFDM interference channel is important to combat the effects of co-channel interference. There are various algorithms in the literature that implement power control, sub-carrier allocation, and rate maximization of the network. Some trade-offs and design challenges exist such as the distributed implementation, computation complexity, signaling overhead, system optimality, and fairness.

In this chapter, we concentrate on a unique perspective of resource allocation from the game theory point of view. The resource allocation problem needs real-time decisions to be made according to the changing channel and resource conditions. This motivated researchers to introduce game theoretic approaches, which are primarily applied in economics, into the field of wireless networks as well. Game theory helps in analyzing and predicting the outcome of the situation based on the decisions of the rational player. Game theory is a recent development in wireless networks, which is applicable to the resource allocation problem. It helps the users in making dynamic decisions about the sharing of the sub-carrier and power adjustments to meet their demand for capacity. Specifically, we analyze performance of cooperative games where users are friends and noncooperative games where users are foes. We investigate the performance in the severe interference channel and moderate interference channel, respectively.

The remaining chapter is organized as follows: Section 20.2 discusses the spectrum management problem formulation followed by a description of the important algorithms developed to optimize this problem. An introduction to game theoretic approaches to spectrum management is given in Section 20.3. Section 20.4.1 focuses mainly on the Nash bargaining solution (NBS) for cooperative games and noncooperative game with a virtual referee scheme is explained in Section 20.4.2. A joint frequency and time division multiplexing scheme for an interference channel in Section 20.4.3. Finally, Section 20.5 summarizes the important points discussed in this chapter.

20.2 Multiuser Dynamic Spectrum Access for OFDM System

Co-channel interference and crosstalk in wireless channels are the major performance degradation factors in the modern wireless communication systems. They are typically 10–20 dB higher than the background noise and form the limiting factor for the design of a network system. Traditional methods to combat crosstalk and co-channel interference consist of a static spectral mask applied to the whole system. The spectral mask is designed for the worst-case scenario (i.e., high-interference condition) and is found to be overtly conservative in weak and medium-interference conditions. Such strict allocation of spectra severely constricts the capacity of the network when operating in weak and medium-interference conditions. To achieve a better spectral efficiency, various dynamic spectrum management (DSM) algorithms have been proposed. Allocation of optimal power, data rate, and channel assignment to the user based on the current channel conditions is done, and the adverse effects of an interference channel are mitigating to a greater extent.

Dynamic spectral management can be categorized into two groups based on the network control architecture. Although it was developed for DSL lines, it is applicable for OFDM wireless networks as well.

1. *Centralized control*: A spectral management center (SMC) monitors and controls the transmit spectra of all the users in the system. It needs a lot of coordination and communication between the users and the SMC. This greatly increases the control signaling overhead but leads to a better performance when compared to the distributed control. Centralized control can be further categorized as follows [1–5]:
   a. *Level 1*: The data rate and transmit power of the user are reported to a central controller, and corresponding control signals are generated to control the rate and transmit power.
   b. *Level 2*: The noise spectra and the received signal spectra are monitored, and the transmit power is controlled by the central controller.
   c. *Level 3*: It allows complete coordination in real-time control of transmit power while monitoring the signals and noise spectra.
2. *Distributed control*: The users have the capability of sensing the channel conditions and adjust their transmit spectra accordingly. This may be efficient from the perspective of signaling and coordination but has the drawback of converging to a suboptimal point. It is also referred to as Level 0 [1,2], where there is no DSM involved and the control is fully distributed.

The control exhibited by the SMC over the users determines the computational complexity of the overall system. A centralized control is efficient but consumes a lot of bandwidth in control messaging between user and base station. On the other hand, distributed algorithms increase the complexity of the receiver of the user.

The algorithms developed for DSM are based on the resource allocation of the available network resources. The resource allocation can be broadly categorized into three areas:

Table 20.1   Notation Table

No.	Symbol	Meaning
1	N	Number of sub-carriers
2	K	Number of users
3	P_k^n	Power of kth user on the nth sub-carrier
4	$\mathbf{y}_k$	The received signal vector on nth sub-carrier $\left[y_k^1, y_k^2, ..., y_k^N\right]$
5	$\mathbf{x}_k$	The input signal vector on nth sub-carrier $\left[x_k^1, x_k^2, ..., x_k^N\right]$
6	$G_{i,k}^n$	The channel gain from ith transmitter to kth receiver on nth sub-carrier
7	σ^2	Additive Gaussian noise variance
8	R_k	The rate or capacity of the kth user
9	λ_k	The Lagrangian multiplier for user k
10	u_i	Utility function of user i
11	r_i	Strategy of user i
12	$\mathbf{r}_i^{-1}$	Strategy vector consisting of strategies of all users except user i

- *Sub-carrier assignment*: The sub-carriers with the best channel gains as seen by the user are allocated to the particular user.
- *Rate allocation*: The data rate is allocated depending on the user application requirements.
- *Power control*: Optimal transmit power is to be allocated to the user, in order to meet its rate requirements while not interfering with the other users.

The optimal allocation of sub-carrier, power, and rate is necessary to maximize the throughput of the network. This chapter discusses the various methods of resource allocation for achieving a better rate and minimizing co-channel interference levels.

Before discussing the subsequent section, the notation table for the variables used in this chapter is given in Table 20.1.

20.2.1 Spectrum Management Problem for Interference Channel

The system model for an interference channel is shown in Figure 20.1:

The received signal vector $\mathbf{y}_k$ can be written as

$$\mathbf{y}_k = \mathbf{G}_k \mathbf{x}_k + \mathbf{z}_k, \tag{20.1}$$

where
  $\mathbf{x}_k$ is the transmitted signal vector
  $\mathbf{z}_k$ is the noise vector
  $\mathbf{G}_k$ is the channel gain vector of user k consisting of the elements $G_k^n \; \forall n$

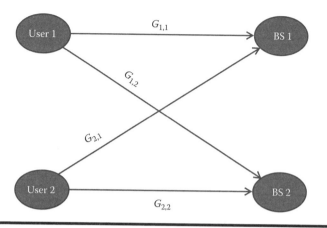

Figure 20.1 2-User interference channel.

For an interference channel, the interference from other users is generally considered as noise. This assumption gives optimal rates for weak and medium interference. So instead of simply using SNR (signal-to-noise ratio, given by $P_k G_{k,k}/\sigma^2$) , we consider the SINR (signal-to-interference-and-noise ratio) to calculate the capacity of the network. Therefore, the rate of user k, R_k is given by

$$R_k = \log_2\left(1 + \frac{P_k G_{k,k}}{\sum_{i\neq k}^{n} P_i G_{i,k} + \sigma^2}\right),\qquad(20.2)$$

where
   P_k is the transmit power of the kth user
   $G_{i,k}$ is the uplink channel gain from user i to base station k
   the term $\sum_{i\neq k} P_i G_{i,k}$ represents the interference caused by other users to user k

Without the loss of generality, we consider the variance of additive gaussian noise as a constant σ^2 for all sub-carriers. The spectrum management problem defines the objective of the network and the various constraints that are to be applied depending on the network capabilities. The spectrum management problem has the following objective and limitations:

- *Objective*: To maximize the overall rate of the network
- *Constraints*: Limited transmit power to achieve the minimum data rate while causing least interference to the other user

Mathematically, it can be expressed as

$$\max_{P\geq 0,\forall n,k} \sum_{k=1}^{K} w_k \sum_{n=1}^{N} \log_2\left(1 + \frac{P_k^n G_{k,k}^n}{\sum_{i\neq k} P_i^n G_{i,k}^n + \sigma^2}\right)$$

$$\text{s.t. } \sum_{n} P_k^n \leq P_k^{\max}.\qquad(20.3)$$

The capacity region of an interference channel is still an open problem. Once we establish the goals of the network, there are various algorithms proposed in the literature that try to achieve the maximum capacity region possible, while adhering to the constraints of maximum transmitter power and minimum target rate of each user. The following section lists out some of the important algorithms that try to find the solution to the above-mentioned optimization problem.

20.2.2 Existing Solutions in the Literature

Researchers have been trying to maximize the capacity region of a network using various algorithms. Listed here are the important algorithms, which will be discussed in this section: Iterative water filling (IWF), optimal spectral balancing (OSB), iterative spectral balancing (ISB), Successive Convex Approximation for Low Complexity (SCALE), autonomous spectral balancing (ASB), and band preference.

20.2.2.1 Iterative Water Filling

The IWF [6] is a level 0–distributed DSM algorithm, which maximizes the individual user's data rate while treating the interference from the other users as Gaussian noise. This is a selfish procedure where each user tries to maximize its own rate through the water-filling procedure for power allocation. This algorithm achieves near optimal values in the low-interference conditions but becomes highly suboptimal in the medium- and high-interference regions. The water filling is done iteratively for each user in turn until convergence to an optimal value of transmit power. The solution to the optimization problem, which is solved by Lagrangian multiplier, is given by

$$P_k^n = \left(\frac{1}{\lambda_k} - \frac{\sum_{j \neq k} P_j^n G_{j,k}^n + \sigma^2}{G_{k,k}^n} \right)^+, \qquad (20.4)$$

where λ_k is the Lagrangian multiplier for user k. To find the optimal λ_k, the above equation is differentiated with respect to P_k^n and equated to zero. The λ_k, which is obtained, is then substituted into the equation to solve for the optimal power allocation P_k^n. This is the simplest of all algorithms and can be implemented for a large number of users in the low-interference condition.

■ *Advantage*: Simple to implement because it is a distributed algorithm, so there is no need of an SMC.
■ *Disadvantage*: In high-interference conditions, the algorithm converges to a suboptimal point, which degrades the performance of the entire system.
■ *Complexity*: $O(KN)$.

20.2.2.2 Optimal Spectral Balancing

OSB [7] is a centralized algorithm where an SMC manages the spectra of all the users in the network in a real-time environment. The SMC is responsible for observing the channel conditions of each user and deciding the optimal transmit power level for each user to mitigate crosstalk among them. This leads to a joint exhaustive search for the optimal power for N users on K sub-carriers simultaneously, which creates an exponential complexity in the number of users and number of sub-carriers.

The spectrum management problem can be solved by using the dual decomposition method, which reduces the joint exponential complexity of number of users and number of sub-carriers. Instead of searching for individual power of each user over each sub-carrier, the dual decomposition method tries to maximize the weighted sum rate of all the users over a particular sub-carrier at a time. Assuming only a two-user case, the weighted sum rate region is given by

$$\max_{P_1, P_2} = wR_1 + (1 - w)R_2, \tag{20.5}$$

where

R_1 and R_2 are the assigned rates
w and $(1 - w)$ are the weights assigned to user 1 and user 2, respectively

The weight w is assigned values between 0 and 1 to ensure convexity [8]. It gives the proportion of the rates assigned to user 1 and user 2 when they use the same sub-carrier for transmission. The Lagrangian for the above optimization problem is defined as follows:

$$L = wR_1 + (1 - w)R_2 - \lambda_1 \sum_k P_k^1 - \lambda_2 \sum_k P_k^2. \tag{20.6}$$

This optimization problem can be now be solved through an unconstrained optimization as follows:

$$\max_{P_1, P_2} L\left(w, \lambda_1, \lambda_2, P_k^1, P_k^2\right). \tag{20.7}$$

This leads to a linear complexity in the number of tones N since $L = \sum_n L^n$. The algorithm steps are briefly listed below for user k on sub-carrier n.

1. The outermost loop contains the iterations to fix the weight w_k.
2. The intermediate loop is for searching λ_1 using bisection search.
3. The innermost loop optimizes λ_2 using bisection search.
4. The Lagranian multipliers λ_1 and λ_2 are optimized such that they satisfy the power constraint as shown below:

$$P_1^n, P_2^n = \arg \max_{P_1^n, P_2^n} L_k\left(P_1^n, P_2^n, w, \lambda_1, \lambda_2\right), \tag{20.8}$$

which is again solved by exhaustive two-dimensional search.

The OSB performance is considered to be the highest when compared to the rest of the algorithms, but the exponential complexity in the number of sub-carriers prevents it from being implemented in practical conditions. So it is sometimes treated as the ideal performance.

- *Advantage*: OSB provides the best weighted rate region compared to all other algorithms. Hence, the resulting rate region is used as a reference for the other algorithms. Also, the weighted sum rate ensures that there is no selfish optimum point for a particular user.
- *Disadvantage*: The OSB algorithm becomes computationally intractable for a large number of sub-carriers.
- *Complexity*: $O(Ke^N)$.

20.2.2.3 Iterative Spectral Balancing

Although OSB achieves the maximum weighted sum rate, it is computationally intractable for a large number of users. So a different approach to maximize the sum rate region is proposed in [9], called the ISB. Instead of a joint search for power as discussed in OSB, the optimal powers are searched in an iterative fashion for each user. Thus, the exhaustive K-dimensional search for powers is replaced by a one-dimensional search for each user.

The steps involved in the algorithm are listed below:

■ The outermost loop contains iterations for users from 1 to K.
■ The inner loop contains iterations for the sub-carriers from 1 to N.
■ By fixing the powers of other users, the weighted rate of current user is maximized by the water-filling method.
■ This process is repeated until convergence, that is, all the users satisfy the minimum rate constraint.

While solving the optimization, the weights and Lagrangian multipliers are updated using gradient descent method as follows:

$$w_k = \left(w_k + \epsilon (R_{k,\text{target}} - R_k) \right)^+,$$

and

$$\lambda_k = \left(\lambda_k + \epsilon (P_{\text{total}} - P_k) \right)^+,$$

where
  w_k is the weight assigned to the rate of user k
  $R_{k,\text{target}}$ is the minimum rate needed to be provided to user K
  ϵ is the gradient descent factor

The power is optimized and updated for a single user at a time while keeping the powers of other users fixed. This may result in a suboptimal operating point because we obtain a local optimum here, rather than a global optimum point as obtained through OSB. The ISB algorithm is the same as the IWF except that instead of maximizing individual user's capacity, the weighted sum rate of all the users is maximized. This ensures that there is a fairness in the allocation of the transmit power of the users and prevents a selfish optimum operating point on the rate region boundary.

■ *Advantage*: It is a computationally tractable algorithm for large number of users and can be implemented in a distributive way.
■ *Disadvantage*: ISB needs the knowledge of all the other user's interference levels. This adds a significant overhead burden on the system.
■ *Complexity*: $O(KN^2)$.

20.2.2.4 SCALE

The SCALE algorithm [10] is a computationally efficient, distributed algorithm, which provides a way to optimize the transmit powers of the users locally and maximize the data rate of each user. The weighted sum rate objective is given in (20.3). The equation forms a non-convex problem because it is non-convex in power, that is, $P_k^n \geq 0, \forall n, k$. Therefore, it becomes an *NP*-hard problem whose global optimum is difficult to compute. In order to simplify the problem, we choose to relax the non-convex problem by using the following expression for lower bound:

$$\alpha \log z + \beta \leq \log(1 + z),\qquad(20.9)$$

where

$$\alpha = \frac{z_0}{1 + z_0}$$

and

$$\beta = \log(1 + z_0) - \frac{z_0}{1 + z_0} \log z_0.$$

Rewriting the original optimization problem using (20.9), we get

$$\max_{P \geq 0} \sum_{k=1}^{K} \sum_{n=1}^{N} w^k \left(\alpha_k^n \log_2 \left(\frac{P_k^n G_{k,k}^n}{\sum_{i \neq k} P_i^n G_{i,k}^n + \sigma^2} \right) + \beta_k^n \right)$$

$$\text{s.t.} \quad \sum_{n} P_k^n \leq P_k^{\max}.\qquad(20.10)$$

We observe that the objective function is still not convex in power. So we define a new variable $\widetilde{P}_k^n = \log\left(P_k^n\right)$. Substituting this expression in Equation 20.10, we get

$$\max_{\widetilde{P}} \sum_{k=1}^{K} \sum_{n=1}^{N} w^k \left(\alpha_k^n \log_2 \left(\frac{\widetilde{P}_k^n G_{k,k}^n}{\sum_{i \neq k} \widetilde{P}_i^n G_{i,k}^n + \sigma^2} \right) + \beta_k^n \right)$$

$$\text{s.t.} \quad \sum_{n} e^{\widetilde{P}_k^n} \leq P_k^{\max}.\qquad(20.11)$$

Now, each term in the expression is concave and the objective function is maximization of concave function. It comprises of log-sum-exponential function, which is convex [8].

$$\log \left(\frac{\widetilde{P}_k^n G_{k,k}^n}{\sum_{j \neq k} P_j^n G_{i,k}^n + \sigma^2} \right) = \log G_{k,k}^n + \widetilde{P}_k^n - \log \left(\sum_{j \neq k} G_{k,j}^n e^{\widetilde{P}_k^n} + \sigma^2 \right).\qquad(20.12)$$

The realization of the SCALE algorithm can be done by formulating dual objective problem using the Lagrangian multipliers:

$$\mathbf{L} = \sum_{k=1}^{K} \sum_{n=1}^{N} w^k \left(\alpha_k^n \log_2 \left(\frac{\widetilde{P}_k^n G_{k,k}^n}{\sum_{i \neq k} \widetilde{P}_i^n G_{i,k}^n + \sigma^2} \right) + \beta_k^n \right) - \sum_{k=1}^{k} \lambda_k \left(\sum_{n=1}^{N} e^{\widetilde{P}_k^n} - P_k^{\max} \right).\qquad(20.13)$$

The Lagrangian is solved by differentiating the above equation with respect to $\widetilde{P}$ and then equating to zero to obtain the minimum value of the Lagrangian multiplier λ_k as

$$\frac{\partial \mathbf{L}}{\partial \widetilde{P}_k^n} = w_k \alpha_k^n - P_k^n \left(\lambda_k + \sum_{j \neq k} w_j \alpha_k^n G_{j,k}^n \right). \tag{20.14}$$

The powers are updated through gradient descent method.

- *Advantage*: The SCALE algorithm uses limited messaging from the SMC, thus creating a low control overhead in the system. It finds the global optimum with a lower complexity when compared to OSB.
- *Disadvantage*: It needs the knowledge of the channels of other users too which may make it impractical for real-time systems with varying channels.
- *Complexity*: $O(K \log N)$.

20.2.2.5 Autonomous Spectral Balancing

ASB is a fully distributed algorithm proposed in [11]. The exponential complexity of OSB and the suboptimal performance of ISB and IW were the motivating factors for the development of this algorithm. It is based on the concept of the "reference line" or the "victim line." The reference line is a statistical average of the worst affected lines. It is generally applied to digital subscriber lines (DSL) since the channel is slowly varying, which helps in gathering the statistics of the reference line. It could also be considered as the farthest modem from the base station and thus suffering the worst channel conditions. The data for the reference line are collected first and are distributed among all the users. Now, each user tries to maximize its own data rate while causing minimum harm to the reference line. In other words, each user tries to maximize the reference line rate while meeting its own minimum rate requirements.

The spectral management problem for ASB becomes

$$\max_{P_k} R_{k,\text{ref}}$$

$$\text{s.t.} \quad \begin{cases} R_k \geq R_{k,\text{target}} \\ \sum_n P_k \leq P_{\text{total}}, \end{cases} \tag{20.15}$$

where

$R_{k,\text{target}}$ is the target data rate
$R_{k,\text{ref}}$ is the rate of the reference line
P_{total} is the total transmit power of user k

In (20.16), the rate of each user is independent of the rates of the other users. Once all the users optimize their rates, the process repeats again because the other user's powers, which are considered as crosstalk noise, are changed. Thus, the optimization continues till convergence. To simplify the analysis of this optimization, we consider the dual of it as shown below:

$$\max_{P_k} w_k R_k + (1 - w_k) R_{k,\text{ref}}$$

$$\text{s.t.} \quad \sum_n P_k \leq P_{\text{total}}. \tag{20.16}$$

This can be solved by using the Lagrangian multiplier method.

$$L_k = (1 - \lambda_k)(w_k R_k + (1 - w_k) R_{k,\text{ref}}) - \lambda_k \sum_n P_k^n, \tag{20.17}$$

which can be solved by the unconstrained optimization problem:

$$\max_{P_k} L_k(w_k, \lambda_k, P_k, \mathbf{P}_{-k}), \tag{20.18}$$

where $\mathbf{P}_{-k}$ represents the power of other users.

The principle difference between ASB and ISB is that ASB is completely autonomous since it assumes the victim line statistics to be constant. This eliminates the control signals between the users informing about the channel variations. Moreover, the optimization does not need the information of the other users, channel gains, it only needs the background noise and the direct channel information. All these factors contribute to the autonomous functioning of this algorithm.

■ *Advantage*: ASB is fully autonomous and has reduced complexity while achieving the rate region close to OSB.
■ *Disadvantage*: This algorithm is designed for DSL lines, which has a slow varying environment and where the channel is considered to be a static channel. Hence, the results may not be the same for a fast varying channel like a wireless OFDMA channel.
■ *Complexity*: $O(KN)$.

20.3 Game Theory Basics

Most of the literature in dynamic resource allocation is dedicated to developing new techniques to improve the rate region under some system constraints. However, the algorithms discussed so far have high computational complexities, which prevents them from being implemented in practical systems. The achievable rates vary considerably over time and the optimal operating point becomes an issue of conflict. In order to resolve such conflicts over the operating point, game theoretic approaches [12,13] can be used to overcome the suboptimality of the interference channel. There are various different approaches or games that can be mainly classified into (a) cooperative games and (b) noncooperative games, which are explained with relevant examples in Section 20.4.1 and 20.4.2, respectively. Before delving into the topic of game theory for an interference channel, we introduce a few terms in this section, which form the basis for the game theory [14].

A game is composed of three main elements: the number of players K, the strategy of each player r, and the utility function or payoff of each player u, which is the interest/preference of a player. A game is called a cooperative game if the players are enforced with a contract before the start of the game to obtain the optimal utility for each player. They form coalitions to achieve the overall goal of the game rather than individual payoffs. On the other hand, a noncooperative game is the one where the players' strategies are independent of each other. Each player tries to maximize his/her payoff. A Nash equilibrium occurs when each player in a noncooperative game chooses the best strategy in response to the strategies of other player. A single player cannot unilaterally deviate from the Nash equilibrium point, otherwise it will degrade his performance. Hence, the Nash equilibrium point is also called the best response point. It can be defined as [15]

$$u_i\left(r_i^*, \mathbf{r}_i^{-1}\right) \geq u_i r_i, \mathbf{r}_i^{-1} \qquad \forall r_i, \tag{20.19}$$

where

r_i^* is the Nash equilibrium point strategy of ith user
r_i is the strategy of user i
u_i is the utility function of user i
$\mathbf{r}_i^{-1}$ is a vector that represents the strategies of the other players

In the context of wireless networks, Nash equilibrium occurs when all the users meet the minimum rate requirement while causing least interference to other users who are transmitting on the same channel. Another equilibrium point that is commonly used in literature is the Pareto optimal point. A Pareto optimal point occurs when a player cannot improve the optimum point without hurting at least one player. If an improvement is made by one player without making the other players worse off, then the situation is called a Pareto improvement.

In a situation where the performance of the game is suboptimal, it is beneficial to go for mixed strategies instead of pure ones. The mixed strategy is a probability assigned to each pure strategy of the player and the player chooses a pure strategy randomly. Sometimes, the opponent benefits by the knowledge of a player's strategy in a pure strategy game whereas in a mixed strategy game, the opponent is kept guessing about the strategy to be played.

Table 20.2 will help in understanding the Pareto optimality condition. This is the "chicken game" where each player wants to play the "dare" rather than "chickening out" of the game. Suppose there are two drivers driving toward each other on a single-lane road. None of them wants to swerve and give way to the other driver. In such a situation, if both of them do not swerve, it would result in a collision with a severe negative payoff. On the other hand, if one of them swerves, he will have a negative payoff because he will be laughed upon by the other player. The other player gains when a driver swerves. If both of them anticipate the collision and swerve, neither of them gets any payoff. The situations with $(-1, 1)$ and $(1, -1)$ are the Pareto optimal points of this game. Also, if the drivers swerve at the last moment to avoid a collision, then it becomes a mixed strategy game because neither of them can anticipate the opponent's next move.

In the wireless network scenario, many a time, the capacity region reaches a suboptimal Nash equilibrium point when the users try to maximize their respective rates. In such a situation, it is beneficial for the users to adopt a mixed strategy to improve the capacity region of the network. The best strategy under the high-interference circumstance is to use a time division multiple access (TDMA) to access the channel.

Table 20.2 Chicken Game

	Drive	Swerve
Drive	$(-100, -100)$	$(1, -1)$
Swerve	$(-1, 1)$	$(0, 0)$

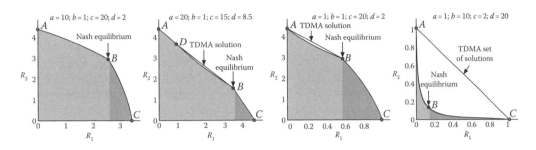

Figure 20.2   2-User rate regions. (a) Low interference. (b) Medium interference. (c) Medium interference. (d) High interference. (From Charafeddine, M. et al., Crystallized rates region of the interference channel via correlated equilibrium with interference as noise, *IEEE International Conference on Communications*, June 14–18, 2009. With permission.)

Figure 20.2 shows the achievable rate regions under different interference conditions. It shows that there is a significant loss in the capacity region in high-interference region. In order to increase the capacity of the network, TDM scheme is applied to maximize the capacity region. The variables a, b, c, and d are the channel gains of the interference channel previously denoted as $G_{1,1}$, $G_{2,1}$, $G_{1,2}$, and $G_{2,2}$ described in Figure 20.1, respectively. The point $A(0, R_{2,max})$ represents the maximum achievable rate of user 2 when it is transmitting alone on the sub-carrier. Similarly, point $C(R_{1,max}, 0)$ represents the maximum achievable rate of user 2 when it is transmitting alone on the sub-carrier. The equilibrium point B is the point where both users share the sub-carrier and transmit at full power given by $(P_{1,max}, P_{2,max})$.

In Figure 20.2a, the achievable rates of both users in the low-interference condition is at point B, which is close to the maximum achievable rates when each user is transmitting alone on the sub-carrier. Thus, it results in a concave rate region. With increase in the interference levels, the rate region tends to become convex, which results in degraded performance. To improve the rate region, TDM is introduced in the region where the performance is low. As we can observe in Figure 20.2b and c, TDM is used between points D and B and between A and B, respectively, to increase the capacity of the region. When high interference is experienced by both the users, the rate region becomes completely convex resulting in the implementation of pure TDM to achieve a significantly improved capacity region as shown in Figure 20.2d.

Thus, the interference level and channel conditions determine the technique to be used to maximize the rate region. In the following section, we discuss interference scenarios where game theory is applied to increase the capacity of the network.

20.4 Friends or Foes Using the Game Theory

In this section, the game theoretic approaches for OFDM interference channels are studied under different interference levels experienced by the users. In a single-cell case, there is high interference among the users while in a multicell case, there is medium interference due to the distance between the cells. For a low-interference case, noncooperative game results in a good performance but the performance degrades considerably with increase in the interference. In high-interference conditions, cooperative games are utilized to allocate the channels to improve network performance and ensure fairness. In medium interference case, noncooperative game outcomes can be improved by introducing a game referee to change the game rules. Finally, the joint consideration of cooperative game and TDM techniques are investigated for the medium-interference case.

20.4.1 Friends in Severe Interference

In a single, user case of OFDMA, the user occupies all the sub-carriers. It then adapts its rate according to the channel conditions on each sub-carrier such that the sum of the rates on all the sub-carriers is at least equal to the target data rate. The real problem arises in a multiuser case when a sub-carrier is good for more than one user. There arises a conflict over which sub-carrier to be allocated to each user in a fair manner. The existing max–min fairness techniques try to improve the worst affected users by penalizing the users with a good channel condition. This results in the reduction of the overall system capacity.

In order to overcome the issue of fairness in the system, cooperative game theory can be applied to the system. Users can negotiate with each other through the base station about the allocation of sub-carriers and achieve a better overall system capacity while satisfying the constraint of minimum rate for each user. In this section, we will focus mainly on the NBS [16] for the cooperative game where the minimum resources are allocated first to each user to satisfy the minimum rate requirements, and then the remaining resources are allocated proportionate to the channel conditions.

Let $R_{i,\min}$ be the minimum payoff given to the user to ensure its participation in the cooperative game and $\mathbf{S}$ be the feasible region of the solution. The basic conditions for r to be an NBS solution $r = \phi(\mathbf{S}, R_{\min})$ are given below:

1. $R_i = \sum r_i^j \geq R_{i,\min} \forall i$. The sum of the rates on all the sub-carriers should be greater than or equal to the minimum target rate.
2. r exists in the feasible region, that is, $r \in \mathbf{S}$. The strategy r, which determines the rate, should lead to a feasible rate.
3. The NBS solution is a Pareto optimal point. Once NBS solution is achieved, any solution that is beneficial to one user cannot be achieved without hurting the other users.
4. The solution can be linearly transformed, that is, any linear operation (addition or scaling) on the NBS solution will lead to an NBS solution again.
5. $\mathbf{S}$ is invariant because it is the set of all possible strategies applicable.

There exists a unique solution for the optimization problem given below, which satisfies all the above conditions too. The problem can be formulated as

$$\max_{\mathbf{A},\mathbf{P}} \mathbf{u} = R_i, \tag{20.20}$$

$$\text{s.t.} \quad \begin{cases} \sum_{i=1}^{K} a_i^j = 1, \ \forall j, \\ \sum_{i=1}^{K} P_i^j \leq P_{\max}, \ \forall i, \\ R_i \geq R_{i,\min}, \ \forall i, \end{cases}$$

where
  $\mathbf{u}$ is the utility function of the system
  $a_{i,j} = 1$ if user i occupies sub-carrier j
  $\mathbf{A}$ is the matrix for allocations of size $K \times N$ having K rows for users and N columns for sub-carriers
  $\mathbf{P}$ is the vector of length K consisting of the powers of K users

The utility function for an NBS system is defined as

$$\mathbf{u} = \prod_{i=1}^{K} (R_i - R_{i,\min}).$$

(20.21)

To maximize this utility function for a fixed $R_{i,\min}$, it is enough to maximize R_i which is the individual achievable rate under the given power constraints. So the problem now reduces to a single user optimization problem for a fixed channel assignment d_i^j.

$$\max_{P_i} \mathbf{u} = R_i$$

(20.22)

$$\text{s.t.} \quad \sum_{j=1}^{N} P_i^j d_i^j \leq P_{\max}, \ \forall i.$$

This can be solved by the water-filling procedure which yields a unique solution for the optimal power assignment under the assumption of fixed channel assignment. If the channel assignment is not fixed, then all the combinations of channel assignment need to be searched to find the optimal channel and power assignment which generates the maximum utility. The solution for this optimization problem has a high computational complexity and may take a long time to converge, so the authors proposed a fast, low-complexity bargaining algorithm in [16].

The basic idea of NBS is that two groups or coalitions are formed among different users in each iteration and the coalitions compete for the sub-carrier. In each iteration, the sub-carriers allocation is optimized to meet the minimum rate requirement of each user. The optimal coalition is formed by the Hungarian method. The steps involved in the proposed algorithm are discussed below for a two user case. The results can be extended to a multiuser case using two coalitions.

1. The sub-carriers are assigned to both the users initially to meet the minimal rate requirement of each user.
2. The sub-carriers are sorted in descending order of the channel gains for each user.
3. For a multiuser case, form coalitions based on Hungarian method and proceed to the next step for bargaining the sub-carriers.
4. There are N iterations for the sub-carriers. Each user follows the water-filling algorithm to occupy the sub-carriers, and the utility function of maximum rate for each user is calculated.
5. Choose the partition of sub-carriers (which produces the maximum rate) is chosen, and the channel allocation is updated.
6. For a two coalition system, the users negotiate to exchange their sub-carriers to obtain the maximum utility. The users regroup and negotiate using the above two-band partition algorithm until the utility can be improved no more.

The fast algorithm described earlier gives the steps for channel assignment. Another important issue in NBS is how to form the coalition among the users, that is, which users should form the coalition at a particular time. The results obtained using NBS when compared to maximal rate algorithm show that although there is slight drop in the overall system capacity region, NBS provides fairness among the users. The Hungarian method, which is used to form the coalition among the users, is described below. It is a more efficient algorithm when compared to the random method of forming coalitions among the users.

20.4.1.1 The Hungarian Method

1. A matrix representing the cost function of each user on each sub-carrier is formed.
2. Now, each element in a row is subtracted by the minimum cost in that row so that there is at least one zero in each row.
3. Similarly, each column is subtracted by the minimum in that column so that there is at least one zero in each column.
4. Cross out the rows and columns that have a zero by drawing a line across them.
5. If the number of the lines is equal to the number of users K, then the channel assignment A is optimal.
6. Otherwise, find the second smallest cost value and subtract it from the rows and columns not covered by the lines.

The simulation results shown in Figure 20.3 gives a comparison of the overall achievable rates obtained by NBS scheme with max–min and maximal rate schemes for a two-user case [16]. The simulations have been conducted for an OFDMA system having 128 sub-carriers spread over 3.2 MHz bandwidth. The maximum transmit power of the users is limited to 50 mW, the acceptable BER is 10^{-2}, the propagation loss factor is taken as 3 and the thermal noise level is fixed at 10^{-11}. The distance between the first user and common base station is fixed at 100 m, and the distance between the second user and the base station is varied from 10 to 200 m. The minimum required rate of each user $R_{i,\min} = 100$ Kbps. The channel conditions vary for user 2 with the varying distance between user 2 and the base station.

Figure 20.4 is a plot for the individual user's rate using NBS, maximal rate, and max–min techniques. Since the distance between user 1 and base station is fixed, the rate of user 1 remains constant while the rate of user 2 decreases with increase in distance D_2 when using the NBS method. So, it shows that NBS supports the fairness among the users, that is, the rate of each user

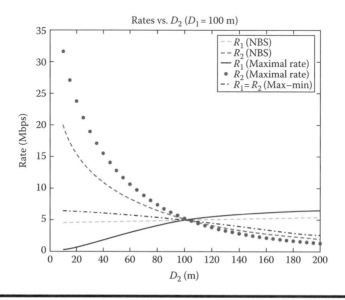

Figure 20.3 **Achievable rate of individual user vs. distance D_2. (From Han, Z. et al., *IEEE Trans. Commun.*, 53, 8, 1366, August 2005. With permission.)**

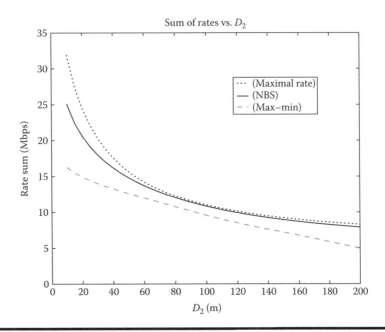

Figure 20.4 Achievable rate vs. reuse factor in multicell case. (From Han, Z. et al., *IEEE Trans. Commun.*, 53, 8, 1366, August 2005. With permission.)

is determined purely by its channel conditions and is not dependent on the interference caused by the other users. Figure 20.3 shows that the rates achieved by the NBS scheme is close to the rate achieved by the max–min scheme and much better than the maximal-rate scheme at longer distances from the base station. This shows that the NBS scheme does not affect the sum rate of the users while maintaining fairness among the users.

Figure 20.5 shows the rate performance of NBS in multiuser case with respect to the number of users. The setup for two-user case is retained except that the minimum target rate of each user is changed to 25 Kbps. The results show that as the number of users increase, the rates obtained are higher. This is due to the diversity provided to the users by the independent fading channels [16]. Also, we observe that the rate performance of the maximal-rate technique is better than NBS, but this gap in performance decreases as the number of users increases because more number of bargaining coalitions can be formed among the users. From the above simulations, we observe that the NBS solution provides a simple, fast, and effective solution to the sub-carrier allocation problem. Also, it achieves a good system performance while ensuring a fair scheme for each user. The fast algorithm for bargaining using the Hungarian method helps in reducing the complexity of the system while maintaining the performance level. Thus, NBS is a viable technique in maximizing the capacity of the system and ensuring fairness of individual users.

20.4.2 Foes in Medium Interference

The multicell, multiuser OFDM resource allocation becomes complicated because of the constant monitoring of the sub-carrier channels by the SMC to ensure an optimal sub-carrier

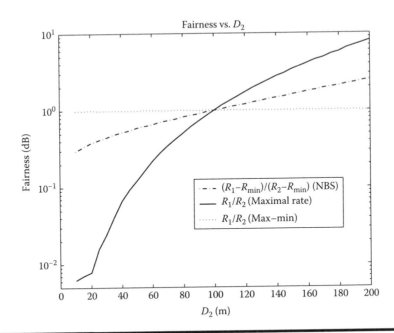

Figure 20.5 Sum rate vs. number of users. (From Han, Z. et al., *IEEE Trans. Commun.*, 53, 8, 1366, August 2005. With permission.)

allocation to each user in each cell. For a real-time system, this adds a considerable amount of delay and messaging overhead on the system. Thus, in order to simplify the resource allocation problem, it can be modeled as a noncooperative game. This ensures a distributive approach among the users rather than a centralized approach that reduces the complexity of the system considerably.

In the proposed noncooperative game, each user meets the target rate requirement while trying to minimize its transmit power. The utility function in this case is the transmit power, which is minimized rather than maximized as seen in most of the cases in literature. If the sub-carriers are carefully assigned to the users, then the transmit powers of the users converge to a unique Nash equilibrium point. This technique can be implemented successfully for a small number of users. But for a large number of users, there maybe multiple Nash equilibrium points because of the multiple local optimal points. This may result in convergence of the overall rate of the network to a suboptimal Nash equilibrium point. In order to overcome this problem, the users causing high interferences need to be prevented from using the particular sub-carrier. This will lead to an improved rate performance of other users using the particular sub-carrier. This is precisely the reason why the concept of the "virtual referee" was introduced in [15].

A referee in any game helps in monitoring the game and punishes or throws out the players who do not obey the rules of the game. Similarly, in the OFDM network, a virtual referee is introduced, which monitors the noncooperative users. If any user causes high interference to other users resulting in an overall performance degradation of the network, then the virtual referee can force the user to stop using the sub-carrier. This helps in improving the Nash equilibrium point to a higher value.

20.4.2.1 Resource Allocation Using Virtual Referee Algorithm

Resource allocation schemes in a network can be classified into two categories:

- *Fixed channel assignment*: Separate groups of sub-carriers are assigned to each cell in the network so that there is no co-channel interference, but this leads to spectral inefficiency due to rigid spectral masks.
- *Iterative water-filling channel assignment*: The users share all the sub-carriers in a noncooperative manner.

The basic task of a virtual referee is to improve the outcome of the game with dynamic channel assignment to a better Nash equilibrium point.

The system objective for a noncooperative game is given in [15] as

$$\min_{\mathbf{A},\mathbf{r}} \sum_{i=1}^{K} \sum_{j=1}^{N} P_i^j. \tag{20.23}$$

$$\text{s.t.} \begin{cases} \sum_{j=1}^{N} r_i^j = R_i \, \forall i, \text{ minimum target rate requirement,} \\ \sum_{j=1}^{N} P_i^j \leq P_{\max} \, \forall i, \text{ maximum power constraint,} \\ r_i^j, P_i^j \geq 0 \, \forall i, j, \text{ nonnegative rate and power,} \end{cases}$$

where
r_i^j is the rate
P_i^j is the transmit power of user i on sub-carrier j

The solution for this optimization problem becomes complicated if a centralized solution is applied. Moreover, it is difficult to obtain the channel information of all the users in the system. So, the noncooperative game approach is applied to retain the simplicity of the system.

The noncooperative game can be mathematically expressed as

$$\min_{\mathbf{r}_i \in \mathbf{S}} u_i = \sum_{n=1}^{N} P_i^n$$

$$\text{s.t.} \quad \sum_{n=1}^{N} r_i^n = R_i, \tag{20.24}$$

where
u_i is the utility function, that is, the transmit power if user i
$\mathbf{S}$ is the set of all strategies r

This can be solved by the iterative water-filling procedure, where the interference from other users is considered as noise.

The iterative water-filling procedure yields a Nash equilibrium point. If the Nash equilibrium point satisfies all the constraints of rate and power for all users, then it is the desired Nash equilibrium point. On the other hand, if it is not the desired Nash equilibrium point, the virtual referee comes into picture. It prevents the users who cause the maximum interference to other users from using the sub-carrier. This leads to a better Nash equilibrium point for the remaining users. The noncooperative game is repeated until the desired Nash equilibrium point is reached.

The iterative water-filling scheme gives desired solutions if the interference level is low, but for high-interference case, the noncooperative game may not converge at all. In such cases, the users, whose maximum transmit powers are achieved yet they have not reached the target rates, play the dual noncooperative game. This improves the feasible convergence region as proved in [15]. Whenever a user reaches his maximum transmit power, he immediately switches to the dual game in order to achieve his respective target data rate. The dual of the noncooperative game is given by

$$
\max_{\mathbf{r}_i \in S} \sum r_i^n
$$
$$
\text{s.t.} \quad \sum P_i^n = P_{\max}. \tag{20.25}
$$

A Nash equilibrium point is achieved when the noncooperative game (20.24) and its dual game (20.25) are played by the different users, although it may not be the desired one. If there is any user who still plays the dual game after convergence, it means that the Nash equilibrium point is not the desired one. Consequently, the virtual referee steps in and the sub-carrier removal process begins in order to bring the game to the desired Nash. Some of the sub-carriers are removed from the other user's transmission group so that the Nash equilibrium point is achieved for all users. The removal of sub-carriers stops when the game converges to a desired Nash equilibrium point.

The flowchart of the game with virtual referee is given in Figure 20.6.

There are certain criteria for the removal of the sub-carrier or the user from the transmission group. The set S_i is the set of transmission sub-carriers used by the user i. The sub-carrier j is dropped by the user i if it offers the highest amount of interference to the user, that is, the lowest SINR. All the users who share the set S_i decide which user should be allowed to use which sub-carrier on the basis of the following conditions.

- Each user is assigned at least one sub-carrier for transmission.
- Each sub-carrier is assigned at-least one user.
- If a user is unable to achieve the target rate even at his maximum power, then the user cannot be removed from the transmission group.

The algorithm is summarized below:

1. This is the initialization step. Each user plays the noncooperative game with a predefined value of total target rate of sub-carriers and occupies all the sub-carriers.
2. If the sum of the powers assigned to the sub-carriers is less than $P_{\max}$ and the target rate has been achieved already, then it is the desired Nash equilibrium point and the virtual referee does nothing. If this condition is not satisfied, go to the next step.
3. If the sum of the powers assigned to the sub-carriers is equal to $P_{\max}$ but the target sum rate of the sub-carriers has not been achieved, then play the dual noncooperative game. If the game converges then go to next step else go back to step 2.
4. This step is for the removal of sub-carrier from the users transmission group by observing which sub-carrier has the lowest gain and highest interference.

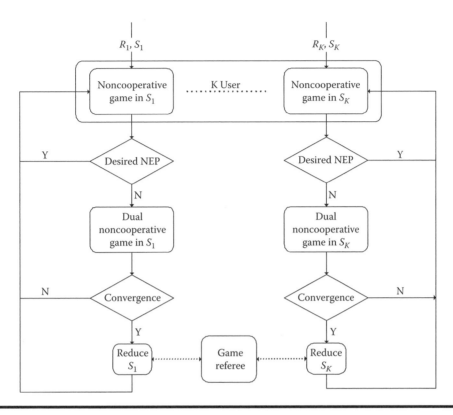

Figure 20.6 Noncooperative game with virtual referee. (From Han, Z. et al., *IEEE J. Selected Areas Commun.*, 25, 6, 1079, August 2007. With permission.)

The virtual referee concept is efficient in improving the Nash equilibrium point while keeping the complexity the same as the water-filling $O(N \log N)$. Moreover, it can be implemented easily in the base stations.

The simulation results prove that the noncooperative game with virtual referee achieves better results when compared to the plain iterative water-filling procedure. Figure 20.7 shows the achievable rate for a two-cell case for a 32 sub-carriers OFDMA system spread over 6.4 MHz bandwidth, with maximum transmit power restricted to 5 mW, acceptable BER of 10^{-3} and the thermal noise power fixed at -70 dBmW. We observe that the achievable rate by virtual referee scheme follows the required rate closely. This proves that the noncooperative game with virtual referee is indeed efficient in the rate performance and is better than the iterative water-filling scheme. For the multicell case, the maximum power is 10 mW, the target rate of each user is 12 Mbps and the other setup values are the same as that of the two-cell case. The reuse factor given here is the distance between the cells. The co-channel interference is smaller for higher reuse factor and vice versa. As shown in Figure 20.8, the achievable rate by the virtual referee scheme is higher than the iterative water-filling scheme in high co-channel interference (i.e., small reuse factor) condition. But, as the distance between the cells increase, the co-channel interference becomes lower and the performance of both the schemes is the same.

Thus, the noncooperative distributive game approach with a virtual referee improves the overall rate of the system while minimizing the transmit power. The performance comparison with

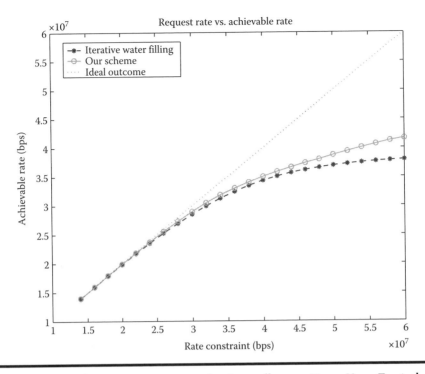

Figure 20.7 Achievable rate vs. target rate for two-cell case. (From Han, Z. et al., *IEEE J. Selected Areas Commun.*, 25, 6, 1079, August 2007. With permission.)

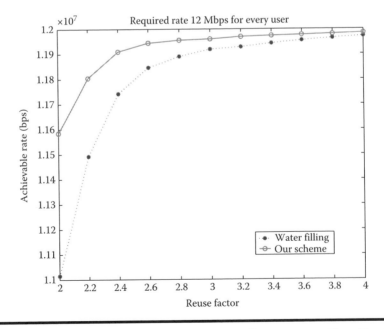

Figure 20.8 Achievable rate vs. reuse factor in multicell case. (From Han, Z. et al., *IEEE J. Selected Areas Commun.*, 25, 6, 1079, August 2007. With permission.)

the iterative water-filling method proves that the virtual referee technique indeed yields a good performance in a multicell environment. The virtual referee can be implemented in a practical system without any significant increase in complexity and provides a simple yet effective way to improve the performance of the network in medium-interference conditions using noncooperative game. The minimization of the utility function also results in lower power consumption, which is desirable for the longevity of battery power of the user.

20.4.3 Joint Scheme for Medium Interference

In this section, a joint FDM/TDM scheme for the frequency selective interference channel is discussed for medium-interference condition. For the medium-interference condition, cooperative games yield a good performance. The NBS for cooperative game, which is discussed in Section 20.4.1, is generally used to ensure fairness among the users, but it does not always converge to a stable Nash equilibrium point. So, an alternative is proposed in [12,13] which takes into account the Power Spectral Density (PSD) mask constraint of a sub-carrier along with the total power constraint while implementing the NBS.

In general, noncooperative games are implemented to ensure a distributive solution while maintaining the simplicity of the system. The utility function of a user in a noncooperative game is rate obtained in the interference channel given by

$$R_k(\mathbf{p}) = \sum_{i=1}^{N} \log_2 \left(1 + \frac{G_{k,k}^n P_k^n}{\sigma^2 + \sum_{j \neq k} G_{j,k}^n P_j^n} \right), \tag{20.26}$$

where $R_k(\mathbf{p})$ is the rate, which can be achieved under the constraint of the PSD mask having a power distribution $\mathbf{p}$. This rate is the achievable rate when all the users are transmitting at their maximum powers under the constraint of the mask that results in a suboptimal performance under medium- and high-interference conditions. Therefore, cooperative games are defined to minimize the effect of interference.

The cooperative game is implemented as a joint FDM/TDM scheme in the system. Each sub-carrier is shared by the users on a TDM basis with each user occupying the sub-carrier for a specific amount of time. For a two-user case, let α_1 and α_2 be the TDM strategies of user 1 and user 2, which is the proportion of the time slot for which they occupy the channel. This eliminates the co-channel interference from the other user because they do not transmit at the same time. Also since $\alpha_2 = 1 - \alpha_1$, there is need for only strategy α_1, which is denoted by α. The utility function for a cooperative game is given by

$$R_k(\boldsymbol{\alpha}) = \sum_{i=1}^{N} \alpha_k^n R_k^n = \sum_{i=1}^{N} \alpha_k^n \log_2 \left(1 + \frac{G_k^{n,n} p_k^n}{\sigma^2} \right), \tag{20.27}$$

where $R_k(\boldsymbol{\alpha})$ is the achievable rate when user k is playing the strategy α_k^n on channel n. Since it is a TDM scheme, there is no interference from the other users during the α_k^n time slot of user k.

The spectrum management problem can now be defined as

$$\max \left(R_1(\alpha) - R_1(\mathbf{p}) \right) \left(R_2(\alpha) - R_2(\mathbf{p}) \right)$$

$$\text{s.t.} \quad R_k(\mathbf{p}) \leq R_k(\boldsymbol{\alpha}). \forall k, n. \tag{20.28}$$

The allocation of the sub-carrier is done in two ways [12,13]:

1. Both users share the channel ($0 \leq \alpha \leq 1$) if

$$\frac{R_1(n)}{R_1(\alpha) - R_1(\mathbf{p})} = \frac{R_2(n)}{R_2(\alpha) - R_2(\mathbf{p})}. \tag{20.29}$$

2. Only one user uses the channel at a time ($\alpha = 1$) if

$$\frac{R_k(n)}{R_k(\alpha) - R_k(\mathbf{p})} > \frac{R_{-k}(n)}{R_{-k}(\alpha) - R_{-k}(\mathbf{p})}. \tag{20.30}$$

In the first case, the channel is shared by the users while the second case is the TDM scheme discussed previously. For simplification, we assume that at least one sub-carrier is shared between the two users represented by n_s, which satisfies Equation 20.29. Then the sub-carriers $1 \leq n < n_s$ are used by user 1 and the sub-carriers $n_s < n \leq N$ are used by user 2. Thus, implementing the TDM technique ensures that the interference is eliminated and leads to maximization of the rate region.

To further improve the rate region, NBS is applied to the sub-carriers. If there is a positive surplus rate given by $R_k(n) - R_k(\mathbf{p})$, then the sub-carriers are subject to the NBS [12,13], that is, the shared frequencies are allocated to an individual user and the resulting rate region is maximized. The simulation results in [12,13] show that the rate region obtained through NBS is close to the optimal rate region.

To find the shared sub-carrier n_s, we begin by defining a few parameters [12,13]. Let $L_n = R_1(n)/R_2(n)$ be the ratio of rates of user 1 and user 2 on sub-carrier n, which is sorted in descending order. Let $\Gamma_n = A_n/B_n$ be the ratio of the surplus rates, where

$$A_n = \sum_{m=1}^{n} R_1(m) - R_1(\mathbf{p}) \tag{20.31}$$

and

$$B_n = \sum_{m=n+1}^{N} R_2(m) - R_2(\mathbf{p}). \tag{20.32}$$

A_n and B_n indicate the surplus of rate of user 1 and user 2 when allocated with sub-carriers $1, 2, \ldots, n$ and $n + 1, \ldots, N$, respectively. We also define $n_{\min} = \arg\min_n A_n \geq 0$ and $n_{\max} = \arg\min_n B_n < 0$. If an NBS exists, then $n_{\min} \leq n_s \leq n_{\max}$. The lemma in [12,13] provides the solution for the TDM strategy α to be followed by the users.

1. If NBS exists and n_s lies in the range of $[n_{\min}, n_{\max})$, then $\alpha(n_s) = \max(0, g)$ where

$$g = 1 + \frac{B_n}{2R_2(n_s)} \left(1 - \frac{\Gamma_{n_s}}{L_{n_s}}\right). \tag{20.33}$$

2. If an NBS exists and n_s is not in the range of $(n_{\min}, n_{\max})$, then $n_s = n_{\max}$ and $\alpha(n_s) = g$.

Thus, the joint FDM/TDM scheme helps improve the rate region by eliminating the co-channel interference through the TDM technique. Also, the cooperative game involving the NBS for the surplus rate of each user results in a rate region that is close to the optimal rate region.

20.5 Summary

This chapter discusses the various approaches to solve the spectrum management problem under interference channel for an OFDM system for maximizing the capacity of the system while adhering to the system constraints. The results of various algorithms along with their complexities were presented, which provide an insight to the recent developments. Game theoretic approaches were analyzed, which provide a novel way to share the sub-carriers to achieve maximum capacity in different interference conditions. Cooperative game based on the forming of coalitions among the users is used to maximize the overall rate of the network in high-interference condition. Although this approach needs limited signaling between the users to decide the coalitions, its performance is much better than the other algorithms in the high-interference conditions. The NBS implemented in the cooperative game ensures fairness among the users while maintaining a good rate performance. For the medium-interference scenario, the noncooperative game approach is simple to implement but the performance can be poor. The network efficiency can be improved by the introduction of the virtual referee, which removes the users from using a sub-carrier if they are causing high interference to other users. Finally, a joint technique employing TDM strategy for the sharing of each sub-carrier is discussed using cooperation under the medium-interference scenario.

References

1. L. Wooyul, K. Youngjae, M. H. Brady, and J.M. Cioffi, Band-preference dynamic spectrum management in a DSL environment, *IEEE Global Telecommunications Conference, 2006 (GLOBECOM '06)*, Istanbul, Turkey, November 27–December 1, 2006, pp. 1–5.

2. K. J. Kerpez, D. L. Waring, S. Galli, J. Dixon, and P. Madon, *Advanced DSL Management*, IEEE Communications Magazine, 41(9), 116–123, September 2003.

3. W. Lee, Y. Kim, M. H. Brady, and J. M. Cioffi, Band-preference dynamic spectrum management in a DSL environment, *IEEE Global Telecommunications Conference (GLOBECOM '06)*, San Francisco, CA, pp. 1–5, November 27–December 1, 2005.

4. J. M. Cioffi, W. Rhee, M. Mohseni, and M. H. Brady, Band preference in dynamic spectrum management, *Eurasip Conference on Signal Processing*, Vienna, Austria, September 2004.

5. M. Charafeddine, Z. Han, A. Paulraj, and J. Cioffi, Crystallized rates region of the interference channel via correlated equilibrium with interference as noise, *IEEE International Conference on Communications*, Dresden, Germany, June 14–18, 2009.

6. W. Yu, W. Rhee, S. Boyd, and J. M. Cioffi, Iterative water-filling for gaussian vector multiple access channels, *IEEE Transactions on Information Theory*, 50(1), 145–152, January 2004.

7. R. Cendrillon, M. Moonen, J. Verlinden, T. Bostoen, and W. Yu, Optimal multiuser spectrum management for digital subscriber lines, *IEEE International Conference on Communications*, vol. 1, pp. 1–5, Paris, France, June 2004.

8. S. Boyd and L. Vandenberghe, *Convex Optimization*, Cambridge University Press, Cambridge, U.K. 2004.

9. R. Cendrillon and M. Moonen, Iterative spectrum balancing for digital subscriber lines, *IEEE Transactions on Communications*, 54(7), 1937–1941, May 2005.

10. J. Papandriopoulos and J. S. Evans, Low-complexity distributed algorithms for spectrum balancing in multi-user DSL networks, *IEEE International Conference on Communications (ICC 2006)*, Istanbul, Turkey, 7, pp. 3270–3275, June 2006.

11. R. Cendrillon, J. Huang, M. Chiang, and M. Moonen, Autonomous spectrum balancing for digital subscriber lines, *IEEE Transactions on Signal Processing*, 55(6), 4241–4257, August 2005.

12. A. Leshem and E. Zehavi, Game theory and the frequency selective interference channel, *IEEE Signal Processing Magazine*, 26(5), 28–40, September 2009.

13. A. Leshem and E. Zehavi, Cooperative game theory and the gaussian interference channel, *IEEE Journal on Selected Areas in Communications*, 26(7), 1078–1088, September 2008.

14. Z. Han and K. J. R. Liu, *Resource Allocation for Wireless Networks*, Cambridge University Press, New York, 2008.

15. Z. Han, Z. Ji, and K. J. R. Liu, Non-cooperative resource competition game by virtual referee in multi-cell OFDMA networks, *IEEE Journal on Selected Areas in Communications*, 25(6), 1079–1090, August 2007.

16. Z. Han, Z. Ji, and K. J. R. Liu, Fair multiuser channel allocation for OFDMA networks using Nash bargaining solutions and coalitions, *IEEE Transactions on Communications*, 53(8), 1366–1376, August 2005.

Chapter 21

Admission Control in IEEE 802.11e Wireless LAN: A Game-Theoretical Approach

Jia Hu, Geyong Min, Weijia Jia, and Mike E. Woodward

Contents

Admission control is an important mechanism for the provisioning of the Quality-of-Service (QoS) in the IEEE 802.11e wireless local area networks (WLANs). In this chapter, we present an efficient admission control scheme based on analytical modeling and noncooperative game theory where the access point (AP) and n new users are the players. The decision of admission control is made by virtue of the strategies to maximize the utilities of the players, which are determined by the QoS performance metrics in terms of the end-to-end delay and frame loss probability. To obtain these required performance metrics, we develop a new analytical model incorporating the contention

483

window (CW) and transmission opportunity (TXOP) differentiation schemes in the IEEE 802.11e protocol under unsaturated working conditions. The efficiency of the proposed admission control scheme is validated via NS-2 simulation experiments. Utilizing the admission control scheme, we investigate the capacity of WLANs under different network configurations and QoS constraints. The numerical results demonstrate that the proposed admission control scheme can maintain the system operation at an optimal point where the utility of the AP is maximized subject to the QoS constraints of both the real-time and non-real-time users. Moreover, this admission control scheme can improve the utility of the AP in WLANs compared to the legacy admission control schemes without the use of game theory.

21.1 Introduction

WLANs based on the IEEE 802.11 standard [10] have been widely deployed in public and residential places for wireless access to the Internet. With the increasing popularity of multimedia applications, the provisioning of QoS in WLANs has drawn considerable research interests. The basic medium access scheme of the IEEE 802.11 WLANs is the distributed coordination function (DCF) [10], which cannot provide any support of QoS. Therefore, an extension of the DCF, called the Enhanced distributed channel access (EDCA), has been proposed in the IEEE 802.11e standard [11]. Although EDCA provides service differentiation among various traffic classes, the QoS constraints of the real-time applications cannot be guaranteed. This problem becomes more serious when the wireless channel is overloaded.

Admission control is an important mechanism to guarantee the user-perceived QoS in WLANs and thus has received significant research efforts [1,2,4,7,13,14,16–19,22,23]. In WLANs, the AP (i.e., service provider) and new users have to cope with a limited radio resource that imposes a conflict of interests. For instance, the AP wants to increase its utility by improving the channel utilization and accommodating more new users. On the other hand, new users want to maximize their own utility by achieving the highest QoS if possible. Since the two objectives are different and often conflict with each other, the AP and new users do not have the apparent incentive to cooperate. Therefore, the noncooperative game theory [8] has been applied to solve the admission control problem in wireless networks from the perspectives of both the service provider and new users.

Cost-efficient analytical models with good accuracy and lower computation complexity compared to simulation experiments can be used in the admission control schemes which need the real-time calculation and estimation of the performance metrics. Therefore, we develop a new analytical model for EDCA and then propose an admission control scheme based on a game-theoretical approach. This model is used to calculate the QoS performance metrics in terms of end-to-end delay and frame loss probability in WLANs, which are adopted by the game-theoretical approach to carry out the admission control. The analytical model accommodates the CW and TXOP differentiation schemes of EDCA. Moreover, it considers the network working conditions where the traffic loads are unsaturated. The admission control is formulated as a noncooperative $(n + 1)$-player game. The AP makes the decision to admit or reject new users (i.e., admission control) according to the strategy that maximizes its utility. On the other hand, new users accept the service offered by the AP only if their own QoS constraints can be guaranteed.

The efficiency of the proposed admission control scheme is validated through NS-2 simulation experiments and is used to carry out the investigation on the capacity of WLANs under different network configurations and QoS constraints. The numerical results demonstrate that our admission control scheme can maintain the system operation at an optimal point where the utility of the AP is maximized with the QoS constraints of various users. Moreover, the admission control scheme can achieve the larger utility of the AP than the legacy admission control schemes without the use of game theory.

The rest of this chapter is organized as follows. Section 21.2 reviews the 802.11 and 802.11e medium access control (MAC) protocols. Section 21.3 describes a detailed survey of the related work on EDCA models and admission control schemes. The analytical model for EDCA and the game-theoretical scheme for admission control are presented in Sections 21.4 and 21.5, respectively. We validate the admission control scheme via simulation results and carry out the performance evaluation in Section 21.6. Finally, the chapter is concluded in Section 21.7.

21.2 Medium Access Control

The DCF is the fundamental channel access scheme in the IEEE 802.11 MAC protocol [10]. A station with backlogged frames first senses the channel before transmission. If the channel is detected idle for a Distributed Inter-Frame Space (DIFS), the station transmits the frame. Otherwise, the station defers until the channel is detected idle for a DIFS, and then starts a backoff procedure by generating a random backoff counter.

The value of the backoff counter is uniformly chosen between zero and CW, which is initially set to CW_{min} and doubled after each unsuccessful transmission until it reaches a maximum value CW_{max}. It is reset to CW_{min} after the successful transmission or if the unsuccessful transmission attempts reach a retry limit. The backoff counter is decreased by one for each time slot when the channel is idle, halted when the channel is busy, and resumed when the channel becomes idle again for a DIFS. A station transmits a frame when its backoff counter reaches zero. Upon the successful reception of the frame, the receiver sends back an ACK frame immediately after a Short Inter-Frame Space (SIFS). If the station does not receive the ACK within a timeout interval [10], it retransmits the frame. Each station maintains a retry counter that is increased by one after each retransmission. The frame is discarded after an unsuccessful transmission if the retry counter reaches the retry limit.

The EDCA [11] was designed to enhance the performance of the DCF and provide the differentiated QoS through assigning different EDCA parameters including AIFS values, CW sizes, and TXOP limits. Specifically, a smaller AIFS/CW results in a larger probability of winning the contention for the channel. On the other hand, the larger the TXOP limit then the longer are channel holding times of the station winning the contention.

In DCF, each station can transmit only one frame once it wins the channel, and then needs to contend for the channel again after a successful transmission. However, the TXOP scheme allows the burst transmission. Specifically, a station gaining the channel can transmit the frames available in its buffer successively provided that the duration of transmission does not exceed the TXOP limit [11]. Each frame is acknowledged by an ACK after an SIFS interval. The next frame is transmitted immediately after it waits for an SIFS upon receiving the ACK. If the transmission of any frame fails, the burst is terminated and the station contends again for the channel to retransmit the failed frame. The TXOP scheme is an efficient method to improve the utilization of scarce wireless bandwidth because the contention overhead is amortized by all the frames transmitted within a burst. Moreover, it can provide service differentiation by assigning different TXOP limits to various applications according to their specific QoS requirements.

21.3 Related Work

There has been much research activity on modeling the performance of the CW and TXOP schemes in EDCA [5,9,15,21,24–26]. Most of these studies have been based on Bianchi's two-dimensional Markov chain [3] for modeling the DCF under the assumption of saturated traffic conditions. For instance, Xiao [26] extended the Markov chain in [3] to account for the CW scheme of EDCA. Tinnirello and Choi [24] compared the saturation throughput of the TXOP scheme coupled by different ACK policies. Vitsas et al. [25] presented a simplified model for calculating the throughput and access delay of the burst transmission scheme under saturated traffic loads. Li et al. [15] analyzed the saturation throughput of the TXOP scheme with the block ACK policy under noisy channel conditions. Peng et al. [21] evaluated the throughput of distinct ACs as a function of various TXOP limits in the saturated WLANs. Since the practical network traffic conditions are always unsaturated, Hu et al. [9] proposed an analytical model for the TXOP scheme under unsaturated traffic loads.

Admission control schemes for WLANs based on analytical modeling have the advantage of computation efficiency and thus have attracted considerable research interests [1,2,4,13,17,18,22,23]. For instance, Pong and Moors [22] proposed to adjust the sizes of CW of different stations to fulfill the goal of admission control. Their scheme was based on the analytical model for DCF proposed in [3] and was limited to saturated traffic conditions. Chen et al. [4] proposed two admission control schemes based on the performance metrics in terms of the average delay and the channel occupancy ratio. However, their model only considered the CW scheme in EDCA. Garroppo et al. [7] presented an admission control approach using the mean channel occupancy time calculated by the EDCA model proposed in [5], which did not take the TXOP scheme into account.

In WLANs, the AP and users have to cope with a limited channel bandwidth that imposes a conflict of interests between them. Therefore, the noncooperative game theory [8] has been applied to solve the admission control problem in wireless networks from the perspectives of both the service provider and users. For example, Kuo et al. [13] used a noncooperative game-theoretic approach for admission control in WLANs where the performance measures required in the game-theoretical approach were obtained from simulations. Lin et al. [16] proposed an integrated admission control and rate control method for Code Division Multiple Access (CDMA) wireless networks based on the noncooperative game theory. Niyato and Hossain [19] proposed a game-theoretical framework for bandwidth allocation and admission control in IEEE 802.16 broadband wireless networks, where the QoS performance metrics were calculated through a queueing model. In this chapter, we aim to present a game-theoretical admission control scheme for the IEEE 802.11e WLANs where the performance measures are obtained from a new analytical model for EDCA.

21.4 Analytical Model for EDCA

We consider a scenario of C classes of stations where Class i ($i = 1, 2, ..., C$) has n_i stations. The arrival traffic at stations in Class i follows a Poisson process with mean arrival rate λ_i. The AIFS differentiation scheme is not considered in this model for the sake of simplicity. In this section, the term *time slot* denotes the time interval between the starts of two consecutive decrements of the backoff counter, while the term *physical time slot* represents a fixed time interval (unit time) specified in the protocol [10]. We extend the analytical model proposed in [3] to derive the probability of burst transmission in a randomly chosen time slot, τ_i, of a station in Class i under non-saturation traffic conditions. The collision probability, p_i, is equal to the probability that at least one of the remaining stations transmits in a considered time slot. p_i can be expressed as

$$p_i = 1 - (1 - \tau_i)^{n_i - 1} \prod_{r \neq i} (1 - \tau_r)^{n_r} \tag{21.1}$$

The burst transmission probability, τ_i, is given by

$$\tau_i = (1 - Pi_0)\tau_i' \tag{21.2}$$

where

  Pi_0 is the probability that the transmission queue of the station in Class i is empty
  τ_i' is the probability that the station transmits given that its transmission queue is nonempty

τ_i' is given by [3]

$$\tau_i' = \frac{2(1 - 2p_i)}{(1 - 2p_i)(W_i + 1) + p_i W_i (1 - (2p_i)^{m_i}} \tag{21.3}$$

where

  m_i represents the maximum backoff stage
  W_i is the minimum CW of the station in Class i

We define the service time as the time interval from the instant that a Head-of-Burst (HoB) frame starts contending for the channel to the instant that the whole data burst is acknowledged following successful transmission. The service time consists of two parts: the channel access delay and burst transmission delay. The former is the time interval from the instant that the HoB frame reaches the head of its transmission queue and starts contending for the channel, until it wins the contention and is ready for transmission. The latter is the time interval of successfully transmitting the data burst. Let $E[S_{vi}]$, $E[A_i]$, and $E[B_v]$ denote the mean service time, channel access delay, and burst transmission delay, respectively, where v represents the number of frames transmitted in a burst and i denotes that the burst is transmitted from a station in Class i. The average channel access delay, $E[A_i]$, can be given by

$$E[A_i] = T_c \varphi_i + \sigma_i' \delta_i \tag{21.4}$$

where

  T_c is the average collision time
  σ_i' is the average length of a time slot

φ_i accounts for the average number of collisions before a successful transmission from the station and is given by

$$\varphi_i = \sum_{j=0}^{\infty} j p_i^j (1 - p_i) = \frac{p_i}{1 - p_i} \tag{21.5}$$

δ_i denotes the average number of time slots that the station defers during backoff stages and can be expressed as

$$\delta_i = \sum_{j=0}^{\infty} \sum_{h=0}^{j} \frac{W_h - 1}{2} p_i^j (1 - p_i) \tag{21.6}$$

where

$p_i^j(1 - p_i)$ is the probability that the HoB frame is successfully transmitted after j collisions

$(W_h - 1)/2$ denotes the mean of the backoff counters generated in the hth backoff stage

The average collision time, T_c, is given by

$$T_c = T_{\text{DIFS}} + T_{\text{DATA}} + T_{\text{SIFS}} + T_{\text{ACK}} + 2\delta \tag{21.7}$$

where

T_{DATA} is the average time required for transmitting the frame

δ is the propagation delay

Let P_{t_i} be the probability that at least one station among the remaining stations transmits in a considered time slot, given that the station in Class i is in the backoff procedure. Similar to Equation 21.1, P_{t_i} can be written as

$$P_{t_i} = 1 - (1 - \tau_i)^{n_i - 1} \prod_{r \neq i} (1 - \tau_r)^{n_r} \tag{21.8}$$

Given that the station in Class i is in the backoff procedure, the probability, $P_{S_i}^r$, that a station in Class r successfully transmits among the remaining stations can be expressed as

$$P_{S_i}^r = \tau_r (1 - \tau_r)^{n_r - 1} (1 - \tau_i)^{-1} \prod_{k \neq r} (1 - \tau_k)^{n_k} \tag{21.9}$$

The average length of a time slot, σ_i', is obtained by considering the fact that the channel is idle with probability $(1 - P_{t_i})$, a successful transmission occurs with probability $\sum_{r=1}^{C} P_{S_i}^r$, and a collision happens with probability $\left(P_{t_i} - \sum_{r=1}^{C} P_{S_i}^r\right)$.

$$\sigma_i' = (1 - P_{t_i})\sigma + \sum_{r=1}^{C} P_{S_i}^r T_S^r + \left(P_{t_i} - \sum_{r=1}^{C} P_{S_i}^r\right) T_c \tag{21.10}$$

where

σ is the duration of a physical time slot

T_S^r is the average time interval for the successful transmission of a burst from the station in Class r

T_S^r is given by

$$T_S^r = \frac{\sum_{v=1}^{K_r} E[B_v] L_{vr}}{1 - Pr_0} \tag{21.11}$$

where

K_r denotes the maximum number of frames that can be transmitted in a TXOP limit of the station in Class r

the denominator $(1 - Pr_0)$ indicates that the occurrence of burst transmission is conditioned on the fact that there is at least one frame in the transmission queue of the station

L_{vr} $(1 \leq v \leq K_r)$ is the probability of having v frames within the burst transmitted from the station in Class r

The transmission queue at the station in Class i can be modeled as an $M/G^{[1,K_i]}/1/N$ queueing system where the superscript $[1, K_i]$ denotes that the number of frames transmitted in a burst ranges from 1 to K_i and N represents the system capacity at each station. The service time of the queueing system is dependent on the number of frames transmitted in a burst and is characterized by an exponential distribution function with mean $E[S_{vi}]$. Thus, the mean service rate, μ_{vi}, is readily given by $1/E[S_{vi}]$.

The state-transition-rate diagram of the underlying Markov chain for the queuing system is shown in Figure 21.1 where each state denotes the number of frames in the queueing system. The transition rate matrix, $\mathbf{G}$, of the Markov chain can be obtained by the state-transition-rate diagram. The steady-state probability vector, $\mathbf{P} = (P_{vi}, \ v = 0, 1, \ldots, N)$, of the Markov chain satisfies the following equations:

$$\mathbf{PG} = 0 \quad \text{and} \quad \mathbf{Pe} = 1 \tag{21.12}$$

Solving these equations yields the steady-state vector as [6]

$$\mathbf{P} = \mathbf{u}(\mathbf{I} - \Re + \mathbf{eU})^{-1} \tag{21.13}$$

where
$\Re = \mathbf{I} + \mathbf{G}/\min\{G(\rho, \rho)\}$
$\mathbf{u}$ is an arbitrary row vector of $\Re$
$\mathbf{e}$ is a unit column vector

Therefore, we can obtain the probability, P_{0i}, that the transmission queue of the station in Class i is empty.

After obtaining the expression of $\mathbf{P}$, we can calculate L_{vi} as

$$\begin{cases} L_{vi} = P_{vi}, & 1 \le v < K_i \\ L_{vi} = \sum_{x=K_i}^{N} P_{xi}, & v = K_i \end{cases} \tag{21.14}$$

Given the loss probability, P_{Ni}, the throughput Γ_i of the station in Class i can be computed by

$$\Gamma_i = \lambda_i E[P](1 - P_{Ni}) \tag{21.15}$$

where
$E[P]$ is the frame payload length
λ_i is the mean traffic arrival rate of the station in Class i

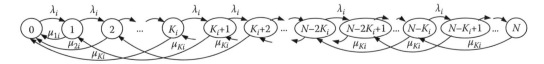

Figure 21.1   State-transition-rate diagram of $M/G^{[1,K_i]}/1/N$ queue.

The end-to-end delay is the time duration from the instant that a frame enters the transmission queue of the station to the instant that the frame is acknowledged after successful transmission. By virtue of Little's theorem [12], the average end-to-end delay, $E[d_i]$, is given by

$$E[d_i] = \frac{E[N_i]}{\lambda_i(1 - P_{Ni})} \tag{21.16}$$

where

$E[N_i] = \sum_{x=0}^{N} xP_{xi}$ is the average number of frames in the queueing system

P_{Ni} is the frame loss probability denoted as LO

$\lambda_i(1 - P_{Ni})$ is the effective arrival rate of the traffic entering into the transmission queue since that the arriving frames are discarded if they find the finite buffer full

After obtaining the performance measures including the end-to-end delay and frame loss probability through the analytical model, the decisions of admission control can be made by virtue of the game-theoretical approach presented in the following section.

21.5 Game-Theoretical Approach for Admission Control

The admission control is formulated as a noncooperative non-zero-sum $(n + 1)$-player game where the AP and n new users requiring services are the players. The game is non-zero-sum [8] because the payoffs of both the AP and new user would increase if the admission of a new user does not degrade the QoS of the ongoing users. In a zero-sum game, an increase of one player's payoff implies a decrease of another player's payoff. Before presenting the admission control scheme, we first introduce the notion of utility (i.e., payoff), which is a measure of the user's satisfaction level and can be modeled as a function of the QoS perceived by the user. We employ the modified sigmoid function [16] to characterize the user's utility. The value of the utility function is between zero and one, which can facilitate the representation of the user's satisfaction from the lowest to highest levels. Moreover, both the slope and the centre of the curve of this function can be adjusted to customize the utility of a specific class of users. Different traffic classes have diverse QoS preferences. For the user with real-time applications such as voice and video, there is a stringent constraint on the end-to-end delay. Therefore, the utility for the real-time user depends on the end-to-end delay, $E[d]$, and can be expressed as

$$U_{re} = 1 - \frac{1}{1 + \exp(-\alpha_r(E[d] - E[d_{tor}]))} \tag{21.17}$$

where

the value of α_r indicates the user's sensitivity to the increase in delay

$E[d_{tor}]$ is the largest delay that the real-time user could tolerate (i.e., delay constraint)

With the increase of delay, α_r determines the rate of the decrease and $E[d_{tor}]$ determines when the utility decreases below 0.5.

For the user with non-real-time applications, there is no strict delay requirement and its QoS satisfaction depends on the frame loss probability, LO. Thus, the utility function for the non-real-time user can be given by

$$U_{nr} = 1 - \frac{1}{1 + \exp(-\alpha_{nr}(LO - LO_{tor}))} \tag{21.18}$$

where the value of α_{nr} indicates the user's sensitivity to the increase of frame loss probability and LO_{tor} is the largest frame loss probability that the non-real-time flow could tolerate (i.e., loss probability constraint). With the increase of frame loss probability, α_{nr} determines the rate of the decrease and LO_{tor} decides when the utility decreases below 0.5.

The admission control scheme is invoked when the AP receives the requests from n new users. The AP then makes a decision which requests to be accepted or rejected. The new users also need to decide whether to accept the service or deny it. The process of making the decisions is modeled as an admission control game that is solved to obtain the best strategies of the players. Let the kth request belong to the user in Class C_k ($1 \leq C_k \leq C$). Based on either admitting or rejecting a request, the AP has 2^n strategies, each of which can be expressed by an n-element vector. The ith strategy for the AP is given by

$$A_i = [a_{i1}, a_{i2}, \ldots, a_{in}] \tag{21.19}$$

where $a_{ik} \in \{0, 1\}$ for $1 \leq i \leq 2^n$ and $1 \leq k \leq n$, $a_{i,k} = 0$ means to accept the kth request while $a_{i,k} = 1$ represents to reject it. Similarly, each new user has two strategies as well. Let B_k be the kth new user's strategy ($1 \leq k \leq n$), then $B_k = 0$ denotes to accept the service provided by the AP and $B_k = 1$ means to deny the service.

The payoff matrices of the AP and the kth new user, **S** and **R**$_k$, are denoted as [16]

$$\mathbf{S} = [s_{A_i B_1 B_2 \cdots B_n}]_{2^n \times 2 \times \cdots \times 2}, \quad \mathbf{R}_k = \left[r^k_{A_i B_1 B_2 \cdots B_n}\right]_{2^n \times 2 \times \cdots \times 2} \tag{21.20}$$

where $s_{A_i B_1 B_2 \cdots B_n}$ and $r^k_{A_i B_1 B_2 \cdots B_n}$ denote the payoff of the AP and kth new user, respectively, if the AP chooses strategy A_i and the jth ($j = 1, 2, \ldots, n$) new user chooses strategy B_j. The values of each element of payoff matrices **S** and **R**$_k$ are given by

$$s_{A_i B_1 B_2 \cdots B_n} = U + \sum_{1 \leq k \leq n \, \cap \, a_{ik} = B_k = 0} (U_k - L_k) \tag{21.21}$$

$$r^k_{A_i B_1 B_2 \cdots B_n} = (1 - a_{ik})((1 - B_k)U_k + B_k(1 - U_k)) \tag{21.22}$$

where
  U represents the total payoff of the ongoing users
  L_k is the decrease in payoff of the ongoing users after admitting the kth new user
  U_k is the payoff of the kth new user after it accepts the service
  $(1 - U_k)$ is the payoff of the new user when it denies the service due to the undesired QoS performance

To obtain the best strategies that maximize the payoffs of the AP and new users, we need to obtain the Nash equilibrium [8], which is a strategy profile where none of the players can increase its own utility by unilaterally changing its strategy. Alternatively, a strategy profile $(A_{i'}, B'_1, B'_2, \ldots, B'_n)$ is said to constitute a Nash equilibrium solution of the game if the following inequalities are satisfied for any other strategies A_i and B_j ($j = 1, 2, \ldots, n$).

$$\begin{cases} S\left(A_{i'}, B'_1, B'_2, \ldots, B'_n\right) \geq S(A_i, B_1, B_2, \ldots, B_n) \\ R_k\left(A_{i'}, B'_1, B'_2, \ldots, B'_n\right) \geq R_k(A_i, B_1, B_2, \ldots, B_n), \quad 1 \leq k \leq n \end{cases} \tag{21.23}$$

Before discussing the solution of Nash equilibrium, we first introduce the concept of dominance in game theory [8].

Definition 21.1   In a noncooperative n-player game, strategy A'_τ of player P_τ ($\tau = 1, 2, \ldots, n$) dominates another strategy A_τ if it obtains the better payoff with strategy A'_τ than with strategy A_τ, for any strategy A_γ of any other player P_γ ($1 \leq \gamma \leq n$ & $\gamma \neq \tau$).

Using the above definition, we can notice that there exists a dominant strategy, A_d, for the AP satisfying

$$s_{A_d B_1 B_2 \cdots B_n} = \max\{s_{A_i B_1 B_2 \cdots B_n}\} \quad \text{for } 1 \leq i \leq 2^n \tag{21.24}$$

which, with Equation 21.21, can be expressed as

$$\sum_{1 \leq k \leq n \, \cap \, a_{dk} = B_k = 0} (U_k - L_k) = \max \left\{ \sum_{1 \leq k \leq n \, \cap \, a_{ik} = B_k = 0} (U_k - L_k) \right\} \quad \text{for } 1 \leq i \leq 2^n \tag{21.25}$$

Similarly, there exists a dominant strategy, B_k, for the kth new user satisfying

$$r^k_{A_i B_1 B_2 \cdots B_n} = \max\{r^k_{A_i B_1 B_2 \cdots B_n}\} \quad \text{for } B_k \in \{0, 1\} \tag{21.26}$$

which, with Equation 21.22, leads to

$$\begin{cases} B_k = 0 & \text{if } U_k \geq 0.5 \\ B_k = 1 & \text{if } U_k < 0.5 \end{cases} \tag{21.27}$$

Therefore, the Nash equilibrium strategy profile can be expressed as $A_d B_1 B_2 \cdots B_n$, if A_d and B_k ($1 \leq k \leq n$) satisfy Equations 21.25 and 21.27.

The proposed game-theoretical admission control algorithm follows the following steps:

- *Step 1*: The new users want to join the network exchange information with the AP to find out the number of active users of each class in the network and the CW and TXOP parameters for them.
- *Step 2*: The user adopts the analytical model to calculate the end-to-end delay (real-time user) or loss probability (non-real-time user) if it is admitted to join the network and then computes the utility U_k using Equations 21.17 and 21.18. Based on U_k, the user determines its strategy on whether to accept or deny the service using the policy shown in Equation 21.27.
- *Step 3*: The AP calculates its payoff with Equation 21.21, if admitting the new user, and then makes the admission control decision based on the policy given in Equation 21.25.

21.6 Validation and Performance Evaluation

This section first investigates the accuracy of the proposed analytical model through extensive NS-2 [20] simulation experiments and then conducts performance evaluation of the TXOP scheme using the model.

21.6.1 Validation

To validate the accuracy of the proposed analytical model, we compare the analytical results to those obtained from NS-2 simulation experiments. We also reveal the superior performance of the proposed admission control scheme. Users in WLANs are classified into the real-time and non-real-time ones with various QoS constraints. The traffic arrival rate of the real-time and non-real-time users is set to 0.2 and 0.1 Mbps, respectively. The sensitivity of the utility function for the real-time and non-real-time users is set to 10 and 5, respectively, that is, $\alpha_r = 10$ and $\alpha_{nr} = 5$. All the users are located in a Basic Service Set (BSS) of 150 m × 150 m rectangular grid. In what follows, without any specification, the system parameters are the same as those given in Table 21.1. The CW_{min} for the real-time and non-real-time user is set to 16 and 32, respectively, and the TXOP limit of the real-time and non-real-time users is set to 2 and 1 (frame), respectively.

Figure 21.2 plots the admission region of the real-time and non-real-time users. The delay and frame loss probability constraints are assumed to be 30 ms and 0.1, respectively. It is shown that the analytical predictions of the maximum number of admitted stations are in a good agreement with the simulation results. Figure 21.3 depicts the performance results of the proposed admission

Table 21.1 System Parameters

Frame payload	8000 bits	PHY header	192 bits
MAC header	224 bits	ACK	112 bits + PHY header
Channel data rate	11 Mbit/s	Maximum backoff stages	5
Basic rate	1 Mbit/s	Buffer size	40 frames
Propagation delay	2 μs	AIFS	50 μs
Slot time	20 μs	SIFS	10 μs

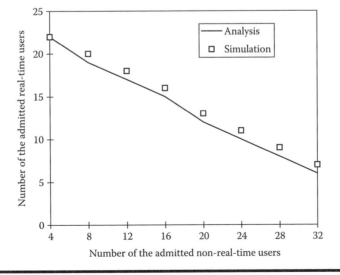

Figure 21.2 Maximum number of admitted real-time users versus that of admitted non-real-time users.

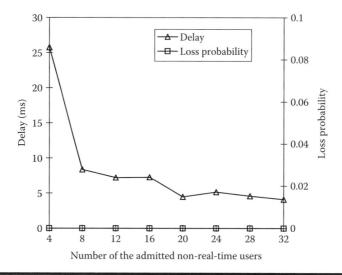

Figure 21.3   Simulation results with the proposed admission control scheme.

control scheme used to regulate the maximum number of admitted users. We can see that the simulation results with the proposed admission control scheme in terms of end-to-end delay and loss probability clearly satisfy the QoS constraints. The above observations demonstrate the effectiveness of the proposed admission control scheme.

21.6.2 Performance Evaluation

Utilizing the admission control scheme, we investigate the capacity of the 802.11e-based WLAN with different QoS requirements and MAC configurations. Figure 21.4 depicts the maximum number of admitted users in each class versus the TXOP limit of the real-time user with the different

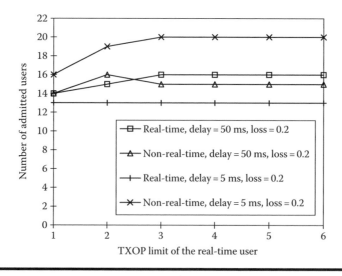

Figure 21.4   Maximum number of admitted users versus the TXOP limit of the real-time user.

delay constraints. The TXOP limit of the non-real-time user is fixed at 1 (frame). In the case that the delay constraint is 50 ms, we observe that the maximum number of both real-time and non-real-time users increases as the TXOP limit increases from 1 to 3. More specifically, the maximum number of the real-time users keeps increasing as the TXOP limit increases from 1 to 3, while the maximum number of the non-real-time users firstly increases as the TXOP limit increases from 1 to 2, and then decreases as the TXOP limit increases from 2 to 3. However, the further increase in the TXOP limit does not have any impact on the system capacity, and, thus, the maximum number of the real-time and non-real-time users that can be admitted stabilizes at 16 and 15, respectively.

We turn next to investigate the impact of the QoS constraints on the system capacity. When the delay constraint of the real-time user becomes more stringent (e.g., $E[d_{tol}] = 5$ ms), the AP admits less real-time users, while it can accept more non-real-time ones. The reason is that the incoming real-time users reject the service since their delay constraint cannot be satisfied while the incoming non-real-time users are admitted as their loss probability requirements can be fulfilled.

In Figure 21.5, we show the utility for the AP and new user, respectively, when the total number of users increases from 5 to 50 in two cases where Case 1 has the more stringent QoS requirements

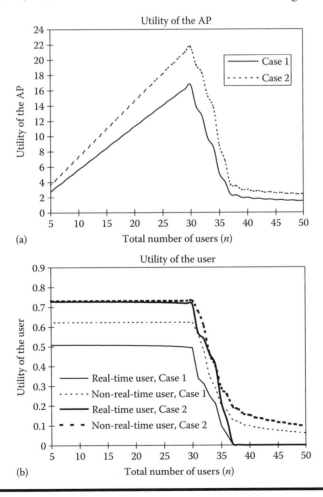

(a)

(b)

Figure 21.5 Utility versus the total number of users. Case 1: $E[d_{tol}] = 5$ ms, $L_{tol} = 0.1$; Case 2: $E[d_{tol}] = 100$ ms, $L_{tol} = 0.2$. (a) Utility of the AP, (b) Utility of the user.

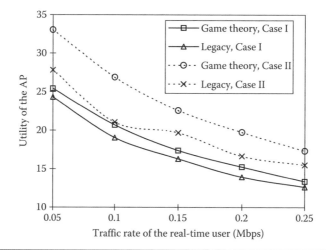

Figure 21.6 Comparison between the game-theoretical admission control scheme and legacy ones. Case I: $\alpha_r = 10$, $\alpha_{nr} = 5$; Case II: $\alpha_r = 30$, $\alpha_{nr} = 15$.

than Case 2. The figure shows that the point for maximizing the utility of the AP is $n = 30$ in both cases. For the new user, the utility decreases as the number of total users increases. The maximum number of users to keep the utility above 0.5 (i.e., the QoS constraints are satisfied) is $n = 26$ and $n = 31$ in Case 1 for the real-time and non-real-time users, respectively, and $n = 32$ in Case 2 for both classes of users. Therefore, the AP could admit 30 users while the new user will not accept the service if the number of total users is larger than 26 in Case 1, which means that the WLAN will accommodate at most 26 users that are decided by the new user in Case 1. On the other hand, in Case 2, the AP can admit 30 users and the new user would only accept the service if the number of total users is not larger than 32. Thus, the WLAN will accept a maximum number of 30 users that are decided by the AP in Case 2. The results demonstrate that the proposed admission control scheme can assure that the system operates under an optimal point where the utility of the AP is maximized with the QoS constraints of both the real-time and non-real-time users.

The legacy admission control schemes of EDCA make the decision of admission control only by virtue of the QoS constraints of users. Therefore, these legacy schemes cannot achieve the goal of maximizing the utility of the AP with the QoS constraints of users. To demonstrate the advantage of the game-theoretical approach, we compare the AP's utility of the proposed game-theoretical admission control scheme to that of the legacy ones in Figure 21.6. We fix the traffic rate of the non-real-time user at 0.2 Mbps and vary the traffic rate of the real-time user from 0.05 to 0.25 Mbps. Figure 21.6 clearly shows that the AP's utility in the game-theoretical scheme is larger than that in the legacy ones. Moreover, the difference of the AP's utility between the game-theoretical scheme and legacy ones rises as the sensitivity of the utility functions for users increases.

21.7 Conclusions

This chapter has presented an admission control scheme for the IEEE 802.11e WLANs based on the analytical model and game theory. The decision of admission control is made on the basis of the strategies maximizing the utilities of the players, which are calculated through the QoS performance measures including the end-to-end delay and frame loss probability derived from an analytical model for the IEEE 802.11e EDCA protocol. The admission control scheme has been validated through NS-2 simulation experiments and has been applied to investigate the network capacity of WLANs under different network configurations and QoS constraints. Specifically, we have evaluated the impact of the TXOP limit and the delay constraint over the network capacity. Moreover, we have demonstrated that the proposed game-theoretical admission control scheme can maintain the system operation at an optimal point where the utility of the AP is maximized with the QoS constraints of various users.

References

1. A. Abdrabou and W. Zhuang, Stochastic delay guarantees and statistical call admission control for IEEE 802.11 single-hop ad hoc networks, *IEEE Transactions on Wireless Communications*, 7, 10, 3972–3981, 2008.
2. C. M. Assi, A. Agarwal, and Y. Liu, Enhanced per-flow admission control and QoS provisioning in IEEE 802.11e wireless lANs, *IEEE Transactions on Vehicular Technology*, 57, 2, 1077–1088, 2008.
3. G. Bianchi, Performance analysis of the IEEE 802.11 distributed coordination function, *IEEE Journal on Selected Areas in Communications*, 18, 3, 535–547, 2000.
4. X. Chen, H. Zhai, X. Tian, and Y. Fang, Supporting QoS in IEEE 802.11e wireless LANs, *IEEE Transactions on Wireless Communications*, 5, 8, 2217–2227, 2006.
5. P. E. Engelstad and O. N. Osterbo, Analysis of the total delay of IEEE 802.11e EDCA and 802.11 DCF, *Proceedings of IEEE ICC'06*, vol. 2, Istanbul, Turkey, pp. 552–559, 2006.
6. W. Fischer and K. Meier-Hellstern, The Markov-modulated poisson process (MMPP) cookbook, *Performance Evaluation*, 18, 2, 149–171, 1993.
7. R. G. Garroppo, S. Giordano, S. Lucetti, and L. Tavanti, A model-based admission control for IEEE 802.11e networks, *Proceedings of IEEE ICC'07*, Glasgow, Scotland, U.K. pp. 398–402, 2007.
8. R. Gibbons, *A Primer in Game Theory*, Prentice Hall, Harlow, U.K., 1992.
9. J. Hu, G. Min, and M. E. Woodward, Analysis and comparison of burst transmission schemes in unsaturated 802.11e WLANs, *Proceedings IEEE GLOBECOM'07*, Washington, DC, pp. 5133–5137, 2007.
10. IEEE, Wireless LAN Medium Access Control (MAC) and Physical Layer (PHY) specifications, IEEE Standard 802.11, 1999.
11. IEEE, Wireless LAN Medium Access Control (MAC) and Physical Layer (PHY) specifications: Medium Access Control (MAC) Quality of Service (QoS) Enhancements, IEEE Standard 802.11e, 2005.
12. L. Kleinrock, *Queueing Systems: Theory*, vol. 1, John Wiley & Sons, New York, 1975.
13. Y. L. Kuo, H. K. Wu, and E. G. Chen, Noncooperative admission control for differentiated services in IEEE 802.11 WLANs, *Proceedings of IEEE GLOBECOM'04*, Dallas, TX, vol. 5, pp. 2981–2986, 2004.
14. F. H. Li, Y. Xiao, and J. Zhang, Variable bit rate VoIP in IEEE 802.11e wireless LANs, *IEEE Wireless Communications*, 15, 1, 56–62, 2008.
15. T. Li, Q. Ni, and Y. Xiao, Investigation of the block ACK scheme in wireless ad-hoc networks, *Wireless Communications and Mobile Computing*, 6, 6, 877–888, 2006.
16. H. Lin, M. Chatterjee, S. K. Das, and K. Basu, ARC: An integrated admission and rate control framework for competitive wireless CDMA data networks using noncooperative games, *IEEE Transactions on Mobile Computing*, 4, 3, 243–258, 2005.
17. L. Lin, H. Fu, and W. Jia, An efficient admission control for IEEE 802.11 networks based on throughput analysis of (un)saturated channel, *Proceedings of IEEE GLOBECOM'05*, St. Louis, MO, vol. 5, pp. 3017–3021, 2005.

18. A. Nafaa and A. Ksentini, On sustained QoS guarantees in operated IEEE 802.11 wireless LANs, *IEEE Transactions on Parallel and Distributed Systems*, 19, 8, 1020–1033, 2008.

19. D. Niyato and E. Hossain, QoS-aware bandwidth allocation and admission control in IEEE 802.16 broadband wireless access networks: A non-cooperative game theoretic approach, *Computer Networks*, 51, 11, 3305–3321, 2007.

20. NS-2 network simulator, http://www.isi.edu/nanam/ns/.

21. F. Peng, H. M. Alnuweiri, and V. C. M. Leung, Analysis of burst transmission in IEEE 802.11e wireless LANs, *Proceedings of IEEE ICC'06*, Istanbul, Turkey, vol. 2, pp. 535–539, 2006.

22. D. Pong and T. Moors, Call admission control for IEEE 802.11 contention access mechanism, *Proceedings of IEEE GLOBECOM'03*, San Francisco, vol. 1, pp. 174–178, 2003.

23. S. Shin and H. Schulzrinne, Call admission control in IEEE 802.11 WLANs using QP-CAT, *Proceedings of IEEE INFOCOM'08*, Phoenix, AZ, pp. 726–734, 2008.

24. I. Tinnirello and S. Choi, Efficiency analysis of burst transmission with block ACK in contention-based 802.11e WLANs, *Proceedings of IEEE ICC'05*, Seoul, South Korea, vol. 5, pp. 3455–3460, 2005.

25. V. Vitsas, P. Chatzimisios, A. C. Boucouvalas, P. Raptis, K. Paparrizos, and D. Kleftouris, Enhancing performance of the IEEE 802.11 distributed coordination function via packet bursting, *Proceedings of IEEE GLOBECOM Workshops'04*, Dallas, pp. 245–252, 2004.

26. Y. Xiao, Performance analysis of priority schemes for IEEE 802.11 and IEEE 802.11e wireless LANs, *IEEE Transactions on Wireless Communications*, 4, 4, 1506–1515, 2005.

Chapter 22

Intelligent Network Selection: Game-Theoretic Approaches

Manzoor Ahmed Khan and Fikret Sivrikaya

Contents

This chapter describes the application of game-theoretic approaches in the design of intelligent network selection and resource allocation in heterogeneous wireless networks. The technical, economical, and business dimensions of the problem are discussed. The chapter justifies the need for a paradigm shift and new business models in future wireless networks. The challenges involved in network selection are outlined and the state of the art is discussed.

22.1 Introduction

We observe an increasingly heterogeneous landscape of wireless access technologies, including UMTS, GSM, WiFi, WiMAX, etc., which are specialized in different environments and user contexts. The development as well as the business cycles of these technologies can assure us that they will be available simultaneously for the years to come. Consequently, there has been significant research activity on the integration and interoperability of these fundamentally different access technologies, which exhibit different service characteristics in terms of bandwidth, coverage, pricing, and Quality of Service (QoS) support. The initial concern for network operators was to increase connectivity by providing diversified methods of access for different types of end devices. However, the emergence of multi-interface terminals has shifted the simple connectivity issue to more rewarding resource allocation problems, whose solutions aimed at increasing the network efficiency and capacity as well as improving users' experience for an ample amount of services such as video on demand, video conferencing, and a variety of other applications.

The possible leveraging of high deployment costs and the possibility to increase revenues have also introduced the concept of network sharing between different operators. The problem of optimal allocation of bandwidth to multimedia applications over different wireless access networks is augmented with the possibility of using the bandwidth of other operators who are willing to share resources for mutual benefits.

In light of these developments, an important aspect that emerges in modern wireless communications is *network selection*, which can be broadly defined as mapping a mobile terminal or application to one of the available wireless access networks. This chapter focuses on the use of game-theoretic approaches for intelligent network selection. The actual decision for network selection may take place on both sides of the wireless channel: on the user terminal side or on the network operator side. To differentiate between the two, the term *interface selection* is used for terminal-based (user-based) network selection, and *resource distribution* for operator-based network selection throughout this chapter.

22.1.1 Generic Network Selection Model

The network selection decision depends on the inputs from different communication entities. The information these entities contain and their objectives vary depending on the level they reside on, as depicted in Figure 22.1. The network selection decision mechanism evaluates different criteria such as service type, user device capabilities, different market segments with varying QoS requirements, network technology conditions, service provider behavior, user location, user speed, user implicit preferences, etc. Evaluation is carried out using some policies or algorithms, and the outcome of decision mechanism is ideally the most suitable network service(s) that can satisfy the users' requirements. The huge number of variables involved in the decision process makes *network selection* an utterly complex problem. Table 22.1 presents a list of some major parameters that influence the network selection decision, together with their characteristics and their point of generation.

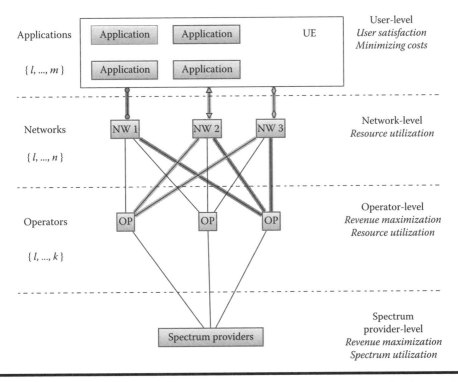

Figure 22.1 A hierarchical diagram of different stakeholders in telecommunications and their typical objectives.

Figure 22.2 depicts the abstract view of network selection process, where the decision mechanism takes network-specific, application-specific, and user-specific requirements as inputs. The decision mechanism is driven by the objective block, which represents the objectives of the user or network operator. The decision of network selection is based on the outcome of objective block that consists of *utility function* and *cost function* blocks. Both of these functions are associated with the stakeholders (i.e., end users and network operators). Let γ represent the set of all parameters for a stakeholder. Then the utility function $u : \gamma \to \mathbb{R}$ ranks each element in the set γ such that if $u(x) \geq u(y)$, then $x \in \gamma$ is preferred to $y \in \gamma$. It is commonly assumed that the utility function is nondecreasing; however, it improves with slower pace near some threshold value of considered parameter, where the player is indifferent to any value higher than the threshold value. For each input parameter, a utility curve or cost curve represents the stake holders' preferences, where the choice of utility or cost is driven by the problem formulation. Entities within objective block in Figure 22.2 may have different definitions for each stakeholder. For example, the utility function for an end user may be defined in terms of available bandwidth, received signal strength, degree of reliability, throughput; and the cost function in terms of bit error rate, transfer delay, service charges, transfer latency, etc. For the sake of simplicity, a common term *payoff* is used interchangeably for utility and cost functions, which refers to the satisfaction level of stakeholders.

There has been significant research activity on decision mechanisms for network selection based on both end users' and network operators' utility/cost functions. This chapter provides a comprehensive overview of both perspectives studied in the literature using game-theoretic approaches.

Table 22.1   Typical Parameters Influencing the Network Selection Decision

Parameter	Dynamic/Static	Source	Factors Involved
User preferences	Static for a call duration	UE	Willingness to pay, security, power, visual quality
Application constraints	Static	UE	QoS constraints, service requirements, application requirements, application context, variety of services, adaptation ability, minimum required bandwidth, maximum loss rate and latency allowed, delay bounds, traffic specification
Reachable APs	Static in a cell	UE	
Device capabilities	Static	UE	CPU speed, memory size, display I/O, transmitted power, network interface, built-in application software platform, supported modalities, maximum object size, screen size, number of colors
Device capabilities	Dynamic	UE	Battery status, interface status
User profile and history	Static	Network	User connectivity pattern, application requests, QoS required
Network capabilities	Static	Network	Network capabilities, network equipment capabilities, access technologies capabilities, access point bandwidth, Uplink and downlink bandwidth, modulation scheme
Network charging model	Static for at least a call duration	Network	Price per unit resource, price per unit time for resource usage
Potential next AP	Static in a cell	Network	Neighboring access points
Network status	Dynamic	Network	Network used capacity, traffic characteristics, maximum saturation throughput of AP, transmitted bandwidth, bandwidth per user, delay, throughput, response time, jitter, bit error rate, burst error, loss, signal strength, available services, average number of connection, connection holding time

Source: Pmamrat, K. et al., Resource management in mobile heterogeneous networks: State of the art and challenges, Inria-00258507, version 4, 2008; Kang, J.M. et al., Towards autonomic handover decision management in 4G networks, *MMNS*, Dublin, Ireland, pp. 145–157, 2006; Jo, J. and Cho, J., A cross-layer vertical handover between mobile Wimax and 3g networks, *IWCMC'08*, Crete, Greece, pp. 644–649, 2008.

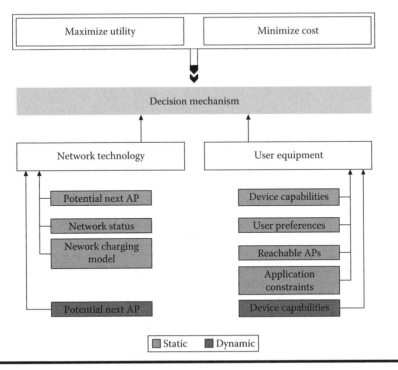

Figure 22.2 Demonstration of a typical decision process and related parameters for network selection.

Section 22.2 considers the user-centric network selection perspective, where the decision mechanism is based on the mobile terminal. Section 22.3 analyzes the network-centric approach, where the focus of game-theoretic network selection is shifted to the network operator side. The next section studies mixed approaches that place the users at the center in terms of their priorities and preferences but delegates the decision making to the network side for better resource utilization. Pricing strategies and their effect on network selection are outlined in Section 22.5. The chapter is concluded with a discussion of open issues and future research directions.

22.2 User-Based Network Selection (Interface Selection)

Modern mobile communication devices are equipped with multiple wireless network interfaces and hence can be associated with different network technologies, potentially belonging to different network operators. The decision mechanism on the user equipment (UE) gets information on relevant factors as inputs, including user preference, application-specific requirements, network conditions, price offers, etc. Upon receiving the information, the decision entity employs various decision-making techniques resulting in an interface selection. Typically the UE selects a single network interface for connectivity of the device, which is used by all user applications. However, with increased multitasking capabilities of mobile terminals, the interface selection can be done on a per-application basis, that is, the interface selection decision associates each application running on the UE separately to a suitable access technology or even multiple access technologies.

In its most generic sense, *user-centric* view considers that the users are free from subscription to any one operator and can instead dynamically choose the most suitable transport infrastructure from the available network providers for their terminal and application requirements [1]. In this approach, the decision of interface selection is delegated to the mobile terminal enabling end users to exploit the best available characteristics of different network technologies and network providers, with the objective of increased satisfaction. The generic term *satisfaction* can be interpreted in different ways, where a natural interpretation would be obtaining a high QoS for the lowest price. In order to more accurately express the user experience in telecommunications, the term QoS has been extended to include more subjective and application-specific measures beyond traditional technical parameters, giving rise to the *Quality of Experience (QoE)* concept. QoE reflects the collective effect of service performances that determines the degree of satisfaction of the end user, for example, what user really perceives in terms of usability, accessibility, retainability, and integrity of the service. The subjective quality perceived by the user has to be linked to the objective, measurable quality, which is expressed in application and network performance parameters resulting in QoE. Feedback between these entities is a prerequisite for covering the user's perception of quality. Nontechnical parameters necessary for evaluating QoE usually require subjective user evaluations that are hard to obtain. Existing methods include utilizing a graphical user interface (GUI) for user feedback [2,3], analytical hierarchy process (AHP) [4], pseudo-subjective quality assessment (PSQA) [5], and learning algorithms to reduce the user-system interaction [6].

Since user requirements span over different parameters of user preferences and application-specific requirements, the simplest and most commonly used methodology to get an overall user policy is a weighted sum of all parameters [7,8]. Letting l denote the number of available parameters, the payoff associated to the selection of an available access network may generally be defined as

$$Payoff\,(network) = \sum_{i=1}^{l} (weight_{parameter(i)} \times g_{parameter(i)}(x)), \tag{22.1}$$

where a careful choice of the cost/utility function $g_{parameter(i)}(x)$ is crucial [9].

To illustrate the fundamental concepts of game theory in network selection, we consider a simple two-player game between a user and a network provider. It is assumed that the user is already associated to a provider and its utility is translated as its satisfaction from the current service of current provider. Similarly, the network provider has a number of users associated to it, and it has some current revenue V prior to game play with player in question, where revenue is the natural utility of the network provider.

22.2.1 Noncooperative Game

Let $G = (P, S, U)$ represent the game, where $P = \{$network provider, end user$\}$ denotes the set of players, S represents the strategy set of players, and U is the set of player utilities. Strategy set of network operator o is $S_o = \{$high QoS/price offer π_h, low QoS/price offer $\pi_l\}$ and user u has the strategy set $S_u = \{$stay connected with current operator (st), handover to network operator o $(ho)\}$.

Let v_c, $(v_e - c)$ be the current and estimated valuation of users from the current operator and the offered service from target operator, respectively, where the valuation is driven by the satisfaction of users from service. Moreover, let $\tilde{c}$ represent the incurring cost by the operator on extending the QoS to users, c the payment by the user for the service, and $\tilde{V} = V + (c - \tilde{c})$ the expected revenue after the user hands over to the network, where $(c - \tilde{c})$ denotes the expected increase in

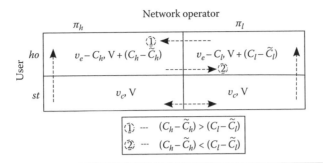

Figure 22.3 A simple two-player game between a user and a network provider.

the revenue of the operator. The utility functions $U_i : S_i \times S_{-i} \to \Re$ for both players are different and conflicting as given in Figure 22.3.

The solution to this problem may be derived by various methods such as strictly/weakly dominant strategies. The solution concept that we adopt in this example utilizes Nash equilibrium using best response function.

Case 1: Assume that the user is indifferent to both offers of the operator (low or high QoS) and that $v_e - c_h > v_c, v_e - c_l > v_c$, which dictates that user is always motivated to handover, as its utility increases following this strategy. Then the best response to this strategy of operator-1 is offer-1 when $(c_h - \tilde{c}_h) > (c_l - \tilde{c}_l)$, and offer-2 when $(c_h - \tilde{c}_h) < (c_l - \tilde{c}_l)$. Since these strategies are mutually best responses to each other, then players have no reason to deviate from given strategy. The efficiency of the solution may be evaluated using the Pareto-optimality concept.

Case 2: Now consider that $(v_e - c_h) < v_c$ in operator first offer (π_h), and that the expected utility of user is more than its current utility, that is, $(v_e - c_l) > v_c$, in the second offer (π_l). In this case, the solution of game turns out to be offer-2 (π_l).

22.2.2 Cooperative Game

It is very difficult to achieve cooperative benefit if it is assumed that players are rational and pursue their self-interest only. The dilemma of achieving cooperation based on selfishness is famously illustrated by the Prisoners' Dilemma Game. In such two-person cooperative games, the players' utilities may be translated in four types (a) temptation, (b) reward (cooperating), (c) sucker, and (d) punishment. However, in the above game to illustrate cooperative nature of users, let $(v_e - c_l) = v_c$ in the second offer (π_l) of operator and also $(c_l - \tilde{c}_l) > (c_h - \tilde{c}_h)$. On the other hand, in the the first offer (π_h), assume that $(v_e - c_h) > v_c$. Then in this case, users can threaten operator to play π_h. If the operator doesn't agree to cooperate, the threat is carried out, and in this case if user plays stay-connected strategy, the operator's expected utility $V + (c_l - \tilde{c}_l)$ reduces to V; therefore, cooperative behavior here guarantees better utilities $(v_e - c_h)$, $V + (c_h - \tilde{c}_h)$ and reduces risk of utility loss.

22.2.3 Noncooperative Interface Selection

Noncooperative game theory studies the strategic choices resulting from the interactions among competing players, where each player chooses its strategy independently for improving its own utility or reducing its costs. A general noncooperative game can be modeled as follows: Let $\Gamma = [N, S_n, U_n]$ represent a game where $N = \{1, \ldots, N\}$ represents the set of players/users, S_n is the strategy set available to each user n, and U_n is the utility function of user n. Each user bases its strategy on maximizing its own utility, that is,

$$\max_{s_n \in S_n} U_k \quad \text{for } N = \{1, \ldots, n\}. \tag{22.2}$$

There exist two main solution concepts for noncooperative games, namely, *Nash equilibrium* and *Pareto optimality*, as defined next.

22.2.3.1 Nash Equilibrium

A pure Nash equilibrium of the game $\Gamma = [N, S_n, U_n]$ is a strategy profile $s^* \in S_n$ such that for every player $i \in N$ there exists

$$U_i\left(s_{-i}^*, s_i^*\right) \geq U_i\left(s_{-i}^*, s_i\right) \quad \text{for all } s_i \in S_i, \tag{22.3}$$

where $s_{-i} = (s_j)_{j \in N \setminus \{i\}}$ denotes the strategy profiles of all players except player i. This indicates that there exists no strategy $s_i \in S_i$ of player i that is preferred over its Nash strategy s_i^* irrespective of the strategy of all other players $j \in N \setminus \{i\}$. The strategy profile $s^* = (s_1, \ldots, s_n)$ is a Nash equilibrium when this holds for every player $i \in N$.

Pareto Optimality is used to assess the efficiency of equilibrium point. A strategy profile s^* is Pareto optimal if there exists no other strategy profile s' that is Pareto superior to s^*. A strategy s' is Pareto superior to strategy s if for any player i, $u_i(s_i', s_{-i}') \geq u_i(s_i, s_{-i})$. In other words, the strategy profile s' is Pareto superior to the strategy profile s, if the payoff of a player i can be increased by changing from s to s' without decreasing the payoff of other players.

As noted earlier, players (i.e., the users) try to form a strategy to maximize their payoff. This results in an important requirement for the appropriate choice of a utility/cost function. The choice of utility/cost function influences the game to a great extent mainly in terms of strategy selection by the players. In the domain of interface selection problem, researchers have contributed various utility/cost functions, which affect end users' strategies to select among available access networks.

One such framework is used in [10] where a *noncooperative game* is formulated among end users, who strive to minimize the cost function. In this scenario, a user plays in a congested WLAN-covered service area, and to maximize its bandwidth, he has the choice to physically traverse a distance to get associated with a less congested access point (AP). The congestion in this case refers to AP load. One of the approaches to model the *payoff function* in this case is maximization of user-received bandwidth. A user is satisfied if (i) it gets associated with an AP that has less number of already associated users, and (ii) the distance to traverse in order to use the target AP is minimum. To capture both these desired characteristics, authors model the cost function as

$$C_{ij} = \alpha x_i + D_{ij}, \tag{22.4}$$

where

C_{ij} is the cost assigned by user i to AP j
x_i is the number of already associated users
D_{ij} is the distance to be traversed to use AP j
α is the relative weight assigned by the user to load and distance with the value between $\{0, \infty\}$

Here α captures user preference relationship between distance and QoS—a higher value of α shows greater concern of user toward target AP load. A myopic algorithm for access selection is suggested, which leads to Nash equilibrium for a simple user exit model (where the users are assumed to exit the system shortly after each other, almost simultaneously). However, the solution is no more stable with more realistic assumptions, that is, when users generate varying load and exit the system dynamically. In this case, the gain that user attains by unilateral change of strategy is proportional to the diversity of users. Thus, a stable system is more likely to result in lower user diversity.

Users' behavior of getting associated with less interfered AP specifically in interference limited network technologies, for example, CDMA, WiFi, etc., can be modeled using *congestion games*, which belong to class of noncooperative games [11]. Such games are well suited to model resource competition where resulting *payoff* is a function of the level of congestion. The generalized form of congestion games is given by $\Gamma = (N, R, (\sigma_i)_{i \in N}, (g_r)_{r \in R})$, where $N = \{1, 2, \ldots, N\}$ denotes a set of users, $R = \{1, 2, \ldots, R\}$ a set of resources, $\sigma_i \subset 2^R$ the strategy space for player i, and $g_r : N \rightarrow Z$ a payoff (cost) function associated with resource r. Cesana et al. [12] model interface selection problem under three different cost functions using congestion games, as explained next.

22.2.3.2 Interference-Based Cost Function

The cost function of user i for selecting the AP j simply depends on the congestion due to x^j interferers connected to the AP j.

$$C_{ij} = C_i(j, x^j) = x^j. \tag{22.5}$$

This is a simplified form of Equation 22.4, with the assumption that all users can select only one AP and that each user has the same effect on AP's congestion. Such a cost function turns the game into single-choice asymmetric congestion games, which admits exact potential function [11].

22.2.3.3 Additive Interference-Rate Cost Function

The cost function is a linear combination of the number of interferers and the reverse of rate perceived by users.

$$C_{ij} = c_i(j, x^j) = \lambda_1 x^j + \lambda_2 T_i^j, \tag{22.6}$$

where
  λ_1 and λ_2 are user-independent parameters
  T_i^j is the reverse of rate perceived by users

The weighing coefficients λ_1 and λ_2 are assumed to be constant for all users. Note that assigning zero to λ_2 reduces the function to the one in (22.5) and that T_i^j replaces the D_{ij} of the cost function in (22.4).

22.2.3.4 Multiplicative Interference Rate Cost Function

The cost function is defined by multiplication of interferers and user-perceived reverse rate.

$$C_{ij} = c_i(j, x^j) = T_i^j x^j. \tag{22.7}$$

The assumption of users' almost simultaneous exit pattern identified earlier in [10] refers to another issue in interface selection. In scenarios such as passengers on a train or bus moving out of the coverage area of current point of attachment, end users tend to handover simultaneously. Simultaneous rational decisions of most users converge, for example, to selection of the operator that has the least load. However, the selected operator/network may no longer remain as the suitable option after the group handover of many users is executed. Such problems fit well in the congestion game branch of game theory. The European Union project C-CAST specifically addresses these issues in the context of mobile multimedia multicasting [13].

Interface selection in the case of *group handover* is discussed in [14], in which end users' satisfaction is translated to minimizing the transfer latency of the application data, that is, a mobile user selects the network that has minimum transfer latency. A network is characterized by its capacity and transfer latency cost, where transfer latency in turn is a function of traffic load and available capacity of the network. Two types of latency functions are considered.

Linear latency function

$$C_{ij} = \frac{w}{a_j},$$ (22.8)

where

  w represents the *traffic* of node
  a_j represents the *available capacity* with the network j

Latency function based on M/M/1 queuing model

$$C_{ij} = \frac{1}{a_j - \min(a_j, w)}.$$ (22.9)

The *interface selection* problem is formulated as K-P model of the congestion games and three algorithms are proposed in [14] for *network selection*: (a) network selection based on Nash equilibrium strategy, (b) network selection with random delay, and (c) network selection with random delay considering handover latency. The algorithm based on the Nash equilibrium requires unrealistic assumptions, for example, end users have the information about all networks and traffic loads of other users, but is used as a reference to compare other proposed algorithms. The algorithms with *random delay* and *random delay with handover latency* carry out the network selection in more realistic scenarios.

User-centric interface selection mainly focuses on the local objective of user satisfaction and usually disregards the global objective of congestion control and optimum network resource utilization. Unlike such approaches described thus far in this section, Niyato et al. discuss the user-driven load balancing approach while network selection is performed on the user side [15]. Competition among groups of users (groups formed on the basis of service class) in different service areas covered by heterogeneous wireless networks is modeled using evolutionary and noncooperative games. The *evolution equilibria solution* based on the replicator dynamics is defined, which is the stable set of fixed points of the replicator dynamics. Replicator dynamics is given by $\bar{x}_j(t) = x_j(t)(\pi_j(t) - \bar{\pi}(t))$, where $\pi(t)$ is the payoff of the users choosing strategy of joining the network j, and $\bar{\pi}(t)$ is the average payoff of the entire population. Evolution equilibria ensure that all users within the same group receive the same payoffs. The evolution game for network selection is implemented using two algorithms: population evaluation and reinforcement learning. Evolutionary equilibria solutions are compared with the Nash equilibrium obtained from the conventional noncooperative games. The payoff of users within evolutionary game is determined by their net utilities.

22.2.4 Cooperative Interface Selection

Cooperative game theory provides analytical tools to study the behavior of rational players when they cooperate. The main branch of cooperative games describes the formation of cooperating groups of players, referred to as coalitions, which can strengthen players' position in a game. There exist various solution concepts for coalition games including *Core*, τ-value, *nucleolus*, and *Shapley value* [16]. Coalition games consist of a set of players, a set of strategies for each coalition, and preferences. Efficiency of a coalition is quantified by a characteristic function. More formally, the coalition games are of the form (N, v), where $N = \{1, 2, \ldots, n\}$ is the set of players and $v : 2^N \to \mathbb{R}$ is the characteristic function that assigns each coalition of players $S \subseteq N$ a worth $v(S)$. It is assumed that $v(\emptyset) = 0$ and $v(S)$ can be defined as the payoff that each member of coalition S can obtain for themselves if they coordinate their actions, independent of the actions taken by the players outside of the coalition S [17].

The strategy set of player $i \in N$ is represented by $S_i = \{T \mid T \subset N \backslash i\}$, where a particular strategy $s_i \in S_i$ represents the set of players with whom player i would like to form coalition. The total utility represented by the characteristic function that coalition achieves can be divided among the players of coalition, and such games are called *transferable utility games*. The preference of users is modeled by their utility functions, similar to the noncooperative case. The outcome of the game consists of a partition of the set of players into groups. To further elaborate this, let us consider an example scenario where the network technologies $w \in N$ cooperate with each other to satisfy the user request by assigning predefined bandwidth (further detailed in Section 22.3.2). Network technologies form coalition $S \subseteq N$, where the payoff of coalition is given by characteristic function v. Let the amount of requested bandwidth received by each network technology is given by a payoff vector $x \in \mathfrak{R}$, which is an imputation for the game if it is efficient and individually rational, that is, (a) $\sum_{i \in N} x_i = v(N)$ and (b) $x_i \geq v(i) \; \forall i \in N$.

To narrow down the imputations and attain stable solution, we illustrate the concept of *core*. The core $C(v)$ of a game is the set $\{x = [x_1, \ldots, x_n] \mid \sum_{i \in S} x_i \geq v(S), \forall S \in 2^N\}$. There are situations when the core is empty. Shapley value and the τ-value are one-point solution concepts. Let us investigate the Shapley value concept for the aforementioned problem. The Shapley value associates each game with a payoff vector in $\mathfrak{R}^n$, which gives the average of the marginal vectors of the game and is denoted by $\phi(v)$:

$$\phi_i(v) = \sum_{S:i \notin S} \frac{|S|!(n - 1 - |S|)!}{n!} (v(S \cup i) - v(S)). \tag{22.10}$$

Here, S indicates the number of elements in the set S. The outcome of above allocates different amount requests to available network technologies, which should be efficient solution.

A cooperative game is usually modeled as a two period structure. Players must first decide whether or not to join coalition, which is done by pairwise *bargaining* [17]. Bargaining problems refer to the negotiation process, modeled using game theory tools, to resolve the conflict that occurs when there are more than one course of actions for all players in a situation. Players involved in the games may try to resolve the conflict by committing themselves voluntarily to a course of action that is beneficial to all of them.

Bargaining problem is modeled as a pair (F, d), where F represents the set of all feasible utility pairs and d represents the *disagreement point*. By definition, players do not participate in a coalition if the utility that they receive is less than the specified disagreement point. Typical bargaining solutions include Nash bargaining solutions, Kalai–Smorodinsky bargaining solutions (KSBS), modified Thomson rule, etc. All such solutions have to satisfy the axioms of (a) individual rationality, (b) Pareto optimality, (c) independence of irrelevant alternative/individual monotonicity, and (d) symmetry.

The most common form of a cooperative relationship in the user-based network selection is network–user cooperation. Recall that users have the objective to increase their satisfaction level by paying less, and network providers strive to increase their revenues, which seemingly indicate conflicting interests for the two entities. On the one hand, it is clear that a user-centric decision alone may not take into account optimal resource utilization or congestion control at the network level, which are key issues for the network providers' objectives. However, on the other hand, when the number of users in a service area increases, the satisfaction level of each individual user decreases. Therefore, the users would be willing to cooperate with the network operators, who in turn are willing to cooperate with users for better utilization of their resources, resulting in a mutual increase in their payoffs [18].

Direct cooperation among users for interface selection is also possible, though they are generally studied within the scope ad hoc networking and falls beyond the scope of this text. In general, most of the cooperative network selection approaches fall into the network-based decision domain, which are covered in Section 22.3.

22.2.5 Auction-Based Interface Selection

Auction is a method to determine the value of a commodity that has an undetermined or variable price. It requires a preannounced methodology, one or more bidders, and an auctioneer to compute the winner bid against the price. An auctioning mechanism checks the bid for validity and updates the set of active bids. Auctions relieve the sellers from fixing the price of items. Dutch, first-price, sealed-bid, and second-price auctions are the four basic auctioning formats [19]. Vickrey introduced the sealed-bid auction type in which highest bid wins, but the price paid is the second highest bid. This mechanism provides an incentive for bidders to declare their true valuation to maximize their payoff. An auction is called efficient if it maximizes the payoff of all the bidders.

Auctioning has been proved to be a powerful tool in networking and communications [20], but its utilization in user-centric network selection is just an emergent topic. Auction in the context of interface selection may be expressed as an auction in which the auctioneer resides on user terminal, bidders are the available network providers, and communication resource requests of users are the auctioned items. Khan et al. [21] use multi-attribute reverse auction for network selection. Here, the item to be auctioned is represented as a tuple $\{(\gamma_q, \theta_k) \mid \gamma_q \in Q, \theta_k \in \theta\}$, where Q is the set of application classes characterized by different amounts of required bandwidth resource, γ_q is the required bandwidth by application class q, and θ defines the vector of various QoS attributes and their expected value ranges for the auctioned item. For any application request, the user announces its intention to acquire the item (γ_q, θ_k) through the auctioneer residing on the user terminal. Auctioneer broadcasts this request in a coverage area a, which is received by the *operator(s)* possessing *network technologies* in that area. Upon receiving this request, those operators submit their bids as offers for providing the *requested bandwidth*. For a given application required bandwidth γ_q and associated attribute vector θ_k, the offer bid by the operator o through network technology w in an area a is given by the tuple

$$b_{w,o}^a(\gamma_q, \theta_k) = \left((\gamma_q, \theta_k), \pi_{w,o}^a\right), \qquad (22.11)$$

where $\pi_{w,o}^a$ is the per unit bandwidth payment (cost). Then the operator utility function is given by

$$U_{w,o}(\gamma, \theta_k) = \begin{cases} (\gamma_q, \theta_k)\pi_{w,o}^a - \sum\limits_{k \in \theta} c_{w,o}(\theta_k) & \text{if } D \le b_w(\theta_k), \\ 0 & \text{if } D \ge b_w(\theta_k), \\ \lambda\left((\gamma_q, \theta_k)\pi_{w,o}^a - \sum\limits_{k \in \theta} c_{w,o}(\theta_k)\right) & \text{if } D = b_w(\theta_k), \end{cases} \qquad (22.12)$$

where

$D = \max_{\{\bar{w}, \bar{o} \ne o\}} b_{\bar{w}, \bar{o}}((\gamma_q, \theta_k), \pi_{\bar{w}, \bar{o}})$ is the maximum suitable bid if operator $o \in O$ does not participate in the game

$c_{w,o}(\theta_k)$ is the cost incurred on a single attribute value

$\sum_{k \in \theta} c_{w,o}(\theta_k)$ is the operators' reservation price for service

$\lambda \in \{0, 1\}$ is a tiebreaking coefficient

The degree of user satisfaction is translated using the *user utility function*, which captures user's preference relationship over various QoS attributes and the behavior toward the amount of offered bandwidth. Since QoS attribute vector is n-dimensional, user preference on every single attribute is represented by assigning weights to those attributes. User terminal evaluates the amount of offered bandwidth and each relevant attribute quoted in the offer from network provider using its utility function (scoring function) and compute the utility of overall bid by using the weighted sum over attributes:

$$U_i\left(b_{w,o}^a(\gamma, \theta_k)\right) = u_i\left(\gamma_q^o\right) \sum_{k \in \theta} \psi_k \zeta_{w,o}(\theta_k) - C_i(\gamma_q, \theta_k), \qquad (22.13)$$

where

ψ_k is the user-assigned weight to attribute k

$u_i\left(\gamma_q^o\right)$ is the utility obtained by the offered bandwidth

The QoS attributes are normalized under two different classes, namely, (a) the smaller the better (e.g., delay, jitter, and packet loss ratio) and (b) the nominal the better (e.g., reliability and security). The cost of frequent handover is reduced by using fuzzy logic approach.

User-based network selection should focus on multi-attribute auctions, in order to capture the negotiation over multiple service attributes. As opposed to the standard auctions, the overall utility of a user depends not only on the price but also on a combination of different attributes that deal with various application-specific QoS attributes, user preferences over these attributes, pricing, etc.

Very little work exists in the research literature in this direction. Suri and Narahari [22] consider splittable applications and propose an auctioning algorithm that is designed as the reverse optimal (REVOPT) mechanism for resource procurement auction problem. In this design, both auction and payment rules are characterized, and it is assumed that network providers are noncooperative.

Mobile users announce the requirements in descending order of importance and network providers submit their bids. Upon receiving the bids, the user needs to employ a selection mechanism for the winning bid with the aim of minimizing price while satisfying the properties of individual rationality and Bayesian incentive compatibility.

22.3 Operator-Based Network Selection (Resource Distribution)

In the world of telecommunications, network operators were traditionally accustomed to have a closed, controlled environment with mostly loyal subscribers. The loyalty of subscribers was generally ensured by long-term contractual agreements, and, hence, users had limited flexibility in their selection and migration among operators. Recent technological developments, advancement toward openness in all areas of computing and communications, and new legislations such as number portability are increasing the users' flexibility. This in turn motivates network operators to increase their service quality and diversity in order to decrease the *churn out rate*, which is defined as the probability of the users' leaving the current service provider and subscribing to another. One such undertaking is to deploy network technologies of different characteristics to have a larger pool of satisfied users and to enable *vertical handovers* among those different technologies in order to maximize the overall network utilization and relieve congestion in the network.

This section focuses on network selection approaches where the decision takes place on the network operator side, which is also termed in this chapter as *resource distribution* (see Figure 22.1). Addressing the network selection problem from a network-centric viewpoint becomes a more complex problem when various network technologies under consideration are owned by different competing operators. In the complex and evolving telecommunication scene, one can observe different scenarios with different levels of cooperation and competition among network operators. Those different scenarios involve unique dynamics and require distinct mechanisms for network selection. This section covers those possible scenarios for the network selection problem, including (a) a single network operator with multiple access technologies (reflecting the case of long-term contracts where the user is bound to a single carrier), (b) multiple network operators and multiple access technologies (where the users are not bound to any operator and can roam freely among different operators), and (c) a mix of the two cases with mobile virtual network operators (MVNO), who may have long-term subscribers and share the resources of other network operators.

The interaction among the entities varies in those scenarios, which can be modeled using different game-theoretic approaches. Similar to the outline of Section 22.2, this section presents those studies under three main subsections; namely, noncooperative, cooperative, and auction-based methods.

22.3.1 Noncooperative Resource Distribution

Consider an area covered by *interference-limited* multiple access technologies, where admitting a new session affects other ongoing sessions. Here, the decision entity on the network operator, which might also be termed as admission controller, should have the goal of not only providing the necessary resources to the new call/session but also maintaining the QoS of existing ones as agreed at the time of admission [23].

The trade-off between the load (number of users serviced) and the offered QoS can be modeled using *game theory*, with the players being network providers and users. Lin et al. [23,24] provide one such approach addressing the resource allocation and call admission control (CAC) issues for a single operator case. Niyato and Hossain [24] formulate a noncooperative game model for bandwidth allocation and CAC in IEEE 802.16 focusing on real-time and non-real-time polling services (rtPS and nrtPS). Upon the arrival of a new connection, the base station invokes the bandwidth allocation and admission control algorithm. The information (requirements) carried by the new call request includes connection type (i.e., rtPS or nrtPS), traffic source parameters (i.e., normal arrival rate, peak rate, and probability of peak rate), and delay and throughput requirements. Base station establishes a set of strategies and computes the expected payoff corresponding to each strategy. Game is played between the new connection (request) and base station, where the objective (utility) of new connection is to have higher throughput in non-real-time polling service and smaller delays in real-time polling services, and the utility of base station is the revenue obtained from the connection. When a new connection is admitted, it is allocated some resources. Therefore, the payoff for the base station becomes the total utility of the ongoing connections after some portion of bandwidth has been taken away (which results in decrease in the total utility for the ongoing connections) plus the utility of the new connection. However, revenue for each connection depends on the perceived performance. Nash equilibrium of the game is computed using best-response function of both players with admissible strategy pair. The allocation of resources to calls are adaptive in a sense that in case of low load cells, ongoing connections receive larger amount of bandwidth, resulting in minimizing the delay and maximizing the throughput.

When this scenario is extended to multiple-operator case with freely roaming users, then unsatisfied users are likely to churn out to other operators. One such scenario is discussed in [23], where operators compete for reducing the churning out of users. CAC problem is formulated as a noncooperative game between a service provider and customer. It is assumed that a user is connected to one operator at a time. The payoff of service provider is a function of the revenue gains from connections and revenue loss as a result of the user churning to other providers. The user has the strategy profiles of staying with the operator or leaving, while the operator has the options of granting the user's request or not. Then the payoff for all possible strategies of the players can be represented by a matrix

$$P = \begin{bmatrix} U + C_k - F - L_k & u + C_k - F \\ U - L_k & U \end{bmatrix}, \tag{22.14}$$

where
U denotes the service provider's current revenue
C_k denotes service provider's average revenue earned from each session of class k customer
F denotes predicted loss due to churning out when call is admitted
L_k denotes service provider's average revenue loss for losing class k customer

Although it seems like an ideal situation, the provider can possibly lose revenue because of other churning users in case of call admission to an already-loaded network. Nash solution to pure strategy is used to determine whether the incoming connection can be accepted or rejected. This model is modified to admit multiple users simultaneously, unlike the one discussed in [24].

22.3.2 Cooperative Resource Distribution

In the simplest scenario of a single network operator, coalitions may be formed among different network technologies of the operator, and the characteristic function/payoff of coalition turns out to be the payoff of network operator. Niyato and Hossain [25] address the resource distribution problem with this perspective using *bankruptcy games*. Bankruptcy game addresses a distribution problem that involves the allocation of a given amount of good among a group of agents, when this amount is insufficient to satisfy the demands of all agents. The available quantity of the good to be divided is usually called *estate* and the agents are called *creditors*. The problem of resource allocation by network technologies (creditors) to users (estate) is mapped to the bankruptcy problem in [25]. With the assumption that end users can connect to multiple network technologies simultaneously, the network selection problem is converted into the distribution of requested bandwidth over available network technologies. A coalition among network technologies is formed with allocated bandwidth as its characteristic function. The decision of what portion of the requested bandwidth should be allocated by which technology is taken by *Shapley Values*, and the stability of allocated bandwidth is ensured by using *Core* [16]. The objective of each network technology within the Core solution is to allocate more bandwidth, that is, to increase the utilization of its own resources.

A solution for resource distribution in a similar scenario is proposed using KSBS in [26]. This work considers a wider perspective of the problem with various combination of different network technologies and mobility of users. However, the scope of this solution is still limited to one-operator scenario. Operator owns technologies of different characteristics in the coverage area. Each network technology is assumed to have two capacity regions, namely, congested and uncongested regions. A network technology is said to be in congested region if its current available bandwidth drops below some threshold value. The congested network behaves different from an uncongested one when allocating resources to any application request, and this behavior is determined by the load balancing factor $\tilde{\psi} = -l_w/(C_w + l_w)$.

Users' applications occupy different amounts of bandwidth capacities of network technologies for some amount of time (holding time), and users leave after staying connected for some random interval and release the resources. To represent the realistic scenario of cellular networks, authors assume that users of applications are mobile and move from one area to the other, resulting in variable number of serving network technologies at different times for the same application request. Players are network technologies $P = \{w | w \in W \vee w \in \tilde{W}\}$, where W and $\tilde{W}$ represent the set of uncongested and congested network technologies, respectively. Strategy set of players include the amount of predefined/offered bandwidth. Predefined offered bandwidth is differentiated from the offered bandwidth in that offered bandwidth can be either more or less depending on the congested/uncongested mode of network technology.

The utility function of network technologies is derived from the bandwidth that these network technologies allocate to any application request. The resource allocation problem is formulated as a bankruptcy problem $\left(r_p^a(q), B^a(q)\right)$, where the application request $r_p^a(q)$ represents estate and the aggregated amount of offered bandwidth $B^a(q)$ takes the role of creditors. Owing to the cooperative behavior of network technologies, the bargaining problem turns out to be 0-associated bargaining problem $\left(r_p^a(q), B^a(q), 0\right)$. The solution to this problem is dictated by taking the maximal element in the feasible set on the line connecting the disagreement point (which in this case is zero) and the ideal point; therefore, it was learnt that recommendations made by KSBS to 0-associated bargaining problem coincides with the recommendations made by proportional distribution rule. Proportional bandwidth distribution algorithm (resource allocated according to offered bandwidth), CAC and area handover algorithms are developed.

In extraordinary situations, for example, during popular events or emergencies, a specific network provider may not be able to meet the bandwidth demand from its user base in a given area. Dropping connections (during a new call or handover request) results in unsatisfied users, which motivates the operator to cooperate with other operators available in that area. On the other hand, the motivation for other operators in the area to cooperate could be idle resource utilization and revenue maximization. One such scenario is addressed in [27], an extension of [26], where the problem of resource distribution is formulated as a cooperative game among different operators and also among network technologies of an operator. Cooperative game therein is played in two stages: (a) at network technology level, termed as the *intra-operator game*, and (b) at network operator level, termed as the *inter-operator game*. User-generated requests are first received by their operator (*home operator*). Then the home operator conducts an intra-operator game, where network technologies cooperate with each other to satisfy the user request and the solution of the game is found using KSBS with a disagreement point of zero.

This game is similar to the one described earlier in [26]. However, if the resources of an operator in a coverage area are not sufficient to satisfy the requested bandwidth, then the inter-operator game is played among different operators over this request. The inter-operator game is also modeled using a cooperative game-theoretic approach and KSBS is applied to obtain the solution. Authors define the following two congestion regions in inter-intra-operator game: (a) RAN congestion (similar to mentioned in [26]) and (b) aggregated congestion: An operator network is said to be in the aggregated congestion region in an area a if the aggregated available bandwidth of the RANs belonging to the operator in this area falls below some threshold value. In intra-operator games, the players, strategy, and utility function are similar to the ones in [26]; however, the players are the network operators in inter-operator games, and the strategy set of players includes the amount offered bandwidth $b_{w,o}$ and adjustment of potential factor μ. In contrast to intra-operator game, the disagreement point in inter-operator game is taken as characteristic function $d_i(q) = \max \left\{ 0; \sum_{k \in O} \bar{r}_k^a(q) - \sum_{i \neq j} b_j^a(q) \right\}$, where $\bar{r}_k^a(q)$ represents the excess bandwidth request for a service class q that an operator k cannot answer and would like to offer in the inter-operator game to other operators. The characteristic function is defined as the amount of bandwidth not covered by offers of other operators. The problem is thus converted to $D(q)$-associated bargaining problem $(S(\bar{Q}^a(q), B^a(q)), D(q))$, where $\bar{Q}^a(q)$ represents the vector of all $\bar{r}_k^a(q)$ requests. It is found that for such $D(q)$-associated bargaining problem, the recommendations made by KSBS coincide with adjusted proportional distribution rule. In other words, bargaining comes up with allocation of requested bandwidth to different operators, x_o^a, which is the member of compact and convex feasibility set. The allocation is dictated by

$$
x_o^a = d_o(q) + \frac{\left(b_o^a(q) - d_o(q) \right)}{\sum_{i \in O} \left(b_i^a(q) - d_i(q) \right)} \cdot \left(\sum_{i \in O} \bar{r}_i^a(q) - \sum_{i \in O} d_i(q) \right),
$$

where $d_i(q) \in D(q)$. This distribution rule is applied by SLA broker. Authors present bandwidth offer algorithm both at intra- and inter-operator level, and the required information exchange between individual RANs in the earlier is implemented by CRRM, whereas in the latter, SLA broker is used for this purpose. The algorithm for the operators turns out to be relatively simpler than the RANs, given the most actual bandwidth offers that RANs made for a specific service class contain a considerable amount of information about status of the operator network in an area. Specifically, the operator sums the most up-to-date bandwidth offers from the RANs in the area for service class. Then this aggregated offer is scaled with motivation factor of the operator $0 \leq \mu_o \leq 1$. By setting this factor, the operator is able to adjust the cooperative nature of its strategy: $b_o^a(q) = \mu_o \sum_{i \in W_o^a} b_{o,i}^a(q)$. There is an incentive for cooperative behavior, as operators can allocate unused bandwidth to increase revenue and utilization.

Let us now consider a mixed scenario of an operator with contracted subscribers and MVNOs extending the operator's resources to nonsubscribers. The resource distribution problem in such a scenario is formulated in [28] using bargaining games. An operator is naturally motivated to sell out its unused capacity to an MVNO, who resells the bandwidth to their user pools. The bargainers in the game are (i) a set of MVNOs and (ii) users having contractual agreements with the operator. The operator acts as the arbitrator. Let the bargaining problem in this case be denoted by (S, d), where S represents the feasible set and d the disagreement point. In this case, the natural selection of disagreement point by both users and MVNOs is *minimum required performance*. The selection of this disagreement point is justified by the assumption that operator provides elastic services, which are characterized by their *minimum* and *maximum* ranges. Bargaining solution can be analyzed using players' preference function. For example, in the case of two players, $v_1 = u_1 + \beta(1 - u_2)$ and $v_2 = u_2 + \beta(1 - u_1)$, where $0 \leq u_1, u_2 \leq 1$, and β is the proportionality factor that measures the trade-off between one's gain and another's loss. The values of $\beta = 0$, $\beta = 1$, and $\beta = -1$ correspond to Nash bargaining, KSBS, and modified Thomson solutions, respectively. The bargaining outcome is expressed as $u* = \max_u(v_1 v_2)$. The resource bargaining problem is formulated as

$$\max_x \ln \left(\prod_{i=1}^{N} \left(x_i - x_i^{\min} + \frac{\beta}{N-1} \sum_{j \neq i} x_j^{\max} - x_j \right) \right) \qquad (22.15)$$

such that $x_i^{\min} \leq x_i \leq x_i^{\max}$ and $\sum_{i=1}^{N} x_i \leq T$, where x_i represents the resource allocated to player i. The utility of player is tied to this allocation, which in turn represents the players' satisfaction. A generic utility function is given by $w_i(x_i) = u_i(x_i) - \pi x_i$. In the case of users, $u_i(x_i)$ represents users' willingness to pay, whereas in the case of MVNO, it is the revenue earned by reselling the acquired bandwidth. Realistically knowing the exact utility of all the players is difficult owing to the fact that operators or MVNOs are reluctant to share commercially sensitive information.

The issue of resource allocation becomes more complex in multiple access and relaying network scenarios, especially for high data rate transmissions [29,30]. Pan et al. [30] model the resource (power and spectrum) allocation problem as a game where OFDMA base station sends out data symbol to UE, and relay stations (RS) retransmit the data symbols by their subcarrier (in different slots and frequency bands). RSs that are designated to the same set of UEs form coalitions. Such grouping results in multiple coalitions, each containing different number of RSs. Similar to the approach discussed in [27], resource distribution is performed at two levels: intra-coalition and inter-coalition. The utility of player RS_i is defined as $u(r_i) = \sum_{n \in N} R(u)$, where N is the set of UEs. *Inter-coalition* problem is modeled using both cooperative and noncooperative games. The objective that motivates the coalitions to cooperate is to maximize the total capacity of UEs while keeping fairness among coalitions. A non-symmetric Nash bargaining solution is used to maintain proportional fairness among coalitions.

The resource sharing problem among relay and multiple nodes is studied in [29]. Assuming that the nodes are assigned some amount of bandwidth, cooperation here can be interpreted as the relay selling out its bandwidth to the source. This in turn defines the source i strategy as the bandwidth size, w_i, that it wants to buy from relay. This cooperation is driven by the pricing at relay, which is given by $p(w) = a + b \left(\sum_{i=1}^{N} w_i \right)$, where a and b are nonnegative constants and w is the strategy adopted by the user. The utility of source is interpreted as the number of data bits successfully received per joule of energy consumed, which is given by

$$U_i = T_{s_i,d_i}(P_i, w - w_i) + \frac{T_{s_i,d_i}^{AF}(p_i, w_i)}{p_i} - c \cdot w_i, \qquad (22.16)$$

where

p_i is the transmit power of source s_i

$T_{s_i,d_i}(P_i, w - w_i)$ is the throughput derived from cooperative transmission help by relay r with bandwidth w_i

$c \cdot w_i$ represents the payment by user i to relay node for resource consumption

Authors present a distributed algorithm to find the Nash equilibrium of the noncooperative bandwidth allocation game in FDMA-based cooperative relay networks.

22.3.3 Auction-Based Resource Distribution

The resource allocation in a decentralized and noncooperative environment requires interaction mechanisms among network providers that can guarantee system-wide properties such as efficiency, stability, and fairness. A suitable means for interaction in this context is provided by auctioning mechanisms. When auctioning is used for operator-based network selection in wireless networks, network providers are the sellers and users are the buyers, that is, bidders. In multi-operator cases, multiple sellers exist; therefore, multiple and possibly simultaneous auctions may take place. The bidding strategy of a user is influenced by the auction format used and the scenario that represents the composition of available network providers and their resources.

In telecommunications, users are traditionally classified as gold, bronze, silver, etc., with respect to the application QoS requirements, price models, etc. This kind of classification holds for a long duration and requires a serious study of application characteristics and network policies. On the other hand, such classes can dynamically emerge for short periods of time (e.g., for a call duration) if auctioning mechanisms are used. If the dynamic class formation is tied to the users' private valuation of the network resources, then a user's bid within every auction reflects the importance of resource to users and operators will be able to approximate the user's willingness to pay. Sallent [31] address the problem of resource allocation using auctioning, where bidders are users, and auctioned items are the radio resources goods (RRG), such as frequency, time slot, etc. RRG of each radio access technology (RAT) is auctioned separately and in each auction several items are auctioned using *multi-unit sealed-bid* auction format.

Most of the schemes discussed so far do not explicitly consider the duration of resource allocation. Normally, the duration of network slots is quite short, but there are situations when network services are needed for relatively larger duration of time. The problem is further complicated when varying user population is considered. "Auction-based THird gEneration Networks resource reservAtion" (ATHENA) addresses such a scenario where a number of mini-auctions of short duration t_a take place [32]. These short auctions are used to attain consistent reservation of resources. Each mini-auction is a sealed-bid auction with atomic bids. Bids are represented by (p, q), where p represents the willingness to pay for quantity q of resource units. In a special case of *UMTS* for service of a specific rate m, $q = m \cdot t_a$. Each mini-auction is a *generalized Vickery auction*, where users' preferences are translated using utility function. Each base station runs mini-auctions by bidding truthfully on behalf of each user. It is assumed that users' value for obtaining the service is the sum of the marginal *sub-utilities* attained due to each successful allocation. These functions reflect the fact that when there are gaps in the resource allocation pattern, not only the amount of resources but also the way these resources are allocated cause a great difference to user satisfaction. The following user types are considered: (a) indifferent to allocation type—volume oriented users, (b) sensitivity to the service continuity—users that prefer watching consistently, and (c) sensitivity of the smoothness of the allocation pattern—users that are sensitive to the jitter of their resource allocation pattern. Results show that users either receive satisfactory service or are not admitted in the network with the presented auctioning scheme [32]. The authors extend their work (ATHENA) in [33] to address multicast traffic in UMTS networks.

The auction over time slots of fading channel is addressed in [34], where individual users compete for resources through bidding for use of the channel. Auctioning is carried out over time slots, where each user has different valuation of a time slot. Users' private valuation in this case is the function of channel quality perceived by them. Within each time slot, (a) users submit their bids for time slots to the transmitter according to channel condition, (b) transmitter computes the winning bid, the bid with the highest price, (c) user pays the second bid price, and (d) transmitter allocates the time slot to the user with the winning bid. It is shown that Nash equilibrium strategy of auction leads to an allocation at which the total throughput is no worse than 75% of the maximum possible throughput when fairness constraints are not imposed.

In a multi-operator scenario, the fact that application flows can be directed over multiple technologies of different operators simultaneously introduces more complexity in terms of winner computation among the bids. Roggendorf et al. discuss one such scenario in [35]. The divisible auction items therein are network flows and the *modified progressive second-price* auction format is used. A buyer (user) can bid for a portion of the item in different auctions being independently and asynchronously run on the various available base stations. Since the user agents strive to increase their utility that is a function of allocated resources, they can exploit the competition among sellers by dynamically selecting the network that has lowest congestion level or bundle the resources from multiple providers to balance the load. In this context, the buyer is interested in finding the best bidding strategy for submitting bids in various auctions. The strategy of submitting the truthful bid works well in a single auction case; however, when it comes to multiple simultaneous auctions, computing the bidding strategy becomes more complex. The complexity is increased if the auction items are heterogeneous and buyers have different behavior toward the auctioned items. Revealing true valuation to such simultaneous auctions may result in overbidding by the user. This complexity is relaxed in [35] by assuming the homogeneity of auctioned items (e.g., existence of a single access technology). In that case, a buyer is truthful to overall market than to a single seller. It is also assumed that auction is a complete information game, in which buyers know the bid profiles from previous rounds and can calculate their utility for a combination of resources from providers. It is shown through simulations that when such strategy is followed by all the users, the solution converges to Nash equilibrium, resulting in efficient resource allocation.

Let us now consider the most general scenario with multiple operators, MVNOs, and multiple technologies. Bodic et al. discuss a market-based framework where different stakeholders are able to trade communication services [36]. Conceptually, the framework refers to a digital marketplace where communication service trade is conducted using software agents. Considered business model contains four groups of actors, namely, service providers, network providers, users, and market providers. Here, *market provider* is a negotiation platform that helps service providers to find network resources for their customers and are responsible for monitoring transactions and penalizing the actors that do not fulfill the contractual agreements. The proposed framework comprises four layers, where the first layer specifies the applications' QoS subjectively, second layer contains the market provider negotiation platform, and third and fourth layers are concerned with the network functions. One of the objectives of this framework is to enable users to select dynamically the serving network infrastructure according to QoS and pricing requirements. Market provider negotiation is carried out by *market agents*. The penalty tag that differentiates between the trusted and non-trusted service provider is maintained by market agent. The penalty imposed on the network influences the reputation of network and is proportional to the number of calls that have been decommitted in the past by the specific network. Owing to the fact that users are not associated with any network before contract/negotiation, *logical market channel (LMC)* is proposed, which is negotiated between market agents and registered network agents at the marketplace initialization. Agents' negotiation here is carried out using auctions where the market provider acts as the *auctioneer*. To overcome the constraints of timing and negotiation strategy requirements, a variant of sealed-bid first-price auction format is used. However, the strategy of service agent is based on service valuation and the user preferences, for example, price, QoS, and network operator reputation. A weight is associated with each parameter as a penalty, $w_{penalty}$, or price, w_{price}, such that $w_{penalty} + w_{price} = 1$. The objective of introducing competition at the call level is also to equalize the overall market supply with the market demand, which is performed by adapting the offered price so as to reserve resources to users who value them the most. Smart applications can exploit the marketplace dynamics by for instance using the radio resources only when the offered price is low [36].

22.4 User-Centric Operator-Based Network Selection

Next-generation networks and the $4G$ vision evolve around user priorities, while the objectives of efficient resource allocation and congestion control can best be achieved when the network selection decision is delegated to network operators. If user policies are taken as a priority but the decision is still made on the network operator side, the two objectives of better resource management and higher user satisfaction may be simultaneously achieved. This sections covers game-theoretic approaches for network selection that place the decision entity on the operator side while taking user preferences or inputs into account.

Cesana et al. [37] capture the interface selection resource distribution by modeling a *bi-level stage game*, where the resource distribution game is dependent on the interface selection game. At the interface selection level, a noncooperative game is played among end users. Since the reference scenario is a *homogeneous networks scenario* (only WiFi APs), the cost function of the player is defined as the function of interference perceived by users, which in turn is defined as the total number of end users covered by the same access network covering the player i and is given by $\{n_i(S_u)\}_{i \in U}$. The game formulated at the user level is defined as

$$\Gamma_u = (U, N, S_u, \{n_i(S_u)\}_{i \in U}), \tag{22.17}$$

where
    U represents the users
    N represents access networks
    S_u represents the strategy set of users that include the reachable access point

Another non-cooperative game at network level is played among network technologies for resource distribution. Strategy set includes the set of frequencies that can be chosen by any network j and payoff $p_j(S_n, \Gamma_u)$ of network j, which is defined as the number of end users that decide to associate to the access network j when it plays the strategy profile S_n. In this *bi-level stage game*, the lower level game leads to Nash equilibria of the network selection with the frequency assignment as a parameter, whereas the upper-level game seeks to allocate the frequencies among access networks given the responses of the end users to these assignments.

$$\Gamma = (N, S_n, \{P_j(S_n, \Gamma_u)\}_{j \in N}). \tag{22.18}$$

Numeric results show that despite the rational behavior of players, this game finds equilibria condition where both end users and networks tend to be fair with respect to the other players.

Feng et al. [38] use a utility function to quantitatively capture the users' satisfaction in multicell CDMA systems. An optimization problem for users and networks is formulated, where user optimizes over the *transmit power* and *base-station assignment*, given by $\max\{U_i(a_i, \rho - \lambda T_i(a_i, P))\}$ for the user utility of $u_i(a_i, P) = T_i(a_i, P)/P_i$. Here $T_i(a_i, P)$ is user-received throughput at base station a_i with transmit power P. Similarly, networks optimize over revenue earned from extending resources $\max_{\lambda \geq 0} \rho(\lambda)$, which is the sum over payment ρ by all users associated to themselves.

In [39–41] different mathematical techniques are used for evaluating the preferred network selection. Although not explicitly mentioned in [39], the format of formulated game among network providers belongs to congestion games, expressed as

$$G = (N, R, S_n, U_n), \tag{22.19}$$

where

the player set N represents network operators
resource R represents the requests from users
S_n represents the strategy set of players, that is, for acquiring resources
U_n is the payoff value tied to a user-assigned value or alternatively to the network technology evaluation process of users

A user evaluates available network technologies for attainable *application QoS*, *pricing*, etc. Different QoS parameters are assigned weights using AHP, which refer to the priority of parameters for service/applications. Given the quantitative relationship among different QoS evaluations (obtained by AHP), Grey relational analysis (GRA) [42] indicates the optimal network for specific application, where the network with higher grey relational coefficient (GRC) value is preferred to the one with a smaller GRC value. Dynamic status of network technologies in terms of available capacity is considered and input as a parameter to the decision mechanism, which computes the preferred network ψ as

$$\psi = GRC \frac{\text{Available bandwidth}}{\text{Required bandwidth}}. \tag{22.20}$$

Therefore, each service request is accompanied by GRC values assigned to different available network technologies. The stability and efficiency of the solution are not discussed.

Antoniou et al. [40,41] formulate the CAC in a similar way using noncooperative games played among operators where common resources are the service requests. Game output is decision of the subset of service requests that is admitted to the network. Players, resource, payoff, and strategy set in this game formulation are similar to the one discussed in [39]. The sequence of process is as follows: (a) providers separately offer access to service request with some predicted value of different QoS parameters, (b) users send service request with associated preference value of available networks to the *4G system*, which in turn is translated into network technology payoffs, and (c) network technologies select the service request that increases their payoff. As different from [39], the suitability of a network technology is computed by dividing the possible user service requests into two broad categories, namely, real-time services (RT) and non-real-time services (NRT). User satisfaction is translated in time (in terms of delay and jitter) and observed packet loss. Since NRT services are not sensitive to delays, users' preference over delay can be a constraint in such a case, whereas in RT traffic application-required delay is a constraint. Assuming that users can not perceive packet losses until they are higher than some threshold, the utility function is given as the function of delay and jitter. Network-predicted parameter values of delay and jitter are mapped to the utility of users, who accordingly assign preference to available networks. Utility function of user in turn decides the payoff of network technology, where the preference value assigned to any particular network motivates the networks to compete for service request to increase their payoffs.

Chen et al. [43] introduce the concept of *arbitration probability* to evaluate the available technologies and compute the preferred technology. It is used to model the utility function and reflects the level of willingness that a user wants to use the service of any available network. The attributes of arbitration probability include (a) satisfaction level of users, (b) relative link quality, and (c) service pricing. User satisfaction level is modeled using *sigmoid function*. Users select the network with the highest arbitration probability value. A noncooperative game $G = [N, B_j, U(\cdot)]$ is modeled to reflect the competition among available networks. Network technologies are players; B_j is the strategy set of network j. Utility function that represents the profit that network may earn by providing resource to the user is modeled using arbitration probability and is given by $U_j = n_j P_j(b_j) A_j(b)$, where n_j is the maximum number of users that network can accommodate at the same time, P_j is the price charged by network j for allocating bandwidth b to request, and $A_j(b)$ represents the arbitration probability value assigned by the user against the offer (containing amount of bandwidth, link quality, and pricing information) of network j. The unique Nash equilibrium is computed as the solution of the modeled game.

22.5 Pricing

Although this chapter focuses on the problem of network selection mainly based on technical parameters and objectives, economic models also play a crucial role in the decision of both users and network operators. Integrating pricing mechanisms into technical aspects of network selection helps capture the problem more realistically. Therefore, we briefly visit pricing issues in this section in a compact tabular form.

Some of the commonly used pricing schemes include (a) flat-rate pricing—services are offered for a fixed period of time with a fixed price irrespective of the amount of service used, (b) usage base pricing—users are charged for the amount of service they use, (c) static pricing—pricing is known prior to service usage, and (d) dynamic pricing—pricing is tuned during service consumption

This section concentrates on dynamic pricing owing to its realistic applicability to next-generation user-centric wireless networks, where optimum resource utilization and congestion control are also key issues. Different game-theoretic approaches in this direction are categorized as *leader–follower*, *multi-stage*, and *auction-based* methods, and are summarized in Tables 22.2 through 22.4. In *leader–follower (Stackelberg)* games [44], the dominant firm, the leader, maximizes its profit subject to all other firms, the followers, being in a competitive equilibrium.

22.6 Conclusion and Open Issues

This chapter focused on network selection problem in heterogeneous wireless networks and presented game-theoretic approaches to address the problem in different scenarios involving various combinations of operators and access technologies as well as different business models and assumptions. Figure 22.1 depicts a generalized scenario where a spectrum provider distributes spectrum to network providers, who allocate spectrum resources to their networks (access technologies), which in turn extend those resources to end users for connectivity. One can observe from the literature survey presented in this chapter that there is a tendency away from static allocation schemes toward

Table 22.2 Leader–Follower Pricing Schemes

Reference	Utility	Details
[36]	**Leader**: Service provider—$U_l(\pi, r)$: Leader sets price π for resource r	Addresses Internet pricing. Takes maximal acceptable response time as QoS metric of user service. It is shown that performance of leader–follower approach does not lie on Pareto boundary
	Followers: Users—$U_f(\pi, r)$: Followers compute their demand function depending on price π	Bargaining solution dictating cooperation among players is used. It is concluded that players are better off when they cooperate. Approach is extended to two ISP cases
[45]	**Leader**: WiMAX	WiMAX uses dynamic pricing scheme to extend services to WiFi owing to elastic nature of applications on WiFi user nodes. WiMAX charges its subscriber stations with fixed prices. Payoff of both WiMAX and WiFi depends on demand functions and accordingly leader sets the prices
	Followers: WiFi (APs)	
	Utility of WiFi k is the difference of revenue earned from its users and cost paid to WiMAX base station: $U_k(R_k - C_k)$	
	Demand function of WiFi k is $b_j(\pi_k) = C_j - d_j\pi_k$, where $b_j(\pi_k)$ is required bandwidth by j for announced price π_k	
[46]	**Leader**: Network	Network decides unit price of resource and accordingly users decide the transmitted power. A joint user and network-centric problem is formulated using noncooperative games among users, where users adjust their powers in a distributed fashion to maximize their utilities. Unique NE in user-centric case is found and solution for revenue maximization is found by multidimensional search on unit prices of each user
	Followers: Users	
	Utility of network n is the revenue earned by payment λ_i of all users $i \in N$, $U_n = \sum_{i=1}^{N} \lambda_i T_i$	
	Utility of the user i is the average number of bits transmitted correctly per joule battery power $U_i = T_i/p_i$	

dynamic interface selection on the user side and dynamic resource distribution on the operator side to maximize user satisfaction and resource usage efficiency.

The increasingly dynamic nature of the telecommunication scene is expected to go beyond the technical domain and also cover business models and socioeconomic aspects of telecommunications. For example, PERIMETER [51], an FP7 (Framework Program 7) project funded by the European Union, aims to establish a new user-centric paradigm in telecommunications where users can switch among different operators and different technologies in real time without being bound to any contractual agreements. Also in [40] and [41], the term *4G system* is used to refer to a scenario where a third party is responsible for collecting user-generated service requests and distributing them

Table 22.3 Multistage Pricing Schemes

Reference	Utility	Details
[38]	Three levels of market vision	The three different levels of market include (a) spectrum level (b) service provider level, and (c) user level. Service providers sitting in the middle layer compete with each other on one hand for spectrum share from spectrum provider and on the other hand tend to increase satisfied user pool. Such multiple competitions are modeled using multi-stage games. In first stage, service providers decide the amount of bandwidth to be bought from spectrum provider. In the second stage, service providers produce services and set prices. Game is solved using backward induction. Existence of Nash equilibrium for two-stage game is proven
	Utility of spectrum provider for bandwidth allocation to service provider 1 (b_1) and provider 2 (b_2) is given by $u_i(b_1, b_2) = \theta_i b_i P(\theta_1 b_1 + \theta_2 b_2) - b_i C(b_1 + b_2)$, where θ_i represents the provider i's spectrum usage efficiency factor	
[47]	Users' utility function is a function of client's intended session length K and the price paid for duration of service used; $f(t, k) - \sum_{i=1}^{t} P_i$. Base stations' utility function is given by $\sum_{i=1}^{t} P_i$	The interaction between a base station and paying client is formulated using a simple two-player game model. At the beginning of each game, AP selects the price and users have options to connect by accepting the price or reject. Players choose between two service providers for (a) web-browsing and (b) file transfer. Browsing formulation leads to a constant price Nash equilibrium but the outcome is inefficient for file transfer users
[48]	Utility of service provider i is the difference between revenue earned from service and price paid for network infrastructure: $U_i(\pi, f) = \pi_i D_i(\pi, f) - R_i(f_i - d_i)$	In this, service providers and network infrastructure owner are decoupled. Competition among service providers is modeled using a noncooperative game. Service demand of the service provider is assumed to be a linear function of vector of prices and QoS of all providers. Two types of games are modeled: (a) keeping QoS of all providers fixed and allowing service prices to vary, and (b) both QoS and pricing are set by providers. Existence of Nash equilibrium is proven analytically under some assumptions
	Cost function is given by $C_i = R_i(\pi, f_i, d_i)$. Aggregated demand function of i is $d_i = D_i(\pi_i, f)$, where $(\pi = $ payment, $R_i = $ revenue$)$	

Table 22.4   Auction-Based Pricing Schemes

Reference	Utility	Details
[26]	Negotiation of QoS for the required service with registered networks is considered. A decision of selecting any suitable network in this connection is a consequence of a trade-off between price and QoS	It provides offered resources at different prices in a dynamic market. Offered price is one of the factors that motivate users to select the service that best suits their preferences in terms of price and reputation of providers. Network resources are sold using a variant of first-price sealed-bid auctioning format. Simulation results exhibit that two-operator case admits equilibrium and offer the same price, whereas any additional operator causes decrease in market prices
[49]	Sellers utility = Revenue: $n \int_{v*}^{\bar{v}}(vF'(v) + F(v) - 1)p(v)dv$	Discusses the effect of pricing on congestion and traffic management. The auction is carried out on calls arriving at *GPRS*-based networks. In case delay is observed, auction-based pricing mechanism is invoked to admit the data calls. Multi-unit Vickery auction is used and auctions are carried out for new calls only. GPRS base station has reserved price v_O that enables it to withdraw auction from bid not covering the operating cost. Reserve price varies in congestion situations v_C, such that $v_C \geq v_O$. Correlation between mean system delay and congestion reservation price is studied
	$v*$ is reserve price, n is number of bidders, $\bar{v}$ is maximum bidders value, and $p(v)$ is probability of a bidder to be one of the auction winners, $F(v) \in \{0, 1\}$	
[50]	User j's utility when connected to base station a_j is modeled as the received QoS and measures the performance in bits per joule	Considers distributed power control in multicell wireless data system. Base station assignment is based on maximum receive signal strength (MRSS), and maximum signal to interference ratio (MSIR) is discussed. Both models are analyzed in the presence of pricing. Two pricing models are introduced as local pricing and global pricing. Nash equilibrium of both models is given

among available network technologies. This entity should also ensure that the network operator delivers the agreed level of service quality to user and takes care of billing and security issues. There are many challenges, both technical and socioeconomic, that need to be addressed for this vision to come true, such as the need for a standardized view of QoE among all stakeholders that should act as a common performance and valuation criterion. Pricing schemes also become very important in such a dynamic environment, which is addressed to some extent by researchers, as listed in Section 22.5. Further research is needed in the direction of mechanism design for such independent entities or alternative solutions for inter-operator arbitration in future dynamic networks.

One of the most static resources in telecommunications has traditionally been the spectrum licences that are allocated to network providers through large auctions. With technological innovations such as *software-defined radio* and with shortened business cycles, the dynamism in other levels of telecommunications may also stretch to the spectrum allocation level, as depicted in Figure 22.1 and studied in the literature. In this context, multidimensional multi-stage dynamic network selection will play an important role for attaining the objectives of all stakeholders. In such dynamic decision mechanisms, timely locating the required information during the network discovery process is a challenging issue. Efficient and light-weight signaling for seamless vertical/session handovers, capturing the accurate preference relationships of different stakeholders, finding the optimal number of comparable metric parameters of different access networks/operators/service providers are additional challenges to be met in order to realize the intelligent network selection concept.

Before we conclude the chapter, let us specify a few open points in the reviewed literature for future research directions. Splittable applications, where a request can be simultaneously associated with multiple network technologies, are considered, for example, in [25,26]. However, there is a lack of studies that consider a realistic mix of different application types (e.g., rigid, elastic, non-partitionable applications). In [25], the predefined offered bandwidth by each technology is decided by a central controller, which fits well into a single-operator scenario. This work may not be extended to multi-operator scenarios unless the resource management mechanism among different technologies of different operators is well defined. Therefore, the criteria or mechanism of defining predefined offered bandwidth for different application classes will be an interesting research area. Although some of these issues are addressed in [27] to some extent, a potential research question that still needs to be addressed is the *mechanism design* at SLA broker level, which drives intelligent and fair decision making for resource allocation achieving the global objectives of both operators and end users.

References

1. T.G. Kanter, Going wireless: Enabling an adaptive and extensible environment, *ACM Journal on Mobile Networks and Applications*, 8, 37–50, 2003.
2. E. Bircher and T. Braun, An agent-based architecture for service discovery and negotiation in wireless networks, *WWIC*, Frankfurt, Germany, pp. 295–306, 2004.
3. E. Adamopoulou, K. Demestichas, and A. Koutsorodi, Intelligent access network selection in heterogeneous networks-simulation results, *IEEE Wireless Communication System Symposium*, Siena, Italy, pp. 279–283, 2005.
4. T.L. Saaty, How to make a decision: The analytical hierarchy process, *European Journal of Operational Research*, 48, 9–26, 1990.
5. A.P.C. Silva, M. Varela, E.S. Sliva, R.M.M. Leao, and G. Rubino, Quality assessment of interactive voice applications, *International Journal of Computer and Telecommunications Networking*, 52, 6, 1179–1192, 2008.
6. V. Menkovski, A. Oredope, A. Liotta, and A. Cuadra Sanchez, Optimized online learning for QoE prediction, *21st Benelux Conference on Artificial Intelligence (BNAIC 2009)*, Eindhoven, the Netherlands, October 2009.
7. B.P. Kafle, E. Kamioka, and S. Yamada, User-centric performance and cost analysis for selecting access networks in heterogeneous overlay systems, *Proceedings of 8th IFIP Conference on MMNS 2005*, Barcelona, Spain, LNCS 3754, pp. 277–288. October 2005.

8. X. Cai, L. Chen, R. Sofia, and Y. Wu, Dynamic and user-centric network selection in heterogeneous networks, *Proceedings of 26th IEEE International Performance Computing and Communications Conference (IPCCC)*, New Orleans, LA, pp. 538–544, 2007.

9. A. Iera, A. Molinaro, C. Campolo, and M. Amadeo, An access network selection algorithm dynamically adapted to user needs and preferences, *Proceedings of 17th International Symposium of Personal, Indoor and Mobile Radio Communications*, Helsinki, Finland, pp. 1–5, 2006.

10. K. Mittal, E.M. Belding, and S. Suri, A game-theoretic analysis of wireless access point selection by mobile users, *Computer Communication*, 31, 10, pp. 2049–2062, 2008.

11. B. Vocking, Congestion games: Optimization in competition, *Proceedings of 2nd Algorithms and Complexity in Durham Workshop*, Durham, U.K., pp. 9–20, 2006.

12. M. Cesana, N. Gatti, and I. Malanchini, Game theoretic analysis of wireless access network selection: Models, inefficiency bounds, and algorithms, *Workshop on Game Theory and Communication Networks (GAMECOMM)*, Athens, Greece, 2008.

13. A. Neto, S. Sargento, E. Logota, J. Antoniou, and F.C. Pinto, Multiparty session and network resource control in the context casting (C-CAST) project, *Second International Workshop on Future Multimedia Networking (FMN 2009)*, Coimbra, Portugal, June 2009.

14. X. Cai and F. Liu, Network selection for group handover in multi-access networks, *Proceedings of IEEE International Conference on Communication*, Beijing, China, pp. 2164–2168, 2008.

15. D. Niyato and E. Hossain, Dynamics of network selection in heterogeneous wireless networks: An evolutionary game approach, *Proceedings of IEEE Transactions on Vehicular Technology*, 58, 4, 2008–2017, 2009.

16. R.B. Myerson, *Game Theory: Analysis of Conflict*, Harvard University Press, Cambridge, MA, September 1991.

17. K. Apt and A. Witzel, A generic approach to coalition formation, *Proceedings of the International Workshop on Computational Social Choice (COMSOC)*, Amsterdam, the Netherlands, December 2006.

18. M. Yoshino, K. Sato, R. Shinkuma, and T. Takhashi, Incentive-rewarding mechanism for user-position control, *IEICE Transaction*, 91-B, 10, 2132–3140, Mobile Services, 2008.

19. R. Boeheim and C. Zulehner, Auctions—A survey, Technical Report, December 1996.

20. S. Frattasi, B. Can, F. Fitzek, and Ra. Prasad, Cooperative services for 4G, *Proceedings of 14th IST Mobile & Wireless Communications*, Dresden, Germany, 2005.

21. M.A. Khan, F. Sivrikaya, and S. Abayrak, Auction based interface selection in heterogeneous wireless networks, *2nd Wireless Days Conference*, Paris, France, December 2009.

22. N.R. Suri and Y. Narahari, An auction algorithm for procuring wireless channel in a heterogenous wireless network, *Wireless and Optical Communications Networks*, 1–5, 2006.

23. H. Lin, M. Chatterjee, S.K. Das, and K. Basu, ARC: An integrated admission and rate control framework for competitive wireless CDMA data networks using noncooperative games, *Journal IEEE Transactions on Mobile Computing*, 1536–1233, 2005.

24. D. Niyato and E. Hossain, A game-theoretic approach to bandwidth allocation and admission control for polling services in IEEE 802.16 broadband wireless networks, *Proceedings of 3rd International Conference on QoS in Heterogeneous Wired/Wireless Networks*, Waterloo, Canada, 2006.

25. D. Niyato and E. Hossain, A cooperative game framework for bandwidth allocation in 4g heterogeneous wireless networks, *Proceedings of IEEE International Conference on Communications ICC 06*, Istanbul, Turkey, vol. 9, pp. 4357–4362, 2006.

26. M.A. Khan, C. Truong, T. Geithner, F. Sivrikaya, and S. Albayrak, Network level cooperation for resource allocation in future wireless networks, *Proceedings of the IFIP Wireless Days Conference '08*, Dubai, UAE.

27. M.A. Khan, A.C. Toker, C. Truong, F. Sivrikaya, and S. Albayrak, Cooperative Game Theoretic Approach to Integrated Bandwidth Sharing and Allocation, *GameNets09*, Istanbul, Turkey, 2009.

28. S.L. Hew and L.B. White, Fair resource bargaining solutions for cooperative multi-operator networks, *International Zurich Seminar on Communication*, Zurich, Switzerland, pp. 58–61, 2006.

29. G. Zhang, L. Cong, L. Zhao, K. Yang, and H. Zhang, Competitive resource sharing based on game theory in cooperative relay networks, *ETRI Journal*, vol. 31, 89–91, 2009.

30. Y. Pan, A. Nix, and M. Beach, A game theoretic approach to distributed resource allocation for OFDMA-based relaying networks, *19th Symposium on Personal, Indoor and Mobile Radio Communications*, Cannes, France, 2008.

31. O. Sallent, J. Perez-Romero, R. Agusti, L. Giupponi, C. Kloeck, I. Martoyo, S. Klett, and J. Luo, Resource auctioning mechanisms in heterogeneous wireless access networks, *Conference of Vehicular Technology*, Melbourne, Australia, 2006.

32. M. Dramitinos, G.D. Stamoulis, and C. Courcoubetis, Auction-based resource reservation in 2.5/3G networks, *Mobile Networks and Applications*, 9, 6, 577–566, 2004.

33. M. Dramitinos, G.D. Stamoulis, and C. Courcoubetis, Auction-based resource reservation in 3G networks serving multicast, *15th IST Mobile Summit*, Myconos, Greece, June 4–8, 2006.

34. J.S. Modiano and E. Lizhong Zheng, Wireless channel allocation using an auction algorithm, *Journal on Selected Areas in Communications*, 24, 2, 1085–1096, 2006.

35. M. Roggendorf and F. Beltran, Flow-based resource allocation in a multiple-access wireless market-setting using an auction, *Proceedings of 26th IEEE International Conference Workshops on Distributed Computing Systems*, Lisbon, Portugal, 2006.

36. G.L. Bodic, D. Girma, J. Irvine, and J. Dunlop, Dynamic 3G network selection for increasing the competition in the mobile communications market, *52nd Vehicular Technology Conference*, vol. 3, Boston, MA, vol. 3, pp. 1064–1071, 2000.

37. M. Cesana, I. Malanchini, and A. Capone, Modelling network selection and resource allocation in wireless access networks with non-cooperative games, *Mobile Ad Hoc and Sensor Systems*, 404–409, 2008.

38. N. Feng, S.C Mau, and N.B. Mandayam, Joint network-centric and user-centric radio resource management in a multicell system, *IEEE Transactions on Communications*, 53, 7, 1114–1118, 2005.

39. D. Charilas, O. Markaki, and E. Tragos, A theoretical scheme for applying game theory and network selection mechanisms in access admission control, *3rd International Symposium on Wireless Pervasive Computing (ISWPC 2008)*, Santorini, Greece, pp. 303–307, 2008.

40. J. Antoniou and A. Pitsillides, 4G converged environment: Modeling network selection as a game, *16th IST Mobile and Wireless Communications Summit*, Budapest, Hungary, pp. 1–5, 2007.

41. J. Antoniou, M. Stylianou, and A. Pitsillides, Ran selection for converged networks supporting IMS signalling and MBMS service, *18th Symposium on Personal, Indoor and Mobile Radio Communications*, Athens, Greece, pp. 1–5, 2007.

42. J.L. Deng, Introduction to grey system theory, *Journal of Grey System*, 1, 1, 1–24, 1989.

43. J. Chen, K. Yu, Y. Ji, and P. Zhang, Non-cooperative distributed network resource allocation in heterogeneous wireless data networks, *IEEE Transactions on Mobile Computing*, 7, 3, 332–345, 2008.

44. H.V. Stackelberg, *The Theory of Market Economy*, Oxford University Press, Oxford, 1952.

45. E. Takahashi and Y. Tanaka, Auction-based effective bandwidth allocation mechanism, *10th International Conference on Telecommunications*, vol. 6, Tahiti, pp. 1046–1050, 2003.

46. X.R. Cao, H.X. Shen, R. Milito, and P. Wirth, Internet pricing with a game theoretical approach: Concepts and examples, *Networking, IEEE/ACM Transactions*, 10, 2, 208–216, 2002.

47. D. Niyato and E. Hossain, Integration of WiMAX and WiFi: Optimal pricing for bandwidth sharing, *IEEE Communication Magazine*, 45, 5, 140–146, 2007.

48. J. Jia and Q. Zhang, Bandwidth and price competitions of wireless service providers in two-stage spectrum market, *International Conference on Communications*, Beijing, China, pp. 4953–5957, 2008.

49. R. El-Azouzi, E. Altman, and L. Wynter, Telecommunications network equilibrium with price and Quality-of-Service characteristics, *Proceedings of ITC*, Charlotte, NC, 2003.
50. J. Musacchio and J. Walrand, Game theoretic modeling of WiFi pricing, in game theoretic modeling of WiFi pricing, *41st Annual Allerton Conference on Communication, Control and Computing*, Monticello, IL, 2003.
51. PERIMETER, EU FP7 Project, http://www.ict-perimeter.eu

Chapter 23

Network Selection and Handoff in Wireless Networks: A Game Theoretic Approach

Josephina Antoniou, Vicky Papadopoulou, Vasos Vassiliou, and Andreas Pitsillides

Contents

Convergence, that is, the integration of various access technologies combining their resources to best serve the increased user requirements, may be supported, through a system architecture where different access networks, terminals, and services coexist. Thus, a new, *user centric* communication paradigm is motivated, that is, the user is no longer bound to only one access network but may indirectly *select* the best available access network(s) to support a service session. Upon a new service request or even any dynamic change, for example, mobility, one (or a group) of the participating access networks needs to be selected in order to support the session. Thus, the converged system architecture must be equipped with a *network selection* mechanism to effectively assign the best access network(s) to handle a service session.

23.1 Introduction

Converged networks allow different access networks, terminals, and services to coexist, bringing forth a new communication paradigm, which is *user-centric* [1], that is, the user is no longer bound to only one access network but may indirectly *select* the *best* available access network(s) to support a service session [2]. Upon a new service request or even any dynamic change affecting the session, (e.g., mobility) one, two, or multiple of the participating access networks need to be selected in order to support the session.

Next-generation communication networks—controlled by an IP core network—need to be equipped with a *network selection mechanism* to assign the *best* access network(s) to handle service activation or any dynamic session change. Such decisions may result in one, two, or multiple access networks handling a service (the cooperation of multiple networks to handle a service is treated in [3]). This chapter studies the resulting interaction between a user and a network when a single network is selected and proposes a payment partition between two networks cooperating to support a session, seeking the best behavior for each entity such that their conflicting interests are overcome and satisfaction is achieved.

Since any communication network, such as a converged network [4] considered in this chapter, is a multi-entity system, decisions are taken by different system entities. The decision-making entities in a converged network are (a) the user and (b) the network operator. Both entities are motivated to make decisions that maximize their own *satisfaction* functions [5,6], which are based on each entity's criteria. The criteria for the user include quality of experience (QoE) and cost; on the other hand, the network operator makes decisions that are driven mainly by one criterion: the network's revenue maximization. Therefore, the network operator sets the *price per user session* for using network resources based on its own revenue-maximization criterion, and the user, given the price set by the network as well as some further parameters that comprise its own satisfaction function, makes the decision whether or not to participate in the particular network, always subject to the network's own admission policy, for example [7,8].

We provide a study of the interaction between a user and a network during network selection, and we define the selection process based on the conclusions obtained from the study of the user–network interaction. The selection decision can be modified to result in a list of prioritized available networks instead of a single network. Thus, for users requesting enhanced service with additional quality guarantees, for example, *premium* service, the two *best* networks on the priority list cooperate to provide these guarantees; such additional guarantees include, for example, service continuity during fast session handoffs that may be required during a session. This situation is considered in this chapter, and we utilize the notion of a Nash bargaining solution (NBS) to compute an optimal partition agreement between the two cooperative networks for the service payment.

Therefore, we consider first the interaction between one user and one access network in order to propose a model for the interaction of each user and each one of the available access networks for network selection. Then, our study turns toward the case of the interaction between two networks cooperating to support a *premium* service, to handle fast session handoffs and ensure service continuity throughout the duration of the session. Such interactions between entities with conflicting interests follow action plans designed by each entity in such a way as to achieve a particular selfish goal and are known as strategic interactions. *Game theory* is a theoretical framework that studies strategic interactions, by developing models that prescribe actions in order for the interacting entities to achieve satisfactory gains from the situation. In this chapter, we utilize game theory in order to model, analyze, and finally propose solutions for the selected interactions. The user–network interaction during network selection is modeled as a game between the user and the network, whereas the second game models the cooperation between two users to support premium service and the user is not involved in the game model. Figure 23.1 presents the relation between the two game models representing the earlier-mentioned situations.

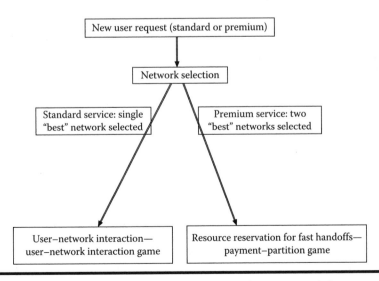

Figure 23.1   Relation between the user–network interaction game and the payment–partition game.

The chapter is organized as follows: Section 23.2 provides an overview of network selection–related research. Section 23.3 defines and analyzes models of the network selection problem when a single network is selected.* In Section 23.4, we introduce an enhanced game, which captures the case where the two best networks cooperate to enable fast session handoffs when the service continuation and the time to handoff are crucial for a specific service.† Next, Section 23.5 provides a simulative evaluation of the strategies for the models presented, and Sections 23.6 and 23.7 offer conclusions for the chapter and a discussion on open research challenges, respectively.

23.2 Related Work

The growing popularity of the next-generation communication networks, which promote technology convergence and allow the coexistence of heterogeneous constituents, for example, various access networks, has pushed toward the efficient resource management (*including network selection*) of the overall converged system.

Earlier works on network selection have explored varying approaches of describing and analyzing this new resource management mechanism. Such approaches included fuzzy logic [10,11], adaptive techniques [1,12], utility-based and game-theoretic models [13,14], technology-specific solutions, especially focusing on the interoperation of cellular systems with wireless LANs [15], as well as architectural models focusing on a more comprehensive architectural view [2]. Decision making in these works is either user controlled [11,15] or network controlled [2,12]; the interaction between them is not really considered and the selection decision mainly involves one entity, in some cases involving the other entity indirectly, that is, network decision considering some user preference [1,14] or user decision considering some network-specific rankings [10,13].

Further, in more recently published research, we observe a popular trend toward both dynamic/automated solutions that are either based on the user or the network as a decision-making entity [16,17], in some cases proposing improvements to the user-perceived quality during network selection [18,19]. Other than the general solutions for network selection, specific solutions handling particular services such as multicast have been recently proposed [20,21].

Given that game theory is a theoretical framework for strategical decision making, it has been a very popular approach among the most recently presented research works. These explore various game-theoretic models such as noncooperative games [22,23] and cooperation schemes where limited resources or need for quality guarantees exists [3,24]. In addition to network selection, game-theoretic models for admission and rate control exist where appropriate admission criteria guard against users leaving the selected network due to, for instance, unsatisfactory service [25]. The game-theoretic study of users having the option to leave their network, that is, churning, is further carried out, in particular, pricing schemes (both noncooperative and cooperative) that maximize the revenues of service providers [26]. Another topic of interest in our work is how to model bargaining situations using game theory. An interesting such work that dives into the two-player bargaining situation, considering a game of incomplete information where players are uncertain about each other's preferences and offer a utility-efficient solution, is found in [27].

An issue that arise in game-theoretic models, which we will also address in this chapter, is that of truthfulness. An interesting and very promising way to guarantee truthfulness of the participating

---

* A detailed study for the case of single network selection is found in [9].
† The case of multiple networks cooperating to support a service request is explored in [3].

networks is through pricing mechanisms [28]. Such mechanisms could penalize a player who turned out to lie, when a revelation of the real value of the particular quantity is possible at a later stage of the procedure. Such a mechanism would enforce the players in a particular game to be truthful. In the worldwide literature, there is a whole research field that is focused on the development, limitations, and capabilities of such pricing mechanisms: the algorithmic mechanism design [28]. Successful paradigms in this context include (combinatorial) auctions [29] and task scheduling [30] using techniques such as the revelation principle [28], incentive compatibility [28], direct revelation [28], and Vickrey–Clarke–Groves mechanisms [31]. We believe that the algorithmic tools and theoretical knowledge that is already developed in the field of algorithmic mechanism design constitute a fruitful pool for extracting algorithmic tools for enforcing players to truthfulness, through pricing mechanisms.

The above-mentioned game-theoretic approaches have inspired our work and the current chapter to explore network selection through a game-theoretic modeling framework. In this chapter, we model aspects of network selection as cooperative game models between interacting entities, capturing the case where decision making is controlled by both the user and the network.

23.3 Single Network Selection

Network selection is an inherently complex decision because it involves considerations of user characteristics, individual access network characteristics, and overall network efficiency. Selecting the network that both satisfies the user and the network may become a challenging task. A careful examination of the user–network interaction prior to defining the proposed network selection model is necessary.

In this section, we consider the interaction between a user and a network, and we seek to investigate the outcomes of their interaction and indicate possible incentives to motivate the cooperation of the two. We first look at the interaction between a user and a network during network selection, to reach a decision that is both user-satisfying and network-satisfying. For a fixed quality level, the interaction may be viewed as an exchange between two entities: the user gives some *compensation* and the network gives a *promise* of the specific quality level. Primarily, we seek the incentives for each entity to select certain strategies, that is, sets of actions, that result in a cooperative selection decision, by which both entities are satisfied; such incentives are usually the payoffs to the entities involved. In the user–network interaction, the payoffs are the following:

- *User's payoff*: the difference between the *perceived satisfaction* and the compensation offered by the user to the network (for a given quality level).
- *Network's payoff*: the difference between the compensation received and the cost of supporting the session, for a given quality level. The cost is analogous to the requested quality by the user quality level.

Prior to modeling the user–network interaction, we make the following general assumptions:

1. The players in the game are heterogeneous, aiming at different payoffs.
2. The players make simultaneous decisions, without knowing the decision of their opponent at the time of their own decision.

3. There is a minimum and a maximum payment as well as a minimum and a maximum quality, considered in the user–network interaction model.
4. There is always a probability that quality degradation will be perceived by the user because of the dynamic nature of the network, even if the network decides to cooperate.
5. For a set quality q, it holds that $p(q) - \kappa > 0$, that is, the difference between the maximum satisfaction perceived by the user and the maximum compensation given by the user is greater than 0. Similarly, the difference between the minimum satisfaction and the minimum compensation is greater than 0, that is, $p(q') - \kappa' > 0$.
6. Both the user and the network have nonnegative payoff functions. This assumption is introduced in order to motivate the players to participate in the interaction (reflecting a selection of an appropriate access network during network selection).

The above assumptions result in the following specifications, which are imposed by the game model:

■ Assumption 6 results in the requirement that the maximum compensation* offered by the user is less than or equal to the satisfaction corresponding to the minimum requested quality. Thus, given that compensation and satisfaction from perceived quality are measurable in comparable units, the maximum compensation offered by the user should be defined to be less than or equal to the satisfaction corresponding to the minimum requested quality. In this way, the user *plays* the game without risking to have a negative payoff, satisfying Assumption 6.
■ Assumption 6 results in the requirement that the minimum compensation offered by the user is greater than or equal to the cost of the maximum requested quality. Thus, given that the compensation and cost of supporting a requested quality are measurable in comparable units, the minimum compensation offered by the user should be defined to be greater than or equal to the cost of the maximum requested quality. In this way, the network *plays* the game without risking to have a negative payoff satisfying Assumption 6.

23.3.1 One-Shot User–Network Interaction Game

We first consider the simpler case of the interaction, where the current actions of the involved entities do not affect future actions, nor are affected by previous ones. Since we aim to extract conclusions on the user–network cooperation, we model this simple scenario as a one-shot game. The game model, considering several quality levels, is illustrated in Figure 23.2; the subsequent analysis considers the quality level requested to be fixed. The players play only once and in which the current outcomes do not affect any future outcomes of the game. In particular, we assume that the user and network interaction is as follows: the user decides on a specific quality level to request and proposes to the network a corresponding compromise for that particular quality level. Then the network decides to accept it or not. In case of acceptance,

---

* Regarding, setting the maximum compensation, if a telecommunications regulator exists, as currently in Europe, the regulator takes the role of setting this maximum compensation, otherwise the platform administrator will set the maximum compensation to encourage the user to remain a customer. The regulator or the administrator has information that prevents exploitation of the users by the operators and based on that information the maximum compensation is set.

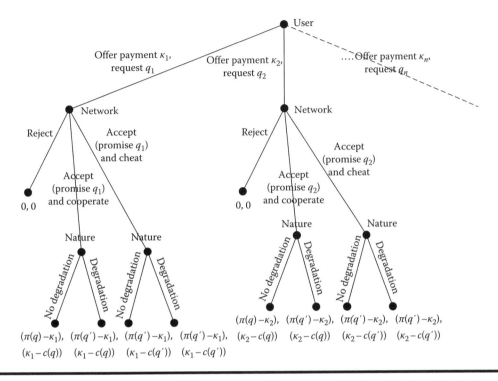

Figure 23.2 **The one-shot sequential moves game of user–network interaction in extensive form.**

the network may then decide to provide the promised quality level or to cheat. Finally, we assume that there exists a random event of quality degradation provided by the user, even in the case of deciding not to cheat. More formally, the user–network interaction game is defined as follows:

Definition 23.1 (The user–network interaction game). Let the user choose an incentive compensation scheme $\kappa_i \in K$ to offer to the network, where $K = \{\kappa_1, \kappa_2, \ldots \kappa_n\}$, in order to encourage the network to make and keep a *promise* of offering $q_i \in Q$, where $Q = \{q_1, q_2, \ldots q_n\}$. The user has the capability to cheat and offer κ_i', such that $\forall\, i \in [n], \kappa_i' < \kappa_i$, but this will not be known to the network until the end of the game when payoffs are collected. The network makes a decision of whether to accept the user's compensation and promise to offer q_i or to reject it. If the network rejects the offer, the interaction terminates and the payoff to both players is 0. If the network accepts the offer, it can decide to keep the promise of supporting q_i or to cheat, by offering q_i', such that $\forall\, i \in [n], q_i' < q_i$. This will be known to the user also at the end of the game, when payoffs are collected.

Regardless of the network's decision, there is a random event, represented by *Nature*, that is, an event that is not controlled by neither the network nor the user. This random event has two outcomes: the quality offered by the network degrades (the user perceives a lower quality $q_i' < q_i$), or the perceived quality indicates no degradation (quality q_i perceived by the user). A quality degradation might be observed by the user, even when the network takes all the necessary measures to keep the promise for the requested quality. Whether the network cheats, or *nature's* event is

Table 23.1   User and Network Play the User–Network Interaction Game, Nature's Play: No Degradation

	Network Accepts and Cooperates	Network Accepts and Cheats	Network Rejects
User cooperates	$\pi(q) - \kappa, \kappa - c(q)$	$\pi(q') - \kappa, \kappa - c(q')$	$0, 0$
User cheats	$\pi(q) - \kappa', \kappa' - c(q)$	$\pi(q') - \kappa', \kappa' - c(q')$	$0, 0$

one of degradation, or whether both occur, the user perceives a quality q_i' without knowing what caused it.

In the rest of the section, we assume a fixed requested quality q and corresponding compensation κ, so we simplify κ_i to κ and q_i to q. Then, the payoffs of the two players are given in Table 23.1, assuming the *nature* outcome is no degradation; for the case of degradation caused by cheating, q is replaced with q'. Let $\pi(q)$ represent the satisfaction resulting from the perceived quality q and $c(q)$ represent the cost for supporting quality q. The payoffs are presented by order pairs: the first term is the user payoff and the second term is the network payoff.

23.3.1.1 Equivalence to Prisoner's Dilemma

Definition 23.2 (Prisoner's dilemma equivalent game) [32]   Consider a one-shot strategic game with two players in which each player has two possible actions: to cooperate with his opponent or to defect from cooperation. Furthermore, assume that the two following additional restrictions on the payoffs are satisfied:

1. The payoffs of the two players for each possible outcome of the game are shown in Table 23.2. For each player $j \in [2]$ and is such that $A_j > B_j > C_j > D_j$.
2. The reward for mutual cooperation is such that each player is not motivated to exploit his opponent or be exploited with the same probability, that is, for each player it holds that $B_j > (A_j + D_j)/2$.

Then, the game is said to be equivalent to a prisoner's dilemma type of game.

Consider now the following outcome for the user–network interaction game: the network accepts the user's offer and the outcome of nature's random event is *no degradation*. We prove that this outcome of the game is equivalent to a *prisoner's dilemma type of game*. In the subsequent

Table 23.2   General Payoffs for the Prisoner's Dilemma

	Player 2 Cooperates	Player 2 Cheats
Player 1 cooperates	B_1, B_2	D_1, A_2
Player 2 cheats	A_1, D_2	C_1, C_2

discussion, we recall Assumptions 5 and 6 from the general assumptions and we generate the following requirements for the user–network interaction game:

1. The difference between the maximum and minimum satisfaction, $\pi(q) - \pi(q')$, is greater than the diifference between the maximum and minimum compensation, $\kappa - \kappa'$, that may be offered by the user, since the user will always prefer that the maximum compensation offered is set as low as it is allowed by the platform administrator, whereas the satisfaction received from the maximum quality may be evaluated as high as possible, since no such constraints exist.
2. The network overall payoff (*compensation cost*) is always positive if accepting, that is, $\kappa - \kappa' > c(q) - c(q')$.

We now show the equivalence between the user–network interaction game and the prisoner's dilemma game:

Proposition 23.1 Consider the user–network interaction game. Assume that the network accepts the user's offer, and the event of nature is no degradation. Then the game is equivalent to a prisoner's dilemma game.

Proof By Definition 23.1 we observe that

Observation 23.1 Both the network and the user have two possible actions: to cooperate and to cheat (i.e., defect from cooperation).

Observation 23.1 combined with Definition 23.2 imply that the actions of the players in the user–network interaction game are the same as the actions of the players of a prisoner's dilemma game. Furthermore, consider the mapping of each player's payoffs (for the cases that the network accepts the user's offer), shown in Table 23.1, to payoffs A_j, B_j, C_j, D_j, where $j \in [2]$, as defined in Definition 23.2. We proceed to prove the following:

LEMMA 23.1 Set A_j, B_j, C_j, D_j according to Table 23.3 in the user–network interaction game. Then condition 1 of Definition 23.2 is satisfied.

Table 23.3 The Mapping between the User and Network Payoffs and the Payoffs in the Prisoner's Dilemma Type of Game

	User Payoffs (j = 1)	*Network Payoffs (j = 2)*
A_j	$\pi(q) - \kappa'$	$\kappa - c(q')$
B_j	$\pi(q) - \kappa$	$\kappa - c(q)$
C_j	$\pi(q') - \kappa'$	$\kappa' - c(q')$
D_j	$\pi(q') - \kappa$	$\kappa' - c(q)$

Proof We prove the claim by showing that $A_j > B_j > C_j > D_j$ for each $j \in [2]$ as required by condition 1 of Definition 23.2, when the network accepts the user's offer and no degradation occurs.

Firstly, consider the user. We verify straightforward that $\pi(q) - \kappa' > \pi(q) - \kappa$; thus, $A_1 > B_1$, and that $\pi(q') - \kappa' > \pi(q') - \kappa$; thus, $C_1 > D_1$, since $\kappa > \kappa'$. Given that $\pi(q) - \pi(q') > \kappa - \kappa'$, it holds that $B_1 > C_1$. It follows that $A_1 > B_1 > C_1 > D_1$ as required by condition 1 of Definition 23.2.

Consider now the network. We verify straightforward that $\kappa - c(q') > \kappa - c(q)$; thus, $A_2 > B_2$, and that $\kappa' - c(q') > \kappa' - c(q)$, since $c(q) > c(q')$; thus, $C_2 > D_2$. Assuming that the network accepts to participate in the interaction in a riskless manner, that is, only if the range of possible compensations exceeds the range of possible costs, then $\kappa - \kappa' > c(q) - c(q')$, and $B_2 > C_2$. It follows that $A_2 > B_2 > C_2 > D_2$ as required by condition 1 of Definition 23.2. $\square$

We now proceed to prove the following

LEMMA 23.2 Set A_j, A_j, B_j, C_j, D_j for each $j \in [2]$ as in Table 23.3. Then, the user–network interaction game satisfies condition 2 of Definition 23.2.

Proof To prove the claim, we must prove that the reward for cooperation is greater than the payoff for the described situation, that is, for each player it must hold that $B_j > (A_j + D_j)/2$.

For the user,

$$\pi(q) - \kappa > \frac{\pi(q) - \kappa' + \pi(q') - \kappa}{2} > \frac{\pi(q) + \pi(q')}{2} - \frac{\kappa' + \kappa}{2} \qquad (23.1)$$

since $\pi(q) > \pi(q')$ and $\kappa > \kappa'$.

For the network,

$$\kappa - c(q) > \frac{\kappa - c(q') + \kappa' - c(q)}{2} > \frac{\kappa + \kappa'}{2} - \frac{c(q) + c(q')}{2} \qquad (23.2)$$

Setting A_2, B_2, D_2 as shown in Table 23.3, we get that $B_j > (A_j + D_j)/2$ as required by condition 2 of Definition 23.2. $\square$

Observation 23.1, Lemmas 23.1, and 23.2 together complete the proof of Proposition 23.1. $\square$

The decision of each player in the user–network interaction game is based on the following reasoning [32]: If the opponent cooperates, defect to maximize payoff (A_j *in Table 23.3*); if the opponent defects, defect (*payoff C_j instead of payoff D_j*). This reasoning immediately implies

---

COROLLARY 23.1 [32] In a one-shot prisoner's dilemma game, a best-response strategy of both players is to defect.

Proposition 23.1 combined with Corollary 23.1 immediately implies

---

COROLLARY 23.2 In the one-shot user–network interaction game, when the event of nature is one of no degradations, a best-response strategy of both players is to cheat.

Given this result, we investigate next how the behavior of both the user and the network changes when the interaction between the two entities takes into account the history of their interaction.

23.3.2 Repeated User–Network Interaction Game

Here, we capture the fact that the interactions between networks and users do not occur only once but are recurring. In such relationships, the players do not only seek the immediate maximization of payoffs but instead the long-run optimal payoff. Such situations are modeled in game theory by finite horizon and infinite horizon repeated game models [33, Chapter 11]. We model the user–network interaction model as an infinite horizon repeated game since the number of new session requests is not known.

Well-known strategies in repeated game are *trigger strategies* [36, Chapter 6], that is, strategies that change in the presence of a predefined trigger. A popular trigger strategy is the *grim strategy* [33, Chapter 11], which dictates that the player participates in the relationship, but if *dissatisfied*, leaves the relationship forever. Another popular strategy used to elicit cooperative performance from an opponent is for a player to mimic the actions of his opponent, giving him the incentive to play cooperatively, since in this way he will be rewarded with a similar mirroring behavior. This strategy is referred to as *tit-for-tat* strategy [33, Chapter 11].

23.3.2.1 Present Value

In order to compare different sequences of payoffs in repeated games, we utilize the idea of the *present value of a payoff sequence* [33, Chapter 11], and we refer to it as the *present value* (PV). PV is the sum that a player is willing to receive now as payoff instead of waiting for future payoffs. Consider the interaction of the network and the user, which takes into account their decisions in the previous periods of their interaction. Let r be the rate by which the current payoffs increase in the next period, for example, in terms of satisfaction since the longer the user and network cooperate the more the satisfaction is for both players. Therefore, if the user (*or the network*) were set to receive a payoff equal to 1 in the next period, today the current payoff the user (*or the network*) would be willing to accept would be equal to $1/(1 + r)$.

Since in each game repetition it is possible that degradation may be perceived, there is always a probability p that the game will not continue in the next period. Then, the probability of no degradation is $1 - p$, and the payoff the user (*or the network*) is willing to accept today, that is, its PV, would be equal to $(1 - p)/(1 + r)$. Let $\delta = (1 - p)/(1 + r)$, where $\delta \in [0, 1]$ and often referred to as the *discount factor* in repeated games [33, Chapter 11]. In order to determine whether cooperation is a better strategy in the repeated game for both the network and the user, we utilize PV and examine for which values of δ, cooperation is a player's best-response strategy to the other player's strategy.

23.3.2.2 Equilibria

We first provide some useful strategies for the repeated game.

Definition 23.3 (Cheat-and-leave strategy). When the user cheats in one period and leaves in the next period to avoid punishment, the strategy is called the cheat-and-leave strategy.

Definition 23.4 (Cheat-and-return strategy)   When the network cheats in one period and returns to cooperation in the next period, the strategy is called the cheat-and-return strategy.

We are now ready to introduce a repeated game modeling the user–network interaction when the history is taken into account in the decisions of the entities:

Definition 23.5 (Repeated user–network interaction game)   Consider a game with infinite repetitions of the one-shot user–network interaction game, where $p \in [0, 1]$ is the probability of degradation, and r is the rate of satisfaction gain of continuing cooperation in the next period. Let PVs of each player be calculated after a history, that is, record of all past actions that the players made [33, Chapter 11], of cooperation from both players in terms of a discount factor $\delta = (1 - p)/(1 + r)$.

The user has a choice between the two following strategies: (i) the *grim* strategy, that is, offer a compensation κ but if degradation is perceived, then leave the relationship forever, and (ii) the *cheat-and-leave* strategy. The network has a choice between (a) the *tit-for-tat* strategy, that is, mimic the actions of its opponent, and (b) the *cheat-and-return* strategy.

The payoffs from each iteration are equal to the payoffs from the one-shot user–network interaction game. The cumulative payoffs (or PVs) for each player from the repeated interaction are equal to the sum of the player's payoffs in all periods; thus, the number of periods, for example, infinite, should be considered.

Then, the game is referred to as repeated user–network interaction game.

A sequence of strategies with which in each iteration a player plays its best-response strategy, that is, giving the player the highest payoff, to the opponent's strategy after every history is called a *subgame perfect strategy* for the player [33, Chapter 11]. To have a subgame perfect strategy, we must show that for every possible iteration of the game, the current choice of each player results in the highest payoff, against all possible actions of the opponent player. In the repeated user–network interaction game, we consider a history of cooperation from both players. Hence, the best-response strategy of each player against all possible actions of the opponent, in the current period, is given in terms of PV assuming such a cooperation history. When both players play their *subgame perfect strategies*, the profile of the game is an equilibrium in the repeated game, known as a *subgame perfect equilibrium* [33, Chapter 11].

When neither of the two players cheats, the sequence of profiles is defined more formally next:

Definition 23.6 (Conditional-cooperation profile)   When the user employs the grim strategy and the network employs the tit-for-tat strategy, the profile of the repeated game is referred to as conditional-cooperation profile of the game.

The following theorem identifies the *conditional-cooperation* profile as a subgame perfect equilibrium for the repeated user–network interaction game.

THEOREM 23.1   In the repeated user–network interaction game, assume $\delta > (c(q) - c(q'))/(\kappa - c(q'))$ and $\delta > (\kappa - \kappa')/(\pi(q) - \kappa')$. Then, the conditional-cooperation profile is a subgame perfect equilibrium for the game.

Proof We assume a history of cooperative moves in the past. We compute the PVs of both the user and the network, and after comparing them, we conclude that the conditional-cooperation profile is a subgame perfect equilibrium.

1. Assume first that the user plays the grim strategy. If the network plays the tit-for-tat strategy, it will cooperate in the current period, whereas, if it plays the cheat-and-return strategy, it may cheat.

If the network cooperates, then

$$\text{PV}^{net}_{coop} = \frac{\kappa - c(q)}{1 - \delta}$$

If the network cheats, then

$$\text{PV}^{net}_{cheat} = \kappa - c(q') + \frac{\delta \cdot 0}{1 - \delta}$$

For the network to be motivated to cooperate, its PV in case of cooperation must be preferable than its PV in case of cheating. Thus

$$\text{PV}^{net}_{coop} > \text{PV}^{net}_{cheat} = \frac{\kappa - c(q)}{1 - \delta} > \kappa - c(q') + \frac{\delta \cdot 0}{1 - \delta}$$

If the user plays the grim strategy, the network is motivated to cooperate when $\delta > (c(q) - c(q'))/(\kappa - c(q'))$.

2. Assume now that the user plays the cheat-and-leave strategy. Considering the network's possible strategies, it could either cooperate or cheat in the current period.

If the network cooperates, then

$$\text{PV}^{net}_{coop} = \kappa' - c(q) + \frac{\delta \cdot 0}{1 - \delta}$$

If the network cheats, then

$$\text{PV}^{net}_{cheat} = \kappa' - c(q') + \frac{\delta \cdot 0}{1 - \delta}$$

If the user plays the cheat-and-leave strategy, the network is not motivated to cooperate since $\text{PV}^{net}_{cheat} > \text{PV}^{net}_{coop}$

3. Assume now that the network plays the tit-for-tat strategy. If the user plays the grim strategy, it will cooperate in the current period, whereas, if it plays the cheat-and-leave strategy, it may cheat.

If the user cooperates, then

$$\text{PV}^{user}_{coop} = \frac{\pi(q) - \kappa}{1 - \delta}$$

If the user cheats, then

$$\text{PV}^{user}_{cheat} = \pi(q) - \kappa' + \frac{\delta \cdot 0}{1 - \delta}$$

For the user to be motivated to cooperate, its PV in case of cooperation must be preferable than its PV in case of cheating. Thus

$$\text{PV}^{user}_{coop} > \text{PV}^{user}_{cheat} = \frac{\pi(q) - \kappa}{1 - \delta} > \pi(q) - \kappa' + \frac{\delta \cdot 0}{1 - \delta}$$

If the network cooperates, the user is motivated to cooperate when $\delta > (\kappa - \kappa')/(\pi(q) - \kappa')$.

4. Assume finally that the network plays the cheat-and-return strategy. Considering the user's possible strategies, it could either cooperate or cheat in the current period.

If the user cooperates, then

$$\text{PV}^{user}_{coop} = \pi(q') - \kappa + \frac{\delta \cdot 0}{1 - \delta}$$

If the user cheats, then

$$\text{PV}^{user}_{cheat} = \pi(q') - \kappa' + \frac{\delta \cdot 0}{1 - \delta}$$

If the network plays the cheat-and-return strategy, the user is not motivated to cooperate since $\text{PV}^{user}_{cheat} > \text{PV}^{user}_{coop}$.

It follows that the conditional-cooperation profile is a subgame perfect equilibrium. □

23.3.3 Introducing Punishment in the Repeated Interaction

We now investigate the repeated user–network interaction game, where we introduce punishment for cheating behavior. The game is modified as follows: the user may employ a strategy such that the punishment imposed on the network for cheating lasts only for one period; namely, let the user be allowed to employ the *leave-and-return* strategy as defined next:

Definition 23.7 (Leave-and-return strategy) When the user cooperates as long as the network cooperates and leaves for one period in case the network cheats, returning in the subsequent period to cooperate again, the user's strategy is called leave-and-return strategy.

Based on the newly defined strategy, we introduce the *one-period punishment* profile of the game.

Definition 23.8 (One-period-punishment profile) When the user employs the leave-and-return strategy and the network employs the tit-for-tat strategy, the profile of the repeated game is called one-period-punishment profile of the game.

In [34], it was proved that the conditions to sustain cooperation with grim trigger strategies, which are the stricter strategies that may be employed in a repeated prisoner's dilemma, are necessary conditions for the feasibility of any form of conditional cooperation. That is, a grim trigger strategy can sustain cooperation in the iterated prisoner's dilemma under the least favorable circumstances of any strategy that can sustain cooperation.

Motivated by the result in [34], we show next that it is easier to impose cooperation in our game under the *conditional-cooperation* profile than under the *one-period-punishment* profile.

THEOREM 23.2   Assume that $\delta > (c(q) - c(q'))/(\kappa - c(q'))$ and $\delta > (\kappa - \kappa')/(\pi(q) - \kappa')$ in the repeated user–network interaction game. Then, the conditional-cooperation profile motivates cooperation of the players. The same conditions on δ are also necessary to motivate cooperation in the one-period-punishment profile.

Proof   We first show that cooperation is motivated under the one-period punishment profile. Given a history of cooperation, we seek the values of δ that can motivate cooperation under the *one-period-punishment* profile.

Assume first that the user cooperates in the current period. Then, the network has two options: to cooperate or to cheat.

If the network cooperates, then

$$\text{PV}^{net}_{coop} = \kappa - c(q) + \delta \cdot (\kappa - c(q)) + \frac{\delta^2 \cdot (\kappa - c(q))}{1 - \delta}$$

If the network cheats, then

$$\text{PV}^{net}_{cheat} = \kappa - c(q') + \delta \cdot 0 + \frac{\delta^2 \cdot (\kappa - c(q))}{1 - \delta}$$

In order for cooperation to be motivated, it must be that

$$\text{PV}^{net}_{coop} > \text{PV}^{net}_{cheat} = \kappa - c(q) + \delta \cdot (\kappa - c(q)) + \frac{\delta^2 \cdot (\kappa - c(q))}{1 - \delta}$$

$$> \kappa - c(q') + \delta \cdot 0 + \frac{\delta^2 \cdot (\kappa - c(q))}{1 - \delta}$$

Simplifying, we get $\delta > (c(q) - c(q'))/(\kappa - c(q))$.

Now, assume that the network cooperates in the current period. Then, the user has two options: to cooperate or to cheat.

If the user cooperates, then

$$\text{PV}^{user}_{coop} = \pi(q) - \kappa + \delta \cdot (\pi(q) - \kappa) + \frac{\delta^2 \cdot (\pi(q) - \kappa)}{1 - \delta}$$

If the user cheats, then

$$\text{PV}^{user}_{cheat} = \pi(q) - \kappa' + \delta \cdot (\pi(q') - \kappa) + \frac{\delta^2 \cdot (\pi(q) - \kappa)}{1 - \delta}$$

Table 23.4   Cooperation Thresholds

	Conditional Cooperation	One-Period Punishment
Network cooperates if	$\delta_{cc}^{net} > \dfrac{c(q) - c(q')}{\kappa - c(q')}$	$\delta_{pun}^{net} > \dfrac{c(q) - c(q')}{\kappa - c(q)}$
User cooperates if	$\delta_{cc}^{user} > \dfrac{\kappa - \kappa'}{\pi(q) - \kappa'}$	$\delta_{pun}^{user} > \dfrac{\kappa - \kappa'}{\pi(q) - \pi(q')}$

For cooperation to be motivated, it must be that

$$\text{PV}_{coop}^{user} > \text{PV}_{cheat}^{user} = \pi(q) - \kappa + \delta \cdot (\pi(q) - \kappa) + \frac{\delta^2 \cdot (\pi(q) - \kappa)}{1 - \delta}$$

$$> \pi(q) - \kappa' + \delta \cdot (\pi(q') - \kappa) + \frac{\delta^2 \cdot (\pi(q) - \kappa)}{1 - \delta}$$

Simplifying, we get $\delta > (\kappa - \kappa')/(\pi(q) - \pi(q'))$.

The cooperation thresholds for both players are summarized in Table 23.4, where the conditional cooperation profile is indicated by the subscript *cc* and the one-period punishment profile is indicated by the subscript *pun*.

Remark 23.1   It holds that $\delta_{cc}^{net} < \delta_{pun}^{net}$ since $c(q') < c(q)$, and also that $\delta_{cc}^{user} < \delta_{pun}^{user}$ since $\pi(q') - \kappa' > 0$; hence, $\pi(q') > \kappa'$. Thus, both players are more motivated to cooperate under the conditional-cooperation profile.

The proof of Theorem 23.2 is now complete.   □

23.3.4 User–Many Networks Interaction

Having examined the generic relationship between a single user and a single network through two different types of strategies for each of the players, we now propose a model for network selection in a converged environment, which is based on the knowledge obtained from the analysis of previous games investigated. The user–network interaction is modeled as a game between one user of the converged environment and the participating access networks that are available to the specific user: the networks play simultaneously as one player called *Networks*. The payoff for the player *Networks* is given as an array of payoffs corresponding to each of the individual access networks.

The situation we model is the following: the user plays first and offers a compensation to *Networks*, and *Networks* examine the compensation and make a decision concerning how many of them to accept and how many to reject the compensation. Any subset of *Networks* could accept or reject the proposed offer, including all accepting and all rejecting. In the latter case, the game terminates with zero payoff to the user and to *Networks*. If one or more networks accept the compensation, then for each network, the user estimates his own satisfaction, given by the function $\pi(q)$, and selects the network that is predicted to best satisfy that value (if only one network accepts, the selection is trivial).

The estimation of $\pi(q)$ by the user is based on user measures of network context and is different for each network. The user's decision to select one of the networks induces the specific network to start interacting with the user, while the rest of the networks do not interact any further with the user in subsequent iterations of the game. From then on, the interaction between the user and the network is as previously described. The payoffs for the networks that are not selected are zero, while the payoffs for the user and the selected network are as previously discussed.

Similarly, the estimation of $\pi(q)$ for selecting a single network by the user–many networks interaction may be used to prioritize the available networks by the user, in order to be able to select more than one networks, for example, two networks. Thus, the two networks ranked first, for example, the two *best* networks, in this list could be the ones chosen to interact in order to support fast session handoffs throughout a session. The subsequent section considers the case that $\pi(q)$ is used to prioritize all available networks and further that the two *best* networks are selected to cooperate for serving a user *premium* service request, that is, providing additional quality guarantees including handoffs throughout a session.

23.4 Support for Fast Session Handoffs

In a converged network, there may exist multiple network operators, each one of them interacting with the participating users indirectly, through an IP-based common management platform, operated by a platform administrator, to achieve user participation and consequently revenue maximization. In this section, we consider the case where enhanced quality demands may require the cooperation of two networks in advance. The two networks are selected by the use of a prioritized list based on the estimation of the satisfaction to be received by the user, $\pi(q)$, as this is calculated by the user. Once the selection is completed, the platform administrator takes over and instructs the networks to cooperate, enforcing a payment partition configuration on the two cooperating networks, to avoid either of the two networks gaining bargaining advantage by handling the partition, and furthermore to ensure that the whole process is transparent to the user, obeying the user-centric paradigm followed in converged, next-generation communication networks.

In particular, the cooperation between the two networks occurs if delay constraints are critical for a service session, and then the second best network also reserves resources, in order to act as immediate backup in case quality degradation is detected and session handoff is necessary. The cooperation of the two networks does not take place only during one handover but further continues throughout the whole duration of the session to act as a "backup," ensuring service continuity, in case that quality degradation is observed in the current network, or that the user performs more than one handoffs. The support for faster session handoffs, enabled by the cooperation of the two best networks, enhances service quality in terms of service continuity and handoff delay, which is a crucial aspect for real-time, critical services (e.g., medical video).

Since a network's satisfaction is represented by its revenue gain, and since two networks must cooperate for a single service, the payment for supporting the service needs to be partitioned between them, in order for the networks to have an incentive to cooperate. Moreover, the partitioning must be done in a way that is satisfying to both networks. This reasoning motivated our research toward modeling this situation through a cooperative game presented next.

We model the *payment partition* as a game of bargaining between the two networks. Firstly, we define the payment partition as a game between the two networks and we show that this is equivalent to the well-known Rubinstein bargaining game [35, Chapter 3], when the agreement is reached in the first negotiation period. Given this equivalence, an optimal solution to the Rubinstein bargaining game would also constitute an optimal solution to the payment partition game. To reach the optimal solution, we utilize the well-known NBS [35, Chapter 2], which applies to Rubinstein bargaining games when the agreement is reached in the first negotiation period and therefore to the payment–partition game.

23.4.1 Cooperative Bargaining Model

Let $q \in Q$ be the quality level for which the two networks negotiate. Consider the payment partition scenario, where two networks want to partition a service payment $\pi_i(q)$ set by the converged platform administrator. Let $c_i(q)$ be the resource reservation cost of network i. Given the cost characteristics of network i, each network seeks a portion:

$$\pi_i(q) = c_i(q) + \phi_i(q) \tag{23.3}$$

where $\phi_i(q)$ is the actual profit of network i, such that

$$\pi_1(q) + \pi_2(q) = \Pi(q) \tag{23.4}$$

where $\Pi(q)$ is the total payment announced by the converged platform administrator. The networks' goal is to find the payment partition, which will maximize the value of $\phi_i(q)$, given the values of $\Pi(q)$ and $c_i(q)$. Definition 23.9 defines the bargaining game between the two networks:

Definition 23.9 (Payment–partition game) Fix a specific quality level $q \in Q$ such that a fixed payment Π is received. Consider a one-shot strategic game with two players corresponding to the two networks. The profiles of the game, that is, the strategy sets of the two players, are all possible pairs (π_1, π_2), where $\pi_1, \pi_2 \in [0, \Pi]$ such that $\pi_1 + \pi_2(q) = \Pi$. All such pairs are called agreement profiles and define set S^a. So, $S^a = \pi_1 \times \pi_2$. In addition, there exists a so-called disagreement pair $\{s_1^d, s_2^d\}$, which corresponds to the case where the two players do not reach an agreement. So, the strategy set of the game is given by $S = S^a \bigcup \{s_1^d, s_2^d\}$. For any agreement point $s \in S^a$, the payoff $U_i(s)$, for player $i \in [2]$, is defined as follows:

$$U_i(s) = \pi_i(q) - c_i(q) \tag{23.5}$$

Otherwise,

$$U_1(s_1^d) = U_2(s_2^d) = 0 \tag{23.6}$$

This game is referred to as the payment–partition game.

Fact 23.1 Let $s^* = (\pi_1^*, \pi_2^*)$ be an optimal solution of the payment–partition game. Then $\phi_i(q) = \pi_i^*(q) - c_i(q)$, where $i \in [2]$, comprises an optimal solution of the payment partition scenario.

23.4.1.1 Equivalence to a Rubinstein Bargaining Game

Initially, we show the equivalence between the *payment–partition* game and a Rubinstein bargaining game, a.k.a., the basic alternating-offer game defined next according to [29]:

Definition 23.10 (Rubinstein bargaining game)   Assume a game of offers and counteroffers between two players, π_{it}^r, where $i \in \{1, 2\}$ and t indicates the time of the offer, for the partition of a cake, of initial size of Π^r. The offers continue until either agreement is reached or disagreement stops the bargaining process.

At the end of each period without agreement, the cake is decreased by a factor of δ_i. If the bargaining procedure *times out*, the payoff to each player is 0. Offers can be made at time slot $t \in \mathcal{N}_0$. If the two players reach an agreement at time $t > 0$, payment $U_i(t)$ of player i receives a share $\pi_{it}^r \cdot t \cdot \delta_i$, where $\delta_i \in [0, 1]$ is a player's discount factor for each negotiation period that passes without agreement being reached. The following equation gives the payment partitions of the two players:

$$\pi_1^r(t) \cdot t \cdot \delta_1 = \Pi^r - \pi_2^r(t) \cdot t \cdot \delta_2 \tag{23.7}$$

So, if agreement is reached in the first negotiation period, the payment partition is as follows:

$$\pi_1^r(t) = \Pi^r - \pi_2^r(t) \tag{23.8}$$

The payoff U_i^r of the players $i, j \in [2]$ if the agreement is reached in iteration t is the following:

$$U_i^r(t) = \pi_i^r(t) = \Pi^r - \pi_j^r(t) \tag{23.9}$$

Such a game is called a Rubinstein bargaining game.

Proposition 23.2   Fix a specific quality q. Then, the payment–partition game is equivalent to the Rubinstein bargaining game, when the agreement is reached in the first negotiation period.

Proof   Assuming that an agreement in the Rubinstein bargaining game is reached in the first negotiation period $t = 1$, then the game satisfies the following:

$$U_1^r(1) + U_2^r(1) = \Pi^r(1)$$

which is a constant.

In the payment–partition game, assuming an agreement profile s, we have

$$U_1(s) + U_2(s) = \pi_1(q) - c_1(q) + \pi_2(q) - c_2(q) = \Pi(q) - c_1(q) - c_2(q)$$

since $\Pi(q) = \pi_1(q) + \pi_2(q)$ and $c_1(q), c_2(q)$ are constants for a fixed quality level. It follows that $U_1(s) + U_2(s)$ is also constant. It follows that the *Rubinstein bargaining* game and the *Payment–partition* game are equivalent.   □

We define

Definition 23.11 (Optimal payment partition)   The optimal partition is when bargaining ends in an agreement profile that gives the highest possible payoff to each player given all possible actions taken by the opponent.

Proposition 23.2 immediately implies:

COROLLARY 23.3   Assume that agreement in a Rubinstein bargaining game is reached in the first negotiation period, that $\Pi^r = \Pi$, and that the corresponding profile s^* is an optimal partition for the Rubinstein game. Then, s^* is also an optimal partition for the payment–partition game.

23.4.2 Payment Partition Based on the Nash Bargaining Solution

Since the Nash bargaining game and the payment–partition game are equivalent when agreement is reached in the first period, we utilize the solution of a Nash bargaining game in order to compute an optimal solution, that is, a partition, which is satisfactory for the two networks in terms of payoffs from the payment–partition game. The solution of the Nash bargaining game, known as the NBS, captures such configuration. Therefore, since disagreement results in payoffs of 0, we are looking for an agreement profile $s = (\pi_1, \pi_2)$ such that the corresponding partition of the players is an optimal payment partition, that is, the partition that best satisfies both players' objectives.

The next theorem proves the existence of an optimal partition of the payment between the two players, given each network's cost c_i.

THEOREM 23.3   There exists an optimal solution for the payment–partition game and is given by the following: $\pi_1(q) = \frac{1}{2}(\Pi(q) + c_1(q) - c_2(q))$ and $\pi_2(q) = \frac{1}{2}(\Pi(q) + c_2(q) - c_1(q))$

Proof   We consider only agreement profiles and thus refer to the partition $\pi_i(q)$ assigned to player i. In any such profile, it holds that $\pi_1(q) + \pi_2(q) = \Pi(q)$. Assuming a disagreement implies that cooperation fails between the two networks and the payoff gained by player i equals to $U_i(s^d) = 0$. Since in any such profile, it holds that $U_i(s^d) > 0$, it follows that the disagreement point is not an optimal solution. Since the payment–partition game is equivalent to the Nash bargaining game (Proposition 23.2), an NBS of the bargaining game is an optimal solution of the payment–partition game between two players, that is, a partition $(\pi_1^*(q), \pi_2^*(q))$ of an amount of goods (such as the payment). According to the NBS properties, it holds that

$$\text{NBS} = (U_1(\pi_1(q)^*) - U_1(s^d))(U_2(\pi_2(q)^*) - U_2(s^d))$$
$$= \max(U_1(\pi_1(q)) - U_1(s^d))(U_2(\pi_2(q)) - U_2(s^d))$$
$$0 \le \pi_1(q) \le \Pi(q), \quad \pi_2(q) = \Pi(q) - \pi_1(q)$$

Since $U_1(s^d) = 0$, $U_2(s^d) = 0$, and $\pi_2(q) = \Pi(q) - \pi_1(q)$,

$$\max(\pi_1(q) - c_1(q))(\Pi(q) - \pi_1(q) - c_2(q)) = (-2\pi_1(q) + \Pi(q) - c_2(q) + c_1(q)) = 0$$

Therefore,

$$\pi_1(q) = \frac{1}{2}(\pi(q) - c_2(q) + c_1(q)), \quad \pi_2(q) = \frac{1}{2}(\pi(q) + c_2(q) - c_1(q)) \tag{23.10}$$

$\square$

Remark 23.2   Note that the probability of degradation does not affect the optimal partition, but it is still part of the networks' payoff functions; therefore, a network with a high estimated probability of degradation will receive much less of a payoff than its actual payment partition.

We proceed to investigate how the solution to the payment partition behaves when we consider the existence of a constant set by the converged platform administrator representing the probability of degradation.

THEOREM 23.4   Assume that the converged platform administrator assigns a constant value p_i^f to network $i \in [2]$ representing the expected quality degradation, based on the particular service and current network conditions. Then, the value of the optimal solution is the same as in Theorem 23.3.

Proof   Our game has the same strategy set as before. Concerning the utility functions of the players in case of disagreement, we have also $U_1(s^d) = 0$, $U_2(s^d) = 0$, and $\pi_2(q) = \Pi(q) - \pi_1(q)$ as before. In case of agreement, we have in addition a constant probability p_i^f in the payoff function of each network:

$$U_i(s) = (1 - p_i^f)(\pi_i(q) - c_i(q))$$

Therefore,

$$\max(1 - p_1^f)(\pi_1(q) - c_1(q))(1 - p_2^f)(\Pi(q) - \pi_1(q) - c_2(q))$$
$$= (1 - p_1^f)(1 - p_2^f)(-2\pi_1(q) + \Pi(q) - c_2(q) + c_1(q)) = 0$$

The optimization removes the probabilities and the optimal solution is the same as Equation 23.10:

$$\pi_1(q) = \frac{1}{2}(\pi(q) - c_2(q) + c_1(q)), \quad \pi_2(q) = \frac{1}{2}(\pi(q) + c_2(q) - c_1(q)) \tag{23.11}$$

$\square$

23.4.3 A Bayesian Form of the Payment–Partition Game

Since the partitioning is based on each network's cost, it is required that the networks are truthful about their costs. Truthfulness is a very important consideration in cooperative situations, especially in bargaining games. The question that arises is whether it would be wise for a player to lie, considering that the player cannot be aware of who the other player is from the original set of available networks and thus cannot guess whether the other player has more or less cost, thus not being able to correctly assess the risk of such an action.

A Bayesian game [36, Chapter 5] is a strategic form game with incomplete information attempting to model a player's knowledge of private information, such as privately observed costs, that the other player does not know. Therefore, in a Bayesian game, each player may have several types of behavior (with a probability of behaving according to one of these types during the game). We use the Bayesian form for the *payment–partition* game, in order to investigate the outcomes of the game, given that each network does not know whether the cost of its opponent is lower or higher than its own.

A Bayesian player begins with a prior knowledge, which is not always precise, and is expressed as the distribution on the "type" parameters of the opponent. The consistency shows whether the updated knowledge on these "type" parameters becomes more accurate as "evidence" data are collected; in fact it is believed that two Bayesian players will ultimately have very close predictive distributions [37]. We allow for both players to have the same "type" parameters and we investigate their behavior according to belief for these types. We study both players in a similar manner adopting the above statement, that both networks will ultimately have very close predictive distributions.

Let each network in the *payment–partition* game have two types: the *lower-cost* type (including networks of equal cost) and the *higher-cost* type. Suppose that each of the two networks has incomplete information about the other player, that is, it does not know the other player's type. Furthermore, each of the two networks assigns a probability to each of the opponent's types according to own beliefs and evaluations. Let p_i^l be the probability according to which network i believes that the opponent is likely to be of type *lower cost*, and $p_i^h = (1 - p_i^l)$ be the probability according to which network i believes that the opponent is likely to be of type *higher cost*.

Since the two players are identical, that is they have the same two types and the same choice of two actions, we will only analyze network i; conclusions also hold for network j, where $i, j \in [2]$, $i \neq j$. Therefore, network i believes that network j is of type *lower cost* with probability p_i^l and of type *higher cost* with probability $1 - p_i^l$. Each network has a choice between two possible actions: to declare its own real costs (D) or to cheat (C), that is, declare higher costs $c_i'(q) > c_i(q)$. The possible payoffs for network 1 are given in Tables 23.5 and 23.6.

Table 23.5 Network i Payoffs when Opponent Is of Type *Lower Cost*

Network i Strategies	Network j Strategies	
	D	C
D	$\frac{1}{2}(\Pi(q) + c_i(q) - c_j(q))$	$\frac{1}{2}(\Pi(q) + c_i(q) - c_j'(q))$
C	$\frac{1}{2}(\Pi(q) + c_i'(q) - c_j(q))$	$\frac{1}{2}(\Pi(q) + c_i'(q) - c_j'(q))$

Table 23.6 Network *i* Payoffs When Opponent Is of Type *Higher Cost*

Network *i* Strategies	Network *j* Strategies	
	D	*C*
D	$\frac{1}{2}(\Pi(q) + c_j(q) - c_i(q))$	$\frac{1}{2}(\Pi(q) + c_j'(q) - c_i(q))$
C	$\frac{1}{2}(\Pi(q) + c_j(q) - c_i'(q))$	$\frac{1}{2}(\Pi(q) + c_j'(q) - c_i'(q))$

LEMMA 23.3 If $p_i^l > p_i^h$, that network j is believed by network i to be of lower cost, then it is more preferable for network i to lie, where $i, j \in [2], i \neq j$.

Proof In Table 23.5, network i has higher or equal costs to network j since network j is of type *lower cost*; thus, $c_i(q) \geq c_j(q)$. When both players play D, that is, they both declare their real costs, and an equal or greater piece of the payment is assigned to network i, since the partition of the payment is directly proportional to the networks' costs. If network i plays C, that is, cheats, while network j plays D, then $c_i'(q) > c_i(q) > c_j(q)$, a profitable strategy for network i, since an even greater piece of the payment will be received. For the cases that network j decides to play C, then the payment partition may or may not favor network j (it depends on the actual amount of cheating and the action of network i). If network i plays C, then it is more likely that $c_i'(q) > c_j'(q)$, and network i will get a greater piece, than it would if it plays D. □

LEMMA 23.4 If $p_i^h > p_i^l$, that network j is believed by network i to be of higher cost, then it is more preferable for network i to lie, where $i, j \in [2], i \neq j$.

Proof In Table 23.6, network i has lower costs compared to network j; thus, $c_i(q) < c_j(q)$. When both players play D, that is, they both declare their real costs, and an equal or greater piece of the payment is assigned to network j. If network i plays C, that is, cheats, then $c_i'(q) > c_i(q)$, so playing C will end up in a higher payoff for network i, and in case network j plays D, i may even get the bigger piece of the partition. If network j plays C, it is still better for network i to play C, since this will end up in network i receiving a greater piece than it would if it plays D when network j plays C, although, more likely, not the greater of the two pieces. □

23.4.3.1 Motivating Truthfulness

The earlier results motivated us to seek suitable conditions in order to motivate networks to be truthful. Next, we enforce truthfulness through the penalty functions of the networks. First note:

COROLLARY 23.5 Two networks playing the Bayesian form of the *payment–partition* game are not motivated to declare their real costs but instead they are motivated to cheat and declare higher costs, that is, $c_i'(q) > c_i(q)$, $1 \in [2]$, in order to get greater payoffs.

Proof Straightforward by Lemmas 23.3 and 23.4. □

In order to motivate the two networks to declare their real costs, there must exist a mechanism that can penalize a player who turns out to lie on its real cost, assuming that it is detectable whether a player has lied or not; we refer to such mechanisms as pricing mechanisms [30]. Let the converged platform administrator be able to detect after the service session has terminated, whether either of the participating networks has lied about its costs. In order to motivate the networks to declare their real costs, we introduce a *pricing mechanism*, that is, a new variable that tunes the resulting payoffs, in the payoff function of each player. The pricing mechanism is a postgame punishment, that is, cheating in a game does not affect the game in which a network cheats but subsequent games. Thus, a state of history of a player's behavior in similar interactions must be kept.

We define a pricing mechanism consisting of variable $\beta_i \in [0, 1]$, which represents the probability of being truthful, and it may adaptively modify the payoffs of a player, according to the player's history of actions. In particular, it is a ratio of the number of times network i has been caught lying, over a selected finite number of periods representing the window of history that the converged platform administrator monitors for each network.*

The less frequently a player cheats, the closer to 1 its β_i is. Thus, the administrator sets the payoff of network i to be

$$\pi_i(q) = \frac{1}{2}(\Pi(q) + \beta_i \cdot c_i(q) - \beta_j \cdot c_j(q)) \tag{23.12}$$

where $i, j \in [2], i \neq j$. The players are motivated to declare their real costs, since any cheating would decrease β_i, affecting any future payoffs from such procedure. An evaluation of the the Bayesian form of the *payment–partition* game including β_i in the players' payoffs and allowing a history of behavior to be collected from repeating the game is given in Section 23.5.

23.5 Evaluating Solutions through Simulations

23.5.1 Evaluation of User–Network Interaction in Single Network Selection

The evaluation presented here is based on a MATLAB® [38] implementation of an iterated user–network interaction game, where user and network strategies are played against each other repeatedly, in order to evaluate each strategy in terms of payoff. The implementation of the user–network interaction game was based on a publicly available MATLAB implementation of the iterated prisoner's dilemma game [39], which has been extended to include all the strategies examined in Section 23.3. In each simulation run, both players play their strategies and get payoffs accordingly; we use numbers that follow the relationships of the payoffs as described in their general case in the repeated game model (Table 23.3). A randomly generated perceived quality and a fixed threshold of expected quality are also implemented in all simulation runs.

For each simulation, a random number of iterations were run to get cumulative user and network payoffs for each combination of a user strategy playing against a network strategy. The user payoffs per strategy and the network payoffs per strategy are eventually added to give the most profitable user and network strategies, respectively, for the total number of iterations of a simulation run; then the average cumulative payoffs from all simulation runs are calculated. Although the number

---

* We consider that a revelation of the real costs of the two access networks is always possible at a later stage of the procedure (e.g., after session termination).

Table 23.7   User Payoffs from All Strategy Combinations

User Payoffs	Network Strategies	
User Strategies	Tit-for-Tat	Cheat and Return
Grim	793.14	4.42
Cheat and leave	6.62	4.65
Leave and return	793.14	252.92

Table 23.8   Network Payoffs from All Strategy Combinations

Network Payoffs	Network Strategies	
User Strategies	Tit-for-Tat	Cheat and Return
Grim	793.14	7.42
Cheat and leave	3.82	4.23
Leave and return	793.14	617.21

of iterations for each simulation is randomly generated, we run the simulation process 100 times, that is, by randomly generating 100 different numbers of iterations, in order to include behaviors when the number of iterations is both small and large.

The payoffs are calculated for an average of 264.38 iterations per simulation run* and presented in Table 23.7 (user payoffs) and in Table 23.8 (network payoffs). The numbers we see are the average payoffs from the 100 simulation runs; however, for each simulation run, a cumulative payoff is given for all the iterations run, because it is important to capture the continuation of behavior in the repeated game by adding the payoffs for all iterations in the same simulation run.

In both tables, we see a score for each strategy combination. The score corresponds to either a user payoff (Table 23.7) or a network payoff (Table 23.8). The results show that the most profitable user strategies, when the network plays with the *tit-for-tat* strategy, are the *grim* and the *leave-and-return* strategies. The most profitable user strategy, when the network employs the *cheat-and-return* strategy, is the *leave-and-return* strategy. The reason is that the *grim* strategy involves the action of leaving forever from the interaction upon detection of cheating from the network, so it receives no more payoff from the interaction upon leaving. On the other hand, the most profitable network strategy when the user employs either the *grim* strategy or the *leave-and-return* strategy is the *tit-for-tat* strategy. When the user plays the *cheat-and-leave* strategy, the results generated a slightly increased payoff for the network's *cheat-and-return* strategy.

The *tit-for-tat* strategy generates better payoffs when the user strategy used involves cooperation as long as the network cooperates, no matter what the punishment is for cheating because, according to the mirroring behavior guided by the *tit-for-tat* strategy, the network keeps cooperating until the end of the game. This is seen in the identical payoffs for both the user and the network generated for the cases when the *tit-for-tat* strategy plays against (a) the *grim* strategy and (b) the *leave-and-return* strategy.

---

* Minimum iterations generated: 8, maximum iterations generated: 1259.

The minimum punishment applied by the *leave-and-return* user strategy appears to have a more positive effect on the payoffs compared to the rest of the user strategies for both network strategies evaluated. However, we must note that the user strategy that motivates cooperation the most, according to the theoretical results, is the *grim* strategy; thus, it is more likely for network to cooperate when the *grim* strategy is employed by the user. The more mild punishment employed by the *leave-and-return* strategy may be less motivating, but for the case that the network cheats, it will return to the interaction after leaving for one period and thus collect more cumulative payoffs. Thus, the *one-period-punishment* profile of the game, that is, when the user plays the *leave-and-return* strategy and the network plays the *tit-for-tat* strategy, is the most profitable.

23.5.2 Evaluation of Truthfulness in Supporting Fast Session Handoffs

Next, we evaluate the Bayesian form of the *payment–partition* game between two networks cooperating to support a service session with strict QoE constraints. The payoffs of the game are based on the NBS as this has been demonstrated in Section 23.4.2 and further include the term β_i (Equation 23.12), in order to motivate the networks to cooperate. The numerical values used for $\Pi(q)$, $c_i(q)$, $c_{-i}(q)$ in case the networks are truthful or lying obey the payoff relations as these are given in Tables 23.5 and 23.6 and the overall model of the payment–partition game as described in Section 23.4.1 and resolved in Section 23.4.2.

The player types are 1 = *lower cost* or 2 = *higher cost* as these are explained in Section 23.4.3. Each network has four strategies per simulation run: (1) always declare its real costs—indicated as *Only D*, (2) always cheat—indicated as *Only C*, (3) randomly declare real costs or cheat with a probability of 0.5 for each option—indicated as 50% C/D, and (4) adapt to the value of β_i, that is, cheat only if $\beta_i \geq 0.9$ else declare real costs—indicated as *Adapt*. Each player's initial value of β_i in the simulations is randomly generated and then adapted to the player's play according to the selected strategy. Overall, we have run 100 simulation runs with different initial values for β_i and with a random number of iterations, indicated as R. Table 23.9 presents the results from the 100 simulation runs as cumulative payoffs for the number of iterations, for each of the two networks partitioning the service payment. We observe that in all simulation runs, the second strategy, that is, always to cheat, is the least profitable strategy for both networks, regardless of their types and initial values of β_i. This is because of the presence of β_i in the payoffs, which punishes the choice of cheating, by detecting such an action after any iteration. On the other hand, the more profitable

Table 23.9 Average Network 1 Payoffs from Payment–Partition Game

Avg. R = 288.59	*Min. R = 4*	*Max. R = 1749*			
Network 1 Payoffs	*Network 2 Strategies*	*Only D*	*Only C*	*50% C/D*	*Adapt*
Network 1 Strategies					
Only D		1462.19	1735.50	1563.25	1481.17
Only C		1179.02	1471.58	1308.93	1212.26
50% C/D		1354.71	1600.62	1457.92	1377.08
Adapt		1441.80	1700.63	1539.63	1460.79

strategies include actions of declaring the real costs, that is, being truthful; specifically, the first strategy of always being truthful is in most cases the most profitable, with the fourth strategy of adapting to β_i generating comparable payoffs.

23.6 Conclusions

This chapter has provided game-theoretic studies for the interactions during the network selection mechanism. In particular, we present the interaction between a user and its serving network, as well as for handoff during a session, specifically for the payment–partition procedure between two cooperating networks. The game-theoretic tools used for these studies included the prisoner's dilemma game model, one-shot and repeated game models, bargaining game models, and Bayesian game models with their corresponding equilibrium notions.

Primarily, we have modeled the interaction between a single user and its serving access network as a one-shot prisoner's dilemma and have concluded that if the actions available to the user and the access network were to either cooperate or to cheat, then both players would be motivated to cheat. Since this was not a desirable behavior, we moved on to model the interaction as a repeated game, which ended up motivating cooperation between the two players. Moreover, it has been shown that cooperation is motivated more easily if the strategies employed by the players incorporate threats for harsh punishments in case the opponent cheats. Based on the study of user–network interaction, a network selection model was provided.

The second game-theoretic study concerns the case where network cooperation is required to support a session handoff with service continuity and minimum handoff delays. The particular aspect we have investigated is the optimization of the payment–partition process between the two cooperating networks. It has been shown that this process is equivalent to the well-known Rubinstein bargaining game when the agreement is reached during the first negotiation period as well as to a Nash bargaining game. Therefore, an optimal payment partition has been derived using the NBS, which indicated the cost of resource reservation for each participating network to be the deciding factor for the optimal payment partition. Furthermore, it has been shown that any constant probability for the expected quality degradation does not affect the optimal payment partition. The payment partition game raises the issue of truthfulness, that is, whether the networks are motivated to be truthful about their resource reservation costs, since lying could in some cases increase their overall partition. Simple reasoning illustrates that lying is worth the risk, so networks are motivated to lie in the absence of pricing mechanisms, that is, mechanisms that punish players for cheating behavior. Hence, to motivate truthfulness, we have proposed such pricing mechanism in the players' payoff, through a Bayesian game model.

Finally, numerical results for both the user-network-interaction game and the Bayesian model of the payment–partition game have been provided (based on MATLAB simulations). In both games, the numerical results provide cumulative payoffs of the players when different strategies are played iteratively. The numerical results verify the conclusions drawn from the theoretical studies.

23.7 Open Issues

We wish to explore further the repeated nature of the interaction between the networks and the user. In particular, we will seek to find optimal solutions of the repeated game and identify players' behaviors as well as the overall system performance. Furthermore, we will try to develop other tools that enable us to capture the effect of the previous outcomes of the interaction between the entities, in any future decisions.

Furthermore, we plan to investigate additional parameters for network selection that can also improve handoffs, such as consideration of reserved resources as an incentive for a user to select a particular network. This could be defined differently for each network and would be an additional factor for the user to consider, prior to evaluating the satisfaction estimation function for each network. Moreover, we plan to perform further studies on the way compensation is set to avoid user exploitation by the network, in the case that the user valuates this privately and such private valuation of service is significantly higher than network-incurring costs.

In this study, we utilized the notion of the discount factor. We would like to analyze further this notion but moreover to discover other such tools that may be able to capture better the interaction of the players. We also plan to further investigate the user–network interaction when more than one networks are involved, modeling it as a repeated game between multiple players. For this enhanced game, we plan to find the necessary and sufficient conditions enforcing the players into behaviors that maximize both local (individual) and global (system) objectives.

Finally, we plan to investigate mechanisms that will enforce the modeled interactions to converge into desired outcomes. Toward this goal, we plan to explore tools from the mechanism design. In particular, we will try to find suitable utilities for the players that the converged network will impose on the involved entities, such that trying to maximize their individual gain will end up in desired behaviors, for example, being truthful.

References

1. A. Iera, A. Molinaro, C. Campolo, and M. Amadeo, An access network selection algorithm dynamically adapted to user needs and preferences, *Proceedings of the IEEE International Symposium on Personal, Indoor and Mobile Radio Communications* (PIMRC '06), Helsinki, Finland, pp. 1–5, IEEE, September 2006.
2. F. Bari and V. Leung, Service delivery over heterogeneous wireless systems: Network selection aspects, *Proceedings of the 2006 International Conference on Wireless Communication and Mobile Computing*, New York, 2006.
3. J. Antoniou, I. Koukoutsidis, E. Jaho, A. Pitsillides, and I. Stavrakakis, Access network synthesis game in next generation networks, *Elsevier Computer Networks Journal*, 53, 15, 2716–2726, October 2009.
4. I. M. Weinstein, Real World Convergence for conferencing: Considerations for hosting IP video on the enterprise data network, Wainhouse Research Issue Paper, July 2005.
5. J. S. Dyer, P. C. Fishburn, R. Estever, J. Wallenius, and S. Zionts, Multiple-criteria decision-making: Multiattribute utility theory—The next ten years, *Management Science* 38, 5, 645–653, April 1992.
6. X. Xiaochun and W. Xiaoyan, A simple rate control algorithm for maximizing total user utility, *Proceedings of the International Conference on Communication Technology* 2003, pp. 135–138, IEEE, April 2003.
7. S. E. Elayoubi, T. Chahed, and G. Hebuterne, Admission control in UMTS in the presence of shared channels, *Computer Communications* 27, 11, 1115–1126, July 2004.
8. D. Gao, J. Cai, and K. N. Ngan, Admission control in IEEE 802.11e wireless LANs, *IEEE Network* 19, 4, 6–13, July 2005.
9. J. Antoniou, V. Papadopoulou, V. Vassiliou, and A. Pitsillides, Cooperative user network interactions in next generation communication networks, Submitted to Elsevier *Computer Networks Journal*, May 2009 (under 2nd review).
10. S. Kher, A. K. Somani, and R. Gupta, Network selection using fuzzy logic, *Proceedings of the 2nd International Conference on Broadband Networks*, Boston, MA, pp. 876–885, IEEE, October 2005.

11. W. Zhang, Handover decision using fuzzy MADM in heterogeneous networks, *Proceedings of the IEEE Wireless Communications and Networking Conference (WCNC '04)*, Atlanta, GA, pp. 653–658, IEEE, March 2004.

12. Q. Song and A. Jamalipour, An adaptive quality-of-service network selection mechanism for heterogeneous mobile networks, *Wireless Communications and Mobile Computing*, 5, 6, 697–708, 2005.

13. H. Chan, P. Fan, and Z. Cao, A utility-based network selection scheme for multiple services in heterogeneous networks, *Proceedings of the IEEE Wireless Networks, Communications and Mobile Computing*, pp. 1175–1180, IEEE, June 2005.

14. J. Antoniou and A. Pitsillides, Radio access network selection scheme in next generation heterogeneous environments, *Proceedings of the IST Mobile Summit 2006*, Mykonos, Greece, IEEE, June 2006.

15. V. Gazis, N. Alonistioti, and L. Merakos, Towards a generic "Always Best Connected" capability in integrated WLAN/UMTS cellular mobile networks, *IEEE Wireless Communications Magazine*, 12, 22–29, 2005.

16. X. Cai, L. Chen, R. Sofia, and Y. Wu, Dynamic and user-centric network selection in heterogeneous networks, *Proceedings of the IEEE International Conference on Performance, Computing and Communications*, New Orleans, LA, pp. 538–544, IEEE, April 2007.

17. F. Bari and V. C. M. Leung, Automated network selection in a heterogeneous wireless network enviroment, *IEEE Network*, 21, 1, 34–40, January 2007.

18. O. Markaki, D. Charilas, and D. Nikitopoulos, Enhancing quality of experience in next generation networks through network selection mechanisms, *Proceedings of the B3G Environment Workshop, PIMRC 2007*, Athens, Greece, IEEE, September 2007.

19. D. Charilas, O. Markaki, D. Nikitopoulos, and M. Theologou, Packet switched network selection with the highest qos in 4G networks, *Computer Networks*, 52, 1, 248–258, 2008.

20. J. Antoniou, M. Stylianou, and A. Pitsillides, RAN selection for converged networks supporting IMS signalling and MBMS service, *Proceedings of the 18th International Symposium on Personal, Indoor and Mobile Radio Communications (PIMRC '07)*, Athens, Greece, June 2008.

21. J. Kim, J. H. Lee, T. Kwon, and Y. Choi, Wireless access network selection for live streaming multicast in future internet, *Proceedings of the 3rd International Conference on Future Internet Technologies*, Seoul, South Korea, June 2008.

22. D. Charilas, O. Markaki, and E. Tragos, A theoretical scheme for applying game theory and network selection mechanisms in access admission control, *Proceedings of the 3rd International Symposium on Wireless Pervasive Computing*, Santorini, Greece, pp. 303–307, May 2008.

23. J. Antoniou and A. Pitsillides, 4G converged environment: Modeling network selection as a game, *Proceedings of the 16th IST Mobile and Wireless Communications Summit 2007*, Budapest, Hungary, IEEE, July 2007.

24. D. Niyato and E. Hossain, A cooperative game framework for bandwidth allocation in 4G heterogeneous wireless networks, *Proceedings of the IEEE International Conference on Communications 2006 (ICC '06)*, Istanbul, Turkey, pp. 4357–4362, IEEE, June 2006.

25. H. Lin and M. Chatterjee, ARC: An integrated admission and rate control framework for competitive wireless CDMA data networks using noncooperative games, *IEEE Transactions on Mobile Computing*, 4, 3, 243–258, 2005.

26. D. Niyato and E. Hossain, Modeling user churning behavior in wireless networks using evolutionary game theory, *Proceedings of IEEE WCNC 2008*, Las Vegas, NV, March 31–April 3, 2008.

27. R. B. Myerson, Two-person bargaining problems with incomplete information, *Econometrica*, 52, 2, 461–487, 1984.

28. N. Nisan and A. Ronen, Algorithmic mechanism design, *Games and Economic Behavior*, 35, 1–2, 166–196, April 2001.

29. P. Cramton, Y. Shoham, and R. Steinberg, editors. *Combinatorial Auctions*. MIT Press, Cambridge, MA, 2006.

30. W. E. Walsh and M. P.Wellman, A market protocol for decentralized task allocation: Extended version. *Proceedings of the Third International Conference on Multi-Agent Systems (ICMAS-98)*, Paris, France, 1998.

31. T. Groves, Incentives in teams, *Econometrica*, 41, 4, 617–631, July 1973.

32. G. Kendall, X. Yao, and S. Y. Chong, *The Iterated Prisoner's Dilemma: 20 Years On*, World Scientific Publishing Co., Singapore, Advances in Natural Computation Book Series, vol. 4, 2009.

33. A. Dixit and S. Skeath, *Games of Strategy*, W.W.Norton & Company, New York, 1999.

34. R. Hegselmann and A. Flache, Rational and adaptive playing: A comparative analysis for all possible prisoner's dilemmas, *Analyse and Kritik*, 22, 1, 75–97, 2000.

35. A. Muthoo, *Bargaining Theory with Applications*, Cambridge University Press, Cambridge, U.K., 2002.

36. H. Gintis, *Game Theory Evolving: A Problem-Centered Introduction to Modeling Strategic Interaction*, Princeton University Press, Princeton, NJ, 2000.

37. S. Ghosal, A review of consistency and convergence of posterior distribution. *Proceedings of Symposium in Bayesian Inference*, Banaras Hindu University, Varanasi, India, 1996.

38. *MATLAB: The Language of Technical Computing*, The Mathworks, version 7.6.0.324 (R2008a), Natick, MA, February 2008.

39. G. Taylor, Iterated Prisoner's Dilemma in MATLAB, Archive for the Game Theory Category http://maths.straylight.co.uk/archives/category/game-theory, March 2007.

Index